W9-BHV-643

PLACE IN RETURN BOX to remove this checkout from your record.
TO AVOID FINES return on or before date due.
MAY BE RE~~~~~~~~~~~~~~~~~~~~~ if requested.

Handbook
of
Functional
Plant Ecology

BOOKS IN SOILS, PLANTS, AND THE ENVIRONMENT

Growth and Mineral Nutrition of Field Crops: Second Edition, N. K. Fageria, V. C. Baligar, and Charles Allan Jones

Fungal Pathogenesis in Plants and Crops: Molecular Biology and Host Defense Mechanisms, P. Vidhyasekaran

Plant Pathogen Detection and Disease Diagnosis, P. Narayanasamy

Agricultural Systems Modeling and Simulation, edited by Robert M. Peart and R. Bruce Curry

Agricultural Biotechnology, edited by Arie Altman

Plant–Microbe Interactions and Biological Control, edited by Greg J. Boland and L. David Kuykendall

Handbook of Soil Conditioners: Substances That Enhance the Physical Properties of Soil, edited by Arthur Wallace and Richard E. Terry

Environmental Chemistry of Selenium, edited by William T. Frankenberger, Jr., and Richard A. Engberg

Principles of Soil Chemistry: Third Edition, Revised and Expanded, Kim H. Tan

Sulfur in the Environment, edited by Douglas G. Maynard

Soil–Machine Interactions: A Finite Element Perspective, edited by Jie Shen and Radhey Lal Kushwaha

Mycotoxins in Agriculture and Food Safety, edited by Kaushal K. Sinha and Deepak Bhatnagar

Plant Amino Acids: Biochemistry and Biotechnology, edited by Bijay K. Singh

Handbook of Functional Plant Ecology, edited by Francisco I. Pugnaire and Fernando Valladares

Additional Volumes in Preparation

Handbook of Plant and Crop Stress: Second Edition, Revised and Expanded, edited by Mohammad Pessarakli

Handbook of Pest Management, edited by John R. Ruberson

Plant Responses to Environmental Stresses: from Phytohormones to Genome Reorganization, edited by H. R. Lerner

Handbook of Functional Plant Ecology

edited by

Francisco I. Pugnaire
Arid Zones Experimental Station
Spanish Council for Scientific Research
Almería, Spain

Fernando Valladares
Center of Environmental Sciences
Spanish Council for Scientific Research
Madrid, Spain

MARCEL DEKKER, INC.   NEW YORK · BASEL

ISBN: 0-8247-1950-6

This book is printed on acid-free paper.

Headquarters
Marcel Dekker, Inc.
270 Madison Avenue, New York, NY 10016
tel: 212-696-9000; fax: 212-685-4540

Eastern Hemisphere Distribution
Marcel Dekker AG
Hutgasse 4, Postfach 812, CH-4001 Basel, Switzerland
tel: 41-61-261-8482; fax: 41-61-261-8896

World Wide Web
http://www.dekker.com

The publisher offers discounts on this book when ordered in bulk quantities. For more information, write to Special Sales/Professional Marketing at the headquarters address above.

Current printing (last digit):
10 9 8 7 6 5 4 3 2 1

PRINTED IN THE UNITED STATES OF AMERICA

Preface

Plants are distributed from the equator to the poles, and from aquatic to the most xeric habitats of the earth. The diversity of plant form and life history in these very different habitats suggests that plant function should underlie this diversity. Consequently, plant functional ecology has a central role in our efforts to understand the community and ecosystem structure and dynamics, and has attracted the interest of ecologists trying to intepret patterns of adaptive specialization in plants.

The *Handbook of Functional Plant Ecology* reflects the diversity of research into functional plant ecology, providing readers with the broadest possible view. Our aim was to include original reviews in a readable style, giving a comprehensive overview of the topic with a historical perspective when needed. The book is intended for a broad audience, from practicing plant ecologists to students, with characteristics of both a textbook and a collection of essays. When commissioning chapters for this book, we were not interested in presentation of new experimental data or novel theoretical intepretations or hypotheses. Instead, we asked the contributors to produce an up-to-date introduction to each topic and to incorporate their opinions on future directions where possible. Given the substantial number of potential topics, deciding the composition of the book was not an easy task; practicality ultimately limited the contents to 24 chapters.

The *Handbook* is divided into five parts with a bottom-up approach, from the more specific, detailed studies focused on plant organs to the broader ecosystem level. The history, aims, and potential of functional approaches are established in the first chapter, which also sets the limits of functional plant ecology, a science centered on the study of whole plants that attempts to predict responses in plant functioning caused by environmental cues. The discipline emphasizes the influence of plants on ecosystem functions, services, and products, and aims to establish patterns and derive functional laws from comparative analyses. The search for these patterns is likely to be most effective if driven by specific hypotheses tested with comparative analyses at the broadest possible scale. Functional laws thus developed may hold predictive power irrespective of whether they represent direct cause–effect relationships. However, the nested nature of the control of functional responses implies uncertainties when scaling functional laws, toward either lower or higher levels of organization.

Part I (Chapters 2–5) deals with plant structure and growth form, a central topic because structure is the basis for function. While limiting the range of possible functions, structure may enhance the efficiency of some of them or at least make them possible. Thus, a functional understanding of plant structures, from tissue anatomy to whole-plant architecture, is essential to fully understand how a plant functions as a whole. These aspects are reviewed for both above- and belowground plant structures, and for the remarkable case of photosynthetic organisms that are able to cope with desiccation: the poikilohydrous autotrophs (unicellular algae, lichens, mosses, and resurrection plants).

Part II (Chapters 6–8) deals with physiological ecology, a field concerned with the performance of plants in their natural environment, and the responses of physiological, biochemical, and molecular attributes to habitat characteristics. Perhaps the most important way in which plants cope with environmental extremes and disturbance is through physiological adaptation, and an understanding of physiology helps to define the limits for plant growth and survival from the individual plant point of view. Physiological ecology widens the scope of physiology by considering the natural scenario for each particular attribute, providing a more realistic account of the efficiency and the ecological performance of different features.

Adaptation to environmental extremes is the focus of Part III (Chapters 9–14), which deals with a number of prominent environments, including Mediterranean, arid, tropics, and high-latitude ecosystems. In these habitats, plant communities are often influenced by singular environmental cues, and plants in similar climate zones in different parts of the globe are characterized by a high proportion of shared adaptations. These plant communities provide examples of convergent evolution with different, distantly related taxa adopting similar forms and functions. Very often, however, different solutions to the same ecological circumstances are apparent in comparisons of co-occurring species, demonstrating the effects of phylogeny in the evolution of functional traits.

Community ecology is covered in Part IV (Chapters 15–20), which discusses plant reproduction and establishment, plant–plant and plant–herbivore interactions, and the role of diversity. In Chapter 17, the role of positive interactions among plants is emphasized, because a large amount of experimental evidence on interspecific positive interactions suggests that conceptual models focused on competition only partly explain the forces that structure plant communities.

Part V (Chapters 21–24) focuses on new approaches to questions about photosynthesis, functional ecology and phylogeny, plant responses to pollution, and plant performance at different scales. The aim of this part is to provide the reader with a sample of new questions in functional plant ecology, as well as insight into how some old topics in plant ecology can be addressed with new techniques.

This volume does not include a chapter on temperate ecosystems, because most advances in theoretical and functional ecology have been developed in these environments and they are comprehensively covered in other texts. Instead, we selected complementary topics to fully illustrate plant functional ecology in different ecosystems and how traits vary between them. Similarly, there is no chapter explicitly devoted to photosynthesis, but included throughout many chapters herein are references to gas exchange and carbon fixation in relation to a host of other subjects such as canopy architecture, water relations, nutrients, environmental changes, and light acclimation. Thus, we consider that aspects of photosynthesis have been covered in several chapters of the book, without needing a specific treatment that would have required focusing more on physiology than ecology. Undoubtedly, individual readers will think that other topics are underrepresented in the book, but we have endeavored to incorporate the widest subject area possible and to establish a fresh approach.

We would like to express our sincere thanks to the authors who contributed to this volume, for their effort, dedication, and enthusiasm, and for meeting the deadlines despite already busy schedules. We also thank our contributing authors who served as reviewers of other chapters, and Carmen Ascaso, Dennis Baldocchi, Matthias Boer, Steve Compton, Pedro Jordano, Sue McIntyre, María Eugenia Ron, Tsvi Sachs, John Willott, and Kell B. Wilson, who also kindly reviewed some of the chapters. Finally, we want to thank Russell Dekker for his support, and the staff of Marcel Dekker, Inc., especially Jeanne McFadden, for their assistance and patience. All of them made possible and greatly improved the quality of this work.

Francisco I. Pugnaire
Fernando Valladares

Contents

II. PHYSIOLOGICAL ECOLOGY

III. HABITATS AND PLANT DISTRIBUTION

IV. POPULATIONS AND COMMUNITIES

Contributors

Luis Balaguer, Ph.D. Associate Professor, Department of Plant Biology I, Complutense University, Madrid, Spain

Jeremy Barnes, Ph.D. Royal Society Research Fellow and Lecturer in Plant Sciences, Air Pollution Laboratory, Department of Agricultural and Environmental Science, University of Newcastle, Newcastle upon Tyne, England

Frank Berendse, Ph.D. Professor, Nature Conservation and Plant Ecology Group, Department of Environmental Sciences, Wageningen Agricultural University, Wageningen, The Netherlands

Wolfram Beyschlag, Ph.D. Professor, Department of Experimental and Systems Ecology, University of Bielefeld, Bielefeld, Germany

Wim G. Braakhekke, Ph.D. Lecturer, Nature Conservation and Plant Ecology Group, Department of Environmental Sciences, Wageningen Agricultural University, Wageningen, The Netherlands

Ragan M. Callaway Division of Biological Sciences, University of Montana, Missoula, Montana

Claire Damesin, Ph.D.* Center for Functional and Evolutionary Ecology, CNRS, Montpellier, France

---

* *Current affiliation*: Assistant Professor, Department of Plant Ecophysiology, University Paris-Sud, Orsay, France.

Alan Davison, Ph.D., M.I.E.E.M. Professor, Air Pollution Laboratory, Department of Agricultural and Environmental Science, University of Newcastle, Newcastle upon Tyne, England

Hans de Kroon, Ph.D. Senior Lecturer, Nature Conservation and Plant Ecology Group, Department of Environmental Sciences, Wageningen Agricultural University, Wageningen, The Netherlands

Carlos M. Duarte, Ph.D. Research Professor, Center for Advanced Studies of Blanes, Spanish Council for Scientific Research, Blanes, Spain

Michael Fenner, Ph.D. Senior Lecturer in Ecology, School of Biological Sciences, University of Southampton, Southampton, England

Helena Freitas, Ph.D. Associate Professor, Department of Botany, University of Coimbra, Coimbra, Portugal

John A. Gamon, Ph.D. Associate Professor, Department of Biology and Microbiology, California State University–Los Angeles, Los Angeles, California

Eric Garnier, Ph.D. Center for Functional and Evolutionary Ecology, CNRS, Montpellier, France

José M. Gómez, Ph.D. Assistant Professor, Department of Animal Biology and Ecology, University of Granada, Granada, Spain

T. G. Allan Green, Ph.D. Department of Biological Sciences, Waikato University, Hamilton, New Zealand

Sarah E. Hobbie, Ph.D. Assistant Professor, Department of Ecology, Evolution, and Behavior, University of Minnesota, St. Paul, Minnesota

José A. Hódar, Ph.D. Assistant Professor, Department of Animal Biology and Ecology, University of Granada, Granada, Spain

William A. Hoffmann, Ph.D. Postdoctoral Fellow, Department of Botany, University of Texas–Austin, Austin, Texas

Robert B. Jackson, Ph.D. Assistant Professor, Department of Botany, University of Texas–Austin, Austin, Texas

Richard Joffre, Ph.D.   Researcher, Center for Functional and Evolutionary Ecology, CNRS, Montpellier, France

Ludger Kappen, Ph.D.   Professor of Botany, Botanical Institute, University of Kiel, Kiel, Germany

Kaoru Kitajima, Ph.D.   Assistant Professor, Department of Botany, University of Florida, Gainesville, Florida

Esteban Manrique-Reol, Ph.D.   Associate Professor, Department of Plant Biology II, Complutense University, Madrid, Spain

Ernesto Medina, Ph.D.   Professor and Senior Researcher, Center of Ecology, Venezuelan Institute for Scientific Investigations, Caracas, Venezuela

Robert W. Pearcy, Ph.D.   Professor, Section of Evolution and Ecology, Division of Biological Sciences, University of California, Davis, California

William T. Pockman, Ph.D.   Postdoctoral Fellow, Department of Botany, University of Texas–Austin, Austin, Texas

Hendrik Poorter, Ph.D.   Department of Plant Ecology and Evolutionary Biology, Utrecht University, Utrecht, The Netherlands

Francisco I. Pugnaire, Ph.D.   Researcher, Arid Zones Experimental Station, Spanish Council for Scientific Research, Almería, Spain

Juan Puigdefábregas, Ph.D.   Researcher, Arid Zones Experimental Station, Spanish Council for Scientific Research, Almería, Spain

Hong-lie Qiu, Ph.D.   Assistant Professor, Department of Geography and Urban Analysis, California State University–Los Angeles, Los Angeles, California

Serge Rambal, Ph.D.   Researcher, Center for Functional and Evolutionary Ecology, CNRS, Montpellier, France

Heather L. Reynolds*   Visiting Assistant Professor, W. K. Kellogg Biological Station, Hickory Corners, Michigan

---

* *Current affiliation*: Assistant Professor, Department of Biology, Kalamazoo College, Kalamazoo, Michigan.

Ronald J. Ryel Department of Range Resources and the Ecology Center, Utah State University, Logan, Utah

Leopoldo G. Sancho, Ph.D. Associate Professor, Department of Plant Biology II, Complutense University, Madrid, Spain

Burkhard Schroeter, Ph.D. Privatdozent, Botanical Institute, University of Kiel, Kiel, Germany

Anna V. Traveset, Ph.D. Research Ecologist, Department of Natural Resources, Mediterranean Institute of Advanced Studies, Spanish Council for Scientific Research, Palma de Mallorca, Spain

Melvin T. Tyree Aiken Forestry Sciences Laboratory, Burlington, Vermont

Fernando Valladares, Ph.D. Ecologist, Center of Environmental Sciences, Spanish Council for Scientific Research, Madrid, Spain

Mark Westoby, Ph.D. Professor, School of Biological Sciences, Macquarie University, New South Wales, Australia

S. Joseph Wright, Ph.D. Research Biologist, Smithsonian Tropical Research Institute, Balboa, Ancon, Republic of Panama

Regino Zamora, Ph.D. Associate Professor, Department of Animal Biology and Ecology, University of Granada, Granada, Spain

1

Methods in Comparative Functional Ecology

Carlos M. Duarte
Center for Advanced Studies of Blanes, Spanish Council for Scientific Research, Blanes, Spain

I. THE DEVELOPMENT OF FUNCTIONAL PLANT ECOLOGY

The quest to describe the diversity of the extant plants and the identification of the basic mechanisms that allow them to occupy different environments has been the center of scientists' attention from the time of ancient Greece to the present. This interest was prompted by two fundamental aims: (1) a pressing need to understand the basic functions and growth requirements of plants, because they provide direct and indirect services to humans; and (2) the widespread belief that the distribution of organisms was not random, for there was essential order in nature, and that the there ought to be a fundamental link between differences in the functions of these organisms and their dominance in contrasting habitats. The notion that differences in plant functions are essential components of their fitness, accounting for their relative dominance in differential habitats, was, therefore, deeply rooted in the minds of early philosophers and, later on, naturalists. Al-

1

though animal functions were relatively easy to embrace from a simple parallel with our own basic functions, those of plants appeared more inaccessible to our ancestors, and the concepts of plant and plant functions have unfolded through the history of biology.

The examination of plant functions in modern science has largely followed a reductionistic path aimed at the explanation of plant functions in terms of the principles of physics and chemistry (Salisbury and Ross 1992). This reductionistic path is linked to the parallel transformation of traditional agriculture science into plant science and the technical developments needed to evolve from the examination of the coarser, integrative functions to those occurring at the molecular level. Although this reductionistic path has lead us toward a thorough catalogue and understanding of plant functions, its limited usefulness to explain and predict the distribution of plants in nature has been a source of frustration. This is largely a result of the multiple interactions that are expected to be involved in the responses of plants to a changing environment (Chapin et al. 1987). Yet the need to achieve this predictive power has transcended the academic arena to become a critical component of our ability to forecast the large-scale changes expected from on-going climatic change. For instance, increased CO_2 concentrations are expected to affect the water and nutrient requirements of plants, but resource availability is itself believed to be influenced by increasing temperatures. Such feedback effects cannot be appropriately predicted from knowledge of the controls that individual factors exert on specific functions. Moreover, the changes expected to occur from climate change are likely to derive mostly from changes in vegetation and dominant plant types rather than from altered physiological responses of the extant plants to the new conditions (Betts et al. 1997).

Failure of plant physiology and plant science to provide reliable predictions of the response of vegetation to changes in their environment likely derive from the hierarchical nature of plants. The response of higher organizational levels are not predictable from the dynamics of those at smaller scales, although these set constraints on the larger-scale responses of hierarchical systems. The component functions do not exist in isolation, as the dominant molecular approaches in modern plant physiology investigate them. Rather, these individual functions are integrated within the plants, which can modulate the responses expected from particular functions, leading to synergism, whether amplifying the responses through multiplicative effects or maintaining homeostasis against external forces.

Recognition of the limitations of modern physiology to provide the needed predictions at the ecological scale led to the advent of plant ecophysiology, which tried to produce more relevant knowledge by the introduction of larger plant components, such as plant organs (instead of cells or organelles), as the units of analysis. Plant ecophysiology therefore represented an effort toward approaching the relevant scale of organization by examining the functions of plant organs. Most often, however, practitioners of the discipline fell somewhere between the molecular approaches dominant in plant physiology and the more integrative approach championed by plant ecophysiology. Because of the strong roots in the tradition of plant physiology, the suite of plant functions addressed by plant ecophysiology still targeted basic functions (e.g., photosynthesis, respiration, etc.) that can be studied through chemical and physical laws (Salisbury and Ross 1992). As a consequence, plant ecophysiology failed to consider more integrative plant functions, such as plant growth, which do not have a single physiological basis, but which are possibly the most relevant function for the prediction of plant performance in nature (see Chapter 3).

Therefore, the efforts of plant ecophysiology proved to be insufficient in predicting how plant function allows the prediction of plant distribution and changes in plant abundance in a changing environment. The realization that the knowledge required to effectively address this question would be best achieved through a more integrative approach led to the advent of a new approach, hereafter referred to as functional plant ecology, which is emerging as a coherent research program (Duarte et al. 1995). Functional plant ecology is centered on whole plants as the units of analysis, and their responses to external forcing are examined in nature or under field conditions. Functional plant ecology attempts to bypass the major uncertainties derived from the extrapolation of responses tested in isolated plant organs maintained under carefully controlled laboratory conditions to nature, and to incorporate the integrated responses to multiple stresses plants display onto the research program.

Although centered in whole plants, functional plant ecology encompasses lower and higher scales of organization, including studies at the organ or cellular level (see Chapter 8), as well as the effect of changes in plant architecture (see Chapters 4 and 5) and the importance of life history traits (see Chapters 15 and 16), interactions with neighbors (see Chapters 17 and 18), and those with other components of the ecosystem (see Chapter 19). In fact, this research program is also based on a much broader conception of plant functions than previously formulated. The plant functions that represent the core of present efforts in functional plant ecology are those by which plants influence ecosystem functions, particularly those that influence the services and products provided by ecosystems (Costanza et al. 1997). Hence, studies at lower levels of organization are conducted with the aim of being subsequently scaled up to the ecosystem level (see Chapter 10).

Because of the emphasis on the prediction of the consequences of changes in vegetation structure and distribution for the ecosystem, functional plant ecology strives to encompass the broadest possible range of functional responses encountered within the biosphere. Yet the elucidation of the range of possible functional responses of plants is not possible with the use of model organisms that characterize most of plant (and animal) physiology. Therefore, functional plant ecology arises as an essentially comparative science concerned with the elucidation of the range of variation in functional properties among plants and the search for patterns and functional laws accounting for this variation (Duarte et al. 1995). Although practitioners of functional plant ecology share the emphasis on the comparative analysis of plant function, the approaches used to achieve these comparisons range broadly. These differences rely largely in the breadth of the comparison and how the subject organisms are described in the analysis. However, the implications of these choices has not been the subject of explicit discussions despite their considerable epistemological implications and their impact on the power of the approach.

II. SCREENING, BROAD-SCALE COMPARISONS AND THE DEVELOPMENT OF FUNCTIONAL LAWS

The success and the limitations of comparative functional plant ecology depend on the choices of approach made, involving the aims and scope of the comparison, as well as the methods to achieve them. The aims of the comparisons range widely, from the compilation of a ''functional taxonomy'' of particular sets of species or floras to efforts to uncover patterns of functional properties that may help formulate predictions or identify possible controlling factors. Many available floras incorporate considerable knowledge, albeit rarely quantitative, on the ecology of the species, particularly as to habitat requirements. An outstanding example is the biological flora of the British Isles published in the *Journal of Ecology*, which incorporates some functional properties of the plants (e.g., Aksoy et al. 1998). The likely reason why ''functional'' floras are still few is the absence of standardized protocols to examine these properties while ensuring comparability of the results obtained. A step toward solving this was provided by Hendry and Grime (1993), who described a series of protocols to obtain estimates of selected basic functional traits of plants in a comparable manner. Unfortunately, while exemplary, those protocols were specifically designed for use within the screening program of the British flora conducted by those investigators (Grime et al. 1988), rendering them of limited applicability for broader comparisons or comparisons of other vegetation types.

If pursued further, the screening approach may generate an encyclopedic catalogue of details on functional properties of different plants. Some ecologist may hold the hope that, once completed, such catalogues will reveal a fundamental order in the functional diversity of the plants investigated, conforming a predictive rule similar to a ''periodic table'' of plant functional traits. Although I do not dispute here that this goal may eventually be achieved, the resources required to produce such catalogues are likely to be overwhelming, since by definition such a screening procedure is of a exploratory nature, where the search for pattern is made from observed facts. Given the number of elements to be screened and the potentially large number of traits to be tested, the cost-effectiveness of the approach is likely to prove suboptimal. Therefore, a screening approach to functional plant ecology is unlikely to improve our predictive power or to uncover basic patterns unless driven by specific hypotheses. Moreover, a hypothesis-driven search for a pattern is likely to be most effective if based on a comparative approach, encompassing the broadest possible relevant range of plants. It is not necessary to test every single plant species to generate and test such general laws.

The comparisons attempted may differ greatly in scope, from comparisons of variability within species to broad-scale comparisons encompassing the widest possible range of phototrophic organisms, from the smallest unicells to trees (Agustí et al. 1994, Nielsen et al. 1996). However, experience shows that the patterns obtained at one level of analysis may differ greatly from those observed at a broader level (Duarte 1990), without necessarily involving a conflict (Reich 1993). The scope of the comparison should depend on the question being posed. However, whenever possible, progress in comparative functional plant ecology should evolve from the general to the particular, thereby evolving from comparisons at the broadest possible scales to comparisons within species or closely related species. In doing so, we shall first draw the overall patterns that yield the functional laws to help identify the constraints of possible functional responses for the organisms.

The simplest possible comparison involves only two subjects, which are commonly enunciated under the euphemism of "contrasting" plant types. Such simple comparisons between one or a few subject plants are very common in the literature. However, these simple comparisons are deceiving because they cannot possibly be conclusive as to the nature of the differences or similarities identified. The implicit suggestion in these contrasts is that the trait on which the contrast is based (e.g., stress resistance vs. stress tolerance) is the cause underlying any observed differences in functional traits. This is, of course, fallacious and at odds with the simplest principles of method in science. Hence, contrasts are unlikely to be an effective approach to uncover regular patterns in plant function, since the degrees of freedom involved are clearly insufficient to venture any strong inferences from the outcome of the comparison.

Therefore, broad-scale comparison involving functional responses across widely different species is the approach of choice when the description of general laws is sought. The formulation of the comparative analyses of plant functions at the broadest possible level has been strongly advocated (Duarte et al. 1995) on the grounds that it will be most likely to disclose the basic rules that govern functional differences among plants. Broad-scale comparisons are most effective when encompassing the most diverse range of plant types possible (Agustí et al. 1994, Niklas 1994). In addition, they are most powerful when the functional properties are examined in concert with quantification of plant traits believed to influence the functions examined, because comparisons based on qualitative or nominal plant traits cannot readily be falsified and therefore remain, unreliable tools for prediction. Hence, the development of broad-scale comparisons requires the careful selection of both the functional property examined and the plant traits that account for the differences in functional properties among the plants to be tested.

Broad-scale comparisons must be driven by sound hypotheses or questions. Yet this approach is of a statistical nature, often involving allometric relationships (Niklas 1994), so that observation of robust patterns is no guarantee of underlying cause and effect relationships, which must be tested experimentally. Nevertheless, the functional laws developed through broad-scale comparative analysis may hold predictive power, regardless of whether they represent direct cause–effect relationships. However, if the law is to have practical application, the independent predictor variable must be more simple than the functional trait examined. There are many examples of such functional laws (Niklas 1994, Agustí et al. 1994, Duarte et al. 1995, Enríquez et al. 1996, Nielsen et al. 1996) that have been generally derived from the compilation of literature data and/or the use of plant cultures in phytotrons or the use of the functional diversity found, for instance, in botanical gardens (Nielsen et al. 1998). This choice of subject organisms is appropriate whenever the emphasis is on the functional significance of intrinsic properties. However, the effect of environmental conditions can hardly be approached in this manner, and the functional ecologist must transport the research to the field, which is the ultimate framework of relevance for this research program.

The comparative approach is also a powerful tool to examine the effect of environmental conditions in situ. Gradient analysis, where functional responses are examined along a clearly defined environmental gradient, has proven a powerful approach to investigate the relationship between plant function and environmental conditions (Vitousek and Matson 1990). Gradient analysis is particularly prone to spurious relationships. The relationship between the gradient property and the functional response may reflect a functional relationship to a hidden factor covarying with that nominally defining the gradient. Therefore, inferences from gradient analysis are also statistical in nature and must be confirmed experimentally to elucidate the nature of underlying relationships.

Broad-scale comparisons often entail substantial uncertainty (typically in the order-of-magnitude range) in their predictions, which is a result of the breadth (typically 4 or more orders of magnitude) in the functions examined. This imprecision limits the applicability of these functional laws and renders their value greatest for the description of the general, large-scale patterns over which the effect of less-general functional regulatory factors, both intrinsic and extrinsic, are superimposed. Hence, the multiple factors that constrain the functional responses of plants are nested in a descending rank of generality, whereby the total number of traits involved in the control is very large, but only a few of them are general across a broad spectrum of plants.

The nested nature of the control of functional responses implies uncertainties when scaling functional laws, either toward lower or higher levels of organization (Duarte 1990). Therefore, there is no guarantee that the patterns observed at the broad-scale level will apply when focusing on particular functional types. Changes in the nature of the patterns when shifting across scales have prompted unnecessary disagreement in the past (Reich 1993). Whenever possible, the thorough investigation of functional properties of plants should include a nested research program, whereby the hypotheses on functional controls examined are first investigated at the broadest possible scale, with subsequent focus on particular subsets of species or functional groups along environmental gradients.

The chapters in this volume provide a clear guide to functional ecology by example, emphasizing the nested nature of the research program both within the chapters and in the manner they have been separated into different parts. The chapters also provide an overview of the entire suite of approaches available to address the goals of functional ecology, thereby providing a most useful tool for the prospective practitioners of the research program. The result is a heuristic description of functional ecology that should serve the dual role of providing a factual account of the achievements of functional ecology, while endowing the reader with the tools to design research within this important research program.

REFERENCES

Agustí, S., S. Enríquez, H. Christensen, K. Sand-Jensen, and C.M. Duarte. 1994. Light harvesting among photosynthetic organisms. Functional Ecology 8: 273–279.

Aksoy, A., J.M. Dixon, and W.H.G. Hale. 1998. *Capsella bursa-pastoris* (L.) Medikus (Thlapsi bursa-pastoris L., *Bursa bursa-pastoris* (L.). Shull, *Bursa pastoris* (L.) Weber). Journal of Ecology 86: 171–186.

Betts, R.A., P.M. Cox, S.E. Lee, and F.I. Woodward. 1997. Contrasting physiological and structural vegetation feedbacks in climate change simulations. Nature 387: 796–799.

Cebrián, J., and C.M. Duarte. 1994. The dependence of herbivory on growth rate in natural plant communities. Functional Ecology 8: 518–525.

Chapin F.S. III, A.J. Bloom, C.B. Field, and R.H. Waring. 1987. Plant responses to multiple environmental factors. Bioscience 37: 49–57.

Costanza R., R. d'Arge, R. de Groo, S. Farber, M. Grasso, B. Hannon, K. Limburg, S. Naeem, R.V. O'Neill, J. Paruelo, R.G. Raskin, P. Sutton, and M. van der Belt. 1997. The value of the world's ecosystem services and natural capital. Nature 387: 253–260.

Duarte, C.M. 1991. Variance and the description of nature. In:Cole, J.J., G. Lovett, and S. Findlay, eds. Comparative Ecology of Ecosystems: Patterns, Mechanisms, and Theories. New York: Springer-Verlag, pp 301–318.

Duarte, C.M., K. Sand-Jensen, S.L. Nielsen, S. Enríquez, and S. Agustí. 1995. Comparative functional plant ecology: rationale and potentials. Trends in Ecology and Evolution 10: 418–421.

Enríquez, S., S.L. Nielsen, C.M. Duarte, and K. Sand-Jensen. 1996. Broad-scale comparison of photosynthetic rates across phototrophic organisms. Oecologia (Berlin) 108: 197–206.

Grime, G.P., J.G. Hodgson, and R. Hunt. 1988. Comparative Plant Ecology. Boston: Unwin Hyman.

Hendry, G.A.F., and J.P. Grime. 1993. Methods in Comparative Plant Ecology. A Laboratory Manual. London: Chapman and Hall.

Nielsen, S.L., S. Enríquez, and C.M. Duarte. 1998. Control of PAR-saturated CO_2 exchange rate in some C_3 and CAM plants. Biologia Plantarum 40: 91–101.

Nielsen, S.L., S. Enríquez, C.M. Duarte, and K. Sand-Jensen. 1996. Scaling of maximum growth rates across photosynthetic organisms. Functional Ecology 10: 167–175.

Niklas, K.J. 1994. Plant Allometry. The Scaling of Form and Process. Chicago: The University of Chicago Press.

Salisbury, F.B., and C.W. Ross. 1992. Plant Physiology. 4th Ed. Belmont, Cal: Wadsworth.

Vitousek, P.M., and P.A. Matson. 1990. Gradient analysis of ecosystems. In:Cole, J.J., G. Lovett, and S. Findlay, eds. Comparative Ecology of Ecosystems: Patterns, Mechanisms, and Theories. New York: Springer-Verlag, pp 287–298.

2
Opportunistic Growth and Desiccation Tolerance: The Ecological Success of Poikilohydrous Autotrophs

Ludger Kappen
University of Kiel, Kiel, Germany

Fernando Valladares
Center of Environmental Sciences, Spanish Council for Scientific Research, Madrid, Spain

I. THE POIKILOHYDROUS WAY OF LIFE

Poikilohydry, or the lack of control of water relations, has typically been a subject studied by lichenologists and bryologists. For many years, much was unknown about poikilohydrous vascular plants, and evidence for their abilities was mostly anecdotal. A small number of these plants were studied by a few physiologists and ecologists who were fascinated by the capability of these ''resurrection plants'' to quickly switch from an anabiotic to a biotic state and vice versa (Pessin 1924, Heil 1925, Walter 1931, Oppenheimer and Halevy 1962, Kappen 1966, Vieweg and Ziegler 1969). Recently, a practical demand has released an unprecedented interest in poikilohydrous plants. The increasing importance of developing and improving technologies for preserving living material in the dry state for breeding and medical purposes has induced tremendous research activity aimed at uncovering the molecular and biochemical basis of desiccation tolerance. Poikilohydrous plants have proven to be very suitable for exploring the basis of this tolerance (Stewart 1989, Oliver and Bewley 1997). Consequently, much of the current literature discusses poikilohydrous plants mainly as a means of explaining basic mechanisms of desiccation tolerance (Bewley and Krochko 1982, Oliver and Bewley 1997, Hartung et al. 1998) instead of exploring their life history and ecology.

    Many new resurrection plants have been discovered during the last 20 years, especially in the southern hemisphere (Gaff 1989), and have provided new insights into the biology of these organisms. In this chapter, structural and physiological features of poikilohydrous autotrophs, and the different strategies in different ecological situations are discussed. The desiccation tolerance itself is covered briefly here because it has been extensively treated in recent reviews (Leopold 1990, Proctor 1990, Oliver and Bewley 1997, Hartung et al. 1998). Our main goal is to assess the ecological success of poikilohydrous autotrophs. We give special attention to the productivity of poikilohydrous autotrophs, how they manage to live in extreme environments, the advantages of their opportunistic growth, and in the case of vascular pokilohydrous plants, what happens to the vascular system during desiccation and resurrection.

A. Poikilohydrous Constitution Versus Poikilohydrous Performance: Toward a Definition of Poikilohydry

According to Walter (1931), poikilohydry in plants can be understood as being analogous to poikilothermy in animals. The latter show variations of their body temperature as a function of ambient temperature, whereas poikilohydrous autotrophs (plants) exhibit variations of their hydration levels as a function of ambient water status (Walter and Kreeb 1970). The term *autotroph* is used here to comprise an extensive and heterogenous list of autotrophic unicellular and multicellular organisms (cyanobacteria, algae, bryophytes, and vascular plants), including the lichen symbiosis. Poikilohydrous performance (from the Greek words *poikilos*, changing or varying, and *hydor*, water) is applied to organisms that passively change their water content in response to water availability (''hydrolabil''; Stalfelt 1939), eventually reaching a hydric equilibrium with the environment. This fact does not necessarily imply that the organism tolerates complete desiccation. There is not a general consensus on the definition of poikilohydrous plants. The greek word *poikilos* also means malicious, which may apply to the difficulty of comprising the outstanding structural and functional heterogeneity of this group of autotrophs.

Nearly all life forms of plants are represented by poikilohydrous species except trees (Table 1). Poikilohydrous or resurrection plants have been found in all continents of the world, although knowledge about these plants in Asia is very poor. Although it is difficult to be precise about the number of poikilohydrous nonvascular species, we can be more exact about vascular plants. According to Gaff (1989), we know approximately 82 pteridophyte species and 106 species of angiosperms belonging to only nine families that can be considered as poikilohydrous. Most of the angiosperm species are monocotyledonous. In a revision of the African Lindernieae (Scrophulariaceae), Fischer (1992) identified at least 17 of a total of 80 species that were poikilohydrous. *Isoetes australis* and seven species of *Borya* (Liliaceae) were recorded by Lazarides (1992) as poikilohydrous plants. Thus, a total of approximately 200 species of vascular plants are known at present to be poikilohydrous. Some families such as the Vellociaceae and the genera *Cheilanthes, Tripogon, Borea*, and *Lindernia* are particularly rich in poikilohydrous species, suggesting an early specialization (Gaff 1989); this might also be the case for some small genera such as *Ramonda, Haberlea*, and *Myrothamnus*.

Nonvascular autotrophs (cyanobacteria, algae, bryophytes, and lichens) are considered constitutively poikilohydrous because they lack means of controlling

Table 1 Desiccation Tolerance of Isolated Chloroplasts and of Poikilohydrous Autotrophs[a]

Species	Degree of desiccation survived	Time of drought survived	Reference
Isolated chloroplasts (*Beta*)	15% RWC	Until equilibrium	Santarius 1967
Nostoc and *Chlorococcum*	Air-dry	73 years	Shields and Durrell 1964
Lichens			
Lichens (about 50 species)	2–9% d.wt.	38–78 wk air-dry	Lange 1953
		24–56 wk P$_2$O$_5$	
Pseudocyphellaria dissimilis	45–65% rh	8–10 hr until equilibrium	Green et al. 1991
Bryophytes			
Ctenidium molluscum	<10% rh in winter	40 hr	Dircksen 1964
Dicranum scoparium	<10% rh in winter	40 hr	Dircksen 1964
Epiphytic mosses (14 species)	0–30% rh	Until equilibrium	Hosokawa & Kubota 1957
Fissidens cristatus	<10% rh in winter	40 hr	Dircksen 1964
Mnium punctatum	<10% rh in winter	40 hr	Dircksen 1964
Rhacomitrium lanuginosum	32% rh	239 d	Dilks and Proctor 1974
Sphagnum sp.	Air-dry	<5 d	Wagner and Titus 1984
Exormotheca holstii	Air-dry	8 mo	Dinter 1921 (s. Hellwege et al. 1994)
Riccia canescens (bulbils)	Air-dry	7 yr	Jovet-Ast 1961
Pteridophytes			
Selaginella lepidophylla	Air-dry	About 1 yr	Eickmeier 1979
Isoetes australis	0% rh	Until equilibrium	Gaff and Latz 1978
Asplenium ruta-muraria	5–10% RWC	1–3 d H$_2$SO$_4$ (winter)	Kappen 1964
Asplenium septentrionale	7–20% RWC	1–3 d H$_2$SO$_4$ (winter)	Kappen 1964
Asplenium trichomanes	7–15% RWC	1–3 days H$_2$SO$_4$ (winter)	Kappen 1964

Species	Water content	Duration	Reference
Camptosorus rhizophyllus (gametophyte)			Picket 1914
Ceterach officinarum	Air-dry (4% RWC)	H_2SO_4 5 mo	Rauschal 1938, Oppenheimer and Halevy 1962
Cheilanthes (8 species)	2–20% rh	Until equilibrium	Gaff and Latz 1978
Notholaena maranthae	6%RWC		Iljin 1931
Paraceterach sp.	15% rh	Until equilibrium	Gaff and Latz 1978
Pellea (2 species)	2–30% rh	Until equilibrium	Gaff and Latz 1978
Pleurosorus rutifolius	2% rh	Until equilibrium	Gaff and Latz 1978
Polypodium polypodioides	3% RWC	50 hr	Stuart 1968
Polypodium vulgare	3–10% RWC	10 d P_2O_5 (winter)	Kappen 1964
Polystichum lobatum	7–10% RWC	24 hr (air) winter	Kappen 1964
Gametophytes of ferns (5 species)	20–65 rh	36 hr	Kappen 1965
Angiosperms			
Dicotyledons			
Ramonda serbica	Air-dry	6–12 mo	Markowska et al 1994
Ramonda myconi	2% RWC	2 d	Kappen 1966
Haberlea rhodopensis			
Boea hygroscopia	0% rh	Until equilibrium	Gaff and Latz 1978
Chamaegigas intrepidus	Dry desert soil	10 mo	Heil 1925
Chamaegigas intrepidus, floating leaves	96% rh	Until equilibrium	Gaff 1971
Chamaegigas intrepidus, submerged leaves	5% rh	4.5 mo	Gaff 1971
Craterostigma (2 species)	0–15% rh	Until equilibrium	Gaff 1971
Limosella grandiflora (corms)	5% rh	4.5 mo (but both species decayed at 100% rh)	Gaff and Giess 1986
Myrothamnus flabellifolius	Air-dry, 0%	Leaves "several yr"	Ziegler and Vieweg 1969, Gaff 1971
Blossfeldia liliputana	18% of initial weight	33 mo	Barthlott and Porembski 1996

Table 1 Continued

Species	Degree of desiccation survived	Time of drought survived	Reference
Monocotyledons			
Poaceae from India (10 species)	0–2 (11)% rh	3 mo	Gaff and Bole 1986
Southern African Poaceae (11 species)	0–5% rh	2–7 mo	Gaff and Ellis 1974
Poaceae from Africa and Kenya (5 species)	0–15% rh	Until equilibrium	Gaff and Latz 1978
Trilepis pilosa (African Inselberg)	8% RWC	Up to 1 yr	Hambler 1961
Coleochloa setifera	Air-dry	5 yr	Gaff 1977
Oropetium sp.	0–15% rh	Until equilibrium	Gaff 1971
Australian Poaceae (6 species)	0–15% rh	Until equilibrium	Gaff and Latz 1978
Australian Liliaceae (2 species)	0–5% rh	Until equilibrium	Gaff and Latz 1978
Borya nitida	air-dry (c. 0% rh)	>4 yr	Gaff and Churchill 1976
Xerophyta (5 species), *Coleochloa* sp	0–15% rh	Until equilibrium	Gaff 1971
Xerophyta squarrosa	Air-dry	5 yr	Gaff 1977
Xerophyta scabrida	Air-dry	5 yr	Csintalan et al 1996
Australian Cyperaceae (1 species)	0–2% rh	Until equilibrium	Gaff and Latz 1978
Southern African Cyperaceae (4 species)	0–5% rh	27 mo	Gaff and Ellis 1975
Cyperaceae from Africa and Kenya (3 species)	Tissues: 0–5% rh		Gaff 1986a
Velloziaceae from Africa and Kenya (2 species)	Mature leaf tissues: 5–30% rh		Gaff 1986a

[a] The list in this table is not exhaustive.
RWC, relative water content; d.wt., dry weight; rh, relative humidity; until equilibrium, until equilibrium between moisture content and ambient air relative humidity.

water relations (Stocker and Holtheide 1938, Biebl 1962, Walter and Kreeb 1970). This is in contrast with vascular plants, which are constitutively homoiohydrous organisms, i.e., able to keep their hydration state within certain limits by such means as roots, conducting tissues, epidermis, cuticles, and stomata. Nonetheless, nearly 200 vascular plant species are presently known to not make use of their homoiohydrous potential, but perform as poikilohydrous plants (Bewley and Krochko 1982). Thus, poikilohydrous performance as an acquired trait can be found in plants that are phylogenetically unrelated. Because poikilohydry is constitutional in nonvascular autotrophs and rare among vascular plants, it is tempting to consider it a primitive property, and to suggest that early terrestrial, photosynthetic organisms based their survival on tolerance instead of avoidance mechanisms. However, poikilohydry is not a primitive trait among vascular plants. Although several pteridophytes are poikilohydrous, there is no known poikilohydrous gymnosperm, and pokilohydry is more frequent in highly derived than in primitive angiosperm families, as was illustrated by Oliver and Bewley (1997). Poikilohydrous performance by vascular plants can be interpreted evolutionarily as an adaptive response to climates with infrequent moist periods.

The term *resurrection* has been commonly used for some species and, in general, matches well with the capability of poikilohydrous plants to quickly reactivate after falling into a period of anabiosis caused by dehydration. It is very appropriate for spikemosses (*Selaginella*) and certain bryophytes and lichens that curl strongly with water loss and unfold conspicuously upon rehydration. Similar behavior can be observed in the dead remnant of the desert annual *Anastatica hierochuntica* and other desert plants. *A. hierochuntica* was called a resurrection plant by some investigators (Wellburn and Wellburn 1976) because of the dramatic change between a curled and/or shrivelled stage in the dry season and the spreading of the branches in the rainy season. Consequently, *resurrection*, in a broad, intuitive sense, could also be applied to certain homoiohydrous desert perennials (e.g., *Aloe*, Mesembryanthemaceae, and certain cacti). On the other hand, shape and appearance of some constitutively poikilohydrous autotrophs, such as terrestrial unicellular algae and crustose lichens, do not visibly change. To add to the confusion, water loss can be dramatic in some homoiohydrous desert plants, whereas it can be minor in constitutively poikilohydrous plants such as *Hymenophyllum tunbridgense* or bryophytes and lichens from moist environments. Therefore, the resurrection phenomenon (visible changes in shape and aspect with hydration) is only part of the poikilohydrous performance and it is not exhibited to the same extent by all poikilohydrous autotrophs.

By shifting between functioning with larger dolichoblasts in the rainy season and small, extremely desiccation-tolerant brachyblasts in the dry season, the small shrub *Satureja gilliesii* can reduce its transpiring leaf surface (Montenegro et al. 1979). The brachyblasts have a mesophytic anatomy and are covered by filamentous trichomes. Among the plants that can live in ephemeral African rock pools, *Aponogeton desertorum* loses its leaves when dry but remains preserved by a desiccation-tolerant rhizome. Similarly, southern African *Limosella* species persist partly with desiccation-tolerant corms (Gaff and Giess 1986), and perhaps other Scrophulariaceae may perform likewise. The performance of *Satureja* and these other taxa may be considered as transiently poikilohydrous, since only a part, albeit major, of the individual acts as a resurrection plant. The difference from homoiohydrous plants is small.

Among the vascular poikilohydrous plants, ferns are dual because they produce constitutively poikilohydrous gametophytes, and a cormophytic sporophyte with the full anatomy of a homoiohydrous plant. Knowledge about the gametophytes is scant. They are usually found in very moist, sheltered habitats where hygric and mesic bryophytes also grow. Previous literature reports on extremely desiccation-tolerant prothalli of the North American *Camptosorus rhizophyllus*, and of *Asplenium platyneuron* and *Ceterach officinarum* (= *Asplenium ceterach*) (Walter and Kreeb 1970). The desiccation tolerance of prothallia of some European fern species varied with species and season (Kappen 1965). They usually overwinter, and their ability to survive low temperatures and freezing is based on increased desiccation tolerance. Prothallia of rock-colonizing species (*Asplenium* species, *Polypodium vulgare*) could withstand 36 hours drying in 40% relative humidity; some species were partly damaged but could regenerate from surviving tissue. Prothallia of other ferns from European forests were more sensitive to desiccation (Table 1).

The poikilohydrous nature of a terrestrial vascular plant is frequently defined by the combination of a passive response to ambient water and a tolerance to desiccation (Gaff 1989), but the emphasis on the different functional aspects involved and the actual limits of poikilohydry are matters of debate. Some poikilohydric species cannot even tolerate a water loss greater than 80% of their maximal water content (Gaff and Loveys 1984), and poikilohydric autotrophs of the transient type can be shown to gain their tolerance only by a preconditioning procedure. Boundaries between poikilohydrous and homoiohydrous plants can be rather blurry, especially if we include examples of xerophytes that can survive extremely low water potentials (Kappen et al. 1972). Another frequently used indicator for poikilohydry is the capacity to survive at very low relative humidities. Again, this is not a useful indicator because a limit of 0–10% relative humidity excludes many nonvascular plants that are undoubtedly poikilohydrous. Considering the low tolerance to desiccation of certain forest lichens (Green et al. 1991) and the fact that lichens can dominate in more xeric habitats while most bryophytes prefer moist environments, Green and Lange (1994) concluded that poikilohydry represents a passive response to ambient water, and that the minimal water potential tolerated is species and environment specific.

The conflict between ecologically based and physiologically or morphologically based criteria cannot be easily solved. However, a compromise can be reached by distinguishing between stenopoikilohydrous (narrow range of water content) and eurypoikilohydrous (broad range of water content) autotrophs. This distinction is especially useful for nonvascular, i.e., for constitutively poikilohydrous autotrophs. For instance, microfungi that spend all their active lifetime within a narrow range of air humidity are stenopoikilohydrous. These xeric species grow in equilibrium with relative humidities as low as 60% (Pitt and Christian 1968, Zimmermann and Butin 1973). Aquatic algae and cyanobacteria as well as the so-called hygric and mesic bryophytes and filmy ferns that are not able to survive drying to less than 60% water content or less than 95% relative humidity also belong to the stenopoikilohydrous type.

All nonvascular and vascular species that are extremely tolerant to desiccation and typically perform as resurrection plants (Gaff 1972, 1977, Proctor 1990) belong to the eurypoikilohydrous group. Because many of these species grow in dry or desert environments, poikilohydry was often associated with xerophytism (Hickel 1967, Patterson 1964, Gaff 1977). Seasonal changes in the tolerance to desiccation can confound this distinction between stenopoikilohydrous and eurypoikilohydrous organisms. These changes have been found in bryophytes (Dilks and Proctor 1976b) and ferns (Kappen 1964) and are very likely to occur in angiosperms. As suggested by Kappen (1964), such plants may be considered as temporarily poikilohydrous. Hence, the number of eurypoikilohydrous bryophyte species cannot be fixed until temporal changes of desiccation tolerance are better studied in mesic species (Proctor 1990, Davey 1997). Most of the available information and, consequently, most of what follows involves eurypoikilohydrous autotrophs. The different groups of poikilohydric autotrophs that can be identified according to the range of water contents experienced in nature or tolerated are summarized later. The classification becomes clearer by considering homoiohydrous plants, which can also be separated in two groups according to their tolerance to changes in their water status.

B. Ecology and Distribution of Poikilohydrous Autotrophs

Nowhere else in the world are poikilohydrous autotrophs more conspicuous than in arid and climatically extreme regions. It is somewhat paradoxical that precisely in habitats with extreme water deficits the dominant organisms are the least protected against water loss. With the exception of some Australian poikilohydrous grasses, sedges, and the Xanthorhoeaceae species *Borya nitida* (Gaff and Churchill 1976, Lazarides 1992), poikilohydrous plants have no xeromorphic properties, but they can compete well with extremely specialized taxa of homoiohydrous plants. But again, the distinction between stenopoikilohydrous and eurypoikilohydrous plants becomes important, because stenopoikilohydrous autotrophs can be very abundant in moist habitats. In fact, mosses and lichens can reach their largest possible dimensions and their fastest growth rates under favorable, moist conditions. For instance, in the moist and misty climate of San Miguel, Azores, *Sphagnum* species are able to grow as epiphytes on small trees. However, eurypoikilohydrous autotrophs, which are capable of enduring prolonged drought and extreme temperatures, represent the most interesting group because they have more specifically exploited the ecological advantages of their opportunistic strategy. The remainder of this chapter presents examples of pokilohydrous autotrophs living under very limiting ecological conditions in many different regions of the earth.

Aerophytic algae, bryophytes, and lichens form the great majority of the poikilohydrous autotrophs in temperate and cold climates. Depending on their habitat, bryophytes can be eurypoikilohydrous or stenopoikilohydrous. Vascular plants are rarely subjected to extreme water stress in such climates. Eurypoikilohydrous vascular plants in Europe are in the genera *Ramonda* and *Haberlea* and in the rock-colonizing fern genera *Ceterach, Cheilanthes*, and *Notholaena*. In contrast to most lichens, bryophytes can live as therophytes (Longton 1988a). Many bryophytes survive the desert by a drought evasion strategy based on prolific sporophyte production. They germinate after heavy rain and then quickly develop gametophytes and sporogons. Some examples of this strategy are *Riella, Riccia*, and species of Sphaerocarpales, Pottiaceae, and Bryobatramiaceae. These annual shuttle species are characteristic of seapage areas and pond margins where the soil remains moist for a few weeks (Volk 1984).

Antarctica, climatically the most extreme continent on earth, is solely inhabited by algae, bryophytes, lichens and fungi, which are constitutively poikilohydrous. Vascular plants are limited to the mildest part of the Antarctic peninsula and surrounding islands (Longton 1988b, Smith 1984). In polar regions and in the most arid deserts, algae, cyanobacteria, lichens, and a few bryophytes colonize clefts and fissures of rocks, or even grow inside the rock as endolithic organisms or hypolithic on the underside of more or less translucent rock particles and stones (Friedmann and Galun 1974, Danin 1983, Scott 1982, Kappen 1988, 1993b, Nienow and Friedmann 1993).

Bryophytes, algae, and lichens are well known as crust-forming elements on desert soils, where they can accumulate considerable biomass (Shields et al. 1957, Rogers and Lange 1971, Scott 1982, Belnap 1994, Lange et al. 1994, 1997). A particular phenomenon in semiarid regions is the occurrence of erratic or vagrant poikilohydrous species such as lichens (Kappen 1973, Perez 1997) and bryophytes (Scott 1982). The coastal Namib desert, with extremely scattered rainfall, consists of wide areas where no vascular plants can be found, but a large cover of mainly lichens forms a prominent vegetation. In rocky places of southern Africa, arid northwest North America, coastal southwest North America, and the Negev desert, lichens and bryophytes coexist with xeromorphic or succulent plants. They occupy rock surfaces and places where vascular plants do not find enough water, or they grow as epiphytes on cacti. Under such extreme conditions, lichens share the habitat with poikilohydrous vascular plants. This is the case with the lichen *Siphula* and the resurrection plant *B. nitida*, which grow on granitic outcrops (Figure 1) with shallow soil cover in southern and western Australia (Gaff and Churchill 1976).

Many poikilohydrous vascular plants exploit temporal ponds or runoffs. Volk and Leippert (1971) described poikilohydrous communities of bryophytes and vascular plants on granitic hills that are temporarily flooded in Namibia. These communities (*Xerophytetum humilis*; Volk 1984) consist of *Riccia* and *Bryum* species, cyanobacteria, *Barbacenia minuta* (=*X. humilis*) and, locally, *Craterostigma plantagineum* associated with some succulents and small grasses. Similar phanerogamic communities of poikilohydrous Scrophulariaceae have been found in savannah regions of Africa (Fischer 1992). *C. plantagineum, C. hirsutum, C. lanceolatum,* and *Lindernia philcoxii* together with *Riccia* and Cyperaceae form a characteristic community (*Craterostigmetum nanolanceolati*) on lateritic crusts in Zaire (Lebrun 1947). Particularly remarkable are poikilohydrous aquatic *Lindernia* species (*L. linearifolia, L. monroi, L. conferta*) and *Chamaegigas (Lindernia) intrepidus* (Heil 1925, Hickel 1967, Gaff and Giess 1986) that

Figure 1   Two very different examples of poikilohydrous autotrophs co-occurring on a shallow depression of a granite outcrop near Armadale, western Australia: the monocotyledonous plant *Borya nitida* (center) and the fruticose lichen *Siphula* sp. (to the right of *B. nitida*) (Photograph by L. Kappen, 1981).

grow in small temporarily water-filled basins of granitic outcrops in Africa (Angola, Zaire, Zimbabwe, South Africa, Namibia). Describing communities of the *Nanocyperion teneriffae*, Volk (1984) states that they are typical of a pioneer vegetation that combines the strategies of therophytes and resurrection plants.

Typical habitats for poikilohydrous vascular plants are rock fissures and cavities. Subfruticose poikilohydrous plants such as *Lindernia crassifolia* (20 cm) and *L. acicularis* (6–10 cm) grow in sheltered rock niches (Fischer 1992). The same is true for the largest (up to 1 m) fruticose poikilohydrous species, *Myrothamnus flabellifolius* from Namibia (e.g., the Erongo mountains; Child 1960, Puff 1978), which is frequently associated with other resurrection plants (*Pellaea viridis, P. calomelanos*). In the wet season, these plants benefit from run-off water that floods the shallow ground (Child 1960). A large number of poikilohydrous rosette plants such as the Gesneriaceae *Ramonda* and *Haberlea* from Europe, and in particular most of the poikilohydrous fern species so far known, are preferentially found in rock cervices, gullies, or sheltered in shady rocky habitats (Gildner and Larson 1992, Nobel 1978). By contrast, the famous resurrection plant *Selaginella lepidophylla* colonizes open plains in Texas (Eickmeier 1979, 1983). Many of the poikilohydrous grass species (< 20 cm high) and sedges (30–50 cm high) are pioneering perennial plants colonizing shallow soil pans in southern Africa (Gaff and Ellis 1974). In Kenya and West Africa, the resurrection grasses, sedges, and Vellociaceae are confined to rocky areas, except *Sporobolus fimbriatus* and *S. pellucidatus. Eragrostis invalida* is the tallest poikilohydrous grass species known with a foliage up to 60 cm (Gaff 1986). *Vellozia schnitzleinia* is a primary mat former following algae and lichens on shallow soils of African inselbergs, persisting during the dry season with brown, purple-tinged rolled leaves that turn green in the wet season (Owoseye and Sandford 1972).

The resurrection flora of North and South America is similar, except for *Vellociaceae*, which do not occur in North America. Gaff (1987) has enumerated 12 fern species and 24 phanerogam species, 20 of which were *Vellociaceae*. One of the most remarkable poikilohydrous vascular plants could be *Blossfeldia liliputana*, a tiny cactus that grows in shaded rock crevices of the eastern Andean chain (Bolivia to northern Argentine) at altitudes between 1200 and 2000 m (Barthlott and Porembski 1996). This plant is unable to maintain growth and appearance during periods of drought, and it persists in the dry state (18% of initial weight) for 12–14 months looking like a piece of paper. When water is again available, it can rehydrate and resume CO_2 assimilation within 2 weeks, being the only known example of a succulent poikilohydrous plant.

From a plant–geographical perspective, southern Africa is the region of the World that has the largest diversity of poikilohydrous vascular plants. Despite the existence of similar potential habitats for poikilohydrous vascular plants in Australia, species are less numerous there than in southern Africa. Lazarides (1992) suggested that this biogeographical difference between Australia and southern Africa is due to the fact that the Australian arid flora has been exposed to alternating arid and pluvial cycles for a shorter geological period of time than the arid flora of southern Africa (the former has experienced this alternation since the Tertiary, whereas the latter has been exposed to dry–wet cycles since the Cretaceous). Ferns, represented by a relatively large number of species [14] and most of the poikilohydrous grasses found in Australia [10] grow in xeric rocky sites (Lazarides 1992). We have no records about poikilohydrous vascular plants from central and eastern Asia, although such type of plant must exist there as well. For instance, the Gesneriaceae *Boea hygrometrica* in China, closely related to the Australian *B. hygroscopia*, may most likely also be poikilohydrous.

C.   Does Poikilohydry Rely on Specific Morphological Features?

Poikilohydrous performance cannot be typified by any one given set of morphological and anatomical features because the heterogeneity of this functional group of photosynthetic organisms. Poikilohydry can be found in autotrophs ranging from those with the most primitive unicellular or thallose organization to those with the most highly derived vascular anatomy. In angiosperms, poikilohydry is mainly considered a secondary development, and their desiccation tolerance is, in general, inversely related to anatomical complexity. It seems that plants can operate either avoidance or tolerance mechanisms at all levels of organization if they are adapted to temporarily dry habitats. Gaff (1977) called resurrection plants "true xerophytic" just because they live in xeric environments. But poikilohydrous angiosperms do not necessarily have xeromorphic traits. Xeromorphic features such as small, leathery leaves (*Myrothamnus*), needle-like leaves (*Borya nitida*; Figure 1), massive sclerenchymatic elements (*Xerophyta* species and *Borya*), hairs, and scales, scleromorphy and succulence are the exception rather than the rule in poikilohydrous vascular plants. Xeromorphic structures would also counteract the potential of rehydration during the wet period.

Most of the approximately 120 resurrection phanerogams do not show anatomical features peculiar to them. Thus, it is hard to decide whether a particular plant is poikilohydrous just from herbarium material or from short-term observations in the field (Gaff and Latz 1978). It is still uncertain, for instance, whether more members of the Lindernieae can be identified as poikilohydrous in studies such as that by Fischer (1992). Many species that grow in shady habitats or that colonize temporarily inundated habitats exhibit a hygromorphic tendency (Markowska et al. 1994, Volk 1984, Fischer 1992). For instance, *Chamaegigas intrepidus* has, like other aquatic plants, aerenchyma and two types of leaves, floating and submerged. *Blossfeldia liliputana*, the only known poikilohydrous Cactaceae, combines a succulent habit with a typically hygromorphic anatomy: very thin cuticle, no thickened outer cell walls, absence of hypodermal layers, and extremely low stomatal density (Barthlott and Porembski 1996). Poikilohydrous vascular plants exhibit, in general, very low stomatal control of transpiration (Gebauer 1986). The leaves of *Satureja gilliesii* even have protruding stomata on the underside (Montenegro et al. 1979).

II. EXPLOITING AN ERRATIC RESOURCE

Water is evasive in many terrestrial habitats, and plants in general have to deal with the changing availability of this crucial resource. This is especially true for poikilohydrous autotrophs, which have successfully explored many different strategies within their general tolerance to water scarcity. However, some of the features that make tolerance to desiccation possible are irreconcilable with those that enhance water use. Poikilohydrous autotrophs, therefore, have had to trade off between surviving desiccation against uptake, transport, and storage of water. Some adaptive conflicts appear, for instance, when a particular feature retards water loss. The important functional problems that arise when the plant has to resume water transport after desiccation might have limited the range of growth forms and plant sizes compatible with poikilohydry.

A. Different Modes of Water Uptake and Transport

Plants must be efficient in acquiring water, particularly in arid regions where rainfall is scarce and sometimes the only available water comes from dew or is in the air in the form of mist or fog. Poikilohydrous plants can outcompete their homoiohydric counterparts in dry habitats if they can rehydrate efficiently. The following section describes the different possibilities for water capture exhibited by poikilohydrous autotrophs, with emphasis on the role of the growth form and of the morphology and anatomy of the structures involved.

   Aerophytic algae and lichens with green-algal photobionts can take up enough water from humid atmospheres to become metabolically active (Bertsch 1996a, b, Blum 1973, Lange 1969b, Lange and Kilian 1985, Lange et al. 1990a). Rehydration in lichens from humid atmospheres may take 1–4 days until equilibrium, whereas mist and dew fall yield water saturation within hours (Kappen et al. 1979, Lange and Redon 1983, Lange et al. 1991). As a consequence, lichens in deserts can survive well with sporadic or even no rainfall (Kappen 1988, Lange et al. 1990c, 1991). Anatomical structures such as long cilia, rhizines, branching, or a reticulate thallus structure are characteristic of lichens from fog deserts (e.g., *Ramalina melanothrix, Teloschistes capensis, Ramalina menziesii*), suggesting that these structures are means for increased water absorption (Rundel 1982a). In lichens, liquid water is absorbed by the entire body (thallus), usually within a few minutes (Blum 1973, Rundel 1982a, 1988). The thallus swells and can unfold lobes or branches. No special water storage structures have been described for lichens, and there is little evidence of a water transport system in these organisms (Green and Lange 1994). However, not all lichens have the same capacity for exploiting the various forms of water from the environment. For example, lichens with cyanobacteria as photobiont cannot exist without liquid water (Lange et al. 1988). For the Australian erratic green-algal *Chondropsis semiviridis*, rainwater is necessary to allow net photosynthetic production because the curled lobes must be unfolded (Rogers and Lange 1971, Lange et al 1990a).

The kinetics of water uptake seems to be similar in lichens and mosses, and the larger the surface area to weight ratio, the more rapid the water uptake (Larson 1981). Rundel (1982a) suggested that thin cortical layers of coastal Roccellaceae in desert regions may be a morphological adaptation to increase rates of water uptake. However, textural features of the upper cortex seem to be more important for water uptake than just thickness (Larson 1984, Valladares 1994a). Valladares (1994a) found that species of Umbilicariaceae that possess the most porous and hygroscopic upper cortex (equal to filter paper) are adapted to live mainly from water vapor (aerohygrophytic), whereas species that have an almost impervious cortex were more frequently exploiting liquid water from the substratum (substratehygrophytic; Sancho and Kappen 1989).

Most bryophytes need a humid environment or externally adhered water to keep a level of hydration high enough for metabolic functions. Many species form cushions, turfs, or mats that aid to keep capillary water around the single shoots (Gimingham and Smith 1971, Giordano et al. 1993). At full saturation, the water content of mosses (excluding external water) can vary between 140 and 250% dry weight (d.wt.) (Dilks and Proctor 1979), which is similar to that of macrolichens. Thallose hygrophytic liverworts require higher levels of hydration, and their maximal water content can be more than 800% d.wt. In shaded or sheltered habitats, hygric and some mesic bryophytes are able to keep their water content relatively constant throughout the year, which is characteristic for a stenopoikilohydrous lifestyle. In more open and exposed sites, the fluctuations in water content are very large (Dilks and Proctor 1979).

The more complex and differentiated morphology and anatomy of bryophytes, in comparison with lichens, allow for more varied modes of water uptake (Proctor 1982, 1990, Rundel 1982b). Bryophytes can take up water vapor to a limited extent and reach only low ($<$ 30% of maximum water content) relative values (Rundel and Lange 1980, Dhindsa 1985, Lange et al. 1986). Dew uptake was shown for *Tortula ruralis* (Tuba et al. 1996a) and for 10 sand-dune mosses (Scott 1982). Leaves of certain desert mosses (e.g., Pottiaceae) act as focus for condensation of water vapor and mist by means of their recurved margins, papillose surfaces, and hair points (Scott 1982). However, the presence of lamellae, filaments, and other outcrops on the adaxial surface of the leaves, which is common in arid zone mosses, may act more as sun shelter rather than as means to enhance water uptake. The role of scales and hyaline structures on the midrib of desert liverworts (e.g., *Riccia, Exormotheca,* and *Grimaldia*), which is inverted and exposed to the open when the thallus is dry, is not clear, but they start absorbing rainwater and swelling to turn down rapidly and may help in storing water (Rundel and Lange 1980). Mosses of the family Polytrichaceae have so-called rhizomes or root-like structures, which are not very efficient for water uptake (Hebant 1977). In general, water uptake of mosses from the soil is poor and needs to be supplemented by external water absorption.

Two main groups of bryophytes have been described according the mode of water transport. The so-called *ectohydrous* resemble lichens because they take up water over all or most of their thallus surface and have no internal water transport system, while the so-called *endohydrous* have various water-proofed surfaces, often well developed near to the gas exchange pores, and have a significant water-transport pathway (Proctor 1984, Green and Lange 1994). Characteristic desert mosses are ectohydrous (Longton 1988a), and water transport in eurypoikilohydrous bryophytes growing in dry environments is predominantly external. Endohydrous mosses and hepatics possess elements similar to those of homoiohydrous plants such as stomata on the sporophytes, a cuticle, and the capacity to conduct water (Hebant 1977). However, they differ from vascular plants in that their conductive structures are not lignified. Water transport can be very different among bryophyte species and, even within a given thallus, water can be carried in various ways.

Proctor (1982) summarized four different pathways or modes by which water moves in a bryophyte: (1) inside elongated conductive cells (hydroids), forming a central strand in the stems of mosses and some liverworts; (2) by the cell walls, which are frequently thickened (in fact, bryophyte cell walls have higher water conductivity than those of vascular plants); (3) through intervening walls and membranes; and (4) by extracellular capillary spaces. The highest internal conduction for water in Polytrichaceae at 70% relative humidity was 67% of the total conduction (Hebant 1977). According to Zamski and Trachtenberg (1976), the stereom (usually considered a supporting tissue) can be an alternative route for the conduction of water. This may be of particular importance for many eurypoikilohydrous (xeric) mosses (Fabroniaceae, Orthotrichiaceae) because they have large masses of stereom tissue.

Water uptake in poikilohydrous vascular plants can be very complex because of interactions between different organs. For instance, in the fern *Cheilanthes fragrans*, water uptake through the leaf surface from a water vapor–saturated atmosphere allows it to reach 80% of its maximal water content within 50 hr (Figure 2a). Petiolar water uptake was also efficient, but only if the leaves were in high air moisture (Figure 2b). Stuart (1968) found that the fern *Polypodium polypodioides* was not able to rehydrate by soil moistening if the air was dry, and the leaves reached only 50% of their maximal water content within 2–

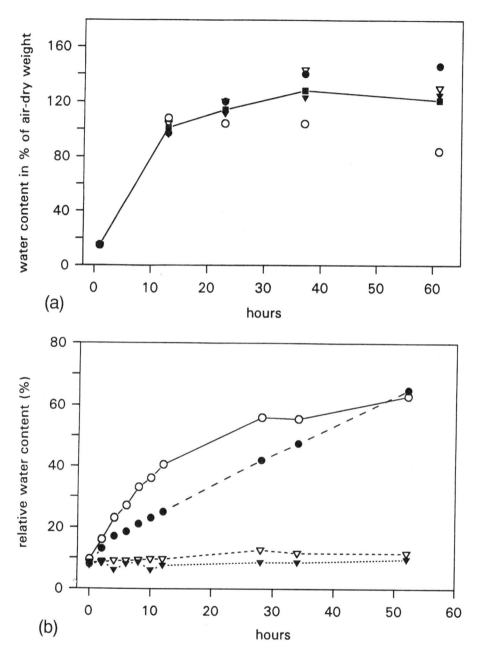

Figure 2 (a) Water vapor uptake of leaves of the fern *Cheilanthes fragrans* with sealed petioles in a moist chamber. The different symbols stand for 4 replicates (L. Kappen, unpublished results). (b) Water uptake of leaves of *Cheilanthes fragrans* placed on filter paper in a moist chamber (open circles); with petiole in a vessel with water and standing in a moist chamber (closed circles), and (open and closed triangles) with petiole in water in a room (approximately 60% rh) (L. Kappen, unpublished results).

3 days, even in a water vapor–saturated atmosphere (Stuart 1968, confirming the results of Pessin 1924). Fronds of the highly desiccation-tolerant *Polypodium virginianum* were, however, not able to absorb water from air as was shown with Deuterium-labeled water (Matthes-Sears et al. 1993). Thus, the capacity of the leaves to take up water vapor varies significantly among species and seems not to be associated with the tolerance to desiccation. In contrast, liquid-water uptake by leaves has been shown to be a common feature in poikilohydrous vascular plants. Detached leaves of *Polypodium polypodioides* regained full saturation within 20–30 minutes if submersed in liquid water (Stuart 1968). However, leaves attached to the rhizome needed 10 times longer for saturation than detached leaves. Stuart explained this by alluding to anaerobic conditions that impede rapid water uptake. Rapid water uptake by leaves was also shown in *Selaginella lepidophylla* (Eickmeier 1979). It seems that, in pteridophytes, water uptake through leaves is an important mechanism for reestablishing water relations of the whole plant and for resuming xylem function. Similarly, rehydration of the whole plant solely by watering the soil in dry air is also incomplete in angiosperm poikilohydrous plants (Gaff 1977). Evidence of partial resaturation of plants from soil water was reported for *Craterostigma, Xerophyta humilis* (Gaff 1977), and *Myrothamnus flabellifolius* (Child 1960).

Water uptake from mist or from saturated atmospheres is insignificant in poikilohydrous angiosperms (Vieweg and Ziegler 1969), as has been shown for isolated leaves of *Ramonda myconi* (Gebauer et al. 1987). In addition, exposure to dewfall could only raise the relative water content to less than 13% in *Craterostigma wilmsii* (Gaff 1977). Foliar water uptake by desert plants has been investigated, particularly with respect to dew uptake (Barthlott and Capesius 1974), but it seems to be insignificant in homoiohydrous plants except in the genus *Tillandsia* (Rundel 1982b). In contrast, foliar water uptake from rain by poikilohydrous vascular plants is important to resume the functioning of the hydraulic system. A brief shower may be very effective in activating these poikilohydrous plants, particularly by run-off effects (Gaff 1977). Leaves of resurrection plants in contact with liquid water can rehydrate within 1–14 hr (Gaff 1977) depending on the species. The quickest uptake was measured in *Chamaegigas intrepidus* (Hickel 1967). The cuticle of vascular plants is generally considered to be an efficient protection against water loss. However, the cuticle of poikilohydrous vascular plants may also enhance water uptake by leaves (e.g., *Borya*; Gaff 1977). The permeability of the cuticle to water was assumed for *Chamaegigas intrepidus* (Hickel 1967). Barthlott and Capesius (1974) suggested that the cuticle of some of these plants seems to be more permeable to water from outside than from inside the leaf. However, this is not clear as some studies attribute permeability to the state of the cuticular layer rather than to the cuticle itself (Schönherr 1982). According to Kerstiens (1996), water uptake through the cuticle is most likely, but evidence needs to be shown.

Hairs and scales can function as auxiliary structures for water uptake because they absorb water more easily than the leaf epidermis. The lower surface of the curled and folded leaflets of ferns like *Ceterach officinarum*, densely covered with scales and trichomes, should enhance water capture (Oppenheimer and Halevy 1962). The so-called hydathodes on the leaves of *Myrothamnus* may actually function as water-absorbing trichomes (Rundel 1982b). However, scales of *Polypodium polypodioides* did not facilitate water uptake, but allowed the water to spread homogeneously on the leaf surface (Pessin 1924, Stuart 1968), and the scales on the leaves of several species of *Ceterach* and *Cheilanthes* retarded water uptake for several hours because of the air that was trapped between the scales (Oppenheimer and Halevy 1962, Gaff 1977, Gebauer 1986). The hairs of the leaves of *Ramonda myconi* (and of other *Ramonda* species) had the same effect (Figure 3). Spraying of the detached hairy leaves resulted in less water uptake than immersion in water or spraying hairless leaves. The retarding effect of scales and hairs suggests that a very rapid water uptake after desiccation could be injurious to the leaf cells.

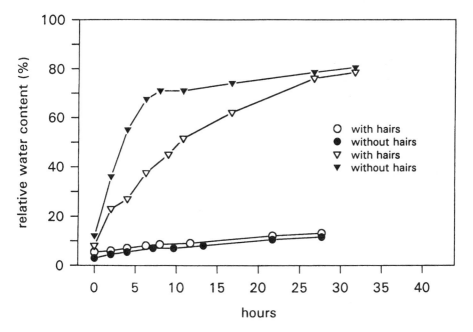

Figure 3 Water uptake of leaves of *Ramonda myconi* (Gesneriaceae) with sealed petioles. Leaves with hairs (▽) and after removing the hairs (▼) soaking from sprayed water; and leaves with hairs (○) and without hairs (●) in a moist chamber. (*Source*: after Gebauer et al. 1987.)

B. The Problems of Resuming Water Transport

Poikilohydrous plants that possess an internal system for water transport (endohydrous bryophytes and vascular plants) are exposed to cavitation (break down of the cohesion in the water column) and embolism (obstruction or occlusion of a vessel by air bubbles) during desiccation, which compromise the functioning of the conducting tissues upon rehydration. However, except for bryophytes, the problems facing poikilohydrous plants when resuming water transport with desiccated conducting tissues have never been discussed. Many studies have been devoted to cavitation and embolism in trees, especially during the last decade, proving that water stress can induce severe cavitation and the appearance of air bubbles in the vessels (Lewis et al. 1994, Kolb et al. 1996, Sperry and Tyree 1988, Hargrave et al. 1994, Tyree and Sperry 1988; Tyree [Chapter 6] this volume). Emboli in a fraction of the conductive elements confines water transport into a diminished number of vessels, which requires an increased tension and further increases the risk of embolism. Embolized conduits can become functional again through bubble dissolution or expulsion, which requires a positive pressurization (Zimmermann and Milburn 1982). Poikilohydrous plants face the dilemma of restoring water transport through their old, embolized tissues or investing in new conducting tissues, which reduces the resources available to be allocated elsewhere in the plant.

   Poikilohydrous vascular plants that can resume water transport only if their shoot tissues have been hydrated by external water uptake perform like the endohydrous bryophytes, where water conduction in the hydroids is supported by lateral and apoplastic water transport. The imbibition of the cell walls of leaf and stem tissues generates the necessary pressure to induce dissolution of emboli in the tracheary tissues. Capillary forces in poikilohydrous plants at 40% of relative humidity and under laboratory conditions could move water to a height of 2–12 cm (Galace 1974, cited in Gaff 1977). These forces are sufficient to rehydrate many of the small poikilohydrous species. Other mechanisms to eliminate emboli are temperature-associated osmosis at the plant apex (Pickard 1989) and generation of a root pressure that is able to dissolve gas bubbles in the conduits of small herbs and grasses (Zimmermann and Milburn 1982). The latter was shown to be able to restore full liquid continuity and was assumed to be important in larger poikilohydrous species like *Xerophyta eglandulosa* (Gaff 1977).

   Earlier literature already records the formation of new adventitious roots in poikilohydrous plants subsequent to rehydration (Walter and Kreeb 1970). This was confirmed recently in *Xerophyta scabrida*, where such roots appeared after the regreening of the plant, replacing the drought-killed original roots (Tuba et al. 1993a). However, there is very little information on the fate of the old roots and on the production of new roots during rehydration of poikilohydrous plants. The slow recovery upon rehydration of whole plants in comparison to detached leaves (Stuart 1968, Gebauer 1986) might be due to the time required to form new roots. This topic has been better explored in homoiohydrous plants from arid environments. For instance, the hydraulic conductivity of roots of *Agave deserti* and *Ferocactus acanthodes*, which decreased dramatically after several days of drought, rapidly recovered when water was again available, not by formation of new roots but by refilling the extant tissues, which were made up of flexible and unlignified vessels (North and Nobel 1991, Ewers et al. 1992). Reversal of almost complete embolism in stems of *Salvia multiflora* was related to the presence of narrow vessels and tracheids, which were better able to refill after embolism than wider conduits (Hargrave et al 1994). It seems likely that these structural properties of the xylem are also important for resuming water transport in poikilohydrous plants. In fact, a flexible structure in the conductive system was identified in the poikilohydrous *Blossfeldia liliputana* (Barthlott and Porembski 1996).

The fact that most of the poikilohydrous vascular plants are herbaceous and smaller than 50 cm (with a few exceptions, e.g., *Xerophyta eglandulosa*; Gaff 1997) might be explained by the difficulties of restoration of conductivity of the xylem. Therefore, it would be very interesting to study the rehydration mechanisms and the recovery from embolized tissues in the few fruticose species (e.g., *Satureja gilliesii* from the Chilean Matorral) and, in particular, in the relatively large (up to 1 m; Puff, 1978) *Myrothamnus flabellifolius*, and in the subfruticose Scrophulariaceae. The same mechanical difficulties of resuming water conductivity of embolized tissues might also be behind the remarkable lack of poikilohydrous species among gymnosperms, which consist only of trees and shrubs. The low flexibility due to the xylem anatomy of gymnosperms was also demonstrated by Ingrouille (1995).

C. Retarding Water Loss

Poikilohydrous plants cannot maintain their water content at a constant level, but they can extend hydration into the dry period by certain, mostly structural, mechanisms. By retarding water loss, these plants can enhance their exploitation of the transient periods of water availability. Retarding water loss could, however, counteract some advantageous aspects of the poikilohydrous strategy. For instance, extending hydration sometimes reduces water capture. Poikilohydry also provides a remarkable tolerance for desiccated plants to other stresses that usually occur with drought, such as heat and excessive light (see Section III). Thus, if metabolic activity is extended into these harmful periods, it could not only reduce overall productivity but also compromise survival.

One way of retarding water loss in lichens is by increasing the water that can be stored within the tissues (Valladares 1994b, Valladares et al. 1998). Anatomical characteristics, such as porous and thick medulla layers and rhizinae, have been suggested as means of increasing water storage in lichens (Snelgar and Green 1981, Valladares et al. 1993, Valladares and Sancho 1995). However, because water and CO_2 share the pathway in lichens, enhanced water storage can hamper CO_2 diffusion and consequently reduce photosynthetic carbon uptake (Green et al. 1985, Lange et al. 1996, Maguas et al. 1997). Thus, again, these plants must reach a compromise between two opposing situations. Large foliose lichens can possibly separate photosynthesis and water storage in space some-what, as certain zones of the thallus optimize gas diffusion (young, actively grow-ing regions), while other zones (old, thick regions) act primarily as water reser-voirs, sacrificing gas diffusion and carbon gain (Green et al. 1985, Valladares et al. 1994). This trade-off between gas diffusion and water storage seems to be flexible, and lichens have been shown to exhibit a remarkable phenotypic plastic-ity in their water storage and retention capacities in response to habitat conditions (Larson 1979, 1981, Pintado et al. 1997). Dry habitats induce increased water storage (Tretiach and Brown 1995), but there are complex interactions with the light availability for photosynthesis. In shaded sites without access to liquid wa-ter, the Antarctic lichen *Catillaria corymbosa* enhanced both water storage and photosynthesis via increased light harvesting by chlorophylls (Sojo et al. 1997), while in exposed sites (dry and receiving high irradiance), the lichen *Ramalina capitata* enhanced photosynthetic utilization of the brief periods of activity via improved gas diffusion at the expense of reducing water storage capacity (Pintado et al. 1997). These problems are not faced by bryophytes, because most of them have rather complex photosynthetic tissues, where the CO_2 exchange surface is separated from water storage volumes (Green and Lange 1994). The capacity of bryophytes to keep high rates of photosynthesis at high water contents is a very likely explanation for the dominance of these organisms in wet habitats (Green and Lange 1994). However, despite mechanistic differences between mosses and lichens, in certain cases overall performance can be similar.

Discussing the role of morphological properties for lichens, Rundel (1982a) concluded that evaporative water loss can be reduced by a decrease of surface/ volume ratio, but such a decrease reduces uptake of water vapor similarly. How-ever, this seems not to be the case for liquid water, since structures on the lower side of the thallus such as the rhizinomorphs of the lichen family Umbilicariaceae enhance capture and storage of water from run-offs without significantly increas-ing water loss by evaporation (Larson 1981, Valladares 1994b, Valladares et al. 1998). In general, selection appears to favor morphological traits that maximize rates of water uptake rather than those that minimize rates of water loss, especially in cool and foggy habitats. Reduction of evaporative water loss by structural traits such as a tomentum on the upper surface (Snelgar and Green 1981), a thick cortex (Larson 1979, Büdel 1990), or a decreased surface/volume ratio is very limited in open habitats, but can be significant in sheltered or humid places or in areas with frequently overcast skies (Kappen 1988) or under the influence of drip water

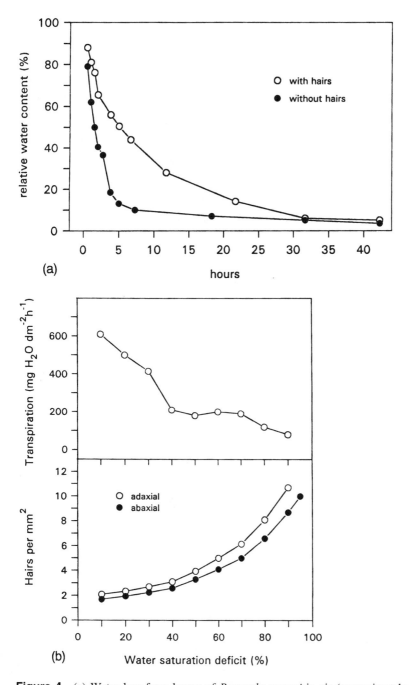

Figure 4 (a) Water loss from leaves of *Ramonda myconi* in air (approximately 50% rh) with hairs and after hairs were removed before the experiment. (*Source*: after Gebauer et al. 1987.) (b) Transpiration rates of leaves of *Ramonda myconi* with increasing water saturation deficit. Hair density on the upper (adaxial) and the lower (abaxial) axial leaf surface increases as the leaf shrinks with increasing water loss. At about 40% saturation deficit, stomatal closure becomes effective. (*Source*: after Gebauer 1986.)

from trees or from antennae on roofs. Bryophytes can also retard water loss by an increased boundary layer resistance and due to growth forms of low surface-to-volume ratios (Gimingham and Smith 1971, Proctor 1982, Giordano et al. 1993).

Comparing water retention of a terricolous moss and a terricolous lichen, Klepper (1968) found that although the moss stored initially more water, the two remained hydrated for the same period of time after the rainfall. Then, regardless of the initial water content, they dried out quickly when the ambient water vanished and the atmosphere became dry. Measurements on desert lichens have clearly shown that the thalli start drying as soon as the sun is rising and their period of hydration depends solely on the evaporative conditions (Kappen et al. 1979, Kappen 1988).

As has been repeatedly demonstrated, the water content of vascular resurrection plants varies with the soil moisture and, like their nonvascular counterparts, they dehydrate within a few hours or days after soil water supply has declined (Gaff and Churchill 1976, Gaff 1977). Since stomatal conductance of resurrection vascular plants is, in general, rather high (Tuba et al. 1994), water loss must be retarded by structural features such as scales, as was shown for leaves of *Ceterach officinarum* (Oppenheimer and Halevy 1962). However, scales were almost ineffective in *Cheilanthes maranthae* (Gebauer 1986, Schwab et al. 1989). A retarding effect of hairs in leaves of *Ramonda* (Figure 4a) has long been known (Bewley and Krochko 1982) and the efficiency of leaf pubescence in retarding water loss is increased when the leaves shrink (Gebauer et al. 1987, Figure 4b). The effect of hair density in reducing the transpiration rate strongly increased when the leaf water saturation deficit went beyond 40%. Reduction of the exposed surface is also a typical mechanism to reduce transpiration. This is the case with poikilohydrous ferns (Pessin 1924); with *Selaginella lepidophylla*, which curls the whole shoot rosette (Lebkuecher und Eickmeier 1993); with *Myrothamnus flabellifolius*, which regularly pleats its fan-like leaves (Vieweg and Ziegler 1969, Puff 1978); and with Velloziaceae, which fold or curl their leaves (Gaff 1977). Leaf or shoot movements of most of these plants are due to differential imbibition of the tissues involved rather than to osmotic phenomena, because they still operate in dead plants.

III. PREVENTING DAMAGE AND TOLERATING STRESSES

A. Desiccation Tolerance

The first, but not the only, stress during desiccation is the lack of water itself, which imposes dramatic structural and physiological changes on the tissues of poikilohydrous organisms. Some poikilohydrous plants exhibit a remarkable tol-

erance to intense desiccation. The moss *Tortula ruralis* survived a water content
as low as 0.008%, which is equivalent to -6000 bars (Schonbeck and Bewley
1981a). In several bryophyte and fern species, the degree of desiccation tolerance
varied seasonally or, for eurypoikilohydrous plants, only occurred during the winter period (Dircksen 1964, Kappen 1964, Dilks and Proctor 1976b). Desiccation
tolerance of eurypoikilohydrous autotrophs can be defined as the capacity to withstand equilibrium with a relative humidity less than 20% (Lange 1953, Biebl
1962, Bertsch 1966b, Gaff 1986). The lowest relative water contents tolerated
by vascular plants were reported to be between 4% and 15%, which is equivalent
to 1.4–8.4% of their dry weight (Kaiser et al. 1985) (see Table 1 for a detailed
list). Detached leaves of resurrection plants proved to be much less tolerant to
severe water loss (Gaff 1980). Table 1 also shows tolerance to extended periods
of anabiosis. *Nostoc* may tolerate 5 years of desiccation, but other cyanobacteria
do not survive desiccation at all (Biebl 1962, Scherer et al. 1986). Most lichens,
bryophytes (Lange 1953, Biebl 1962, Proctor 1982), and poikilohydrous vascular
plants (Hallam and Gaff 1978, Lazarides 1992) survive dry periods that last for
a few months. Although Marchantiales were reported to be exposed to 6–8
months of drought in their habitat in Namibia (Volk 1979), such extremely long
periods of desiccation are usually rare for this kind of bryophyte. Several vascular
resurrection plants have been shown to tolerate air-dry periods of 2–5 years
(Hickel 1967, Gaff and Ellis 1974, Gaff 1977), and some lichens can become
photosynthetically active again after a period of 10 years frozen in the dry state
(Larson 1988).

Bryophytes comprise taxa that have an intrinsic capacity of desiccation
tolerance, as well as taxa that need acclimation. Although the so-called xerophytic
moss species such as *Syntrichia ruralis, Rhacomitrium canescens, Neckera
crispa*, and others always immediately survive extreme desiccation, mesophytic
and even hygrophytic species such as *Bryum caespititium, Plagiothecium platy-
phyllum, Pohlia elongata*, or *Mnium seligeri* become extremely desiccation tolerant only if they were pretreated at a relative humidity of 96% for 24 hours (Abel
1956, Biebl 1962). Most remarkable is that the water moss *Fontinalis squamosa*
is as tolerant as a xerophytic moss, whereas *Fontinalis antipyretica* is drought
sensitive even when pretreated at 96% relative humidity. The intrinsic desiccation
tolerance of eurypoikilohydrous bryophytes can be modified by the rapidity of
desiccation processes (Gaff 1980). Desiccation tolerance was observed to vary
seasonally in many bryophytes (Dilks and Proctor 1976a), in ferns, and also in
Borya nitida as they can acclimate to frost desiccation in winter or to summer
drought (Gaff 1980). Several bryophytes (*Dicranum scoparium* and *Mnium pun-
catum*) and ferns (*Asplenium* and *Polypodium vulgare*) gained an extremely high
desiccation tolerance in winter (Dircksen 1964, Kappen 1964). Cultivation under
moist conditions for 2 weeks can decrease the desiccation tolerance of most eury-

poikilohydrous plants (algae and lichens; Kappen 1973, Farrar 1976b; bryophytes: Schonbeck and Bewley 1981b, Hellwege et al. 1994; vascular plants: Gaff 1977, 1980). Like seeds, air-dry lichens (Lange 1953), bryophytes (Hosokawa and Kubota 1957, Gaff 1980, Proctor 1990), and vascular resurrection plants can persist longer if stored at humidities lower than 30% relative humidity (Leopold 1990). This was explained by the fact that intermediate-to-low water contents allow some enzyme activity and lead to respiratory carbon loss, destructive processes, and infections (Gaff and Churchill 1976 [Borya nitida], Proctor 1982). Most of our current knowledge of the effects of environmental conditions before and during desiccation on the tolerance to desiccation and related stresses comes from experiments under controlled conditions. Slow drying of poikilohydric plants probably occurs under real field conditions, but whether this is the rule, and the ecological significance of hardening and preconditioning phenomena remain to be investigated.

More important for eurypoikilohydrous plants is their capacity to withstand repeated changes between dry and moist states. Lichens of deserts and Mediterranean regions, for instance, oscillate regularly between periods of a few hours of activity and anabiosis for the rest of the day (Kappen et al. 1979, Redon and Lange 1983, Kappen 1988, Lange et al. 1991, Sancho et al. 1997). The capacity to tolerate several changes between dry and wet states was tested experimentally for different moss species by Dilks and Proctor (1976a, 1979). *Tortula ruralis*, as an eurypoikilohydrous species, performed well during up to 63 changes within a period of 18 months, whereas *Rhytidiadelphus lorius* was killed when continually dry for 18 months or when the oscillation phase was 1 day wet/1 day dry, but it retained 50% of its normal net photosynthetic rate if the wet periods were longer (6 or 7 days) or the dry period was longer. Mosses such as *Tortula ruralis* and the Xanthorhoeaceae *Borya nitida* (Schonbeck and Bewley 1981b) were actually able to increase their desiccation tolerance if drying and rehydration were repeated. In contrast, continuous hydration over several days decreased the tolerance to desiccation in *Tortula ruralis* (Schonbeck and Bewley 1981b), a result also found for some lichens (Ahmadjian 1973). Apparently, in the absence of contrasting oscillations in moisture content, the algal partner of lichens grows excessively, altering its symbiotic relation with the mycobiont (Farrar 1976b). Repeated drying seems to be essential for the internal metabolic balance of lichens (McFarlane and Kershaw 1982), and also in poikilohydrous vascular plants, as indicated by the fact that cultivation of *Myrothamnus flabellifolius* is only successful if the plants dry occasionally (Puff 1978). For several desert cyanobacteria, hydration is a very rare event. Pleurococcoid green algae (e.g., *Apatococcus lobatus*) and many epilithic lichens are water repellent (Bertsch 1966a). Species of the genera *Chrysothrix, Lepraria*, and *Psilolechia* growing under overhanging rocks never receive liquid water during their lifetime (Wirth 1987).

B. Cellular and Physiological Changes During Desiccation

Great attention has been given in the last 20 years to the investigation of the ultrastructural changes and the biochemical processes that take place during dehydration and rehydration of poikilohydrous plants. Rather than providing a detailed account here, we refer the interested reader to some recent reviews (Gaff 1980, 1989, Stewart 1989, Leopold 1990, Bewley and Krochko 1982, Proctor 1990, Bewley and Oliver 1992, Ingram and Bartels 1996, Hartung et al. 1998). A few notes may illustrate the principal aspects.

Nonvascular plants are desiccation tolerant if they can retain cellular integrity and limit cellular damage during drying. To accomplish this, certain bryophytes increase ribonuclease activity and exhibit minor qualitative changes in the types of proteins synthesized during drying (Bewley 1995). As extensively shown with the moss *Tortula ruralis*, this tolerance is based on inherent ultrastructural and metabolic properties that can operate without activation of specific protective substances or mechanisms (Bewley and Oliver 1992, Oliver and Bewley 1997), i.e., sugar content does not change significantly, and no membrane protective mechanism is detectable. The mechanism of lichens for dealing with desiccation remains broadly unknown but is strongly related to their high content of polyols (Farrar 1976a). Membrane leakage in lichens as a consequence of repeated drying and wetting (Farrar 1976a) is harmless but, for less desiccation-tolerant bryophytes, irreversible changes of membranes and metabolism can occur during dehydration (Bewley 1979).

Poikilohydrous vascular plants exhibit a great variety of down- and upregulation of cellular processes, which can be retained at very low water potentials (Leopold 1990). The nucleus always remains intact (Bartley and Hallam 1979) and contains a dense mass of chromatin (Hetherington and Smillie 1982). Structural changes are usually small in the *homoiochlorophyllous* species (species that stay green during desiccation and rehydration), and thylacoid membranes and associated chlorophyll complexes remain widely preserved (Owoseye and Sandford 1972, Hallam and Cappicchiano 1974, Gaff and McGregor 1979, Gaff 1989). Preservation of polysomes and of RNA also enable protein synthesis after drought (Bewley 1973). Changes in the chloroplast structure were, however, reported for *Myrothamnus flabellifolius* (Wellburn and Wellburn 1976) and *Talbotia elegans*. In the latter, the mitochondria were reduced to membrane-bound sacks (Hallam and Gaff 1978). Membrane leakage has been reported for a few species. Vacuoles, fragmented into numerous vesicles, and lysosomes seem to be maintained intact (Gaff 1980, 1989, Hartung et al. 1998). Kaiser and Heber (1981) and Schwab and Heber (1984) state that the lens shape of the chloroplast permits dehydration without greater surface aerea reduction, and in vivo rupture of chloroplasts during desiccation is rare.

In contrast, a loss of virtually all thylacoids and most of their carotenoid content, and also of most cristae of the mitochondria is common in the *poikilochlo-rophyllous* species (plants that destruct their chlorophyll during desiccation [Gaff 1980, Hetherington and Smillie 1982, Tuba et al. 1996b]). However, respiration continued in desiccated plants in some of these species (*Xerophyta scabrida*) and was still measurable far below -3.2 MPa (Tuba et al. 1996b). Obviously, the destruction of the chlorophyll is structured, as grana are retained (Bartley and Hallam 1979). The ultrastructural changes in poikilochlorophyllous plants may be even greater than in desiccation-sensitive species at a comparable dehydration level (Gaff 1989). Thus, rather than being deleterious, these organelle changes involve an organized remobilization of cell resources in these resurrection plants.

C. Synthesis of Proteins and Protective Substances

On the one hand, desiccation induces down-regulation of genes, particularly of those that code for enzymes relevant to photosynthesis, both in vascular plants and in mosses (Ingram and Bartels 1996, Bernacchia et al. 1996, Oliver and Bewley 1997). In general, the decline of total protein is smaller than in drought-sensitive plants. Loss of water-insoluble proteins is common in resurrection plants, especially in the poikilochlorophyllous species, probably because of degradation of the lipoproteins of the membrane (Gaff 1980). In *Selaginella lepidophylla*, 74% of the activity of nine enzymes of the carbohydrate metabolism was conserved during drying. The conservation of photosynthetic enzymes was lower than that for respiratory enzymes (Eickmeier 1986).

On the other hand, the polysome content rises significantly with water loss (*Xerophyta villosa*; Gaff 1989), and incorporation of amino acids increases (Tymms et al. 1982) together with protein synthesis (Bartley and Hallam 1979, Hallam and Luff 1980, Eickmeier 1988). Obviously, a network of genes with presumably different functions are activated by water stress. Hartung et al. (1998) estimated that 800–3000 genes could be involved in the response of plants to desiccation. The so-called late embryogenesis abundant (LEA) proteins seem to be involved because they commonly occur in plant tissues under water stress (Ingram and Bartels 1996, Hartung et al. 1998) and they were found in some resurrection plants (Schneider et al. 1993, Ingram and Bartels 1996). Many novel proteins are synthesized during desiccation, most of which are specific for extremely desiccation-tolerant plants (Bartels et al. 1990, Piatkowski et al. 1990, Bartels et al. 1993, Kuang et al. 1995). Nevertheless, certain polypeptides, such as those found in desiccated *Polypodium virginianum*, are not exclusive to the desiccation regime (Reynolds and Bewley 1993b). Desiccation-related proteins (*dehydrins*) are cytosolic and are believed to protect desiccation-sensitive enzymes during dehydration (Schneider et al. 1993, Bernacchia et al. 1996, Hartung et al. 1998).

In most resurrection plants, including the aquatic species *Chamaegigas intrepidus*, abscisic acid is strongly accumulated and is involved in attaining desiccation tolerance and in stimulating the synthesis of dehydrins (Gaff 1980, 1989, Gaff and Loveys 1984, Reynolds and Bewley 1993a, Hellwege et al. 1994, Schiller et al. 1997). It is hypothesized that abscisic acid (ABA) coordinates the activation of special genes upon drying (Bartels et al 1990, Nelson et al. 1994, Oliver and Bewley 1997). Abscisic acid was not detected in the moss *Tortula ruralis* (Reynolds and Bewley 1993a), but it was present in the moss *Funaria hygrometrica* (Werner et al. 1991), as well as in the eurypoikilohydrous *Exormotheca holstii* and other hepatic species (Hellwege et al. 1994, Hartung personal communication). Abscisic acid was accumulated during drought, disappeared during wet cultures, and induced hardiness when applied to dehardened thalli. This suggests that ABA is involved in a hardening process that may not be necessary for *T. ruralis*. The role of ABA recently discovered in lichens is unclear, as this hormone is produced by the fungal biont and, as opposed to its activity in plants, increases as a response to water uptake (Dietz and Hartung. 1998). Leaves of *Myrothamnus flabellifolia* and *Borya nitida* did not survive dehydration if they were dried so rapidly that ABA could not be accumulated (Gaff and Loveys 1984). Abscisic acid accumulation obviously can only occur in leaves attached to the whole plant (Hartung et al. 1998).

A common phenomenon in drought stress is the accumulation of organic compatible solutes because they stabilize proteins and membranes (Levitt 1980, Crowe and Crowe 1992). Lichens are permanently rich in sugar alcohols, which are assumed to be the basis of their remarkable desiccation tolerance (Kappen 1988). Cowan et al. (1979) have demonstrated that the synthesis of amino acids and sugar alcohols was active in lichens in equilibrium with humidities as low as 50%. In contrast, desiccation-tolerant bryophytes contain a low amount of sugars, mainly sucrose, and show no or very little increase in sugar content during drying (Bewley and Pacey 1978, Santarius 1994). Strong sugar accumulation during desiccation has been demonstrated in many resurrection grasses, whereas other resurrection plants such as *Ceterach* and *Craterostigma* already contained comparatively high amounts of sugar in the leaves when turgid (Schwab and Gaff 1986). In these and other species (e.g., *Myrothamnus flabellifolius*), changes in the sugar composition were observed during dehydration (Bianchi et al. 1991, Hartung et al. 1998). Unusual sugars that appear in the turgid leaves are storage products, but they are converted mainly into sucrose during the drying process (Bianchi et al. 1991, 1993, Albini et al. 1994). Sucrose significantly increased during desiccation in the Gesneriaceae *Boea hygroscopia* and two poikilohydrous *Sporobolus* (Poaceae) species, but this tendency was also noticed in a desiccation-sensitive *Sporobolus* species (Kaiser et al. 1985). Proline concentration in many plant species associated with water stress was comparatively high in *Ceterach* and *Craterostigma*, but did not significantly increase during dehydration (Schwab and Heber 1984). Application of proline had no effect on detached leaves of *Borya nitida* and *Myrothamnus flabellifolius* (Gaff 1980).

One of the internal hazards of desiccation is the increase of oxidative processes, which occurs in plants exposed to a wide range of environmental stresses (Smirnoff 1995). An increase in or a high level of defense enzymes of the ascorbate/glutathione cycle was associated with the protection of the membrane lipids in *Sporobolus stapfianus* during drying (Sgherri et al. 1994a). Oxidized glutathione was much lower in slowly dried (unimpaired) samples than in rapidly (injured) dried samples of *Boea hygroscopica* (Sgherri et al. 1994b), and glutathione was shown to play the primary role in maintaining the sulfhydryl groups of thylacoid proteins in reduced state during desiccation (Navarri-Izzo et al. 1997). In *Craterostigma plantagineum*, lipoxigenase, which catalyzes lipid peroxidation at membranes, becomes increasingly inhibited during drying (Smirnoff 1993). Similar processes may also operate in desiccation-tolerant bryophytes (Dhindsa and Matowe 1981). Lipid peroxidation during drought is low in eurypoikilohydrous mosses (Seel et al. 1992a). Oxidative processes can take place both in the presence and the absence of light, and light can exacerbate oxidation. This situation—oxidative stress caused or accentuated by light—is discussed in the following section.

D. Photoprotection of the Photosynthetic Units

If plants absorb more light than can be used in photosynthesis, they are exposed to the risk of photo-oxidative destruction of their photosynthetic apparatus (Long et al. 1994). Photo-oxidative stress can therefore be an important limiting factor for poikilohydrous plants that live in open habitats. In fact, stenopoikilohydrous mosses are more damaged by drying at high irradiance than at low irradiance (Seel et al. 1992a). Hence, the poikilochlorophyllous strategy of destructing the photosynthetic apparatus during desiccation might be very adaptive in open habitats, because photo-oxidative potentials are avoided during drying (Smirnoff 1993). Nevertheless, tolerance to strong light can be enhanced by acclimation. Cyanobacterial mats taken from exposed habitats proved to be highly tolerant to high irradiance, whereas cyanobacteria from shaded sites were very sensitive (Lüttge et al. 1995). In addition, bryophytes and algae restricted to shady habitats were shown to have limited photoprotective capacities (Öquist and Fork 1982b).

It has long been considered that the photosynthetic apparatus cannot be affected by strong irradiance when poikilohydrous plants are dry because it undergoes a functional dissociation between light harvesting complexes and photosystem II during desiccation (Bilger et al. 1989, Smirnoff 1993). However, this tolerance to high irradiances when dry might only be partial, at least in certain autotrophs and under particular environmental conditions. For instance, some air-dried lichens exhibited damage after exposure to high light (Valladares et al. 1995, Gauslaa and Solhaug 1996). The most important risks of photodamage occur while the plants are active and especially during the process of desiccation. Down-regulation of photosynthetic processes and the so-called dynamic or recoverable photoinhibition (that is, inhibition of photosynthesis by light) has been observed in a number of poikilohydrous plants and lichens (Demmig-Adams et al. 1990a, Seel et al. 1992b, Eickmeier et al. 1993, Valladares et al. 1995, Leisner et al. 1996). This down-regulation of photosynthesis can avoid photo-oxidation by energy transfer or dissipation (Demmig-Adams et al. 1990b, Öquist and Fork 1982a, Wiltens et al. 1978). One of the mechanisms suggested to be involved in this photoprotection is energy dissipation via the carotenoids of the xanthophyll cycle (zeaxanthin, violaxanthin, and anteraxanthin; Demmig-Adams and Adams 1992). This mechanism has been found in different poikilohydrous organisms (Demmig-Adams et al. 1990a, Eickmeier et al. 1993, Valladares et al. 1995, Calatayud et al. 1997). Energy dissipation via carotenoids operates in cyanobacteria by means of different carotenoids to those of the xanthophyll cycle characteristic of green algae and vascular plants (Demmig-Adams et al. 1990b, Leisner et al. 1994). The zeaxanthin-associated mechanism of photoprotection was increasingly activated during drying in *Selaginella lepidophylla*, efficiently protecting photosystem II during water loss (Casper et al. 1993, Eickmeier et al. 1993). Thus, this poikilohydrous pteridophyte did not differ from homoiohydrous xerophytic plants (e.g., *Nerium oleander*: Demmig et al. 1988; CAM Clusia: Winter and Königer 1989). In *Xerophyta scabrida*, 22% of the carotenoids were still preserved in the dry leaves. Although the photosynthetic apparatus was destroyed, carotenoids seemed to be protective or essential when the chloroplasts reorganized during rehydration (Tuba et al. 1993b).

Isolated *Trebouxia* as well as green-algal lichens resisted photostress in the field by energy dissipation and by a quick desiccation under high irradiances (Öquist and Fork 1982b, Leisner et al. 1996). Water content has been proved to influence both dynamic and chronic photoinhibition of lichens (Valladares et al. 1995, Calatayud et al. 1997). Negative effects of strong light were observed in tropical, temperate (Coxson 1987a, b), and Mediterranean lichens (Manrique et al. 1993, Valladares et al. 1995), but since they dry out quickly under intense radiation, the ecological significance of photostress may be low. Field studies have revealed, for instance, that the cyanobacterial lichen *Peltigera rufescens* was only photoinhibited under certain conditions in winter (Leisner et al. 1996) and that lichen species in Antarctica such as *Umbilicaria aprina* and *Leptogium puberulum* were very resistant to the combination of low temperatures and high irradiance while the thallus was photosynthetically active (Schlensog et al. 1997, Kappen et al. 1998a, Valladares et al. in preparation).

Poikilohydrous autotrophs exhibit various photoprotective mechanisms in addition to the energy dissipation via carotenoids. In the case of lichens, filtering or screening effects of the upper cortex and of certain secondary compounds have been shown to be potentially important in reducing the risk of photodamage of the photosynthetic units (Büdel 1987, Kappen et al. 1998a, Solhaug and Gauslaa 1996). Leaf and shoot curling during drought, despite not being very effective in certain bryophytes (Seel et al. 1992a), can confer photoprotection. For instance, leaf curling by everting the reflectant abaxial leaf surface was effective in *Polypodium polypodioides*, a species sensitive to strong light under water stress (Muslin and Homann 1992). Additionally, photorespiration and light-activated photosynthetic enzyme activity were not affected by intense radiation in *Selaginella lepidophylla* due to leaf curling (Lebkuecher and Eickmeier 1991, 1993). The protective role of curling is important for plants growing in open and exposed habitats and, although not yet thoroughly explored, it may be relevant for lichens such as *Parmelia convoluta* and *Chondropsis semiviridis*, or for thallose desert liverworts that evert a whitish underside (Lange et al. 1990a).

E. Tolerance to Extreme Temperatures: A Property Linked to Poikilohydry

Plants capable of surviving desiccation are usually tolerant to other environmental stresses such as extreme temperatures. Resistance to extreme temperatures is greater once the plant is desiccated, and poikilohydrous species differ in their tolerance to freezing and heat while they are wet and active. Some lichen species, for example, are extremely resistant to freezing both in hydrated and dehydrated states. The freezing tolerance of hydrated lichens exceeds by far the temperature stresses that occur in winter or in polar environments (Kappen 1973). Bryophytes are less freezing-tolerant, but many species are well adapted to live and persist in cold environments (Richardson 1981, Proctor 1982, Longton 1988b). Fern gametophytes and vascular resurrection plants from the temperate zone have a moderate (-9 to $-18°C$) freezing tolerance in winter (Kappen 1964, 1965). Most resurrection vascular plants come from subtropical climates and are sensitive to freezing. For example, *Borya nitida* can only survive temperatures between -1 and $-2°C$ in winter (Gaff and Churchill 1976), and other grass species do not survive temperatures below $0°C$ (Lazarides 1992). Desiccated leaves, fronds, or other vegetative parts may become highly tolerant to cold depending on the remaining water content. Experiments with *Ramonda myconi*, *Polypodium vulgare*, and *Myrothamnus flabellifolius* revealed that leaves resisted $-196°C$ if they were desiccated to a relative water content of 6% (Kappen 1966, Vieweg and Ziegler 1969). Thus, the tolerance of these vascular poikilohydrous plants to low temperatures was not different from that of dry algae, lichens, and bryophytes (Levitt 1980), which indicates water content as a crucial factor in the freezing tolerance of organisms in general.

Heat tolerance of plants can also be increased if they are desiccated (Kappen 1981). This fact becomes ecologically relevant if we consider that dry plants can easily heat up to over 50°C in their natural habitats. Thallus temperatures in dry moss turfs and dry lichens can reach 60–70°C under field conditions (Lange 1953, Lange 1955, Richardson 1981, Proctor 1982, Kershaw 1985). A similarly high temperature was reported for desert soil around *Selaginella lepidophylla* (Eickmeier 1986). The temperature in the rock pools where *Chamaegigas intrepidus* lives can reach 41°C (Gaff and Giess 1986). Although some lichens such as *Peltigera praetextata* and *Cladonia rangiferina* did not survive temperatures higher than 35°C, even in the desiccated state (Tegler and Kershaw 1981), tolerance to 70°C and even to 115°C was recorded for many other dry lichen and bryophyte species (Lange 1953, 1955, Proctor 1982). Heat tolerance of bryophytes and vascular plant species varies with season. The maximal temperatures tolerated in the turgescent state by the temporarily poikilohydrous fern *Polypodium vulgare* and by *Ramonda myconi* were highest in winter (approximately 48°C), and by decreasing water content, the heat tolerance could be increased to approximately 55°C (Kappen 1966, Figure 5). *Chamaegigas intrepidus* was able to thrive after being subjected to 60°C temperatures (Heil 1925), and dry leaves of *Myrothamnus flabellifolius* were reported to have resisted at 80°C (Vieweg and Ziegler 1969). Although Eickmeier (1986) could demonstrate an increase

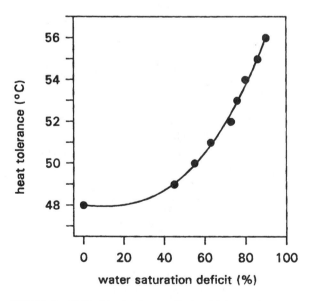

Figure 5 Shift of heat tolerance limit with increasing water saturation deficit in leaves of *Ramonda myconi*. (*Source*: after Kappen 1966.)

in photosynthetic repair capacity of *Selaginella lepidophylla* if the plants were subjected when dry to 45 and 65°C, he found that desiccation tolerance decreased with increasing temperatures (25, 45, 65°C). However, since the water content of the ''dry'' plants was not defined, it still remains open whether metabolic disturbance at the higher temperature weakened the plants.

IV. CARBON BUDGET

Photosynthetic features such as speed of reactivation upon rehydration and maximal rates of carbon uptake determine the productivity of poikilohydrous plants. Water content of poikilohydrous plants is frequently a crucial factor for gas exchange and, at low water contents, the relationship between hydration and CO_2 exchange is very close. The carbon economy of these organisms is influenced by the frequency and duration of the periods of metabolic activity and by repair and other processes that delay maximal net photosynthesis once the plant enters in the biotic state by rehydration. In fact, poikilohydrous species exhibit significant differences in their speed of recovery from dehydration and in their photosynthetic performance while they are active.

A. Photosynthesis

While cyanobacteria, algae, lichens, and bryophytes can reach maximal photosynthetic rates within 10–30 minutes after rehydration, poikilohydrous vascular plants usually require much longer (Table 2). Generally, the threshold for respiratory activity is lower than that for net photosynthesis (Lange and Redon 1983). The capacity of chloroplasts and mitochondria to function at low water potentials is remarkable in aerophytic green algae and green-algal lichens (-50 MPa and -38 MPa, respectively; Bertsch 1966a, Nash et al. 1990), and also in homoiohydrous vascular plants (-19 MPa in the mesophyll of *Valerianella chlorenchyma*, Bertsch 1967; beet chloroplasts, Santarius 1967). However, each particular poikilohydric organism has a suite of morphological and functional traits that exert specific influences on photosynthetic yield and overall performance during the periods of water availability. These organisms are covered in the following sections.

1. Lichens and Bryophytes

Lichens can be photosynthetically active at water contents as low as 20% dry weight (Lange 1969b, Lange et al. 1990b), under high saline stress (Nash et al. 1990), and at temperatures as low as $-20°C$ (Kappen 1989, 1993a, Schroeter et al. 1994; see also Chapter 15). Schroeter and Scheidegger (1995) demonstrated

Table 2 Time Required by Different Poikilohydrous Autotrophs To Resume Photosynthetic Activity (Net Photosynthesis) After Rehydration with Water Vapor (>90% Relative Humidity) or with Liquid Water

Species	Water source	Time	Reference
Parmelia hypoleucina	Water vapor	<1 hr	Lange and Kilian 1985
Cladonia portentosa	Water vapor	6 hr	Lange and Kilian 1985
Caloplaca regalis	Water vapor	24 hr	Lange and Kilian 1985
Lobaria pulmonaria	Liquid	<30 sec[a]	Scheidegger et al. 1997
Hypnum cupressiforme	Water vapor	12 hr	Lange 1969
Tortula ruralis	Liquid	90 min	Scott and Oliver 1994
Exormotheca holstii	Liquid	2 hr[a]	Hellwege et al. 1994
Polypodium virgineanum	Liquid	6 hr[b]	Reynolds and Bewley 1993
Selaginella lepidophylla	Liquid	24 hr[b]	Eickmeier 1979
Craterostigma plantagineum	Liquid	15 hr[b]	Bernachia et al. 1969
Ramonda serbica	Liquid	24 hr	Markowska et al. 1997
Xerophyta scabrida	Liquid	72 hr	Tuba et al. 1993b

[a] Chlorophyll a fluorescence positive.
[b] Chloroplasts reconstituted.

that net photosynthesis was still positive when the algal cells were shriveled and the fungal hyphae cavitated due to low water content of the thallus. At a given low water potential, different lichen species exhibited different fractions of collapsed photobiont cells and different photosynthetic rates (Figure 6). It has therefore been suggested that an increase of net photosynthesis is, in general, proportional to an increase in the amount of functional photobiont cells (Figure 7). However, the levels of hydration at which algal cells reach maximal and minimal photosynthetic rates are unknown. The fungal part of the lichens proved to be totally impotent in aiding the algal cells to absorb water vapor, as was shown by comparing lichens and isolated photobionts (Lange et al. 1990b, Kappen 1994). Most importantly, these lichens are able to start photosynthesis by water vapor uptake at low water potentials, as was observed in the laboratory (Lange and Bertsch 1965, Lange and Kilian 1985, Scheidegger et al. 1995) as well as in the field (Kappen 1993a, Schroeter et al. 1994).

Lichens with cyanobacteria as photobionts are not capable of initiating any significant photosynthetic activity when exposed to humid air. As Figure 8 shows, cyanolichens need liquid water for activation of their photobionts. Their water content for starting photosynthesis (85–100% dry weight) is 3–6 times higher than that of green-algal lichens (15–30% dry weight; Lange et al. 1988). The deficiency of cyanobacteria to become photosynthetically activated by water vapor cannot be compensated by the symbiosis with the fungus (Kappen 1994).

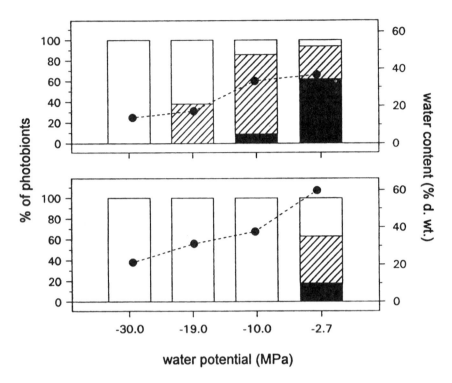

Figure 6 Amount of photobionts (open bars, heavily collapsed; hatched bars, partially folded; solid bars, globular) and water content (solid circles, % dry weight) of thalli of the lichens *Ramalina maciformis* (top) and *Pseudevernia furfuracea* (bottom) at an equilibrium with water potentials between −30 and −2.7 MPa. (*Source*: after Scheidegger et al. 1995.)

The explanation for this inability to resume turgidity by water vapor uptake remains unclear, but the possibility that diffusion resistance by the gelatinose sheath of cyanobacteria could impede water vapor absorption has been excluded (Büdel and Langel 1991).

Quick recovery of CO_2 exchange during the daily wetting and drying cycles has been demonstrated in lichens from various habitats (Kappen et al. 1979, Lange and Redon 1983, Lange et al. 1990c, 1991, Bruns-Strenge and Lange 1991). However, in some cases, CO_2 exchange of lichens and bryophytes revealed only respiratory CO_2 release, frequently at high rates, during short periods of soaking with water. This release was interpreted as being due to recovery processes (Ried 1960, Hinshiri and Proctor 1971, Smith and Molesworth 1973, Dilks and Proctor 1974, 1976a, Farrar 1976b). These recovery periods, which range

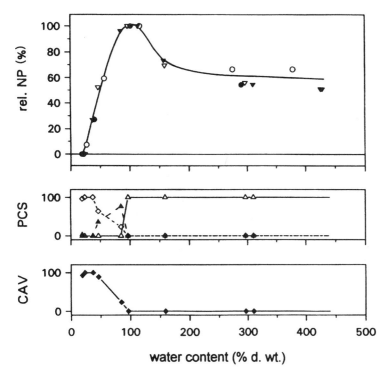

Figure 7 Relative net photosynthetic rates (rel. NP as % of the maximum rate) in the lichen *Lobaria pulmonaria*, changes of photobiont cell shape (PCS) given as percentage of globular (△), slightly indented (▲), and heavily collapsed cells (◇), and percentage of cavitated hyphae (CAV) in the upper cortex in relation to water content (% dry weight). (*Source*: after Scheidegger et al. 1995.)

from 15 minutes to 3 hours, cause a significant carbon loss (Lechowicz 1992). Slow recovery was shown by species of moist habitats (stenopoikilohydrous species), whereas quick recovery was found in species subjected to frequent and severe drying (eurypoikilohydrous species).

Most bryophytes require liquid water for activation of photosynthesis, although some species (e.g., *Hypnum cupressiforme*) are able to activate photosynthesis by uptake of water vapor (Lange 1969a). Thalli of the desert moss *Barbula aurea* showed high respiration rates in a water-saturated atmosphere, but could achieve an almost negligible net photosynthetic rate when irradiated (Rundel and Lange 1980). In bryophytes as well as in lichens, net photosynthetic rate becomes maximal within 15–30 minutes if the dry thallus is moistened with liquid water. Photosynthetic rates of mosses decreased with increasing length of the drought

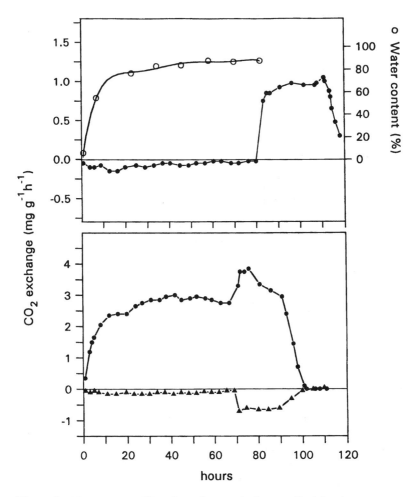

Figure 8 Water content (○) and net photosynthetic rates (●) following water vapor and then liquid water uptake of the lichen species *Nephroma resupinatum* with cyanobacterial photobionts (top) and *Ramalina menziesii* with green-algal photobionts (bottom). In *N. resupinatum*, water vapor uptake (○) resulted in maximum thallus water content of 85% dry weight but induced only a weak respiratory response in light, whereas spraying activated photosynthetic CO_2 uptake. In *R. menziesii*, both dark respiration (▲) and net photosynthesis (●) were activated within the first 30 minutes of water vapor uptake. (*Source*: after Lange et al. 1986.)

period before hydration. Repeated dry–wet cycles caused a greater decrease of net photosynthesis in mesic stenopoikilohydrous mosses than in xeric species (Davey 1997).

As mentioned in Section II, photosynthetic CO_2 uptake can be severely impeded by excessive water contents. This has been intensively studied in lichens (Lange 1980, Lange and Tenhunen 1981, Lange et al. 1993, Green et al. 1994), where the major resistance to CO_2 diffusion seems to result from the soaking of the cortex (Cowan et al. 1992), but the effect is specific to species and can also be absent. According to Lange et al. (1993), four different types of net photosynthetic response to water content can be discerned in lichens (Figure 9): (1) net photosynthesis follows a saturation curve and does not decrease at high water content; (2) photosynthesis is maximal over a wider range of water contents but decreases slightly at high water content; (3) similar to (2) but net photosynthesis becomes zero or even negative at high water contents; and (4) the range of water content for high rates of net photosynthesis is narrow, with a clear optimum followed by a sharp decrease at high water contents where a low, but positive net photosynthesis occurs. Surprisingly, these different photosynthetic responses to hydration could not be related with the environmental conditions under which

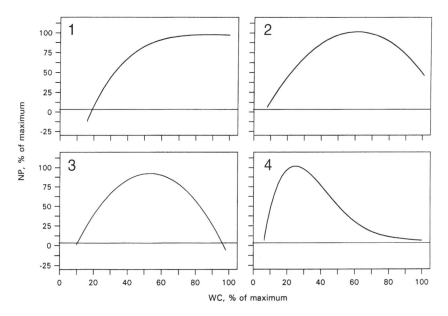

Figure 9   The four types of response of net photosynthesis (NP) to thallus water content (WC). Schematic curves based on measurements on 11 different lichen species from forests in New Zealand. (*Source*: after Green et al. 1994.)

each species was found, but species from xeric habitats have been found to have their maximal photosynthetic rates at relatively low water contents (60–120% dry weight) (Lange 1980, Lange and Matthes 1981, Kappen 1983).

In bryophytes (mainly endohydrous species; Green and Lange 1994), a similar effect can be identified (Dilks and Proctor 1979, Rundel and Lange 1980, Tuba et al. 1996a). However, the effect of diffusion resistance is diminished in ectohydrous bryophyte species as their growth form and anatomy allow a separation of the water storing sites from the photosynthetic tissues and as they have ventilated structures (Proctor 1990, Green and Lange 1994). It is remarkable that photosynthetic response curves of intertidal algae show a depression at high water contents (Quadir et al. 1979, Bewley and Krochko 1982). This indicates that increased CO_2 resistance with hydration is not only due to free water blocking gaseous pathways within the thallus, but also to tissue swelling, which seems to be a widespread phenomenon among nonvascular plants.

Maximal net photosynthetic rates of lichens are generally low, although there are rare exceptions (Table 3). The most productive lichens are epiphytes from shady forests (Green and Lange 1994). The photobionts of lichens from open habitats are frequently screened by a pigmented or thick upper cortex (Büdel 1987, Kappen et al. 1998a). Thus, although the chlorophyll content of lichens is usually enough to harvest more than 80% of the incident light (Valladares et al 1996), the photosynthetic units of lichens experience light limitations even in high-light environments (Büdel and Lange 1994). This fact, together with some additional structural and physiological features (thylakoid structure, low light compensation point, large amounts of light-harvesting protein complexes), has led to the suggestion that lichens resemble ''shade plants'' (Green and Lange 1994).

2. Homoiochlorophyllous Vascular Plants

Resurrection vascular plants strictly depend on liquid water to recover their photosynthetic activity after desiccation, and they need longer periods of time than nonvascular autotrophs to regain full activity (Table 2). Ferns and allies require 4.5–48 hours to gain full or reduced photosynthetic rates (Oppenheimer and Halevy 1962, Stuart 1968), whereas in other cases, up to 4 days of hydration are necessary to recover net photosynthesis (e.g., *Myrothamnus flabellifolius*; Hoffmann 1968, Vieweg and Ziegler 1969). In contrast, dark respiration of most poikilohydrous plants is quickly activated in a saturated atmosphere. Plants of *Selaginella lepidophylla* that were dry during a whole year recovered full photosynthetic capacity within 20 hours whether the plants were rehydrated in light or in darkness (Eickmeier 1979). During the first 6 hours, the plants exhibited only respiration, and the photosynthetic recovery was linked to the regaining of Rubisco (ribulose bis-phosphate carboxylase [RuBP]) carboxylase activity,

Table 3 Maximal Rate of Net Photosynthesis, Photosynthetic Light Compensation, and Saturation Points, and Minimum Water Content Allowing Photosynthetic Activity in Different Poikilohydrous Autotrophs

Organism	Maximum rate of net photosynthesis (different units)	Light compensation point (μmol photons m^{-2}s^{-1})	Light saturation point (μmol photons m^{-2}s^{-1})	Minimum water content for net photosynthesis	Source
Fucus distichus (high intertidal zone)	6.3 mg CO$_2$ g^{-1}h^{-1}			80% desiccation	Quadir et al. 1979
Lichens (in general)	0.2–5.0 mg CO$_2$ g dwt^{-1}h^{-1}				Kallio and Kärenlampi 1975, Green et al. 1984
Lecidella crystallina	5.9 μ mol CO$_2$ m^{-2}s^{-1}	28–43		0.13 mg m^{-2}	Lange et al. 1994
Barbula aurea	1.4 mg CO$_2$g dwt^{-1}h^{-1}	20	200	23% dwt	Rundel and Lange 1983
Tortula ruralis	4.8 mg CO$_2$ g dwt^{-1}h^{-1}			10% dwt	Tuba 1987
Polypodium virginianum	3.5 μ mol CO$_2$ m^{-2}s^{-1}	20–30	100–200		Gildner and Larson 1992
Cheilanthes maranthae	3.48 mg CO$_2$g dwt^{-1}h^{-1}	5–50	500–750		Gebauer 1986
Cheilanthes maderensis	4.04 mg CO$_2$ g dwt^{-1} h^{-1}	5–25	100–500		Gebauer 1986
Notholaena parryi	5.5 μmol CO$_2$ m^{-2} s^{-1}		150		Nobel 1978
Selaginella lepidophylla	2.44 mg CO$_2$ g dwt^{-1} h^{-1}		>2000		Eickmeier 1979
Ramonda myconi	2.39 mg CO$_2$ g dwt^{-1} h^{-1}	30	400–500	20% RWC	Gebauer 1986
Xerophyta scabrida	4.0 μmol CO$_2$ m^{-2} s^{-1}				Tuba et al. 1994
Homoiohydrous sclerophylls of dry regions	7.9–23.8 μmol CO$_2$ m^{-2} s^{-1}				Larcher 1980
Homoiohydrous sun plants	10–20 μmol CO$_2$ m^{-2} s^{-1}				Lüttge et al. 1988

dwt, dry weight; RWC, relative water content.

which was conserved by 60% during the long abiotic period (Eickmeier 1979). *Selaginella* species differed in the rapidity of recovering photosynthesis and in the length of the initial period of respiration. While *S. lepidophylla* is a characteristic eurypoikilohydrous species, *S. pilifera*, living in more mesic habitats, behaves as a stenopoikilohydrous species (Eickmeier 1980).

Long periods of drought, too-frequent cycles of drying and wetting, and too-rapid drying can decrease productivity of poikilohydrous plants by affecting the maximal rates of net photosynthesis. In *Selaginella*, rates of net photosynthesis after hydration decreased with increasing length of the period of desiccation (Eickmeier 1979). In *Ramonda myconi*, net photosynthesis decreased to one third after the fifth cycle of drying and wetting (Gebauer 1986). In *S. lepidophylla*, intermediate drying speeds (52–94 hours until complete curling of the plant was reached) led to maximal recovery, whereas either very rapid (5.5 hours) or very slow (175 hours) drying was associated with significantly reduced photosynthetic rates (Eickmeier 1983). Rapid drying caused increased membrane dysfunction, whereas slow drying caused retarded de novo protein synthesis.

Although C_3 photosynthesis is by far the most common photosynthetic pathway among poikilohydrous vascular plants, all of the other pathways have also been found in these plants. In many resurrection grasses of the genera *Eragrostidella, Sporobolus*, and *Tripogon* from Australia, Kranz-type anatomy indicated that they were C_4 plants (Lazarides 1992). Certain poikilohydrous vascular plants such as *Ramonda serbica* and *Haberlea rhodopensis*, with slightly succulent leaves when soaked, exhibited a crassulacean acid metabolism (CAM) under water stress (Kimenov et al. 1989, Markowska et al. 1994, 1997). This was confirmed by the diel CO_2 exchange pattern and by phosphoenol pyruvate (PEP) carboxylase activity. CAM activity was interpreted as a mechanism to delay periods of metabolic inactivity and to recycle CO_2. As this enhances reductive capacity, photo- and drought-induced oxidative processes can be ameliorated. The only known truly succulent resurrection plant, *Blossfeldia liliputana*, is a CAM plant according to its carbon isotopic and carbon discrimination values (Barthlott and Porembski 1996). The presence of CAM activity in poikilohydrous plants demonstrates that water conservation mechanisms can be combined with a desiccation-resurrection strategy.

3. Poikilochlorophyllous Vascular Plants

Poikilochlorophyllous plants have only been found among the monocots (e.g., *Borya nitida*, Vellociaceae, Cyperaceae, and Poaceae; see Gaff and Hallam 1974, Gaff et al. 1976). In the dry state, leaves of poikilochlorophyllous plants are yellow-brown and, apparently, some energy transfer from the plant is necessary for regreening because detached young leaves need longer to recover than attached ones. Their photosynthetic activity after rehydration logically depends on the rate of resynthesis of photosynthetic pigments, which are lost during desiccation (Gaff 1989). Chloroplasts have to be resynthesized de novo on rehydration from desiccoplasts (Tuba et al. 1993b). In poikilochlorophyllous plants, desiccation affects not only the functioning but also the ultrastructure of the photosynthetic system. The breakdown of the chloroplasts requires a relatively slow desiccation. For instance, if the leaves of *Borya nitida* were dried very quickly, their chloroplasts remained preserved, whereas, if they were slowly dried in a humid atmosphere, most of their chlorophylls were lost within 22 hours. During regreening upon rehydration, photosystem I activity appeared to recover faster than that of the photosystem II (Gaff and Hallam 1974, Hetherington and Smillie 1982).

During slow desiccation (at 96% relative humidity), net photosynthesis of *Xerophyta scabrida* declined to the compensation point within 8 hours, dark respiration rate was decreased by 50%, and it took 4 days to reach an 85% reduction of the chlorophyll content (Tuba et al. 1996b). Drying at higher relative humidities maintained net photosynthesis over 4 days but extended the period of respiratory carbon loss to 14 days (Tuba et al. 1997). Respiration of *X. scabrida* was activated within 20 minutes of rehydration and reached full rates within 6 hours, even before turgor was restored in the cells (Tuba et al. 1994). However, chlorophyll resynthesis started only after 12 hours of rehydration and was not complete until 36 hours had elapsed. The time required to fully restore photosynthetic capacity upon rehydration was not different whether the plants were exposed to air with 350 or 700 ppm of CO_2, but down-regulation of photosynthetic rates was found at 700 ppm (Tuba et al. 1996b).

Although the reconstruction of the photosynthetic apparatus during rehydration of poikilocholorphyllous plants may reduce the period of photosynthetic activity and their overall productivity, it seems to enhance tolerance to environmental stresses like excessive irradiance. Comparative studies between homoio- and poikilochlorophyllous species may reveal more about the ecophysiological advantages of this extreme poikilohydric strategy.

B. Processes that Delay Recovery of Carbon Gain

Among the many possible structural and functional processes that cause a delay of the CO_2 exchange upon rehydration, those occurring in chloroplasts and mitochondria have received special attention (Bewley and Krochko 1982). Intrinsic "repair-based" mechanisms, as have been investigated in *Tortula ruralis*, can enable a quick start of CO_2 exchange (Bewley and Oliver 1992, Gaff 1989). The rate of protein synthesis on rehydration is considerably lower in rapidly than in slowly dried mosses. The accumulation of oxidized glutathione during drying causes an initial inhibition of protein synthesis, which can be prevented by glutathione reductase activity (Dhindsa 1991). In contrast to lichens and mosses, repair processes in vascular plants during rehydration involve dramatic ultrastructural changes, which may last more than 2–8 days in species of *Ramonda* and *Haberlea* (Gaff 1989, Markowska et al. 1995). Two main mechanisms are involved: one that down-regulates the processes leading to desiccation tolerance during dehydration, and another one that contributes to full metabolic recovery (Bernacchia et al. 1996). Protein synthesis is necessary not only for protection against desiccation and associated stresses (see Section III), but also for repair of the plastid membranes, for de novo synthesis of chlorophylls, and for the restitution of cell organelles once the plant becomes active again. Transient synthesis of rehydration proteins seems to be required for increased photosynthetic activity in certain species (e.g., chlorophyll a/b binding protein; Bernacchia et al. 1996). Recovery of thylakoid function is a major cause for the delay of reactivation of CO_2 uptake, provided that the plasmalemma is intact (Schwab et al. 1989). Thylakoid membrane proliferation and chloroplast polysome function in *Selaginella lepidophylla* starts only after 8–12 hours of hydration with subsequent grana stacking during the next 6–10 hours, a process that requires organelle and cytoplasmic-directed de novo protein synthesis (Eickmeier 1982). In the African resurrection fern *Pellaea calomelanos*, chloroplasts remain intact (Gaff 1989), and in plants capable of a relatively quick resurrection (bryophytes and some pteridophytes), mechanisms that permit conservation of structures during desiccation seem to be more important than repair (Eickmeier 1982). Among the most resistant organelles to desiccation are the mitochondria, which explains why respiration can start so quickly in comparison to photosynthetic processes (Smith and Molesworth 1973, Dilks and Proctor 1974, Hallam and Capicchiano 1974, Eickmeier 1979, 1980, Schwab et al. 1989).

Not surprisingly, the restoration processes are more complex and time consuming in poikilochlorophyllous resurrection plants than in their homoiochlorophyllous counterparts. The fine structure of the chloroplast is radically changed during desiccation: grana disappear and most thylakoids are lost forming the so-called desiccoplasts (Bartley and Hallam 1979, Gaff 1989, Tuba et al. 1993b). The regeneration of chloroplast structure may take 24 hours or more in plants such as *Talbotia elegans* (Hallam and Gaff 1978). In *Xerophyta scabrida*, the degree of stacking in thylakoids was still increasing after 72 hours of rehydration and continued until the thylakoid system was fully reconstituted (Tuba et al. 1993b). When most of the water was taken up by leaves, modifications in the protein profile still took place in *Craterostigma plantagineum* (Bernacchia et al. 1996). Within the first 3–6 hours of rehydration, no new messenger RNA (mRNA) was available, and changes in the translatable mRNA populations could not be detected earlier than 15 hours after the beginning of the rehydration (Hsiao 1973, Kaiser 1987, Stewart 1989). Only then did the synthesis of the CO_2-fixing enzyme Rubisco start (Bernacchia et al. 1996).

Available studies on poikilohydrous plants have focused on the cellular and molecular mechanisms underlying inactivation and reactivation of photosynthesis and respiration. With a few exceptions, seemingly relevant aspects of the recovery of plant CO_2 exchange such as stomatal responses (Schwab et al. 1989, Lichtenthaler and Rinderle 1988, Tuba et al. 1994), leaf performance (Stuart 1968, Matthes-Sears et al. 1993), and whole plant behavior (Eickmeier et al. 1992) have been neglected. Field studies of the CO_2 exchange during the dehydration/rehydration cycles, such as those conducted with lichens (see Section V.A), are necessary to reach a better ecological understanding of the carbon economy of poikilohydrous plants.

C. Opportunistic Metabolic Activity

The photosynthetic activity of lichens has been monitored in situ in various climatic regions (Kappen 1988). Figure 10 depicts diurnal courses for lichens in two xeric habitats, the Namib desert in spring (Lange et al. 1990c), and south-exposed rocks of southern Norway in winter (Kappen et al. 1996). As can be seen in this figure, frozen water can maintain CO_2 exchange in winter if irradiance is sufficient.

Different types of diurnal courses of gas exchange of lichens result from the existence of several possible sources of moisture (dew, fog, and rainfall, each one alone or in combination). Lösch et al. (1997) have proposed a simplified scheme for the seasonal variation of the periods of photosynthetic activity of nonvascular autotrophs with regard to the climatic conditions: in subpolar and polar regions, the most productive and extended periods of activity occur in spring, summer, and fall; in temperate regions, the periods of activity are regular but rather short and occur during all seasons, although they are somewhat limited in summer; in hot arid regions, the periods of activity are brief but frequent, since they potentially occur throughout the year or every day during the wet season, depending on the region (Figure 11). Examples of field studies in different environments are the studies by Hahn et al. (1989, 1993), Bruns-Strenge and Lange (1991), Sancho et al. (1997), and Green et al. (Chapter 15 of this volume). The daily periods of moistening may last from half of the night to up to 3 hours in sunlight if dewfall is involved. In arid regions, the length of the period of lichen hydration is directly influenced by the exposure and orientation of the location. In warm and temperate climates, shaded habitats allow the longest periods of hydration, but in frigid climates, hydration is combined with insolation of the habitat (Kappen et al. 1980, Kappen 1982, 1988, Nash and Moser 1982, Pintado et al. 1997).

A long-term investigation of the lichen *Ramalina maciformis* in the Negev desert revealed that thalli were active on most days of the year, with dew causing 306 days of metabolic activity (of which carbon balance was positive on 218 days and negative on 88 days), whereas rainfall produced activity on only 29 days per year (Kappen et al. 1979). In high mountain and polar regions, melting snow can extend the productive period of lichens and bryophytes for many days

Photosynthesis

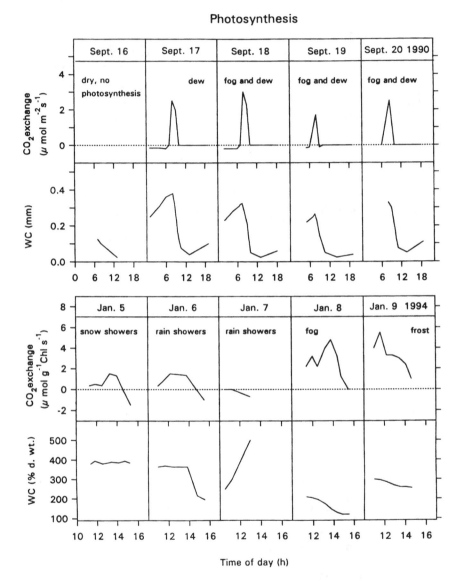

Figure 10 Series of diurnal courses of lichen water content (WC) and CO_2 exchange of *Lecidella crystallina* (on an area basis) from southern Africa in spring (Sept. 16–20, 1990) and of *Umbilicaria spodochroa* (on chlorophyll basis) from Norway in winter (Jan. 5–9, 1994). (*Source*: after Lange et al. 1994 and Kappen et al. 1996.)

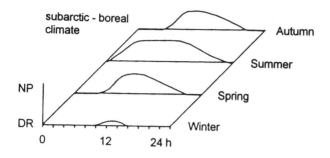

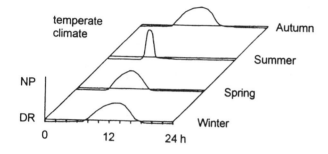

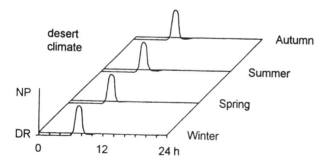

Figure 11 Schematic description of typical diurnal photosynthetic activity in the seasonal course for bryophytes and lichens in different climatic regions. NP, net photosynthesis; DR, dark respiration. (*Source*: after Lösch et al. 1997.)

Table 4 Number of Days with Metabolic Activity Within 1 Year in Various Regions

	Active	Positive CO_2 balance	Negative CO_2 balance	Source
Negev 1971/1972, *Ramalina maciformis*	306	218	88	Kappen et al. 1979
Maritime Antarctic 1992/1993, *Usnea aurantiaco-atra*	268	150	118	Schroeter et al. 1997
Continental Antarctic 1985/1988, Cryptoendolithic lichen community	120	120(?)	0(?)	Friedmann et al. 1993

or weeks (Kappen et al. 1995a, Kappen et al. 1998b). In Antarctica, the lichen *Usnea aurantiaco-atra* was active for a total of 3359 hours within 268 days in 1 year in the wet Maritime region (Schroeter et al. 1997), whereas the period of activity was reduced to one fifth of this value within about 120 days in the Antarctic Dry Valleys for cryptoendolithic microorganisms (see Table 4) (Friedmann et al. 1993, Nienow and Friedmann 1993).

Diurnal cycles of photosynthetic activity of bryophytes have been recorded in the field in temperate (Lösch et al. 1997) and subarctic and polar habitats (Hicklenton and Oechel 1977, Oechel and Sveinbjörnson 1978), demonstrating the dependency of the periods of activity on weather conditions and habitat factors, as shown in lichens. In polar regions and high mountains, the photosynthetic activity of the typically shade-adapted bryophytes may be depressed by temporarily high insolation combined with snow melt (Valanne 1984, Kappen et al. 1989).

For poikilohydrous vascular plants, we can only refer to one diurnal course measured in *Notholaena parryi* of the western Colorado desert showing full light-driven CO_2 exchange (Nobel 1978), and to a study on *Chamaegigas intrepidus* by Gaff and Gies (1986). In the latter investigation, the authors found 11 events of activity, each lasting from 2–23 days over a period of 1.5 years (Figure 12).

V. LIMITS AND SUCCESS OF POIKILOHYDRY

A. The Different Strategies

The life strategy of poikilohydrous plants in contrast to the homoiohydrous plants can be characterized by the hydration range rapidity with which they enter into the anabiotic state during desiccation and with which they recover normal activity during rehydration (Figure 13). According to the features discussed throughout

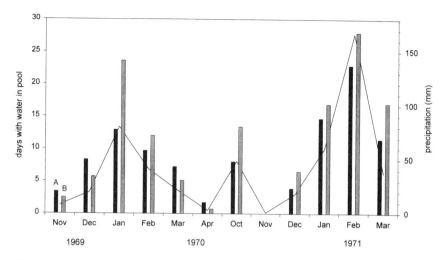

Figure 12 Amount of rainfall (line) and periods of filling two rockpools (columns) in Fritz Gaerdes Reserve, Okahandja, Namibia, where *Chamaegigas intrepidus* (Scrophulariaceae) grows. (*Source*: after Gaff and Giess 1986.)

this chapter, we can discern four kinds of strategies among the eurypoikilohydrous organisms.

The *Ready type* is exhibited by constitutively poikilohydrous nonvascular species (cyanobacteria, algae, lichens, and some bryophytes), which can easily lose and absorb water, switching their metabolism on and off very quickly. Their cellular structures are highly strengthened and do not need extensive, time-consuming repair processes, which allows them to oscillate very frequently (daily or even within a few hours) between anabiosis and active state.

Certain bryophytes and vascular plants need variable periods of time (hours in the case of the former, days in the case of the latter) to recover completely from desiccation and therefore represent a *Repair type*. The most clear example of this strategy are the poikilochlorophyllous plants. Ecologically, the partial destruction of organelles and photosynthetic pigments during desiccation seems to be disadvantageous, since it reduces the period of activity and limits photosynthetic carbon gain. However, this response type may prevent oxidative membrane deterioration, particularly in plants that grow in open-exposed habitats. Such plants need rather long and continuous periods of activity and oscillate between dry and wet states only a few times per year (Gaff and Gies 1986).

Certain bryophytes and vascular plants are not capable of tolerating extreme desiccation without a previous hardening or preconditioning, and are therefore typical of the *Prepare type*. Tolerance to desiccation is increased if they are

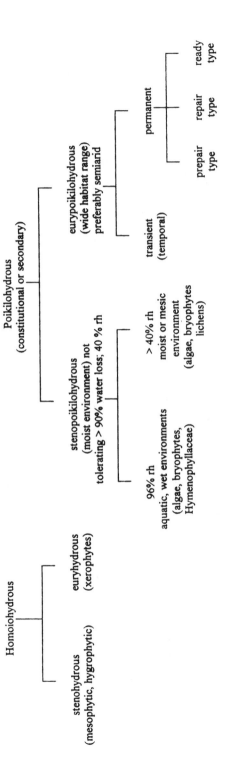

Figure 13 Water status–related plant performance.

exposed to a slow water loss, or if water loss occurs under a low vapor pressure deficit (e.g., *Bryum caespititium* and *Pohlia elongata*). Such plants either make use of structural features that retard water loss or grow in sheltered habitats where the evaporative potential is low (rock colonizing ferns, *Borya nitida*).

The *Transient type* includes certain bryophytes and ferns that acquire a eurypoikilohydrous character only temporarily as their fronds become extremely desiccation tolerant during the winter or the dry season (*Polypodium vulgare, Asplenium* species, and *Satureja gilliesii*), and *Limosella* and *Aponogeton* that stay poikilohydrous with a part of the vegetative organs such as rhizomes through the dry season.

B. A Place in Plant Communities

Their remarkable tolerance to climatically extreme conditions, their capacity to colonize both the exterior and the interior of rocks, and their relative success as epiphytes allow many poikilohydrous autotrophs to survive and grow in the absence of strong competition by homoiohydrous plants. Nonvascular and vascular poikilohydrous species are successful pioneers and they become dominant as long as the site is not well-suited for the establishment of homoiohydrous plants. For instance, a transition from dominance to exclusion of poikilohydrous vascular plants was found in southern Africa as a function of soil depth (Gaff 1977). Initial stages are characteristic with poikilohydrous water plants (*Chamaegigas intrepidus* in Namibia or *Craterostigma monroi* in Zimbabwe) and lead to grass-herb communities over several steps to, finally, a pluvio-therophytic grass vegetation or associations between *Xerophyta* species and perennial succulent life-forms. Although they grow slowly, they are also very successful in cracks and rock ledges with changing water supply once they are established. In fact, most of the poikilohydrous vascular desert plants preferably colonize sheltered, semi-shady habitats in temperate as well as in semi-arid climates (Nobel et al. 1978, Markowska et al. 1994). This is also evident with most eurypoikilohydrous bryophytes and lichens that benefit from preserved moisture in temporarily shaded conditions (exposure effect; Kappen 1982, 1988, Rundel 1982a).

In many habitats, poikilohydrous communities are interspersed among the vegetation. Lichens, bryophytes, and vascular poikilohydrous plants form permanent communities in xeric habitats on shallow soils (Gaff 1977, Volk 1984, Müller 1985, Lazarides 1992). In these places, they can outcompete therophytes, as they are able to defend their place against invasion of seeds and seedlings. On open, exposed soils, many vascular poikilohydrous species form herds, turfs, or mats, whereas cryptogamic species typically form crusts in desert lands. Both forms, but especially the latter, are ecologically crucial because they reduce erosion and contribute to the preservation of hydric, chemical, and physical properties of the soil (Danin and Garty 1983, Belnap 1994).

Algae, lichens, and bryophytes grow potentially everywhere due to their small and easily transportable vegetative and generative propagules (Kappen 1995). The evolution of a great variety of asexual means of reproduction has made nonvascular plants extremely successful in the colonization of remote and difficult sites. We have only limited knowledge about reproductive strategies and success of eurypoikilohydrous vascular plants. Among the few exceptions is *Myrothamnus flabellifolius*, which forms perianthless inconspicuous monoecic flowers that are most likely anemogamous (Puff 1978, Child 1960). According to Puff (1978), pollen tetrades may increase the fertility on a receptaculum (success of mating) in this species. The seeds are extremely small and can be dispersed over long distances by the wind, but we have no records about periods and conditions of flowering. The flowers of poikilohydrous Scrophulariaceae do not differ much from those of the homoiohydrous members of the family (Heil 1925, Hickel 1967, Smook 1969, Gaff 1977). However, the rapidity of producing flowers is remarkable in species such as *Chamaegigas intrepidus*, as flower buds appeared simultaneously with the floating levels. As expected, the reproductive phase occurred predominantly during the wet season, and a general requirement was a period of photosynthetic activity before reproduction, as was shown for *Vellozia schnitzleinia* (Nigeria; Owoseye at Sandford 1972) and poikilohydrous grass species in Australia (Lazarides 1972). Nevertheless, a few grass species can reproduce during the dry period. Clearly, more studies of the reproductive biology of poikilohydrous vascular plants are required to interpret their ecological strategy and to compare their relative success with that of homoiohydrous co-occurring species.

C. Primary Production of Poikilohydrous Autotrophs

A measure of the success of a given plant is its rate of biomass production and its primary production integrated throughout a growth period. As previously discussed, photosynthetic and growth rates of poikilohydrous plants are low in comparison with homoiohydrous species (Table 3). Thus, the time period during which the plant is hydrated becomes a crucial factor for the primary production of poikilohydrous plants in relation to that of the potentially more productive homoiohydrous plants. To our knowledge, no data are available regarding the primary production of poikilohydrous vascular plants. Studies with lichens and bryophytes have revealed the existence of different diel cycles of activity depending on the season, the habitat, and the weather (Figure 11), each combination resulting in very different carbon balances. In the Namib desert, fog resulted in an annual carbon gain of 20% of the standing carbon mass for the lichens studied (Lange et al. 1990c). Annual carbon gain was 20–30% in the maritime Antarctic and northern Europe (Schroeter et al. 1995, Kallio and Kärenlampi 1975) and went up to 79% in a coastal habitat with high precipitation in northern Germany

(Bruns-Strenge and Lange 1992). In contrast, the estimated annual carbon gain was only 3.8% in the harsh conditions of the continental Antarctic (Kappen 1985).

Biomass of lichens can be surprisingly high in tundra ecosystems, especially when compared with that of vascular plants. For instance, in the Alaskan tundra (Atkasook), lichen biomass was 76 g dry matter m^{-2}, which represented 26% of the total aboveground biomass of the system, and it reached 1372 g m^{-2} near Anatuvuk Pass in the Brooks Range (Lechowicz 1981). In Scotland, the annual increase in dry biomass of the lichen *Cladonia portentosa* alone was 47 g m^{-2} (Prince, quoted by Bruns-Strenge and Lange 1992). Lichen primary production is also surprising in deserts, where they locally exhibit similar or higher values to those of the homoiohydrous vascular plants. A field study of the desert lichen *Ramalina maciformis* found an annual carbon gain of 60–195 mg CO_2 g^{-1} dry weight, which is equivalent to a net carbon gain of 4.5–15.0% (Kappen et al. 1980). In the poor soils of the Namib desert, the lichen crust has an annual carbon gain of 16 g C m^{-2} (Lange et al. 1991), and the lichen *Teloschistes capensis* forms dense turfs of a biomass of 250 g m^{-2} (Kappen 1988). Similar values were found in the Negev desert, where the lichen biomass was within the range of that of the vascular plant vegetation (145–323 g m^{-2}; Kappen et al. 1980).

Bryophytes are frequent in deserts, but they are not very productive in these arid environments. They rarely surpass a 0.3–8% cover and a biomass of 2 g m^{-2} (Nash et al. 1977). However, in tundra ecosystems they can be as productive and abundant as lichens and vascular plants. For example, in the Alaskan Arctic, bryophytes exhibit an average cover of almost 60%, and they contribute 36% of the net aboveground production of the ecosystem (Webber 1978). In certain areas of the maritime Antarctic, where cryptogamic phytomass reaches values of 1900 g m^{-2}, bryophytes are dominant (Kappen 1993b). However, most of them are stenopoikilohydrous moss species that, in warmer and moist habitats, form up to 4750 g m^{-2} measured in the uppermost 10 cm of the profile (Longton 1988b). *Sphagnum* species, for instance, may cover 1%, and lichens were estimated to dominate in total approximately 8% of the land surface of the globe (Clymo 1970, Larson 1987, Ahmadjian 1995).

ACKNOWLEDGMENTS

The authors wish to thank Professors Carmen Ascaso, T. G. Allan Green, Wolfram Hartung, O. L. Lange, and Rainer Lösch for kindly providing information and discussion on poikilohydrous plants. Allan Green also helped to edit the manuscript. Stefan Pannewitz (Kiel, Germany) helped to acquire the literature and organize the reference list. F.V. was supported by the Spanish Ministry of Education and Culture and CSIC, within the project DGICT PB 95-0067.

REFERENCES

Abel, W.O. 1956. Die Austrocknungsresistenz der Laubmoose. Sitzungsber. d. Wiener Akademie d. Wissenschaften. Math. Nat. Kl. Abt. I, 165: 619–707.

Ahmadjian, V. 1973. Resynthesis of lichens. In: Ahmadjian, V., and M.E. Hale, eds. The Lichens. New York: Academic Press, pp 565–580.

Ahmadjian, V. 1995. Lichens are more important than you think. BioScience 45: 124.

Albini, F.M., C. Murelli, G. Patritti, M. Rovati, P. Zienna, and P.V. Finzi. 1994. Low-molecular weight substances from the resurrection plant Sporobolus stapfianus. Phytochemistry 37: 137–142.

Bartels, D., R. Alexander, K. Schneider, R. Elster, R. Velasco, J. Alamillo, G. Bianchi, D. Nelson, and F. Salamini. 1993. Dessication-related gene products analysed in a resurrection plant and barley embryos. In: Close, T.J., and E.A. Bray, eds. Plant Responses to Cellular Dehydration During Environmental Stress. Current Topics in Plant Physiology, vol. 10. New York: American Society of Plant Physiology, pp 119–127.

Bartels, D., K. Schneider, G. Terstappen, D. Piatkowski, and F. Salamini. 1990. Molecular cloning of abscisic acid-modulated genes which are induced during desiccation of the resurrection plant Craterostigma plantagineum. Planta 181: 27–34.

Barthlott, W., and I. Capesius. 1974. Wasserabsorption durch Blatt- und Sproßorgane einiger Xerophyten. Z. Pflanzenphysiol. 72: 443–455.

Barthlott, W., and S. Porembski. 1996. Ecology and morphology of Blossfeldia liliputana (Cactaceae): a poikilohydric and almost astomate succulent. Botanica Acta 109: 161–166.

Bartley, M., and N.D. Hallam. 1979. Changes in the fine structure of the desiccation tolerant sedge Coleochloa setifera (Ridley) Gilly under water stress. Australian Journal of Botany 27: 531–545.

Beckett, R.P. 1995. Some aspects of the water relations of lichens from habitats of contrasting water status studied using thermocouple psychrometry. Annals of Botany 76: 211–217.

Belnap, J. 1994. Potential role of cryptogamic soil crusts in semiarid rangelands. In: Monson, S.B., and S.G. Kitchen, eds. Proceedings, Ecology and Management of Annual Rangeland. USDA Forest Service, General Technical Report INT-GTR 313: 179–185.

Bernacchia, G.S, Salamini, F., Bartels, D. 1996. Molecular characterisation of the rehydration process in the resurrection plant Craterostigma plantagineum. Plant Physiology 111: 1043–1050.

Bertsch, A. 1966a. Über den CO2-Gaswechsel einiger Flechten nach Wasserdampfaufnahme. Planta 68: 157–166.

Bertsch, A. 1966b. CO_2-Gaswechsel und Wasserhaushalt der aerophilen Grünalge Apatococcus lobatus. Planta 70: 46–72.

Bertsch, A. 1967. CO_2-Gaswechsel, Wasserpotential und Sättigungsdefizit bei der Austrocknung epidermisfreier Blattscheiben von Valerianella. Naturwissenschaften 8: 204.

Bewley, J.D. 1973. Polyribosomes conserved during desiccation of the moss Tortula ruralis are active. Plant Physiology 51: 285–288.

Bewley, J.D. 1979. Physiological aspects of desiccation tolerance. Annual Review of Plant Physiology 30: 195–238.

Bewley, J.D. 1995. Physiological aspects of desiccation tolerance—a retrospect. International Journal of Plant Science 156: 393–403.

Bewley, J.D., and J.E. Krochko. 1982. Desiccation tolerance. In: Pirson, A., and M.H. Zimmermann, eds. Encyclopedia of Plant Physiology. New Series Vol. 12b. Berlin-Heidelberg: Springer-Verlag, pp 325–378.

Bewley, J.D., and M.J. Oliver. 1992. Desiccation-tolerance in vegetative plant tissues and seeds: protein synthesis in relation to desiccation and a potential role for protection and repair mechanisms. In: Osmond, C.B., and G. Somero, eds. Water and Life: A Comparative Analysis of Water Relationships at the Organismic, Cellular and Molecular Levels. Berlin: Springer Verlag, pp 141–160.

Bewley, J.D., and J. Pacey. 1978. Desiccation-induced ultrastructural changes in drought-sensitive and drought-tolerant plants. In: Crowe, J.W., and J.S. Clegg, eds. New York: Academic Press, pp 53–73.

Bianchi, G., A. Gamba, C.R. Limiroli, N. Pozzi, R. Elster, F. Salamini, and D. Bartels. 1993. The unusual sugar composition in leaves of the resurrection plant *Myrothamnus flabellifolia*. Physiologia Plantarum 87: 223–226.

Bianchi, G., A. Gamba, C. Murelli, F. Salamini, and D. Bartels. 1991. Novel carbohydrate metabolism in the resurrection plant *Craterostigma plantagineum*. Plant Journal 1: 355–359.

Biebl, R. 1962. Protoplasmatische Ökologie der Pflanzen, Wasser und Temperatur. Protoplasmatologia XII, 1. Wien: Springer-Verlag, 344 pp.

Bilger, W., S. Rimke, U. Schreiber, and O.L. Lange. 1989. Inhibition of energy-transfer to photosystem II in lichens by dehydration: different properties of reversibility with green and blue-green phycobionts. Journal of Plant Physiology 134: 261–268.

Blum, O.B. 1973. Water relations. In: Ahmadjian, V., and M.E. Hale, eds. The Lichens. New York: Academic Press, pp 381–400.

Bruns-Strenge, S., and O.L. Lange. 1991. Photosynthetische Primärproduktion der Flechte *Cladonia portentosa* an einem Dünenstandort auf der Nordseeinsel Baltrum I. Freilandmessungen von Mikroklima, Wassergehalt und CO2-Gaswechsel. Flora 185: 73–97.

Bruns-Strenge, S., and O.L. Lange. 1992. Photosynthetische Primärproduktion der Flechte *Cladonia portentosa* an einem Dünenstandort auf der Nordseeinsel Baltrum III. Anwendung des Photosynthesemodells zur Simulation von Tagesabläufen des CO_2-Gaswechsels und zur Abschätzung der Jahresproduktion. Flora 186: 127–140.

Büdel, B. 1987. Zur Biologie und Systematik der Flechtengattung *Heppia* und *Peltula* im südlichen Afrika. Bibliotheca Lichenologica 23: 1–105.

Büdel, B. 1990. Anatomical adaptions to the semiarid/arid environment in the lichen genus *Peltula*. Bibliotheca Lichenologica 38: 47–61.

Büdel, B., and O.L. Lange. 1991. Water status of green and blue-green phycobionts in lichen thalli after hydration by water vapor uptake: do they become turgid? Botanica Acta 104: 345–404.

Büdel, B., and O.L. Lange 1994. The role of the cortical and epineral layers in the lichen genus *Peltula*. Cryptogamic Botany 4: 262–269.

Calatayud, A., V.I. Deltoro, E. Barreno, and S. del Valle-Tascón. 1997. Changes in *in vivo* chlorophyll fluorescence quenching in lichen thalli as a function of water content and suggestion of zeaxanthin-associated photoprotection. Physiologia Plantarum 101: 93–102.

Casper, C., W.G. Eickmeier, and C.B. Osmond. 1993. Changes of fluorescence and xanthophyll pigments during dyhydration in the resurrection plant *Selaginella lepidophylla* in low and medium light intensities. Oecologia 94: 528–533.

Child, G.F. 1960. Brief notes on the ecology of the resurrection plant *Myrothamnus flabellifolia* with mention of its water-absorbing abilities. Journal of South African Botany 26: 1–8.

Clymo, R.S. 1970. The growth of *Sphagnum*: methods and measurments. Journal of Ecology 58: 13–49.

Cowan, D.A., T.G.A. Green, A.T. Wilson. 1979. Lichen metabolism I. The use of tritium labelled water in studies of anhydrobiotic metabolism in *Ramalina celestri* and *Peltigera polydactyla*. New Phytologist 82: 489–503.

Cowan, I.R., O.L. Lange, and T.G.A. Green. 1992. Carbon-dioxide exchange in lichens: determination of transport and carboxylation characteristics. Planta 187: 282–294.

Coxson, D.S. 1987a. Photoinhibition of net photosynthesis in Stereocaulon virgatum and S. tomentosum, a tropical-temperate comparison. Canadian Journal of Botany 65: 1707–1715.

Coxson, D.S. 1987b. The temperature dependance of photoinhibition in the tropical basidiomycete lichen *Cora pavonia* E. Fries. Oecologia 73: 447–453.

Crowe, J.H., and L.M. Crowe. 1992. Membrane integrity in anhydrobiotic organisms: towards a mechanism for stabilizing dry cells. In: Somero, G.N., C.B. Osmond, and C.L. Bolin, eds. Water and Life. A Comparative Analysis of Water Relationships at the Organismic, Cellular and Molecular Levels. Berlin: Springer Verlag, pp 87–113.

Csintalan, Z., Z. Tuba, H. Lichtenthaler, and J. Grace. 1996. Reconstitution of photosynthesis upon rehydration in the desiccated leaves of the poikilochlorophyllus shrub *Xerophyta scabrida* at elevated CO_2. Journal of Plant Physiology 148: 345–350.

Danin, A., and J. Garty. 1983. Distribution of cyanobacteria and lichens on hillsides of the Negev Highlands and their impact on biogenic weathering. Zeitschrift für Geomorphologie N.F. 27: 423–444.

Davey, M.C. 1997. Effects of continuous and repeated dehydration on carbon fixation by bryophytes from the maritime Antarctic. Oecologia 110: 25–31.

Demmig, B., K. Winter, K. Krüger, and F.-C. Czygan. 1988. Photoinhibition and the heat dissipation of excess light energy in *Nerium oleander* exposed to a combination of high light and water stress. Plant Physiology 87: 17–24.

Demmig-Adams, B., and W.W. Adams. 1992. Photoprotection and other responses of plants to high light stress. Annual Review of Plant Physiology and Plant Molecular Biology 43: 599–626.

Demmig-Adams, B., W.W. Adams III, F.-C. Czygan, U. Schreiber, and O.L. Lange. 1990b. Differences in the capacity for radiationless energy dissipation in the photochemical apparatus of green and blue-green algal lichens associated with differences in carotenoid composition. Planta 180: 582–589.

Demmig-Adams, B., C. Máguas, W.W.I. Adams, A. Meyer, E. Kilian, and O.L. Lange. 1990a. Effect of light on the efficiency of photochemical energy conversion in a variety of lichen species with green and blue-green phycobionts. Planta 180: 400–409.

Dhindsa, R.S. 1985. Non-autotrophic CO_2 fixation and drought tolerance in mosses. Journal of Experimental Botany 36: 980–988.

Dhindsa, R.S. 1991. Drought stress, enzymes of glutathion metabolism, oxidation injury, and protein synthesis in *Tortula ruralis*. Plant Physiology 95: 648–651.

Dhindsa, R.S., and W. Matowe. 1981. Drought tolerance in two mosses: correlated with enzymatic defence against lipid peroxidation. Journal of Experimental Botany 32: 79–91.

Dietz, S., and W. Hartung 1998. Abscisic acid in lichens: variation, water relations and metabolism. New Phytologist 138: 99–106.

Dilks, T.J.K., and M.C.F. Proctor. 1974. The pattern of recovery of bryophytes after desiccation. Journal of Bryology 8: 97–115.

Dilks, T.J.K., and M.C.F. Proctor. 1976a. Effects of intermittent desiccation on bryophytes. Journal of Bryology 9: 249–264.

Dilks, T.J.K., and M.C.F. Proctor. 1976b. Seasonal variation in desiccation tolerance in some British bryophytes. Journal of Bryology 9: 239–247.

Dilks, T.J.K., and M.C.F. Proctor. 1979. Photosynthesis, respiration and water content in bryophytes. New Phytologist 82: 97–114.

Dirksen, A. 1964. Vergleichende Untersuchungen zur Frost-, Hitze- und Austrocknungsresistenz einheimischer Laub- und Lebermoose unter besonderer Berücksichtigung jahreszeitlicher Veränderungen. Doctor Thesis, Universität Göttingen.

Eickmeier, W.G. 1979. Photosynthetic recovery in the resurrection plant *Selaginella lepidophylla* after wetting. Oecologia 39: 93–106.

Eickmeier, W.G. 1980. Photosynthetic recovery of resurrection spike mosses from different hydration regimes. Oecologia 46: 380–385.

Eickmeier, W.G. 1982. Protein synthesis and photosynthetic recovery in the resurrection plant *Selaginella lepidophylla*. Plant Physiology 69: 135–138.

Eickmeier, W.G. 1983. Photosynthetic recovery of the resurrection plant *Selaginella lepidophylla*: effects of prior desiccation rate and mechanisms of desiccation damage. Oecologia 58: 115–120.

Eickmeier, W.G. 1986. The correlation between high-temperature and desiccation tolerances in a poikilohydric desert plant. Canadian Journal of Botany 64: 611–617.

Eickmeier, W.G. 1988. The effects of desiccation rate on enzyme and protein—synthesis dynamics in the desiccation-tolerant pteridophyte *Selaginella lepidophylla*. Canadian Journal of Botany 66: 2574–2580.

Eickmeier, W.G., C. Casper, and C.B. Osmond. 1993. Chlorophyll fluorescence in the resurrection plant *Selaginella lepidophylla* (Hook & Grev) spring during high-light and desiccation stress, and evidence for zeaxanthin-associated photoprotection. Planta 189: 30–38.

Eickmeier, W.G., J.G. Lebkuecher, and C.B. Osmond. 1992. Photosynthetic water oxidation and water stress in plants. In: Somero, G.N., G.N. Osmond, and C.L. Bolis, eds. Water and Life: Comparitive Analysis of Water Relationships at the Organismic,

Cellular, and Molecular Levels. Berlin-Heidelberg-New York: Springer Verlag, pp 223–239.

Ewers, F.W., G.B. North, and P.S. Nobel. 1992. Root-stem junctions of a desert monocotyledon and a dicotyledon: hydraulic consequences under wet conditions and during drought. New Phytologist 121: 377–385.

Farrar, J.F. 1976a. The lichen as an ecosystem: observation and experiment. In: Brown, D.H., D.L. Hawksworth, and R.H. Bailey, eds. Lichenology: Progress and Problems. New York: Academic Press, pp 385–406.

Farrar, J.F. 1976b. Ecological physiology of the lichen *Hypogymnia physodes* I. Some effects of constant water saturation. New Phytologist 77: 93–103.

Fischer, E. 1992. Systematik der afrikanischen Lindernieae (Scrophulariaceae). Franz Steiner Verlag, Stuttgart.

Friedmann, E.I., and M. Galun. 1974. Desert algae, lichens, and fungi. In: G.W.J. Brown, ed. Desert Biology New York: Academic Press, pp 166–212.

Friedmann, E.I., L. Kappen, M.A. Meyer, and J.A. Nienow. 1993. Long-term productivity in the cryptoendolithic microbial community of the Ross Desert, Antarctica. Microbial Ecology 25: 51–69.

Gaff, D.F. 1971. Desiccation-tolerant flowering plants in southern Africa. Science 174: 1033–1034.

Gaff, D.F. 1972. Drought resistance in *Welwitschia mirabilis* Hook. fil. Dinteria 7: 3–19.

Gaff, D.F. 1977. Desiccation tolerant vascular plants of southern Africa. Oecologia 31: 95–109.

Gaff, D.F. 1980. Protoplasmic tolerance of extreme water stress. In: Turner, N.C., and P.J. Kramer, eds. Adaptation of Plants to Water and High Temperature Stress. New York: Wiley, pp 207–230.

Gaff, D.F. 1986. Desiccation tolerant ''resurrection'' grasses from Kenya and west Africa. Oecologia 70: 118–120.

Gaff, D.F. 1987. Desiccation tolerant plants in South America. Oecologia 74: 133–136.

Gaff, D.F. 1989. Responses of Desiccation Tolerant ''Resurrection'' Plants to Water Stress. The Hague, The Netherlands: SPB Academic Publishing.

Gaff, D.F., and P.V. Bole. 1986. Resurrection grasses in India. Oecologia 71: 159–160.

Gaff, D.F., and D.M. Churchill. 1976. *Borya nitida* Labill.—an Australian species in the Liliaceae with desiccation-tolerant leaves. Australian Journal of Botany 24: 209–224.

Gaff, D.F., and R.P. Ellis. 1974. Southern African grasses with foliage that revives after dehydration. Bothalia 11: 305–308.

Gaff, D.F., and W. Giess. 1986. Drought resistance in water plants in rock pools of southern Africa. Dinteria 18: 17–36.

Gaff, D.F., and N.D. Hallam. 1974. Resurrecting desiccated plants. In: Bieleski, R.L., A.R. Ferguson, and M.M. Cresswell, eds. Wellington: The Royal Society of New Zealand, pp 389–393.

Gaff, D.F., and P.K. Latz. 1978. The occurrence of resurrection plants in the Australian flora. Australian Journal of Botany 26:485–492.

Gaff, D.F., and B.R. Loveys. 1984. Abscisic-acid content and effects during dehydration of detached leaves of desiccation tolerant plants. Journal of Experimental Botany 35: 1350–1358.

Gaff, D.F., and G.R. McGregor. 1979. The effect of dehydration and rehydration on the nitrogen content of various fractions from resurrection plants. Biologia Plantarum (Praha) 21: 92–99.

Gaff, D.F., S.-Y. Zee, and T.P. O'Brien. 1976. The fine structure of dehydrated and reviving leaves of *Borya nitida* Labill.—a desiccation-tolerant plant. Australian Journal of Botany 24: 225–236.

Gauslaa, Y., and K.A. Solhaug. 1996. Differences in the susceptibility to light stress between epiphytic lichens of ancient and young boreal forest stands. Functional Ecology 10: 344–354.

Gebauer, R. 1986. Wasserabgabe und Wasseraufnahme poikilohydrer höherer Pflanzen im Hinblick auf ihre physiologische Aktivität. Diploma thesis, Universität zu Kiel, 97 pp.

Gebauer, R., R. Lösch, and L. Kappen. 1987. Wassergehalt und CO_2-Gaswechsel des poikilohydren Kormophyten *Ramonda myconi* (L.) Schltz. während der Austrocknung und Wiederaufsättigung. Verhandlungen Gesellschaft für Ökologie. XVI: 231–236.

Gildner, B.S., and D.W. Larson. 1992. Seasonal changes in photosynthesis in the desiccation-tolerant fern *Polypodium virginianum*. Oecologia 89: 383–389.

Gimingham, C.H., and R.I.L. Smith. 1971. Growth form and water relations of mosses in the maritime Antarctic. British Antarctic Survey Bulletin 25: 1–21.

Giordano, S., C. Colacino, V. Spagnuolo, A. Basile, A. Esposito, and R. Castaldo-Cobianchi. 1993. Morphological adaptation to water uptake and transport in the poikilohydric moss *Tortula ruralis*. Giornale Botanico Italiano 127: 1123–1132.

Green, T.G.A., W.P. Snelgar, and A.L. Wilkins. 1985. Photosynthesis, water relations and thallus structure of Stictaceae lichens. In: Brown, D.H., ed. Lichen Physiology and Cell Biology. New York: Plenum Press, pp 57–75.

Green, T.G.A., E. Kilian, and O.L. Lange. 1991. *Pseudocyphellaria dissimilis*: a desiccation-sensitive, highly shade-adapted lichen from New Zealand. Oecologia 85: 498–503.

Green, T.G.A., and O.L. Lange. 1994. Photosynthesis in poikilohydric plants: a comparison of lichens and bryophytes. In: Schulze, E.-D., and M.M. Caldwell, eds. Ecophysiology of Photosynthesis. Ecological Studies 100. Berlin Heidelberg-New York: Springer-Verlag, pp 320–341.

Green, T.G.A., O.L. Lange, and I.R. Cowan. 1994. Ecophysiology of lichen photosynthesis: the role of water status and thallus diffusion resistances. Cryptogamic Botany 4: 166–178.

Hahn, S., D. Speer, A. Meyer, and O.L. Lange. 1989. Photosynthetic primary production of epigean lichens growing in local xerothermic steppe formations in Franconia. I. Diurnal time course of microclimate, water content and CO_2 exchange at different seasons. Flora 182: 313–339.

Hahn, S., D. Tenhunen, P.W. Popp, A. Meyer, and O.L. Lange. 1993. Upland tundra in the foothills of the Brooks Range, Alaska: diurnal CO_2 exchange patterns of characteristic lichen species. Flora 188: 125–143.

Hallam, N.D., and P. Cappicchiano. 1974. Studies on dessication tolerant plants using nonaqueous fixation methods. Proceedings Eighth International Congress on Electronic Microscopy, Canberra 2: 612–613.

Hallam, N.D., and D.F. Gaff. 1978. Regeneration of chloroplast structure in *Talbotia elegans*: a desiccation-tolerant plant. New Phytologist 81: 657–662.

Hallam, N.D., and S.E. Luff. 1980. Fine structure changes in the leaves of the desiccation-tolerant plant *Talbotia elegans* during extreme water stress. Botanical Gazzette 141: 180–187.

Hambler, D.J. 1961. A poikilohydrous, poikilochlorophyllous angiosperm from Africa. Nature 191: 1415–1416.

Hargrave, K.R., K.J. Kolb, F.W. Ewers, and S.D. Davis. 1994. Conduit diameter and drought-induced embolism in *Salvia mellifera* Greene (Labiatae). New Phytologist 126: 695–705.

Hartung, W., P. Schiller, and K.-J. Dietz. 1998. The physiology of poikilohydric plants. Progress in Botany 59: 299–327.

Hébant, C. 1977. The Conducting Tissues of Bryophytes. Vaduz: Gantner Verlag.

Heil, H. 1925. *Chamaegigas intrepidus* Dtr., eine neue Auferstehungspflanze. Beiträgezum Botanischen Zentralblatt 41: 41–50.

Hellwege, E.M., K.J. Dietz, O.H. Volk, and W. Hartung. 1994. Abscisic acid and the induction of desiccation tolerance in the extremely xerophilic liverwort *Exormotheca holstii*. Planta 194: 525–531.

Hetherington, S.E., and R.M. Smillie. 1982. Humidity-sensitive degreening and regreening of leaves of *Borya nitida* Labill. as followed by changes in chlorophyll fluorescence. Australian Journal of Plant Physiology 9: 587–599.

Hickel, B. 1967. Zur Kenntnis einer xerophilen Wasserpflanze: *Chamaegigas intrepidus* Dtr. aus Südwestafrika. Hydrobiologie 52: 361–400.

Hicklenton, P.R., and W.C. Oechel. 1977. The influence of light intensity and temperature on the field carbon dioxide exchange of *Dicranum fuscescens* in the subarctic. Arctic and Alpine Research 9: 407–419.

Hinshiri, H.M., and M.C.F. Proctor. 1971. The effect of desiccation on subsequent assimilation and respiration of the bryophytes *Anomodon viticulosus* and *Porella platyphylla*. New Phytologist 70: 527–538.

Hoffmann, P. 1968. Pigmentgehalt und Gaswechsel von Myrothamnus-Blättern nach Austrocknung und Wiederaufsättigung. Photosynthetica 2: 245–252.

Hosokawa, T., and H. Kubota. 1957. On the osmotic pressure and resistance to desiccation of epiphytic mosses from a beech forest, South-west Japan. Journal of Ecology 45: 579–591.

Hsiao, T.C. 1973. Plant responses to water stress. Annual Review of Plant Physiology 24: 519–570.

Ilgin, W.S. 1931. Austrocknungsresistenz des Farnes *Notholaena maranthae* R. Br. Protoplasma 10: 379–414.

Ingram, J., and D. Bartels. 1996. The molecular basis of dehydration tolerance in plants. Annual Review of Plant Physiology and Plant Molecular Biology 47: 377–403.

Ingrouille, M. 1995. Diversity and Evolution of Land Plants. London: Chapman and Hall.

Jovet-Ast, S. 1969. Le caryotype des Ricciaceae. Revue Bryologic Lichenologique 36: 573–689.

Kaiser, K., D.F. Gaff, and W.H. Outlaw Jr. 1985. Sugar contents of leaves of desiccation-tolerant plants. Naturwissenschaften 72: 608–609.

Kaiser, W.M. and U. Heber. 1981. Photosynthesis under osmotic stress. Effect of high solute concentrations on the permeability properties of the chloroplast envelope and the activity of stoma enzymes. Planta 153: 429–432.

Kaiser, W.M. 1987. Effect of water deficit on photosynthetic activity. Physiologia Plantarum 71: 142–149.

Kallio, P., and L. Kärenlampi. 1975. Photosynthesis in mosses and lichens. In: Cooper, J.P., ed. Photosynthesis and Productivity in Different Environments. Cambridge: Cambridge University Press, pp 393–423.

Kappen, L. 1964. Untersuchungen über den Jahreslauf der Frost-, Hitze- und Austrocknungsresistenz von Sporophyten einheimischer Polypodiaceen (Filicinae). Flora 155: 123–166.

Kappen, L. 1965. Untersuchungen über die Widerstandsfähigkeit der Gametophyten einheimischer Polypodiaceen gegenüber Frost, Hitze and Trockenheit. Flora 156: 101–115.

Kappen, L. 1966. Der Einfluß des Wassergehaltes auf die Widerstandsfähigkeit von Pflanzen gegenüber hohen und tiefen Temperaturen, untersucht an Blättern einiger Farne und von *Ramonda myconi*. Flora 156: 427–445.

Kappen, L. 1973. Response to extreme environments. In: Ahmadjian, V., and M.E. Hale, eds. The Lichens. New York: Academic Press, pp 311–380.

Kappen, L. 1981. Ecological significance of resistance to high temperature. In: Lange, O.L., P.S. Nobel, C.B. Osmond, and H. Ziegler, eds. Physiological Plant Ecology, vol. I. Responses to the Physical Environment. Berlin-Heidelberg: Springer-Verlag, pp 439–473.

Kappen, L. 1982. Lichen oases in hot and cold deserts. Journal of the Hattori Botanical Laboratory 53: 325–330.

Kappen, L. 1983. Ecology and physiology of the Antarctic fructicose lichen *Usnea sulphurea* (Koenig) Th. Fries. Polar Biology 1: 249–255.

Kappen, L. 1985. Water relations and net photosynthesis of *Usnea*. A comparison between *Usnea fasciata* (maritime Antarctic) and *Usnea sulphurea* (continental Antarctic). In: Brown, D.H., eds. Lichen physiology and cell biology. New York-London: Plenum Press, pp 41–56.

Kappen, L. 1988. Ecophysiological relationships in different climatic regions. In: Galun, M., eds. Handbook of Lichenology, vol. 2. Boca Raton, FL: CRC Press, pp 37–100.

Kappen, L. 1989. Field measurements of carbon dioxide exchange of the Antarctic lichen *Usnea sphacelata* in the frozen state. Antarctic Science 1: 31–34.

Kappen, L. 1993a. Plant activity under snow and ice, with particular reference to lichens. Arctic 46: 297–302.

Kappen, L. 1993b. Lichens in the Antarctic region. In: Friedmann, E.I. ed. Antarctic Microbiology. New York: Wiley-Liss, pp 433–490.

Kappen, L. 1994. The lichen, a mutualistic system? Some mainly ecophysiological aspects. Cryptogamic Botany 4: 193–202.

Kappen, L. 1995. Terrestrische Mikroalgen und Flechten der Antarktis. In: Hausmann, K., and B.P. Kremer, eds. Extremophile. Mikroorganismen in ausgefallenen Lebensräumen. Weinheim, New York, Basel, Cambridge, Tokyo: VCH Verlagsgesellschaft mbH, pp 3–25.

Kappen, L., O.L. Lange, E.-D. Schulze, M. Evenari, and U. Buschbom. 1972. Extreme water stress and photosynthetic activity of the desert plant *Artemisia herba-alba* Asso. Oecologia 10: 177–182.

Kappen, L., O.L. Lange, E.-D. Schulze, M. Evenari, and U. Buschbom. 1979. Ecophysiological investigations on lichens of the Negev desert. VI. Annual course of photosynthetic production of *Ramalina maciformis* (Del.) Bory. Flora 168: 85–108.

Kappen, L., O.L. Lange, E.-D. Schulze, U. Buschbom, and M. Evenari. 1980. Ecophysiological investigations on lichens of the Negev desert. VII. The influence of the habitat exposure on dew imbition and photosynthetic productivity. Flora 169: 216–229.

Kappen, L., R.I. Lewis Smith, and M. Meyer. 1989. Carbon dioxide exchange of two ecodemes of *Schistidium antarctici* in continental Antarctica. Polar Biology 9: 415–422.

Kappen, L., B. Schroeter, T.G.A. Green, and R.D. Seppelt. 1998a. Chlorophyll a fluorescence and CO_2 exchange of *Umbilicaria aprina* under extreme light stress in the cold. Oecologia 113: 325–331.

Kappen, L., B. Schroeter, T.G.A. Green, and R.D. Seppelt 1998b. Microclimatic conditions, meltwater moistening, and the distributional pattern of *Buellia frigida* on rock in a southern continental Antarctic habitat. Polar Biology 19: 101–106.

Kappen, L., B. Schroeter, G. Hestmark, and J.B. Winkler. 1996. Field measurements of photosynthesis of umbilicarious lichens in winter. Botanica Acta 109: 292–298.

Kappen, L., M. Sommerkorn, and B. Schroeter. 1995a. Carbon acquisition and water relations of lichens in polar regions-potentials and limitations. Lichenologist 27: 531–545.

Kerstiens, G. 1996. Cuticular water permeability and its physiological significance. Journal of Experimental Botany 47: 1813–1832.

Kershaw, K.A. 1985. Physiological Ecology of Lichens. Cambridge: Cambridge University Press.

Kimenov, G.P., Y.K. Markovska, and T.D. Tsonev. 1989. Photosynthesis and transpiration of *Haberlea rhodopensis* Friv. in dependence on water deficit. Photosynthetica 23: 368–371.

Klepper, B. 1968. A comparison of the water relations of a moss and a lichen. Nova Hedwigia 15: 13–20.

Kolb, K.J., J.S. Sperry, and B.B. Lamont. 1996. A method for measuring xylem hydraulic conductance and embolism in entire root and shoot system. Journal of Experimental Botany 47: 1805–1810.

Kuang, J., D.F. Gaff, R.D. Gianello, C.K. Blomstedt, A.D. Neale, and J.D. Hamill. 1995. Changes in *in vivo* protein complements in drying leaves of the desiccation-tolerant grass *Sporobolus stapfianus* and the desiccation-sensitive grass *Sporobolus pyramidalis*. Australian Journal of Plant Physiology 22: 1027–1034.

Lange, O.L. 1953. Hitze- und Trockenresistenz der Flechten in Beziehung zu ihrer Verbreitung. Flora 140: 39–97.

Lange, O.L. 1955. Untersuchungen über die Hitzeresistenz der Moose in Beziehung zu ihrer Verbreitung. I. Die Resistenz stark ausgetrockneter Moose. Flora 142: 381–399.

Lange, O.L. 1969. CO_2 Gaswechsel von Moosen nach Wasserdampfaufnahme aus dem Luftraum. Planta 89: 90–94.

Lange, O.L. 1969b. Experimentell-ökologische Untersuchungen an Flechten der Negev-Wüste. I. CO_2-Gaswechsel von Ramalina maciformis (Del.) Bory unter kontrollierten Bedingungen im Laboratorium. Flora 158: 324–359.

Lange, O.L. 1980. Moisture content and CO_2 exchange of lichens. I. Influence of temperature on moisture-dependent net photosynthesis and dark respiration in *Ramalina maciformis*. Oecologia 45: 82–87.

Lange, O.L., J. Belnap, H. Reichenberger, and A. Meyer. 1997. Photosynthesis of green algal soil crust lichens from arid lands in southern Utah, USA: role of water content on light and temperature responses of CO_2 exchange. Flora 192: 1–15.

Lange, O.L., and A. Bertsch. 1965. Photosynthese der Wüstenflechte *Ramalina maciformis* nach Wasserdampfaufnahme aus dem Luftraum. Naturwissenschaften 52: 215–216.

Lange, O.L., B. Büdel, U. Heber, A. Meyer, H. Zellner, and T.G.A. Green. 1993. Temperate rainforest lichens in New Zeeland: high thallus water content can severely limit photosynthetic CO_2 exchange. Oecologia 95: 303–313.

Lange, O.L., T.G.A. Green, and H. Ziegler. 1988. Water status related photosynthesis and carbon isotope discrimination in species of the lichen genus *Pseudocyphellaria* with green or blue-green phycobionts and in photosymbiodemes. Oecologia 75: 494–501.

Lange, O.L., T.G.A. Green, H. Reichenberger, and A. Meyer. 1996. Photosynthetic depression at high water contents in lichens: concurrent use of gas exchange and fluorescence techniques with cyanobacterial and green algal *Peltigera* species. Botanica Acta 109: 43–50.

Lange, O.L., and E. Kilian. 1985. Reaktivierung der Photosynthese trockener Flechten durch Wasserdampfaufnahme aus dem Luftraum: Artspezifisch unterschiedliches Verhalten. Flora 176: 7–23.

Lange, O.L., E. Kilian, and H. Ziegler. 1986. Water vapor uptake and photosynthesis of lichens: performance differences in species with green and blue-green algae as phycobionts. Oecologia 71: 104–110.

Lange, O.L., E. Kilian, and H. Ziegler. 1990a. Photosynthese von Blattflechten mit hygroskopischen Thallusbewegungen bei Befeuchtung durch Wasserdampf oder mit flüssigem Wasser. Bibliotheca Lichenologica 38: 311–323.

Lange, O.L., and U. Matthes. 1981. Moisture-dependent CO_2 exchange of lichens. Photosynthetica 15: 555–574.

Lange, O.L., A. Meyer, I. Ullmann, and H. Zellner. 1991. Mikroklima, Wassergehalt und Photosynthese von Flechten in der küstennahen Nebelzone der Namib-Wüste: Messungen während der herbstlichen Witterungsperiode. Flora 185: 233–266.

Lange, O.L., A. Meyer, H. Zellner, and U. Heber. 1994. Photosynthesis and water relations of lichen soil crusts: field measurements in the coastal fog zone of the Namib desert. Functional Ecology 8: 253–264.

Lange, O.L., A. Meyer, H. Zellner, I. Ullmann, and D.C.J. Wessels. 1990c. Eight days in the life of a desert lichen: water relations and photosynthesis of *Teloschistes capensis* in the coastal fog zone of the Namib desert: Madoqua 17: 17–30.

Lange, O.L., H. Pfanz, E. Kilian, and A. Meyer. 1990b. Effect of low water potential on photosynthesis in intact lichens and their liberated algal components. Planta 182: 467–472.

Lange, O.L., and J. Redon. 1983. Epiphytische Flechten im Bereich einer chilenischen "Nebeloase" (Fray Jorge) II. Ökophysiologische Charakterisierung von CO_2-Gaswechsel und Wasserhaushalt. Flora 174: 245–284.

Lange, O.L., and J.D. Tenhunen. 1981. Moisture content and CO_2 exchange of lichens. II. Depression of net photosynthesis in *Ramalina maciformis* at high water content is caused by increased thallus carbon dioxide diffusion resistance. Oecologia 51: 426–429.

Larcher, W. 1980. Ökologie der Pflanzen. Stuttgart: Verlag Eugen Ulmer.

Larson, D.W. 1979. Lichen water relations under drying conditions. New Phytologist 82: 713–731.

Larson, D.W. 1981. Differential wetting in some lichens and mosses: the role of morphology. Bryologist 84: 1–15.

Larson, D.W. 1984. Habitat overlap/niche segregation in two *Umbilicaria* lichens: a possible mechanism. Oecologia 62: 118–125.

Larson, D.W. 1987. The absorption and release of water by lichens. Bibliotheca Lichenologica 25:351–360.

Larson, D.W. 1988. The impact of ten years at $-20°C$ on gas exchange in five lichen species. Oecologia 78: 87–92.

Lazarides, M. 1992. Resurrection grasses (*Poaceae*) in Australia. In: Chapman, G.P., ed. Desertified Grasslands: Their Biology and Management. London: Academic Press, pp 213–234.

Lebkuecher, J.G., and W.G. Eickmeier. 1991. Reduced photoinhibition with stem curling in the resurrection plant *Selaginella lepidophylla*. Oecologia 88: 597–604.

Lebkuecher, J.G., and W.G. Eickmeier. 1993. Physiological benefits of stem curling for resurrection plants in the field. Ecology 74: 1073–1080.

Lebrun, J. 1947. La vegetation de la pleine alluviale au sud du Lac Edouard. Exploration du Parc National Albert. Mission J. Lebrun (1937–1938). Bruxelles: Institut des Parc nationaux du Congo Belge, Bruxelles, 100 pp.

Lechowicz, M.J. 1981. The effects of climatic pattern on lichen productivity: *Cetraria cucullata* (Bell.) Ach. in the Arctic tundra of northern Alaska. Oecologia 50: 210–216.

Lechowicz, M.J. 1992. The niche at the organismal level: lichen photosynthetic response. In: Carroll, G.C., and D.T. Wicklow, eds. The Fungal Community. New York: Marcel Dekker, pp 29–42.

Leisner, J.M.R., W. Bilger, F.-C. Czygan, and O.L. Lange. 1994. Light exposure and the composition of lipophilous carotenoids in cyanobacterial lichens. Journal of Plant Physiology 143: 514–519.

Leisner, J.M.R., W. Bilger, and O.L. Lange. 1996. Chlorophyll fluorescence characteristics of the cyanobacterial lichen *Peltigera rufescens* under field conditions. Flora 191: 261–273.

Leopold, A.C. 1990. Coping with desiccation. In: Alscher, R.G., ed. Stress Responses in Plants: Adaptation and Acclimation Mechanisms. New York: Wiley-Liss, pp 36–56.

Levitt, J. 1980. Responses of Plants to Environmental Stress. Chilling, Freezing, and High Temperature Stresses. New York: Academic Press.

Lewis, A.M., V.D. Harnden, and M.T. Tyree. 1994. Collapse of water-stress emboli in the tracheids of *Thuja occidentalis* L. Plant Physiology 106: 1639–1646.

Lichtenthaler, H.K., and U. Rinderle. 1988. The role of chlorophyll fluorescence in the detection of stress conditions in plants. CRC Critical Review Analytical Chemistry 19: 29–85.

Long, S.P., S. Humphries, and P.G. Falkowski. 1994. Photoinhibition of photosynthesis in nature. Annual Review of Plant Physiology and Plant Molecular Biology 45: 633–662.

Longton, R.E. 1988a. Life-history strategies among bryophytes of arid regions. Journal of the Hattori Botanical Laboratory 64: 15–28.

Longton, R.E. 1988b. Biology of Polar Bryophytes and Lichens. Cambridge: Cambridge University Press.

Lösch, R., R. Pause, and B. Mies. 1997. Poikilohydrie und räumlich-zeitliche Existenznische von Flechten und Moosen. Bibliotheca Lichenologica 67: 145–162.

Lüttge, U., B. Büdel, E. Ball, F. Strube, and P. Weber. 1995. Photosynthesis of terrestrial cyanobacteria under light and desiccation stress as expressed by chlorophyll fluorescence and gas exchange. Journal of Experimental Botany 46: 309–319.

Lüttge, U., M. Kluge, and G. Bauer. 1988. Botanik, ein grundlegendes Lehrbuch. Weinheim: Verlag Chemie.

MacFarlane, J.D., and K.A. Kershaw. 1982. Physiological-environmental interactions in lichens. XIV. The environmental control of glucose movement from alga to fungus in *Peltigera polydactyla, P. rufescens* and *Collema furfuraceaum*. New Phytologist 91: 93–101.

Maguas, C., F. Valladares, and E. Brugnoli. 1997. Effects of thallus size on morphology and physiology of foliose lichens: new findings with a new approach. Symbiosis 23: 149–164.

Manrique, E., L. Balaguer, J. Barnes, and A.W. Davison. 1993. Photoinhibition studies in lichens using chlorophyll fluorescence analysis. Bryologist 96: 443–449.

Markovska, Y., T. Tsonev, and G. Kimenov. 1997. Regulation of CAM and respiratory recycling by water supply in higher poikilohydric plants—*Haberlea rhodopensis* Friv. and *Ramonda serbica* Panc. at transition from biosis to anabiosis and *vice versa*. Botanica Acta 110: 18–24.

Markovska, Y.K., T.D. Tsonev, G.P. Kimenov, and A.A. Tutekova. 1994. Physiological changes in higher poikilohydric plants—*Haberlea rhodopensis* Friv. and *Ramonda serbica* Panc. during drought and rewatering at different light regimes. Journal of Plant Physiology 144: 100–108.

Markovska, Y.K., A.A. Tutekova, and G.P. Kimenov. 1995. Ultrastructure of chloroplasts of poikilohydric plants *Haberlea rhodopensis* Friv. and *Ramonda serbica* Panc. during recovery from desiccation. Photosynthetica 31: 613–620.

Masuch, G. 1993. Biologie der Flechten. Heidelberg Wiesbaden: UTB Quelle und Meyer.

Matthes-Sears, U., P.E. Kelly, and D.W. Larson. 1993. Early spring gas exchange and uptake of deuterium-labelled water in the poikilohydric fern *Polypodium virginianum*. Oecologia 95: 9–13.

Montenegro, G., A.J. Hoffmann, M.E. Aljaro, and A.E. Hoffmann. 1979. *Satureja gilliesii*: a poikilohydric shrub from the Chilean Mediterranean vegetation. Canadian Journal of Botany 57: 1206–1213.

Müller, M.A.N. 1985. Gräser Südwestafrika/Namibias. Windhoek, Südwestafrika/Namibia: John Meinert (Pty.) Ltd.

Muslin, E.H., and P.H. Homann. 1992. Light as a hazard for the desiccation-resistant "resurrection" fern *Polypodium polypodioides* L. Plant, Cell and Environment 15: 81–89.

Nash, T.H.I., and T.J. Moser. 1982. Vegetational and physiological patterns of lichens in North American deserts. Journal of the Hattori Botanical Laboratory 53: 331–336.

Nash, T.H.I., A. Reiner, B. Demmig-Adams, E. Kilian, W.M. Kaiser, and O.L. Lange. 1990. The effect of atmospheric desiccation and osmotic water stress on photosynthesis and dark respiration of lichens. New Phytologist 116: 269–276.

Nash, T.H.I., S.L. White, and J.E. Marsh. 1977. Lichen and moss distribution and biomass in hot desert ecosystems. Bryologist 80: 470–479.

Navari-Izzo, F., S. Meneguzzo, B. Loggini, C. Vazzana, and C.L.M. Sgherri. 1997. The role of the glutathione system dehydration of *Boea hygroscopica*. Physiologia Plantarum 99: 23–30.

Nelson, D., F. Salamini, and D. Bartels. 1994. Abscisic acid promotes novel DNA-binding activity to a desiccation-related promotor of *Craterostigma plantagineum*. The Plant Journal 5: 451–458.

Nienow, J.A., and E.I. Friedmann. 1993. Terresetrial lithophytic (rock) communities. In: Friedmann, E.I., ed. Antarctic Microbiology. New York: Wiley-Liss. pp 343–412.

Nobel, P.S. 1978. Microhabitat, water relations, and photosynthesis of a desert fern, *Notholaena parryi*. Oecologia 31: 293–309.

Nobel, P.S., D.J. Longstreth, and T.L. Hartsock. 1978. Effect of water stress on the temperature optima of net CO_2 exchange for the two desert species. Physiologia Plantarum 44: 97–101.

North, G.B., and P.S. Nobel. 1991. Changes in hydraulic conductivity and anatomy caused by drying and rewetting roots of Agave-deserti (Agavaceae). American Journal of Botany 78: 906–915.

Oechel, W.C., and B. Sveinbjörnsson. 1978. Primary production progresses in the arctic bryophytes at Barrow, Alaska. In: Tieszen, L.L., ed. Ecological Studies 29: Vegetation and Production Ecology of an Alaskan Arctic Tundra. New York-Heidelberg-Berlin: Springer-Verlag, pp. 269–299.

Oliver, M.J., and J.D. Bewley. 1997. Desiccation-tolerance of plant tissues: a mechanistic overview. Horticultural Reviews 18: 171–213.

Oppenheimer, H.R., and A.H. Halevy. 1962. Anabiosis of *Ceterach officinarum*. Bulletin of the Research Council Israel 11: 127–164.

Öquist, G., and D.C. Fork. 1982a. Effects of desiccation on the exitation energy distribution from phycoerythrin to the two photosystems in the red alga *Porphyra perforata*. Physiologia Plantarum 56: 56–62.

Öquist, G., and D.C. Fork. 1982b. Effect of desiccation on the 77 Kelvin fluorescence properties of the liverwort *Porella navicularis* and the isolated lichen green alga *Trebouxia pyriformis*. Physiologia Plantarum 56: 63–68.

Owoseye, J.A., and W.W. Sanford. 1972. An ecological study of *Vellozia schnitzleinia*, a drought-enduring plant of northern Nigeria. Journal of Ecology 60: 807–817.

Patterson, P.M. 1964. Problems presented by bryophytic xerophytism. Bryologist 67: 390–396.

Perez, F.L. 1997. Geoecology of erratic globular lichens of *Catapyrenium lachneum* in a high Andean Paramo. Flora 192: 241–259.

Pessin, L.J. 1924. A physiological and anatomical study of the leaves of *Polypodium polypodioides*. American Journal of Botany 11: 370–381.

Piatkowski, D., K. Schneider, F. Salamini, and D. Bartels. 1990. Characterization of five abscisic acid-responsive cDNA clones isolated from the desiccation-tolerant plant *Craterostigma plantagineum* and their relationship to other water-stress genes. Plant Physiology 94: 1682–1688.

Pickard, W.F. 1989. How might a tracheary element which is embolized by day be healed by night? Journal of Theoretical Biology 141: 259–279.

Picket, F.L. 1914. Ecological adaptations of certain fern prothallia. American Journal of Botany I: 477–498.

Pintado, A., F. Valladares, and L.G. Sancho. 1997. Exploring phenotypic plasticity in the lichen *Ramalina capitata*: morphology, water relations and chlorophyll content in North- and South-facing populations. Annals of Botany 80: 345–353.

Pitt, J.I., and J.H.B. Christian. 1968. Water relations of xerophilic fungi isolated from prunes. Applied Microbiology 16: 1853–1858.

Proctor, M.C.F. 1982. Physiological ecology: water relations, light and temperature responses, carbon balance. In: Smith, A.J.E., ed. Bryophyte Ecology. London-New York: Chapman and Hall, pp 333–381.

Proctor, M.C.F. 1984. Structure and ecological adaptation. In: Dyer, A.F., and J.Q. Duckett, eds. The Experimental Biology of Bryophytes. London: Academic Press, pp 9–37.

Proctor, M.C.F. 1990. The physiological basis of bryophyte production. Botanical Journal of the Linnean Society 104: 61–77.

Puff, C. 1978. Zur Biologie von *Myrothamnus flabellifolius* Welw. (Myrothamnaceae). Dinteria 14: 1–20.

Quadir, A., P.J. Harrison, and R.E. Dewreede. 1979. The effects of emergence and subemergence on the photosynthesis and respiration of marine macrophytes. Phycologia 18: 83–88.

Redon, J., and O.L. Lange. 1983. Epiphytische Flechten im Bereich einer chilenischen ''Nebeloase'' (Fray Jorge) I. Vegetationskundliche Gliederung und Standortsbedingungen. Flora 174: 213–243.

Reynolds, T.L., and J.D. Bewley. 1993a. Absicic acid enhances the ability of the desiccation tolerant fern *Polypodium virginianum* to withstand drying. Journal of Experimental Botany 44: 1771–1779.

Reynolds, T.L., and J.D. Bewley. 1993b. Characterization of protein synthetic changes in a desiccation tolerant fern, *Polypodium virginianum*. Comparison of the effects of drying and rehydration, and absicic acid. Journal of Experimental Botany 44: 921–928.

Richardson, D.H.S. 1981. The Biology of Mosses. Oxford: Blackwell Scientific.

Ried, A. 1960. Thallusbau und Assimilationshaushalt von Laub- und Krustenflechten. Biol. Zentralblatt 79: 129–151.

Rogers, R.W., and R.T. Lange. 1971. Lichen populations on arid soil crusts around sheep watering places in south Australia. Oikos 22: 93–100.

Rundel, P.W. 1982a. The role of morphology in the water relations of desert lichens. Journal of the Hattori Botanical Laboratory 53: 315–320.

Rundel, P.W. 1982b. Water uptake by organs other than roots. In: Lange, O.L., P.S. Nobel, C.B. Osmond, and H. Ziegler, eds. Physiological Plant Ecology, vol. II. Water Relations and Carbon Assimilation. Berlin-Heidelberg-New York: Springer-Verlag, pp 111–134.

Rundel, P.W. 1988. Water relations. In: Galun, M., ed. Handbook of Lichenology, vol. 2. Boca Raton, FL: CRC Press, pp 17–36.

Rundel, P.W., and O.L. Lange. 1980. Water relations and photosynthetic response of a desert moss. Flora 169: 329–335.

Sancho, L.G., and L. Kappen. 1989. Photosynthesis and water relations and the role of anatomy in Umbilicariaceae (lichens) from central Spain. Oecologia 81: 473–480.

Sancho, L.G., B. Schroeter, and F. Valladares. 1997. Photosynthetic performance of two closely related *Umbilicaria* species in central Spain: temperature as a key factor. Lichenologist 29: 67–82.

Santarius, K.A. 1967. Das Verhalten von CO_2-Assimilation, NADP- und PGS-Reduktion und ATP-Synthese intakter Blattzellen in Abhängigkeit vom Wassergehalt. Planta 73: 228–242.

Santarius, K.A. 1994. Apoplasmic water fractions and osmotic water potentials at full turgidity of some Bryidae. Planta 193: 32–37.

Scheidegger, C., B. Frey, and B. Schroeter. 1997. Cellular water uptake, translocation and PSII activation during rehydration of desiccated *Lobaria pulmonaria* and *Nephroma bellum*. Bibliotheca Lichenologica 67: 105–117.

Scheidegger, C., B. Schroeter, and B. Frey. 1995. Structural and functional processes during water vapor uptake and desiccation in selected lichens with green algal photobionts. Planta 197: 399–409.

Scherer, S., T.-W. Chen, and P. Böger. 1986. Recovery of adenine-nucleotide pools in terrestrial blue-green algae after prolonged drought periods. Oecologia 68: 585–588.

Schiller, P., H. Heilmeier, and W. Hartung 1997. Absisic acid (ABA) relations in the aquatic resurrection plant *Chamaegigas intrepidus* under naturally fluctuating environmental conditions. New Phytologist 136: 603–611.

Schlensog, M., B. Schroeter, L.G. Sancho, A. Pintado, and L. Kappen. 1997. Effect of strong irradiance and photosynthetic performance of the melt-water dependent cyanobacterial lichen *Leptogium puberulum* Hue (Collemataceae) from the maritime Antarctic. Bibliotheca Lichenologica 67: 235–246.

Schneider, K., B. Wells, E. Schmelzer, F. Salamini, and D. Bartels. 1993. Desiccation leads to the rapid accumulation of both cytosolic and chloroplastic proteins in the resurrection plants *Craterostigma plantagineum* Hochst. Planta 189: 120–131.

Schonbeck, M.W., and J.D. Bewley. 1981a. Responses of the moss *Tortula ruralis* to desiccation treatments. I. Effects on minimum water content and rates of dehydration and rehydration. Canadian Journal of Botany 59: 2698–2706.

Schonbeck, M.W., and J.D. Bewley. 1981b. Responses of the moss *Tortula ruralis* to desiccation treatments. II. Variations in desiccation tolerance. Canadian Journal of Botany 59: 2707–2712.

Schönherr, J. 1982. Resistance of plant surfaces to water loss: transport properties of Cutin, Suberin, and associated lipids. In: Lange, O.L., P.S. Nobel, C.B. Osmond, and H. Zeigler, eds. Physiological Plant Ecology, vol. II. Water Relations and Carbon Assimilation. Berlin-Heidelberg-New York: Springer-Verlag, pp 153–180.

Schroeter, B., T.G.A. Green, L. Kappen, and R.D. Seppelt. 1994. Carbon dioxide exchange at subzero temperatures. Field measurements on *Umbilicaria aprina* in Antarctica. Cryptogamic Botany 4: 233–241.

Schroeter, B., L. Kappen, and F. Schulz. 1997. Long-term measurements of microclimatic conditions in the fruticose lichen *Usnea aurantiaco-atra* in the maritime Antarctic. In: Cacho, J., and D. Serrat, eds. Actas del V Simposio de Estudios Antárcticos. Madrid: CICYT, pp 63–69.

Schroeter, B., M. Olech, L. Kappen, and W. Heitland. 1995. Ecophysiological investigations of *Usnea antarctica* in the maritime Antarctic. I. Annual microclimatic conditions and potential primary production. Antarctic Science 7: 251–260.

Schroeter, B. and C. Scheidegger 1995. Water relations in lichens at subzero temperatures: structural changes and carbon dioxide exchange in the lichen *Umbilicaria aprina* from continental Antarctic. New Phytologist 131: 273–285.

Schwab, K.B., and D.F. Gaff. 1986. Sugar and ion contents in leaf tissues of several drought tolerant plants under water stress. Journal of Plant Physiology 125: 257–265.

Schwab, K.B., and U. Heber. 1984. Thylakoid membrane stability in drought tolerant and drought sensitive plants. Planta 161: 37–45.

Schwab, K.B., U. Schreiber, and U. Heber. 1989. Response of photosynthesis and respiration of resurrection plants to desiccation and rehydration. Planta 177: 217–227.

Scott, G.A.M. 1982. Desert bryophytes. In: Smith, A.J.E., ed. Bryophyte Ecology. London, New York: Chapman and Hall, pp 105–122.

Seel, W.E., N.R. Baker, and J.A. Lee. 1992b. Analysis of the decrease in photosynthesis on desiccation of mosses from xeric and hydric environments. Physiologia Plantarum 86: 451–458.

Seel, W.E., G.A.F. Hendry, and J.A. Lee. 1992a. The combined effects of desiccation and irradiance on mosses from xeric and hydric habitats. Journal of Experimental Botany. 43: 1023–1030.

Sgherri, C.L.M., B. Loggini, A. Bochicchio, and F. Navari-Izzo. 1994b. Antioxidant system in *Boea hygroscopica*: changes in response to desiccation and rehydration. Phytochemistry 37: 377–381.

Sgherri, C.L.M., B. Loggini, S. Puliga, and F. Navari-Izzo. 1994a. Antioxidant system in *Sporobolus stapfianus*: changes in response to desiccation and rehydration. Phytochemistry 35: 561–565.

Shields, L.M. 1957. Algal and lichen floras in relation to nitrogen content of certain volcanic and arid range soils. Ecology 38: 661–663.

Smirnoff, N. 1993. Tansley review No. 52: the role of active oxygen in the response of plants to water deficit and desiccation. New Phytologist 125: 27–58.

Smirnoff, N. 1995. Antioxidant systems and plant response to the environment. In: Smirnoff, N., ed. Environment and Plant Metabolism, Flexibility and Acclimation. Oxford: Bios Scientific Publishers, pp 217–244.

Smith, D.C., and S. Molesworth. 1973. Lichen physiology. XIII. Effects of rewetting dry lichens. New Phytologist 72: 525–533.

Smith, R.I.L. 1984. Terrestrial plant biology of the sub-Antarctic and Antarctic. In: Laws, R.M., ed. Antarctic Ecology. London: Academic Press, pp 61–162.

Smook, L. 1969. Some observations on *Lindernia intrepidus* (Dinter) Oberm. (=*Chamacgigas intrepidas* Dinter). Dinteria 2: 13–21.

Snelgar, W.P., and T.G.A. Green. 1981. Ecologically-linked variation in morphology, acetylene reduction, and water relations in *Pseudocyphellaria dissimilis*. New Phytologist 87: 403–411.

Sojo, F., F. Valladares, and L.G. Sancho. 1997. Structural and physiological plasticity of the lichen *Catillaria corymbosa* in different microhabitats of the maritime Antarctic. Bryologist 100: 171–179.

Solhaug, K.A., and Y. Gauslaa. 1996. Parietin, a photoprotective secondary product of the lichen *Xanthoria parietina*. Oecologia 108: 412–418.

Sperry, J.S., and M.T. Tyree. 1988. Mechanism of water stress-induced xylem embolism. Plant Physiology 88: 581–587.

Stålfelt, M.G. 1939. Vom System der Wasserversorgung abhängige Stoffwechselcharaktere. Botaniska Notiser 40: 176–192.

Stewart, G.R. 1989. Desiccation injury, anhydrobiosis and survival In: Jones, H.G., F.J. Flowers, and M.B. Jones, eds. Plant Under Stress. Cambridge: Cambridge University Press, pp 115–130.

Stocker, O., and W. Holdheide. 1938. Die Assimilation Helgoländer Gezeitenalgen während der Ebbezeit. Z. Botany 32: 1–59.

Stuart, T.S. 1968. Revival of respiration and photosynthesis in dried leaves of *Polypodium polypodioides*. Planta 83: 185–206.

Tegler, B., and K.A. Kershaw. 1981. Physiological-environmental interactions in lichens. XII. The variation of the heat stress response of *Cladonia rangiferina*. New Phytologist 87: 395–401.

Tennant, J.R. 1954. Some preliminary observations on the water relations of a mesophytic moss, *Thamnium alopecurum* (Hedw.) B. & S. Transactions of the British Bryological Society 2: 439–445.

Trachtenberg, S., and E. Zamski. 1979. The apoplastic conduction of water in *Polytrichum juniperinum* Willd. gametophytes. New Phytologist 83: 49–52.

Tuba, Z. 1987. Light, temperature and desiccation responses of CO_2-exchange in the desiccation-tolerant moss, *Tortula ruralis*. Symposia Biologica Hungarica 35: 137–149.

Tuba, Z., Z. Csintalan, and M.C.F. Proctor. 1996a. Photosynthetic responses of a moss, *Tortula ruralis*, ssp, and the lichens *Cladonia convoluta* and *C. furcata* to water deficit and short periods of desiccation, and their ecophysiological significance: a baseline study at present-day CO_2 concentration. New Phytologist 133: 353–361.

Tuba, Z., H.K. Lichtenthaler, Z. Csintalan, Z. Nagy, and K. Szente. 1994. Reconstitution of chlorophylls and photosynthetic CO_2 assimilation upon rehydration of the desiccated poikilochlorophyllous plant *Xerophyta scabrida* (Pax) Th. Dur. et Schinz. Planta 192: 414–420.

Tuba, Z., H.K. Lichtenthaler, Z. Csintalan, Z. Nagy, and K. Szente. 1996b. Loss of chlorophylls, cessation of photosynthetic CO_2 assimilation and respiration in the poikilochlorophyllous plant *Xerophyta scabrida* during desiccation. Physiologia Plantarum 96: 383–388.

Tuba, Z., H.K. Lichtenthaler, Z. Csintalan, and T. Pocs. 1993a. Regreening of desiccated leaves of the poikilochlorophyllous *Xerophyta scabrida* upon rehydration. Journal of Plant Physiology 142: 103–108.

Tuba, Z., H.K. Lichtenthaler, I. Maroti, and Z. Czintalan. 1993b. Resynthesis of thylakoids and chloroplast ultrastructure in the desiccated leaves of the poikilochlorophyllous plant *Xerophyta scabrida* upon rehydration. Journal of Plant Physiology 142: 742–748.

Tuba, Z., N. Smirnoff, Z. Csintalan, K. Szente, and Z. Nagy. 1997. Respiration during slow desiccation of the poikilochlorophyllous desiccation tolerant plant *Xerophyta scabrida* at present-day CO_2 concentration. Plant Physiology and Biochemistry 35: 381–386.

Tymms, M.J., D.F. Gaff, and N.D. Hallam. 1982. Protein synthesis in the desiccation tolerant angiosperm *Xerophyta villosa* during dehydration. Journal of Experimental Botany 33: 332–343.

Tyree, M.T., and J.S. Sperry. 1988. Do woody plants operate near the point of catastrophic xylem dysfunction caused by dynamic water stress? Plant Physiology 88: 574–580.

Valanne, N. 1984. Photosynthesis and photosynthetic products in mosses. In: Dyer, A.F., and J.G. Duckett, eds. The Experimental Biology of Bryophytes. London: Academic Press, pp. 257–273.

Valladares, F. 1994a. Texture and hygroscopic features of the upper surface of the thallus in the lichen family Umbilicariaceae. Annals of Botany 73: 493–500.

Valladares, F. 1994b. Form-functional trends in Spanish Umbilicariaceae with special reference to water relations. Cryptogamie, Bryologie et Lichénologie 15:117–127.

Valladares, F., C. Ascaso, and L.G. Sancho. 1994. Intrathalline variability of some structural and physical parameters in the lichen genus *Lasallia*. Canadian Journal of Botany 72: 415–428.

Valladares, F., A. Sanchez-Hoyos, and E. Manrique. 1995. Diurnal changes in photosynthetic efficiency and carotenoid composition of the lichen *Anaptychia ciliaris*: effects of hydration and light intensity. Bryologist 98: 375–382.

Valladares, F., and L.G. Sancho. 1995. Medullary structure of Umbilicariaceae. Lichenologist 27: 189–199.

Valladares, F., L.G. Sancho, and C. Ascaso. 1996. Functional analysis of the intrathalline and intracellular chlorophyll concentration in the lichen family Umbilicariaceae. Annals of Botany 78: 471–477.

Valladares, F., L.G. Sancho, and C. Ascaso. 1998. Water storage in the lichen family Umbilicariaceae. Botanica Acta 111: 1–9.

Valladares, F., J. Wierzchos, and C. Ascaso. 1993. Porosimetric study of the lichen family Umbilicariaceae: anatomical interpretation and implications for water storage capacity of the thallus. American Journal of Botany 80: 263–272.

Vieweg, G.H., and H. Ziegler. 1969. Zur Physiology von *Myrothamnus flabellifolia*. Berichte der Deutschen Botanischen Gesellschaft 82: 29–36.

Volk, O.H. 1979. Beiträge zur Kenntnis der Lebermoose (Marchantiales) aus Südwestafrika (Namibia). I. Mitteilungen der Botanischen Staatssammlungen München 15: 223–242.

Volk, O.H. 1984. Pflanzenvergesellschaftungen mit *Riccia*-Arten in Südwestafrika (Namibia). Vegetatio 55: 57–64.

Volk, O.H., and H. Leippert. 1971. Vegetationsverhältnisse im Windhoeker Bergland, Südwestafrika. Journal S.W.A. wiss. Ges. 25: 5–44.

Wagner, D.J., and J.E. Titus. 1984. Comparative desiccation tolerance of two *Sphagnum* mosses. Oecologia 62: 182–187.

Walter, H. 1931. Die Hydratur der Pflanze. Jena: Gustav Fischer-Verlag.

Walter, H., and K. Kreeb. 1970. Die Hydration und Hydratur des Protoplasmas der Pflanzen und ihre ökophysiologische Bedeutung. Wien, New York: Springer-Verlag.

Webber, P.J. 1978. Spatial and temporal variation of the vegetation and its production, Barrow, Alaska. In: Tieszen, L.L., ed. Ecological Studies 29: Vegetation and Production Ecology of an Alaskan Arctic Tundra. New York, Heidelberg, Berlin: Springer-Verlag, pp 37–112.

Wellburn, F.A.M., and A.R. Wellburn. 1976. Novel chloroplasts and unusual cellular ultrastructure in the "resurrection" plant *Myrothamnus flabellifolia* Welw. (Myrothamnaceae). Botanical Journal of the Linnean Society 72: 51–54.

Winter, K., and M. Königer. 1989. Dithiothreitol, an inhibitor of violaxanthin de-epoxidation, increases the susceptibility of leaves of *Nerium oleander* L. to photoinhibition of photosynthesis. Planta 180: 24–31.

Wirth, V. 1987. Die Flechten Baden-Württembergs, Verbreitungsatlas. Stuttgart: Eugen Ulmer Verlag.

Zamski, E., and S. Trachtenberg. 1976. Water movement through hydroids of a moss gametophyte. Israel Journal of Botany 25: 168–173.

Zimmermann, G., and H. Butin. 1990. Untersuchungen über die Hitze- und Trockenresistenz holzbewohnender Pilze. Flora 162: 393–419.

Zimmermann, M.H., and J.A. Milburn. 1982. Transport and storage of water. In: Lange, O.L., P.S. Nobel, C.B. Osmond, and H. Ziegler, eds. Physiological Plant Ecology, vol. II. Water Relations and Carbon Assimilation. Berlin, Heidelberg, New York: Springer-Verlag, pp 135–152.

3

Ecological Significance of Inherent Variation in Relative Growth Rate and Its Components

Hendrik Poorter
Utrecht University, Utrecht, The Netherlands

Eric Garnier
Center for Functional and Evolutionary Ecology, CNRS, Montpellier, France

I. INTRODUCTION

There is an amazing number of higher plant species on earth, with estimations greater than 250,000 (Wilson 1992). These species are not randomly distributed, but often can be found in rather specific habitats. What is the reason that some species flourish in a desert and others in the tundra? Clearly, a certain degree of specialization must have taken place. It is the aim of functional ecology to explain the distribution of species (or genotypes within a species) from their functional attributes. These attributes can be related to the physiological, morphological, anatomical, and/or chemical characteristics of a plant species, but they could also depend on life history characteristics such as seed longevity, flowering time, and life form. Since the question is why an individual of species A is performing better in a given environment than an individual of species B, a comparative approach is needed (Bradshaw 1987). The strength of such an approach is not only that we gain insight into the processes that determine a plant's success or failure in a given habitat, but it also enables us to categorize the wide variety of species into a more limited number of functional groups. This may be an avenue

Table 1 Terms, Abbreviations, and Units Used in This Chapter, As Well As Normal Ranges Found in Herbaceous Species[a]

Abbreviation	Meaning	Preferred units	Normal range
LAR	Leaf area ratio	m^2 leaf kg^{-1} plant	5–50
LD	Leaf density	kg m^{-3}	100–400
LTh	Leaf thickness	μm	100–600
LMF	Leaf mass fraction	g leaf g^{-1} plant	0.15–0.60
LR_m	Rate of leaf respiration	nmol CO_2 g^{-1} leaf s^{-1}	5–50
MRT	Mean residence time	year	0.2–20
NUR	Net nitrogen uptake rate	mmol N g^{-1} root day^{-1}	1–8
NP	Nitrogen productivity	g increase mol^{-1} N day^{-1}	20–100
NUE	Nitrogen use efficiency	kg increase mol^{-1} N taken up or lost	0.5–15
PCC	[C] in the plant	mmol C g^{-1} plant	25–33
PNC	[N] in the plant	mmol N g^{-1} plant	1–4
PS_a	Rate of photosynthesis	μmol CO_2 m^{-2} leaf s^{-1}	3–30
PS_m	Rate of photosynthesis	nmol CO_2 g^{-1} leaf s^{-1}	100–600
RGR	Relative growth rate	mg increase g^{-1} plant day^{-1}	40–400
RMF	Root mass fraction	g root g^{-1} plant	0.15–0.60
RR_m	Rate of root respiration	nmol CO_2 g^{-1} root s^{-1}	10–80
SLA	Specific leaf area	m^2 leaf kg^{-1} leaf	8–80
SMF	Stem mass fraction	g stem g^{-1} plant	0.05–0.30
SR_m	Rate of stem respiration	nmol CO_2 g^{-1} stem s^{-1}	?
ULR	Unit leaf rate	g increase m^{-2} leaf day^{-1}	2–20

[a] All mass-based parameters are expressed per unit dry mass.

toward simplification of a complex reality, with a great number of species in a given habitat. On the basis of functional groups, we are probably better able to predict effects of environmental changes on vegetations (Hobbs 1997).

This chapter uses the comparative approach to analyze the characteristics and distribution of species varying in the maximum relative growth rate (RGR_{max}; see Section II for definition and Table 1 for a listing of abbreviations and units used throughout this chapter) they can achieve. Plant species grown under uniform and more or less optimum conditions in the laboratory differ several-fold in RGR_{max} (100–400 mg g^{-1} d^{-1} for herbaceous species; 10–150 mg g^{-1} d^{-1} for woody species). Over the last three decades, evidence has accumulated that RGR_{max} is linked to the characteristics of the habitat from which the species originated (Parsons 1968, Chapin 1980, Lambers and Poorter 1992; see also Section III). This linkage is intriguing, and leads to a number of questions that will be addressed in this chapter. After introducing the different analytical concepts that will be used and providing evidence of a relationship between RGR_{max} and habitat characteristics, the physiological, morphological, and anatomical attributes that lead to variation in RGR_{max} between species are explored. We will show that, whatever the evolutionary forces have been, fast- and slow-growing species grown under laboratory conditions show consistent suites of traits. To have an ecological meaning, these sets of traits should not only be found in the laboratory, but also in the field. After testing whether this is indeed the case, we analyze the ecological implications of interspecific differences in growth rate and of variation in the underlying parameters. This leads us to suggest that selection has acted on components of RGR, rather than on RGR itself.

II. ASSESSING THE GROWTH POTENTIAL OF A SPECIES

This section focuses on the analysis of inherent differences in RGR (Appendix 1). The concept of RGR was first introduced by Blackman (1919), who recognized that the increase in plant biomass over a given period of time was proportional to the biomass present at the beginning of this period. He saw a parallel with money in a bank account accumulating at compound interest. In the case of plants, newly formed biomass is immediately deployed to fix new carbon and take up extra nutrients and water, thus leading to an accelerated biomass increase. Borrowing from economic theory, he derived the following equation:

$$M_2 = M_1 e^{RGR(t_2 - t_1)}$$

where M_1 and M_2 are the plant masses at time t_1 and t_2, respectively. The RGR in this equation indicates the dry mass increment per unit dry mass already present in the plant and per unit time. For a more detailed discussion on the background of RGR, see Appendix 1.

Originally, Blackman thought of RGR as a physiological constant, characteristic for a given species under given conditions. However, a constant RGR implies that plants grow exponentially throughout their entire life (Figure 1). In reality, plants hardly ever show a true exponential growth phase, as RGR changes continuously with ontogeny (Hunt and Lloyd 1987, Robinson 1991, Poorter and Pothmann 1992). During germination there is a gradual transition from growth being dependent on seed reserves to complete autotrophy. When plants get older and larger, the upper leaves start to shade lower leaves. Moreover, larger plants have to allocate more resources away from the assimilating parts of leaves and roots, and invest more in support tissue, especially in stems. Consequently, RGR decreases with size and/or time (Figure 1). Does this imply that the concept of RGR can only be used in the seedling stage, during what is often termed the

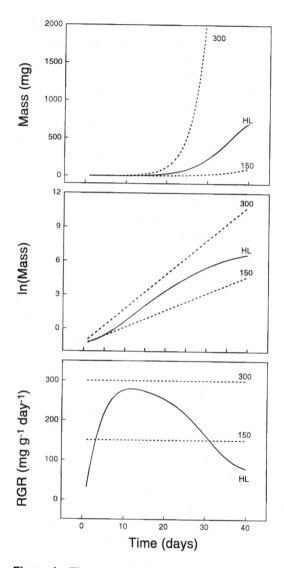

Figure 1 Time course in (top) total plant mass, (middle) ln-transformed values of total plant mass, and (bottom) RGR of a theoretical plant population growing continuously with an RGR of 150 or 300 (dotted lines) mg g^{-1} day^{-1}, and experimental data on a population of *Holcus lanatus* (continuous line marked HL; adapted from Hunt and Lloyd 1987). All populations had a similar starting mass at day 0.

"exponential growth phase"? Mathematically, there is no requirement for RGR to be constant because it is a parameter that can be used as a quantification of growth at any point in time, even if growth is not strictly exponential. It can also be used as an average over a given time period (Evans 1972) or a given mass trajectory. Therefore, as long as one is convinced that the growth of the plants under study is somehow proportional to the plant biomass already present, RGR is the most appropriate parameter to use. However, it is not a parameter fully independent of plant size.

In the field, where plants experience a fluctuating environment, growth is restricted by a continuously changing array of abiotic factors (light, temperature, nutrients, and water) and affected by biotic interactions (competitors, herbivores, pathogens, but also symbionts). In comparing species (or genotypes), it is of interest to know their genetic potential, or growth achieved in the absence of constraining factors. Such a goal is difficult to achieve. It would require knowledge for each species about the exact combination of factors that enables fastest growth. Even if such a goal could be technically achieved, it would have the drawback of comparing species that had been grown in more or less different environments. The practical solution has been to choose a set of conditions that is close to optimal for growth of most species and technically achievable (Grime and Hunt 1975). Growth rate is then measured for relatively small and young plants over rather short time intervals (10–20 days) and without interference from other plants. The RGR value obtained in this way is considered to be the RGR_{max}. These values are not absolute because they depend on ontogeny as well as growth conditions. However, with the exception of very low nutrient levels (Shipley and Keddy 1988) or light levels (Mahmoud and Grime 1974), RGR ranking remains rather similar (see Poorter et al. 1995, Biere et al. 1996 for nutrients; Hunt and Cornelissen 1997, Poorter and Van der Werf 1998 for light). Therefore, ranking of species for RGR_{max} does not change strongly across experiments and can be used in a relative way to order species on the fast-slow continuum. However, there is variability in such relative rankings across experiments, with correlation coefficients approximately 0.6 (Table 2). Part of this variation is probably caused by imprecisions related to RGR determinations, especially in larger screening programs with a limited number of plants harvested per species (Poorter and Garnier 1996).

What do RGR_{max} values obtained in the laboratory tell about plant growth in the field? With respect to light, field-grown plants generally experience stronger fluctuations in instantaneous irradiance and higher levels of total quantum input, when considered over the whole growing season (Garnier and Freijsen 1994). However, RGR does not strongly depend on the total daily quantum input above $20 \, mol \, m^{-2} \, d^{-1}$ (Poorter and Van der Werf 1998), a value often reached in growth chambers. Temperature is often lower in the field, especially during vegetative growth in temperate climates. With respect to nutrients, conditions are generally far more limiting in the field. Moreover, plants in the field encounter competition

Table 2 Correlation Coefficients Between RGR_{max} Values of Herbaceous Species Shared by Some Larger-Scale Comparative Experiments and Number of Species in Common on Which the Correlation Coefficient Is Based[a]

	P89	P&O	H97	V98
G75	0.77 (n = 8)	0.55 (n = 23)	0.66 (n = 30)	0.57 (n = 33)
P89		0.98 (n = 7)	—	—
P&O			0.73 (n = 13)	0.76 (n = 13)
H97				0.62 (n = 13)

[a] Only those correlations are given when seven or more species were in common.

Sources: Data from Grime and Hunt 1975 (G75; 130 species); Poorter 1989 (P89; 9 species); Poorter and Remkes 1990, supplemented with data from Van der Werf et al. 1993, Den Dubbelden and Verburg 1996, and J.J.C.M. van Arendonk (unpublished results), which were all grown under identical conditions (P&O; 47 species); Hunt and Cornelissen 1997 (H97; 43 species); and Van der Werf et al. 1998 (V98; 71 species).

and other biotic interactions. Consequently, there is a large difference between the RGR_{max} as measured in the laboratory and RGR of plants in the field (Garnier and Freijsen 1994). Although we expect some relationship between the relative ranking in laboratory and field, data are too scarce for a proper evaluation. In this chapter, we treat RGR_{max} more as a representation of a complex of traits than as a predictor of growth rate under natural conditions. This aspect is discussed further in Section VII.

III. RGR_{max} AND PLANT ECOLOGY

A. RGR_{max} and Plant Distribution

Bradshaw et al. (1964) were among the first to establish a relationship between the RGR_{max} of wild species as measured under laboratory conditions and the characteristics of the habitat they originated from. Others followed, but generally the number of species was rather low (< 10) to infer strong conclusions. Grime and Hunt (1975) determined the RGR_{max} of 130 species from England and classified them according to habitat. Fast-growing species were found relatively more often in fertile habitats, whereas species with a low potential growth rate tended to occupy infertile habitats. It is not always easy and straightforward to quantitatively classify a specific habitat along a fertility scale. A semiquantitative approach has been used by Ellenberg (1988), who assigned so-called "N-numbers" to a wide range of species from central Europe. The higher the value, the higher the fertility of the habitats in which such a species would generally occur. Plotting RGR_{max} data of herbaceous perennials against the N-number of Ellenberg generally yields positive relationships (Figure 2A). A positive relation between RGR_{max} and nutrient availability also is likely for woody species (Cornelissen et al. 1998). However, annuals seem to have a high RGR_{max} independent of soil fertility (Fichtner and Schulze 1992).

Is inherent variation in potential RGR of species also related to other environmental gradients? Evidence is less well documented than in the case of nutrient availability. Alpine species have lower RGR_{max} under laboratory conditions than lowland species (Figure 2B). Dry (Rozijn and Van der Werf 1986; Figure 2C) or saline habitats (Figure 2D) harbor species that grow more slowly under optimal conditions, and the same relation between RGR_{max} and plant occurrence was found for sites with heavy metal pollution (Wilson 1988, Verkley and Prast 1989). Disturbance regime may also be a source of variation in RGR_{max}: annuals that have to complete their life cycle in periodically disturbed habitats display a higher RGR_{max} than congeneric perennials from more stable habitats (Garnier 1992). The intensity of trampling has also been identified as selecting for species with different RGR_{max}: plants from trampled places have a lower RGR_{max} than those from nontrampled sites (Figure 2E). Finally, when determined at relatively high light levels, species from strongly shaded habitats have a lower RGR than those from light-exposed environments. In most of the cases shown in Figure 2B–F, however, it seems that differences in RGR_{max} between species from favorable and unfavorable habitats are not as clear as in the case of species adapted to habitats differing in fertility.

B. RGR_{max} and Plant Strategies

We have shown above that in a variety of cases, there is a link between the potential growth rate of a species and its occurrence in a given habitat. As such, RGR_{max} forms one of the cornerstones in the plant strategy theory formulated by Grime (1979). According to this theory, plant strategies are shaped by the possible combinations of two factors experienced by plants: stress and disturbance. *Stress* in this sense is defined as the extent to which a combination of environmental variables retards growth (e.g., low nutrient availability, low or high temperature, low water availability). *Disturbance* is defined as the degree of physical disruption of the plant's biomass (e.g., grazing, trampling). Species from habitats with a high degree of stress and a low degree of disturbance are called "stress tolerators." They are generally perennials with a low RGR_{max} (Figure 3). Species from sites with a high disturbance but with low stress are called "ruderals." They are

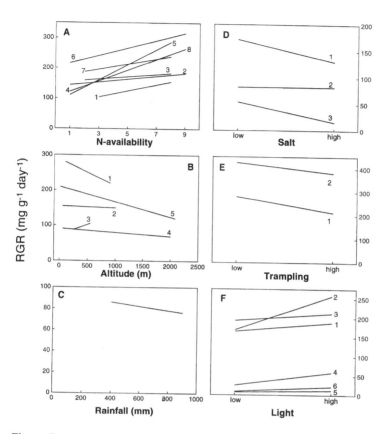

Figure 2 RGR$_{max}$ of species originating from habitats differing in: (A) nutrient availability, as indicated by the N-number of Ellenberg; (B) altitude, in meters above sea level; (C) rainfall; (D) salt concentration; (E) trampling intensity; and (F) light intensity. Each study is represented by a regression line. *Sources*: numbers refer to the following studies: A: data from (1) Rorison 1968, (2) Grime and Hunt (1975), (3) Boorman (1980), (4) Poorter (1989), (5) Poorter and Remkes (1990), (6) Reiling and Davidson (1992), (7) Stockey and Hunt (1994), and (8) Van der Werf et al. (1998); B: data from (1) Woodward (1979), (2) Woodward (1983), (3) Graves and Taylor (1986), (4) Atkin and Day (1990), and (5) Atkin et al. (1996a); C: data from Mooney et al. (1978); D: data from (1) Ball (1988), (2) Van Diggelen (1988), and (3) Schwarz and Gale (1994); E: data from (1) Dijkstra and Lambers (1989) and (2) Meerts and Garnier (1996); F: data from (1) Pons (1977), (2) Corré (1983a), (3) Corré (1983b), (4) Kitajima (1994), (5) Osunkoya et al. (1994), and (6) L. Poorter, unpublished results.

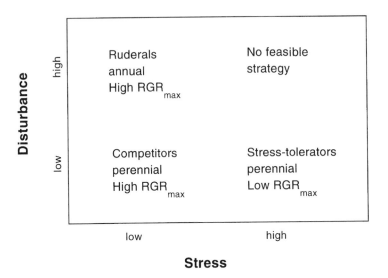

Figure 3 Relation between stress and disturbance and the growth strategy, life form, and RGR of the three types of strategies. (*Source*: Adapted from Grime 1979.)

mostly annuals that have a high RGR_{max}, enabling them to complete their life cycle quickly. This would ensure that seeds are produced before a disturbance event takes place that kills the plant. Habitats in which both stress and disturbance are low are favorable for plant growth. According to the plant strategy theory, these are sites where a strong competition between plants is expected; consequently, species that thrive here are called "competitors." They are generally perennials, and also have a high RGR_{max}. Sites with a high level of both stress and disturbance (vulcanos, nutrient-poor and strongly drifting sand dunes) do not bear plants because there is no feasible strategy to cope with such an environment.

Given the importance that is generally attached to biomass gain and the relative fitness of a plant (McGraw and Garbutt 1990), it may be hypothesized that there has been a selection pressure in fertile (and favorable, even if only temporarily) habitats toward plant species with a high RGR, and toward a low potential RGR in places that are unfavorable (Grime 1979, Chapin 1980). However, RGR is a parameter that is the result of a combination of many physiological, morphological, anatomical, and biochemical traits. Alternatively, it could well be that it is one or more of these traits underlying RGR that has been the target of selection, rather than RGR itself (Lambers and Dijkstra 1987, Grime 1979, Coley 1983). RGR would then merely be a by-product of selection. Before being able to evaluate these contrasting hypotheses, we first must investigate the components underlying RGR.

IV. COMPONENTS UNDERLYING RGR$_{max}$

A. Growth Parameters

Growth is more than photosynthesis. It is the balance between the carbon gain per unit leaf area and the carbon losses in the plant (which depend on the respiration rate but also on the relative proportion of the assimilatory and nonassimilatory organs), corrected for the C-concentration of the newly formed biomass. Evaluating the relative importance of each of these factors requires a top-down approach, in which RGR is broken down into components. A common way to do so is factorizing RGR into the increase in mass per unit leaf area and time (ULR) and the leaf area per unit plant mass (LAR). LAR can be factorized further into the components leaf area:leaf mass (SLA) and leaf mass:plant mass (LMF). A definition of these components of RGR, as well as an explanation of the concept of growth parameters underlying RGR, is given in Appendix 2.

To what extent is inherent variation in RGR$_{max}$ caused by variation in the components ULR and LAR? A wide variety of results has been published; some experiments found ULR to be the factor determining growth, others found LAR to be the cause of variation in growth, whereas others found intermediate results. Variation may be due to the choice of the species as well as growth conditions used and the experimental procedure followed (e.g., duration of the experiment and number of harvests). To enable a quantitative analysis of the cause of variation in RGR within a given experiment, we use the growth response coefficient (GRC). This coefficient can be calculated after determining the linear regression between the growth parameter X (which can be ULR, LAR, SLA, or LMF) as the dependent variable and RGR as the independent variable. GRC$_X$ then is defined as the relative increase in growth parameter X divided by the relative increase in RGR:

$$GRC_X = \frac{dX/X}{dRGR/RGR}$$

The sum of GRC$_{ULR}$ and GRC$_{LAR}$ for any experiment should be 1, and this is also the case for the sum of GRC$_{ULR}$, GRC$_{SLA}$, and GRC$_{LMF}$. A value of 1 for GRC$_{ULR}$ indicates that species variation in RGR within an experiment is completely due to variation in ULR, whereas a value of 0 indicates no effect of this parameter at all.

What is the overall picture that emerges from the literature? Poorter and van der Werf (1998) analyzed a total of 111 experiments on herbaceous C$_3$ species, and calculated the average GRC for the various growth parameters. They found that, on average, variation in SLA is by far the most dominant factor explaining variation in inherent RGR. ULR is second, and LMF is, on average, the quantitatively least important variable (Table 3). For a more elaborate analysis on GRC and the literature compilation, see Poorter and van der Werf (1998).

Table 3 Growth Response Coefficients for ULR, LAR, SLA, and LMF[a]

	GRC
ULR	0.26
LAR	0.74
SLA	0.63
LMF	0.11

[a] Average values from a literature survey on causes of inherent variation in RGR. A total of 111 articles were compiled. Only those reports were considered in which RGR differences between species or genotypes were at least 40 mg g^{-1} day^{-1}. More details are given in Poorter and Van der Werf (1998). See Table 1 for definitions of abbreviations.

B. Physiological Parameters

The aforementioned technique of growth analysis has the advantage of being simple. Moreover, technical requirements to conduct such an experiment are low. A drawback is that ULR is a parameter that integrates various aspects of plant functioning (photosynthesis, respiration, allocation, chemical composition) and therefore cannot be related directly to a specific physiological process. One may assume that a considerable part of the variation in ULR is due to differences in the area-based rate of photosynthesis (Konings 1989). To obtain more insight into the physiological basis of variation in RGR, however, a more mechanistic approach must be followed, in which growth is analyzed in terms of the plant's carbon (C) economy. This requires knowledge of the C-gain of the plant in photosynthesis, and C-losses in leaf, stem, and root respiration, all integrated over the day. The net increase in C over the day can be converted into a dry mass increase if the C-concentration of the newly formed material is known. A formula to relate the C-fluxes to RGR is presented in the second part of Appendix 2.

Given that 85–95% of plants consist of carbon-based compounds (for a review, see Poorter and Villar 1997), it is beyond doubt that almost all newly formed biomass is fixed during the photosynthetic process. However, that does not necessarily imply that variation in RGR must be due to variation in the rate of photosynthesis, measured per unit leaf area and per unit of time (PS$_a$). Only a few attempts have been made to quantify both whole-shoot photosynthesis at growth conditions (rather than determining photosynthetic capacity for the youngest expanded leaf) and RGR. In a number of cases, no relationship at all has been observed (Dijkstra and Lambers 1989, Poorter et al. 1990, Van der Werf et al. 1993, Atkin et al. 1996a), but in others a positive relation was found (Garnier et al., unpublished results). If indeed the growth parameter ULR is correlated well with the rate of photosynthesis per unit leaf area (Konings 1989, Poorter and Van der Werf 1998), we might derive from the GRC_{ULR} value in Table 3 that, in the average experiment, there is a modestly positive relationship between PS_a and RGR.

It may well be that PS_a (and ULR) is not the best variable to consider if one seeks to understand the physiological basis of growth. Traditionally, in physiological research, the rate of photosynthesis is expressed per unit leaf area. In this way, the flux of C can be related to the flux of incoming photons and the efflux of water. In analyzing growth, however, it may be more relevant to consider how much C is fixed by 1 g of leaf biomass, which is the photosynthetic rate per unit leaf mass (PS_m). This would relate much better to RGR, which is the increase in biomass per unit of biomass and time. PS_m is the product of PS_a and SLA (Appendix 2), and this parameter is strongly associated with RGR (Poorter et al. 1990, Reich et al. 1992, Van der Werf et al. 1993, Walters et al. 1993, Atkin et al. 1996a, Garnier et al. unpublished results).

Both shoot and root respiration are generally positively correlated with RGR (Poorter et al. 1990, Van der Werf et al. 1993, Walters et al. 1993; but see Atkin et al. 1996a). This is partly a reflection of the higher rate of growth of the fast-growing species, which requires extra amounts of energy (ATP) and reducing power (NADH) per unit dry mass. As far as roots are concerned, the higher respiration rate is also a reflection of a much higher uptake rate of ions (Van der Werf et al. 1994).

An alternative approach to the carbon balance decomposition just described can be followed, focusing on the nitrogen economy of the plant. RGR can then be expressed as the product of the root mass fraction and the nitrogen uptake rate divided by the mean plant nitrogen concentration (Garnier 1991, Lambers and Poorter 1992). Breaking down the RGR of fast- and slow-growing species in this way shows that faster growth is associated with a relatively lower allocation of biomass to roots, a higher plant nitrogen concentration, and a higher nitrogen uptake rate (NUR). In fact, NUR is the parameter that shows the widest variation and the strongest correlation with RGR in the fast-slow continuum (Garnier 1991).

C. Chemical and Anatomical Parameters

There is a wide range of other parameters for which fast- and slow-growing species differ under more or less optimal growth conditions. Most notably, fast-growing species have higher concentrations of reduced nitrogen in all organs, as well as higher amounts of minerals (Poorter and Bergkotte 1992, Reich et al. 1992, Van der Werf et al. 1993, Garnier and Vancaeyzeele 1994, Atkin et al. 1996b). They tend to have slightly lower concentrations of C (but see Garnier and Vancaeyzeele 1994 and Atkin et al. 1996a), but these differences are marginal for species within the same life form. Differences in C-concentration may become substantial when the growth differences between herbaceous and woody species are analyzed (Poorter 1989).

Given the importance of SLA in explaining variation in RGR_{max} (Table 3), it is appropriate to factorize this parameter further. SLA, or rather its inverse, 1/SLA (expressed in g m^{-2}), is the product of leaf thickness and leaf density (see Appendix 3). Generally, the lower SLA of slow-growing species is due more to a higher leaf density (LD) than to a higher leaf thickness (Van Arendonk and Poorter 1994, Garnier and Laurent 1994; but see Körner and Diemer 1987, Shipley 1995). The density of a leaf or root is strongly related to its water content per unit dry mass (Garnier and Laurent 1994). This can easily be understood by envisaging a cell as a box. A higher density is often caused by an extra deposition of cell wall material (lignin, cellulose). A doubling of the amount of cell wall material will hardly affect the cell size or the amount of water in the cell (the volume of the box). However, the amount of water relative to the dry mass will decrease, whereas the density, the dry mass per unit volume will increase. It can therefore be expected that fast-growing species, with a low density of tissues, will have a high water content per unit biomass. This turns out to be the case and not only applies to leaves but also to stems and roots (Garnier 1992, Ryser 1996). Another factor that will strongly affect the density is the relative volume of intercellular spaces.

V. RGR AND ITS COMPONENTS: A SYNTHESIS

It is quite clear from the previous section that inherent differences in RGR_{max} among species are associated with differences in a large array of plant traits (Lambers and Poorter 1992, Chapin et al. 1993). To what extent are these parameters related to each other? To answer this question, a principal component analysis (PCA) was conducted on two sets of data obtained from wild herbaceous species for which a large number of variables was measured. The first one (*Herbs*) concerns 11 grasses and 13 dicotyledonous species; the second (*Grasses*) is for 12 wild grasses. The results are shown in Figure 4.

The patterns confirm the differences between the fast- and slow-growing species as presented in Section IV: in both data sets, a high RGR is strongly related to a high uptake capacity of both leaves (photosynthesis, PS_m) and roots (nitrogen uptake rate [NUR]) expressed per unit mass of leaves and roots. The correlation between RGR and PS_a (and ULR) is looser. Shoot respiration and root respiration (only available for *Herbs*) are also positively associated with RGR. Species with a high RGR also have a high SLA, a high nitrogen concentration in the leaves, and a high concentration of minerals in their tissues (only available for *Herbs*). All these parameters can be found at the righthand side of

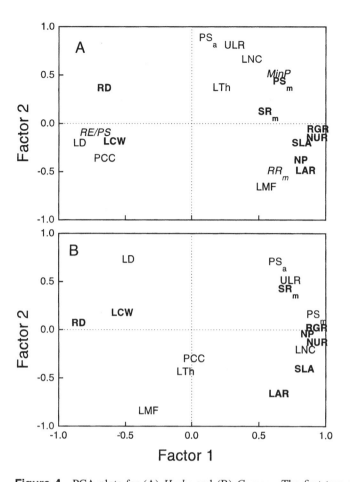

Figure 4 PCA plots for (A) *Herbs* and (B) *Grasses*. The first two axes explain 66% and 71% of the overall information for *Herbs* and *Grasses*, respectively. In both analyses, axis 1 can be interpreted as a biomass production axis; axis 2 appears to be mainly determined by gas exchange properties (photosynthesis and transpiration) per unit leaf area. Variables printed in bold are on similar places in the plane. Variables printed in italics were determined for *Herbs* only. Abbreviations not given in Table 1 are: LCW, proportion of cell walls in the leaves; MinP, mineral concentration of the plant; RE/PS, the fraction of daily fixed photosynthesis that is respired again. In this analysis, two new variables (factor 1 and factor 2) are computed out of a combination of all original variables. For each of these variables it is calculated whether they contribute positively (close to 1.0), negatively (close to −1.0), or not (close to 0.0) to factor 1. The amount of variance thus explained is taken out of the data and the procedure is repeated with the remaining variance. The result is somewhat comparable to a two-dimensional electrophoresis. Variables that are close together (like RGR and NUR) are generally positively correlated, variables that are at the opposite parts of the graph (like RGR and LCW) are negatively correlated, and variables that have values close to 1 or −1 for one factor and values close to 0 for the other axis [like RGR and PS$_a$ in (A)] are generally not correlated at all. *Source*: Data for *Herbs* are from Poorter and Remkes (1990), Poorter et al. (1990), and Poorter and Bergkotte (1992), and concern 11 grasses and 13 dicotyledoneous species. Data for *Grasses* are from Garnier (1992), Garnier and Laurent (1994), Garnier and Vancaeyzeele (1994), and Garnier et al. (unpublished results) and comprise 12 grass species.

Figure 4. At the lefthand side are the variables associated with slow-growing species. They have a high density in leaves and roots, a high proportion of leaf biomass in cell walls, and a large fraction of the total C fixed daily in photosynthesis is spent in respiration (only available for *Herbs*).

A PCA analysis can also be used to gain insight into how well the various other parameters of the plant are correlated with each other, and probably also mechanistically associated. For example, PS_a and ULR are strongly associated. The rates of shoot and root respiration (only available for *Herbs*) have a similar value in common for factor 1, but at the axis of factor 2 they are separated from each other. This may indicate that, although they are both highly correlated with RGR, the reason for being correlated is—at least to some extent—different. Part of the respiration is necessary for growth and can be expected to be high in both organs for fast-growing species. Another part of the shoot respiration may have to do with a high transport rate of sugars to the sink, whereas part of the root respiration is related to nutrient uptake. It is this type of information we need if we want to understand the trade-offs that play a role in determining the functioning of the whole plant.

From the above analyses, suites of traits can be associated with fast and slow growth: high RGR_{max} is achieved in species exhibiting high rates of resource acquisition, made possible, among other things, through high light interception per unit leaf mass (high SLA), high concentration of enzymatic machinery (reflected by a high nitrogen concentration), and low density of tissues (i.e., high water content). The latter two are an indication that fast-growing plants display a high concentration of protoplasmic elements (except for starch; see also Niemann et al. 1992). Opposite traits are found in slow-growing species, a low RGR_{max} being associated with a high amount of cell walls and starch—which are metabolically inactive and lead to low acquisition rates per unit mass. In Section VIII, we put forward the hypothesis that these contrasting suites of traits can be interpreted as a functional trade-off between high biomass productivity and efficient conservation of nutrients.

VI. DO LABORATORY FINDINGS APPLY TO THE FIELD?

As stated in Section II, plants growing in the field certainly do not achieve their potential growth rate. But to bear ecological significance, differences in the aforementioned suite of traits between species, as found under laboratory conditions, should at least reflect to some extent the differences between the same species growing in their natural habitat. Is there evidence that this is the case? This question can be approached by comparing plant traits that have been measured for the same species both in the laboratory and in the field. This will be done here for SLA and leaf nitrogen concentration on a leaf mass basis, which are positively

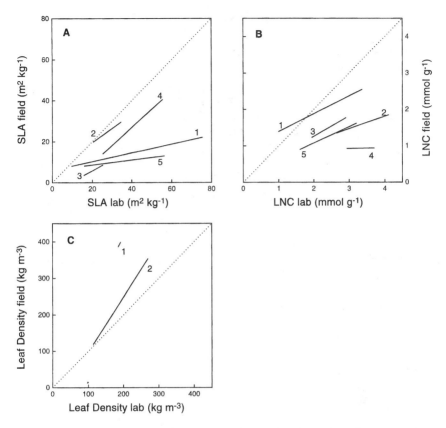

Figure 5 (A) SLA, (B) leaf nitrogen concentration, and (C) leaf density measured on the same species in field and laboratory conditions. Each study is represented by a regression line and contains several species. *Sources*: numbers refer to the following studies: A: data from (1) J.H.C. Cornelissen et al. (unpublished results, laboratory and field), (2) Garnier and Laurent (1994, laboratory) and Garnier et al. (1997, field), (3) Mooney et al. (1978, laboratory and field), (4) Poorter and Remkes (1990, laboratory) and H. Poorter and R. de Jong (unpublished results, field), (5) Walters et al. (1993) and P.B. Reich (unpublished results, laboratory), and Reich et al. (1992, field). B: data from (1) Cornelissen et al. (1997), (2) Garnier and Vancaeyzeele (1994, laboratory) and Garnier et al. (1997, field), (3) Hull and Mooney (1990, laboratory and field), (4) Mooney et al. (1978, laboratory and field), (5) P.B. Reich (unpublished results, laboratory) and Reich et al. (1992, field). C: data from Garnier and Laurent (1994, laboratory) and E. Garnier and J.-L. Cordonnier (unpublished results, field), (2) Poorter and Bergkotte (1992, laboratory) and H. Poorter and R. de Jong (unpublished results, field).

related to RGR_{max}, and leaf density, which was negatively correlated with RGR_{max} for plants grown under optimum conditions in the laboratory (Sections IV and V). How are these attributes when plants grow in the field, and is the ranking of species the same under both conditions?

Several data sets are available to address this question, comprising both herbaceous and woody plants. For the 86 species for which such data are available, SLA is on average 50% lower in the field than in the laboratory. Nonetheless, there is a good relationship between SLA measured in the field and in the laboratory for four of the five data sets taken individually (Figure 5A). Leaf nitrogen concentration is 34% lower in field-grown plants, and the relationship between field- and laboratory-growing plants is significant for three of the five data sets (Figure 5B). Finally, leaf density is approximately 40% higher in the field than in the laboratory, and the relationship between density measured in the field and in the laboratory is highly significant for the only data set with a substantial amount of data points (11 species; Figure 5C). Differences between laboratory- and field-measured leaf traits are probably caused by a combination of a higher light intensity, higher wind speed, and lower nutrient availability in the latter environment.

Although there are differences in plant traits measured under field and laboratory conditions, these results show that the ranking of species for such important traits as SLA and leaf density are maintained under a wide range of growing conditions. This may be less so for leaf nitrogen concentration. We may therefore expect that the suite of traits discussed in the previous section is likely to be maintained, at least partially, under field conditions.

VII. SELECTION FOR RGR OR UNDERLYING COMPONENTS?

A. Selection for RGR

As discussed in the previous section, RGR is a parameter that is associated with a suite of traits. Based on the negative correlations between RGR_{max} and the harshness of the environment (Figure 2), it has been suggested that RGR was the target of selection (Grime 1977, Chapin 1988). Alternatively, it may have been one or more of the components of RGR that has been selected for (Grime 1977, Coley 1983, Lambers and Dijkstra 1987).

What are the arguments in favor of selection for a high RGR? For ruderal and other annual species, a fast completion of the life cycle seems of paramount importance (Grime 1979). A high RGR may be of help to reach the size required for high seed production (Benjamin and Hardwick 1986, McGraw and Garbutt 1990, Garnier and Freijsen 1994). Competitive species, be it woody or herbaceous, seem to have an advantage by quickly occupying the available space within the vegetation. The acquisition of resources, both above and below ground, will depend on how much of the volume of soil and air is occupied by roots and leaves. A high RGR may enable this.

What would be the adaptive value of a low RGR? Some explanations have been offered, focusing on plant species from nutrient-poor habitats (Chapin 1980, 1988). However, these suggestions are questionable (Poorter 1989, Lambers & Poorter 1992):

1. *If plants grow slowly, they are less likely to deplete the available nutrient resources early in the season.* This does not seem to be an evolutionarily stable strategy. As soon as one species or genotype starts to take up as much nutrients as possible early in the season, the others miss out.

2. *Plants in nutrient-poor areas will never grow fast. If they have a low potential RGR, they will be closer to their "optimum" in the field.* For all plants it applies that their physiological optimum differs from the conditions they experience in the field (see Section II). However, a clear difference between physiological and ecological optimum does not necessary imply a disadvantage. Fast- as well as slow-growing species have a large flexibility, morphological as well as physiological, to cope with varying conditions (Reynolds and D'Antonio 1996). At low nutrient supply, for example, it is found that species with a high RGR_{max} still grow faster or at similar rates than those with a low RGR_{max} (Shipley and Keddy 1988, Poorter et al. 1995).

3. *If plants grow slowly, they can accumulate sugars and nutrients during favorable times, so as to enable growth in later times when nutrients are less easily available.* As far as sugars are concerned this explanation does not hold. Plants experiencing nutrients stress are limited more in growth than in photosynthesis. They fix more C than can be used in growth, resulting in accumulation of nonstructural sugars (mainly starch; Poorter and Villar 1997). Therefore, it does not seem necessary to store sugars for times with a low nutrient availability. For nutrients, however, this reasoning could be valid. A prerequisite is that nutrients become available in flushes. Such flushes have been shown, for example, during freeze-thaw events that lyse microbial cells or during drought-wet cycles (Bilbrough and Caldwell 1997). However, although these processes have been shown to occur, there is at present not enough evidence to conclude that species found in sites of different nutrient availability (or those with contrasting RGR_{max}) differ in this respect. For plants grown hydroponically in the laboratory at a nonlimiting nutrient supply, we found fast-growing species to accumulate 4–5 times more NO_3 than slow-growing species.

4. A last hypothesis does not explain why plants in low-resource environments do have a low RGR, but rather why they do not have a high potential RGR: *A high RGR cannot be realized in low-resource environments and therefore RGR is a selectively neutral trait.* Although a very high RGR would not be reached, a genotype with a slightly higher RGR could still occupy some extra space and consequently would be able to acquire some extra nutrients. Thus, in those cases RGR would not necessarily be a selectively neutral trait.

B. Selection for Components of RGR

Is there evidence that components of RGR have been selected for, rather than RGR itself? This is not a question that can be easily answered. If all of the components scale positively with RGR, it will be impossible to separate between the two alternatives. Moreover, it will be difficult to single out one component if the different traits have not been selected independently, or if they are functionally related. For example, a high rate of photosynthesis will be achieved with a high concentration of protein. A high concentration of protein may imply a high rate of protein turnover, and therefore a high maintenance respiration. In addition, a high rate of photosynthesis may result in a high rate of export to the phloem, which will also cause an increase in shoot respiration (Figure 4). Given these mechanistic interrelations, it will not be easy to break the correlation between photosynthesis and respiration.

Nevertheless, we believe that there are good reasons to think that selection in a low-resource environment has been for components related to a low SLA, rather than for RGR per se. Generally, SLA is the most important growth parameter associated with inherent variation in RGR_{max} (Sections IV and V). Moreover, differences in SLA observed in the laboratory are preserved in the field (Section VI). What are the arguments in favor for this alternative hypothesis?

C. Selection for SLA-Related Traits in Adverse Environments

Plants from extremely nutrient-poor environments are often evergreens, with a high leaf longevity. As was discussed in Section VII.B, availability of photosynthates is not a growth-limiting factor in these environments, but the availability of nutrients is. A first problem of plants in a low-nutrient environment is to acquire nutrients; a second problem is to use them efficiently. Berendse and Aerts (1987) showed that an efficient use of, for example, N (nitrogen use efficiency [NUE]; gram of growth per unit N taken up or lost by the plant) depends on two components: the biomass increase per unit N and per unit of time (the mean annual nitrogen productivity [NP]) and the average time that a unit of N stays in the plant (mean residence time [MRT] of nitrogen; see Appendix 4). They argue that there is a trade-off between these two components: a plant cannot achieve both a high NP and a high MRT. Such a trade-off has indeed been shown in several cases (e.g., Eckstein and Karlsson 1997; Figure 6), but is less evident in other experiments (Garnier and Aronson 1998). The putative trade-off between NP and MRT is most likely due to the suite of characters discussed in Section V: a high nitrogen productivity is strongly correlated with both a high SLA and a low density of organs (characteristics found at the righthand side in Figure 4). Plants with those characteristics are mainly directed toward attaining a high rate of resource capture (carbon, nutrients). This strategy is discussed further in Section VII.D. The other extreme is formed by the species with a high MRT. Species with such a strategy can be found in nutrient-poor environments. Why is this so? In nutrient-poor environments, it may be a disadvantage for an individual plant to lose acquired nutrients, as it is very questionable whether that individual plant will be able to take them up again once they have entered the nutrient cycling process. Therefore, there is a premium to increase the residence time of nutrients. Theoretically, this can be achieved in two ways (Appendix 4): either a plant resorbs nutrients from senescing organs very efficiently, or it restricts the turnover of organs and thus the loss of biomass per unit of biomass present (Aerts 1990, Garnier and Aronson 1998). Although there is variation in resorption efficiency

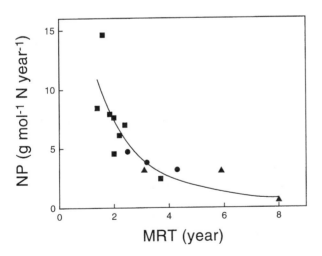

Figure 6   The relation between nitrogen productivity (NP; on an annual basis) and the mean residence time (MRT) of nitrogen of 14 species grown in the field. Squares, herbaceous species; circles, deciduous shrubs; triangles, evergreen shrubs. (*Source*: data from Eckstein and Karlsson 1997.)

among species, it does not differ strongly among life-forms or species originating from habitats differing in fertility (Aerts 1996). Thus, the way species from nutrient-poor habitats achieve a high MRT is by a long life span of both leaves and roots (Aerts 1995, Ryser 1996, Eckstein and Karlsson 1997, Garnier and Aronson 1998).

What determines the life span of a leaf or root? For perennial species, there is an effect of the environment, with increases in leaf life span under relatively predictable low-resource conditions (low temperature, low irradiance). However, most of the variation in leaf life span is due to inherent differences between species (Reich 1998). Leaves of some species live for less than 2 months, whereas leaves of others have been reported to function for more than 20 years (Ewers and Schmid 1981). The physiological mechanism determining the life span of a leaf is unknown. However, it is evident that a leaf can only become long-lived if it can withstand adverse periods. Therefore, compared with a leaf with a short life span, extra investments have to be made to survive periods of drought or coldness. It also should be less attractive to herbivores, and not be so frail that it is damaged in storms. Finally, in a nutrient-poor environment, one could expect extra investments of plants to prevent nutrients leaching out of the leaf. Drought tolerance may be achieved by leaf hairs, a thick cuticle, an increase in lipid concentration, and possibly small cell sizes (Jones 1983). Cold resistance may be acquired by increases in osmotic solutes. Herbivory may be counteracted by the accumulation of phenolics or other secondary compounds. In all of these cases, extra lignification may occur. Extra lignification and thicker cell walls can also be expected for plants that have adapted to a high degree of mechanical disturbance, such as trampling or strong winds. Nutrient leaching could be prevented by extra investment in wax layers.

Compared with a "basic" leaf with a short life span, all of these additional investments will increase the biomass per unit leaf area and thus decrease SLA. Therefore, we might expect a negative relation between SLA (and the suite of traits positively associated with SLA) and leaf life span. This has indeed been shown in an analysis of plant species across a wide range of habitats (Reich et al. 1992, 1997; Figure 7). Data on the life span of roots are far less abundant (Eissenstat and Yani 1997). As far as data are available, they seem to indicate that species with a higher density of root tissue may have longer life spans (Ryser 1996). It is this connection between the nutrient economy (in the form of a high MRT through a long leaf or root life span) and the carbon economy (in the form of a low SLA) that may explain the success of species with a low RGR_{max} in nutrient-poor environments.

Up to now, we have focused the discussion on RGR components of plants from habitats low in nutrients. Basically, similar considerations could work for plants in other environments adverse to growth. In all cases where the production

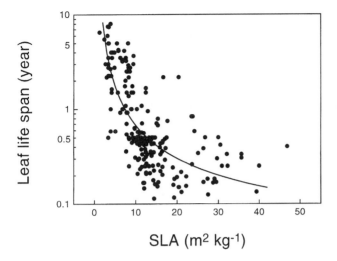

Figure 7 Leaf longevity as a function of SLA for a wide range of trees, shrubs, and herbaceous C₃ species found in six biomes of contrasting climates. (*Source*: after Reich et al. 1997.)

of biomass is difficult, one may expect a premium on maintaining existing biomass rather than replacing lost leaves or roots.

D. Selection for SLA-Related Traits in Favorable Environments

What about a possible selection for RGR components in an environment favorable for plant growth? Vegetation will be dense there, with a fast leaf area development during the growing season. Consequently, there is a strong competition for light, necessitating a fast increase in stem height and an efficient light interception per unit leaf biomass. In simulations of competition under agricultural conditions, it has been shown how important SLA is during the period before canopy closure: high-SLA plants have a higher light interception per unit leaf biomass and therefore faster growth than competing plants with a low SLA (Spitters and Aerts 1983, Gutschick 1988). In a closed canopy, maximum photosynthetic carbon gain of the vegetation as a whole is highest at lower values of SLA (Gutschick 1988). However, also under those conditions there may be a selection pressure toward an increase in SLA. Using game theory, Schieving (1998) analyzed mathematically the total carbon gain of two putative genotypes, which were similar in all traits, except for SLA. Simulating competition in a dense stand on the basis of functions for the distribution of light, nitrogen, and leaf area over the canopy, he found that the genotype with the higher SLA replaced the lower-SLA genotype.

VIII. EVIDENCE FOR THE IMPORTANCE OF SLA-RELATED TRAITS

As outlined in Section VII, it is difficult to really prove that selection has been for SLA-related traits rather than for RGR itself. At best, we can show that SLA and the suite of leaf and root characteristics associated with it vary along environmental gradients, or that high-SLA species that have been introduced in new continents displace low-SLA species in favorable habitats, but not in harsher environments. In analyzing these data, one should realize that SLA of a given species is not very sensitive to differences in nutrient availability, but rather sensitive to light.

A. Nutrient Availability

The most obvious difference in SLA is between woody deciduous and woody evergreen species. Evergreens have much lower SLAs and are generally found in nutrient-poor habitats (Monk 1966, Small 1972). Comparing productivity of various woody deciduous and evergreen species at the stand scale, a positive correlation has been found between SLA and annual productivity per unit leaf mass (Reich et al. 1997). Additionally, for herbaceous vegetations, there is a general trend that productive sites bear species with a higher SLA (H. Poorter, unpublished results). However, there is considerable variability around this trend, as only half of the total variation in SLA between species was explained by differences between sites, the other half due to differences in species within sites. This is in line with data of Van Andel and Biere (1989), who found a large variation in RGR_{max} and SLA for species co-occurring in the same habitat.

Species replacement has been observed in Venezuela, where two introduced C_4 grasses have outcompeted the original C_4 grass in most areas, but not in drier sites with low fertility (Baruch et al. 1985, Baruch 1996). Conforming to our expectations, the native species has a lower SLA, a higher proportion of sclerenchyma, and a lower concentration of N (Table 4).

A controlled experimental test to analyze the relation between growth parameters and field performance was conducted by Biere et al. (1996). They selected a range of families of *Lychnis flos-cuculi* and performed a growth analysis for each of these families in the glasshouse. At the same time, seeds of these families were sown in the field, along a fertility gradient. The above-ground bio-

Table 4 Differences in Leaf Characteristics Between Introduced and Native Species[a]

	H. rufa (introduced)	*T. plumosus* (native)	*A. stolonifera* (introduced)	*A. maggelanica* (native)
SLA (m^2 kg^{-1})	34	21	33	13
Leaf thickness (μm)	140	190	170	330
Leaf density (kg m^{-3})	210	250	180	235
Volume (%)				
Epidermis	36	38	24	20
Mesophyll	53	39	72	62
Vascular	11	17	3	3
Sclerenchyma	1	5	2	16
Concentrations (mg g^{-1})				
N	20	13	21	16

[a] The first comparison is a C_4 grass introduced in Venezuela (*Hyparrhenia rufa*) with a native C_4 grass species (*Trachypogon plumosus*; data from Baruch et al. 1985). The second comparison is an *Agrostis* species introduced on subantarctic islands (*A. stolonifera*) with a native species from the same genus (*A. maggelanica*; data from Pammenter et al. 1986).

mass of all field plants at the end of the growing season, a good predictor of next year's fecundity, was estimated and correlated with the growth parameters determined in the glasshouse. At the poorest site, families with a high RGR_{max}, LAR, and SLA fared worst (Table 5). However, the higher the site fertility, the more these parameters gained importance, and in the most productive site families with a high SLA (and LAR) were those that attained the highest biomass.

Table 5 Correlations Between Plant Mass in the Field, Determined at Four Sites Differing in Productivity, and a Number of Growth-Related Parameters, Determined in the Glasshouse[a]

Site productivity	Seed mass	Emergence	RGR	ULR	LAR	SLA	LMF
Low	**0.77**	0.07	**−0.70**	0.11	**−0.70**	**−0.76**	0.49
Low–intermediate	0.25	−0.02	−0.06	0.15	−0.15	−0.11	−0.09
Intermediate–high	0.12	−0.27	0.04	0.03	0.02	−0.02	0.12
High	−0.10	**−0.45**	0.21	−0.27	**0.38**	**0.41**	−0.02

[a] Data are family means (n = 56) of aboveground plant mass (ln-transformed) of plants of *Lychnis flos-cuculi* sown in the field, and seed mass, time to emergence, and growth parameters as determined in the glasshouse; after Biere et al. (1996), slightly simplified and reworked for this table. Values in bold are significant at $P < 0.05$.

B. Water Availability

Mooney et al. (1978) investigated *Eucalyptus* species along a rainfall gradient, and showed that, both for plants growing in the field and in the laboratory the SLA of the low-rainfall species was lower than those of species growing at sites with higher rainfall. The high-SLA species had the highest concentration of N per mass. In this case, the correlation between seedling RGR and rainfall was not very tight, whereas the correlation with SLA was. Similarly, the SLA of sun-lit leaves from *Eucalyptus* canopies sampled along a climatic gradient in Australia was found to be inversely related to the potential evaporation at the various sites studied (Specht and Specht 1989). Finally, *Encelia farinosa* plants from dry places are reported to have lower SLA than those from wetter places, and lower SLA than *Encelia frutescens* plants growing nearby but tapping in on deeper water (Ehleringer and Cook 1984, Ehleringer 1988). The lower SLA of *E. farinosa* from the driest sites is completely due to a thick layer of leaf hairs, which reflect the sunlight and may take up 50% of the biomass of the leaf.

C. Trampling and Wind Damage

A very clear case in which a low SLA is of survival value is trampling. Dijkstra and Lambers (1989) studied two subspecies of *Plantago major*. One is an annual, growing on occasionally flooded river banks, whereas the other is a perennial, occurring in frequently mown and trampled lawns. The leaves of the first subspecies have a high SLA and are erect. Those of the second are prostrate and have a low SLA, as well as a higher proportion of biomass in cell walls. An experiment showed that the subspecies with the low SLA survived trampling better (Table 6). Similarly, Meerts and Garnier (1996) found that *Polygonum aviculare* genotypes from trampled habitats show a lower SLA under laboratory conditions than those from nontrampled places.

   Leaf destruction can also take place by strong winds. Pammenter et al. (1986) analyzed differences in leaf anatomy between two *Agrostis* species that occur at subantarctic islands. *A. magellanica* is native to these islands; *A. stolonifera* has been introduced recently. The latter species has displaced *A. magellanica* in wind-sheltered areas, but not in more open terrain. This is most likely explained by the fact that the leaves of the introduced species are relatively thin and fragile. *A. magellanica*, with an unusual high fraction of the leaf volume occupied by sclerenchyma, a high leaf thickness, and a much lower SLA (Table 4), seems better able to withstand wind damage.

   Alpine species have been found to have lower SLA than lowland species, both in the field (Körner and Diemer 1987) and in the laboratory (Atkin et al. 1996a). Under both sets of conditions, this was at least partly due to thicker leaves. It has been suggested that the increased wind speed measured at higher

Table 6 Leaf Characteristics, Chemical Composition
and Tramping Survival Determined for Laboratory-
Grown Plants of Two Subspecies of *Plantago major*[a]

	P. major	
	River bank	Trampled lawn
SLA (m^2 kg^{-1})	39	29
Leaf thickness (μm)	280	280
Leaf density (kg m^{-3})	90	125
Concentrations (mg g^{-1})		
N	42	42
Minerals	190	130
Cell walls	200	240
Trampling survival (%)	6	45

[a] One subspecies is an annual, occurring on irregularly flooded
riverbanks; the other is a perennial from a frequently mown
and trampled lawn (Dijkstra and Lambers 1989).

elevations could play a role here as well. In an experiment with an upland, low-
SLA species and a lowland high-SLA species grown in a wind tunnel at high
windspeed, Woodward (1983) found much greater leaf damage in the high-SLA
species.

D. Herbivory

Herbivory by insects and mammals may have a large impact on the vegetation
and strongly damage individual plants. Species that have a low attractivity to
herbivores are those with a low water content, a low organic nitrogen concentration, high concentrations of lignin and other cell wall components, a high concentration of secondary compounds like tannins, and tough leaves in general (Scriber
1977, Grubb 1986, Coley and Barone 1996). These are all traits associated with
a low SLA.

An interesting case in which species with an inherently high SLA perform
less is in the understory of tropical forests. Plants generally acclimate to a low-
light environment by an increase in SLA, enabling them to capture more light per
unit leaf biomass. Therefore, one might expect at first to find high-SLA species in
the understory. Indeed, high-SLA pioneer species have a higher RGR at low light
intensities than shade-tolerant species (Veneklaas and Poorter 1998). Notwith-
standing their initially higher growth rate, these individual plants suffer from a
high mortality rate under these conditions compared to the shade-tolerant seed-

lings with low SLA (Kitajima 1994). Susceptibility to herbivory may play a role here as well.

IX. CONCLUSIONS

We have shown that there is a relationship between the potential RGR of a species, as measured under optimal conditions in the laboratory, and the distribution of species in the field. Fast-growing species are found in habitats favorable for plant growth, either on the short term (annuals, ruderals) or in the longer term (competitors according to the classification of Grime). Species found in harsh environments generally have a lower potential RGR (Section III). These interspecific differences in RGR_{max} are largely due to inherent differences in SLA (Sections IV and V). We have shown evidence that selection in the field may have acted, at least partly, on parameters related to SLA. An inherently low SLA (and the suite of traits associated with it) diminishes losses of nutrients or biomass due to grazing, trampling, or leaf turnover, and may be advantageous for plants growing in harsh environments. If correct, this implies that a low RGR_{max} is merely a side effect of the low SLA.

Up to now, we have focused on the "average" fast- and slow-growing species as representatives of two groups of functional types of species. However, it would be naive to think that any categorization into functional types can account for more than a modest amount of the variation that we encounter in the field. Moreover, it is evident that the plant traits discussed here are not the only that are shaped by evolution. To capture more of the variability, it would be desirable to include some other plant characteristics that are to a large extent independent of SLA and RGR. Westoby (1998) presented an interesting scheme, proposing to categorize species on the basis of SLA, seed mass, and maximum height achieved in the canopy. Seed mass is an important predictor of seed output per square meter of canopy cover, and a good indicator of seedling survival (Westoby 1998). The maximum height a plant can achieve is important for the type of vegetation in which it can survive. Such a characterization, which is still untested, could be a promising avenue to further increase our understanding of the success or failure of species in a given habitat.

APPENDIX 1: RELATIVE GROWTH RATE

The problem of how to express the growth rate of a plant can best be illustrated by the following example. Suppose there are two plants, A and B, whose dry masses are 0.1 and 1.0 g, respectively. Given that both increase in biomass by 0.1 g in 24 hours, can they be considered to grow at the same rate? One way to express growth is to consider the absolute growth rate (AGR), which is defined as the increase in plant mass, M, over a period of time, t:

$$AGR = \frac{dM}{dt} \tag{1}$$

Plant mass M_2 at time t_2 can be calculated for a given AGR when mass M_1 at time t_1 is known:

$$M_2 = M_1 \cdot AGR \cdot (t_2 - t_1) \tag{2}$$

In the above example, plants A and B have the same AGR. However, A achieved this increase with far less starting material than B. To take this into account, the rate of biomass increase can be defined relative to the mass of the plant already present. This is the relative growth rate (RGR):

$$RGR = \frac{1}{M} \cdot \frac{dM}{dt} \tag{3}$$

For a given RGR and plant mass M_1 at time t_1, M_2 at time t_2 can be calculated by:

$$M_2 = M_1 \cdot e^{RGR \cdot (t_2 - t_1)} \tag{4}$$

By taking the natural logarithm of both sides of eq. 4, and a little rearranging, we obtain a formula by which RGR can be calculated from experimental data:

$$RGR = \frac{\ln M_2 - \ln M_1}{t_2 - t_1} \tag{5}$$

Note that RGR can be interpreted graphically as the slope of the line that connects the ln-transformed dry masses of plants at several harvesting times (Figure 1B). Because of the ln-transformation, the numerator has no dimension; therefore, RGR has the unit "d^{-1}." As RGR values were based on plant masses rather than on leaf area or another parameter of plant size, units could be expressed as $g\ g^{-1}\ d^{-1}$ (gram increase per gram dry mass present and per unit of time). These numbers are generally low, and therefore we use $mg\ g^{-1}\ d^{-1}$ as the unit of expression.

In the previously given example, plant A has an RGR of 693 $mg\ g^{-1}\ d^{-1}$, while B has an RGR of 95 $mg\ g^{-1}\ d^{-1}$. We thus conclude that plant A is more efficient in terms of growth than B. (It may be surprising at first sight that plant A, which doubles its mass in 24 hours, has an RGR less than 1000 $mg\ g^{-1}\ d^{-1}$. The reason for this is that the assumption underlying eq. 4 is that extra biomass, produced during the beginning of the day, is immediately deployed to fix new C and minerals, leading to compounding growth.)

There are two additional points to which we want to draw attention. First, the dry mass of the whole plant is generally taken as the basis for the RGR calculation. However, RGRs have also been calculated on the basis of fresh weight, shoot weight, leaf area, or leaf number. As long as plant growth is in steady state, i.e., as long as there are no changes in the dry mass:fresh mass ratio, allocation, morphology, leaf size, etc., the RGR values expressed in several ways should be equal. Second, plants with a low RGR can still achieve a large biomass when they start with a high seed mass or grow over prolonged periods of time (see eq. 4). For a discussion on various approaches for the experimental design in growth analysis and details on calculations, see Causton and Venus (1981), Hunt (1982), and Poorter and Garnier (1996).

APPENDIX 2: COMPONENTS OF RELATIVE GROWTH RATE

A simple framework to factorize RGR was developed at the beginning of the 20th Century (Blackman 1919, West et al. 1923). The basic assumption underlying this framework is that plant growth is dependent on photosynthesis and that leaf area is the plant variable driving total C-gain. RGR is then factorized into two components: unit leaf rate (ULR) and leaf area ratio (LAR) (Hunt 1982). ULR is defined as the increase in biomass per unit time and leaf area:

$$ULR = \frac{1}{A} \cdot \frac{dM}{dt} \tag{6}$$

where A is the total leaf area of the plant, dM the increase in mass over period dt. LAR is defined as the total leaf area per unit total plant mass:

$$LAR = \frac{A}{M} \tag{7}$$

and consequently

$$ULR \cdot LAR = \frac{1}{A} \cdot \frac{dM}{dt} \cdot \frac{A}{M} = RGR \tag{8}$$

By determining leaf mass and stem and root mass separately, one is also able to break down LAR into two components: specific leaf area (SLA) and leaf mass fraction (LMF). SLA is the amount of leaf area per unit leaf mass (M_L):

$$SLA = \frac{A}{M_L} \tag{9}$$

and LMF is the fraction of total plant mass that is invested in the leaves:

$$LMF = \frac{M_L}{M} \tag{10}$$

and consequently

$$SLA \cdot LMF = \frac{A}{M_L} \cdot \frac{M_L}{M} = LAR \tag{11}$$

The advantage of this approach is that it requires only data on progressions in leaf area and plant mass to obtain a good indication about the causes of variation in growth rate. This is because differences in ULR are often due to differences in the area-based rate of photosynthesis. However, ULR is, in fact, the net balance of total plant carbon gain and carbon losses, expressed per unit leaf area and corrected for the C-concentration of the newly formed biomass. If one really wants to obtain insight in the relation between the various C-fluxes and RGR, the following formula can be used:

$$RGR = \frac{PS_a \cdot SLA \cdot LMF - LR \cdot LMF - SR \cdot SMF - RR \cdot RMF}{PCC} \tag{12}$$

where PS_a is the amount of C fixed per unit leaf area integrated over a 24-hour period, and LR, SR, and RR are the rates of leaf, stem, and root respiration, expressed as C lost per unit organ mass and integrated over a day as well. SMF and RMF are the fractions of total plant biomass invested in stems and roots, respectively. In this equation, all rates of C-gain and C-losses per organ are converted to a rate per unit total plant mass. The numerator is thus the net amount of C fixed per unit of biomass and per day. The denominator converts this net amount of C into a biomass increase (Poorter and Welschen 1993). It this equation, C-losses due to exudation or leaf and root turnover are considered to be negligible.

APPENDIX 3: COMPONENTS OF SPECIFIC LEAF AREA

Specific leaf area (SLA; leaf area per unit leaf mass) is a parameter that scales linearly and positively with RGR and is, in this respect, an easy parameter to use in growth analyses. However, if one would like to analyze what factors play a role in determining SLA, it is more appropriate to consider its inverse, 1/SLA, for which many other terms have been used (SLW, SLM, LMA). This is because components of the leaf, like leaf thickness, or anatomical and biochemical features, increase linearly with the inverse of SLA. Witkowski and Lamont (1991) factorized 1/SLA into two components: leaf thickness (LTh) and leaf density (LD). Leaf density is defined as the mass of a leaf per unit leaf volume:

$$LD = \frac{M_L}{LTh \cdot A} \tag{13}$$

and therefore,

$$\frac{1}{SLA} = LD \cdot LTh = \frac{M_L}{LTh \cdot A} \cdot LTh \tag{14}$$

A leaf can also be separated into its underlying anatomical tissues: epidermis, mesophyll, sclerenchyma, vascular tissue, and intercellular spaces. The $1/SLA$ is then the sum of the densities of the various tissues i, weighted by the volume per unit leaf area taken by each tissue (Garnier and Laurent 1994):

$$\frac{1}{SLA} = \sum_{i=1}^{5} \frac{V_i}{A} \cdot \frac{M_i}{V_i} \tag{15}$$

Leaf biomass can also be separated into its various biochemical fractions. A simple grouping of the wide range of compounds is: lipids, lignin, soluble phenolics, protein, structural carbohydrates (cellulose, hemicellulose, pectin), nonstructural carbohydrates (glucose, fructose, sucrose, starch), organic acids, and minerals (Poorter and Villar 1997). $1/SLA$ is then the sum of the masses of each of the 8 classes of compounds expressed per unit area:

$$\frac{1}{SLA} = \sum_{i=1}^{8} \frac{M_i}{A} \tag{16}$$

The three ways of breaking down $1/SLA$ are interrelated: a high leaf density, for example, can be the result of a high proportion of sclerenchyma, which shows up in the biochemical analysis as a high concentration of cell walls (lignin and structural carbohydrates).

APPENDIX 4: COMPONENTS OF NITROGEN USE EFFICIENCY

In Appendix 2, RGR was factorized into components based on the assumption that leaf area is the important plant variable driving photosynthesis, and thus growth. An alternative approach is to consider plant (organic) N as the driving variable, as proteins play a vital role in C-fixation, nutrient uptake, and in most other physiological processes of the plant. RGR is then factorized into the components NP (nitrogen productivity) and PNC (plant nitrogen concentration; preferably restricted to organic nitrogen, as NH_4^+ and NO_3^- play less of a physiological role). NP is defined as the increase in biomass per unit time and plant nitrogen (Ingestad 1979):

$$NP = \frac{1}{N} \cdot \frac{dM}{dt} \qquad (17)$$

where N is the total amount of nitrogen in the plant. PNC is simply the concentration of nitrogen in the plant:

$$PNC = \frac{N}{M} \qquad (18)$$

and consequently

$$NP \cdot PNC = \frac{1}{N} \cdot \frac{dM}{dt} \cdot \frac{N}{M} = RGR \qquad (19)$$

Considered over a short and fixed period of time, plants use their internal nitrogen efficiently if they have a high NP. Considered over a longer time scale, an alternative way to make efficient use of a unit of nitrogen taken up is to increase the time this unit remains in the plant. This time span is called the mean residence time (MRT; expressed in years). Under steady-state conditions (i.e., when the amount of nutrients taken up by the plant equals that lost by leaf and root shedding, herbivory, etc.), MRT is equal to the ratio between the average amount of nitrogen in the plant (N) and that taken up or lost over a given period of time (dN/dt; see, e.g., Frissel 1981). Berendse and Aerts (1987) have proposed that it is the product of NP and MRT that determines the nutrient use efficiency (NUE):

$$NP \cdot MRT = \frac{1}{N} \cdot \frac{dM}{dt} \cdot N \cdot \frac{dt}{dN} = \frac{dM}{dN} = NUE \qquad (20)$$

NUE is therefore the total amount of biomass produced per unit N taken up or lost. Note that in Eq. 19, NP is generally determined over a short time scale (days to weeks), whereas in Eq. 20 it is determined on an annual basis.

Under steady-state functioning, MRT depends on two parameters in the following way (Garnier and Aronson 1998):

$$MRT = \frac{1}{1 - R_{eff}} \cdot \frac{1}{e} \qquad (21)$$

where R_{eff} is the nitrogen resorption efficiency during senescence (defined as the reduction in the amount of nutrient between mature and senesced organs, relative to the amount in mature organs (Aerts 1996, Killingbeck 1996), and e is the rate of biomass lost by the plant, which depends on the life span of organs. These two parameters are central to the definition of the nutrient conservation strategy of the various species.

ACKNOWLEDGMENTS

We thank Arjen Biere, Hans Cornelissen, Lourens Poorter, Peter Reich, Jeroen van Arendonk, and Adrie van der Werf for generously providing published and unpublished data. Lourens Poorter, Owen Atkin, and an anonymous reviewer made helpful remarks on a previous version of the manuscript.

REFERENCES

Aerts, R. 1990. Nitrogen use efficiency in evergreen and deciduous species from heathlands. Oecologia (Berlin) 84: 391–397.

Aerts, R. 1995. The advantages of being evergreen. Trends in Ecology and Evolution 10: 402–407.

Aerts, R. 1996. Nutrient resorption from senescing leaves of perennials: are there general patterns? Journal of Ecology 84: 597–608.

Atkin, O.K., and D.A. Day. 1990. A comparison of the respiratory processes and growth rates of selected Australian alpine and related lowland plant species. Australian Journal of Plant Physiology 17: 517–526.

Atkin, O.K., B. Botman, and H. Lambers. 1996a. The causes of inherently slow growth in alpine plants: an analysis based on the underlying carbon economies of alpine and lowland *Poa* species. Functional Ecology 10: 698–707.

Atkin, O.K., B. Botman, and H. Lambers. 1996b. The relationship between the relative growth rate and nitrogen economy of alpine and lowland *Poa* species. Plant, Cell and Environment 19: 1324–1330.

Ball, M.C. 1988. Salinity tolerance in mangroves *Aegiceras corniculatum* and *Avicennia marina*. I. Water use in relation to growth, carbon partitioning, and salt balance. Australian Journal of Plant Physiology 15: 447–464.

Baruch, Z. 1996. Ecophysiological aspects of the invasion by African grasses and their impact on biodiversity and function of neotropical savannas. In: Solbrig, O.T., E. Medina, and J. Silva, eds. Biodiversity and Savanna Ecosystem Processes: A Global Perspective. Ecological Studies Vol. 121. Berlin: Springer-Verlag, pp 73–93.

Baruch, Z., M.M. Ludlow, and R. Davis. 1985. Photosynthetic responses of native and introduced C4 grasses from Venezuelan savannas. Oecologia 67: 388–393.

Benjamin, L.R., and R.C. Hardwick. 1986. Sources of variation and measures of variability in even-aged stands of plants. Annals of Botany 58: 757–778.

Berendse, F., and R. Aerts. 1987. Nitrogen-use-efficiency: a biologically meaningful definition? Functional Ecology 1: 293–296.

Biere, A. 1996. Intra-specific variation in relative growth rate: impact on competitive ability and performance of *Lychnis flos-cuculi* in habitats differing in soil fertility. Plant and Soil 182: 313–327.

Bilbrough, C.J., and M.M. Caldwell. 1997. Exploitation of springtime ephemeral N pulses by six great basin plant species. Ecology 78: 231–243.

Blackman, V.H. 1919. The compound interest law and plant growth. Annals of Botany 33: 353–360.

Boorman, L.A. 1982. Some plant growth patterns in relation to the sand dune habitat. Journal of Ecology 70: 607–614.

Bradshaw, A.D. 1987. Functional ecology: comparative ecology? Functional Ecology 1: 71.

Bradshaw, A.D., M.J. Chadwick, D. Jowett, and R.W. Snaydon. 1964. Experimental investigations into the mineral nutrition of several grass species. IV. Nitrogen level. Journal of Ecology 52: 665–676.

Causton, D.R., and J.C. Venus, 1981. The Biometry of Plant Growth. E. Arnold: London.

Chapin, F.S. 1980. The mineral nutrition of wild plants. Annual Review of Ecology and Systematics 11: 233–260.

Chapin, F.S. 1988. Ecological aspects of plant mineral nutrition. Advances in Mineral Nutrition 3: 161–191.

Chapin, F.S., K. Autumn, and F. Pugnaire. 1993. Evolution of suites of traits in response to environmental stress. American Naturalist 142: S78–92.

Coley, P.D. 1983. Herbivory and defense characteristics of tree species in lowland tropical forest. Ecological Monographs 53: 209–233.

Coley, P.D., and J.A. Barone. 1996. Herbivory and plant defenses in tropical forests. Annual Review of Ecology and Systematics 27: 305–335.

Corré, W.J. 1983a. Growth and morphogenesis of sun and shade plants. I. The influence of light intensity. Acta Botanica Neerlandica 32: 49–62.

Corré, W.J. 1983b. Growth and morphogenesis of sun and shade plants. II. The influence of light quality. Acta Botanica Neerlandica 32: 185–202.

Cornelissen, J.H.C., M.J.A. Werger, P. Castro-Diez, J.W.A. van Rheenen, and A.P. Rowland. 1997. Foliar nutrients in relation to growth, allocation and leaf traits in seedlings of a wide range of woody plant species and types. Oecologia 111: 460–469.

Cornelissen, J.H.C., P. Castro-Díez, and A.L. Carnelli. 1998. Variation in relative growth rate among woody species. In: Lambers, H., H. Poorter, and M. van Vuuren, eds. Inherent Variation in Plant Growth. Physiological Mechanisms and Ecological Consequences. Leiden: Backhuys Publishers, pp 363–392.

Den Dubbelden, K.C., and R.W. Verburg. 1996. Inherent allocation patterns and potential growth rates of herbaceous climbing plants. Plant and Soil 184: 341–347.

Dijkstra, P., and H. Lambers. 1989. Analysis of specific leaf area and photosynthesis of two inbred lines of *Plantago major* differing in relative growth rate. New Phytologist 113: 283–290.

Eckstein, R.L., and P.S. Karlsson. 1997. Above-ground growth and nutrient use by plants in a subarctic environment: effects of habitat, life-form and species. Oikos 79: 311–324.

Ehleringer, J.R. 1988. Comparative ecophysiology of *Encelia farinosa* and *Encelia frutescens*. I. Energy balance considerations. Oecologia 76: 553–561.

Ehleringer, J.R., and C.S. Cook. 1984. Photosynthesis in *Encelia farinosa* Gray in response to decreasing leaf water potential. Plant Physiology 75: 688–693.

Eissenstat, D.M., and R.D. Yanai. 1997. The ecology of root lifespan. Advances in Ecological Research 27: 1–60.

Ellenberg, H. 1988. Vegetation Ecology of Central Europe. Cambridge: Cambridge University Press.

Evans, G.C. 1972. The Quantitative Analysis of Plant Growth. London: Blackwell Scientific.

Ewers, F.W., and R. Schmid. 1981. Longevity of needle fascicles of *Pinus longeava* (Bristlecone Pine) and other North American pines. Oecologia 51: 107–115.

Fichtner, K., and E.-D. Schulze. 1992. The effect of nitrogen nutrition on growth and biomass partitioning of annual plants originating from habitats of different nitrogen availability. Oecologia 92: 236–241.

Frissel, M.J. 1981. The definition of residence times in ecological models. In: Clark, F.E., and T. Rosswall, eds. Terrestrial Nitrogen Cycles. Processes, Ecosystem Strategies and Management Impacts. Stockholm: Swedish Natural Science Research Council, pp 117–122.

Garnier, E. 1991. Resource capture, biomass allocation and growth in herbaceous plants. Trends in Ecology and Evolution 6: 126–131.

Garnier, E. 1992. Growth analysis of congeneric annual and perennial grass species. Journal of Ecology 80: 665–675.

Garnier, E., and J. Aronson. 1998. Nitrogen-use-efficiency from leaf to stand level: clarifying the concept. In: Lambers, H., H. Poorter, and M. van Vuuren, eds. Inherent Variation in Plant Growth. Physiological Mechanisms and Ecological Consequences. Leiden: Backhuys Publishers, pp 515–538.

Garnier, E., and A.H.J. Freijsen. 1994. On ecological inference from laboratory experiments conducted under optimum conditions. In: Roy, J., and E. Garnier, eds. A Whole Plant Perspective on Carbon-Nitrogen Interactions. The Hague: SPB Academic Publishing, pp 267–292.

Garnier, E., and G. Laurent. 1994. Leaf anatomy, specific mass and water content in congeneric annual and perennial grass species. New Phytologist 128: 725–736.

Garnier, E., and S. Vancaeyzeele. 1994. Carbon and nitrogen content of congeneric annual and perennial species: relationships with growth. Plant, Cell and Environment 17: 399–407.

Garnier, E., P. Cordonnier, J.-L. Guillerm, and L. Sonié. 1997. Specific leaf area and leaf nitrogen concentration in annual and perennial grass species growing in Mediterranean old-fields. Oecologia 111: 490–498.

Graves, J.D., and K. Taylor. 1986. A comparative study of *Geum rivale* L. and *Geum urbanum* L. to determine those factors controlling their altitudinal distribution. I. Growth in controlled and natural environments. New Phytologist 104: 681–691.

Grime, J.P. 1977. Evidence for the existence of three primary strategies in plants and its relevance to ecological and evolutionary theory. American Naturalist 111: 1169–1194.

Grime, J.P. 1979. Plant Strategies and Vegetation Processes. Chichester: Wiley.

Grime, J.P., and R. Hunt. 1975. Relative growth rate: its range and adaptive significance in a local flora. Journal of Ecology 63: 393–422.

Grubb, P.J. 1986. Sclerophylls, pachyphylls and pycnophylls: the nature and significance of hard leaf surfaces. In: Juniper, B., and R. Southwood, eds. Insects and Plant Surface. London: Edward Arnold, pp 137–150.

Gutschick, V.P. 1988. Optimization of specific leaf mass, internal CO_2 concentration, and chlorophyll content in crop canopies. Plant Physiology and Biochemistry 26: 525–537.

Hobbs, R.J. 1997. Can we use plant functional types to describe and predict responses to environmental changes? In: Smith, T.M., H.H. Shugart, and F.I. Woodward, eds.

Plant Functional Types. Their Relevance to Ecosystem Properties and Global Change. Cambridge: Cambridge University Press, pp 66–90.

Hull, J.C. and H.A. Mooney. 1990. Effects of nitrogen on photosynthesis and growth rates of four California annual species. Acta Oecologica 11: 453–468.

Hunt, R. 1982. Plant Growth Curves. The Functional Approach to Plant Growth Analysis. London: E. Arnold.

Hunt, R., and J.H.C. Cornelissen. 1997. Components of relative growth rate and their interrelations in 59 temperate plant species. New Phytologist 135: 395–417.

Hunt, R., and P.S. Lloyd. 1987. Growth and partitioning. New Phytologist 106: 235–249.

Ingestad, T. 1979. Nitrogen stress in birch seedlings. II. N, K, K, P, Ca and Mg nutrition. Physiologia Plantarum 45: 149–157.

Jones, H.G. 1983. Plants and Microclimate. Cambridge: Cambridge University Press.

Killingbeck, K.T. 1996. Nutrients in senesced leaves: keys to the search for potential resorption and resorption proficiency. Ecology 77: 1716–1727.

Kitajama, K. 1994. Relative importance of photosynthetic traits and allocation patterns as correlates of seedling shade tolerance of 13 tropical trees. Oecologia 98: 419–428.

Konings, H. 1989. Physiological and morphological differences between plants with a high NAR or a high LAR as related to environmental conditions. In: Lambers, H., M.L. Cambridge, H. Konings, and T.L. Pons, eds. Causes and Consequences of Variation in Growth Rate and Productivity in Higher Plants. The Hague: SPB Academic Publishing BV, pp 101–123.

Körner, C., and M. Diemer. 1987. In situ photosynthetic response to light, temperature and carbon dioxide in herbaceous plants from low and high altitude. Functional Ecology 1: 179–194.

Lambers, H., and P. Dijkstra. 1987. A physiological analysis of genotypic variation in relative growth rate: can growth rate confer ecological advantage? In: van Andel, J., J.P. Bakker, and R.W. Snaydon, eds. Disturbance in Grasslands. Dordrecht: Junk Publishers, pp 237–251.

Lambers, H., and H. Poorter. 1992. Inherent variation in growth rate between higher plants: a search for physiological causes and ecological consequences. Advances in Ecological Research 23: 187–261.

Mahmoud, A. and Grime, J.P. 1974. An analysis of competitive ability in three perennial grasses. New Phytologist 77: 431–435.

McGraw, J.B., and K. Garbutt. 1990. The analysis of plant growth in ecological and evolutionary studies. Trends in Ecology and Evolution 5: 251–254.

Meerts, P., and E. Garnier. 1996. Variation in relative growth rate and its components in the annual *Polygonum aviculare* in relation to habitat disturbance and seed size. Oecologia 108: 438–445.

Monk, C.D. 1966. An ecological significance of evergreenness. Ecology 47: 504–505.

Mooney, H.A., P.J. Ferrar, and R.O. Slatyer. 1978. Photosynthetic capacity and carbon allocation patterns in diverse growth forms of *Eucalyptus*. Oecologia 36: 103–111.

Niemann, G.J., J.B.M. Pureveen, G.B. Eijkel, H. Poorter, and J.J. Boon. 1992. Differences in relative growth rate in 11 grasses correlate with differences in chemical composition as determined by pyrolysis mass spectrometry. Oecologia 89: 567–573.

Osunkoya, O.O., J.E. Ash, M.S. Hopkins, and A.W. Graham. 1994. Influence of seed size and ecological attributes on shade-tolerance of rainforest tree species in northern Queeensland. Journal of Ecology 82: 149–163.

Pammenter, N.W., P.M. Drennan, and V.R. Smith. 1986. Physiological and anatomical aspects of photosynthesis of two *Agrostis* species at a sub-antarctic island. New Phytologist 102: 143–160.

Parsons, R.F. 1968. The significance of growth rate comparisons for plant ecology. American Naturalist 102: 595–597.

Pons, T.L. 1977. An ecophysiological study in the field layer of ash coppice. II. Experiments with *Geum urbanum* and *Cirsium palustre* in different light intensities. Acta Botanica Neerlandica 26: 29–42.

Poorter, H. 1989. Interspecific variation in relative growth rate: on ecological causes and physiological consequences. In: Lambers, H., M.L. Cambridge, H. Konings, and T.L. Pons, eds. Causes and Consequences of Variation in Growth Rate and Productivity of Higher Plants. The Hague: SPB Academic Publishing, pp 45–68.

Poorter, H., and M. Bergkotte. 1992. Chemical composition of 24 wild species differing in relative growth rate. Plant, Cell and Environment 15: 221–229.

Poorter, H., and E. Garnier. 1996. Plant growth analysis: an evaluation of experimental design and computational methods. Journal of Experimental Botany 47: 1343–1351.

Poorter, H., and P. Pothmann. 1992. Growth and carbon economy of a fast-growing and a slow-growing grass species as dependent on ontogeny. New Phytologist 120: 159–166.

Poorter, H., and C. Remkes. 1990. Leaf area ratio and net assimilation rate of 24 wild species differing in relative growth rate. Oecologia 83: 553–559.

Poorter, H., and A.K. Van der Werf. 1998. Is inherent variation in RGR determined by LAR at low irradiance and NAR at high irradiance? A review of herbaceous species. In: Lambers, H., H. Poorter, and M. van Vuuren, eds. Inherent Variation in Plant Growth. Physiological Mechanisms and Ecological Consequences. Leiden: Backhuys Publishes, pp 309–336.

Poorter, H., and R. Villar. 1997. The fate of acquired carbon in plants: chemical composition and construction costs. In: Bazzaz, F.A., and J. Grace, eds. Plant Resource Allocation. New York: Academic Press, pp 39–72.

Poorter, H., and R.A.M. Welschen. 1993. Variation in RGR, the underlying carbon economy. In: Hendry, G.A.F., and J.P. Grime, eds. Methods in Comparative Plant Ecology. London: Chapman & Hall, pp 107–110.

Poorter, H., C. Remkes, and H. Lambers. 1990. Carbon and nitrogen economy of 24 wild species differing in relative growth rate. Plant Physiology 94: 621–627.

Poorter, H., C.A.D.M. van de Vijver, R.G.A. Boot, and H. Lambers. 1995. Growth and carbon economy of a fast-growing and a slow-growing grass species as dependent on nitrate supply. Plant and Soil 171: 217–227.

Reich, P.B. 1998. Variation among plant species in leaf turnover rates and associated traits: implications for growth at all life stages. In: Lambers, H., H. Poorter, and M. van Vuuren, eds. Inherent Variation in Plant Growth. Physiological Mechanisms and Ecological Consequences. Leiden: Backhuys Publishers, pp 467–487.

Reich, P.B., C. Uhl, and D.S. Ellsworth. 1991. Leaf lifespan as a determinant of leaf structure and function among 23 amazonian tree species. Oecologia 86: 16–24.

Reich, P.B., M.B. Walters, and D.S. Ellsworth. 1992. Leaf life-span in relation to leaf, plant, and stand characteristics among diverse ecosystems. Ecological Monographs 62: 365–392.

Reich, P.B., M.B. Walters, and D.S. Ellsworth. 1997. From tropics to tundra: global convergence in plant functioning. Proceedings of the National Academy of Sciences (USA) 94: 13730–13734.

Reiling, K., and A.W. Davison. 1992. The response of native, herbaceous species to ozone: growth and fluorescence screening. New Phytologist 120: 29–37.

Reynolds, H.L., and C. D'Antonio. 1996. The ecological significance of plasticity in root weight ratio in response to nitrogen: opinion. Plant and Soil 185: 75–97.

Robinson, D. 1991. Strategies for optimising growth in response to nutrient supply. In: Porter, J.R., and D.W. Lawlor, eds. Plant Growth. Interactions with Nutrition and Environment. Cambridge: Cambridge University Press, pp 177–199.

Rorison, I.H. 1968. The response to phosphorus of some ecologically distinct plant species. I. Growth rates and phosphorus absorption. New Phytologist 67: 913–923.

Rozijn, N.A.M.G., and D.C. van der Werf. 1986. Effect of drought during different stages in the life cycle on the growth and biomass of two *Aira* species. Journal of Ecology 74: 507–523.

Ryser, P. 1996. The importance of tissue density for growth and life span of leaves and roots: a comparison of five ecologically contrasting grasses. Functional Ecology 10: 717–723.

Schieving, F. 1998. Plato's Plant: On the Mathematical Structure of Simple Plants and Canopies. Leiden: Backhuys Publishers.

Schläpfer, B., and P. Ryser. 1996. Leaf and root turnover of three ecologically contrasting grass species in relation to their performance along a productivity gradient. Oikos 75: 398–406.

Schwarz, M., and J. Gale. 1984. Growth response to salinity at high levels of carbon dioxide. Journal of Experimental Botany 35: 193–196.

Scriber, J.M. 1977. Limiting effects of leaf-water content on the nitrogen utilization, energy budget and larval growth of *Hyalophora cecropia* (Lepidoptera: Saturniidae). Oecologia 28: 269–287.

Shipley, B. 1995. Structures interspecific determinants of specific leaf area in 34 species of herbaceous angiosperms. Functional Ecology 9: 312–319.

Shipley, B., and P.A. Keddy. 1988. The relationship between relative growth rate and sensitivity to nutrient stress in twenty-eight species of emergent macrophytes. Journal of Ecology 76: 1101–1110.

Small, E. 1972. Photosynthetic rates in relation to nitrogen cycling as an adaptation to nutrient deficiency in peat bog plants. Canadian Journal of Botany 50: 2227–2233.

Specht, R.L., and A. Specht. 1989. Canopy structure in *Eucalyptus*-dominated communities in Australia along climatic gradients. Acta Oecologica, Oecologia Plantarum 10: 191–213.

Spitters, C.J.T., and R. Aerts. 1983. Simulation of competition for light and water in crop-weed associations. Aspects of Applied Biology 4: 467–483.

Stockey, A., and R. Hunt. 1994. Predicting secondary succession in wetland mesocosms on the basis of autecological information on seeds and seedlings. Journal of Applied Ecology 31: 543–559.

Van Andel, J., and A. Biere 1989. Ecological significance of variability in growth rate and plant productivity. In: Lambers, H., M.L. Cambridge, H. Konings, and T.L. Pons, eds. Causes and Consequences of Variation in Growth Rate and Productivity of Higher Plants. The Hague: SPB Academic Publishing, pp 257–267.

Van Arendonk, J.J.C.M., and H. Poorter. 1994. The chemical composition and anatomical structure of leaves of grass species differing in relative growth rate. Plant, Cell and Environment 17: 963–970.

Van der Werf, A., M. van Nuenen, A.J. Visser, and H. Lambers. 1993. Effects of N-supply on the rates of photosynthesis and shoot and root respiration of inherently fast- and slow-growing monocotyledonous species. Physiologia Plantarum 89: 563–569.

Van der Werf, A., H. Poorter, and H. Lambers. 1994. Respiration as dependent on a species' inherent growth rate and on the nitrogen supply to the plant. In: Roy, J., and E. Garnier, eds. A Whole Plant Perspective on Carbon-Nitrogen Interactions. The Hague: SPB Academic Publishing, pp 91–110.

Van der Werf, A., R.H.E.M. Geerts, F.H.H. Jacobs, H. Korevaar, M.J.M Oomes, and W. de Visser. 1998. The importance of relative growth rate and associated traits for competition between species. In: Lambers, H., H. Poorter, and M. van Vuuren, eds. Inherent Variation in Plant Growth. Physiological Mechanisms and Ecological Consequences. Leiden: Backhuys Publishers, pp 489–502.

Van Diggelen, J. 1988. A Comparative Study on the Ecophysiology of Salt Marsh Halophytes. PhD Thesis, Free University, Amsterdam.

Veneklaas, E.J., and L. Poorter. 1998. Growth and carbon partitioning of tropical tree seedlings in contrasting light environments. In: Lambers, H., H. Poorter, and M. van Vuuren, eds. Inherent Variation in Plant Growth. Physiological Mechanisms and Ecological Consequences. Leiden: Backhuys Publishers, pp 337–361.

Verkleij, J.A.C., and J.E. Prast. 1989. Cadmium tolerance and co-tolerance in *Silene vulgaris* (Moench.) Garcke (= *S. cucubalis* (L.) Wib.). New Phytologist 111: 637–645.

Walters, M.B., E.L. Kruger, and P.B. Reich. 1993. Relative growth rate in relation to physiological and morphological traits for northern hardwood tree seedlings: species, light environment and ontogenetic considerations. Oecologia 96: 219–231.

West, C., G.E. Briggs, and F. Kidd. 1920. Methods and significant relations in the quantitative analysis of plant growth. New Phytologist 19: 200–207.

Westoby, M. 1998. A leaf-height-seed (LHS) plant ecology strategy scheme. Plant and Soil, in press.

Wilson, E.O. 1992. The Diversity of Life. Cambridge, MA: Harvard University Press.

Wilson, J.B. 1988. The cost of heavy-metal tolerance. Evolution 42: 408–413.

Witkowski, E.T.F., and B.B. Lamont. 1991. Leaf specific mass confounds leaf density and thickness. Oecologia 88: 486–493.

Woodward, F.I. 1979. The differential temperature response of the growth of certain plant species from different altitudes. I. Growth analysis of *Phleum alpinum* L., *P. bertolonii* D.C., *Sesleria albicans* KIT. and *Dactylis glomerata* L. New Phytologist 82: 385–395.

Woodward, F.I. 1983. The significance of interspecific differences in specific leaf area to the growth of selected herbaceous species from different altitudes. New Phytologist 96: 313–323.

4

Architecture, Ecology, and Evolution of Plant Crowns

Fernando Valladares
Center of Environmental Sciences, Spanish Council for Scientific Research, Madrid, Spain

I. INTRODUCTION

Plants exhibit a striking diversity of forms and structures, and ever since the ancient Egyptian or Greek naturalists, the observer has struggled to interpret the variations in plant form. The functional approach to the study of plant form emerged as a separate discipline at the beginning of the 20th century (Raunkiaer 1905, Werger et al. 1988) with the first classifications of growth forms in relation to climate and with tentative ecophysiological studies of plant responses to the environment (Waller 1991). However, the lack of precise and handy instrumentation forced botanists and plant ecologists to concentrate on purely descriptive approaches to the study of plant form until the middle of the 20th century. Toward the end of this century, the increasing interest in testing of causal hypotheses and in quantification of ecologically relevant processes involving plant structure, together with the increasing availability of sophisticated field instruments and computers, accelerated the progress in the understanding of the functional aspects of plant form and their evolutionary relevance. Physiological ecology now makes detailed predictions on how physical and physiological characteristics affect plant photosynthesis, while plant population ecology translates patterns of growth into fitness of individuals and populations. And plant structure remains an essential tool for these exercises of interpreting plant performance in natural habitats and for scaling from leaves to ecosystems (Ehleringer and Field 1993, Schulze et al. 1994).

    Plant performance, like the performance of any given organism, can be understood as the crucial link between its phenotype and ecological success (Bock and Won Wahlert 1965, Koehl 1996). Consequently, the form becomes ecologically and evolutionary relevant when it affects the performance of the organism. It is important to consider that misconceptions can arise from studies in which selective advantages of particular structures are not made with a mechanistic understanding of how the structural traits affect performance. Koehl (1996) showed that the relationship between morphology and performance can be nonlinear, context-dependent, and sometimes surprising. Remarkably, new functions and novel ecological consequences of morphological changes can arise simply as the result of changes in size or habitat.

While all would agree that structure is intrinsically coupled with function, the impetus is often stronger to investigate physiological mechanisms rather than the functional implications of plant form. This is not to say that functional plant architecture has been ignored. For instance, the role of canopy architecture in competition for light has been addressed in several works after the keystone study by Horn (1971). But the architectural constraints of plant success, i.e., of plant persistence or expansion in the community, has not been explored extensively. Plant architecture involves the manner in which the foliage is positioned in different microenvironments and determines the flexibility of a shoot system to take advantage of unfilled gaps in the canopy, to allocate and utilize photoassimilates, and to recover from herbivory or mechanical damage (Caldwell et al. 1981, 1983, Küppers 1989, 1990). Except in particular or very extreme environments, plant physiology alone does not explain ecological success, since growth and competition have been clearly related to structural features (Küppers 1994). In agreement with Tomlinson (1987), the study of plant morphology is an integrative discipline rather than a subject restricted to the comparison of anatomical details of plant life cycles. This chapter attempts to demonstrate that plant morphology in general and plant architecture in particular belongs more rightly within the fields of plant ecophysiology and plant population biology.

This chapter deals with plant shape from a functional point of view. The basic plant design and the many interpretations and implications of its modular nature are presented here as an indispensable starting point to enter into discussions on function and adaptive value of crown structural features. Models of plant growth and classifications of the architecture of the crown are also presented in the first sections of the chapter. After the introductory sections, the chapter discusses the relationships between plant shape and light capture. Plants depend on the light energy that they capture by photosynthesis, and solar radiation is the major driving force affecting not only photosynthetic activity, but also leaf temperature, leaf water status, and many other physiological processes of the plant. Plant architecture is crucial for light capture and the distribution of light to each particular photosynthetic unit of the crown, but it must also serve several other functions. The architectural design of a given plant must provide safety margins to cope with gravity and wind; therefore, biomechanical constraints must be taken into account when assessing the influence of morphology and architecture on plant performance. The structural basis of light capture by plant crowns is explored here from the leaf level to the community level, with special attention to leaf angle, phyllotaxis, branching patterns, and crown shape. Examples of plant architecture in extreme light environments are included, where the functional implications for light capture of a range of structural features can be better seen. In the analysis of plant shape at the community level, two main functional concepts involving plant interactions are discussed: the occupation of the space and the shading of neighbors. The chapter then discusses the complex but fascinating world of plasticity, in particular the adaptive plastic response of plants to environmental heterogeneities and changes. Current limits in our understanding of phenotypic plasticity and a brief analysis of the potential ecological and evolutionary implications of a plastic plant response to the environment are included. The chapter finishes with a quick look to old and new concepts in evolution, with reference to the poorly known field of modes and causes of plant evolution.

With this review of functional crown architecture an attempt is made to analyze the importance of adaptations to the light environment for plant survival and evolution with strong emphasis on structural features, which are not as frequent in the literature as their physiological counterparts. The presentation of some structural, biomechanical, and genetic constraints to adaptation to the light environment helps to illustrate the intrinsic complexity of understanding the ecology and evolution of the plant crown, a compromise of usually conflicting strategies.

II. PLANT DESIGN

The shape of a given plant is determined by the shape of the space that it fills, but most plants attain a characteristic shape when grown alone in the open due to an inherited developmental program (Horn 1971). This developmental program usually implies the reiterative addition of a series of structurally equivalent subunits (branches, axes, shoots, leaves), which confers plants a modular nature. This developmental program is the result of plant evolution under some general biomechanical constraints. For instance, the shape of the crown of a tree is constrained by the fact that the cost of horizontal branches is greater than that of vertical branches (Mattheck 1991). This section explores the functional implications of these two general aspects, the modular nature of plants and the biomechanical constraints of shape, which, in addition to the environment where the plant grows, determine plant architecture.

A. The Basic Architecture of Terrestrial Plants

Terrestrial vascular plants must combine the structural requirements of water-conduction and gas-exchange systems with the problems of mechanical support of aerial structures and light capture by the photosynthetic surfaces. Many different solutions to these frequently opposing problems have been found during plant evolution (Niklas and Kerchner 1984, Speck and Vogellehner 1988, Niklas 1990, 1997b). The diversification into trees, shrubs, and herbs occurred relatively rapidly (Raven 1986), and by the end of the Devonian, many alternative plant designs were successfully tested in most terrestrial systems. From primitive cylindrical or flat, two-dimensional photosynthetic surfaces restricted to liquid environments, terrestrial plants evolved complex three-dimensional arrangements of the photosynthetic units, which required stomata for control over water loss preventing embolism (Woodward 1998), lignified fibers for support, and a specialized root system for efficient competition for belowground resources (see Chapter 5). But because no one design dominated in all environments, specialization for efficiency in any given environment involved structural trade-offs that made the same plant less competitive in other environments (Waller 1991). Most of what follows in this chapter aims to explore the ecological implications and the trade-offs involved in the various and varying architectural designs of extant plants.

B. The Modular Nature of Plants

In the crown of most vascular plants, it is easy to recognize a hierarchical series of subunits. The largest subunit is the branch, which is made up of modules (Porter 1983, 1989). *Module* is a general term (see Appendix 1) that refers to a shoot with its leaves and buds, and the term can be applied to either determinate (structures whose apical meristem dies or produces a terminal inflorescence) or indeterminate shoot axes (Waller 1991). Modules are, in turn, made up of smaller subunits consisting of a leaf, its axillary buds, and the associated internode. These small subunits have been called metamers (White 1984). Since plants have many redundant modules or organs that have similar or identical functions (e.g., leaves or shoots transforming absorbed light into biomass), plants have been seen as metapopulations (White 1979). Such redundant modules are not fully dependent on one another, and, in fact, individual modules continue to function when neighbor organs are removed (Novoplansky et al. 1989, Sachs et al. 1993). The existence of hundreds of redundant subunits within a single plant raises the question: To what degree are these structural subunits (e.g., shoots) responding independently to the environment?

1. Scaling Up and Down

The study of functional modularity of plants can be tackled at different scales. The smallest end is the so-called nutritional or physiological unit, comprising a unit of foliage, the axillary bud, and the corresponding portion of stem (Watson 1986). As pointed out by Sprugel et al. (1991), the opposite end of the spectrum would be the clonal herbs (see Section V), in which each module (ramet) contains all of the structural parts necessary for independent existence. The branch is an intermediately scaled unit, which is very convenient because it is large enough to integrate most relevant physiological processes, but small enough to be used in ecophysiological experiments. For this reason, branches have been used extensively by ecologists and ecophysiologists to scale from leaf-level measurements to the whole plant or to the plant community (Ehleringer and Field 1993, Gartner 1995). Additionally, branches allow researchers to reduce the complex crown of a tree to a much simpler experimental unit.

All branches within a plant are structurally and physiologically connected to one another, but the mutual interactions are not always easy to elucidate. To make reliable scaling and generalization exercises, branch autonomy must be investigated thoroughly. Branch autonomy depends on the resource, i.e., carbon, water, or nutrients. The most clear aspect of branch autonomy is that related with carbon budget, since most branches fix all the carbon they need, and usually fix more, becoming exporters or sources of carbon in contrast with roots or reproductive structures, which are important carbon sinks (Geiger 1986). Although branches cannot be completely autonomous with respect to water and nutrients, which come from the roots via the stem, they exhibit different levels of uncoupling with the rest of the branches of the crown, i.e., different levels of relative autonomy (Zimmerman 1983; also see Chapter 6). In most species, branches are somewhat hydraulically isolated from the rest of the plant; thus, in words of Tyree (1988), "branches can be treated as small, independent seedlings rooted in the main bole." Nevertheless, branches are imperfect substitutes for studies on whole plants, especially when exceptions to the general branch autonomy can be expected (Sprugel et al. 1991).

2. The Ecology of Branch Autonomy

Branch autonomy has two major ecological advantages: (1) control of stress and damage, and (2) a more efficient exploitation of heterogeneous environments (Hardwick 1986). A compartmentalized plant may be less vulnerable to pathogens or herbivores than an integrated plant. It is well known that trees are capable of walling off injured or too-old branches, which provides an efficient protection against spreading of infections and against a net energy drain on the organism, respectively (Zimmermann 1983, Sprugel et al. 1991). A similar argument on the advantages of branch autonomy can be built for the prevention of runaway cavitation, i.e., the formation of gas bubbles when transpiration rate exceeds water transport that block xylem vessels or tracheids (Zimmermann 1983; also see Chapter 6).

Because the different aerial parts of a plant (e.g., branches and leaves) are generally in different light environments, plants frequently face the problem of distributing limited resources in a way that would optimize the performance of units exposed to heterogenous light conditions. Although plants do not forage in the classical sense of moving around to different prey locations, they do exhibit a foraging behavior (Bell 1984, Hutchings and de Kroon 1994). Plants forage because they must spend energy producing the leaves and the associated supporting structures necessary to harvest light, and their fitness is increased if this energy is spent efficiently, i.e., if leaves are arranged appropriately to maximize light capture. A plant that puts the new leaves in high-light areas of the crown will have an advantage over one that remains symmetric and puts out leaves equally in all possible locations. Branch autonomy with respect to carbon budget enhances the efficiency of light foraging because branches exposed to high light will grow bigger, shaded branches will stop growing, and no energy will be wasted in producing leaves in shaded areas (Sprugel et al. 1991). However, this is the case only for woody plants with indeterminate or multiple flushing growth patterns, where photosynthate for new leaves at the top of a shoot has been shown to come primarily from the older leaves of the same shoot (Fujimori and Whitehead 1986). In woody plants with determinate, single-flush growth patterns, efficient light foraging is not achieved via branch autonomy but rather via increased bud production in high-light areas; the buds will draw on reserves throughout the tree in the next growing season (Sprugel et al. 1991).

3. Modularity Versus Integrity

Despite the ecological relevance and the functional evidence of a certain autonomy of the different modules of a given plant, many different studies suggest that a plant is more than just a population of redundant organs because it responds to the environment as an integrated individual and not as a simple colony with limited mutual aid (Novoplansky et al. 1989, Sprugel et al. 1991, Sachs et al. 1993). The simplistic, albeit tempting, concept that a single plant is not a unit but a collection of independent subunitary parts became widespread during the 19th century and persisted until modern times (White 1979). It is reminiscent of the assumption that organismal structure and function can be understood by studying the cells, since cells have been considered the building blocks of organism form since the publication of the cell theory in 1838 (Kaplan and Hagemann 1991). In the advocacy of plant integrity, plants have been considered "metapopulations" (White 1979), not in the sense of Wilson (1975), but in the context of the classical etymology of *meta-* as "sharing." Therefore, plants are referred to as metapopulations when the shared elements that make up the morphological structure of an individual are to be emphasized. Under controlled conditions, plants have been shown to do more than respond locally to the degree to which they are damaged: interactions and mutual support between branches allowed treated plants for the comparison of available branches, and for the diversion of resources so as to increase the chances of greatest overall success (Sachs and Hassidim 1996). These processes that are essential for plant organization and for the control of plant morphogenesis seem to be mediated by phytohormones (Sachs 1988b, 1993).

Nevertheless, two interesting lines of evidence support the notion that the modules of a plant are functionally independent of one another, at least to some extent: (1) independent patterns of phenology between branches, and (2) competitive interactions between modules for limited resources as a consequence of a eustelic arrangement. Each module undergoes a complete life cycle of birth, growth, maturation, senescence, and death; therefore, a plant can be studied as a dynamic population of modules with a distinct age structure following rigorous demographic analyses (Harper 1977, Room et al. 1994). Individual leaves and foliage units are manifestly not all the same due to the two simple facts that they are not of the same age and that they are borne in different positions relative to each other (Harper 1989). In this sense, and considering the remarkable genetic variability of the different modules of a plant and the fitness differentials between modules, individual plants can be tackled as colonies of evolutionary individuals (Gill 1991). Leaving aside the integrated response of the many modules of a single plant, it is by focusing on plants as assemblages of metameric units that the ecology of their canopy shape will be better understood.

C. Plant Biomechanics: Coping with Gravity and Wind

Among the many interpretations of plant form, a crucial one is that plants are units of constructional engineering where shapes of their branches, their elasticity, and resistance to strain, are constrained by well-known mechanical principles (McMahon and Kronauer 1976, Niklas 1992). Because aerial shoots face the obvious forces of gravity and wind, a fraction of the biomass must be devoted to support. And this fraction increases quickly with plant size. For instance, the strength of a column (e.g., a branch or a stem) scales with the square of its diameter, while its mass increases with diameter squared times length. For any given plant, the mechanical costs associated with its crown geometry must be balanced with the photosynthetic benefits associated with its light-capture efficiency.

The height to which a plant should grow depends on the environment and on what the other individuals in the vicinity are doing, becoming a text-book example of an evolutionarily stable strategy (Waller 1991, see Appendix 6). The taller a plant becomes in its competition for light, the more light it needs to support its pre-existing biomass and to achieve growth. In fact, it has been shown that the maximum height of the tree *Liriodendron tulipifera* was determined by the point at which carbon gain in full sunlight is evened by construction and maintenance costs of crown and roots (Givnish 1988). Although small-stature plants have much smaller growth maintenance requirements per unit of light-absorbing machinery than large plants, for most plant shapes, the higher the plant, the more light it intercepts during the course of the day (Jahnke and Lawrence 1965, King 1981, 1990). Thus, there is a payback of investing in height that can be especially relevant under situations of strong competition for light. Plant height in light-limited environments interacts not only with competition, but also with growth form and leaf phenology (Givnish 1982, 1986).

Mechanical stability imposes the minimum amount of tissue required to support the crown and its units. The most likely mode of stem failure is elastic toppling, not failure under the weight of the crown. Accordingly, stem diameter scales with height, obeying a law that prevents elastic toppling and not compressive failure (McMahon 1973). For most plants, height varies with trunk diameter in such a way that there is a margin of safety against buckling (Niklas 1994). When trees grow in the open with little competition, their size and shape is conservative, being only one quarter of their theoretical buckling height (McMahon 1975, McMahon and Kronauer 1976). But when competition in a forest is strong, trees cannot afford large safety margins, especially when they have not reached the canopy. Based on this, Givnish (1995) predicted that shade-intolerant pioneer species should have lower mechanical safety margins than shade-tolerant species of similar stature. High wood density, usually reached in long-lived species with slow tissue turnover, provides resistance against mechanical failure and against attack by fungi and insects (King 1986). However, it adds extra weight for a given height or length of the stem or branch, so the biomechanical advantages of a stronger building material are frequently neutralized by the additional load. Structural costs are minimized by constructing stems of low-density wood, and for this reason softwoods can grow faster than hardwoods (Horn 1971). Hence, pioneer trees are expected to have light, energetically inexpensive wood, whereas late successional trees should have dense, highly lignified wood. Most studies in temperate and tropical forests confirm this trend (Horn 1971, Givnish 1995). The different biomechanics and associated costs of the crowns of hardwoods and softwoods can be crucial depending on the sign and intensity of factors such as frequency of storms, stability of the substrate, competition for light, or availability of water and nutrients.

Another important biomechanical aspect of the crown is the branching pattern. Branching angles should minimize both structural costs and leaf overlap to achieve optimal plant growth. However, these two features are mutually exclusive because branching patterns and leaf arrangements that reduce leaf overlap often require more investment in supporting tissues (Givnish 1995). Plants segregate in the cost and benefit trade-offs that their crown design entails in a given environment. In general, tree mechanisms concentrate on a good mechanical design only if light capture is sufficient (Mattheck 1991, 1995), but the biomechanical theory of crown design is still insufficient for integrated comparisons of the particular advantages of each crown architecture.

Although gravity leads to static loading of a plant based on the weight of individual parts, the dynamic loading caused by wind is often transitory (Grace 1977, King 1986, Speck et al. 1990). However, the wind exerts permanent modifications of the overall shape of plants and affects the anatomy and density of the wood, inducing biomechanical changes at architectural and anatomical levels (Coutts and Grace 1995, Ennos 1997). The greatest effects of strong winds on trees are seen near the tree line, where most species exhibit the so-called krummholz form (Ennos 1997). *Krummholz* refers to environmentally dwarfed trees, in which the crown is a prostrate cushion that extends leeward from the short trunk (Arno and Hammerly 1984). Despite the fact that light harvesting can be decreased by the krummholz habit, carbon gain is enhanced in comparison with upright trees in equivalent environmental conditions due to the increased photosynthetic rates exhibited by the leaves, which are deep in the boundary layer and warmed more by the sun (James et al. 1994). Another interesting, albeit little explored, aspect of plant biomechanics and wind is the dynamic reconfiguration of crown shape while the wind is blowing. Branches and foliage bend away with the wind, which reduces drag. It has been suggested that drag reduction should lead to flexible twigs in windy environments (Vogel 1981), and also to pinnate or lobed leaves due to the great degree of reconfiguration of these leaves in comparison with that of simple leaves (Vogel 1989). Increasing evidence is pointing to the existence of two main strategies regarding the wind as an ecological factor: (1) pioneer trees in windy habitats with flexible branches and pinnate or lobed leaves to reduce aerodynamic drag; and (2) late-successional trees or species from sheltered sites with simple leaves and rigid branches to maintain optimal light interception (Vogel 1989, Ennos 1997). A similar reasoning was given for woody plants that dwell along shores of streams and torrents: flexible twigs and narrow, willow-like leaves should prove adaptive since they reduce pressure drag during flash floods (Van Steenis 1981, Vogel 1981).

Unusual growth forms pose specific biomechanical problems, and precise studies are required to interpret certain plant designs. For instance, in most species of *Opuntia* (Cactaceae), shoots are formed as a sequence of short, flattened stem segments called cladodes. Cladodes have an elliptical base that supports the greatly enlarged upper portion and joins over only a small portion of their periphery so that there is considerable flexing at the cladode–cladode junctions (Nobel and Meyer 1991). Despite the fact that the contact between cladodes is only 20% of that occurring in a similar stem of constant width, the resulting shoot structure is rigid and resistant to typical wind and gravity loadings. The remarkable strength of this cladode–cladode junction cannot, be fully explained from a biomechanical point of view (Nobel and Meyer 1991). Another interesting study case are palm trees. Their lack of secondary thickening exposes them to a risk of toppling that increases with crown height. Mechanical safety of certain palm trees seems to be maintained by increasing the tissue density over time, and by proliferation of existing tissues that leads to an increase in actual cross-section of the stem (Rich 1986, 1987).

III. THE DEVELOPMENT OF A CROWN SHAPE

Despite the fact that most plants exhibit an indefinite growth, which produces a remarkable variability in their final size, they have a recognizable form. The many meristems of a plant are integrated into a galaxy of possible but not random morphologies. Understanding the mechanisms behind the production, arrangement, and turnover of plant modules led morphologists to group plants in a small number of "architectural models," allowing plant form to be linked not only to taxonomy but also to functional ecology. However, there is some controversy regarding the ecological implications of these architectural models followed by different plants during their ontogeny, and architectural models do not suffice for the complete description of plant form.

A. Crown Architecture and Models of Growth

Plants exhibit an extraordinary variety of branching patterns and foliage arrangements. The luxuriance of structural details of a forest canopy or the diversity of morphologies displayed by the herbs of a subalpine meadow can be overwhelming. For this reason, botanists and plant ecologists have looked at the developmental organization (architecture) of plants with a reductionist approach, slimming the complexity of plant shape to a sequence of simpler processes, but retaining the holistic features that determine plant construction (Tomlinson 1987). The questions of how many possible ways there are to build a plant and how many architectural models are exhibited by real plants have led to several classifications of plant shape. One of the best-known detailed classifications of plant architecture was reported by Hallé and Oldeman (1970), and resulted from an extensive, comparative study of the ontogenetic changes of the shape of tropical trees (Appendix 2). In fact, most systematic descriptions and cataloguing of architectural patterns have been based on trees. The most interesting features of these classifications are (1) a revival of the notion of modular construction and its importance in the generation of plant shape, and (2) an emphasis on understanding the mechanisms behind the dynamics of the arrangement, production, and turnover of plant modules and subunits (Porter 1989). This sort of information has made possible the realistic reconstruction of "virtual plants" (see Appendix 4), which is leading to in-depth understanding of plant growth in response to the environment and to promising orientations for plant breeding and pests and pathogens management thanks to the potential of "virtual experimentation" (Room et al. 1994, 1996, Prusinkiewicz and Lindenmayer 1996).

1. The Framework of Crown Shape: Branching Patterns

Branching complexity ranges from plants with a single axis to large trees with many orders of branching in three-dimensional space. However, the overall complex shape of a tree can be determined by surprisingly few parameters since a new branch is geometrically determined by just two parameters: branching angle and branch length (Honda et al. 1997). Repetition of the branching generates the distinctive complexity of plant crowns. The relative simplicity of the process has resulted in the generation of numerous computer models that simulate branching and growth of plants with remarkable realism (Waller and Steingraeber 1985, Fisher 1992, Prusinkiewicz and Lindenmayer 1996, Room et al. 1996). Moreover, the realism of tree models has increased with the modern development of fractal geometry (Lorimer et al. 1994). Basic notions and key references regarding fractals and modeling of plant growth are given in Appendices 3 and 4, respectively.

Although some trees have a single axis (e.g., most palms) and some have many similar branching axes, most species of trees have two or more types of axes that can be distinguished by their primary orientation, symmetry, or form. In general, leader axes are radially symmetrical, whereas lateral branch axes are dorsiventrally symmetrical (Fisher 1986). Differences in initial vigor of lateral branches results in a well-defined main axis, which is established commonly in a regular, alternating zigzag pattern (Fisher 1986).

The branching and consequent growth of trees and shrubs can be characterized by vertical or longitudinal, and horizontal or lateral symmetries (Strasburger et al. 1991). Vertical symmetry is characterized by growth of branches at the top (acrotony) or at the base (basitony), whereas lateral symmetry is characterized by branch growth at the upper or lower side of the lateral branch (epitony and hypotony, respectively; see Appendix 1 for architectural terms). Logically, shrubs exhibit a basitonic branching, whereas trees are characterized by acrotonic branching. Analogously, while typical trees exhibit a hypotonic branching, most shrubs and small trees exhibit epitonic branching. However, there are many exceptions to these rules. For instance, the pyramidal shape of the crown of many conifers is due to the combination of basitonic branching (typically a shrub pattern) with a monopodial growth of the bole (Strasburger et al. 1991). The dominance of branch development when branch originates from buds on the upper side of stems or main branches (epitonic shoots) appears to be important in shrub competition for space, since hypotonic branching confers the capacity of extending laterally but not overtopping an existing canopy (Schulze et al. 1986b). More implications of branching patterns in the way shrubs and trees occupy and compete for space are discussed in Section IV.

Relative number of branches has been examined in trees using the Strahler ordering technique, which begins at the edge of the canopy (first-order branches) and works its way toward the trunk, incrementing the order of a branch each time it intersects the junction of two similarly ordered branches (Waller 1991). The bifurcation ratio, an index of the degree of branching from one order to the next, was initially related to the successional status of the tree (Whitney 1976). However, later studies have shown that it varies within a given species (Steingraeber et al. 1979, Boojh and Ramakrishnan 1982, Veres and Pickett 1982) and even within a given crown (Steingraeber 1982, Borchert and Slade 1984). The ratio between terminal and subterminal branches can be of ecological interest, but higher-order bifurcation ratios are difficult to interpret (Steingraeber et al. 1979, Steingraeber 1982).

2. Arranging the Leaves

Fisher (1986) distinguished five different factors that determine the position of leaves. Among them, only two (phyllotaxis, which will be addressed in this section, and secondary leaf reorientation by internode twisting, petiole bending, or pulvinus movement, which will be addressed in Section IV) apply to the leaves themselves. The other three concern the branching pattern and the position of the leaf-bearing branches. For instance, internode length affects the longitudinal distribution of leaves along the axis, or the existence of short and long shoots determines whether leaves would be produced every year or not, since, in general, only short shoots continue to produce leaves after one growing season.

Phyllotaxis is responsible for the morphological contrast between plants with leaves along the sides of horizontal twigs, forming horizontal sprays of foliage, and those with leaves spiraling around erect twigs. Phyllotaxis, as the sequence of origin of leaves on a stem, has a great impact not only on the shape of a crown (it affects the position of axillary buds or apical meristems and thus determines branching patterns), but also in many functional aspects of the crown since it affects the interception of light (see Section IV) and the patterns of assimilate movement (Watson 1986). With regard to leaves, there can be one per node (as in all monocotyledons and in some dicotyledons) or more than one per node (as in many dicotyledons). Leaves that lie directly above one another at different nodes form vertical ranks called orthostichies. When there is only one leaf per node, the phyllotaxis can be monostichous, distichous, tristichous, or spiral if the stem has one, two, three, or more than three orthostichies, respectively (Figure 1). Monostichous is a very rare phyllotaxis and is usually accompanied by a slight twist of the stem that arranges the leaves in a shallow helix; the corresponding phyllotaxis is called spiromonostichous (Bell 1993; Figures 1 and 7). In a distichous foliage, the two rows of leaves are 180° from each other, whereas in a tristichous foliage, leaves are in three rows with 120° between rows. Spiral phyllotaxy results when each leaf is at a fixed angle from its predecessor in such a way that a line drawn through successive leaf bases forms a spiral (the genetic spiral) around the stem. This widespread phyllotaxis, also called disperse due to the apparent lack of geometrical pattern, has led to a mathematical formulation of functional implications (see Section IV). Spiral phyllotaxis can be mathematically described as a fraction in which the denominator is the number of leaves that develop before a direct vertical overlap between two leaves occurs, and the numerator is the number of turns around the stem before this happens (Figure 2). This fraction times 360° is a measure of the angle around the stem between insertion of any two successive leaves (e.g., for a tristichous phyllotaxis, the fraction is 1/3, meaning that 3 leaves are developed before vertical overlap between two leaves, and this overlap happens in one turn around the stem, and the 120° between the orthostichies or between two successive leaves results from

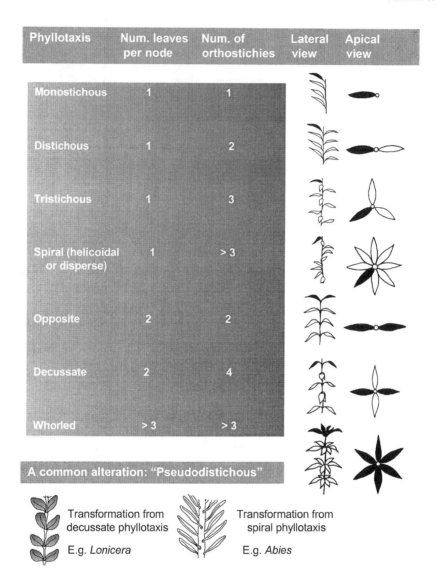

Phyllotaxis	Num. leaves per node	Num. of orthostichies	Lateral view	Apical view
Monostichous	1	1		
Distichous	1	2		
Tristichous	1	3		
Spiral (helicoidal or disperse)	1	> 3		
Opposite	2	2		
Decussate	2	4		
Whorled	> 3	> 3		

A common alteration: "Pseudodistichous"

Transformation from decussate phyllotaxis
E.g. *Lonicera*

Transformation from spiral phyllotaxis
E.g. *Abies*

Figure 1 Main patterns of leaf arrangement (phyllotaxis) in plants. In the sketches, the black leaf (or leaves) represents the uppermost one in the shoot. A common alteration that results in a phyllotaxis that looks distichous (pseudodistichous), which has been generally interpreted as an adaptation to avoid self-shading, is shown in the lower part of the figure.

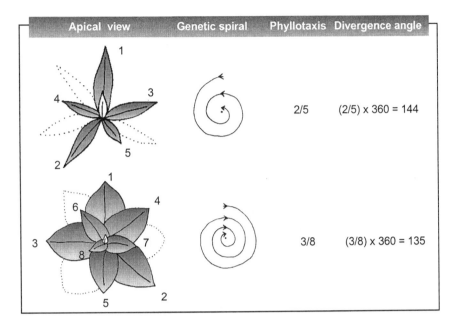

Figure 2   Two examples of spiral phyllotaxis. Spiral phyllotaxis can be described as a fraction in which the denominator is the number of leaves that develop before a direct vertical overlap between two leaves occurs, and the numerator is the number of turns around the stem before this happens. This fraction times 360° is a measure of the angle around the stem between insertion of any two successive leaves. The genetic spiral is the virtual line connecting successive leaves. Numbers in the two sketches of the left-hand side indicate leaf order in the genetic spiral.

1/3 times 360°). When the phyllotactic fraction of plants with spiral phyllotaxis was calculated and ordered, the following series was obtained: 1/2, 1/3, 2/5, 3/8, 5/13, 8/21, etc. (Figure 2). Interestingly, in this series both numerators and denominators form Fibonacci series since each number is the sum of the preceding two numbers. When multiplied by 360°, this series converges toward 137.5° (Fibonacci angle), which is the divergence angle between two successive leaves in most plants with spirally arranged leaves (Leigh 1972, Bell 1993). This and other interesting geometrical properties of phyllotaxes have been reviewed in the literature (Erickson 1983, Prusinkiewicz and Lindenmayer 1996).

Other phyllotaxes can be observed when more than one leaf is present on each node. The simplest case is the opposite foliage, with two leaves 180° apart at each node, forming two orthostichies. A common variation is the decussate phyllotaxis, which has four orthostichies due to the fact that successive pairs of leaves are orientated 90° to each other (Figure 1). A more complex variation is the bijugate or spiral decussate phyllotaxis, where successive leaf pairs are less than 90° apart, leading to a double spiral (Bell 1993).

The ease with which the phyllotactic fraction is measured in a given plant is frequently confounded by internode twisting or leaf primordium displacement. It is relatively frequent that several phyllotaxes converge in an apparent distichous foliage. For example, the needles of some *Abies* are in two rows and look distichous, but their real phyllotaxis is spiral, as indicated by the petiole insertion (Figure 1). A similar case was found in the shade shoots of the chaparral shrub *Heteromeles arbutifolia*, which exhibited a pseudodistichous phyllotaxis instead of the characteristic spiral phyllotaxis of the species (Valladares and Pearcy 1998). Decussate phyllotaxis might also look distichous, as observed in horizontal shoots of *Lonicera* (Figure 1). Spiral and distichous leaf arrangements are also sometimes found in the same plant species. For instance, certain plants may first, as seedlings, set leaves spirally around an erect stem, and then, as mature individuals, develop a distichous foliage on the horizontal branches produced in the axes of the initial leaves. This seems to be the case for the tropical forest understory herb *Dichorisandra hexandra* (see Section IV and Figure 10).

3.   Classifying Crown Architectures

The first, and possibly the best known, classification of tree architecture was reported by Hallé and Oldeman (1970; improved for the English version: Hallé et al. 1978). Basic features of this classification were dichotomic characteristics of the tree crown, such as monopodial or sympodial branching, basitonic or acrotonic branching, orthotropic or plagiotropic shoots, etc. (see Appendix 1 for terms and concepts and Appendix 2 for a key for these classic architectural models). From the practical point of view, this classification can be very difficult to use with certain species because the researcher must know the way by which the shape of the crown is achieved during the ontogeny of the tree from the seedling to sexual maturity, something that exceeds the time frame of most field studies dealing with long-lived plants. In addition, certain species exhibit "architectural ambiguities," shifting from one model to another during their ontogeny or under different environmental conditions. Leigh (1990, 1998) modified Hallé and Oldeman's classification, simplifying it by merging some models that cannot be easily distinguished.

As previously mentioned, the most widespread architectural classification has been developed for trees, but it can be used with other plants such as shrubs. However, characteristics such as multiple versus single stems should be considered in more detail within the classification to better differentiate the several designs of shrubs. The multiple-stemmed characteristic results from the growth of buds from the belowground level that escape apical dominance to form new stems or modules (Wilson 1995). Multiple-stemmed shrubs exhibit not only a different shape than single-stemmed shrubs, but also a different tolerance to perturbations (e.g., fire and pests). Multiple-stemmed shrubs can survive indefinitely as a clone by producing new stems, whereas single-stemmed shrubs die when the stem dies. The maintenance of an apical control; the tendency of the stems to bend toward the horizontal, producing vigorous vertical shoots in a series of arching segments; or the location of the underground buds of multiple-stemmed shrubs (on the basis of the shoot, along rhizomes, layered branches) are also important features to consider in the description of shrub architecture (Wilson 1995).

Architectural models are a convenient starting point for interpreting plant form, but there is a series of variations and exceptions to each program of development that complicates classification and suggests the search of additional descriptions of crown shapes. For instance, *Arbutus* sp. exhibit two different architectural patterns depending on the light environment, and *Acer pseudoplatanus*, as with many other woody plants, undergo significant changes of branching patterns during the ontogeny, switching from one model to another (Bell 1993). There are also many examples of metamorphosis (abrupt change from plagiotropic to orthotropic disposition of a branch) and intercalation of shoots infringing the rules of each model (Bell 1993).

Nevertheless, architectural models are useful to predict the form that a plant will assume in the absence of unusual external forces or when affected by the common circumstance of losing a structural subunit (e.g., a branch) through injury. The modules that regrow when a tree loses a subunit usually mirror the architecture of the whole crown of the tree in a process called "reiteration" (Hallé et al. 1978, Hallé 1995). As the tree grows, the number of reiterated units tends to increase, but their size tends to diminish, and ultimately only parts of the architectural unit are reiterated in a so-called "partial reiteration" (Hallé 1995). This reiteration process that occurs during the growth of a large tree reinforces the idea that most trees are colonies, the elementary individual being not the bud, but the architectural unit. This idea of a plant as a colony (discussed in Section II) dates back to eighteenth century: botanists such as de la Hire, Bradley and von Goethe (see references in White 1979), and Charles Darwin (Darwin 1839) and his grandfather Erasmus Darwin (Darwin 1800) thought that coloniality existed in trees. In convergence with these interpretations, the architectural models developed since 1970 provided morphological evidence for coloniality in trees, which was considered a corollary of their sessile way of life. Although reiterated units have largely been considered as leafy branch systems, Hallé (1995) went one step beyond, posing the hypothesis that these units comprise their own root system, and thus the bole is made up of the aggregated root systems of all the reiterated units forming the tree crown. Needless to say, this hypothesis is controversial and may be somewhat heretical to certain readers, as acknowledged by Hallé himself (Hallé 1995, p. 41).

B.   Functional Insights into Architectural Classifications

Since the shape of the crown influences important aspects of growth and survival of plants, such as light interception and competition for space, the adaptive significance of the architectural models of Hallé and Oldeman (1970) has interested many ecologists dealing with plant form. While all investigators agree that crown shape is generally adaptive, there is no consensus regarding the ecological and evolutionary implications of these architectural models (Porter 1989). On the one hand, as observed by Porter (1989), fossil plants exhibit only three of 23 possible architectural models originally described by Hallé and Oldeman (1970), mostly due to the remarkable lack of fossil examples of sympodial branching. This clumping of fossil trees among Hallé and Oldeman models suggests that some plant forms may have paid an evolutionary penalty for their mode of whole plant development, that is, the limited number of ancestral architectures may have limited the number of architectural models that have survived. On the other hand, Ashton (1978) pointed out that in West Malaysia, certain models were very rare in shady habitats, whereas a very plastic type of organization (Troll's model) was very widespread. The relatively small number of models found in temperate deciduous forests (the conifer forests of the boreal regions have even fewer models) suggests that some models are selected against in some regions (Ingrouille 1995).

However, three arguments have been given to support the notion that these architectural models are not adaptive (Fournier 1979): (1) all models coexist in lowland tropical rainforests, so a single ecological region has not favored some models at the expense of others; (2) the same model exists at different levels in the forest canopy, despite the remarkable vertical gradients of light, predation, and nutrients; and (3) the same model exists in different growth forms from very tall trees to small herbs, which clearly do not share the same ecology. Actually, as shown in Figure 3, developmentally different models can produce functionally similar crown shapes (ecological convergence). And a single model shared by different plant species can produce functionally divergent crowns due to differences in factors such as the relative elongation of axes and the exact arrangements of leaves (Fisher and Hibbs 1982). There is a wide plasticity allowable within one model of Hallé et al., so these models may lead to unequivocal ecological predictions only for the simplest crowns (Porter 1989, Waller 1991). Additionally, efficiency of leaf display, which is crucial in the ecological strategy of most

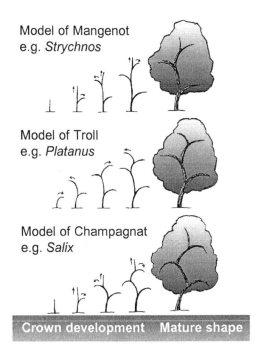

Model of Mangenot
e.g. *Strychnos*

Model of Troll
e.g. *Platanus*

Model of Champagnat
e.g. *Salix*

Crown development   Mature shape

Figure 3   Example of how three developmentally different architectural models can produce functionally similar crown shapes. Arrows indicate direction of meristem growth or
branch bending. (*Source*: Adapted from graphs in Waller 1991 and Bell 1993.)

species (see Section IV), is not included in the parameters used to define the
architectural models (Tomlinson 1987).

Because developmental plasticity is an intrinsic characteristic of plant form
(see Section V), any attempt to classify the architectural patterns of plants should
include the structural response of each species to different environments or perturbations. And in agreement with Sachs (1995), a response is a quantitative process,
which would make the separation of species into discrete models very difficult.
Tomlinson (1982) has emphasized the complex balance between design and
chance in plant construction. In many plants, and especially in long-lived trees,
it is a challenge to distinguish the genetically determined structure from environmental damage and phenotypic plasticity (Fisher 1992). Consequently, searching
for a single ecological classification of plant architecture seems a vain endeavor.
The critical parameters for the classification of plant shape must vary depending
on the problem at hand (Sachs 1995).

C. Real Crowns: Imperfect Architectures or Controlled Variability?

In contrast to human designs such as buildings, the final shape of the crown of a plant expresses a remarkable variability, which is evident even in comparisons of two halves of the very same individual (Sachs and Novoplansky 1995). However, there is a characteristic design or architectural pattern for each plant species. Therefore, the general shape of a crown is rather constant for a given species under a given environment, whereas many aspects of branch growth and survival do not follow a strict program, exhibiting an apparently stochastic behavior. At least the following three parameters have been shown to introduce variability in the shape of a plant: (1) the location and number of developing apices; (2) the developmental rates of individual apices; and (3) the shedding of branches (Sachs and Novoplansky 1995). Is this variability in the crown shape due to a malfunction of the genetic program that determines the development of the shape of a plant? How could the general form of a tree be more predictable than the individual events (e.g., production and shedding of branches) that lead to it? Variability in the final shape is not characteristic of primitive or maladapted plants, and it is not the result of errors in the developmental program. On the contrary, it has a crucial ecological role in changing and heterogeneous environments (see Section V). On the other hand, predictable mature structures can result from selection (the so-called epigenetic selection) of the most appropriate developmental events from an excess of possibilities that are genetically equivalent (Sachs 1988a). In this way, the final shape or pattern is genetically specified, but the development of the crown gravitates toward this final shape without a detailed genetic program. This tendency toward the final shape is accomplished by means of internal systems that control the variability in the aforementioned parameters, but allow for developmental plasticity. These control systems that constrain development variability were explained by Sachs (1995), and they include internal correlative interactions between branches, responses to local shading, and programmed limitations of successful branches. In conclusion, although the architecture of a plant limits its range of possible shapes, a plant's architectural model does not determine its final shape.

IV. STRUCTURAL DETERMINANTS OF LIGHT CAPTURE

Plants survive the freezing temperatures of high mountains, the desiccating effects of the desert, and the gas diffusion problems of the aquatic environment (Crawford 1990), but with the exception of parasitic life-forms, no plant survives without light. Photosynthesis in a canopy proceeds at a rate that depends not only on the physiology of the process, but also on how photons are distributed over individual elements of the foliage (Russel et al. 1989, Attridge 1990). The amount of leaf present is the most basic structural property that affects the fraction of the available radiation that is absorbed, although the distribution and arrangement of leaves within a crown obviously affects their efficiency. Because three-dimensional arrangement of leaves in a crown is difficult to measure, it is often assumed in models of light interception and canopy photosynthesis that the leaves are randomly distributed throughout the canopy volume (Russel et al. 1989; also see Chapter 22). This section discusses the different structural features of plant crowns that affect light interception from the individual level to the community level.

A. Shaping the Foliage: The Single-Crown Level

The shape of the crown of a given plant and the arrangement of its foliage units are the two most basic parameters affecting the efficiency of light capture. From a photosynthetic point of view, the most efficient canopy is achieved when all of the leaves are evenly illuminated at intermediate light flux densities. However, this is found in nature only very rarely. Different crown shapes and different dispositions of the leaves within the crown result in different diurnal and seasonal patterns of light interception at both the single-leaf and the whole-crown levels. Leaves at the uppermost positions of the canopy are frequently exposed to higher irradiances than lower leaves, but the light available for the latter is affected not only by the amount of neighboring leaves, but also by the general form of the crown and the angle and orientation of the surrounding units of the foliage. In addition, the level of incident photon irradiance can be regulated by diurnal leaf movements, that is, by a crown of a changing geometry. All of these factors and their interactions with latitude, season, and time of day are explored in the following sections.

1. Crown Shape

Is there a perfect crown shape that maximizes light interception in a given environment? In spite of the fact that many studies have addressed the influence of crown shape on light interception, only a few have tried to answer this question (Jahnke and Lawrence 1965, Horn 1971, Terjung and Louie 1972, Oker-Blom and Kellomäki 1982, Kuuluvainen 1992, Chen et al. 1994). The most stimulating attempts have been conducted by considering crown shape across two different kinds of environmental gradients: the variation of solar elevation and radiation with latitude (Kuuluvainen 1992, Chen et al. 1994) and the variation of available light with forest succession (Horn 1971).

The shape of the crown can be described by three parameters: the absolute size, the ratio of height to width, and the convexity or shape of its contour. Based

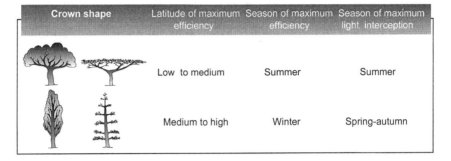

Crown shape	Latitude of maximum efficiency	Season of maximum efficiency	Season of maximum light interception
	Low to medium	Summer	Summer
	Medium to high	Winter	Spring-autumn

Figure 4 Latitude and season of maximum efficiency of light interception, and season of maximum light interception for two main types of crown shape: flat and broad versus thin and tall.

on latitudinal changes of solar elevation, it has been postulated that narrow, vertically extended crowns are more efficient intercepters of direct light than flat, horizontally extended crowns at high latitudes, whereas the reverse is true at low latitudes (Figure 4). The occurrence of tall and thin conifers at high latitudes and of acacia-like trees at low latitudes seemed to confirm the different adaptive value of these two general crown shapes in regions of different latitude. However, Chen et al. (1994) showed that the latitudinal variation of potential sunlight interception by different crown shapes does not reflect very well the existing latitudinal gradient of crown shapes because (1) light is not the only factor affecting the shape of the crown, and (2) light interception is not only controlled by the shape of the crown but also by the geometry and distribution of the foliage; therefore, crowns of different shapes can intercept a similar fraction of the available light. In addition, the latitude of maximum light interception efficiency is not fixed for a given crown shape. Depending on the conical or elliptical outline of the crown, a vertically extended crown can intercept light very efficiently at high latitudes, and its efficiency can be similar to that of a broad crown at low latitudes. Cones are more efficient sunlight intercepters than ellipsoids at high-sun conditions (i.e., at low and medium latitudes or during the summer), and, in general, the higher the cone, the larger fraction of irradiance captured (Jahnke and Lawrence 1965). For a given latitude, either very small or very large values of the height-to-width ratio result in maximum direct light interception (Chen et al. 1994). Another important fact to be considered in the functional analysis of the crown shape is the total radiation available and the time of the year of maximum irradiance and light interception. For instance, although the maximum efficiency of light interception by narrow-crowned trees is achieved during the winter, they intercept more light in spring and autumn. In the case of broad-crowned trees, both light interception efficiency and amount of light intercepted reach their maximum values during the summer. The fraction of available radiation made up by diffuse radiation is also an important factor affecting the efficiency of a crown capturing light. In cloudy regions, for example, a large fraction of radiation is received from all angles (diffuse radiation), and as a result, the role of the shape of the crown for light interception becomes less relevant. To understand the relation between the latitudinal gradient in crown shape and the latitudinal variation of the light regime, not only is a theoretical analysis of crown shape and light interception required (like the one by Chen et al. 1994), but also more case studies of the real light environment experienced by trees of different shapes growing at different latitudes, and a thorough exploration of the interference of other factors and constraints, such as water, snow shedding, and gravity, that also affect the shape of the crown.

The height-to-width ratio of the crown must reach a balance between growth in height to reach the brighter areas of the canopy, and growth in width to intercept light and occupy enough space (Horn 1971). In addition, the greater the convexity of a crown, the greater the irradiance intercepted (at most common latitudes), but also the greater the amount of supporting and conductive tissues. Horn (1971) predicted the optimal shape of trees of three different successional strategies: early successionals, late successionals, and early successionals that persist in the mature forest (Figure 5). Because early succession is a race to form a canopy, fast-growing softwoods are favored over stronger hardwoods,

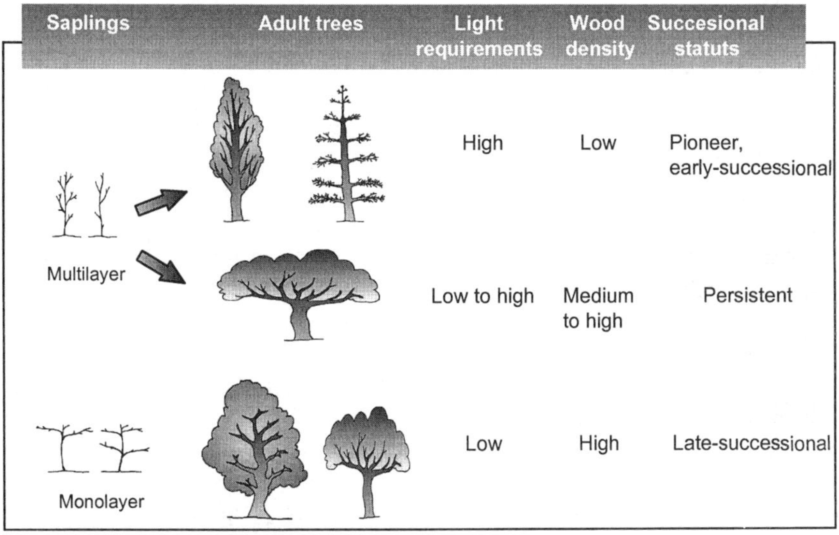

Saplings	Adult trees	Light requirements	Wood density	Succesional statuts
Multilayer		High	Low	Pioneer, early-successional
		Low to high	Medium to high	Persistent
Monolayer		Low	High	Late-successional

Figure 5 Crown shape as sapling and adult, light requirements, and wood density predicted for trees of different successional status. (*Source*: Adapted from Horn 1971.)

and growth in height is favored over growth in width (Horn 1971). Because softwood is weak, lateral growth must be relatively scant in early successional trees. Saplings of early-successional, shade-sensitive species should be multilayered, whereas those of late-successional, shade-tolerant species should be monolayered. Some multilayered trees persist by invading small openings in the forest and should have a mixed strategy. Because these species must initially race to the canopy, they must be tall, thin, multilayered, and made of softwood. Once they reach the canopy, they should spread out and dominate the forest gap. The height-to-width ratio decreases with age, and their wood should become harder to provide lateral support. These predictions of the crown shape in relation to the successional strategy of the species ar e simplistic and represent an incomplete theory, as Horn himself acknowledged (Horn 1971, p. 121), because they are based on a single factor (light). However, they provide an explicit list of testable assumptions, and with certain exceptions, such as the large-leaved, early-successional monolayer species of the wet tropics, they agree reasonably well with study cases such as those of North American temperate forests (Horn 1971).

2. Geometry of the Foliage

The distribution and arrangement of leaf area in space defines the probability that a beam of light will pass through a gap in the foliage to the lower leaves. The same leafiness of a crown, measured by its leaf area index (LAI: total leaf area divided by the total ground area where it stands) can lead to different efficiencies in light capture (Hunt 1990). A given branching pattern, for instance, can minimize the overlapping of leaf clusters on a horizontally spreading branch (Honda and Fisher 1978, Takenaka 1994). The geometry of the *Terminalia* type of branching (Aubréville's architectural model; see Appendix 2) has been shown to optimize effective leaf area both in theoretical studies of the architecture by computer simulations (Honda and Fisher 1978, Fisher and Honda 1979a) and in surveys of real trees (Fisher and Honda 1979b). A branch system with a high bifurcation ratio is theoretically more efficient in terms of the amount of stem tissue required to display a given area of leaves for the columnar crowns characteristic of high-light environments, whereas low bifurcation ratios are more efficient for the planar leaf display characteristic of low-light sites (Leopold 1971, Whitney 1976, Canham 1988).

In addition to the branching pattern, light interception efficiency of the crown is mainly determined by two topological aspects of the foliage: (1) proximity (degree of clustering) and (2) angle and orientation of the leaves. These two features significantly influence both the overlap of the leaves as seen from the sunpath, that is, the capability of a foliage unit to intercept direct sunlight, and the transmission of diffuse light to lower layers of foliage. Leaf angle and orientation alone, that is, without considering self-shading effects, can generate a complex diurnal pattern of light interception. Steep leaves project a small fraction of their area to the sun during the central hours of the day, but the effect is not the same for leaves with different orientations (Figure 6). To reduce leaf area by 50% at midday in spring, north-facing leaves must be positioned at an angle of 40°, whereas south-facing leaves must be positioned at an angle of 80° (at a latitude of 38° N). Steep east- and west-facing leaves exhibit an increased light interception during the early morning and late evening, respectively, with a marked decline during the central hours of the day, whereas it is almost the reverse for south- and north-facing leaves. Thus, steep leaf angles can have very different effects on leaves of different orientations, and although they always reduce light interception at the individual leaf level, this reduction can vary from very limiting for photosynthetic carbon fixation to negligible, and can generate maximum interception at very different times of the day (Figure 6).

Since leaves are frequently small units of large and complex crowns, the influence of leaf angle on light interception must be scaled up to the level of the whole crown in spite of the fact that three-dimensional distribution and arrangement of leaves in these crowns is difficult to measure. For plants with an LAI less than approximately 3, differences in leaf angle usually have a very small effect on whole-plant light interception and photosynthesis (Duncan 1971). However, the angle of the leaves has a strong influence on whole-plant photosynthesis for plants with a dense foliage (LAI > 3). In particular, nonrandom distribution of leaf angles in the plant crown, with leaves near the top of the crown more vertically inclined than leaves near the bottom of the crown, has been seen as a way of maximizing whole-plant photosynthesis (Herbert 1991, Herbert and Nilson 1991). A change from more near vertical to more nearly horizontal leaves lower down the crown leads to effective beam penetration and a more even distribution of light. In such cases, large LAIs can be sustained because only a few leaves are then light-saturated, and leaves at the base of the crown receive enough light for photosynthesis (Russel et al. 1989). Horizontal leaves at the top of the crown exhibit their maximum light interception efficiency at times of the day and the year (midday and summer, respectively) at which irradiance in sunny environments is well above the light saturation point for photosynthesis; therefore, their superior light capture usually translates into a negligible increase of potential carbon gain (Figure 7). For these reasons, erectophile crops have a marked yield advantage over those with horizontal leaves, especially at high LAIs and at high solar elevations (Isebrands and Michael 1986). However, light interception by steep leaves themselves is poor; therefore, if they represent a large fraction of the foliage or if their angle is too steep and the leaf blades are too close to each other (see computer images in Figure 7), light interception and potential carbon gain by the whole plant decrease. In a simulation of light interception and potential carbon gain by shoots of *Heteromeles arbutifolia* with leaves set at different angles, vertical foliages absorbed 20–30% less photosynthetic photon flux density (PPFD) and had 30% lower daily carbon gain than normal shoots (average leaf angle = 71°; Valladares and Pearcy 1998). The spa-

SINGLE LEAVES
(no selfshading)

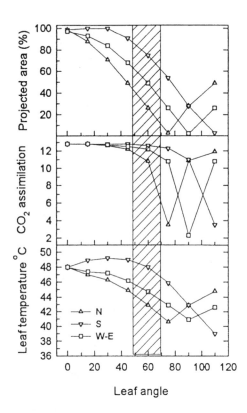

Figure 6 Leaf area projected to the sun, net photosynthetic rate, and leaf temperature at midday as a function of leaf angle for leaves of different orientations (north, south, west–east). Leaf projected area and photosynthesis were calculated using the three-dimensional YPLANT model (Pearcy and Yang 1996). Leaf temperature was calculated entering real data of stomatal conductance, windspeed, and air temperature in an energy balance model. Measurements and calculations were carried out for a clear day of spring in a Californian chaparral. No self-shading effects were considered. The hatched bars indicate the natural range of leaf angle observed in the chaparral shrub *Heteromeles arbutifolia*. (*Source*: Data from Valladares and Pearcy, unpublished.)

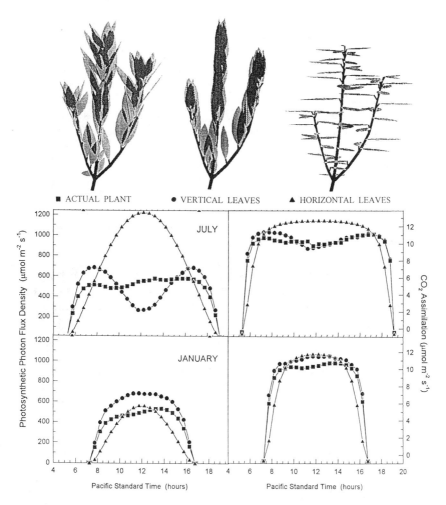

Figure 7 Diurnal course of interception of photosynthetically active radiation and CO_2 assimilation calculated for whole shoots of the chaparral shrub *Heteromeles arbutifolia* on a clear day of winter (lower graphs) and summer (upper graphs). Data were calculated for real shoots and for the same shoots with either vertical or horizontal leaves. Simulations were performed using the three-dimensional YPLANT model (Pearcy and Yang 1996). (*Source*: Data from Valladares and Pearcy, unpublished.)

tial packing and inclination of leaves are interrelated, and variations in these two factors can produce a set of equivalent solutions to the problem of maximizing light interception and total plant photosynthesis (Herbert 1996).

A little-explored aspect of leaf angle is its influence on light harvesting and utilization at the chloroplast level. Leaf inclination affects the amount of light that is received by the abaxial side of the leaf, typically in a light environment of lower intensity than that of the adaxial side. Photosynthetic tissues of the two sides of the leaf of both dorsiventral and isolateral leaves have been shown to acclimate to the particular light environment of each side, which is dramatically different in horizontal leaves or in artificially inverted leaves, but can be very similar in vertical or steep leaves (Evans et al. 1993, Poulson and DeLucia 1993; Valladares and Pearcy, in preparation). Large differences in mesophyll and chloroplast photosynthetic properties between the two sides of the leaves seem to depend on a complex interaction among light environment, leaf anatomy, and leaf angle (Myers et al. 1997).

Mutual shading of leaves is a complex phenomenon that is determined by relative distances among foliage units, leaf size and shape, and features of the light environment of the plant such as elevation angle of the sun and the relative importance of diffuse light. The sun disk is completely blocked by a leaf for a theoretical distance of 108 times the leaf diameter (umbra; Figure 8). An object farther than this distance from the leaf is lit by at least part of the sun (penumbra; Figure 8). Empirically, this distance is approximately 50–70 times the leaf diameter with the sun at the zenith on a clear day (Horn 1971). On a cloudy day, light (diffuse skylight) comes from 180 degrees, and the shadow of the leaf vanishes at a distance equal to its diameter. Therefore, the minimum distance between layers of an optimal multilayer crown is very large in high-light environments, and significantly smaller in low-light environments; in other words, the optimal size of a leaf is small in a sunny climate and large in a cloudy climate or in the shade (Horn 1971). Mutual shading among leaves is also affected by the orientation of target leaves and their position in the crown. For instance, in a California chaparral shrub, self-shading had a larger impact than steep leaf angles in the reduction of the leaf area displayed to the sun in south-facing leaves, whereas the reverse was true for leaves with other orientations (Figure 9).

Certain patterns of leaf arrangements (phyllotaxis) has been seen as genetically specified ways of enhancing light interception by avoiding self-shading. Leaves along the sides of horizontal twigs (distichous phyllotaxis; Figure 1) are commonplace in conditions where avoidance of excessive mutual shading among leaves is presumably crucial (Givnish 1995, Leigh 1998). Spirally arranged leaves are also common under low-light conditions. Plants with spiral leaf arrangements invariably exhibit phyllotactic fractions, forming Fibonacci series (see Section II.A), and converge on a divergence angle of 137° 30′ 28″, at which point no leaf will exactly overlap another. However, the influence of phyllotaxis on light interception has been evaluated quantitatively only very rarely because of the

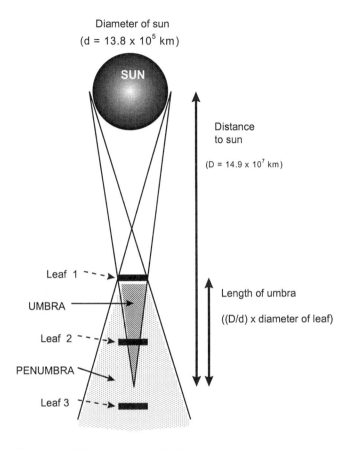

Diameter of sun
(d = 13.8 x 10⁵ km)

SUN

Distance to sun

(D = 14.9 x 10⁷ km)

Leaf 1

UMBRA

Length of umbra

((D/d) x diameter of leaf)

Leaf 2

PENUMBRA

Leaf 3

Figure 8 Below a leaf that blocks the sunrays (leaf 1) there are two zones: one that receives no direct sunlight (umbra) and one that receives direct sunlight (penumbra). However, sunlight in the penumbra zone is not full sunlight because the zone is illuminated by only a fraction of the solar disk. Since the umbra has a definite length (theoretically 108 times the diameter of leaf 1, empirically 50–70 times the leaf diameter), a leaf far enough away from leaf 1 can receive direct sunlight (leaf 3), despite being aligned with leaf 1, while leaves at intermediate distances from leaf 1 (e.g., leaf 2) will have parts completely shaded and parts lit by part of the sun. (*Source*: Adapted from Horn 1971.)

complicating effects of branching patterns and leaf fluttering (Niklas 1988). Changes in leaf shape, size, and orientation, and in petiole or stem length can compensate for the negative effects of leaf overlap produced by certain phyllotaxes (Niklas 1988, Pearcy and Yang 1998). In large crowns, the effect of phyllotaxis and other geometrical properties of the foliage on light interception is also obscured by the fact that leaves are not randomly distributed, but instead are

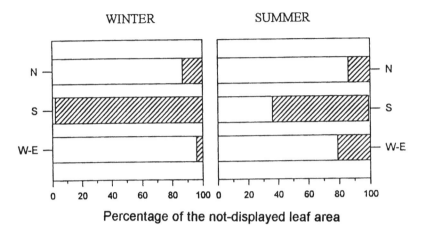

Figure 9   Influence of season and orientation of the leaf (north, south, west–east) in the relative importance of selfshading effects (hatched bars) and of steep leaf angles effects (open bars) in reducing leaf area displayed to the sun in the shrub *Heteromeles arbutifolia.* Data are expressed as percentage of the leaf area that is not displayed to the sun at midday, which was, on average, 66% of the total leaf area. Data were calculated for real shoots growing in a patch of chaparral (see Valladares and Pearcy 1998) using the three-dimensional YPLANT model (Pearcy and Yang 1996). (*Source*: Data from Valladares and Pearcy, unpublished.)

most frequently grouped into whorls of branches, and within whorls, they are frequently grouped around stems. Russell et al. (1989) pointed out the relevance of leaf grouping for light interception, calculating that if leaves were randomly distributed, an LAI greater than 6 could not be sustained due to excessive self-shading, whereas grouping allows an LAI of 10 or more.

Most studies of the interplay of plant architecture, light interception, and photosynthetic efficiency are either two-dimensional or too simplistic because they calculate overlapping of leaf silhouettes as viewed from the sun or estimate light penetration applying extinction coefficients that require random distribution of leaf angles and orientations and no leaf grouping. These studies miss the vertical component of the crown and the nonrandom geometrical features of the foliage units, which have a strong impact on the diffuse light penetration and distribution (Valladares and Pearcy 1998). Diffuse light, frequently neglected in simulations of light interception by plant canopies, is an important component of incident radiation (Gutschick and Wiegel 1988, Herbert 1991). The orientation of the crown and its leaves in the vicinities of forest gaps frequently respond to diffuse light rather than direct light (Ackerly and Bazzaz 1995, Clearwater and Gould 1995). The geometry of the foliage, basically proximity of leaves and distribution of leaf angle throughout the canopy, affect the transport of diffuse light to lower layers, which can be relevant for whole-plant photosynthesis not only in low-light environments, but also in open and sunny sites (Valladares and Pearcy 1998).

3. Changing Geometries: Leaf Movements

Leaves from a number of species do not remain in a fixed position during the day, but move so the leaf blade remains either perpendicular or parallel to the direct rays of the sun. These leaf movements result in a partial regulation of the intensity of the incident photon irradiance, especially when individuals are widely spaced and LAI values are lower than 1.5 (Ehleringer and Forseth 1989). When the leaf remains perpendicular to the sunrays throughout the day (these movements and the leaves that exhibit them are called diaheliotropic), that is, when there is leaf solar tracking, light interception can be increased by as much as 35% when compared with a fixed leaf with a horizontal position (Ehleringer and Forseth 1980, Ehleringer and Werk 1986). When the leaf remains parallel to the sunrays (paraheliotropic leaves or movements), it is exposed to much lower light levels and heat loads than leaves that remain in fixed positions. Leaf solar tracking occurs in herbaceous species and is most common in annuals (Ehleringer and Werk 1986). In drier sites, the frequency of leaf solar-tracking species seems to be inversely related to the length of the growing season, reaching values as high as 75% of the flora in the summer annuals of the Sonoran Desert (Ehleringer and Forseth 1980). However, leaf solar tracking poses a physiological dilemma when photosynthesis is impaired at midday by water stress or heat: diaheliotropic leaves can intercept excessive radiation under these far-from-optimum conditions. Certain species, such as the desert annual *Lupinus arizonicus*, avoid the dilemma, exhibiting either diaheliotropic or paraheliotropic leaf movements depending on the availability of water (Ehleringer and Werk 1986). The effects of leaf movements on canopy productivity depend on the foliage density (LAI). When LAI is low, leaf solar tracking enhances canopy productivity since leaves absorb photons that would otherwise pass through the sparse canopy. However, when the LAI is greater than 4, leaf solar tracking reduces canopy productivity because the bulk of the canopy photosynthesis is restricted to the leaves of the upper parts of the crown (Ehleringer and Forseth 1989). Since leaf movements require a high ratio of direct to diffuse components of the solar radiation, few of the inner leaves can maintain diurnal movements; therefore, in dense or very large crowns, leaf movements are restricted to the external layer of leaves (Ehleringer and Forseth 1989). For the same reason, leaf movements are not expected to occur in habitats with a high incidence of overcast days or in understory habitats. However, it has been shown that leaves do move into a vertical orientation as a result of high temperatures, independent of the direction of light. This movement to a vertical position due to increasing temperatures could produce the appearance of the leaf movement leading or lagging sun position (Herbert 1996). These leaf movements, which translate into a lead of 38° in the morning and a lag of 44° in the afternoon in *Phaseolus vulgaris* (Shell et al. 1974), still require full understanding and experimental verification of the mechanisms involved and of the implications for total plant carbon gain (Herbert 1996).

B. Crown Architecture in Extreme Light Environments

Light can be a limiting resource in dark environments or when plants are subject to strong neighborhood competition, and light can be excessive and even harmful in open environments where plant metabolism is impaired by environmental stresses. Plant shape and size have been shown to change as a function of the light environment, and plants are capable of orienting their light-gathering surfaces in different ways to increase or decrease the leaf surface area projected in the direction of ambient light (Ellison and Niklas 1988). For a more physiological approach to plant adaptation and acclimation to sun and shade, see Chapter 7. In this section, two case studies of crown architecture adaptation to extreme light environments are presented. The first case explores the influence of growth form and leaf arrangement on light interception in plants growing in the dark understory of a tropical rainforest, while the second case examines the photoprotective role of plant architecture in Mediterranean-type ecosystems.

1. When Light Is Scarce

When plants grow in dense stands or in the understory, the resource of radiant energy becomes scarce, unreliable, and patchy, and evolution has led to two principal approaches to survival under these conditions: avoid or tolerate the shade. Angiosperms, in particular, have evolved an impressive capacity to avoid shade. The so-called shade avoidance syndrome involves accelerated extension growth, strengthened apical dominace, and retarded leaf and chloroplast development, among other processes (Smith and Whitelam 1997). Since the avoidance of shade involves a certain degree of structural plasticity, this issue will be addressed again in Section V. Here the focus is on the functional aspects of the crown of plants that tolerate shade and on the structural features that are relevant for such tolerance, with some preliminary conclusions from a study of plant architecture conducted in the understory of a lowland tropical rainforest on Barro Colorado Island, Panama (Valladares et al., in preparation).

Tropical rainforests exhibit an outstanding diversity of plant species and growth forms (see Chapters 11 and 12). Despite the extremely low levels of irradiance experienced in the understory of mature rainforests, a relatively large number of shrubs, herbs, and seedlings can be found within a few hectares. These plants suffer shading not only from the forest canopy and neighboring plants, but also from the leaves of their own crowns. The efficiency of light capture of

24 understory species differing in their habit and growth form was compared, and the influence of phyllotaxis and leaf size and shape in the avoidance of self-shading was explored in the field study in Barro Colorado Island. The species studied included understory palms, saplings of canopy trees, shrubs, and a wide variety of monocots of contrasting architectures. Plant size and total leaf surface area also varied significantly among the species considered. Most of the phyllo-taxes shown in Figure 1 were represented, and leaf size ranged from a few to several hundred square centimeters. Light harvesting efficiency was calculated with the three-dimensional plant architecture model YPLANT (Pearcy and Yang 1996). The most remarkable result of this study was the functional convergence of the different plant species co-occurring in the forest understory of Barro Colorado Island: most of the species intercepted between 80% and 90% of the available radiation, and mutual shading of the leaves during the brightest hours of the day was little, approximately 10% of the foliage area in most cases (Figure 10). Since the amount of available radiation varied among different sites, the total PPFD intercepted by each plant in a clear day exhibited more differences than the pa-rameters of efficiency, ranging from 0.26–0.75 mol m^{-2} day^{-1}. Thus, the rare spiromonostichous phyllotaxis of *Costus pulverulentus* (Figure 10), apparently a unique solution to avoid self-shading, was no better for this purpose than the spiral phyllotaxis of the saplings of *Thevetia ahouai* or the pseudodistichous fo-liage of the shrub *Hybanthus prunifolius*. Nevertheless, significant differences among species were found when the fraction of the plant biomass invested in support was considered in the analysis of the efficiency of the different architec-tures. Monocots, with a lower investment in dry weight, generally reached a more favorable compromise in this simple cost-benefit analysis of plant architecture under limiting light conditions. The drawbacks of the monocot strategy are a reduced survival to mechanical damage, and in some cases, a shorter plant lon-gevity and a limited capacity to reach the forest canopy.

In some cases, spiral and distichous phyllotaxis are linked in low-light envi-ronments: certain plants first set leaves spirally around an erect stem, and then produce horizontal branches bearing distichous leaves (Leigh 1998). This combi-nation of two phyllotaxes has been interpreted as a way of minimizing leaf over-lap. *Dichorisandra hexandra* exhibited this combination of spiral leaves around vertical stems and distichous leaves around horizontal branches (Figure 10), but leaf overlap was as reduced as in other species with different leaf arrangements. It seems more likely that this combination of two phyllotaxes is an efficient way of filling the space with leaves while growing in height.

Where there are many leaves in one spiral, long petioles in older leaves or narrow leaf bases in certain species can minimize leaf overlap (Leigh 1998). In the redwood forest understory plant *Adenocaulon bicolor*, which exhibits a spiral phyllotaxis with a mean divergence angle of 137° (phyllotactic fraction of 8/21), leaf overlap was reduced by particular combinations of leaf size and petiole length

Costus pulverulentus   *Dichorisandra hexandra*

0.07	Fraction of leaf area self-shaded during the central hours	0.06
0.84	Fraction of leaf area displayed during the central hours	0.85
0.49	Intercepted PPFD in a clear day of spring (mol m-2 day-1)	0.75
0.90	Intercepted PPFD in a clear day of spring (fraction of available)	0.93

Hybanthus prunifolius

Thevetia ahouai

0.12	Fraction of leaf area self-shaded during the central hours	0.07
0.77	Fraction of leaf area displayed during the central hours	0.81
0.30	Intercepted PPFD in a clear day of spring (mol m-2 day-1)	0.26
0.86	Intercepted PPFD in a clear day of spring (fraction of available)	0.87

at successive nodes (both increasing initially and then decreasing). The petiole length observed in this plant corresponded to the optimal petiole length obtained in simulations of the dependence of light absorption efficiency on petiole length (Pearcy and Yang 1998).

In the search for light, the crown of certain plants becomes thin instead of broad and flat in the shade. Light interception is not favored by such transformation, which usually represents an escape strategy of shade-intolerant species. In some cases, the whole developmental sequence of the plant is changed in the shade. Shrubs such as *Arbutus* switch from a sympodial growth in the open (Leeuwemberg architectural model; see Appendix 2) to a monopodial trunk (according to the model of Scarrone) in low-light environments (Bell 1993). Many plants accommodate their structure to the light environment, enhancing light interception efficiency under low-light conditions. This is the case of the chaparral shrub *Heteromeles arbutifolia*, which changes from orthotropic stems with spirally arranged leaves in the open to plagiotropic stems with pseudodistichous foliage when exposed to the moderate shade of a *Quercus* woodland. This structural change, in contrast to the escape strategy of more shade-intolerant species, significantly enhances light interception on a leaf area basis (Valladares and Pearcy 1998).

Considering the geometry of leaf self-shading, Horn (1971) predicted that the optimal distribution of leaf area in a monolayer crown is a regular spacing of large leaves, while the leaves in a multilayer crown can be randomly distributed as long as the leaves are small or lobed enough. Although leaves of many species tend to be larger and arranged in a monolayer or in plagiotropic shoots when the plant grows in the shade, in our study of the understory of the Panamanian forest, we found species with relatively small leaves in a multilayer (e.g., *Piper cordulatum*) and no clear correlation between leaf size and a monolayer versus multilayer crown.

Figure 10   Four plant species co-occurring in the understory of a tropical rainforest (Barro Colorado Island, Panama). *Costus pulverulentus* and *Dichorisandra hexandra* are monocot herbs, *Hybanthus prunifolius* is an understory shrub, and the individual of *Thevetia ahouai* (a canopy tree) presented is a 2-m-high sapling. Beneath each photograph, two computer images at dawn (left) and at noon (right) of a representative of each species are provided. A lighter gray in the computer images indicates overlap between two or more leaves as seen from the sunpath. For each species, the fraction of the total leaf area that is either self-shaded or displayed during the central hours of the day, and the photosynthetic photon flux density (PPFD) intercepted in a clear day of spring (both as daily total and as a fraction of available) were calculated using the three-dimensional YPLANT model (Pearcy and Yang 1996). (*Source*: Data from Valladares et al., in preparation.)

2. When Light is Excessive

Plants in open environments are exposed to high irradiance, which frequently leads to a decline in the efficiency of photosynthesis (photoinhibition), particularly under adverse conditions (Long et al. 1994). High irradiance can also lead to leaf overheating, especially in warm regions or in places where transpirational cooling is reduced due to water deficits, as in arid or Mediterranean-type environments. Under these circumstances, plants exhibit two general strategies: one physiological and the other structural or architectural. In the first strategy, the photosynthetic units are protected against light intensities in excess to those that can be utilized in photosynthesis by physiological and biochemical processes (Horton et al. 1996). The second strategy is based on the avoidance of excessive irradiance by structural features that reduce the leaf area directly exposed to the sun. Plants exposed to high light combine these two strategies, but with different emphasis on each depending on the species and the environmental conditions. However, it has been shown that structural avoidance of excessive irradiance can be crucial for survival under extreme conditions, even in plants capable of remarkable physiological adjustment to stress (Valladares and Pearcy 1997). Stress itself can also have a direct effect on crown architecture because it can modify the allocation pattern and the developmental processes of the plant. In fact, these deviations from the normal architectural program of a plant can be used to detect stress (see Appendix 5).

Shoot architecture of the Chaparral shrub *Heteromeles arbutifolia* was very different in the open and in the understory of a *Quercus* woodland. Sun shoots exhibited a remarkable structural photoprotection. Photosynthetic photon flux density intercepted on a leaf surface area basis was reduced by orthotropic stems with steeply inclined leaves arranged at short internodes (Figure 11). Despite having seven times more PPFD available, sun shoots intercepted only four times more and had potential daily carbon gains only double of those of shade shoots (Valladares and Pearcy 1998). The resulting fraction of leaf area that was displayed during the central hours of a typical day of spring was only one third of the total leaf area of the shoot. Leaf angle, the most plastic character in the response of *H. arbutifolia* shoots to high light, played a key role in achieving an efficient compromise between maximizing carbon gain while minimizing the time that the leaf surfaces were exposed to PPFDs in excess of those required for light saturation of photosynthesis, and therefore potentially photoinhibitory (Figure 6). For relatively simple canopies, leaf angle and orientation are the main structural photoprotective features (Werk and Ehleringer 1984, Smith and Ullberg 1989), but mutual shading among leaves can be even more important in complex, multilayered canopies (Roberts and Miller 1977, Caldwell et al. 1986). In *H. arbutifolia*, 27% of the foliage was self-shaded during the central hours of a clear spring day (Figure 11), but this percentage was far higher for leaves of certain

The following table appears within the figure:

	Fraction of leaf area self-shaded during the central hours	
0.27	0.25	0.20
	Fraction of leaf area displayed during the central hours	
0.34	0.16	0.37
	Intercepted PPFD in a clear day of spring (mol m-2 day-1)	
18.1	16.3	30.0
	Intercepted PPFD in a clear day of spring (fraction of available)	
0.33	0.30	0.55

Figure 11   Three plant species from open, dry environments. *Heteromeles arbutifolia* is an evergreen sclerophyll of the California chaparral, *Stipa tenacissima* is a tussock grass frequent in the driest regions of the Iberian Peninsula, and *Retama sphaerocarpa* is a leguminous, leafless shrub also frequent in dry and warm areas of the Iberian Peninsula. Beneath each photograph, two computer images of a representative of each species are provided. A lighter gray in the computer images indicates overlap between two or more leaves as seen from the sunpath. For each species, the fraction of the total leaf area that is either self-shaded or displayed during the central hours of the day, and the photosynthetic photon flux density (PPFD) intercepted in a clear day of spring (both as daily total and as a fraction of available) were calculated using the three-dimensional YPLANT model (Pearcy and Yang 1996). (*Source*: Data from Valladares and Pearcy 1998, Valladares and Pugnaire 1998.)

orientations, such as those facing south (Figure 9). A steeply oriented foliage and moderate self-shading that reduces the photosynthetic surface area displayed during the central hours of the day were also characteristic structural features of the crowns of two other plants from high-light environments: *Stipa tenacissima*, a tussock grass, and *Retama sphaerocarpa*, a leguminous, leafless shrub (Valladares and Pugnaire 1998). These two species exhibited similar leaf display and PPFD interception efficiencies to those of *H. arbutifolia* (Figure 11). However, the differences in growth form and general crown architecture among these three species translated into certain differences in their light interception patterns. The open crown of *R. sphaerocarpa* exhibited less self-shading and higher PPFD interception than the other two species. This species has a remarkable root system, that allows for an extended period of photosynthetic activity during the summer drought, while the other two species, with a less extended root system, exhibit a marked decline of photosynthetic carbon gain during the drought as a result of stomatal closure. Consequently, *R. sphaerocarpa* can make better use of the available PPFD during the critical dry season than co-occurring species such as *S. tenacissima*, and the structural photoprotection is less adaptive in this species since it comes at the cost of missing opportunities for carbon gain (Valladares and Pugnaire 1998). Actually, the costs in terms of missed opportunity for carbon gain (comparing plant crowns with equivalent horizontal photosynthetic surfaces) for these two species were similar to those imposed by the summer drought (approximately 50% of the potential carbon gain), the main limiting factor for plant survival in semiarid environments. This elevated cost of structural photoprotection emphasizes the ecological relevance of avoidance of high irradiance stress in these species.

Stress factors, such as those occurring in high-light environments in addition to high light itself (heat, water deficit), favor increased inclination angles of leaves (Shackel and Hall 1979, Ehleringer and Forseth 1980, Comstock and Mahall 1985, Lovelock and Clough 1992). It has been shown that leaves move into a vertical orientation as a result of high temperatures, independent of the direction of light (Herbert 1996). Since different stresses co-occur at certain times of the day or during certain seasons, a protective strategy that is triggered by one type of stress (heat, water deficit, or excessive light) but also increases protection or tolerance to other types of stress is very adaptive. This situation with regard to physiological features (down-regulation of photosynthesis, heat tolerance) was observed in plants from Mediterranean-type climates such as those presented in this section, and it was found to be very efficient when the plants were exposed to multiple stresses during the summer (Valladares and Pearcy 1997).

High leaf inclination angles favor increased plant photosynthesis in high-light environments (Herbert 1991, Herbert and Nilson 1991). Steep leaf angles facilitate the penetration of light to lower layers of foliage, especially diffuse light. Diffuse light in an open environment can represent only 5% of the total PPFD absorbed by the leaves, but it can be responsible for more than 20% of the total daily carbon gain of a whole shoot (Valladares and Pearcy 1998). Small leaves have a similar effect to steep leaf angles on crown photosynthesis due to penumbral effects (Figure 8), and for that reason Horn (1971) predicted small leaves in sunny climates. However, it must be taken into account that because both steep leaf angles and small leaf sizes have other ecophysiological implications such as those regarding boundary layer and heat dissipation, predictions and generalizations based only on light harvesting for photosynthesis might not hold under some circumstances.

The comparison in this section of *H. arbutifolia*, *S. tenacissima*, and *R. sphaerocarpa*, which represent three different growth forms of high-light environments, indicates that while canopy geometries of sympatric species can differ dramatically, the differences can be compensated by the spatial distribution of the canopy elements so that light interception, particularly during the critical season (summer in Mediterranean-type climates), can be similarly low. A similar conclusion was reached in a study of two structurally contrasting shrub species from the Mediterranean-type climate region of central Chile (Roberts and Miller 1977). Steep photosynthetic surfaces have proved an efficient and extended solution to ameliorate high light stresses. All of these are remarkable examples of functional convergence in high-light environments, with different distantly related taxa adopting a similar architecture from the point of view of light interception.

C. Occupying Space and Casting Shade: The Community Level

Whenever plants grow in close proximity there is competition for light and space. Branching poses an ecological dilemma to many plants, because a wide and low crown with a lot of branches is highly efficient in terms of light capture versus construction costs, but is easily overtopped, whereas a tall crown with little branches represents the opposite trade-off. Certain plants have reached an interesting compromise by building compound leaves whose long rachises act as "throw-away branches," extending the photosynthetic surface of the crown without investing in permanent and expensive support tissue (Givnish 1978). Probably one of the most extreme examples of this strategy is that of the Devil's walking stick (*Aralia spinosa*), which avoids branching by producing very long, light leaves, allowing for a very fast growth of the trunk. To support the same leaf area, a co-occurring tree, the flowering dogwood (*Cornus florida*), has to invest 7–15 times more in wood (White 1984).

Height growth is extended in the forest understory because light increases exponentially toward the canopy surface, while the costs of structural tissues escalate less rapidly with plant height (Givnish 1982; see Section II for further analyses of the economics of plant height). Competition between trees in its most general sense is competition to fill space, quickly in early succession, where r-selection seems to be dominant, but completely in late succession, where K-selection seems to dominate (Horn 1971). The dynamic nature of a forest canopy, with numerous gaps and clearings caused by treefalls, contributes to the coexistence of many more species of trees than could be expected in a theoretical analysis of crown architecture and monopolization of light; therefore, species of different successional status and crown shapes can coexist (Figure 5).

Forests are vertically stratified, from the towering emergent trees to the herbs on the forest floor, each strata comprising a distinct suite of plant species adapted for the conditions at each particular level, mainly light conditions. In the initial description of this idea of vertical strata in forests, Paul W. Richards did not provide an explicit mechanism that would account for stratification (see discussion in Terborgh 1992). One of the questions that remained unanswered in this description was why certain tree species cease their upward growth when they attain a given height, unlike canopy species that pursue an upward trajectory until they reach the open sky or die. Is there an optimum height for a midstory tree? Terborgh (1992) suggested an explanation, considering how direct sunlight passes through the holes of the forest canopy to the lower layers and eventually to the forest floor. As the sun progresses across the sky, sunlight penetrates into the forest over a wide range of angles. In a simplified and regular canopy, the sunlight passing into the forest interior through a single gap forms a triangular area on the way to the ground (Figure 12). At the upper parts of the triangle, the number of hours of sunlight is larger than at the lower parts. These triangular areas of direct sunlight spread out below the canopy so that areas from adjacent gaps overlap. Some points receive direct sunlight twice a day (intersection of two areas), while others lower down in the forest receive sunlight from an increasing number of gaps, although for increasingly briefer periods of time (briefer "sunflecks"; for more details on sunflecks see Chapter 7). This generates a spatially uniform light field near the ground (Figure 12), and it was predicted that a midstory tree must grow as high as the higher limit of this field because above this point, at least part of the tree crown might not receive enough light to pay its costs, and the whole construction and maintenance costs of the tree would be increased at the expenses of other functions such as reproduction (Terborgh 1992). This prediction was found to be true (midstory trees were of the expected height) in a mature temperate forest in North America, but not in the more complex and irregular forests of the tropics. In addition to the fact that the canopy of tropical forests is uneven and complex, Terborgh (1992) pointed to the shape of the crown of the canopy trees as another factor to explain the lack of a predictable, uniform light field. The shape of their crown determines the size of the triangular area of sunlight beneath a gap (Figure 12). Crown shape tends to vary with latitude, with mushroom-like trees in the tropics and conical crowns in boreal regions (see Figure 4 and Section IV), which allows for either generous shafts of direct sunlight or very little sunlight reaching the floor, respectively (Figure 12). Thus, although the forest has plenty of understory plants in the tropics, it is nearly

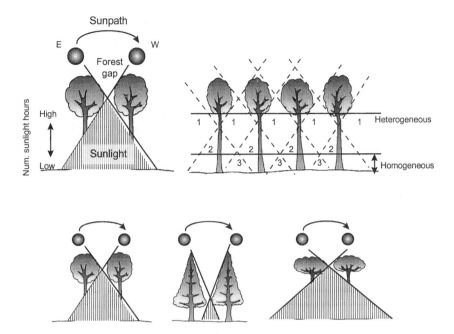

Figure 12 A gap in the forest canopy allows direct sunlight to reach the understory (upper left). The number of sunlight hours increases from the ground to the canopy. When the sunlight passing through more than one gap is considered, a more complex pattern is found (upper right), with understory areas affected by one, two, three, or more neighbor gaps (indicated by numbers). Where the cones of several gaps intersect, a spatially uniform light field is produced. Both the distance between the trees forming the limits of the gap, and the shape of the crown of these trees determine the duration of direct sunlight in the understory (lower graphs). Pyramidal crowns allow little sunlight to reach the understory, while the reverse is true for flat and broad crowns. (*Source*: Adapted from Terborgh 1992.)

devoid of them in the boreal regions; temperate forests represent an intermediate situation, with their rather simple canopy structure being very suitable for Terborgh's theoretical description of vertical light gradients and for the corresponding predictions of optimal height of understory trees. Another prediction regarding crown architecture resulting from the thesis that forests are vertically stratified is that crown shape varies systematically with vertical position. This was found to be true in tropical forests with more than two plant strata: while emergent trees possessed crowns that were more broad than deep, those of trees immediately below were more deep than broad (for the rationale, see Terborgh 1992).

In their search for light, understory plants are not only exposed to the vertical gradient of light, but to other physical factors that interact and influence their architecture. If height growth in a low-light environment has the risk of too-

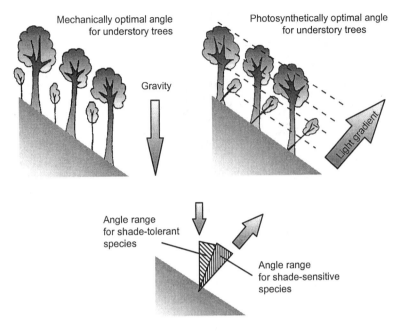

Figure 13   Understory trees that grow on slopes are exposed to the dilemma of growing vertically, which is mechanically optimal, or with their trunks inclined downward, that is, parallel to the light gradient occurring from the ground to the upper canopy, which shortens the distance of their foliage from the canopy surface. Depending on their light requirement or shade tolerance, species are expected to exhibit two ranges of trunk angle, as shown in the lower figure. (*Source*: Adapted from Ishii and Higashi 1997, Alexander 1997.)

expensive construction and maintenance costs, the situation becomes riskier or at least more complicated when the ground is not even, as is the case with hillsides. Since the lines of equal light intensity from the canopy to the ground run parallel to the ground (Horn 1971), the most efficient height growth occurs at right angles to the ground (Figure 13). However, to do this on a slope, trees should lean outward (Alexander 1997). Trunk inclination on slopes has been shown to be adaptive (Ishii and Higashi 1997), but the greater the angle of lean, the stronger the trunk of the tree needs to be for biomechanical reasons (Mattheck 1991, 1995), which entails additional costs. Under low-light conditions, leaning trees cannot grow a trunk as tall as it could if it were vertical, so their optimal angle on a slope is neither vertical nor perpendicular to the forest floor (Alexander 1997). Ishii and Higashi (1997) constructed a model to explore tree coexistence on a slope and to predict how tree survival is affected by trunk inclination. The predictions were that survival rate increases with slope angle more sharply under poorer light conditions. These predictions were supported by the understory tree *Rhododendron tashiroi*, which exhibited sharper trunk inclination and coexisted more successfully on steeper slopes with the dominant canopy trees (Ishii and Higashi 1997). Trunk inclination also seems to be affected by the shade tolerance of the species, with the relationship between slope and trunk inclination being more marked in shade-sensitive trees (Figure 13). This model provides an explanation based on optimizing processes of evolution by natural selection for the common observation that the trunks of trees on a slope often incline downward. This explanation is more complete and convincing than previous ones alluding to landslides or wind (King 1981, Del Tredici 1991, Mattheck 1991). Another way to enhance light capture on slopes is with an asymmetrical crown, with more branches on the downhill side (Halle et al. 1978, Alexander 1997). Although this architectural solution does not require trunk inclination, it must imply additional construction costs if the trees are not to snap or fall over.

Another interesting study case regarding understory plants and slopes is that of leaning herbs that arrange their leaves in a distichous array along an arching stem (e.g., *Disporum, Polygonatum, Smilacina, Streptopus, Uvularia* in the Liliaceae, and *Renealmia* in the Zingiberaceae). Such crown architecture is mechanically less efficient than an umbrella-like arrangement that supports leaves at the same height on a vertical stem. Their competitive advantage derives from their tendency to orient strongly downslope (Givnish 1982, 1986). Above a critical slope inclination, leaning shoots become mechanically more efficient than other herb architectures and tend to supplant them. The correspondence between their observed and predicted distribution relative to slope inclination provides support for their competitive ability on slopes (Givnish 1986).

Hedgerows, linear arrangements of woody species that follow property boundaries between fields, are an interesting case of plant communities with strong competition for light and space. The interactions between carbon relationships, growth, and plant architecture have been thoroughly studied in these systems (Küppers 1984, 1985a, 1985b, 1985c, 1989, 1994, Schulze et al. 1986a). An important conclusion from these studies was that plant architecture in combination with carbon input can lead to a better understanding of plant success since net primary production by itself is not sufficient to explain competitive relationships between woody species (Schulze et al. 1986). Branching and leaf exposure were principal mechanisms in competition: short internodes and thorns were important in early successional species, whereas shading of neighbors with a minimum of self-shading (capacity for occupying new, higher aerial space coupled with maintenance of a closed leaf cover above the occupied space) provided a competitive advantage to late successional species (Schulze et al. 1986). Particular branching patterns and lower costs of space occupation permitted late successional species to grow a crown quickly and then outcompete shade-intolerant pioneers (Küppers 1989, 1994).

Using a stochastic model of plant growth, Ford (1987) reached some interesting conclusions on the implications of crown architecture for plant competition and spatial interference. Growth rates depended on branch probability and also on angle of branching in sympodial plants but not in monopodial plants. In sympodial plants, an optimal branching angle of approximately 30° was found when the ratio of the interference distance to the internode distance was 1.5:5. Sympodial plants with a branching angle of 30° outcompeted monopodial plants (Ford 1987). Promising research into the role of plant architecture for competitive interactions among plants will follow the development and application of models like MADEIRA (Küppers and List 1997), which combines morphological information with carbon gain and allocation, MAESTRO (Wang and Jarvis 1990a, 1990b, Wang et al. 1991), LIGNUM (Perttunen et al. 1996), and others (de Reffye 1981, de Reffye et al. 1995, Kellomäki and Strandman 1995; see also Appendix 4).

Plant cover can change very rapidly. The light environment of a plant can vary due to the activity of nearby vegetation. Sensing their neighbors, perceiving the light opportunities, and responding in a timely fashion is crucial for plant survival at the community level (Ballaré 1994). An adaptive way of coping with competition is by a plastic response to the environmental changes caused by neighboring plants, an issue addressed in the next section.

V. PLASTICITY IN RESPONSE TO ENVIRONMENTAL HETEROGENEITY

Plants exhibit a remarkable within-species and within-individual variability in their structural features. While in some cases it can be due to a malfunction of the genetic program, in most cases this variability is a plastic response to local conditions. For instance, branching pattern of trees is not stationary, and it has been shown that the variation of branching pattern can be the result of developmental–phenotypic interactions (Steingraeber 1982). Structural plasticity of plants enables a fine tuning with environmental changes so that the efficiency of the limiting processes at each stage is maximized. A common environmental change experienced by plants is the decreasing availability of light with the advance of succession. It has been shown for the succulent halophyte *Salicornia europaea* that morphological changes in the branching patterns during succession maximized light interception (Ellison and Niklas 1988). However, even taxonomically closely related plants differ in their capacity for a plastic response to the light environment (Valladares et al. 1998). For instance, certain species that exhibit an architecture suited to high irradiance conditions do not change significantly when grown in the shade. That was the case for mangroves in Malaysia: architecture and allometry of shaded mangroves were consistently more similar to those of unshaded mangroves than to shaded, broad-leaved, evergreen, rainforest trees (Turner et al. 1995). Light is possibly the most spatially and temporally heterogeneous environmental factor affecting plant survival and growth. Although spatial and temporal heterogeneities are frequently considered independently, they are highly correlated (i.e., small spatial scales correspond with rapid temporal scales. Causes, scales, and plant responses to both spatial and temporal heterogeneities are covered in Chapter 7).

Shade avoidance, a feature that angiosperms have developed to a remarkable extent, is based on a plastic response to light, which in turn responds to signals that anticipate that shade is going to change (via changes in the red–far red ratio). The so-called shade avoidance syndrome involves a strong elongation of internodes and petioles in the shade, the production of less dry matter, larger and thinner leaves, a higher shoot–root ratio, and a series of remarkable physiological changes (Smith and Whitelam 1997). The response of plants to shade is multiple and is mediated by multiple phytochromes (Smith and Whitelam 1997). The variety of morphogenic programs triggered to move the photosynthetic area toward better-lit canopy regions, and all of the morphological and physiological adjustments to the light environment, is called "foraging for light" (Ballaré 1994, Ballaré et al. 1997). However, until recently the effect of shade avoidance responses on performance and fitness (Appendix 6) had not been investigated explicitly (Schmitt 1997).

Phenotypic plasticity in plants has three main functional roles: maintenance of homeostasis, foraging for resources, and defense. Although plant plasticity has been a commonplace observation, its ecological and evolutionary consequences are only beginning to be thoroughly explored. This is due to the fact that most ecological and ecophysiological studies of plant form to date have focused on adaptive, habitat-based specialization, and plasticity has been interpreted as a feature of generalists. The supposed superiority of specialized ecotypes or taxa over generalists has led biologists to focus on evolutionary specialization and neglect the ecological and evolutionary implications of plastic responses of the phenotype to the environment (Schlichting 1986, Sultan 1992). However, phenotypic plasticity is essential for survival in heterogeneous and variable environments, especially in sessile photosynthetic organisms (Bradshaw 1965, Sultan 1992, Pintado et al. 1997). Although the adaptive implications of plasticity for relative allocation of biomass or other fitness-related parameters have been widely recognized, plasticity has been traditionally viewed as an alternative to specialization. The evidence that plastic, and thus generalist, species are less able to compete with specialized species is weak at best (Niklas 1997b). A new approach to this question has postulated that plasticity in some plant traits may, in fact, represent a product of specialization (Lortie and Aarssen 1996). The new predictions for plasticity of specialized genotypes were proposed to depend on whether specialization is associated with the more or the less favorable end of an environmental gradient (Lortie and Aarssen 1996). Specialization to the more favorable end was proposed to increase plasticity, whereas specialization to the less favorable end was proposed to decrease plasticity. Certain studies have found greater photosynthetic plasticity in gap-dependent compared with shade-tolerant species (Bazzaz and Carlson 1982, Strauss-Debenedetti and Bazzaz 1991, Bazzaz 1996, Strauss-Debenedetti and Bazzaz 1996). In a study of 16 species of the genus *Psychotria* occurring on Barro Colorado Island, Panama, we found that the mean phenotypic plasticity was significantly higher for the high-light species than for the low-light species (Valladares et al. 1998). Selection for greater plasticity may be stronger in the high-light species because forest gaps (high-light environments) exhibit a relatively predictable decrease in PPFD for which this plasticity could be adaptive. In contrast, the low-light species experience relatively unpredictable changes in light caused by infrequent gap formation. Under these conditions, phenotypic stability may have higher adaptive value. The results are consistent with the view that plasticity, rather than being an alternative to specialization, is indeed a specialization to the high-light end of the light gradient in tropical forests. On the other hand, the relative stability exhibited by the low-light species is consistent with a stress-tolerant syndrome, with its low potential maximum growth rates, low maximum photosynthetic rates, and low leaf turnover (Valladares et al. 1998). In this study, we also found that plasticity was generally greater for the physiological than for the structural characteristics.

The definition of phenotypic plasticity is not always obvious, and some controversy exists regarding whether variations in certain physiological or structural features can be considered expressions of phenotypic plasticity (Schlichting and Pigliucci 1998). However, it seems clear that changes in stomatal conductance or enzymatic activity (regulatory responses) do not fall within the current concept of phenotypic plasticity, and neither do the phenotypic changes due to stress injury (in contrast to changes that ameliorate the effects of the stress). Phenotypic plasticity is restricted to phenotypic changes that enhance plant performance (see Chapter 7).

Because plasticity increases the variance of the mean value of a given plant feature, it complicates experiments, predictions, and generalizations. But since plasticity itself conveys a message regarding plant adaptation, it must be considered a subject of study. Plant ecologists should not be embarrassed by a large standard deviation in their data, but must try to interpret it instead. Phenotypic plasticity is not always easy to study because the genetic variability comprised in the samples is unknown; therefore, the differences among the individuals compared cannot be undoubtedly ascribed to a plastic response of the phenotype to the environment. This contingency seems to be very difficult to avoid, for instance, in the case of lichens, due to the problematic handling of these organisms in laboratory studies and their slow growth, which forces those interested in exploring phenotypic plasticity of lichens to compare populations or individuals growing under different conditions in the field and accept some potential noise in the observed differences introduced by an unknown genetic variability (Pintado et al. 1997, Sojo et al. 1997). In the analysis of plant plasticity, choice of which subunits to count is essential, since within a plant there is clearly a "hierarchy of plasticities," and not all structures exhibit the same degree of plastic response (White 1979). For instance, the range of variation of a phenotype (the norm of reaction) for vegetative structures tends to be broader than that for reproductive traits (Niklas 1997b). The adaptive significance of the plasticity of a character becomes clear only when it is scaled up to the performance of the whole plant. Usually, if a plastic response in a feature improves performance, the result at the next level up is an enhanced homeostasis (see Chapter 7).

A remarkable number of studies on plant phenotypic plasticity have been conducted with clonal plants. These plants are excellent study material because the same genotype produces new phenotypes again and again, each time under a potentially different environment, so that a large plant is a living norm of reaction if the environments under which each ramet was produced is known. There are four types of morphological plasticity exhibited by these plants (in a broad sense, all modular organisms can exhibit these four types): (1) switching between types of units; (2) changing in size or shape of the units; (3) creating more units; and (4) moving the relative positions of the units (Alpert and Stuefer 1997). However, clonal plants differ greatly in architectural plasticity, since, for instance, some vary the spacing between ramets according to resource levels, whereas other species do not (de Kroon and van Groenendael 1997). A plastic feature that makes clonal plants a fascinating ecological case is the division of labor among ramets or units, with resource sharing across ramets (Stuefer et al. 1996, Alpert and Stuefer 1997). For more information on plastic response and division of labor in clonal plants, see Chapter 7.

Since phenotypic plasticity is adaptive due to the heterogeneous nature of most environments, the question of why all plants are not equally (and maximally) plastic is a very pertinent one. It is rare to find scenarios in which a plastic response to the environment could be maladaptive. Some examples have been pointed out for plants growing under extreme physiological conditions and usually in the absence of strong competition (Chapin 1991, Chapin et al. 1993). Such plants tend to show a conservative pattern involving slow, steady growth, even when conditions are temporarily favorable (Waller 1991). Under ideal growing conditions, these plants are more likely to store nutrients than to accelerate their growth (Chapin 1980, Chapin et al. 1986) to avoid the production of a plant that is too large or structures that are too expensive to be sustained once conditions deteriorate. Specialization to a low-resource environment seems to start with a modification in a key growth-related character, which results in a cascade of effects that triggers the entire "stress resistance syndrome" (low rates of growth, photosynthesis, and nutrient absorption, high root:shoot ratios, low rates of tissue turnover, and high concentrations of secondary metabolites; Chapin et al. 1993).

Plasticity is more likely included in traits that can experience reversible changes, a possible reason for the common observation that physiological features are considered more plastic than structural traits (Peter Alpert, personal communication, 1998). Sometimes, failure to respond to environmental conditions or cues may reflect, not merely the constraints of unsophisticated physiology, but selection for conservatism. For instance the roots of *Quercus* are genetically programmed to ignore the signals of the environment (water) and to grow in depth, a safe strategy in arid places but useless in humid ones. In considering when is it good to be plastic, Alpert (1998) distinguished six cases: (1) in heterogeneous habitats; (2) in predictable habitats; (3) when resources are high; (4) when plants can respond faster than the environmental changes (anticipation and speed); (5) when resources come in pulses; and (6) when decisions can be made late in development. To understand the ecological and evolutionary implications of phenotypic plasticity, it is necessary to consider not only the benefits but also the costs (direct costs: construction and maintenance; and indirect costs: opportunity), which are usually very difficult to assess. Theory predicts that the evolution of phenotypic plasticity may be constrained by strong genetic correlations between traits across or within environments (Schmitt 1997).

VI. EVOLUTION AND PLANT FORM

The key mechanism leading to evolutionary change is natural selection of heritable variations (see Darwinism and neo-Darwinism vs. Lamarckism in Appendix 6). However, very few unequivocal examples of natural selection have been observed in nature (Ingrouille 1995). It is very difficult to understand adaptation because the unit of selection is not the character or suite of characters under study but rather the whole individual, that is, the sum of all adaptations. In addition, the assessment of the adaptive value of morphological variation is even more complicated than that of physiological features (Ingrouille 1995). Adaptation has been assessed by measuring fitness (Appendix 6). But although Darwinian fitness is conceptually straightforward, it is extraordinarily difficult to measure directly (Niklas 1997b). The fitness of an individual can be measured only in relative terms, and usually through indirect parameters such as longevity, size, or growth, which are frequently correlated with the Darwinian fitness. However, it is not clear which fraction of the variation observed among individuals is nonadaptive (with no effect on survival or reproduction). Convergent evolution provides evidence for the adaptive nature of some characters. The convergence of the crown shape of unrelated tree species that overlap in latitude or habitat (see Section IV. A) and the functional convergence of the crown of different understory plants co-occurring in a tropical forest (see Section IV.B) are examples of structural traits that may not be fully understood, but are surely adaptive. Therefore, ecologists must rely on comparative methods to explain adaptation (Harvey and Purvis 1991) and to find evidence for functionality (see Chapter 1). A fundamental criticism of adaptationist hypotheses is that they do not allow the possibility that many variations are not adaptive in a Darwinian sense, so they are essentially teleological, seeking to find the purpose of every biological structure. It must be taken into consideration that not all differences between species are necessarily adaptive. Because the environment is constantly changing, in the long run, evolutionarily successful species are those that manage to preserve as many variations (potential adaptations) within the species as possible, whether they are presently adaptive or not (Ingrouille 1995).

Tempos and modes or patterns of plant evolution are extremely diverse, and there are three general approaches to determining them (Niklas 1997). The first approach is by molecular comparisons among living species, which gives the rate and direction of genotypic change. The second approach relies on morphological or anatomical comparisons among species, and the rate and direction of phenotypic evolution is inferred from the number of differences that have occurred since species diverged from one another. The third approach is to catalog and study the birth and death of species, that is, the historical fates of phenotypes in their entirety (paleontology). The lack of a direct relationship between genotypic and phenotypic change (i.e., significant reorganizations of a genome may have little effect on the appearance of the plant, and dramatic phenotypic changes may result from small genic alterations) makes the challenge of understanding evolutionary tempos and patterns difficult. Since natural selection acts on the phenotype, not on individual genes or individual structural features, the study of the birth and death of species is the most powerful approach to study long-term evolutionary rates and patterns. However, the fossil record is not always complete. As concluded by Niklas (1997), the debate over rates and modes of speciation or extinction likely will continue for years to come. Readers interested in plant evolution simulated through computer models, assuming that fitness is based on a reduced number of biological tasks (light interception being one of the most conspicuous), are encouraged to follow the adaptive walks through fitness landscapes explored by Niklas (Niklas 1997a, 1997b).

The shape of the crown is an adaptive compromise of conflicting strategies. Multiple functions of an architectural design and functional convergence of alternative architectural features make assessment of optimal design difficult. For instance, phyllotaxis is not strictly a developmental constraint because different phyllotaxes can be functionally equivalent (e.g., in terms of light interception efficiency [Niklas 1988]; also see the case of understory plants discussed in Section IV.B). The same argument can be built for branching patterns or for the general shape of the crown (Figure 3), and these structural features may provide a paradigm for other features in plant evolution. This is the most likely reason why it has proved impossible to describe the many tree architectural models as adaptations (see Section III. B). However, there are particular cases in which the several functions of a given structure do not appear to be a constraint in interpreting its functional optimization, such as the analysis of petiole length versus light capture in the understory herb *Adenocaulon bicolor* (Pearcy and Yang 1998). Certain genetic or developmental constraints can be overcome from a functional point of view by changes in other, more plastic structural features, such as petiole and leaf shape, size, and orientation. Consequently, evolution of architectural features cannot be interpreted without a minimum knowledge of the plastic response to the environment of the involved traits.

Many analyses of adaptations in plant form have assumed that natural selection favors traits that tend to maximize the growth rate of a given plant (Givnish 1986). One important objection to this approach is the lack of consideration of the role of competitors. As Givnish (1986) has shown, certain features of plant architecture, such as leaf height, are examples of a trait in which the strategy that maximizes growth in the absence of competitors is the one that does most poorly in their presence. In addition, certain structural features are not very efficient for their primary function, but provide a competitive advantage. For instance, the generous amount of leaves within certain crowns, well above what could be strictly needed, appears to be competitively more effective by intercepting light that competitors might use than in providing photochemical energy (Margalef 1997). Thus, the evolution of plant form must be interpreted considering not only the immediate function of the organs and structures involved (e.g., light interception), but also its effect on the efficiency with which competitors exploit the resources.

VII. CONCLUSION

This chapter has summarized the main structural and functional rules that determine or at least influence the architecture of plant crowns. Special emphasis was placed on adaptive responses to the light environment. In addition to understanding all mechanisms involved in each particular function of the crown, the real challenge for ecologists is now to integrate the available knowledge, to test the resulting functional hypotheses regarding plant architecture, and to make predictions at both the ecological and the evolutionary levels.

APPENDIX 1: TERMS AND CONCEPTS IN PLANT
ARCHITECTURE*

Acrotony: branching pattern in which apical or distal branches are more developed than basal or proximal ones. Opposed to *basitony*.
Architecture: a static concept of plant shape alluding to the visible morphological

---

* Data from Hallé et al. 1978, White 1979, Küppers 1989, Porter 1989, Bell 1993, Room et al. 1994.

expression of the genotype of a given plant at any one time. It usually implies a more idealized view of the arrangement of the plant crown than the term *plant form*, because it excludes unusual alterations of the normal shape of the plant.

Architectural model: a dynamic concept of plant shape expressing the normal growth program that determines the successive architectural phases of a plant derived from a seed, undisturbed by extrinsic forces (such as pruning, herbivory). The concept excludes *reiterations*.

Architectural unit: the complete set of *axis* types and their relative arrangements found in a plant species. It cannot be seen until an individual is old enough to have expressed its *architectural model* in full.

Axis: a sequence of units of growth in the same general direction from one (*monopodial*) or more (*sympodial*) meristems.

Basitony: branching pattern in which basal or proximal branches are more developed than apical or distal ones. Opposed to *acrotony*.

Branch: an *axis* other than the main stem plus any subordinate axes it bears.

Caulomer: a segment of a module showing morphologically distinct growth increments; a unit of extension. Also used as synonym of *module*.

Clone: a collection of plants derived by vegetative propagation (natural or artificial) from a single *genet*. Although they may remain attached to the ''parental'' *genet*, they are physiologically partially or completely independent.

Epinasty: the effect of one branch on another, which determines their final orientation.

Epitony: branching pattern in which second-order branches grow at the upper side of a lateral branch. Opposed to *hypotony*.

Genet: the plant, of any size and subdivided or propaged in any way, derived from a single seed. Equivalent to *ortet*.

Growth form: genetically determined general growth pattern of a plant, expressed by terms such as tree, shrub, vine, etc.

Hapaxanthic: an axis that is determined by terminal flowering. Opposed to *pleonanthic*.

Hypotony: branching pattern in which second-order branches grow at the lower side of a lateral branch. Opposed to *epitony*.

Life-form: similar to *growth form*, but usually implying species' reproductive and life history strategy.

Metamer: a generic term for a unitary part of a shoot. One of the several segments that shares in the construction of a shoot or into which a shoot may be conceptually resolved: an internode, the axillary bud(s) at its proximal end, and the leaf or leaves at its distal end, but not any shoots resulting from growth of axillary buds. For certain investigators, *caulomers, phytomers*, or *modules* are particular cases of metamers. However, a *module* may contain many metamers if the broad definition of *module* is used.

Module: a monopodial shoot terminated by an inflorescence, spine, or tendril, or by parenchymatization of the apical meristem. A broader definition is the product of one meristem. In shoots, it is a set of *metamers* originating from one axillary/apical bud. It is also the smallest unit of morphology capable of producing daughter units and/or seeds, and can be referred to as a sympodial unit, article, or *caulomer*.

Monolayer: a crown with leaves densely packed in a single layer or in a reduced, discrete number of horizontal layers. Opposed to *multilayer*.

Monopodial: a way of plant growth in which the axis of a *shoot* is built up by the vegetative extension of one apical meristem; in other words, growth by the continued activity of a single meristem. Opposed to *sympodial*.

Multilayer: a crown with leaves loosely scattered among many horizontal layers. Opposed to *monolayer*.

Orthotropic: a vertical axis resulting from a particular gravitational response of shoots. Opposed to *plagiotropic*.

Plagiotropic: an oblique or horizontal axis.

Plant form: particular shape resulting from the expression of a specific genotype in a certain environment, that is, one specific phenotype expression out of a set available to a taxon-specific *growth form*.

Pleonanthic: an axis that is not determined by flowering (flowers or inflorescences are laterally positioned). Opposed to *hapaxanthic*.

Phytomer: a particular kind of *metamer*, defined as a node and its attached single leaf, together with the internode just below it. Equivalent to *phyton*.

Ramet: a useful but not morphologically well-defined reference to a single module or shoot complex of a *genet* that is conveniently enumerated. It may be attached to the *genet* or become detached and independent. It is the unit of clonal growth, capable of an independent existence if severed from the parent plant.

Reiteration: recapitulation of rejuvenescence of the whole or part of the architectural model within the same *genet*, as a result of improved environmental conditions or damage to the initial model. It is usually initiated from dormant meristems.

Shoot: colloquial reference to an aerial axis with associated appendages (mostly leaves) and lateral meristems.

Shoot complex: a sequence of modules linked sympodially, forming a coherent structure; applied to plants with distinct erect or lateral shoot complexes.

Sympodial: a way of plant growth in which the axis of a *shoot* is built up by a linear series of shoot units, each new distal shoot unit developing from an axillary bud on the previous shoot unit; in other words, growth from successive lateral meristems. Opposed to *monopodial*.

APPENDIX 2: KEY FOR THE CLASSICAL TREE ARCHITECTURAL MODELS

Architectural models of Hallé and Oldeman (1970, Hallé et al. 1978) are given. Each model is named after a botanist. Closed circles indicate determinate growth (terminal inflorescence), and open triangles indicate indeterminate growth. (Adopted from Bell 1993 and Leigh 1998.)

I. Unbranched

    A. Flowers once at apex and then dies (monocarpic); *Holttum model #1*: ↑ → |

    B. Flowers repeatedly, laterally, under leaves; *Corner model #2*: ↑ → ⥉

II. Branched

    A. Branches underground

        1. Only underground branching; *Tomlinson model #3*: ↲ → ↲↺

        2. Both underground and above ground branching (horizontal branches with distichous leaves); *McClure model #4*: ↲ → ↲↺

    B. Branches only aboveground

        1. Modular trunk

            a. Trunk built of leaning segments, each segment sprouting where its predecessor bends

                i. Tips of the leafy twigs ascending, with spiral or decussate leaves; *Champagnat model #5*: ↑ → ⤳

                ii. Tips of the leafy twigs horizontal, with distichous leaves

                    a. Branch leans smoothly and succesor sprouts from the bend; *Troll model #6*: ↶ → ⤳

                    b. Horizontal branch formed sharply from erect trunk segment; *Mangenot model #7*: ↑ → ⤳

            b. Trunk not built of leaning segments

                i. Trunk composed of a succession of segments (sympodial), each segment originating just below tip of predecessor *a.* Each module produces a single, unbranched segment; flowers at top; *Chamberlain model #8:*

    b. Each module ends in a whorl of branches

        i. Branches upturned with leaves spiralled or decussate around their twig; *Prévost model #9:*

        ii. Branches horizontal with distichous leaves; *Nozeran model #10:*

  ii. Trunk forks

    a. Each module flowers at its tip and forks

        i. Modules fork symmetrically (no dominant branch); *Leeuwenberg model #11:*

        ii. Modules fork asymmetrically (one branch is larger than others); *Kwan Koriba model #12:*

    b. Trunk forks without flowering; *Schoute model #13:*

2. Nonmodular trunk

  a. Monopodial trunk with segment branching

    i. Each lateral branch flowers and forks; orthotropic branches

        a. Primary branches bunched along the trunk; *Scarrone model #14:*

        b. Primary branches distributed continuously along the trunk; *Stone model #15:*

    ii. Lateral branches with horizontal sprays of leaf rosettes; plagiotropic branches

        a. Flowers appear below or among leaves; primary branches bunched along the stem; *Aubreville model #16:*

       b. Flowers at the end of each branch

          i. Primary branches bunched or in layers along the trunk; *Fagerlind model #17:*

          ii. Primary branches distributed continuously along the trunk; *Petit model #18:*

    b. Straight monopodial trunk

      i. Leaves spiral or decussate around erect stems; orthotropic branches

         a. Branches in bunches or in whorls along branches of higher order; *Rauh model #19:*

         b. Branches distributed continuously along branches of higher order (in spiral or decussate arrangement); *Attims model #20:*

      ii. Leaves distichous; plagiotropic branches

         a. Branches in bunches or whorls along the trunk (forming layers); *Massart model #21:*

         b. Branches distributed continuosly along the trunk

            i. Primary branches shed as units (as they were palm fronds); *Cook model #22:*

            ii. Primary branches not shed as units; *Roux model #23:*

APPENDIX 3: FRACTALS AND PLANT FORM

Plant form can be very complex due to the combination of regular and irregular pattern formation processes. While Euclidean geometry is very useful for studying linear, continuous, or regular structural properties of the objects, fractal geometry is a powerful tool to analyze nonlinear, discontinuous, or irregular structural properties, which are characteristic of plants (Hasting and Sugihara 1993, Lorimer et al. 1994). For this reason, fractal geometry has many applications in ecology (Frontier 1987, Sugihara and May 1990). Fractal objects have four interrelated properties:

Self-similarity. The shape or geometry of the object does not change with the magnification or scale (Gefen 1987). The reiteration of a branching pattern in trees is a good example of this property, which was qualitatively described and used in the classification of architectural models of trees before fractals became popular (Halle et al. 1978).

Fractal dimension. The fractal dimension describes the degree with which an object fills the Euclidean space in which it is embedded. The dimension of the object is not an integer, but a fractional number. For instance, a coastline rich in twists and turns is not a simple one-dimensional line or a two-dimensional plane. Its dimension is intermediate between one and two (Mandelbrot 1967). And the same is true for the contour of leaves (Vicek and Cheung 1986). Trees extend in three dimensions, but their branches are never compact in the occupation of space, and the proposed fractal dimensions fall between 1.1 and 1.9 (Margalef 1997).

Scale independence. For certain domains or regions of scale, the pattern for the phenomenon of interest does not change. The fractal dimension indexes the scale dependency of the pattern, and a change in the fractal dimension signals a change in the process (Lorimer et al. 1994).

Complexity. Spatial data are usually irregular and complex. The fractal dimension is a measure of complexity, but the point on the continuum between complex and simple systems cannot be marked (Phillips 1985). Randomness (noise, the uncertainty in complex systems) also contributes to the complexity of fractals (Mandelbrot 1983, Lorimer et al. 1994).

Plants, and plant architecture in particular, have many fractal properties (Mandelbrot 1983, Prusinkiewicz and Lindenmayer 1996). A tree can be modeled as a fractal, and many functional aspects, such as efficiency of occupation of space by the leaves, total wood volume, stem surface area, and number of branch tips, can be calculated with more accuracy by using fractals rather than Euclidean geometry (Lorimer et al. 1994). Foliage damage by pathogens or herbivores is also fractal (Loehle 1983). However, many natural objects, such as forests, tree branches, plant crowns, or compound leaves, are most likely *multifractals* because they are not strictly self-similar at every scale, that is, not exactly the same at all magnifications (Stewart 1988a, Stewart 1988b). This concept is clearly homologous to the *partial reiteration* concept that Hallé et al. (1978) and Hallé (1995) used in their classification of the architecture of trees.

APPENDIX 4: MODELING PLANTS AND GROWTH

Because symmetries and elegant geometric features of plants have always attracted mathematicians, models of plant shape and growth have received considerable attention, especially when the advantages of reasonably accurate simulations and virtual experimentation became evident. Models can be classified in two main groups: morphological and process-based models (Perttunen et al. 1996). However, the ideal model is a morphological model that deals with physiological processes or a process-based model that incorporates morphological information (Kurth 1994). Models vary greatly in scope and resolution (Room et al. 1994, Remphrey and Prusinkiewicz 1997), but very simple models can mimic response of real plants because the complex integrated growth patterns seem to be emergent properties of a simple system (Cheeseman 1993). Metamer dynamics have been simulated using the tools of population dynamics, which have rather simple mathematical formulation; however, this approach ignores structure and allows little scope for geometric analyses (Room et al 1994). In geometric models, the spatial position and orientation of each structural component is considered, which allows the accurate simulation of interception of light by leaves (Pearcy and Yang 1996), of bending of branches due to gravity, and of collision between branches (Room et al. 1994). Geometric models also provide the information necessary to produce realistic images of plants (see Figures 10 and 11; Prusinkiewicz and Lindenmayer 1996), which has additional applications in education, entertainment, and art. For more examples of models and their applications, see reports by Kellomäki and Strandman (1995), Perttunen et al. (1996), and Küppers and List (1997). From the many models available to simulate plant growth, there are two systems that have the widest potential application for plant ecologists and physiologists: L-systems (initiated by Lindenmayer 1968) and AMAP (Atelier pour la Modélisation de l'Architecture des Plantes) originated by de Reffye (1981). AMAP uses stochastic mechanisms, and L-systems, although initially deterministic, can incorporate stochastic mechanisms as well. Despite the fact that AMAP has remarkable utility in agronomy by giving a central role to the structure of plants (Godin et al. 1997), L-systems are inherently more versatile and hold greater promise (Room et al. 1994). L-systems make good use of the relative simplicity of plant developmental rules and of the regularly repeated appearances of certain structures in a phenomenon called self-similarity (a fractal feature; see Appendix 3). Actually, the fact that plants can be considered as the result from iterations of a basic structural module that is continually re-expressed is used in a wide variety of models of plant growth, such as those aimed at exploring the influence of resource acquisition and utilization on the whole plant morphology and function (Colasanti and Hunt 1997). In L-systems, different plant components are represented by different letters. A sequence of letters, with branches delimited by brackets, represents an entire plant. The order of letters indicates the topological connections in the structure. Development is simulated by rewriting, with every letter in the sequence being replaced according to a specific rule. Rewriting is repeated until a desired stage of development is reached (Room et al. 1994, Prusinkiewicz and Lindenmayer 1996, Prusinkiewicz et al. 1997). Although an ideal growth model will take both internal and external factors into account, models to date focus on either one or the other. Room et al. (1994) summarized all of the internal and external parameters that affect metamer dynamics and that should be considered in modeling plant growth. The ability to simulate metamer dynamics and to handle three-dimensional information on the development and growth of individual plants will provide new insights into relationships between morphology, environment, and productivity (Room et al. 1996). New optimal ideotypes could be suggested to plant breeders. Virtual plants will have applications not only within developmental biology or physiological ecology, but also within agronomy, horticulture, and forestry, providing a new way of exploring plant competition for resources and space, and the action of pests, pathogens, weeds, and herbivores on plant structure and growth (Room 1996). The advance of plant growth modeling is challenged by the difficulties of making virtual plants responsive to the environment and to neighboring plants in real time, and devising efficient methods of measuring plant structure, which is crucial information for the models that is usually hard to obtain (Room et al. 1994).

APPENDIX 5: STRESS AND CROWN ARCHITECTURE

Stress dissipates energy that is otherwise available for homeostatic processes such as those controlling development and pattern formation (Parsons 1993, Escós et al. 1995). It has been shown that growth pattern and shape-related features are more sensitive to this energy dissipation caused by stress than just size (Alados et al. 1994). Stress introduces variation in structure, which is not adaptive, in contrast to plastic phenotypic responses to environmental changes. This kind of increased variability is known as developmental instability (Waddington et al. 1957). Studies on developmental stability in plants include those of Paxman (1956) and Freeman et al. (1993). In the case of bilaterally symmetric structures, random variations during development under stress aggravate right–left asymmetry. The study of the so-called fluctuating asymmetry has been used to detect disruption of homeostasis in this type of structure (Alados et al. 1994, Alados et al. 1995). Stress also causes deviations in radial symmetry and in symmetry of scale, that is, in self-similarity at different spatial scales (Alados et al. 1994). There are other forms of asymmetry in plants. Since both the distance between two successive leaves (internode length) and the internode diameter scale with node order, dispersion about the regression line between length or diameter and node order becomes a measure of the departure from perfect translational symmetry, which is another form of developmental instability (Escós et al. 1997). Fractals (see Appendix 3) have been successfully used to explore the effects of stress on complex structural features of plant crowns. The fractal dimension of the branching pattern of the shrub *Anthyllis cytisoides* decreased at heavy grazing stress (Alados et al. 1994, Escós et al. 1997). Moderate grazing on this shrub promoted growth, but also enhanced stability of vegetative structures, as reflected by the homeostatic maintenance of its translatory symmetry and by the increased complexity of its fractal structures (Escós et al. 1997). The decrease in branching complexity with stress is likely to translate into a poorer occupation of the space and a less-efficient light capture. Internode length follows a simple self-similar sequence, and the internode order from the stem base can be considered a scaling factor. Different plant species fit different general equations. A decline in accuracy of the curve fit is a good estimator of scale asymmetry, and it has been proven to be a good descriptor of grazing stress in the shrub *Chrysothamnos greenii* (Alados et al. 1994). A similar study of the effects of stress on plant architecture based on the same reasoning of developmental instability and increased error of curve fit was conducted for the relationship between branch diameter and branch length with branch order in *Anthyllis cytisoides* (Alados et al. 1994, Escós et al. 1997). The effects of heterozygosity and habitat characteristics on developmental instability have been explored using morphometric measures of vegetative and reproductive structures in the Mediterranean chamaephyte *Teucrium lusitanicum* (Alados et al. 1998).

APPENDIX 6: CONCEPTS IN EVOLUTION*

Despite the importance of evolutionary concepts in ecological studies and the fact that most of them were established a long time ago, there are important points in dispute, and new approaches to the same old questions are constantly being suggested. During this century, the Darwinian theory of evolution has been almost universally accepted, mainly because the Lamarckian inheritance of acquired characteristics has grown more and more remote with advances in molecular genetics. Evolutionary biologists have assumed that the genetic system is the only provider of heritable variation. However, new theories have been developed not to supplant selectionist theories of adaptation, but to complement them. In certain cases, Lamarckian arguments should not be completely forgotten (Jablonka and Lamb 1995). Lamarckian mechanisms such as adaptive mutation and epigenetic inheritance, which allow the inheritance of induced characters,

---

* Definitions from Marrow et al. 1996 and Niklas 1997a, 1997b.

need to be reconsidered in the light of new cellular and molecular findings (Jablonka et al. 1998). Discrepancies arise regarding the motor of evolution. While some investigators support the notion that natural selection is the unifying theory for biology (Dawkins 1996), others argue that historical accidents determine disparate courses of evolution (Gould 1989). Recent studies have shown that adaptive radiation in similar environments can overcome historical contingencies to produce strikingly similar evolutionary outcomes (Harvey and Partridge 1998), that is, selection seems to be stronger than chance, at least some of the time (Vogel 1998). And neo-Darwinism itself has been questioned as a simplistic, excessively mechanistic view of life evolution (Margulis 1997). Lynn Margulis asked evolutionary biologists to think physiologically in general and to recognize the principles of autopoiesis (from Greek; *auto* = self, *poiesis* = to make). Autopoiesis, a set of six principles developed by Humberto Maturana et al. to define the living, is presented as an alternative to neo-Darwinism, which is seen as a product of the physics-centered philosophy of mechanism (Margulis 1997). However, neo-Darwinism is still the most extended scientific school of biological evolution.

Adaptive radiation: rapid *divergence* of closely related species.

Adaptive walk: the sequence of phenotypes leading from the ancestor to the most fit descendant. The expression, which was coined by Sewall Wright, has been successfully applied to plants by Karl J. Niklas, who has simulated several adaptive walks through *fitness landscapes* to interpret why plant evolution proceeded the way it did.

Convergence: a long-term phenotypic trend resulting from *directional selection*, operating on two or more unrelated species or populations that diverge from their respective ancestors and evolve analogous features as a result of identical selection pressures.

Darwinism: theory of evolution proposed by C. Darwin, where the mechanism leading to evolutionary change was natural selection of heritable variations. It included the pangenesis theory to explain the origin and behavior of inherited variation.

Diffusive evolution: diversification unfettered by selection pressure.

Directional selection: a mode of natural selection that results when individuals expressing a particular trait are preferentially subjected to the greatest selection, shifting the frequency distribution of phenotypes to one end of the spectrum by favoring rare phenotypes.

Disruptive selection: a mode of natural selection that results when the intermediate phenotypes have a selective disadvantage relative to those of both ends of the spectrum. Also called *diversifying selection*.

Divergence: a long-term phenotypic trend resulting from either geographic isolation or *disruptive selection* operating on a single taxon, which produces two very different phenotypes. On a larger scale, divergent evolution is called *adaptive radiation.*

Evolutionary stable strategy (ESS): a phenotypic strategy becomes evolutionarily stable when, in addition to being at *Nash equilibrium* (no other strategy gains a higher payoff when playing against it), no other strategy has exactly the same payoff. Thus, when an ESS is fixed in the population, it cannot be invaded by any alternative strategy.

Fitness: a measure of relative adaptation of a species to its environment in terms of reproductive success. Darwinian fitness is defined as the measure of an individual's relative contribution to the gene pool of the next generation.

Fitness landscape: the universe of all conceivable phenotypes or variants within the population with a relative fitness assigned to each of them.

Lamarckism: theory of evolution proposed by J.B. Lamarck that included many aspects (spontaneous generation of simple forms, inherent tendency of living matter to become more complex), the most controversial being that the modifications brought about by use and disuse of parts in response to environment are inherited (inheritance of acquired characters). Recent studies reveal the existence of ''Lamarckian'' mechanisms in Darwinian evolution (see earlier).

Long-term evolution: a theory of evolution based on the repeated invasion and replacement of mutants (emphasis on phenotype and game theory). See *short-term evolution.*

Morphospace: the hypothetical universe of all conceivable morphological or anatomical phenotypic possibilities.

Nash equilibrium: a phenotypic strategy is at Nash equilibrium when no other strategy gains a higher payoff when playing against it.

Neo-Darwinism: theory of evolution resulting from the synthesis of genetics and Darwinian ideas in the 1930s and 1940s. Evolutionary change results from natural selection of the random genetic variations generated by mutation and recombination. Pangenesis and inheritance of acquired characteristics included in the original theory of Darwin were purged.

Norm of reaction: range of phenotypic variation resulting from a range of environmental conditions.

Parallelism: a long-term phenotypic trend resulting from *stabilizing selection* operating on two or more related species or populations.

Phenotypic plasticity: the ability of a given genotype to produce more than one phenotype or alternative form of morphology, physiology, and/or behavior in response to environmental conditions.

Short-term evolution: a theory of evolution based on changes in the frequency of existing alleles over time (emphasis on genotype and population genetics). See *long-term evolution.*

Stabilizing selection: a mode of natural selection that results when interactions

between a population and its environment remain constant and natural selection removes the extreme phenotypes, reducing the phenotypic variation
in the population.

ACKNOWLEDGMENTS

Tsvi Sachs has been most generous with his time and expertise in commenting
on a first draft of this chapter. Thanks to Luis Balaguer and Maria Eugenia Ron
for fitting criticisms, and to Lucía Ramirez for her patience. This chapter is based
on research supported by the Spanish Ministry of Education and Science (postdoctoral fellowship and research contract within the project DGICYT PB95-
0067), and by the Mellon Science Fundation, Smithsonian Institution.

REFERENCES

Ackerly, D.D., and F.A. Bazzaz. 1995. Seedling crown orientation and interception of
diffuse radiation in tropical forest gap. Ecology 76: 1134–1146.
Alados, C.L., J. Escós, and J.M. Emlen. 1995. Fluctuating asymmetry and fractal dimension of the sagittal suture as indicators of inbreeding depression in dama and dorcas
gazelles. Canadian Journal of Zoology 73: 1967–1974.
Alados, C.L., J.M. Escós, and J.M. Emlen. 1994. Scale asymmetry: a tool to detect developmental instability under the fractal geometry scope. In: Novak, M.M., ed. Fractals
in the Natural and Applied Sciences. North Holland: Elsevier, pp 25–36.
Alados, C.L., T. Navarro, B. Cabezudo, J.M. Emlen, and C. Freeman. 1998. Developmental instability in gynodioecious *Teucrium lusitanicum*. Evolutionary Ecology
12: in press.
Alexander, R.M. 1997. Leaning trees on a sloping ground. Nature 386: 327–329.
Alpert, P. 1998. Fixity versus plasticity in clonal plant characteristics: when is it good to
adjust? Phenotypic plasticity in plants: consequences of non-cognitive behaviour.
International Workshop, Israel, March 15–19, 1998, p 15.
Alpert, P., and J.F. Stuefer. 1997. Division of labour in clonal plants. In: de Kroon, H.,
and J. van Groenendael, eds. The Ecology and Evolution of Clonal Plants. The
Hague: Backhuys Publishers, pp 137–154.
Arno, S.F., and R.P. Hammerly 1984. Timberline: Mountain and Arctic Frontiers. Seattle,
WA: The Mountaineers.
Ashton, P.S. 1978. Crown characteristics of tropical trees. In: Tomlinson, P.B., and M.H.
Zimmermann, eds. Tropical Trees as Living Systems. Cambridge: Cambridge Uni-
versity Press, pp 66–84.
Attridge, T.H. 1990. Light and Plant Responses. Kent: Edward Arnold.
Ballaré, C.L. 1994. Light gaps: sensing the light opportunities in highly dynamic canopy
environments. In: Caldwell, M.M., and R.W. Pearcy, eds. Exploitation of Environ-

mental Heterogeniety by Plants: Ecophysiological Processes Above- and Be-lowground. San Diego: Academic Press, pp 73–110.

Ballaré, C.L., A.L. Scopel, and R.A. Sanchez. 1997. Foraging for light: photosensory ecology and agricultural implications. Plant, Cell and Environment 20: 820–825.

Bazzaz, F.A. 1996. Plants in Changing Environments: Linking Physiological, Population, and Community Ecology. Cambridge: Cambridge University Press.

Bazzaz, F.A., and R.W. Carlson. 1982. Photosynthetic acclimation to variability in the light environment of early and late successional plants. Oecologia (Berlin) 54: 313–316.

Bell, A.D. 1984. Dynamic morphology: a contribution to plant population ecology. In: Dirzo, R., and J. Sarukhán, eds. Perspectives on Plant Population Ecology. Sunder-land, MA: Sinauer, pp 48–65.

Bell, A.D. 1993. Plant Form. New York: Oxford University Press.

Bock, W.J., and G. Won Wahlert. 1965. Adaptation and the form-function complex. Evo-lution 19: 269–299.

Boojh, R., and P.S. Ramakrishnan. 1982. Growth strategy of trees related to successional status. I. Architectural and extension growth. Forest Ecology Management 4: 359–374.

Borchert, R., and N.A. Slade. 1984. Bifurcation ratios and the adaptive geometry of trees. Botanical Gazzette 142: 394–401.

Bradshaw, A.D. 1965. Evolutionary significance of phenotypic plasticity in plants. Ad-vances in Genetics 13: 115–155.

Caldwell, M.M., T.J. Dean, R.S. Nowak, R.S. Dzurec, and J.H. Richards. 1983. Bunch-grass architecture, light interception, and water-use efficiency: assessment by fiber optic point quadrats and gas exchange. Oecologia 59: 178–184.

Caldwell, M.M., H.P. Meister, J.D. Tehunen, and O.L. Lange. 1986. Canopy structure, light microclimate and leaf gas exchange of *Quercus coccifera* L. in a Portugese macchia: measurements in different canopy layers and simulations with a canopy model. Trees 1: 25–41.

Caldwell, M.M., J.H. Richards, D.A. Johnson, and D.C. Dzurec. 1981. Coping with herbi-vory: photosynthetic capacity and resource allocation in two semiarid *Agropyron* bunchgrasses. Oecologia 50: 14–24.

Canham, C.D. 1988. Growth and canopy architecture of shade-tolerant trees: response to canopy gaps. Ecology 69: 786–795.

Chapin, F.S. III. 1980. The mineral nutrition of wild plants. Annual Review of Ecology and Systematics 11: 233–260.

Chapin, F.S. 1991. Effects of multiple environmental stresses on nutrient availability and use. In: Mooney, H.A., W.E. Winner, and E.J. Pell, eds. Responses of Plants to Multiple Environmental Stresses. San Diego: Academic Press, pp 67–88.

Chapin, F.S., K. Autumn, and F. Pugnaire. 1993. Evolution of suites of traits in response to environmental stress. American Naturalist 142: S78–S92.

Chapin, F.S., A.J. Bloom, C.B. Field, and R.H. Waring. 1986. Plant responses to multiple environmental factors. BioScience 37: 49–57.

Cheeseman, J.M. 1993. Plant growth modelling without integrating mechanisms. Plant, Cell and Environment 16: 137–147.

Chen, S.G., R. Ceulemans, and I. Impens. 1994. Is there a light regime determined tree ideotype? Journal of Theoretical Biology 169: 153–161.

Clearwater, M.J., and K.S. Gould. 1995. Leaf orientation and light interception by juvenile *Pseudopanax crassifolius* (Cunn.) C. Koch in a partially shaded forest environment. Oecologia 104: 363–371.

Colasanti, R.L., and R. Hunt. 1997. Resource dynamics and plant growth: a self-assembling model for individuals, populations and communities. Functional Ecology 11: 133–145.

Comstock, J.P., and B.E. Mahall. 1985. Drought and changes in leaf orientation for two California chaparral shrubs: *Ceanothus megacarpus* and *Ceanothus crassifolius*. Oecologia 65: 531–535.

Coutts, M.P., and J. Grace. 1995. Wind and Trees. Cambridge: Cambridge University Press.

Crawford, R.M.M. 1990. Studies in plant survival. Ecological case histories of plant adaptation to adversity. Suffolk: Blackwell Scientific Publications.

Darwin, C. 1839. Journal of Researches into Geology and Natural History of the Various Countries Visited by H.M.S. Beagle, Under the Command of Captain Fitz Roy from 1832 to 1836. London.

Darwin, E. 1800. Phytologia; or the Philosophy of Agriculture and Gardening. London: Johnson.

Dawkins, R. 1996. Climbing Mount Improbable. London: Viking.

de Kroon, H., and J. van Groenendael. 1997. The Ecology and Evolution of Clonal Plants. The Hague: Backhuys Publishers.

de Reffye, P. 1981. Modèle mathématique aléatoire et simulation de la croissance et de l'architecture de *Caffeir robusta*. 4e partie. Programmation sur micro-ordinateur du trace en trois dimensions de l'architecture d'un arbre. Application au caféier. Café Caccao Thé 27: 3–20.

de Raffye P., F. Houllier, F. Blaise, D. Barthelemy, J. Dauzat, and D. Auclair. 1995. A model simulating above- and below-ground tree architecture with agroforestry applications. Agroforestry Systems 30: 175–197.

Del Tredici, P. 1991. Natural regeneration of *Ginkgo biloba* from downward growing cotyledonary buds (Basal Chichi). American Journal of Botany 79: 522–530.

Duncan, W.G. 1971. Leaf angles, leaf area, and canopy photosynthesis. Crop Science 11: 482–485.

Ehleringer, J., and I. Forseth. 1980. Solar tracking by plants. Science 210: 1094–1098.

Ehleringer, J.R., and C.B. Field. 1993. Scaling Physiological Processes. San Diego: Academic Press.

Ehleringer, J.R., and I.N. Forseth. 1989. Diurnal leaf movements and productivity in canopies. In: Russell, G., B. Marshall, and P.G. Jarvis, eds. Plant canopies: their growth, form and function. Cambridge: Cambridge University Press, pp 129–142.

Ehleringer, J.R., and K.S. Werk. 1986. Modifications of solar-radiation absorption patterns and implications for carbon gain at the leaf level. In: Givnish, T.J., ed. On the Economy of Plant Form and Function. Cambridge: Cambridge University Press, pp 564–571.

Ellison, A.M., and K.J. Niklas. 1988. Branching patterns of *Salicornia europaea* (Chenopodiaceae) at different successional stages: a comparison of theoretical and real plants. American Journal of Botany 75: 501–512.

Ennos, A.R. 1997. Wind as an ecological factor. Trends in Ecology and Evolution 12: 108–111.

Erickson, R.O. 1983. The geometry of phyllotaxis. In: Dale, J.E., and F.L. Milthrope, eds. The Growth and Functioning of Leaves. Cambridge: Cambridge University Press, pp 108–111.

Escós, J., C.L. Alados, and J.M. Emlen. 1997. The impact of grazing on plant fractal architecture and fitness of a Mediterranean shrub *Anthyllis cytisoides* L. Functional Ecology 11: 66–78.

Escós, J.M., C.L. Alados, and J.M. Emlen. 1995. Fractal structures and fractal functions as disease indicators. Oikos 74: 310–314.

Evans, J.R., I. Jakobsen, and E. Ögren. 1993. Photosynthetic light-response curves .2. Gradients of light absorption and photosynthetic capacity. Planta 189: 191–200.

Fisher, J.B. 1986. Branching patterns and angles in trees. In: Givnish, T.J., ed. On the Economy of Plant Form and Function. New York: Cambridge University Press, pp 493–524.

Fisher, J.B. 1992. How predictive are computer simulations of tree architecture? International Journal of Plant Science 153: S137–S146.

Fisher, J.B., and D.E. Hibbs. 1982. Plasticity of tree architecture: specific and ecological variations found in Aubréville model. American Journal of Botany 69: 690–702.

Fisher, J.B., and H. Honda. 1979a. Branch geometry and effective leaf area: a study of *Terminalia*-branching pattern. I. Theoretical trees. American Journal of Botany 66: 633–644.

Fisher, J.B., and H. Honda. 1979b. Branch geometry and effective leaf area: a study of *Terminalia*-branching pattern. II. Survey of real trees. American Journal of Botany 66: 645–655.

Ford, H. 1987. Investigating the ecological and evolutionary significance of plant growth form using stochastic simulation. Annals of Botany 59: 487–494.

Fournier, A. 1979. Is Architectural Radiation Adaptive? Montpellier, France: Université des Sciences et Techniques du Languedoc.

Freeman, D.C., J.H. Graham, and J.M. Emlen. 1993. Developmental stability in plants: symmetries, stress and epigenesis. Genetica 89: 97–119.

Frontier, S. 1987. Applications of fractal theory to ecology. In: Legendre, P., ed. Developments in Numerical Ecology. NATO Series. Berlin: Springer Verlag, pp 55–76.

Fujimori, T., and D. Whitehead. 1986. Crown and Canopy Structure in Relation to Productivity. Ibaraki, Japan: Forestry and Forestal Products Research Institute.

Gartner, B.L. 1995. Plant Stems. Physiology and Functional Morphology. San Diego: Academic Press.

Gefen, Y. 1987. Fractals. McGraw-Hill Encyclopedia of Science and Technology 7.

Geiger, D.R. 1986. Processes affecting carbon allocation and partitioning among sinks. In: Cronshaw, J., W.J. Lucas, and R.T. Giaquinta, eds. Phloem Transport. New York: Alan R. Liss, pp 375–388.

Gill, D.E. 1991. Individual plants as genetic mosaics: ecological organisms versus evolutionary individuals. In: Crawley, M.J., ed. Plant Ecology. Wiltshire: Blackwell Scientific, pp 321–344.

Givnish, T.J. 1978. On the adaptive significance of compound leaves, with particular reference to tropical trees. In: Tomlinson, P.B., and H. Zimmermann, eds. Tropical Trees as Living Systems. Cambridge: Cambridge University Press, pp 351–380.

Givnish, T.J. 1982. On the adaptive significance of leaf height in forest herbs. American Naturalist 120: 353–381.

Givnish, T.J. 1986. Biomechanical constraints on crown geometry in forest herbs. In: Givnish, T.J., ed. On the Economy of Plant Form and Function. New York: Cambridge University Press, pp 525–584.

Givnish, T.J. 1988. Adaptation to sun and shade: a whole-plant perspective. Australian Journal of Plant Physiology 15: 63–92.

Givnish, T.J. 1995. Plant stems: biomechanical adaptation for energy capture and influence on species distributions. In: Gartner, B.L., ed. Plant Stems: Physiology and Functional Morphology. San Diego: Academic Press, pp 3–49.

Godin, C., Y. Guedon, E. Costes, and Y. Caraglio. 1997. Measuring and analysing plants with the AMAPmod software. In: Michalewicz, M.T., ed. Plants to Ecosystems. Advances in Computational Life Sciences. Collingwood, Australia: CSIRO, pp 53–81.

Gould, S.J. 1989. Wonderful Life. New York: Norton.

Grace, J. 1977. Plant Response to Wind. New York: Academic Press.

Gutschick, V.P., and F.W. Wiegel. 1988. Optimizing the canopy photosynthetic rate by patterns of investment in specific leaf mass. American Naturalist 132: 67–86.

Hallé, F. 1995. Canopy architecture in tropical trees: a pictorial approach. In: Lowman, M.D., and N.M. Nadkarni, eds. Forest Canopies. London: Academic Press, pp 27–44.

Hallé, F., and R. Oldeman. 1970. Essai sur l'architecture et la dynamique de croissance des arbres tropicaux. Paris: Masson.

Hallé, F., R.A.A. Oldeman, and P.B. Tomlinson. 1978. Tropical Trees and Forests. An Architectural Analysis. Berlin-Heidelberg: Springer-Verlag.

Hardwick, R.C. 1986. Physiological consequences of modular growth in plants. Philosophical Transactions of the Royal Society of London, Biology 313: 161–173.

Harper, J.L. 1977. Population Biology of Plants. London: Academic Press.

Harper, J.L. 1989. Canopies as populations. In: Russell, G., B. Marshall, and P.J. Jarvis, eds. Plant Canopies: Their Growth, Form and Function. Cambridge: Cambridge University Press, pp 105–128.

Harvey, P.H., and L. Partridge. 1998. Different routes to similar ends. Nature 392: 552–553.

Harvey, P.H., and A. Purvis. 1991. Comparative methods for explaining adaptations. Nature 351: 619–624.

Hasting, H.M., and G. Sugihara. 1993. Fractals: A User's Guide for the Natural Sciences. Oxford: Oxford University Press.

Herbert, T. 1996. On the relationship of plant geometry to photosynthetic response. In: Mulkey, S.S., R.L. Chazdon, and A.P. Smith, eds. Tropical Forest Plant Ecophysiology. New York: Chapman and Hall, pp 139–161.

Herbert, T.J. 1991. Variation in interception of the direct solar beam by top canopy layers. Ecology 72: 17–22.

Herbert, T.J., and T. Nilson. 1991. A model of variance of photosynthesis between leaves and maximization of whole plant photosynthesis. Photosynthetica 25: 597–606.

Honda, H., and J.B. Fisher. 1978. Tree branch angle: maximizing effective leaf area. Science 199: 888–889.

Honda, H., H. Hatta, and J.B. Fisher. 1997. Branch geometry in *Cornus kousa* (Cornaceae): computer simulations. American Journal of Botany 84: 745–755.

Horn, H.S. 1971. The Adaptive Geometry of Trees. Princeton, NJ: Princeton University Press.

Horton, P., A.V. Ruban, and R.G. Walters. 1996. Regulation of light harvesting in green plants. Annual Review of Plant Physiology and Plant Molecular Biology 47: 655–684.

Hunt, R. 1990. Basic Growth Analysis. London: Unwin Hyman.

Hutchings, M.J., and H. de Kroon. 1994. Foraging in plants: the role of morphological plasticity in resource acquisition. Advances in Ecological Research 25: 159–238.

Ingrouille, M. 1995. Diversity and Evolution of Land Plants. London: Chapman and Hall.

Isebrands, J.G., and D.A. Michael. 1986. Effects of leaf morphology and orientation on solar radiation interception and photosynthesis in Populus. In: Fujimori, T., and D. Whitehead, eds. Crown and Canopy Structure in Relation to Productivity. Ibaraki, Japan: Forestry and Forestal Products Research Institute, pp 155–168.

Ishii, R., and M. Higashi. 1997. Tree coexistance on a slope: an adaptive significance of trunk inclination. Proceedings of the Royal Society of London, Biology 264: 133–140.

Jablonka, E., and M.J. Lamb. 1995. Epigenetic Inheritance and Evolution. Oxford: Oxford University Press.

Jablonka, E., M.J. Lamb, and E. Avital. 1998. ''Lamarckian'' mechanisms in Darwinian evolution. Trends in Ecology and Evolution 13: 206–210.

Jahnke, L.S., and D.B. Lawrence. 1965. Influence of photosynthetic crown structure on potential productivity of vegetation, based primarily on mathematical models. Ecology 46: 319–326.

James, J.C., J. Grace, and S.P. Hoad. 1994. Growth and photosynthesis of *Pinus sylvestris* at its altitudinal limit in Scotland. Journal of Ecology 82: 297–306.

Kaplan, D.R., and W. Hagemann. 1991. The relationship of cell and organism in vascular plants—are cells the building blocks of plant form. Bioscience 41: 693–703.

Kellomäki, S., and H. Strandman. 1995. A model for the structural growth of young Scots pine crowns based on light interception by shoots. Ecological Modelling 80: 237–250.

King, D. 1981. Tree dimensions: maximizing the rate of height growth in dense stands. Oecologia 51: 351–356.

King, D.A. 1986. Tree form, height growth, and susceptibility to wind damage in *Acer saccharum*. Ecology 67: 980–990.

King, D.A. 1990. The adaptive significance of tree height. American Naturalist 135: 809–828.

Koehl, M.A.R. 1996. When does morphology matter? Annual Review of Ecology and Systematics 27: 501–542.

Küppers, M. 1984. Carbon relations and competition between woody species in a central European hedgerow. I. Photosynthetic characteristics. Oecologia 64: 332–343.

Küppers, M. 1985a. Carbon relations and competition between woody species in a central European hedgerow. II. Stomatal responses, water use, and the conductivity to liquid water in the root/leaf pathway. Oecologia 64: 344–354.

Küppers, M. 1985b. Carbon relations and competition between woody species in a central European hedgerow. III. Carbon and water balance on the leaf level. Oecologia 65: 94–100.

Küppers, M. 1985c. Carbon relations and competition between woody species in a central European hedgerow. IV. Growth form and partitioning. Oecologia 66: 343–352.

Küppers, M. 1989. Ecological significance of above-ground architectural patterns in woody plants: a question of cost-benefit relationships. Trends in Ecology and Evolution 4: 375–379.

Küppers, M. 1990. Ecological significance of aboveground architectural patterns in woody plants. Trends in Ecology and Evolution 10: 1–17.

Küppers, M. 1994. Canopy gaps: competitive light interception and economic space filling—a matter of whole plant allocation. In: Caldwell, M.M., and R.W. Pearcy, eds. Exploitation of Environmental Heterogeneity by Plants. San Diego: Academic Press, pp 111–144.

Küppers, M., and R. List. 1997. MADEIRA—a simulation of carbon gain, allocation, canopy architecture in competing woody plants. In: Jeremonidis, G., and J.F.V. Vincent, eds. Plant Biomechanics. Reading: Centre for Biomimetics, The University of Reading, pp 321–329.

Kuuluvainen, T. 1992. Tree architectures adpated to efficient light utilization: is there a basis for latitudinal gradients? Oikos 65: 275–284.

Leigh, E.G. 1990. Tree shape and leaf arrangement: a quantitative comparison of montane forests, with emphasis on Malaysia and south India. In: Daniel, J.C., and J.S. Serrao, eds. Conservation in Developing Countries: Problems and Prospects. Bombay, India: Oxford University Press, pp 119–174.

Leigh, E.G. 1998. Tropical Forest Ecology: A View from Barro Colorado Island. New York: Oxford University Press.

Leigh, E.G.J. 1972. The golden section and spiral leaf-arrangement. In: Deevey, E.S., ed. Growth by Intussusception. Hamden, CT: Archon Books, pp 163–176.

Leopold, L.B. 1971. Trees and streams: the efficiency of branching patterns. Journal of Theoretical Biology 31: 339–354.

Lindenmayer, A. 1968. Mathematical models for cellular interactions in development. Parts I and II. Journal of Theoretical Biology 18: 280–315.

Loehle, C. 1983. The fractal dimension and ecology. Speculations in Science and Technology 6: 131–142.

Long, S.P., S. Humphries, and P.G. Falkowski. 1994. Photoinhibition of photosynthesis in nature. Annual Review of Plant Physiology and Plant Molecular Biology 45: 633–662.

Lorimer, N.D., R.G. Haight, and R.A. Leary. 1994. The fractal forest: fractal geometry and applications in forest science. General Technical Report NC-170. St. Paul, MN: U.S. Department of Agriculture, Forest Service, North Central Forest Experiment Station.

Lortie, C.J., and L.W. Aarssen. 1996. The specialization hypothesis for phenotypic plasticity in plants. International Journal of Plant Sciences 157: 484–487.

Lovelock, C.E., and B.F. Clough. 1992. Influence of solar radiation and leaf angle on leaf xanthophyll concentrations in mangroves. Oecologia 91: 518–525.

Mandelbrot, B.B. 1967. How long is the coast of Britain? Statistical self-similarity and fractional dimension. Science 156: 636–638.

Mandelbrot, B.B. 1983. The Fractal Geometry of Nature. New York: W.H. Freeman.

Margalef, R. 1997. Our Biosphere. Excellence in Ecology 10. Wurzburg, Germany: Ecology Institute of Germany.

Margulis, L. 1997. Big trouble in biology. Physiological autopoiesis versus mechanistic neo-Darwinism. In: Margulis, L., and D. Sagan, eds. Slanted Truths. New York: Springer Verlag, pp 265–282.

Marrow, P., R.A. Johnstone, and L.D. Hurst. 1996. Riding the evolutionary streetcar: where population genetics and game theory meet. Trends in Ecology and Evolution 11: 445–446.

Mattheck, C. 1995. Biomechanical optimum in woody stems. In: Gartner, B.L., ed. Plant Stems: Physiology and Functional Morphology. San Diego: Academic Press, pp 3–49.

Mattheck, G.C. 1991. Trees. The Mechanical Design. New York: Springer Verlag.

McMahon, T. 1975. The mechanical design of trees. Scientific American 233: 92–102.

McMahon, T.A. 1973. Size and shape in biology. Science 179: 1201–1204.

McMahon, T.A., and R.F. Kronauer. 1976. Tree structures: deducing the principle of mechanical design. Journal of Theoretical Biology 59: 443–466.

Myers, D.A., D.N. Jordan, and T.C. Vogelmann. 1997. Inclination of sun and shade leaves influences chloroplast light harvesting and utilization. Physiologia Plantarum 99: 395–404.

Niklas, K.J. 1988. The role of phyllotactic pattern as a ''developmental constraint'' on the interception of light by leaf surfaces. Evolution 42: 1–16.

Niklas, K.J. 1990. Biomechanics of *Psilotum nudum* and some early paleozoic vascular sporophytes. American Journal of Botany 77: 590–606.

Niklas, K.J. 1992. Plant Biomechanics: An Engineering Approach to Plant Form and Function. Chicago: The University of Chicago Press.

Niklas, K.J. 1994. Interspecific allometries of critical buckling height and actual plant height. American Journal of Botany 81: 1275–1279.

Niklas, K.J. 1997a. Adaptive walks through fitness landscapes for early vascular plants. American Journal of Botany 84: 16–25.

Niklas, K.J. 1997b. The Evolutionary Biology of Plants. Chicago: The University of Chicago Press.

Niklas, K.J., and V. Kerchner. 1984. Mechanical and photosynthetic constraints on the evolution of plant shape. Paleobiology 10: 79–101.

Nobel, P.S., and R.W. Meyer. 1991. Biomechanics of cladodes and cladode-cladode junctions for *Opuntia ficus-indica* (Cactaceae). American Journal of Botany 78: 1252–1259.

Novoplansky, A., D. Cohen, and T. Sachs. 1989. Ecological implications of correlative inhibition between plant shoots. Physiologia Plantarum 77: 136–140.

Oker-Blom, P., and S. Kellomäki. 1982. Theoretical computations on the role of crown shape in the absorption of light by forest trees. Mathematical Biosciences 59: 291–311.

Parsons, P.A. 1993. The importance and consequences of stress in living and fossil populations: from life-history variation to evolutionary change. American Naturalist 142: S5–S20.

Paxman, G.J. 1956. Differentiation and stability in the development of *Nicotiana rustica*. Annals of Botany 20: 331–347.

Pearcy, R.W., and W. Yang. 1996. A three-dimensional shoot architecture model for assessment of light capture and carbon gain by understory plants. Oecologia 108: 1–12.

Pearcy, R.W., and W. Yang. 1998. The functional morphology of light capture and carbon gain in the redwood-forest understory plant, *Adenocaulon bicolor* Hook. Functional Ecology 12: 543–552.

Perttunen, J., R. Sievanen, E. Nikinmaa, H. Salminen, H. Saarenmaa, and J. Vakeva. 1996. Lignum: a tree model based on simple structural units. Annals of Botany 77: 87–98.

Phillips, J.D. 1985. Measuring complexity of enviornmental gradients. Vegetatio 64: 95–102.

Pintado, Á., F. Valladares, and L.G. Sancho. 1997. Exploring phenotypic plasticity in the lichen *Ramalina capitata*: morphology, water relations and chlorophyll content in North- and South-facing populations. Annals of Botany 80: 345–353.

Porter, J.R. 1983. A modular approach to analysis of plant growth. I. Theory and principles. New Phytologist 94: 183–190.

Porter, J.R. 1989. Modules, models and meristems in plant architecture. In: Russell, G., B. Marshall, and P.J. Jarvis, eds. Plant Canopies: Their Growth, Form and Function. Cambridge: Cambridge University Press, pp 143–159.

Poulson, M.E., and E.H. DeLucia. 1993. Photosynthetic and structural acclimation to light direction in vertical leaves of *Silphium terebinthinaceum*. Oecologia 95: 393–400.

Prusinkiewicz, P., J.S. Hanan, M. Hammel, and R. Mech. 1997. L-systems: from theory to visual models of plants. In: Michalewicz, M.T., ed. Plants to Ecosystems. Advances in Computational Life Sciences. Collingwood, Australia: CSIRO, pp 1–27.

Prusinkiewicz, P., and A. Lindenmayer. 1996. The Algorithmic Beauty of Plants. New York: Springer Verlag.

Raunkiaer, C. 1905. Types biologiques pour la geographie Botanique. Bull. Acad. Sc. Danemark.

Raven, J.A. 1986. Evolution of plant life forms. In: Givnish, T.J., ed. On the Economy of Plant Growth and Function. New York: Cambridge University Press, pp 421–492.

Remphrey, W.R., and P. Prusinkiewicz. 1997. Quantification and modelling of tree architecture. In: Michalewicz, M.T., ed. Plants to Ecosystems. Advances in Computational Life Science. Collingwood, Australia: CSIRO, pp 45–52.

Rich, P.M. 1986. Mechanical architecture of arborescent rainforest palms. Principles 30: 117–131.

Rich, P.M. 1987. Developmental anatomy of the stem of *Welfia georgii, Iriartea gigantea*, and other arborescent palms: implications for mechanical support. American Journal of Botany 74: 792–802.

Roberts, S.W., and P.C. Miller. 1977. Interception of solar radiation as affected by canopy organization in two mediterranean shrubs. Oecologia Plantarum 12: 273–290.

Room, P., J. Hanan, and P. Prusinkiewicz. 1996. Virtual plants: new perspectives for ecologists, pathologists and agricultural scientists. Trends in Plant Science 1: 33–38.

Room, P.M., L. Maillette, and J.S. Hanan. 1994. Module and metamer dynamics and virtual plants. Advances in Ecological Research 25: 105–157.

Russel, G., P.G. Jarvis, and J.L. Monteith. 1989. Absorption of radiation by canopies and stand growth. In: Russell, G., B. Marshall, and P.J. Jarvis, eds. Plant Canopies: Their Growth, Form and Function. Cambridge: Cambridge University Press, pp 21–40.

Sachs, T. 1988a. Epigenetic selection: an alternative mechanism of pattern formation. Journal of Theoretical Biology 134: 547–559.

Sachs, T. 1988b. Internal controls of plant morphogenesis. In: Greuter, W., and B. Zimmer, eds. Proceedings of the XIV International Botanical Congress, Königstein/Taunus, Koeltz, 1998.

Sachs, T. 1993. The role of auxin in plant organization. Acta Horticulturae 329: 162–168.

Sachs, T., and M. Hassidim. 1996. Mutual support and selection between branches of damaged plants. Vegetatio 127: 25–30.

Sachs, T., and A. Novoplansky. 1995. Tree form: architectural models do not suffice. Israel Journal of Plant Sciences 43: 203–212.

Sachs, T., A. Novoplansky, and D. Cohen. 1993. Plants as competing populations of redundant organs. Plant, Cell and Environment 16: 765–770.

Schlichting, C.D. 1986. The evolution of phenotypic plasticity in plants. Annual Review of Ecology and Systematics 17: 667–693.

Schlichting, C.D., and M. Pigliucci. 1998. Phenotypic Evolution. A Reaction Norm Perspective. Sunderland: Sinauer Associates Inc.

Schmitt, J. 1997. Is photomorphogenic shade avoidance adaptive? Perspectives from population biology. Plant, Cell and Environment 20: 826–830.

Schulze, E.D., F.M. Kelliher, C. Körner, J. Lloyd, and R. Leuning. 1994. Relationships among maximum stomatal conductance, ecosystem surface conductance, carbon assimilation rate, and plant nitrogen nutrition: a global scaling exercise. Annual Review of Ecology and Systematics 25: 629–660.

Schulze, E.D., M. Küppers, and R. Matyssek. 1986a. The roles of carbon balance and branching pattern in the growth of woody species. In: Givnish, T.J., ed. On the Economy of Plant Form and Function. New York: Cambridge University Press, pp 585–602.

Schulze, E.-D., M. Küppers, and R. Matyssek. 1986b. The roles of carbon balance and branching pattern in the growth of woody species. In: Givnish, T.J., ed. On the Economy of Plant Form and Function. New York: Cambridge University Press, pp 585–602.

Shackel, K.A., and A.E. Hall. 1979. Reversible leaflet movements in relation to drought adaptation of cowpeas, *Vigna unquiculata* (L.) Walp. Australian Journal of Plant Physiology 6: 265–276.

Shell, G.S.G., A.R.G. Lang, and P.J.M. Sale. 1974. Quantitative measures of leaf orientation and heliotropic response in sunflower, bean, pepper and cucumber. Agricultural Meteorology 13: 25–37.

Smith, H., and G.C. Whitelam. 1997. The shade avoidance syndrome: multiple responses mediated by multiple phytochromes. Plant, Cell and Environment 20: 840–844.

Smith, M., and D. Ullberg. 1989. Effect of leaf angle and orientation on photosynthesis and water relations in *Silphium terebinthinaceum*. American Journal of Botany 76: 1714–1719.

Sojo, F., F. Valladares, and L.G. Sancho. 1997. Structural and physiological plasticity of the lichen *Catillaria corymbosa* in different microhabitats of the Maritime Antarctica. Bryologist 100: 171–179.

Speck, T., and D. Vogellehner. 1988. Biophysical examinations of the bending stability of various stele types and the upright axes of early ''vascular'' land plants. Botanica Acta 101: 262–268.

Speck, T.H., H.C. Spatz, and D. Vogellehner. 1990. Contributions to the biomechanics of plants. I. Stabilities of plant stems with strengthening elemnts of different cross-sections against weight and wind forces. Botanica Acta 103: 111–122.

Sprugel, D.G., T.M. Hinckley, and W. Schaap. 1991. The theory and practice of branch autonomy. Annual Review of Ecology and Systematics 22: 309–334.

Steingraeber, D.A. 1982. Phenotypic plasticity of branching pattern in sugar maple (*Acer saccharum*). American Journal of Botany 69: 1277–1282.

Steingraeber, D.A., L.J. Kascht, and D.H. Franck. 1979. Variation of shoot morphology and bifurcation ratio in sugar maple (*Acer saccharum*). American Journal of Botany 66: 441–445.

Stewart, I. 1988a. The beat of a fractal drum. Nature 333: 206–207.

Stewart, I. 1988b. A review of the science of fractal images. Nature 336: 289.

Strasburger, E., F. Noll, H. Schenck, A.F.W. Schimper, P. Sitte, H. Ziegler, F. Ehrendorfer, and A. Bresinsky. 1991. Lehrbuch der Botanik. Stuttgart: Gustav Fischer Verlag.

Strauss-Debenedetti, S., and F. Bazzaz. 1996. Photosynthetic characteristics of tropical trees along successional gradients. In: Mulkey, S.S., R.L. Chazdon, and A.P. Smith, eds. Tropical Forest Plant Ecophysiology. New York: Chapman and Hall, pp 162–186.

Strauss-Debenedetti, S., and F.A. Bazzaz. 1991. Plasticity and acclimation to light in tropical moraceae of different successional positions. Oecologia 87: 377–387.

Stuefer, J.F., H. de Kroon, and H.J. During. 1996. Exploitation of environmental heterogeneity by spatial division of labour in a clonal plant. Functional Ecology 10: 328–334.

Sugihara, G., and R.M. May. 1990. Applications of fractals in ecology. Trends in Ecology and Evolution 5: 79–86.

Sultan, S.E. 1992. What has survived of Darwin's theory? Phenotypic plasticity and the neo-Darwinian legacy. Evolutionary Trends in Plants 6: 61–71.

Takenaka, A. 1994. A simulation model of tree architecture development based on growth response to local light environment. Journal of Plant Research 107: 321–330.

Terborgh, J. 1992. Diversity and the Tropical Forest. New York: Scientific American Library.

Terjung, W.H., and S.S.F. Louie. 1972. Potential solar radiation on plant shapes. International Journal of Biometeorology 16: 25–43.

Tomlinson, P.B. 1982. Chance and design in the construction of plants. In: Sattler, R, ed. Axioms and Principles of Plant Construction. The Hague: Junk, pp 77–93.

Tomlinson, P.B. 1987. Architecture of tropical plants. Annual Review of Ecology and Systematics 18: 1–21.

Turner, I.M., W.K. Gong, J.E. Ong, J.S. Bujang, and T. Kohyama. 1995. The architecture and allometry of mangrove saplings. Functional Ecology 9: 205–212.

Tyree, M.T. 1988. A dynamic model for water flow in a single tree: evidence that models must account for hydraulic architecture. Tree Physiology 4: 195–217.

Valladares, F., and R.W. Pearcy. 1997. Interactions between water stress, sun-shade acclimation, heat tolerance and photoinhibition in the sclerophyll *Heteromeles arbutifolia*. Plant, Cell and Environment 20: 25–36.

Valladares, F., and R.W. Pearcy. 1998. The functional ecology of shoot architecture in sun and shade plants of *Heteromeles arbutifolia M. Roem.*, a Californian chaparral shrub. Oecologia 114: 1–10.

Valladares, F., and F.I. Pugnaire. 1998. Tradeoffs between irradiance capture and avoidance in two plants of contrasting architecture co-occurring in a semiarid environment. Annals of Botany, in press.

Valladares, F., S.J. Wright, E. Lasso, K. Kitajima, and R.W. Pearcy. 1998. Plastic phenotypic response to light of 16 rainforest shrubs (*Psychotria*) differing in shade tolerance and leaf turnover. Ecology, in press.

Van Steenis, C.G.G.J. 1981. Reophytes of the World: An Account of the Flood-Resistant Flowering Plants and Ferns and the Theory of Autonomous Evolution. Alphen aan den Rijn, The Netherlands: Sijthoff and Noordhoof.

Veres, J.S., and S.T.A. Pickett. 1982. Branching patterns of *Lindera benzoin* beneath gaps and closed canopies. New Phytologist 91: 767–772.

Vicek, J., and E. Cheung. 1986. Fractal analysis of leaf shapes. Canadian Journal of Forest Research 16: 124–127.

Vogel, G. 1998. For island lizards, history repeats itself. Science 279: 2043.

Vogel, S. 1981. Life in Moving Fluids. Boston: Williard Grant Press.

Vogel, S. 1989. Drag and reconfiguration of broad leaves in high winds. Journal of Experimental Botany 40: 941–948.

Waddington, J.L., M.J. MacCulloch, and J.E. Sambrooks. 1957. The Strategy of Genes. London: Allen and Uniwin.

Waller, D.M. 1991. The dynamics of growth and form. In: Crawley, M.J., ed. Plant Ecology. Wiltshire: Blackwell Scientific, pp 291–320.

Waller, D.M., and D.Q. Steingraeber. 1985. Branching and modular growth: theoretical models and empirical patterns. In: Jackson, J.B.C., L.W. Buss, and R.E. Cooks, eds. Population Biology and Evolution of Clonal Organisms. New Haven, CT: Yale University Press, pp 35–70.

Wang, Y.P., and P.G. Jarvis. 1990a. Description and validation in an array model-MAESTRO. Agricultural and Forest Meteorology 51: 257–280.

Wang, Y.P., and P.G. Jarvis. 1990b. Influence of crown structural properties on PAR absorption, photosynthesis and transpiration in Sitka spruce-application of a model (MAESTRO). Tree Physiology 7: 297–316.

Wang Y.P., P.G. Jarvis, and C.M.A. Taylor. 1991. Par absorption and its relation to above-ground dry matter production of Sitka spruce. Journal of Applied Ecology 28: 547–560.

Watson, M.A. 1986. Integrated physiological units in plants. Trends in Ecology and Evolution 1: 111–123.

Werger, M.J.A., P.J.M. Van der Aart, H.J. During, and J.T.A. Verhoeven. 1988. Plant Form and Vegetation Structure. The Hague: SPB Academic Publishing.

Werk, K.S., and J. Ehleringer. 1984. Non-random leaf orientation in *Lactuca serriola* L. Plant, Cell and Environment 7: 81–87.

White, J. 1979. The plant as a metapopulation. Annual Review of Ecology and Systematics 10: 109–145.

White, J. 1984. Plant metamerism. In: Dirzo, R., and J. Sarukhán, eds. Perspectives on Plant Population Ecology. Sunderland, MA: Sinauer, pp 176–185.

Whitney, G.G. 1976. The bifurcation ratio as an indicator of adaptive strategy in woody plant species. Bulletin of the Torrey Botanical Club 103: 67–72.

Wilson, B.F. 1995. Shrub stems: form and function. In: Gartner, B., ed. Plant stems. Physiology and functional morphology. San Diego: Academic Press, pp 91–103.

Wilson, E.O. 1975. Sociobiology: The New Synthesis. Cambridge, MA: Harvard University Press.

Woodward, F.I. 1998. Do plants really need stomata? Journal of Experimental Botany 49: 471–480.

Zimmerman, M.H. 1983. Xylem Structure and the Ascent of Sap. Berlin: Springer Verlag.

5

The Structure and Function of Root Systems

Robert B. Jackson, William T. Pockman, and William A. Hoffmann
University of Texas–Austin, Austin, Texas

I. INTRODUCTION

The study of root structure and functioning is centuries old (Hales 1727, reprinted 1961). Although great progress has been made (Brouwer et al. 1981), our knowledge is limited by the difficulties in studying roots in situ. These limitations color our perception of plants. A typical layperson knows that forests can grow 50–100 m in height, but rarely recognizes that root systems can grow to similar depths (Canadell et al. 1996). The individual may also never consider the functional consequences of roots that typically spread well beyond the canopy line of most plants (Lyford and Wilson 1964). Just as there is a quiet bias in maps of the world that consistently present the northern hemisphere ''on top,'' our perception of plants would change if they were drawn upside down—roots on top and shoots underneath. A small shrub such as *Prosopis glandulosa* would suddenly appear as majestic as a tree, and many trees would suddenly seem shrubby. The bias of human perception shadows our view of the plant world.

Root and shoot functioning are often studied separately, in part because the techniques and equipment needed can differ substantially. In reality, roots and shoots are functionally integrated. This integration is evident in patterns of standing biomass and allocation. With the exception of forests, most natural systems have root: shoot ratios (the ratio of root to shoot biomass) between 1 and 7, including tundra, deserts, and grasslands (Jackson et al. 1996, 1997). Most forest systems typically have root:shoot ratios of approximately 0.2, with the majority of biomass stored as woody biomass in the boles of trees. Even for forests, half or more of annual primary production is usually allocated below ground (e.g., 60% for a *Liriodendron tulipfera* forest; Reichle et al. 1973). This is not to imply that roots are more important to the functioning of plants than shoots, just to demonstrate that they are no less important.

The purpose of this chapter is to provide an introduction to the structure and functioning of root systems. We begin by outlining the basics of root morphology and development. We next examine four broad categories of root functioning: anchoring, resource uptake, storage, and sensing the environment. Information on more specialized root functions such as reproduction and aeration is available elsewhere (Drew 1997). We discuss two important root symbioses: mycorrhizal associations and symbiotic nitrogen fixation. We end by examining global patterns of root distributions for biomes and plant functional types. Interested readers will find references in each section that provide comprehensive detail on each topic.

II. ROOT MORPHOLOGY AND DEVELOPMENT

We begin by reviewing the generalized structure of primary roots, including the main tissue types within roots and the changes that occur with secondary growth and tertiary root morphology. A more detailed discussion of these and other structural features can be found in complete botanical and anatomical texts (Esau 1977, Mauseth 1988).

A. Primary Root Anatomy

As in other plant organs, cell division during primary root growth occurs in the apical meristem, giving rise to the undifferentiated cells of the protoderm, ground meristem, and procambium (Figure 1A). Root length increases as these newly produced cells elongate and differentiate in the region immediately behind the root tip. As a result, root anatomy and functioning change with distance from the tip, and all developmental stages may be present in a single root.

1. Epidermis

The epidermis is derived from the protoderm and generally consists of a single layer of cells forming the outermost root tissue (Figure 1B). A key feature of the epidermis is root hairs, elongated cells projecting into the surrounding soil (Hofer 1991). Root hair density is typically greatest in the most distal region of the primary root behind the root tip. This so-called root hair zone has been widely regarded as the location of most water and nutrient uptake (but see Section II.A.3). Although root hairs persist in some species, their distribution is generally restricted to the distal portion of the root because the oldest root hairs are lost and new ones are produced only near the root tip. Root hair density ranges from 20–2500 per cm^2 (Dittmer 1937, Kramer 1983) and can more than double the root surface area in contact with the soil, resulting in a greater accessible soil volume (Kramer 1969). In many but not all species, nutrient absorption increases in proportion to root hair density (Bole 1973, Itoh and Barber 1983).

2. Cortex

The cortex is derived from the ground meristem, comprised largely of parenchyma cells. It lies between the epidermis to the outside and the vascular cylinder at the center of the root (Figure 1B). The cortex may develop large air canals that increase oxygen availability to root cells, and it can be an important site for carbohydrate storage (Mauseth 1988). Perhaps the most studied feature of the cortex is the endodermis, a single cell layer defining the interior edge of the cortex. The central feature of the endodermis is the casparian band formed by the deposition of suberin in the primary cell wall and middle lamella of each adjoining endodermal cell. The result is a continuous suberized barrier that prevents passage of soil solutes from the cortex into the vascular cylinder without crossing a cell membrane (Clarkson 1993, Weatherley 1982). Recent studies suggest that a more complex model of water and solute uptake may be appropriate but support the importance of the casparian band in providing control over solute

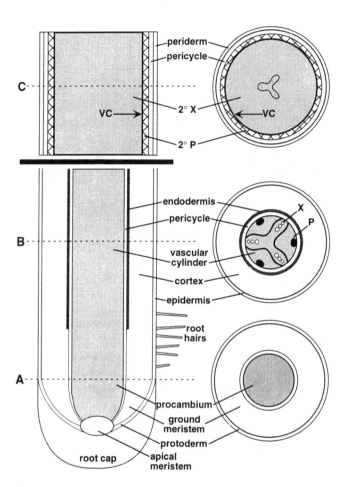

Figure 1 Longitudinal and cross-sectional schematics of generalized root anatomy. Cross-sections are indicated by broken lines through the longitudinal section at left. Drawing is not to scale to allow illustration of the key features of the root. (A) Undifferentiated cells of the protoderm, ground meristem, and procambium produced by divisions of the root apical meristem located immediately behind the root cap. (B) Differentiated primary tissues: epidermis (from protoderm), cortex, and endodermis (from ground meristem), and xylem (X), phloem (P), and pericycle (from procambium). (C) Root anatomy after substantial secondary growth. Differentiation of cells between xylem and phloem bundles (B) has produced the vascular cambium (VC). Divisions of the cambial initials give rise to secondary xylem (2° X) and phloem (2° P). The cortex, and with it the endodermis, has ruptured and sloughed off with growth of the vascular cylinder and is not visible at this stage. The suberized lignified periderm, ultimately derived from the pericycle, has developed and assumed the function of sealing the root from the surrounding soil. The primary xylem is visible at the center of the secondary xylem. (*Source*: Adapted from Raven et al. 1986.)

flow into roots (Steudle 1994, Steudle and Meshcheryakov 1996). Many taxa also form an exodermis, an anatomically similar cell layer located at the outer edge of the cortex (Perumalla et al. 1990, Peterson and Perumalla 1990).

3. Vascular Cylinder

The vascular cylinder develops from the procambium in the center of the root and is delimited by a single layer of parenchyma cells called the pericycle (Figure 1B). In cross-section, the primary xylem is arranged in finger-like projections from the center of the root toward the pericycle. Phloem bundles occur between these xylem projections. The smallest xylem conduits, the protoxylem, occur at the tips of these projections, whereas the larger conduits of the metaxylem are located more centrally and mature later. Root uptake of water and nutrients requires axial transport through functional xylem conduits, which are dead at maturity. Although the root hair zone is often cited as the site of maximum uptake, studies of maize and soybean suggest that water absorption in the root hair zone may be restricted because the largest metaxylem conduits are still alive and nonconducting (McCully and Canny 1988). Future work addressing the relative timing of maturity of the xylem, endodermis, and root hairs across taxa will improve our understanding of the contribution of different development stages of the root to resource uptake (McCully 1995).

4. Lateral and Adventitious Roots

The production of lateral roots is an important feature determining the tertiary (three-dimensional) structure of the root system and the distribution of surface area for resource uptake. Lateral roots occur in gymnosperms and dicots and arise from root primordia in the pericycle and, less commonly, the endodermis (Peterson and Peterson 1986). As a root primordium elongates, it passes through the cortex and epidermis. Vascular tissues differentiate within the developing root and are connected with the plant's vascular system at the base of the primordium. Many cortical cells are crushed during lateral root emergence, although anticlinal divisions of the endodermis (in which the cell plate forms perpendicular to the nearest tissue surface) minimize disruption of the casparian band around the emerging lateral root. Nevertheless, lateral root growth may provide a pathway for the flow of water in or out of the vascular cylinder that is not controlled by the casparian band (Caldwell et al. 1998, Kramer and Boyer 1995).

Adventitious roots originate from aerial or underground plant stems. They occur in most plant taxa and are particularly important in the monocotyledons, comprising most of the root system (Davis and Haissig 1994). Adventitious roots develop from root primordia that can arise in most tissues of plant stems. As in lateral roots, these primordia differentiate into cell types typical of a root, and vascular connections are formed with existing xylem and phloem at the root origin.

B. Secondary Growth

Secondary growth in roots is clearly important for the ability of plants to become "woody" and perennial. It occurs commonly among gymnosperms, to varying degrees among dicotyledonous plants, and is absent among the monocots (Mauseth 1988). The initiation of secondary growth is preceded by the formation of a vascular cambium derived from undifferentiated procambium and parts of the pericycle (Figure 1C). The vascular cambium forms a ring of meristematic cells between the phloem and xylem. These cambial initials give rise to xylem and phloem cells by periclinal divisions (the cell plate forms parallel to the nearest tissue surface) and accommodate increases in root diameter by occasional anticlinal divisions that maintain the continuity of the cambium. Likewise, the pericycle undergoes both periclinal and anticlinal divisions, giving rise to the phellogen, a meristematic tissue that produces the periderm. These changes are not accompanied by further divisions of the cortex, which becomes fractured and lost as the root increases in diameter. The periderm, which includes the suberized cells of the cork, becomes the outer surface of the root and assumes the protective function formerly provided by the epidermis and endodermis.

C. Tertiary Root Morphology

Root development among seed plants begins with the elongation of the taproot. For gymnosperms and dicots, the three-dimensional structure of the taproot and its branched lateral roots define the morphology of the root system (Figure 2). In contrast, the early demise of the taproot and the subsequent growth of adventitious roots in monocots result in a fibrous root system emanating from the base of the stem (limiting the rooting depth of monocots). This difference, combined with the lack of secondary growth in roots of monocots, defines an important functional difference between taxa that possess these very different root systems (Caldwell and Richards 1986). Although the absence of secondary growth limits the rooting depth of monocots, they often have very high root length densities in the soil volume they explore (Glinski and Lipiec 1990). In contrast, the taproot system of dicots and gymnosperms is often capable of exploring a soil volume that extends both laterally and vertically beyond the reach of many monocots. Despite their different structural characteristics, both fibrous and taproot systems are capable of differential root proliferation in response to resource patches in the soil (Bilbrough and Caldwell 1995, Drew 1975). The exploitation of such resources is addressed in Section III.B.

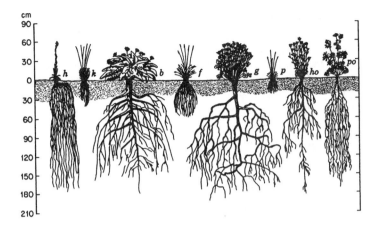

Figure 2   Differences in rooting system morphology among prairie grasses (family Gramineae) and herbs: *h, Hieracium scouleri* (Compositae); *k, Koeleria cristata* (Gramineae); *b, Balsamina sagittata* (Balsaminaceae); *f, Festuca ovina ingrata* (Gramineae); *g, Geranium viscosissimum* (Geraniaceae); *p, Poa sandbergii* (Gramineae); *ho, Hoorebekia racemosa* (Compositae); *po, Potentilla blaschkeana* (Rosaceae). (*Source*: Kramer and Boyer 1995.)

III. ROOT FUNCTIONS

Root systems have at least four broad functions: anchoring plants, capturing re-sources, storing resources, and sensing the environment. Such distinctions are arbitrary but provide a useful framework for examining root functioning.

A. Anchoring

Probably the most fundamental root function is to hold plants in place. This need is most obvious in protecting trees from windthrow, but shrubs and herbaceous vegetation are also exposed to the vagaries of wind, trampling, and herbivores. Resistance to toppling has economic importance for crop species, too, whose root systems tend to be fairly shallow (Brady 1934).

In general, there are three kinds of mechanical failure in plants: uprooting, stem failure, and root failure. The biomechanics of root anchoring can be studied by uprooting plants mechanically and recording the resistance with a strain gauge (Mattheck 1991, Somerville 1979). Results from these and other experiments show that resistance to windthrow has two primary components: the resistance of leeward laterals to bending and the resistance of windward sinkers and taproots to uprooting. Bending tests on the leeward laterals of a deep-rooted larch species showed that they provided approximately 25% of tree anchorage support (Crook and Ennos 1996). Consequently, approximately three quarters of the stability in that system came from tap roots and windward sinkers. Where there is a prevailing wind direction, there is often an asymmetrical development of structural roots—for example, greater root development on the leeward side compared with the windward side of a tree (Nicoll and Ray 1996). The cross-sectional structure of individual roots can also differ depending on the location of the roots. Secondary growth above the center of a root can lead to a T-beam structure (swollen on top and thinner on the bottom). This type of thickening is more common on roots relatively close to the trunk (<1 m away), particularly on the leeward side of trees. Roots with an I-beam structure can be more prevalent at greater distances, especially on the windward side. Such roots resist vertical flexing (Nicoll and Ray 1996). Stokes et al. (1996) developed a theoretical model of anchoring, resistance to uprooting, and root branching patterns. Not surprisingly, deep roots were especially important.

B. Resource Uptake

Leaves and roots play analogous roles for plants. Leaves are the structures primarily responsible for carbon and energy uptake, and fine roots take up most of the water and nutrients acquired by plants. This dichotomy of structure and function is useful conceptually because above- and belowground resources are generally separated. In practice, however, it is difficult to disentangle above- and belowground processes (Donald 1958, Jackson and Caldwell 1992). Light availability powers the enzymes responsible for phosphate transport; adequate root surface area depends on the amount of CO_2 taken up by shoots. In turn, carbon and energy uptake require soil water for the maintenance of turgor and stomatal conductance and nitrogen to build photosynthetic proteins such as ribulose bisphosphate (RuBP) carboxylase. This interdependence has led to many perspectives on the "balance" of root and shoot processes (Brouwer 1963). In the following discussion of resource uptake by roots, we focus on the uptake of water, nitrogen, and phosphorus. A more detailed discussion can be found in Nye and Tinker (1977), Marschner (1995), and Casper and Jackson (1997), on which much of the following section is based. Additional perspective can be found in Chapter 9.

Soil resources typically reach the surface of roots by three processes: root interception, mass flow of water and nutrients, and diffusion (Marschner 1995). Root interception occurs as a root grows through the soil, physically displacing soil particles and clay surfaces and acquiring water and nutrients. This process typically accounts for less than 10% of the resources taken up by roots. Mass flow, which is driven by plant transpiration, depends on the rate of H_2O movement to the root and the concentration of dissolved nutrients in the soil solution. Nutrient diffusion toward the root occurs when nutrient uptake by the root exceeds the supply by mass flow and a depletion zone around the root is created. The supply of nutrients by diffusion is especially important for those nutrients with large fractions bound to the soil matrix, such as K^+ and $H_2PO_4^-$. In nature, mass flow and diffusion work in concert to supply N, P, and K, and are difficult to separate in the field (Nye and Tinker 1977).

Water moves into and out of roots passively based on the water potential gradient in the soil–plant system. In contrast, nutrient uptake is generally an enzymatic process that follows apparent Michaelis–Menten kinetics:

$$V = V_{max} \, C_1(C_1 + K_m)$$

where V is the flux of ion into the root per unit time, V_{max} is the maximum influx rate, C_1 is the soil solution concentration at the root surface, and K_m is the soil solution concentration where influx is 50% of V_{max} (Nye and Tinker 1977). The equation sometimes includes a C_{min} term, the soil solution concentration at which net influx into the root is zero (Barber 1984). Because nutrient uptake is generally enzymatic, it is sensitive to reductions in photosynthesis by shoots (Jackson and Caldwell 1992).

The belowground competitive ability of plants is often directly proportional to the size of their root systems. This is in contrast to shoot systems, where a relatively small portion of leaves can overtop a canopy and acquire most of the available light. There are many examples where root systems with the highest densities and occupying the most space are the strongest competitors (Aerts et al. 1991, Casper and Jackson 1997). Consequently, a plant may grow higher root densities or extend the volume of soil explored to acquire more water and nutrients. Plants can also increase resource uptake by selective foraging. Plants frequently respond to enriched patches of soil water and nutrients by proliferating roots, selectively growing roots in the zone of enrichment (Duncan and Ohlrogge 1958; see also Chapter 9). Proliferated roots tend to be smaller in diameter and greater in density than those found in the background soil. A second related factor that may increase resource uptake is a change in fine-root demography. In a Michigan hardwood forest, not only did roots proliferate in response to water and nitrogen patches, but the new roots lived significantly longer than new roots in control patches (Pregitzer and Hendrick 1993). Architectural adjustment (changes in root topology, length, or branching angles) is another type of morphological plasticity that can increase nutrient uptake. Fitter (1994) examined the architectural attributes of 11 herbaceous species and showed that roots in relatively high-nutrient patches typically had a more herringbone branching pattern than roots in low-nutrient patches, concentrating higher order lateral roots in the patches and increasing the efficiency of nutrient uptake.

Physiological plasticity can selectively increase nutrient uptake by altering enzyme attributes or other physiological traits. A species with more enzymes per root surface area (greater V_{max}), a higher ion affinity of enzymes (smaller K_m), or with a greater ability to draw nutrients down to a low level (smaller C_{min}) will be at a competitive advantage (ignoring other factors). Plants in the laboratory and in the field have been shown to increase V_{max} and decrease K_m in response to localized nutrients (Drew and Saker 1975, Jackson et al. 1990). For example, Jackson et al. (1990) showed that grass and shrub species in the field were able to selectively increase physiological rates of phosphate uptake in portions of their root system in fertilized soil patches (Figure 3). For water uptake, osmoregulation can lower cell water potential and maintain net uptake in the face of drying soils (Kramer and Boyer 1995). It is the suite of morphological and physiological attributes that determines resource uptake by plants. Although their genetic make-up plays a fundamental role in the type of root system plants possess, there is often great flexibility in how those genes are expressed based on environmental cues.

Extensive overviews of morphological and physiological plasticity and resource capture can be found in Hutchings and de Kroon (1994) and Robinson (1994) (see also Chapter 9).

C. Storage

Plants must cope with variable environments in which the timing of resource availability and uptake may not coincide with demand by the plant. The storage of carbohydrates, nutrients, and, to a lesser extent, water when resources are abundant provides insurance for future periods of high demand. Although many plant organs are involved in storage, roots are the most important storage site for many species. Roots play a particularly important role in cases where complete regeneration of aerial biomass is necessary. Perennial and biennial herbaceous species enduring seasonal environments or intense herbivory must rely heavily on belowground reserves. Woody plants in fire-prone environments such as Mediterranean-type ecosystems and savannas also depend heavily on roots for nutrient and carbohydrate storage (Bell et al. 1996, Bowen and Pate 1993, Miyanishi and Kellman 1986). Even in environments not typically subjected to fire, woody roots are an important site for storage, with carbohydrate concentrations often exceeding those in stems (Loescher et al. 1990).

In many species, belowground storage occurs in modified stems such as rhizomes, bulbs, corms, stem tubers, lignotubers, and burls (de Kroon and Bobbink 1997). These organs are functionally similar to roots with respect to storage and will be included in our discussion. Among true roots, storage can occur in all size classes of roots, but specialized, large-diameter roots such as root tubers and taproots often play the most important role.

Most of the mineral nutrients required by plants are stored in roots (Pate and Dixon 1982), and the majority of studies have focused on carbon, nitrogen, and phosphorus. Consequently, most of our discussion will be devoted to these elements. A large number of chemical compounds are involved in their storage.

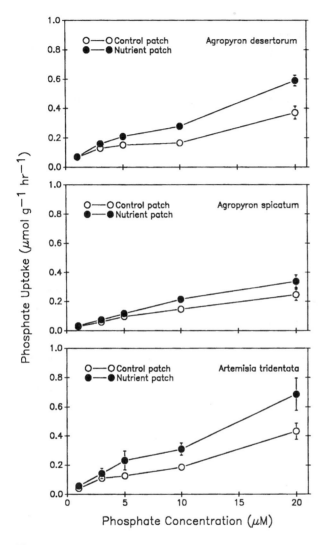

Figure 3 The rate of phosphate uptake for roots from enriched and control soil patches as a function of solution phosphate concentration (mean ± SEM; n = 6–8). Soil patches on opposite sides of plants in monoculture field plots were treated with 750 ml of nutrient solution or distilled water, and samples of the soil patches were cored 1 week after treatment. Roots from each core were subsampled and immersed in ^{32}P solutions. Results were similar for 3-day experiments as for the week-long experiments shown here. (*Source*: Jackson et al. 1990.)

In roots, starch is the most important form of carbon storage, although other polysaccharides can be important (Lewis 1984). In particular, fructan, a polymer of fructose, is common in many monocots and a few dicot families (Pollock 1986). Sucrose, monosaccharides, sugar alcohols, and lipids can also be prevalent (Glerum and Balatinecz 1980, Lewis 1984, Dickson 1991). Nitrogen is stored as specialized storage proteins, amino acids, amides, or nitrate (Tromp 1983, Staswick 1994). Phosphorus is stored primarily as phosphate, phytic acid, and polyphosphate (Bieleski 1973).

Within the cell, the vacuole is the most important site for the storage of sugars, phosphate, and nitrogen (Bieleski 1973, Willenbrink 1992), whereas starch storage occurs in plastids (Jenner 1992). In roots, storage occurs primarily in parenchyma cells (Bell et al. 1996, Bieleski 1973, Jenner 1992).

Plant storage of materials can be classified as accumulation or reserve storage (Chapin et al. 1990). Accumulation occurs when the uptake of a resource is greater than the plant's immediate capacity to use the resource. The plant would not be able to use the resource for other functions, so its storage does not compete with growth and maintenance. In contrast, reserve formation occurs at a time when the resource could otherwise be used for growth. Because reserve formation competes directly with growth and maintenance, there is a large cost in forming reserves.

Both accumulation and reserve formation are strongly influenced by resource availability. Shading reduces root carbohydrate concentrations (Bowen and Pate 1993, Jackson and Caldwell 1992), whereas elevated CO_2 can increase carbohydrate storage (Chomba et al. 1993). Nitrogen and phosphorus accumulation is typical under high nutrient availability (Chapin 1980), a response typically referred to as luxury consumption.

Low availability of one resource often increases the storage of other resources. For example, root carbohydrate storage has been found to be greater under water stress because tissue growth was more inhibited than photosynthesis (Busso et al. 1990). Low nutrient availability can also increase root carbohydrates (Jackson and Caldwell 1992) as has been shown for leaves (McDonald et al. 1991, Waring et al. 1985), presumably because low nutrient availability results in lower tissue production, reducing demand for photosynthate. These cases represent accumulation rather than reserve formation, because the storage results from low demand for the resource within the plant.

Demand for resources within the plant also largely determines the timing of storage. In herbaceous perennials and deciduous woody plants, root reserves are retranslocated at the beginning of the growing season to supply developing leaves, stems, and flowers. As a result, lowest reserve concentrations occur around the end of leaf flush (Chapin et al. 1986, Daer and Willard 1981, Keller and Loescher 1989, Wan and Sosebee 1990, Woods et al. 1959). After leaf maturation, the root undergoes a transition from carbohydrate source to carbohydrate sink, and replenishment of carbohydrate reserves begins. Root carbohydrate concentration reaches a peak around the time of leaf-fall (Chapin et al. 1986, Daer and Willard 1981, Keller and Loescher 1989, Wan and Sosebee 1990, Woods et al. 1959). In contrast to deciduous species, evergreen plants show lower seasonal fluctuations in the concentration of stored carbohydrates (Chapin et al. 1986, Dickson 1991).

Similar trends are observed in woody plants subjected to fire, but replenishment of the reserves often takes longer (Bowen and Pate 1993, Miyanishi and Kellman 1986). This slower replenishment may result from the greater amount of aerial biomass that must be replaced, or may represent an adaptation to a less frequent form of disturbance. On the other hand, grasses tolerant to frequent grazing exhibit rapid recovery of carbohydrate reserves after grazing occurs (Oesterheld and McNaughton 1988, Orodo and Trlica 1990, Owensby et al. 1970).

If disturbance recurs before adequate replenishment is complete, further recovery will be compromised (Kays and Canham 1991, Miyanishi and Kellman 1986). Overall, the cost associated with rapid replenishment of reserves can be large if replenishment competes with plant growth (Chapin et al. 1990).

D. Producing Hormones and Sensing the Environment

In addition to anchoring the plant and taking up and storing resources, roots sense their environment (Mahall and Callaway 1991) and convey information on the balance between root and shoot functioning in the form of hormones. Plants produce a variety of hormones, including auxin, cytokinins, gibberrelins, ethylene, and abscisic acid (ABA). Depending on how broad the definition of a hormone, other compounds such as jasmonates and polyamines may also be considered. Plant responses to hormones depend on changes in their concentration and in the sensitivity of tissues responding to the hormone (Trewavas 1981).

There are two groups of hormones where roots are the dominant site of synthesis: cytokinins and ABA. Cytokinins, named for their role in stimulating cytokinesis, are produced primarily in roots and are then transported in the xylem to leaves where they retard senescence and maintain metabolic activity (Torrey 1976). Cytokinins affect numerous plant processes, including cell division and morphogenesis. They affect protein synthesis and stimulate chlorophyll development. Cytokinin production also helps to coordinate root and shoot activity in plants. Roots become active in the spring and produce cytokinins that are transported to the shoot and activate dormant buds (Mauseth 1995). Cytokinins help to balance total leaf area relative to the root system by affecting rates of leaf expansion. Roots that produce cytokinins delay the senescence of shoot tissue. Cytokinins also provide a link from N uptake and the N status of roots to the synthesis of proteins. There are many other examples of how cytokinins help integrate the functioning of roots and shoots.

Unlike most plant hormones, ABA is primarily a growth inhibitor. ABA and gibberellins are both synthesized from mevalonic acid, the pathway that produces carotenoids and the general class of compounds known as terpenes. First characterized in 1963, ABA was proposed to play a role in the abscision of cotton bolls (Ohkuma et al. 1963). Subsequent evidence suggested that ethylene rather than ABA is a more important controller of abscision. Instead, ABA affects such plant processes as dormancy, senescence, stress responses (including water, freezing, and salt stress), and water uptake and stomatal regulation.

The role of roots in ABA synthesis is particularly important for stomatal closure and water stress. Changes in leaf cell turgor were originally thought to be the sole cause of stomatal closure in response to soil dehydration. However, when leaf cell turgor was maintained by pressurizing the root system, stomatal closure still occurred when part of the root system was in dry soil (Gollan et al. 1986). These data suggested that a chemical signal was produced in response to decreasing soil water status. Subsequent studies have identified ABA as the most important such signal (Davies and Zhang 1991). In leaves, ABA causes stomata to close at low concentrations (< 1 µM) by inhibiting K^+ influx to guard cells and activating K^+ efflux; its concentration can increase 50-fold in leaves experiencing water stress (Salisbury and Ross 1985, Walton and Li 1995). Experimental evidence suggests that ABA production in the roots is stimulated by reduced cell turgor and ABA moves to the shoots in the xylem stream (Tardieu et al. 1992). Further evidence for the importance of ABA in regulating plant water status is that mutants unable to produce ABA wilt permanently (Taiz and Zeiger 1991).

The relative contributions of water potential and ABA signals in determining leaf responses to water stress is a matter of continuing research. Reports that stomata often begin to close before any damage or detectable change in leaf water status occurs (Davies and Zhang 1991) have been interpreted as evidence that ABA signals from the roots provide a feed-forward mechanism for avoiding water stress. Tardieu et al. (1991, 1992) showed that there was a strong correlation between ABA in the xylem and leaf stomatal conductance (g_s) in maize plants, but little relationship between g_s and leaf turgor or water potential. The time lag between ABA production in the root and its arrival at the leaf suggests that root signals may not be well suited for stomatal regulation over short time scales (Kramer 1988). Seasonal studies showed that ABA signals were correlated with maximum stomatal conductance during progressive drying over many days but not on a diurnal basis (Wartinger et al. 1990). Such data suggest that water potential and ABA root signals combine to affect stomatal responses to soil water status (Tardieu and Davies 1993).

A rich literature on the synthesis and effects of ABA and cytokinins is available for the interested reader (Addicott 1983, Davies 1995, Davies and Jones 1991). In addition, Davies (1995) provides an excellent physiological overview of the production and function of plant hormones.

IV. ROOT SYMBIOSES

There are numerous symbioses in nature. Arguably, two of the most important for higher plants are mycorrhizae and symbiotic nitrogen fixation. Mycorrhizae (literally "fungus-root") are mutualistic associations between plant roots and soil fungi. In general, the fungus acts as a fine root matrix, exploring the soil and transporting nutrients and water back to the plant. In return, the plant typically supplies the fungus with a dependable carbon source. Fossil evidence indicates that mycorrhizae are as old as vascular land plants; they have been hypothesized as necessary for plant colonization of land in "soils" that would have been extremely nutrient poor (Pirozynski and Mallock 1975). Excellent introductory texts for the functioning and importance of mycorrhizae are those by Harley and Smith (1983) and Allen (1991).

Mycorrhizae are critical components of the rhizosphere in all terrestrial biomes, and greater than 90% of forest and grassland species are typically mycorrhizal (Brundrett 1991). The direct and indirect benefits of mycorrhizae to plants include increased phosphorus status and, to a lesser extent, improved nitrogen and water uptake (Fitter 1989). Mycorrhizae have been shown to supply 80% of the P taken up by plants and 25% of the N (Marschner and Dell 1984). Increased resource uptake occurs primarily through the greater efficiency with which fungi explore the soil compared with plant roots, but there is also evidence that the fungi mobilize resources otherwise unavailable to the plant. Such potential benefits come with pronounced carbon costs, estimated as 15% of net primary productivity for ectomycorrhizae (Vogt et al. 1982, for a Pacific fir forest) and 7–17% of the energy translocated to roots for arbuscular mycorrhizae (Harris and Paul 1987).

There are four broad classes of mycorrhizae: arbuscular mycorrhizae (AM), ectomycorrhizae (ECM), ericoid mycorrhizae, and orchid mycorrhizae. Arbuscular mycorrhizae are a type of endomycorrhizae, with hyphae, arbuscules (exchange organs), and often storage vesicles produced within the cortex of root cells. Fungi in AM are limited to Zygomycetes. Because only 150 or so fungal species have been shown to participate with vascular plants in AM, there is little specificity between fungi and host plant (although this apparent lack of specificity may also reflect present limitations in fungal taxonomy). Ectomycorrhizae are characteristic of certain woody plants, particularly those in the pine, willow, and beech families. They are characterized by a mantle or hyphal sheath around highly branched roots and by a hyphal network called the Hartig net that grows between cortical root cells. In contrast to AM, ECM do not generally penetrate living root cells. There are thousands of fungal species that form ECM, usually Basidiomycetes, and there is much greater host–fungus specificity than for AM.

By far, the majority of the world's mycorrhizal associations are AM or ECM. The absence of mycorrhizae is apparently limited primarily to early successional annuals of such relatively "advanced" families as the Brassicaceae, Chenopodiaceae, Amaranthaceae, and Zygophyllaceae (Allen 1991). Two other less prevalent but important forms of mycorrhizae are those of plants in the Ericaceae and Orchidaceae. Ericaceous mycorrhizae are intermediate between AM and ECM. They sometimes form a sheath around the root, but they also penetrate the outer root cortical cells with hyphal coils. Ericaceous plants often grow in acidic, nutrient-poor soils such as heathlands. The fungus can constitute 80% of their extensive mycorrhizal association by mass (Raven et al. 1986). These specialized, expensive mycorrhizae seem to be particularly beneficial for mobilizing and taking up organic nitrogen that would otherwise be unavailable to the plant (Read 1983).

Orchids also have a unique mycorrhizal relationship. Their seeds will not germinate in nature in the absence of the mycorrhizal fungus, typically a Basidiomycete in such genera as *Rhizoctonia*. Orchids also have an early nonphotosynthetic stage of growth in which carbon is obtained solely from the mycorrhizal fungus. Mature orchids receive more typical resources such as nutrients and water from mycorrhizae.

A second important root symbiosis is the fixation of nitrogen by plants and bacteria. Biological nitrogen fixation adds approximately 150×10^{12} g N to terrestrial ecosystems each year, roughly 1 g m^{-2} on average for all of the earth's land (Burns and Hardy 1975, Vitousek et al. 1997). Functionally, the reciprocal benefits of symbiotic N fixation are similar to those for mycorrhizae; the plant supplies carbon and energy to the symbiont and the symbiont supplies a resource to the plant. The two symbioses differ in that biological fixation supplies only nitrogen and the symbiont is a bacterium rather than a fungus. In addition to sugars, the plant also supplies an environment conducive to N fixation in which bacterial enzymes are protected from atmospheric oxygen and light. A two-protein, enzyme complex called nitrogenase catalyzes the fixation of N through the following chemical reaction:

$$N_2 + 8H^+ + 6e^- + XATP \rightarrow 2NH_4^+ + XADP + XP_i$$

where the value for the number of ATPs, ADPs, and P_is is approximately 15. The ammonium is converted to amino acids or ureides before being transported to the plant.

The most important N-fixing bacteria are those in the genus *Rhizobium*, which colonize the roots of legumes. The bacterium enters the root through root hairs that curl in response to a chemical secreted by the bacterium. A bacterial infection thread then penetrates the root cortex and induces formation of the nodule where N fixation occurs. A second important group of symbionts are actinomycete bacteria; they fix N as in roots of such plant genera as *Alnus, Myrica*, and *Ceanothus*.

In addition to the symbioses of *Rhizobium*, actinomycetes, and higher plants, there are also rhizosphere bacteria such as *Azospirillum* that fix nitrogen on the surface of roots of many plant species. It is debatable whether these relationships constitute a true symbiosis. There is clearly not the tight coupling that is seen with *Rhizobium* and legumes, and the quantities of N fixed are much smaller. Nevertheless, this source of N is important for a number of grasses and crop species.

V. GLOBAL PATTERNS OF ROOT DISTRIBUTIONS

Despite tremendous natural variation in the soil around plants and in the plasticity individual plants show to such variation (Jackson and Caldwell 1993, Snaydon 1962), there are broad patterns in the distributions of roots observed for biomes and plant functional types. Jackson et al. (1996, 1997) constructed a global database of climate, soil, and root attributes from more than 250 literature studies. The data were separated by terrestrial biomes and plant functional types and fitted to an asymptotic model of vertical root distribution:

$$Y = 1 - \beta^d$$

where Y is the cumulative root fraction (a proportion between 0 and 1) from the soil surface to depth d (in cm) and β is the fitted extinction coefficient (Gale and Grigal 1987). β provides a simple numerical index of rooting distributions. Low β values (e.g., 0.90) correspond to proportionally more roots near the surface than do high β values (e.g., 0.98).

Examining the data for plant functional types, typical β values were 0.957 for grasses and 0.972 and 0.980 for trees and shrubs, respectively (Jackson et al. 1996). Cumulative root biomass in the top 30 cm of soil varied from 75% for a typical grass to 45% for an average shrub. Shrubs had 87% of their root biomass in the top meter of soil, whereas grasses had almost 99%. These estimates for total root biomass are slightly deeper than those in Jackson et al. (1996) because only profiles to $\approx$ 2 m soil depth or greater were used in this revised analysis. Terrestrial biomes also showed clear patterns for root distributions. Deserts, temperate coniferous forests, and savannas had some of the deepest distributions ($0.970 \leq \beta \leq 0.980$), whereas tundra, boreal forests, and temperate grasslands had the shallowest profiles (β = 0.913, 0.943, and 0.943, respectively). Tundra typically had 60% of roots in the upper 10 cm of soil, whereas deserts had approximately 20% of roots in the same depth increment.

The database was also used to examine global patterns of root biomass and annual belowground net primary production (NPP) (Jackson et al. 1997, Jackson et al. 1996). Average root biomass ranged from <1 kg m^{-2} for deserts and croplands to 4–5 kg m^{-2} for most forest systems. When root biomass estimates were combined with the extent of each biome, global root biomass was estimated to be approximately 290 $\times$ 10^{15} g (or 140 $\times$ 10^{15} g C, equivalent to $\approx$20% of atmospheric C). The standing biomass of live and dead fine roots (<2 mm diameter) was approximately 80 $\times$ 10^{15} g. Assuming conservatively that fine roots turn over once per year on average, they represent one third of global NPP for plants, approximately 20 $\times$ 10^{15} g C yr^{-1} (Jackson et al. 1997). The average C:N:P ratio was 450:11:1 for fine roots and 850:11:1 for more coarse roots (2 $\leq$ $\times$ $\leq$ 5 mm).

The upper meter of soil contains the majority of root biomass in most systems. Nevertheless, what constitutes the functional rooting depth of an ecosystem is an important and difficult question. For woody plants, much of surface root biomass is in large-diameter roots that play a strong role in anchoring and transport but not in resource uptake. Furthermore, even where fine-root biomass distributions are known with depth, root functioning is often not proportional to root biomass. As an example, Gregory et al. (1978) showed that winter wheat had only 3% of its root system by mass below 1 m soil depth; this small fraction of roots supplied almost 20% of the water transpired by the wheat canopy during midsummer. The importance of relatively deep roots may frequently be underestimated because few studies examine root abundance and functioning below 1 m soil depth (Jackson et al. 1996). Some of the root distributions estimated above are undoubtedly too shallow. There are also a number of other uncertainties in this type of analysis, particularly seasonal and spatial dynamics that are masked by pooling data across space and time.

Although 2 or 3 m is "deep" for the typical ecological study, roots clearly grow much deeper. Of the 255 species examined for maximum rooting depth by Canadell et al. (1996), almost 10% grew roots below 10 m depth. At least eight woody species have been shown to grow roots below 40 m (Table 1). The functional significance of such roots can be large. In the Brazilian cerrado and in Amazonian rainforests, roots have been found at least 18 m deep in the soil (Nepstad et al. 1994, Rawitscher 1948). More than 75% of transpired water in these systems can come from below 2 m soil depth, particularly during the dry season (Nepstad et al. 1994). Where resources, particularly water, are available at depth in the soil, deep roots can be disproportionately important for resource uptake. The depth to which roots grow and their distribution in the soil are also

Table 1 The Ten Deepest Records for Rooting Depth

Species	System	Maximum rooting depth (m)	Reference
Boscia albitrunca	Kalahari desert	68	Jennings (1974)
Juniperus monosperma	Colorado Plateau	61	Cannon (1960)
Eucalyptus sp.	Australian forest	61	Jennings (1971)
Acacia erioloba	Kalahari desert	60	Jennings (1974)
Prosopis juliflora	Arizona desert	54	Phillips (1963)
Eucalyptus calophylla	Australian forest	45	Campion (1926)
Medicago sativa	Agricultural field	40	Meinzer (1927)
Eucalyptus marginata	Jarrah forest	40	Dell et al. (1983)
Acacia raddiana	Niger desert	35	Anonymous (1974)
Quercus douglassii	California woodland	24	Lewis and Burgy (1964)

Source: Generated in part from data in Stone and Kalisz (1991) and Canadell et al. (1996).

likely to be important for the maintenance of leaf area in evergreen systems and in determining the boundaries between evergreen and deciduous vegetation.

VI. CONCLUSIONS

Roots function in anchoring the plant, capturing soil resources, sensing the environment, and storing reserves. The physical difficulties in studying roots and soil represent both a frustrating barrier and a tremendous opportunity for the creative scientist. There are numerous unanswered questions of root functioning in need of novel approaches (Jackson 1999). What is the relationship of root biomass to root functioning? In what systems are the phenologies of roots and shoots tightly coupled? Are the distribution of microbes and soil fauna in the soil coupled to plant rooting depth? How prevalent are deep roots and are they important only for water uptake? Do general patterns exist for such questions globally?

Fortunately, such tools as minirhizotrons and stable isotopes are improving our understanding of the turnover and functioning of root systems. Future progress will also be made by combining new approaches in ecological studies with the mechanistic insights of molecular biology. For example, Zhang and Forde (1998) recently showed that a nitrate-inducible gene in *Arabidopsis* (ANR1) is a key determinant of developmental plasticity in roots. Perhaps most of all, it is important to remember that the functioning of roots is intimately linked to that of shoots. This integration occurs through the mutual exchange of resources and plant signals. Attempts to understand root or shoot functioning must eventually take this integration into account.

ACKNOWLEDGMENTS

We wish to thank the National Science Foundation, the Andrew W. Mellon Foundation, and the Department of Energy's National Institute for Global Environmental Change for support of this work. L. Anderson, C. Bilbrough, H. de Kroon, and E. Jobbagy provided helpful comments on the manuscript.

REFERENCES

Addicott, F., ed. 1983. Abscisic Acid. New York: Praeger.
Aerts, R., R. Boot, and J. van der Aart., 1991. The relation between above- and belowground biomass allocation patterns and competitive ability. Oecologia 87: 551–559.
Allen, M. 1991. The Ecology of Mycorrhizae. Cambridge: Cambridge University Press.
Anonymous. 1974. The tree of the Tenere desert is dead. Bios For. Trop. 153: 61–65.
Barber, S. 1984. Soil Nutrient Bioavailability. New York: Wiley.
Bell, T., J. Pate, and K. Dixon. 1996. Relationship between fire response, morphology, root anatomy, and starch distribution in south-west Australian Epacridaceae. Annals of Botany 77: 357–364.
Bieleski, R. 1973. Phosphate pools, phosphate transport, and phosphate availability. Annual Review of Plant Physiology 24: 225–252.
Bilbrough, C., and M. Caldwell. 1995. The effects of shading and N status on root proliferation in nutrient patches by the perennial grass Agropyron desertorum in the field. Oecologia 103: 10–16.
Bole, J. 1973. Influence of root hairs in supplying soil phosphorus to wheat. Canadian Journal of Soil Science 53: 169–175.
Bowen, B., and J. Pate. 1993. The significance of root starch in post-fire recovery of the resprouter Stiringia latifolia R.Br. (Proteaceae). Annals of Botany 72: 7–16.
Brady, J. 1934. Some factors influencing lodging in cereals. Journal of Agricultural Science 72: 273–280.
Brouwer, R. 1963. Some aspects of the equilibrium between overground and underground plant parts. Jaarb Int Bio Scheik Onderz Landbgewass 1963: 31–39.
Brouwer, R., O. Gasparíková, J. Kolek, and B. Loughman, eds. 1981. Structure and Function of Plant Roots. The Hague: Martinus Nijhoff.
Brundrett, M. 1991. Mycorrhizas in natural ecosystems. Advances in Ecological Research 21: 171–313.
Burns, R., and R. Hardy. 1975. Nitrogen Fixation in Bacteria and Higher Plants. Berlin: Springer-Verlag.

Busso, C., J. Richards, and N. Chatterton. 1990. Nonstructural carbohydrates and spring regrowth of two cool-season grasses: interaction of drought and clipping. Journal of Range Management 43: 336–343.

Caldwell, M., T. Dawson, and J. Richards. 1998. Hydraulic lift: consequences of water efflux from the roots of plants. Oecologia 113: 151–161.

Caldwell, M.M., and J.H. Richards. 1986. Competing root systems: morphology and models of absorption. In: Givnish, T.J., ed. On the Economy of Plant Form and Function, Cambridge: Cambridge University Press, pp 251–273.

Campion, W.E. 1926. The depth attained by roots. Australian Journal of Forestry 9:128.

Canadell, J., R.B. Jackson, J.R. Ehleringer, H.A. Mooney, O.E. Sala, and E.-D. Schulze. 1996. Maximum rooting depth of vegetation types at the global scale. Oecologia 108: 583–595.

Cannon, H.L. 1960. The development of botanical methods of prospecting for uranium on the Colorado Plateau. US Geological Survey Bulletin 1085-A, 50 pp.

Casper, B., and R. Jackson. 1997. Plant competition underground. Annual Review of Ecology and Systematics 28: 545–570.

Chapin, F.S. III. 1980. The mineral nutrition of wild plants. Annual Review of Ecology and Systematics 11: 233–260.

Chapin, F.S. III, J. Kendrick, and D. Johnson. 1986. Seasonal changes in carbon fractions in Alaskan tundra plants of differing growth forms: implications for herbivory. Journal of Ecology 74: 707–731.

Chapin, F.S. III, E.D. Schulze, and H.A. Mooney. 1990. The ecology and economics of storage in plants. Annual Review of Ecology and Systematics 21: 423–447.

Chomba, B.M., R. Guy, and H. Werger. 1993. Carbohydrate reserve accumulation and depletion in Engelman spruce (*Picea engelmannii* Parry): effects of cold storage and pre-storage CO_2 enrichment. Tree Physiology 13: 351–364.

Clarkson, D.T. 1993. Roots and the delivery of solutes to the xylem. Philosophical Transactions of the Royal Society London B 341: 5–17.

Crook, M., and A. Ennos. 1996. The anchorage mechanics of deep rooted larch, *Larix europea* x *L. japonica*. Journal of Experimental Botany 47: 1509–1517.

Daer, T., and E. Willard. 1981. Total nonstructural carbohydrate trends in bluebunch wheatgras related to growth and physiology. Journal of Range Management 34: 377–379.

Davies, P., ed. 1995. Plant Hormones: Physiology, Biochemistry and Molecular Biology. Dordrecht: Kluwer.

Davies, W., and H. Jones. 1991. Abscisic Acid: Physiology and Biochemistry. Oxford: Bios Scientific.

Davies, W., and J. Zhang. 1991. Root signals and the regulation of growth and development of plants in drying soil. Annual Review of Plant Physiology and Molecular Biology 42: 55–76.

Davis, T., and B. Haissig. 1994. Biology of Adventitious Root Formation. New York: Plenum Press.

De Kroon, H., and R. Bobbink. 1997. Clonal plant dominance under elevated nitrogen deposition, with special reference to *Brachypodium pinnatum* in chalk grassland. In: de Kroon H., and J. van Groenendael, eds. The Ecology and Evolution of Clonal Plants, Leiden: Backhuys Publishers, pp 359–379.

Dell, B., J.R. Bartle, and W.H. Tracey. 1983. Root occupation and root channels of jarrah forest subsoils. Australian Journal of Botany 31: 615–627.

Dickson, R. 1991. Assimilate distribution and storage. In: Raghavendra, A., ed. Physiology of Trees. New York: Wiley.

Dittmer, H. 1937. A quantitative study of the roots and root hairs of a winter rye plant (Secale cereale). American Journal of Botany 24: 417–419.

Donald, C. 1958. The interaction of competition for light and nutrients. Australian Journal of Agricultural Research 9: 421–435.

Drew, M. 1975. Comparison of the effects of a localized supply of phosphate, nitrate, ammonium and potassium on the growth of the seminal root system. New Phytologist 75: 479–490.

Drew, M. 1997. Oxygen deficiency and root metabolism: injury and acclimation under hypoxia and anoxia. Annual Review of Plant Physiology and Plant Molecular Biology 48: 223–250.

Drew, M., and L. Saker. 1975. Nutrient supply and the growth of the seminal root system in barley. II. Localized, compensatory increases in lateral root growth and rates of nitrate uptake when nitrate supply is restricted to only part of the root system. Journal of Experimental Botany 26: 79–90.

Duncan, W., and A. Ohlrogge. 1958. Principles of nutrient uptake from fertilizer bands II. Root development in the band. Agronomy Journal 50: 605–608.

Esau, K. 1977. Anatomy of Seed Plants. 2nd ed. New York: Wiley.

Fitter, A. 1989. The role and ecological significance of vesicular-arbuscular mycorrhizas in temperate ecosystems. Agric Ecosystems Environ 29: 137–151.

Fitter, A. 1994. Architecture and biomass allocation as components of the plastic response of root systems to soil heterogeneity. In: Caldwell, M., and R. Pearcy, eds. Exploitation of Environmental Heterogeneity by Plants. San Diego, CA: Academic Press, pp 305–323.

Gale, M., and D. Grigal. 1987. Vertical root distributions of northern tree species in relation to successional status. Canadian Journal of Forest Research 17: 829–834.

Glerum, C., and J. Balatinecz. 1980. Formation and distribution of food reserves during autumn and their subsequent utilization in jack pine. Canadian Journal of Botany 58: 40–54.

Glinski, J., and J. Lipiec. 1990. Soil Physical Conditions and Plant Roots. Boca Raton: CRC Press.

Gollan, T., J.B. Passioura, and R. Munns. 1986. Soil water status affects the stomatal conductance of fully turgid wheat and sunflower leaves. Australian Journal of Plant Physiology 13: 459–464.

Gregory, P., M. McGowan, and P. Biscoe. 1978. Water relations of winter wheat. 2. Soil water relations. J Agric Sci, Cambridge 91: 103–116.

Hales, S. 1727, reprinted 1961. Vegetable Staticks. London: London Scientific Book Guild.

Harley, J., and S. Smith. 1983. Mycorrhizal Symbiosis. New York: Academic Press,

Harris, D., and E. Paul. 1987. Carbon requirements of vesicular-arbuscular mycorrhizae. In: Safir, G., ed. Ecophysiology of VA Mycorrhizal Plants. Boca Raton: CRC Press, pp 93–105.

Hofer, R.-M. 1991. Root Hairs. In: Waisel, Y., A. Eshel, and U. Kafkafi, eds. Plant Roots: The Hidden Half. New York: Marcel Dekker, pp 129–148.

Hutchings, M., and H. de Kroon. 1994. Foraging in plants: the role of morphological plasticity in resource acquisition. Advances in Ecological Research 25: 159–238.

Itoh, S., and S. Barber. 1983. Phosphorus uptake by six plant species as related to root hairs. Agronomy Journal 75: 457–461.

Jackson, R. 1999. The importance of root distributions for hydrology, biogeochemistry, and ecosystem functioning. In: Tenhunen, J., and P. Kabat, eds. Integrating Hydrology, Ecosystem Dynamics and Biogeochemistry in Complex Landscapes. Chichester: Wiley, in press.

Jackson, R., and M. Caldwell. 1992. Shading and the capture of localized soil nutrients: nutrient contents, carbohydrates, and root uptake kinetics of a perennial tussock grass. Oecologia 91: 457–462.

Jackson, R., J. Manwaring, and M. Caldwell. 1990. Rapid physiological adjustment of roots to localized soil enrichment. Nature 344: 58–60.

Jackson, R., H. Mooney, and E.-D. Schulze. 1997. A global budget for fine root biomass, surface area, and nutrient contents. Proceedings of the National Academy of Sciences, USA 94: 7362–7366.

Jackson, R.B., and M.M. Caldwell. 1993. Geostatistical patterns of soil heterogeneity around individual perennial plants. Journal of Ecology 81: 683–692.

Jackson, R.B., J. Canadell, J.R. Ehleringer, H.A. Mooney, O.E. Sala, and E.-D. Schulze. 1996. A global analysis of root distributions for terrestrial biomes. Oecologia 108: 389–411.

Jenner, C. 1992. Storage of starch. In: Loewus, F., and W. Tanner, eds. Plant Carbohydrates I: Intracellular Carbohydrates, volume 13A. Berlin: Springer-Verlag, pp 700–747.

Jennings, C.M.H. 1974. The hydrology of Botswana. PhD dissertation. University of Natal, South Africa.

Jennings, J.N. 1971. Karst. MIT Press, Cambridge, MA, 252 pp.

Kays, J., and C. Canham. 1991. Effects of time and frequency of cutting on hardwood root reserves and sprout growth. Forest Science 37: 524–539.

Keller, J., and H. Loescher. 1989. Nonstructural carbohydrate partitioning in perennial parts of sweet cherry. Journal of the American Society for Horticultural Science 114: 969–975.

Kramer, P. 1969. Plant and Soil Water Relationships. New York: McGraw-Hill.

Kramer, P.J. 1983. Water Relations of Plants. New York: Academic Press.

Kramer, P., and J. Boyer. 1995. Water Relations of Plants and Soils. San Diego; Academic Press.

Kramer, P.J. 1988. Changing concepts regarding plant water relations. Plant Cell and Environment 11: 565–568.

Lewis, D.C., and R.H. Burgy. 1964. The relationship between oak tree roots and groundwater in fractured rock as determined by tritium tracing. J Geophys Res 69: 2579–2588.

Lewis, D.H. 1984. Occurrence and distribution of storage carbohydrates in vascular plants. In: Lewis, D., ed. Storage Carbohydrates in Vascular Plants. Cambridge: Cambridge University Press, pp 1–52.

Loescher, W., T. McCamant, and J. Keller. 1990. Carbohydrate reserves, translocation, and storage in woody plant roots. Hortscience 25: 274–281.

Lyford, W., and B. Wilson. 1964. Development of the root system of *Acer rubrum* L. Harvard Forest Paper 10.

Mahall, B.E., and R.M. Callaway. 1991. Root communication among desert shrubs. Proceedings of the National Academy of Science USA 88: 874–876.

Marschner, H. 1995. Mineral Nutrition of Higher Plants. London: Academic Press.

Marschner, H., and B. Dell. 1984. Nutrient uptake in mycorrhizal symbiosis. Plant and Soil 159: 89–102.

Mattheck, C. 1991. Trees: The Mechanical Design. Berlin: Springer-Verlag.

Mauseth, J. 1995. Botany: An Introduction to Plant Biology. Philadelphia: Saunders.

Mauseth, J.D. 1988. Plant Anatomy. Menlo Park: Benjamin/Cummings.

McCully, M. 1995. How do real roots work? Some new views of root structure. Plant Physiology 109:1–6.

McCully, M.E., and M.J. Canny. 1988. Pathways and processes of water transport and nutrient movement in roots. Plant and Soil 111: 159–170.

McDonald, A., T. Ericsson, and T. Ingestad. 1991. Growth and nutrition of tree seedlings. In: Raghavendra, A., ed. Physiology of Trees. New York: Wiley, pp 199–220.

Meinzer, O.E. 1927. Plants as indicators of ground water. U.S. Geological Survey, Water Supply Paper 577, 95 pp.

Miyanishi, K., and M. Kellman. 1986. The role of root nutrient reserves in regrowth of two savanna shrubs. Canadian Journal of Botany 64: 1244–1248.

Nepstad, D.C., C.R. de Carvalho, E.A. Davidson, P.H. Jipp, P.A. Lefebvre, G.H. Negreiros, E.D. da Silva, T.A. Stone, S.E. Trumbore, and S. Vieira. 1994. The role of deep roots in the hydrological and carbon cycles of Amazonian forests and pastures. Nature 372: 666–669.

Nicoll, B., and D. Ray. 1996. Adaptive growth of tree root systems in response to wind action and site conditions. Tree Physiology 16: 891–898.

Nye, P., and P. Tinker. 1977. Solute Movement in the Soil-Root System. Berkeley: Blackwell.

Oesterheld, M., and S. McNaughton. 1988. Intraspecific variation in the response of *Themeda triandra* to defoliation: the effect of time of recovery and growth rates on compensatory growth. Oecologia 77: 181–186.

Ohkuma, K., J. Lyon, F. Addicott, and O. Smith. 1963. Abscisin II, an abscission-accelerating substance from young cotton fruit. Science 142: 1592–1593.

Orodo, A., and M. Trlica. 1990. Clipping and long-term grazing effects on biomass and carbohydrates of Indian ricegrass. Journal of Range Management 43: 52–57.

Owensby, C., G. Paulsen, and J. McKendrick. 1970. Effect of burning and clipping on big bluestem reserve carbohydrates. Journal of Range Management 23:358–360.

Pate, J.S., and K.W. Dixon. 1982. Tuberous, Cormous and Bulbous Plants. Nedlands: University of Western Australia Press.

Perumalla, C., C. Peterson, and D. Enstone. 1990. A survey of angiosperm species to detect hypodermal Casparian bands. I. Roots with a uniseriate hypodermis and epidermis. Botanical Journal of the Linnean Society 103: 93–112.

Peterson, C.A., and C.J. Perumalla. 1990. A survey of angiosperm species to detect hypodermal Casparian bands: II. Roots with a multiseriate hypodermis or epidermis. Botanical Journal of the Linnean Society 103: 113–125.

Peterson, R., and C. Peterson. 1986. Ontogeny and anatomy of lateral roots. In: M. Jackson, M., ed. New Root Formation in Plants and Cuttings. Dordrecht: Martinus Nijhoff Publishers, pp 1–30.

Phillips, W.S. 1963. Depth of roots in soil. Ecology 44: 424.

Pirozynski, K., and D. Mallock. 1975. The origin of land plants: a matter of mycotrophism. Biosystems 6: 1533–1164.

Pollock, C. 1986. Fructans and the metabolism of sucrose in vascular plants. New Phytologist 104: 1–24.

Pregitzer, K., and R. Hendrick. 1993. The demography of fine roots in response to patches of water and nitrogen. New Phytologist 125: 575–580.

Raven, P., R. Evert, and S. Eichhorn. 1986. Biology of Plants. 4th ed. New York: Worth.

Rawitscher, F. 1948. The water economy of the vegetation of the ''Campos Cerrados'' in southern Brazil. Journal of Ecology 36: 237–268.

Read, D. 1983. The biology of mycorrhiza in the Ericales. Canadian Journal of Botany 61: 985–1004.

Reichle, D., B. Dinger, N. Edwards, W. Harris, and P. Sollins. 1973. Carbon flow and storage in a forest ecosystem. In: Woodwell, G., and E. Pecan, eds. Carbon and the Biosphere. US Atomic Energy Commission, pp 345–365.

Robinson, D. 1994. The responses of plants to non-uniform supplies of nutrients. New Phytologist 127: 635–674.

Salisbury, F., and C. Ross. 1985. Plant Physiology. 3rd ed. Belmont: Wadsworth.

Snaydon, R. 1962. Micro-distribution of *Trifolium repens* L and its relation to soil factors. Journal of Ecology 50: 133–143.

Somerville, A. 1979. Root anchorage and root morphology of Pinus radiata on a range of ripping treatments. N Z J For Sci 9: 294–315.

Staswick, P. 1994. Storage proteins in vegetative plant tissues. Annual Review of Plant Physiology and Plant Molecular Biology 45: 303–322.

Steudle, E. 1994. Water transport across roots. Plant and Soil 167: 79–90.

Steudle, E., and A.B. Meshcheryakov. 1996. Hydraulic and osmotic properties of oak roots. Journal of Experimental Botany 47: 387–401.

Stokes, A., J. Ball, A. Fitter, P. Brain, and M. Coutts. 1996. An experimental investigation of the resistance of model root systems to uprooting. Annals of Botany 78: 415–421.

Stone, E., and P. Kalisz. 1991. On the maximum extent of tree roots. For Ecol Manage 46: 59–102.

Taiz, L., and E. Zeiger. 1991. Plant Physiology. Redwood City: Benjamin/Cummings.

Tardieu, F., and W.J. Davies. 1993. Integration of hydraulic and chemical signalling in the control of stomatal conductance and water status of droughted plants. Plant Cell and Environment 16: 341–349.

Tardieu, F., N. Katerji, O. Bethenod, J. Zhang, and W. Davies. 1991. Maize stomatal conductance in the field: its relationship with soil and plant water potentials, mechanical constraints and ABA concentration in the xylem sap. Plant Cell and Environment 14: 121–126.

Tardieu, F., J. Zhang, N. Katerji, O. Bethenod, S. Palmer, and W.J. Davies. 1992. Xylem

ABA controls the stomatal conductance of field-grown maize subjected to soil compaction or soil drying. Plant Cell and Environment 15: 193–197.

Torrey, J. 1976. Root hormones and plant growth. Annual Review of Plant Physiology 27: 435–459.

Trewavas, A. 1981. How do plant growth substances work? Plant Cell and Environment 4: 203–228.

Tromp, J. 1983. Nutrient reserves in roots of fruit trees, in particular carbohydrates and nitrogen. Plant and Soil 71: 401–413.

Vitousek, P., J. Aber, R. Howarth, G. Likens, P. Matson, D. Schindler, W. Schlesinger, and D. Tilman. 1997. Human alteration of the global nitrogen cycle: sources and consequences. Ecological Applications 7: 737–750.

Vogt, K., C. Grier, C. Meier, and R. Edmonds. 1982. Mycorrhizal role in net primary production and nutrient cycling in *Abies amabilis* ecosystems in western Washington. Ecology 63: 370–380.

Walton, D., and Y. Li. 1995. Abscisic acid biosynthesis and metabolism. In: P. Davies, P., ed. Plant Hormones: Physiology, Biochemistry and Molecular Biology. Dordrecht: Kluwer, pp 140–157.

Wan, C., and R.E. Sosebee. 1990. Relationship of photosynthetic rate and edaphic factors to root carbohydrate trends in honey mesquite. Journal of Range Management 43: 171–176.

Waring, R., A. McDonald, S. Laarson, T. Ericsson, E. Wiren, A. Arwidsson, A. Ericsson, and T. Lohammer. 1985. Differences in chemical composition of plants grown at constant relative growth rates with stable mineral nutrition. Oecologia 66: 157–160.

Wartinger, A., H. Heilmeier, W. Hartung, and E.-D. Schulze. 1990. Daily and seasonal courses of leaf conductance and abscisic acid in the xylem sap of almond trees (*Prunus dulcis*) under desert conditions. New Phytologist 116: 581–587.

Weatherley, P. 1982. Water uptake and flow in roots. In: Lange, O., P. Nobel, C. Osmond, and H. Ziegler, eds. Encyclopedia of Plant Physiology, volume 12B. Berlin: Springer-Verlag, pp 79–109.

Willenbrink, J. 1992. Storage of sugars in higher plants. In: Loewus, F.A., and W. Tanner, eds. Plant Carbohydrates I: Intracellular Carbohydrates, volume 13A. Berlin: Springer Verlag, pp 684–699.

Woods, F., H. Harris, and R. Caldwell. 1959. Monthly variations of carbohydrates and nitrogen of sandhill oaks and wiregrass. Ecology 40: 292–295.

Zhang, H., and B.G. Forde. 1998. An *Arabidopsis* MADS Box gene that controls nutrient-induced changes in root architecture. Science 279: 407–409.

6

Water Relations and Hydraulic Architecture

Melvin T. Tyree
Aiken Forestry Sciences Laboratory, Burlington, Vermont

I. INTRODUCTION

Water relations of plants is a large and diverse subject. This chapter is confined to some basic concepts needed for a better understanding of the role of water relations in plant ecology. Readers seeking more details should consult Slatyer (1967) and Kramer (1983).

Water movement in plants is purely passive. In contrast, plants are frequently involved in active transport of substances across membranes. Such transport involves membrane-bound proteins (enzymes) that impart metabolic energy to transport and increase the energy of the substance being transported. Although there have been claims of active water movement in the past, no claim of active water transport has ever been proved.

Passive movement of water (like passive movement of other substances or objects) still involves forces, but passive movement is defined as spontaneous movement in a system that is already out of equilibrium in such a way that the system tends toward equilibrium. Active movement, by contrast, requires the input of biological energy and moves the system further away from equilibrium or keeps it out of equilibrium despite continuous passive movement in the counterdirection.

The basic equation that describes passive movement is Newton's law of motion on earth where there is friction:

$$v = (1/f)\ F \tag{1}$$

where v is velocity of movement (m s^{-1}), F is the force causing the movement (N), and f is the coefficient of friction (N s m^{-1}). In the context of passive water or solute movement in plants, it is more convenient to measure moles moved per s per unit area, which is a unit of measure called a flux density (J). Fortunately, there is a simple relationship between J, v, and concentration (C, mol m^{-3}) of the substance moving: $J = Cv$. In addition, in a chemical/biological context, it is easier to measure the energy of a substance and how the energy changes as it moves than it is to measure the force acting on the substance. Passive movement of water or a substance occurs when it moves from a location where it has high energy to one where it has lower energy. The appropriate energy to measure is called the chemical potential, μ, and it has units of energy per mol (J mol^{-1}).

The force acting on the water or solute is the rate of change of energy with distance, hence, $F = -(d\mu/dx)$, which has units of $J\ m^{-1}\ mol^{-1}$ or $N\ mol^{-1}$ (because $J = N\ m$). Therefore, replacing F with $-(d\mu/dx)$ and J with v gives:

$$J = -K(d\mu/dx) \tag{2}$$

where K is a constant equal to C/f. Equation 2 or some variation on it is used to describe water movement in soils and plants. The variations on Eq. 2 generally involve measuring J in kg or m^3 of water rather than mol and measuring μ in pressure units rather than energy units. These changes can be accommodated by incorporating the conversion factor for unit changes into K in the variant equations as shown elsewhere in this chapter.

II. WATER RELATIONS OF PLANT CELLS

The water relations of plant cells can be described by the equation that gives the energy state of water in cells and how this energy state changes with water content, which can be understood through the Höfler diagram. First, the factors that determine the energy state of water in a cell will be considered.

The energy content of water depends on temperature, height in the earth's gravitational field, pressure, and mole fraction of water (X_w) in a solution. For practical purposes, we evaluate the chemical potential of water in a plant cell in terms of how much it differs from pure water at ground level and at the same temperature as the water in the cell, i.e., we measure

$$\Delta\mu = \mu - \mu_0 \tag{3}$$

where μ_0 is the chemical potential of water at ground level at the same temperature as the cell. It has become customary for plant physiologists to report $\Delta\mu$ ($J\ mol^{-1}$) in units of J per m^3 of water because this has dimensions equal to a unit of pressure: Pa ($= N\ m^{-2}$), and this new quantity is called water potential (Ψ). The conversion involves dividing $\Delta\mu$ by the partial molal volume of water ($\overline{V}_w$). i.e.,

$$\Psi = \frac{1}{\overline{V}_w}\Delta\mu. \tag{4}$$

In general, Ψ is given by

$$\Psi = P + \pi + \rho gh \tag{5}$$

where P is the pressure potential (the hydrostatic pressure), $\pi = RT/\overline{V}_w \ln X_w$ is called the osmotic potential (sometimes called osmotic pressure) and is approximated by $\pi = -RTC$ where C is the osmolal concentration of the solution, R is the gas constant, T is the Kelvin temperature, ρ is the density of water, g is the acceleration due to gravity, and h is the height above ground level.

Water will flow into a cell whenever the water potential outside the cell (Ψ_o) is greater than the water potential inside the cell (Ψ_c). Let us consider the water relations of a cell at ground level, i.e., how water moves in and out of the cell in the course of a day. Water flows though plants in xylem conduits (vessels or tracheids), which are nonliving pipes, and the walls of the pipes are made of cellulose. Water can freely pass in and out of the conduits through the cellulose walls (Figure 1A). The water potential of the water in the xylem conduit is given by:

$$\Psi_x = P_x + \pi_x \qquad (6)$$

During the course of a day, P_x might change from slightly negative values at sunrise (e.g., -0.05 MPa) to more negative values by early afternoon (e.g., -1.3 MPa) and then might return again by the next morning, as explained in Section VI. Because the concentration of solutes in xylem fluid is usually very low (i.e., plants transport nearly pure water in xylem), π_o is not very negative (e.g., -0.05 MPa). Ψ_x might change from -0.1 to -1.35 MPa in the example given in Figure 1B. This daily change in Ψ_x will cause a daily change in Ψ_c as water flows out of the cell as Ψ_x falls and into the cell as Ψ_x rises.

The two primary factors that determine the water potential of a cell at ground level (Ψ_c) are turgor pressure P_t ($=P$ inside the cell) and π of the cell sap (π_c).

$$\Psi_c = P_t + \pi_c \qquad (7)$$

Plant cells generally have π_c values in the range of -1 to -3 MPa; consequently, P_t is often a large positive value whenever $\Psi_c = \Psi_x$. There are two reasons for cells having P_t greater than 0: (1) the protoplasm of living cells is enclosed inside a semipermeable membrane (the plasmalemma membrane) that permits relatively

---

Figure 1   (A, inset) A living cell with plasmalemma membrane adjacent to a xylem conduit (shown in longitudinal section). The living cell is surrounded by a soft cell wall composed of cellulose that retards expansion of the cell and does not prevent the collapse of the cell as P_t falls. The xylem conduit is surrounded by a woody cell wall (cellulose plus lignin) that strongly retards both expansion and collapse when P_x becomes negative. (A) A Höfler diagram showing how the cell water potential (Ψ_c), the cell osmotic potential (π_c), and the turgor pressure (P_t) changes with cell volume. (B) A representative daily time course of how cell or xylem water potential (Ψ) changes during the course of a day. The component water potentials shown are xylem pressure potential (P_x), xylem osmotic potential (π_x), cell turgor pressure (P_t), and cell osmotic potential (π_c). Also shown is how cell volume changes with time. See text for more details.

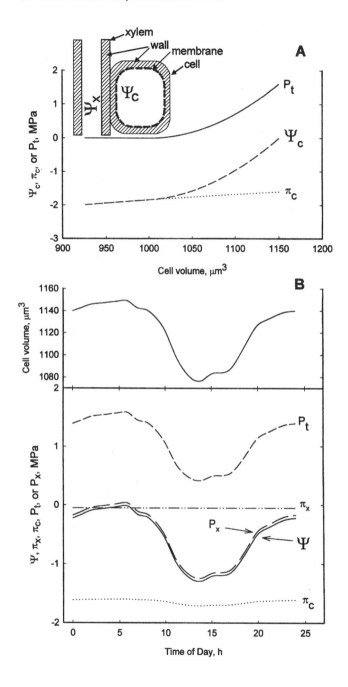

rapid transmembrane movement of water and relatively slow transmembrane movement of solutes; therefore, the solutes inside the cell making π_c negative cannot move out to the xylem to make π_x more negative; and (2) The membrane-bound protoplasm is itself surrounded by a relatively rigid elastic cell wall. The cell wall must expand as water flows into the cell to accommodate the extra volume; the stretch of the elastic wall place the cell contents under a positive pressure (much like a tire pumped up by air puts the air in the tire under pressure). The rise in P_t raises Ψ_c until it reaches a value equal to Ψ_x, at which point water flow stops.

The effect of water movement into or out of a cell is described by a Höfler diagram (Figure 1A). Entry of water into the cell has two effects: it causes a dilution of cell contents, hence π_c becomes slightly less negative, and the turgor pressure, P_t, rises very rapidly with cell volume. The net effect of the increase in π_c and P_t is an increase in Ψ_c toward zero. Conversely, a loss of water makes Ψ_c, π_c, and P_t fall to increasingly negative values. If Ψ_c falls low enough, then the P_t falls to zero and remains at zero in cells in soft tissue. In woody tissues, i.e., in cells with lignified cell walls, the lignification will prevent cell collapse and P_t can fall to negative values. The information in the Höfler diagram can be used to understand the water relations of cells in the course of a day.

A representative time course is shown in Figure 1B. Suppose the sun rises at 6 AM and sets at 6 PM. Radiant energy falling on leaves will enhance the rate of evaporation above the rate at which roots can replace evaporated water, hence, both Ψ_c and Ψ_o will fall to the most negative values in early afternoon (indicated by the solid line marked by Ψ in Figure 1B). As the afternoon progresses, the light intensity diminishes and the rate of water loss from the leaves falls below the rate of uptake of water from the roots, hence Ψ increases. Overnight, the value of Ψ will return to a value near 0 in wet soils or more negative values in drier soils; in either case, Ψ reaches a maximum value just before dawn. The value of Ψ before sunrise (called the predawn water potential) is often taken as a valid measure of soil dryness ($=\Psi_{soil}$) in the rooting zone of the plant. In the xylem, the osmotic potential (π_x) remains more or less constant and only slightly negative during the day; therefore, all change in Ψ_x is brought about by a large change in P_o, which closely parallels changes in Ψ. Similarly, in a living cell, changes in Ψ_c are brought about by large changes in P_t, whereas the cell osmotic potential (π_c) changes only slightly and remains a large negative value.

III. WATER RELATIONS OF WHOLE PLANTS

A. Regulation of Water Loss by Leaves

The water relations of a whole plant can be understood in terms of the fundamental physiological role of the leaf. The leaf is an organ designed to permit CO_2

uptake at a rate needed for photosynthesis while keeping water evaporation from leaves at a reasonably low rate. The roots have the function of extracting water from the soil to replace water evaporated from leaves. The leaf structure of a typical plant is illustrated in Figure 2. The upper and lower epidermis of leaves are covered with a waxy cuticle that reduces water loss from the leaf to negligible level. All gas exchange into and out of the leaf occurs via stomata. The guard cells of the stomata are capable of opening and closing air passages that provide pathways for diffusion of CO_2 into the leaf and for loss of water vapor from the leaf. Photosynthesis occurs primarily in the palisade and mesophyll cells of the leaf (Figure 2B). Leaves are thin enough (approximately 0.1 mm thick) to permit photon penetration to chloroplasts (shown as dots in Figure 2B). The chloroplasts absorb the energy of the photon and through a photochemical process use the energy to convert CO_2 and water into sugar. The consumption of CO_2 in the chloroplasts lowers the concentration of CO_2 in the cells and thus sets up a concentration gradient for the diffusion of more CO_2 into the cells via the stomatal pores and mesophyll air spaces. Because mesophyll and palisade cell surfaces are wet, water will continuously evaporate from these surfaces and water vapor will continuously diffuse out of the leaf by the same pathway taken by CO_2. Evaporated water is continuously replaced by water flow through veins in the leaves. The veins contain xylem for water uptake and phloem for sugar export.

The physiology of guard cells has evolved to optimize photosynthesis when conditions are right for the photochemical reaction and to minimize water loss when conditions are wrong for photosynthesis. Stomata remain closed or only partly open when soils are dry; this is an advantage to the plant because roots are unable to extract water fast enough from dry soils to keep up with evaporation from leaves. Roots send a chemical signal to leaves, abscisic acid (ABA), which mediates stomatal closure (Davies and Zhang 1991). When soil water is not limited, and when the internal CO_2 concentrations fall below ambient levels in the atmosphere, stomata open in sunlight. In approximately half of the known species, stomata have the additional capability of sensing the relative humidity (RH) of the ambient air and tend to progressively close as RH in the air adjacent to the leaves decreases. Rapid stomatal opening is mediated by movement of K^+ from the epidermal cells to the guard cells (Figure 2C–F). The movement of K^+ makes π_c less negative in the epidermal cells and more negative in the guard cells; consequently, water flows from the epidermal cells to the guard cells, P_t falls in the epidermal cells, and P_t rises in the guard cells. The mechanical effect of this water movement and change in P_t causes the guard cells to swell into the epidermal region and open a stomatal pore.

The evaporative flux density (E) of water vapor from leaves is ultimately governed by Fick's law of diffusion of gases in air. The control exercised by the plant is to change the area available for vapor diffusion through the opening and closing of stomatal pores. The value of E (mmol water/s/m^2 of leaf surface) is

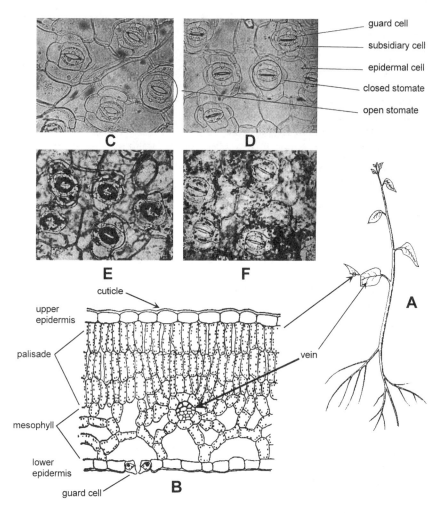

Figure 2 (A) A typical plant. (B) The lower leaf has been cut and the cross-section is shown enlarged approximately 750×. (C–F) Photographs of the lower leaf surface enlarged approximately 1000×. (E) and (F) have been stained to show location of K^+. (C) and (E) illustrate the open state of stomata when the leaf is exposed to light. *Stoma* is from the Greek and means mouth (plural, stomata); the two guard cells for a structure looks much like a mouth. (D) and (F) are similar leaves in light but also exposed to abscisic acid (ABA), causing stomata to remain closed in light. Note that in open stomata, K^+ is concentrated in the guard cells (E) and the guard cells have moved apart to form an air passage for gas exchange (the stomatal pore). Note that the K^+ is located in the epidermal cells when the stomata are closed (F). (*Source*: Adapted from Kramer 1983 and Bidwell 1979.)

given by:

$$E = g_L(X_i - X_o) \tag{8}$$

where g_L is the diffusional conductance of the leaf (largely controlled by the stomatal conductance, g_s), X_i is the mole fraction of water vapor at the evaporative surface of the palisade and mesophyll cells, and X_o is the mole fraction of water vapor in the ambient air surrounding the leaf. The mole fraction is defined as $X = n_w/N$, where n_w is the number of moles of water vapor and N is the number of moles of all gas molecules, including water vapor, the most abundant gas molecules being N_2 and O_2. The dependence of g_L on some environmental and physiological variables is illustrated in Figure 3.

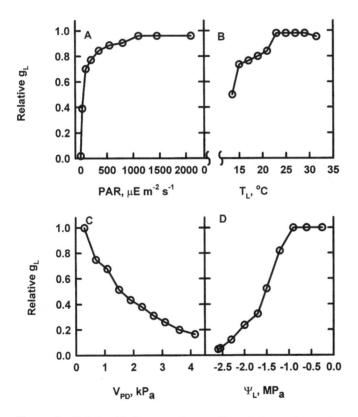

Figure 3 Relationship between change of g_L relative to the maximum value, $g_L/g_{L,max}$, and various environmental factors measured on *Acer saccharum* leaves. The environmental factors were: (A) photosynthetically active radiation (PAR); (B) leaf temperature (T_L); (C) vapor pressure deficit (V_{PD}); and (D) leaf water potential (Ψ_L). (*Source*: Adapted from Yang et al. 1998.)

The maximum value of X occurs when RH is 100%, i.e., when the air is saturated with water vapor; the maximum value of X increases exponentially with the Kelvin temperature of the air. The air at the evaporating surface of leaves is at the temperature of the leaf (T_L), and X_i is taken as the maximum value of X at saturation, which can be symbolized as $X_i = X[T_L]$. The value of X_o depends on the microclimate near the leaf, i.e., the air temperature and RH. However, the microclimate of the leaf is strongly influenced by the behavior of the plant community surrounding the leaf. Therefore, although the leaf has direct control over the value of g_s and g_L, it has less control over E than might appear from Eq. 8.

The qualitative aspects of how leaves influence their own microclimate is easily explained. When the sun rises in the morning, the radiant energy load on the leaf increases. This has two effects: T_L rises as the sun warms the leaves, and hence, X_i rises and g_L increases as stomata open. But the increased evaporation from the leaves causes X_o to rise as all the water vapor is added to the ambient air. Even changes of g_L under constant radiant energy load causes less change in E than might be expected from Eq. 8. When g_L doubles, E also doubles, but only temporarily. The increased E lowers T_L because of increased evaporative cooling, and hence lowers X_i. The increased E from all of the leaves in a stand eventually increases X_o, hence $X_i - X_o$ declines, causing a decline in E. Consequently, Eq. 8 is not very useful in predicting the value of E at the level of plant communities. Leaf-level behavior can be extrapolated to community-level equations if we take into account leaf-level solar energy budgets, i.e., equations that describe light absorption by leaves at all wavelengths and the conversion of this energy to temperature and heat fluxes. Studies of solar energy budgets have been conducted at both the leaf and stand level (Slatyer 1967, and Chang 1968; see also Section VII).

Most of the changes in E at the leaf level can be explained in terms of net radiation absorbed at the stand level, which is relatively easy to measure. This relationship is illustrated in Figure 4, where daily values of E at the leaf level ($kg\ m^{-2}/day$) are correlated with daily values of net radiation (MJ/m^2 of ground per day) measured with an Eppley-type net radiometer. Equations presented elsewhere to describe stand level E in terms of net radiation and other factors are validated by relationships such as shown in Figure 4. The other factors most commonly included account for plant control over E through leaf area index (leaf area per unit ground area), the effect of drought on g_L, and ambient temperature (see Section VII).

B. Tissue–Water Relations (Pressure–Volume Curves)

From the preceding section, the reader might falsely conclude that plants can lose water from leaves without any negative impact on the growth and survival.

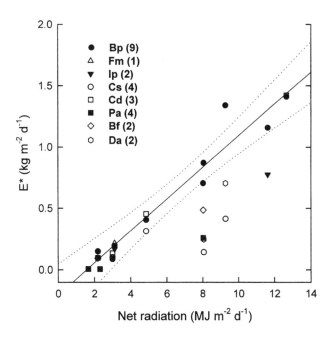

Figure 4 Correlation of daily water use of leaves (E*) and net radiation (NR). Data are from potometer experiments with nine woody cloud forest species in Panama. The number of days per species is given in parenthesis. The regression (with 95% confidence intervals) is for Bp: *Baccharis pendunculata* ($E* = -0.20 + 1.29$ NR, $r^2 = 0.92$). Even when the data for all species are pooled, NR proved to be a very good predictor of daily water use ($E* = -0.2 + 1.04$ NR, $r^2 = 0.72$). Cm, *Citharexylum macradenium*; Cd, *Croton draco*; Fm, *Ficus macbridei*; Ip, *Inga punctata*; Pa, *Parathesis amplifolia*; Bf, *Blakea foliacea*; Cs, *Clusia stenophylla*; and Da, *Dendropanax arboreus*. (*Source*: From Zotz et al. 1998.)

But there is more to water loss than is apparent from stomatal physiology and the energy interactions between leaves and their immediate environment. Whenever leaves lose water faster than the rate of water uptake by roots, the water potential in the xylem (Ψ_x) and of the leaf cells (Ψ_c) must also fall (see Section II). Most of the decline in Ψ_x is due to a drop in P_x, and when P_x becomes too negative, cavitations occur that prevent water flow though vessels (see Section VI). Most of the decline in Ψ_c is due to a drop in P_t, and when P_t becomes too small, cell growth stops. Growing cells are surrounded by relatively ridged "wooden boxes" consisting of cellulose cell walls. Cell walls must be stretched plastically to grow large, and the motive force on plastic stretch is the force of P_t against the cell walls. As leaves lose water, P_t and P_x must fall for the plant to extract water from the soil at a rate approximately equal to the rate of water loss. As soils become dry, values of P_t and P_x must fall even more for plants to extract water from soils. From Eq. 7 it can be seen that $P_t = \Psi_c - \pi_c$, hence plants have some control over P_t through "adjustments" in π_c. Some species have evolved to have lower values of π_c than others, and some species can make π_c more negative in response to drought by increasing the osmolal concentration (C) of solutes in their cells (because $\pi_c = -RTC$).

The relationship between Ψ and its components (P and π) in leaves can be described by a leaf-level Höfler diagram (Figure 2A). Many studies have reported the comparative physiology of tissue–water relationship of leaves and have discussed how differences in these relationships might explain ecological adaptation of plants. Readers interested in learning more should consult the references contained in Tyree and Jarvis (1982). The following is a brief overview of how Höfler diagrams are measured in leaves and some ecological applications of this information.

The fastest way to derive a Höfler diagram is to measure the water potential of leaves (Ψ_{leaf}) versus water loss using a Scholander-Hammel pressure bomb (Scholander et al. 1965, Tyree and Hammel 1972). This is done by obtaining a series of balance pressure (P_B) versus weight loss of leaves or shoots; the P_B is an approximate measure of the Ψ_{leaf}. The pressure bomb is a metal chamber into which is placed an excised leaf or shoot (Figure 5A) that is at an unknown water potential, Ψ_{leaf}. When gas pressure (P_{gas}) is applied to the leaf surface, the pressure of the fluid in the cells is increased by an equal amount so that $\Psi_{leaf} = \Psi_c = \pi_c + P_t + P_{gas}$. The xylem water potential and cell water potential are at equilibrium, $\Psi_c = \Psi_x$; from this it follows that $\pi_x + P_x = \pi_c + P_t + P_{gas}$ or $P_x = \pi_c + P_t + P_{gas} - \pi_x$. When the P_{gas} is at the balance pressure (P_B), P_x has risen to zero and xylem sap is squeezed out of the end of the branch or petiole protruding outside

---

Figure 5 Scholander-Hammel pressure bomb and analysis of pressure bomb data. (A) A pressure bomb is shown in the inset. Open circles (right axis): original data of balance pressure (P_B) versus weight loss of shoot in the bomb. Weight loss is induced by removing the shoot from the bomb and allowing water to evaporate. Closed circles (left axis): Transformed data of $1/P_B$ versus weight loss. The linear portion of the curve has been extrapolated to two points, A = y-intercept = $-1/\pi_0$, where π_0 is the average osmotic pressure of the symplast at full hydration when $\Psi = 0$; B = x-intercept = weight of water in the symplast when $\Psi = 0$ [this value can also be computed from $(-1/\pi_0)/\text{slope}$]; C = total water content of the shoot = maximum weight loss when oven-dried. (B) The Höfler diagram for the whole shoot derived from the data in (A). The x-axis is relative water content of the shoot and the y-axis is π = osmotic potential of the symplasm (solid triangles), P_t = turgor pressure of symplasm (open circles), and Ψ = total water potential of the symplasm (solid circles).

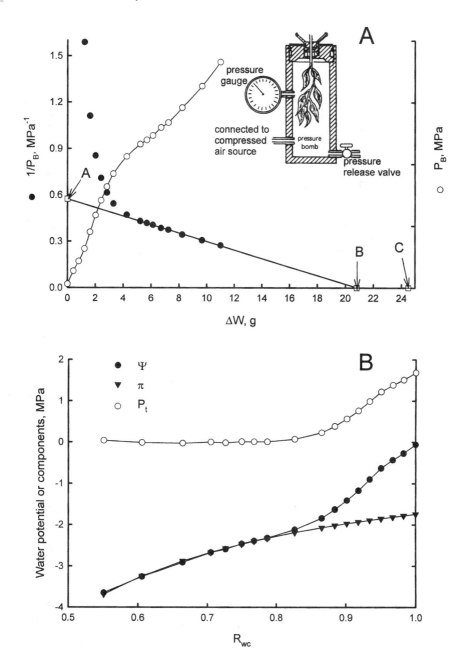

the pressure bomb (Figure 5). Therefore, we have $P_B = -(\pi_c + P_t) + \pi_x = -(\Psi_{leaf},$ when the branch or leaf was outside the bomb) $+ \pi_x$. P_B is usually approximated by $-\Psi_{leaf}$ since π_x is usually much smaller than Ψ_{leaf}.

A pressure–volume curve is obtained by slowly dehydrating a shoot and obtaining a series of weights, W, versus P_B values. If W_o was the original weight, then the cumulative weight loss is $\Delta W = W_o - W$. The pressure–volume curve is a plot of $1/P_B$ versus ΔW. Strictly speaking, the pressure–volume curve should be called a pressure–weight curve, but if ΔW is given in g, then that is the same as volume in mL since 1 mL of water weighs 1 g. The pressure–volume curve has a curved region for small ΔW values and a linear region for larger ΔW values (Figure 5B). When the linear portion of the plot is extrapolated back $\Delta W = 0$, the y-intercept (the point marked A in Figure 5A) is equal to $-1/\pi_o$, where π_o is the solute potential of the living cells at zero water potential. The point marked B in Figure 5A is the turgor loss point (Ψ_{tlp}), i.e., the value of Ψ_{leaf} when P_t reaches zero. The x-intercept (point C) is the volume of water contained in the symplast (W_s = total water in the protoplasm and vacuoles of all living cells), and the difference in x-values (D–C) is the amount of water in the apoplast (W_a = total water in xylem and cell walls). Höfler diagrams for shoots or leaves are usually plots of $\Psi, \pi,$ and P_t versus relative water content of the shoot or leaf. Relative water content (R_{WC}) is defined as (the current water content)/(the maximum water content at full hydration), $R_{WC} = (W_o - \Delta W)/(W_o - W_d)$, where W_d is the dry weight. Values of π at different R_{WC} are calculated from $\pi_o W_s/(W_o - \Delta W)$, values of Ψ are equated to $-P_B$, and values of P_t are calculated from $P_B - \pi_o/R_{WC}$. The justification for these relationships is given in a report by Tyree and Hammel (1972).

As plants dry, changes in π can be caused by changes in symplastic water content, W_s, or in the number of moles of solute in the symplasm, N_s, because $\pi = RTN_s/W_s$. Considerable emphasis has been placed on demonstrating changes in π as a result to changes in N_s (Turner and Jones 1980). A change in π caused by a change in N_s is called an osmotic adjustment. Diurnal changes in π ranging from 0.4–1.6 MPa have been reported for some plants; the amount of change resulting from diurnal changes in N_s is in the range of 0.2–0.8 MPa. Medium-term changes in π induced by slow soil dehydration have also been attributed to osmotic adjustment in drought-stressed versus unstressed plants. Osmotic adjustments of 0.1–1 MPa have been reported over periods of 3 days to 3 weeks. Long-term or seasonal changes in π range from 0.2–1.8 MPa; some of the largest changes are recorded during the onset of winter in temperate plants and appear to be correlated to changes in frost tolerance. The degree of diurnal, medium-term, and long-term osmotic adjustment varies widely between species. There are some species that have shown little or no adjustment (Tyree and Jarvis 1982).

Low values of π in plants should enhance the ability of plants to take up soil water under dry or saline conditions (Tyree 1976). This advantage is probably marginal in sandy soils because the available water reserves at soil water potentials less than -0.4 MPa are very small; therefore, the plant's ability to grow deep roots (Section IV) is probably of greater advantage. In clay soils, however, there are considerable water reserves at water potentials less than -0.4 MPa, so that low leaf and root values of π may be as important as root growth in assisting water uptake. Low values of π in leaves also enable P_t to remain above zero at lower values of Ψ than otherwise would be possible as Ψ falls. This allows the maintenance of open stomata with larger apertures and high stomatal conductances and higher net rates of photosynthesis down to lower values of Ψ than would be the case if π were higher (less negative). Osmotic adjustments and/or lower π values also enable maintenance of turgor pressure for growth, since it has been shown that the rate of volume growth $(r = dV/dt)$ of a cell is given by $m(P_t - Y)$, where m is the growth rate constant of a cell and Y is the yield point of the cell (Green et al. 1971).

Some attention in the past has been focused on the slope of the P_t line in the Höfler diagram in which the x-axis is R_{WC}. The bulk modulus of elasticity of a tissue is defined as:

$$\epsilon = \frac{\Delta P_t}{\Delta R_{wc}} R_{wc} \qquad (9)$$

The bigger the value of ϵ is, the bigger the slope. A large value of ϵ is seen as an advantage for water conservation. For a plant to extract water from the soil, it must first lose some water so that its Ψ_{leaf} falls below the soil water potential. Because most of the change in Ψ_{leaf} is due to the change in P_t, a large value of ϵ means that a plant would have to lose less water to lower Ψ_{leaf} than would a plant with a small value of ϵ.

Although values of ϵ have been reported to range from 0.5–20 MPa, the adaptive advantages of large versus small ϵ have never been clearly established (Tyree and Jarvis 1982) because the ecological advantage of a large ϵ may be rather marginal. All plants lose water during the day and regain most or all of the lost water during the night. Therefore, the advantage of less water loss means more constant concentrations of biochemical substrates during the day. Leaves can lose anywhere from 1–20% of their water content during the day, and this water loss could cause a corresponding 1–20% diurnal variation in substrates. Because the concentration of reactants and products would normally change during the day, even without the influence of leaf dehydration, it is not clear how much is gained by keeping relatively constant cell volume because of large ϵ.

IV. WATER ABSORPTION BY PLANT ROOTS

The primary factor affecting the pattern of water extraction by plants from soils is the rooting depth. Rooting depths can be extremely variable depending on soil conditions and species of plant producing the roots (Figure 6A). Many of the early studies of rooting depth and branching pattern of roots were performed in

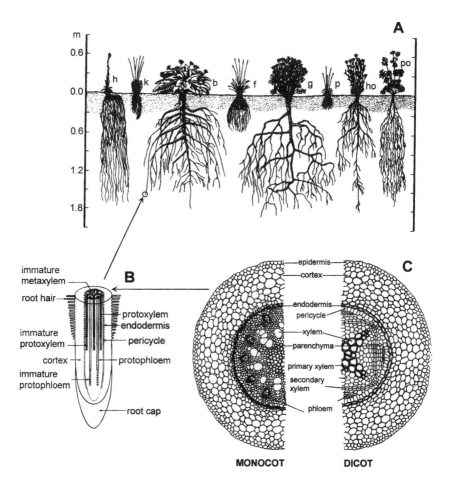

Figure 6 (A) Differences in root morphology and depth of root systems of various species of prairie plants growing in a deep, well-aerated soil. Species shown are: h, *Hieracium scouleri*; k, *Loeleria cristata*; b, *Balsamina sgittata*; f, *Festuca ovina ingrata*; g, *Geranium viscosissimum*; p, *Poa sandbergii*; ho, *Hoorebekia racemosa*; and po, *Potentilla blaschkeana*. (B) Enlargement of a dicot root tip enlarged approximately 50×. (C) Cross-section of monocot and dicot roots enlarged approximately 400×. (*Source*: Adapted from Kramer 1983, Steward 1964, and Bidwell 1979.)

the 1920s and 1930s in deep, well-aerated prairie soils, where roots penetrate to great depths. At the extreme, roots have been traced to depths of 10–25 m, e.g., alfalfa (10 m), longleaf pine, (17 m) (Kramer 1983), and drought-evading species in the California chaparral (25 m) (S. Davis, personal communication, April 1991). The situation is very different for plants growing in heavy soils, where 90% of the roots can be found in the upper 0.5–1.0 m.

In seasonally dry regions, e.g., central Panama, the majority of the roots may be located in the upper 0.5 m, but it is far from clear if the majority of water absorption occurs in the upper 0.5 m during the dry season. Water use by many evergreen trees is higher in the dry season than in the wet season, although the upper 1 m of the soils is much drier than the leaves of the trees ($\Psi_{soil} < \Psi_{leaf}$) (M. T. Tyree, personal observation, February 1989). Hence, the role of shallow versus deep roots of woody species deserves more study.

In addition, deeply rooted species may contribute to the water supply of shallow-rooted species through a process called hydraulic lift (Richards and Caldwell 1987). In one study, it was found that shallow-rooted species growing within 1–5 m of the base of maple trees were in a better water balance than the same species growing more than 5 m away (Dawson 1993). Each night, the Ψ_{soil} at a depth of 0.5 m increased underneath maple trees, but Ψ_{soil} did not increase at distances greater than 5 m from the trees. This indicated that the deep maple roots were in contact with moist soil and were capable of transporting water from deep maple roots to shallow maple roots overnight. Since the water potential of the shallow maple roots exceeded the water potential of the adjacent soil, water flow from the shallow maple roots to the adjacent soil contributed to the overnight rehydration of shallow soils.

Roots absorb both water and mineral solutes found in the soil, and the flow of solutes and water interact with each other. The mechanism and pathway of water absorption by roots is more complex than in the case of a single cell (Section II). Water must travel first radially from the epidermis to cortex, endodermis, and pericycle before it finally reaches the xylem vessels, from which point water flow is axial along the root (Figures 6B and 6C). The radial pathway (typically 0.3 mm long in young roots) is usually much less conductive than the axial pathway (>1 m in many cases); therefore, whole-root conductance is generally proportional to the root surface area. The radial pathway can be viewed as a composite membrane separating the soil solution from the solution in the xylem fluid. The composite membrane consists of serial and parallel pathways made up of plasmalemma membranes, cell wall membranes, and plasmodesmata (pores $<$ 0.5 μm diameter) that connect adjacent cells. The composite membrane is rather leaky to solutes; therefore, differences in osmotic potential between the soil (π_s) and the xylem (π_x) have less influence on the movement of water. At any given point along the axis, the water flux density across the root radius (J_r) will be given by

$$J_r = L_r [(P_s - P_x) + \sigma(\pi_s - \pi_x)] \tag{10}$$

where L_r is the radial root conductance to water and σ is the solute reflection coefficient. For an ideal membrane in which water but not solutes may pass, $\sigma = 1$. For the composite membrane of roots, σ is usually between 0.1–0.8. The system of equations that describes water transport in roots is complex when all of the factors are taken into account, e.g., axial and radial conductances, the fact that each solute has a different σ, and that the rate of water flow is influenced by the solute loading rate (J_s). Water and solute flow in roots can be described by a standing gradient osmotic flow model (readers interested in the details may consult Tyree et al. 1994b and Steudle 1992).

Fortunately, the equations describing water flow become simple when the rate of water flow is high. The concentration of solutes in the xylem fluid is equal to the ratio of solute flux to water flux ($J_s : J_w$) during steady-state flux. Solute flux tends to be more or less constant with time, but water flux increases with increasing transpiration. When water flow is high, the concentration of solutes in the xylem fluid becomes small and approaches values comparable to that in the soil solution, and pressure differences become quite large, hence ($P_s - P_x$) $\gg \sigma(\pi_s - \pi_x)$. Only at night or during rainy periods can values of ($P_s - P_x$) approach those of $\sigma(\pi_s - \pi_x)$. Therefore, water flow (J_w, kg s^{-1}) through a whole root system during the day can be approximated by

$$J_w \cong k_r (P_s - P_{x,b}) \tag{11}$$

where $P_{x,b}$ is the xylem pressure at the base of the plant and k_r is the total root conductance (combined radial and axial conductances).

V. HYDRAULIC ARCHITECTURE AND PATHWAY OF WATER MOVEMENT IN PLANTS

Van den Honert (1948) quantified water flow in plants in a classical paper in which he viewed the flow of water in the plant as a catenary process, where each catena element is viewed as a hydraulic conductance (analogous to an electrical conductance) across which water (analogous to electric current) flows. Thus, van den Honert proposed an Ohm's law analog for water flow in plants. The Ohm's analog leads to the following predictions: (1) the driving force of sap ascent is a continuous decrease in P_x in the direction of sap flow; and (2) evaporative flux density from leaves (E) is proportional to negative of the pressure gradient ($-dP_x/dx$) at any given point (cross-section) along the transpiration stream. Thus, at any given point of a root, stem, or leaf vein, we have:

$$-dP_x/dx = AE/K_h + \rho g \, dh/dx \tag{12}$$

where A is the leaf-area–supplied water by a stem segment with hydraulic conductivity K_h and $\rho g \, dh/dx$ is the gravitational potential gradient, where ρ is density of water, g is acceleration due to gravity, and dh/dx is height gained, dh, per unit distance, dx, traveled by water in the stem segment.

In the context of stem segments of length (L) with finite pressure drops across ends of the segment, we have:

$$\Delta P_x = LAE/K_h + \rho g \Delta h \tag{13}$$

Figure 7 illustrates water flow through a plant represented by a linear catena of conductance elements near the center and a branched catena of conductance ele-

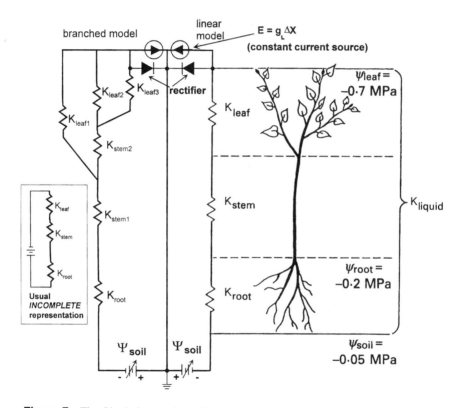

Figure 7 The Ohm's law analogy. The total conductance is seen as resultant conductance (K) of the root, stem, and leaf in series and parallel. Water flow is driven by evaporation of water from leaves, which creates a difference in water potential between the soil, Ψ_{soil}, and the water potential at the evaporating surface, Ψ_{evap}. On the right is the simplest Ohm's law analogy with conductances in series. On the left is a more complex conductance catena in which some conductance elements are in series and some in parallel.

ments on the left. The number and arrangement of catena elements are dictated primarily by the spatial precision desired in the representation of water flow through a plant; a plant can be represented by anywhere from one to thousands of conductance elements. A recent review of the hydraulic architecture of woody plants can be found in a report by Tyree and Ewers (1991).

The usual way of representing the Ohm's law circuit is incomplete. The usual but incomplete representation of the Ohm's law circuit is shown in the boxed inset in the lower left side of Figure 7. The battery represents the water potential drop from root to leaf, but it is an incomplete representation of reality because no ground point is shown in the circuit. The ground point represents zero water potential. Water potential is always measured relative to pure water at the same temperature and pressure as the plant. The incomplete representation gives the correct drop in water potential across each conductance element but not the correct water potential relative to ground (the zero reference). The complete representation has a variable battery that gives the soil water potential (Ψ_{soil}) and has some analog components to represent the evaporation from the leaves. In Figure 8, the evaporation is shown by a constant current source and a rectifier in parallel. The amount of current, E, in the constant current source can be approximated by Eq. 8. When E = 0 we expect leaf water potential to equal soil water potential and that there should be no water current flowing through the circuit; this condition is achieved in the circuit by the introduction of the rectifier that prevents backward current flow when E = 0 and Ψ_{soil} < 0. In some experimental conditions, Ψ_{soil} can be raised to positive values by placing a potted root system in a pressure chamber with the shoot outside the chamber. When E = 0 and Ψ_{soil} > 0 we could expect gutation (exudation of water from leaves); this forward direction of current flow is permitted by the rectifier. When E > 0 and Ψ_{soil} is positive enough, gutation can occur simultaneously with evaporation when Ψ_{leaf} > 0; the rectifier permits this additional current flow.

An alternative representation of evaporation from leaves would be achieved by including an additional conductance element for the vapor phase conductance, K_{vapor} (analogous to $g_L \cong g_s$), with an additional water potential drop from the leaf to the ambient air outside the leaf (Ψ_{air}). K_{vapor} would be orders of magnitude smaller than the whole plant conductance (K_{liquid}), and hence would dominate and control E and behave like a constant current source. There are two disadvantages to using this representation. First, g_L is never measured in the same units as K_{liquid}, making it difficult to compare values with those in the literature. Second, g_L measured in the units of K_{liquid} would no longer be a constant, i.e., its value in theory would change with Ψ in the vapor phase. This can be seen immediately by looking at Eq. 2 and 4, from which we can get:

$$J_{vapor} = -\frac{C\overline{V}_w}{f}\frac{d\Psi}{dx} \tag{14}$$

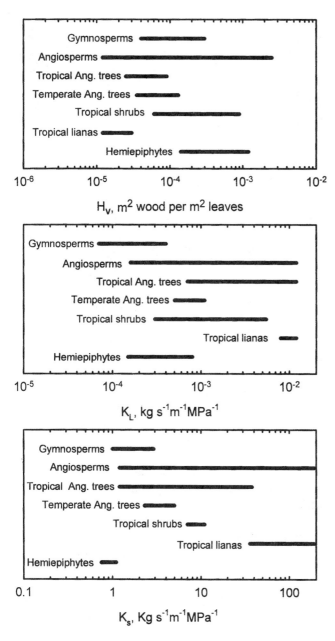

Figure 8 Ranges of hydraulic parameters by phylogeny or growth form. Horizontal bars demark the ranges read from the bottom axis of each parameter, where H_v is the Huber value, K_L is the leaf-specific conductivity, and K_s is the specific conductivity. Values represent ranges for 48 species. (*Source*: Adapted from Patiño et al. 1995.)

where C is the concentration of water. In the liquid phase, C is nearly a constant, but in the vapor phase, the concentration of water vapor can easily change by a factor of 10 from the evaporating surface in a leaf to the bulk air. The value of g_L in Eq. 8 is superior in that it depends only on the geometry of the stomatal pores; this follows because in the vapor phase

$$\frac{d\Psi}{dx} = \frac{RT}{V_w C}\frac{dC}{dx} \qquad (15)$$

hence

$$J_{vapor} = \frac{RT}{f}\frac{dC}{dx}, \qquad (16)$$

which is Fick's law for water vapor transport in air. The coefficient RT/f is independent of dC/dx. Equation 8 can be viewed as derived from Fick's law with account taken of the specific geometry of water diffusion through stomates and replacing dC/dx with the gradient in mole fraction of water vapor.

A. Parameters and Concepts To Describe Hydraulic Architecture

The hydraulic architecture of a plant can be defined as a quantitative description of the plant in terms of the Ohm's law analog using a simple linear model of conductance elements or a complex branched catena of a few or even thousands of conductance elements (Figure 7). The conductance elements are quantified by measurements made on excised stem segments for branched catena models and by measurements on whole roots and shoots for simple linear models.

Measurements on stem segments are performed with a conductivity apparatus. Excised stem segments are fitted into water-filled lengths of plastic tubing. One end of the segment of length (L, m) is connected via tubing to an upper reservoir of water and the other end to a lower reservoir. The height difference between the reservoirs is usually set at 0.3–1 m to create a pressure drop ΔP of 3–10 kPa across the stem segment. The lower reservoir is usually placed on top of a digital balance to measure water flow rate, F, in kg s^{-1}. The fundamental parameter measured is the hydraulic conductivity (K_h), defined as:

$$K_h = F/(\Delta P/L). \qquad (17)$$

Values of K_h are usually measured for stems of different diameter, D, and regressions are used to obtain allometric relationships of K_h versus diameter of the form

$$K_h = A\,D^B \qquad (18)$$

where A and B are regression constants.

Stem segments can be viewed as bundles of conduits (vessels or tracheids) with a certain diameter and number of conduits per unit cross-section. If all of the conduits were of the same diameter and number per unit cross-section, then B = 2, because K_h would increase with the number of conduits in parallel, which would increase with cross-section, which is proportional to diameter squared. Usually, B is found to be more than 2 but less than 3 because the diameter of conduits tends to increase with stem diameter in most plants. According to the Hagen-Poiseuille law, the K_h of a single conduit of diameter, d, increases in proportion to d^4; therefore, the conductance of a stem segment with N conduits would be proportional to Nd^4. Although you cannot pack as many big-diameter conduits into a stem segment as small-diameter conduits, there is a net gain in K_h for a stem segment to have bigger diameter conduits, although fewer would fit into that available space. To prove this, let us imagine a stem segment of 1-mm diameter with 1000 conduits of d_1 = 0.01-mm diameter each and another stem segment of 1-mm diameter with 250 conduits of d_2 = 0.02-mm diameter each. The cross-sectional area of each conduit is $\pi d^2/4$. Both segments would have the same cross-sectional area of conduits since 1000 $\pi d_1^2/4$ = 250 $\pi d_2^2/4$, but each conduit in the latter would be 16 times more conductive than the former because d_1^4 = 16 d_1^4; therefore, although there are only one fourth as many conduits of diameter d_2 than d_1, the stem will be four times more conductive.

Since K_h of stem segments depends on stem cross-section, one useful way of scaling K_h is to divide it by stem cross-section to yield specific conductivity, K_s. Specific conductivity is a measure of the efficiency of stems to conduct water. The efficiency of stems increases with the number of conduits per unit cross-section and with their diameter to the fourth power. In large woody stems, the central core is often nonconductive heartwood. It is often better to calculate K_s from K_h/A_{sw}, where A_{sw} is the cross-sectional area of conductive sapwood.

Leaf specific conductivity (K_L), also known as LSC (Tyree and Ewers 1991), is equal to K_h divided by the leaf area distal to the segment (A_L, m^2). This is a measure of the hydraulic sufficiency of the segment to supply water to leaves distal to that segment. If we know the mean evaporative flux density (E, $kg\ s^{-1}\ m^{-2}$) from the leaves supplied by the stem segment and we ignore water storage capacitance, then the pressure gradient through the segment (dP/dx) = E/K_L. Therefore, the higher the K_L is, the lower the dP/dx required to allow for a particular transpiration rate.

In simple linear Ohm's law models, whole shoot and whole root conductances (k_{sh} and k_r, respectively) are measured using a high-pressure flowmeter (HPFM; see later). These conductances are usually defined as the ratio of flow, F, across the whole root or shoot divided by the pressure drop, ΔP; hence, it differs from K_h, K_s, and K_L in that root or shoot length is not taken into account. The word *conductivity* is usually used when L is taken into account in the calculation, and *conductance* is used when L is not used in the calculation. Because a large plant or root will be more conductive than small plants, some suitable means of normalization for plant size is needed. One way to do this is to calculate leaf specific conductances, i.e., $K_{sh} = K_{sh}/A_L$ and $K_r = k_r/A_L$. The advantage of this versus other kinds of scaling is discussed in Section V.C.

The Huber value (H_v) is defined as the sapwood cross-section (or sometimes the stem cross-section) divided by the leaf area distal to the segment. Because, the H_v is in units of m^2 stem area per m^2 leaf area, it is often written without dimension. It is a measurement of the investment of stem tissue per unit leaf area fed. It follows from the aforementioned definitions that $K_L = H_v K_s$.

Figure 8 summarizes the known ranges of hydraulic architecture parameters in 48 taxa covering a range of growth forms and phylogonies. Comparisons of mean values between species are difficult because K_L, K_s, and H_v often change significantly with stem diameter, D. For example, K_L and K_s can be 10–100 times greater when measured in the bole of trees (D = 300 mm) than when measured in young branches (D = 3 mm). Sometimes, difference between species in K_L or K_s measured at D = 6 mm may be reversed at D = 60 mm; furthermore, stem morphologies may be such that the smallest segments bearing leaves were 20 mm in diameter in some species but just 3 mm in another. The mean values in Figure 8 were computed from regression values at D = 15 mm, except for three tropical species with large stems for which we used D = 45 mm because smaller branches did not exist.

B. Patterns of Hydraulic Architecture of Shoots

Values of K_L have been applied to complex "hydraulic maps" in which the aboveground portion of trees were represented by hundreds to thousands of conductance elements (Figure 7). Using known values of K_L and $\dot{E}$, it is possible to calculate P_x versus path length from the base of a tree to selected branch tips (Figure 9). Some species are so conductive (large K_L values) that the predicted drop in P_x is little more than needed to lift water against gravity (see *Schefflera* in Figure 9). In other species, gradients in P_x become very steep near branch tips (see *Thuja* in Figure 9). The predicted gradients of P_x become steeper in small branches of trees because K_L is often lower in small-diameter branches than in big-diameter branches. In 20 species of angiosperms, the value of K_L was found to be proportional to stem diameter, D, to the power of 0.5–2.0 depending on the species (Tyree and Ewers 1991, Patiño et al. 1995, Zotz et al 1998). The increase in K_L with diameter could be due to an increase in H_v or K_s since $K_L = H_v K_s$, but in most cases, H_v was found to be approximately constant with D so the change was due to an increase in K_s with D. The exceptions seem to be gymnosperms with strong apical dominance, where H_v increases toward the apex, i.e., decreases with increasing D (Ewers and Zimmermann 1984a, 1984b). In one

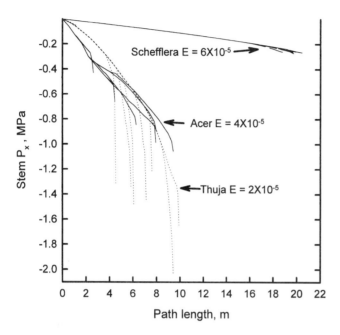

Figure 9 Pressure profiles in three large trees, i.e., computed change in xylem pressure (P_x) versus path length. P_x values were computed from the base of each tree to a few randomly selected branch tips. The drooping nature of these plots near the apices of the branches is caused by the decline of leaf-specific conductivity (K_L) from base to apex of the trees. The pressure profiles do not include pressure drops across roots or leaves. In some species, pressure drop across leaves can be more than shown. The pressure drop across roots is generally equal to that across the shoots (including leaves). E, evaporative flux density in kg s^{-1} m^{-2}.

angiosperm, *Ficus dugandii*, H_v was found to increase with D to the power of 1.0 (Patiño et al. 1995).

There appear to be no hydraulic constrictions at branch junctions in woody plants (Tyree and Alexander 1993), i.e., the hydraulic conductance of water passing through branch junctions is approximately the same as in equal lengths of stem segments above and below branch junctions. This is contrary to earlier reports of major constrictions in branch junctions, but the number of measurements in these reports were too few to draw statistically provable conclusions (Ewers and Zimmermann 1984a, 1984b, Zimmermann 1978). Unfortunately, a myth has arisen regarding the hydraulics of branch junctions because these earlier, preliminary studies have been frequently cited.

C. Root, Shoot, and Leaf Hydraulic Conductances

Most studies have focused on the hydraulic architecture of large woody tropical plants, e.g., trees, shrubs, and vines (Tyree and Ewers 1996, Patiño et al. 1995, Ewers et al. 1991). Parameters measured generally have been confined to hydraulic conductance of woody shoots. But relatively little is known about total root conductance of plants relative to shoot conductances, and little is known about the conductivity of leaves relative to shoots. In large woody plants, the shoots of rapidly growing trees appear to be more conductive than slowly growing trees.

The recent development of a HPFM (Tyree et al. 1995, Yang and Tyree 1994) allows the rapid measurement of root and shoot hydraulic conductance of seedlings to saplings with stem diameters from 1–50 mm. Thus, it is now possible to study the growth dynamics of seedlings while monitoring dynamic changes in hydraulic conductance of roots and shoots. This ability raises the question of how best to scale conductance parameters to reveal ecological adaptation to light regimes. Because more thought has gone into scaling of shoot parameters (Tyree and Ewers 1996) than root parameters, a review of arguments for roots seems appropriate.

Root conductance (k_r) can be defined as water flow rate (kg s^{-1}) per unit pressure drop (MPa) driving flow though the entire root system. Values of k_r could be scaled by dividing by some measure of root size (root surface area, total root length, or mass) or by dividing k_r by leaf surface area. Division by root surface area (A_r) is justified by an analysis of axial versus radial resistances to water flow in roots. In the radial pathway, water flows from the root surface to the xylem vessels through nonvascular tissue. In the axial pathway, water flow is predominately through vessels. The resistance of the radial path is usually more than the axial path (French and Steudle 1989, North et al. 1992). Most water uptake is presumed to occur in fine roots ($<$ 2 mm diameter) and fine root surface area is usually greater than 90% of the total root surface area (M. T. Tyree, personal observation). Root uptake of water would appear to be limited by root surface area, and hence it is reasonable to divide k_r by A_r, yielding a measure of root efficiency. Some roots are more efficient than others. Division of k_r by total root length (L) is not as desirable, but is justified because A_r and L are correlated approximately and L can be estimated by a low-cost, line-intersection technique rather than a high-cost, image-analysis technique.

Scaling by root mass is justified by consideration of the cost of resource allocation. Plants must invest a lot of carbon into roots to grow and maintain them. The benefit derived from this carbon investment is enhanced scavenging for water and mineral nutrient resources. Total root dry weight (TRDW) is a measure of carbon investment into roots. Thus, the carbon efficiency of roots might be measured in terms of K_r/TRDW, A_r/TRDW, or L/TRDW. Scaling by TRDW provides information of ecological rather than physiological importance.

Scaling of k_r by leaf surface are a (A_L) provides an estimate of the sufficiency of the roots to provide water to leaves. The physiological justification of scaling k_r to the leaf surface area (A_L) comes from an analysis of the Ohm's law analog for water flow from soil to leaf (van den Honert 1948). The Ohm's law analog describes water flow rate $(F, kg\ s^{-1})$ in terms of the difference in water potential between the soil (Ψ_{soil}) and the leaf (Ψ_L):

$$\Psi_{soil} - \Psi_L = (1/k_{soil} + 1/k_r + 1/k_{sh})F \tag{19a}$$

where k_{soil} is the hydraulic conductance of the soil. Because it is usually assumed that $k_{soil} \gg k_r$ and k_{sh} except in dry soils, $1/k_{soil}$ can be ignored. Leaf water potential is approximated by:

$$\Psi_L \cong \Psi_{soil} - (1/k_r + 1/k_{sh})F \tag{19b}$$

Or, if we wish to express Eq. 19b in terms of leaf area and average evaporative flux density (E), we have:

$$\Psi_L \cong \Psi_{soil} - (1/k_r + 1/k_{sh})A_L E \tag{20}$$

This equation also can be rewritten so that root and shoot conductances are scaled to leaf surface areas, i.e., to give leaf-specific shoot and root conductances, $K_{sh} = k_{sh}/A_L$ and $K_r = k_r/A_L$, respectively:

$$\Psi_L \cong \Psi_{soil} - (1/K_r + 1/K_{sh})E \tag{21}$$

Meristem growth and gas exchange are maximal when water stress is small, i.e., when Ψ_L is near zero. From Eq. 21 it can be seen that the advantage of high K_r and K_{sh} is that Ψ_L will be closer to Ψ_{soil}. Because leaf-specific stem-segment conductivities, K_L, are high in adult pioneer trees, the water potential drop from soil to leaf is much smaller than in old-forest species (Machado and Tyree 1994). This may promote rapid extension growth of meristems in pioneers compared with old-forest species. In addition, stomatal conductance (g_s) and therefore net assimilation rate are reduced when Ψ_L is too low. During the first 60 days of growth of *Quercus rubra* L. seedlings, there was a strong correlation between midday g_s and leaf-specific plant conductance, $G = k_p/A_L$, where $k_p = k_r k_{sh}/(k_r + k_{sh})$ (Ren and Sucoff 1995). This suggests that whole-seedling hydraulic conductance is limiting g_s though its effect on Ψ_L. There is also reason to believe that whole-shoot conductance limits g_s in mature trees of *Acer saccharum* Marsh (Yang and Tyree 1993). Thus, high values of K_r and K_{sh} may promote both rapid extension growth and high net assimilation rates in pioneers.

Scaling is always necessary to normalize for plant size. As seedlings grow exponentially in size, we would expect an approximately proportional increase in k_r and k_{sh}. Since roots and shoots both supply water to leaves and since an increase in leaf area means an increase in rate of water loss per plant, we would expect k_r and k_{sh} to be approximately proportional to A_L.

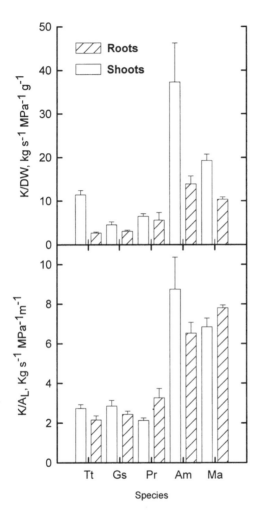

Figure 10 Hydraulic conductances of shoots and roots scaled to dry weight or leaf area (top) k_r per unit total root dry weight (TRDW) and k_{sh} per unit shoot dry weight. (Bottom) k_r and k_{sh}, both scaled to leaf area (A_L). Error bars are SEM, n = 23–36. Species are: Ma, *Miconia argentea*; Am, *Apeiba membranacea*; Pr, *Pouteria reticulata*; Gs, *Gustavia superba*; and Tt, *Trichilia tuberculata*. Ma, Am, and Pr are old-forest species and Pr and Gs are pioneer species. Root and shoot means for Am and Ma were significantly different from corresponding root and shoot means for Tt, Pr, and Gs in both (top) and (bottom) (Tukey test, $P \le .05$). (*Source*: From Tyree et al. 1998.)

The major resistance (= inverse conductance) to liquid water flow in plants resides in the nonvascular pathways, i.e., the radial pathway for water uptake in young roots and in the cells between leaf veins and the evaporating surfaces of leaves. Nonvascular resistances can be 60–90% of the total plant resistance to water flow, although the total path length may be less than 1 mm. Although water flows through distances of 1–100 m in the vascular pathways from root vessel to stems to leaf vessels, the hydraulic resistance of this pathway rarely dominates. The biggest resistance to water flow (liquid and vapor) from the soil to the atmosphere resides in the vapor transport phase. The rate of flow from the leaf to the atmosphere is determined primarily by the stomata, as discussed in Section II.A.

Early successional (pioneer) species grow more rapidly and require higher light levels to survive than late successional (old-forest) species. A pattern is emerging that shows that the dry matter cost of root and shoot conductance is less in pioneer versus old-forest species, and the leaf specific root and shoot conductances are higher. Hence, scaling hydraulic parameters by leaf area is ecologically meaningful. In Figure 10 (top), the root and shoot conductances are scaled by root and shoot dry weights, respectively. In Figure 10 (bottom), the root and shoot conductances are scaled by dividing both by leaf area. In both cases, the adaptive advantages of the pioneer species become evident. All pioneer species differed significantly from the nonpioneer species, although all species were growing in the same light regime. Figure 10 illustrates two advantages of pioneers versus other species in this study. The higher values of K_r and K_{sh} mean that the pioneer species can maintain less negative leaf water potentials than the other species at any given transpiration rate. This might lead to higher rates of extension growth and net assimilation. The higher values of K_r/DW and K_{sh}/DW in pioneers means that pioneers spend less carbon to provide efficient hydraulic pathways than do the other species. Both of these advantages (Figure 10, top and bottom) mean that pioneers can be more competitive in gap environments than old forest species.

VI. THE COHESION-TENSION THEORY AND XYLEM DYSFUNCTION

The Cohesion-Tension (C-T) theory was proposed more than 100 years ago by Dixon and Joly (1894), and some aspects of the C-T theory were put on a quantitative basis by van den Honert (1948) with the introduction of the Ohm's law analog of sap flow in the soil–plant–atmosphere continuum.

A. Water Transport Under Negative Pressure

According to the C-T theory, water ascends plants in a metastable state under tension, i.e., with xylem pressure (P_x) more negative than that of a perfect vacuum. The driving force is generated by surface tension at the evaporating surfaces of the leaf, and the tension is transmitted through a continuous water column from the leaves to the root apices and throughout all parts of the apoplast in every organ of the plant. Evaporation occurs predominately from the cell walls of the substomatal chambers due to the much lower water potential of the water vapor in air. The evaporation creates a curvature in the water menisci of apoplastic water within the cellulosic microfibril pores of cell walls. Surface tension forces consequently lower P_x in the liquid directly behind the menisci (the air–water interfaces). This creates a lower water potential, Ψ, in adjacent regions, including adjoining cell walls and cell protoplasts. The lowering of Ψ is a direct consequence of P_x being one of the two major components of water potential in plants, the other component being solute potential, π (see Eq. 5–7). The energy for the evaporation process ultimately comes from the sun, which provides the energy to overcome the latent heat of evaporation of the water molecules; i.e., the energy to break hydrogen bonds at the menisci.

This tension (negative P_x) is ultimately transferred to the roots, where it lowers Ψ of the roots below the Ψ of the soil–water. This causes water uptake from the soil to the roots and from the roots to the leaves to replace water evaporated at the surface of the leaves. The Scholander-Hammel pressure bomb (Scholander et al. 1965) is one of the most frequently used tools for estimating P_x. Typically, P_x can range down to -2 MPa (in crop plants) or to -4 MPa (in arid-zone species), and in some cases -10 MPa (in California chaparral species). Readers interested in learning more about the C-T theory of sap ascent should refer to the report by Tyree (1997).

B. Cavitation, Embolism, and Stability of Bubbles

Water in xylem conduits is said to be in a metastable condition when P_x is below the pressure of a perfect vacuum, because the continuity of the water column, once broken, will not rejoin until P_x rises to values above that of a vacuum. Metastable conditions are maintained by the cohesion of water to water and by adhesion of water to walls of xylem conduits. Both cohesion and adhesion of water are manifestations of hydrogen bonding. Although air–water interfaces can exist anywhere along the path of water movement, the small diameter of pores in cell walls and the capillary forces produced by surface tension within such pores prevent the passage of air into conduits under normal circumstances. When P_x becomes negative enough, the continuity of the water column in the conduit is rapidly broken, which is called a cavitation event. Water is drawn out of cavitated conduits by surrounding tissue, leaving a void filled with water vapor. Eventually, air diffuses into the void; when this happens, the conduit is said to be embolized.

A cavitation event in xylem conduits ultimately results in dysfunction. A cavitation occurs when a void of sufficient radius forms in water under tension. The void is filled with gas (water vapor and some air) and is inherently unstable, i.e., surface tension forces will make it spontaneously collapse unless the water is under sufficient tension (negative pressure) to make it expand.

The chemical force driving the collapse is the energy stored in hydrogen bonds, the intermolecular force between adjacent water molecules. In ice, water is bound to adjacent water molecules by four hydrogen bonds. In the liquid state, each water molecule is bound by an average of 3.8 hydrogen bonds at room temperature. In the liquid state, hydrogen bonds are forming and breaking all the time, permitting more motion of molecules than in ice (Slatyer 1968). However, when an interface between water and air is formed, some of those hydrogen bounds are broken and the water molecules at the surface are at a higher energy state because of the broken bonds. The force (N = Newtons) exerted at the interface as hydrogen bonds break and reform can be expressed in pressure units (Pa) because pressure is dimensionally equal to energy (J = Joules) per unit volume of molecules, i.e., $J\ m^{-3} = N{\cdot}m\ m^{-3} = N\ m^{-2} = Pa$. Stable voids in water tend to form spheres because spheres have the least surface area per unit volume, and thus a spherical void has the minimum number of broken hydrogen bonds per unit volume of void. The pressure tending to make a void collapse is given by $2\tau/r$, where r is the radius of the spherical void and τ is the surface tension of water (= 0.072 Pa m at 25°C).

For a void to be stable, its collapse pressure ($2\tau/r$) must be balanced by a pressure difference across its surface or meniscus $= P_v - P_w$, where P_w is the absolute pressure (= P_x + the atmospheric pressure) of the water and P_v is the absolute pressure of the void.

$$P_v - P_w = 2\tau/r \tag{22}$$

P_v is always above absolute zero pressure (= perfect vacuum) since the void is usually filled with water vapor and some air. Relatively stable voids are commonplace in daily life, e.g., the air bubbles that form in a cold glass of water freshly drawn from a tap. An entrapped air bubble is temporarily stable in a glass of water because P_w is a relatively constant 0.1 MPa, and P_v is determined by the ideal gas law, $P_v = nRT/V$, where n is the number of moles of air in the bubble, R is gas constant, T is absolute temperature, and V is the volume of the bubble. The tendency of the void to collapse ($2\tau/r$) makes V decrease, which causes P_v to increase according to the ideal gas law because P_v is inversely proportional to V. The rise in P_v provides the restoring force across the meniscus needed for stability. However, an air bubble in a glass of water is only temporarily stable, because according to Henry's law, the solubility of a gas in water increases with the pressure of the gas. Therefore, the increased pressure exerted by $2\tau/r$ makes the gas in the bubble more soluble in water, and it slowly collapses as the air dissolves, i.e., as n decreases.

Air bubbles are rarely stable in xylem conduits because transpiration can draw P_w to values less than zero. As P_w falls toward zero, the bubble expands according to the ideal gas law, but because V can never grow larger than the volume of the conduit, P_v can never fall to or below zero to permit $P_v - P_w$ to balance $2\tau/r$ without a decline in P_w. Once the bubble has expanded to fill the lumen, the conduit is dysfunctional and no longer capable of transporting water. Fortunately for the plant, a dynamic balance at the meniscus in cell walls is ultimately achieved. This stability will be discussed first in the context of a vessel and its pit membranes.

As the air bubble is drawn up to the surface of the pit membrane in vessel cell walls, the pores in the pit membrane break the meniscus into many small menisci at the opening of each pore. As the meniscus is drawn through the pores, the radius of curvature of the meniscus, r_m, falls toward the radius of the pores, r_p. As long as r_m exceeds r_p, the necessary conditions for stability are again achieved, i.e.,

$$P_v - P_w = 2\tau/r_m \tag{23}$$

Usually, a dysfunctional conduit will eventually fill with air at atmospheric pressure (as demanded by Henry's law); therefore, P_v eventually approaches 0.1 MPa as gas diffuses through water to the lumen and comes out of solution. When P_v equals 0.1 MPa, the conduit is said to be fully embolized. As P_w rises and falls as dictated by the demands of transpiration, r_m adjusts at the pit-membrane pores to achieve stability. When the conduit is fully embolized, both sides of Eq. 23 can be expressed in terms of xylem pressure potential,

$$P_x = -(P_v - P_w) = -2\tau/r_m. \tag{24}$$

The minimum P_x that can be balanced by the meniscus is given when r_m equals the radius of the biggest pit-membrane pore bordering the embolized conduit. If the biggest pore is 0.1 or 0.05 μm, then the minimum stable P_x is -1.44 or -2.88 MPa, respectively. The porosity of the pit-membrane is therefore critical to preventing dysfunction of vessels adjacent to embolized vessels (Sperry and Tyree 1988). When P_x falls below the critical value, the air bubble is sucked into an adjacent vessel, seeding a new cavitation.

Consequently, the genetics that determines pit morphology and pit-membrane porosity must be under strong selective pressure. A safe pit membrane will be one with very narrow pores and one thick enough and thus strong enough to sustain substantial pressure differences without rupturing.

The situation for tracheids of conifers is different because air movement from an embolized tracheid to an adjacent tracheid is prevented by the sealing (aspiration) of the torus against the overarching border of the pit. The porosity of the margo that supports the torus is too large to prevent meniscus passage at pressure differences exceeding 0.1 MPa in most cases (Sperry and Tyree 1990). However, because the margo is elastic, a pressure difference of just 0.03 MPa is sufficient to deflect the torus into the sealed position. Air bubbles pass between tracheids when the pressure difference becomes large enough to rip the torus out of its sealed position (Sperry and Tyree 1990).

C. Vulnerability Curves and the Air-Seeding Hypothesis

Water movement can occur at night or during rain when P_x is positive in some plants due to root pressure, i.e., osmotically driven flow from roots. Water flow under positive P_x is often accompanied by gutation, i.e., the formation of water droplet at leaf margins. However, water is normally under negative pressure (tension) as it moves through the xylem toward the leaves during sunny days. The water is thus in a metastable condition and vulnerable to cavitation due to air entry into the water columns. Cavitation results in embolism (air blockage), thus disrupting the flow of water (Tyree and Sperry 1989). Cavitation in plants can result from water stress, and each species has a characteristic vulnerability curve, which is a plot of the percent loss k_h in stems versus the xylem pressure potential, P_x, required to induce the loss. Vulnerability curves are typically measured by dehydrating large excised branches to known P_x. Stem segments are then cut under water from the dehydrated branches; the air bubbles remain inside the conduits for the most part. An initial conductivity measurement is made and compared with the maximum K_h after air bubbles have been dissolved. The vulnerability curves of plants, in concert with their hydraulic architecture, can give considerable insight to drought tolerance and water relations "strategies."

The vulnerability curves for a number of species are illustrated in Figure 11 (Tyree et al. 1994a). These curves represent the range of vulnerabilities observed thus far in more than 60 species; 50% loss K_h occurs at P_x values ranging from -0.7 to -11 MPa. Many plants growing in areas with seasonal rainfall patterns (wet and dry periods) appear to be drought evaders, and others are drought tolerators. The drought evaders evade drought (low P_x and high percent loss K_h) by having deep roots and a highly conductive hydraulic system; alternatively, they evade drought by being deciduous. The other species frequently reach very negative P_x for part or all of the year and are shallow rooted or grow in saline environments. These species survive because they have a vascular system highly resistant to cavitations.

When air bubbles are sucked into xylem conduits from adjacent embolized conduits, the cavitation event is said to be air-seeded. Plants will always have some embolized conduits to seed embolism into adjacent conduits. Embolisms are the natural consequence of foliar abscission, herbivory, wind damage, and other mechanical fates that might befall a plant. It is now appropriate to ask if

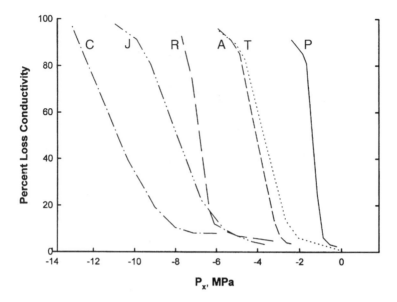

Figure 11 Vulnerability curves for various species. Y-axis is percent loss of hydraulic conductivity induced by the xylem pressure potential, P_x, shown on the x-axis. C, *Ceanothus megacarpus*; J, *Juniperus virginiana*; R, *Rhizophora mangle*; A, *Acer saccharum*; T, *Thuja occidentalis*; P, *Populus deltoides*.

all emboli are seeded from adjacent conduits or if some other mechanism occurs in some or most of the cases.

Four mechanisms for the nucleation of cavitations in plants have been proposed; these are illustrated in Figure 12, which shows for each mechanism the sequence of events that might occur as P_x declines in the lumen of a conduit. Readers are referred to Zimmermann (1983), Tyree et al. (1994a), and Pickard (1981) for a detailed discussion of the four mechanisms. Other air-seeding mechanisms have been proposed that apply to SCUBA divers (Yount 1989), and such mechanisms might occur in plants when gas solubility decreases in xylem water as it warms. However, little is known of the importance of this mechanism in plants. We need to be concerned only about which mechanism occurs most frequently in plants.

Experiments can discriminate between the air-seeding mechanism and the other three mechanisms (Figure 12; Yount 1989). All mechanisms predict cavitation when xylem fluid is under tension, but the air-seeding mechanism predicts that air can be blown into vessels while the fluid is under positive pressure. The air-seeding mechanism requires only a pressure differential $(P_a - P_w)$ where P_a

INCREASING TENSION ⟶

WALL ▨   AIR ☐   WATER ☐

Air Seeding through pore

Air Seeding through hydrophobic crack

Homogeneous nucleation

Hydrophobic adhesion failure

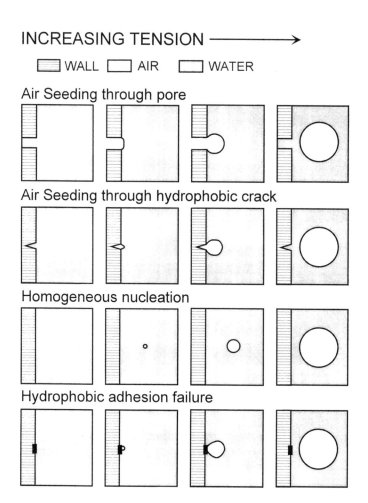

Figure 12 Four possible mechanisms of cavitation induction. (1) Air seeding through a pore occurs when the pressure differential across the meniscus is enough to allow the meniscus to overcome surface tension and pass through the pore. (2) Air seeding through a hydrophobic crack occurs when a stable air bubble resides at the base of a hydrophobic crack in the wall of a xylem conduit. When the P_x becomes negative enough, the bubble is sucked out of the crack. (3) Homogeneous nucleation involves the spontaneous generation of a void in a fluid. It is a random process requiring thermal motion of the water molecules. The hydrogen bonds at a specific locus are broken when all water molecules randomly move away from any locus at the same instant with sufficient energy to break all hydrogen bonds between water molecules. As the tension in the water increases, the hydrogen bonds are stretched and weakened so that the energy needed to break the bonds decreases, making a homogeneous nucleation more likely. (4) Hydrophobic adhesion failure is similar to homogeneous nucleation, except that hydrogen bonds are broken between water and a hydrophobic patch in the wall where the energy of binding between water and wall are reduced.

is the air pressure outside and P_w is the fluid pressure inside. It makes no difference if P_a is 0.1 and P_w is -3.0 or if P_a is 3.1 and P_w is 0.1. Experiments have shown that the same vulnerability curve results whether P_w is reduced by air dehydration or P_a is increased in a pressure bomb (Cochard et al. 1992). A vulnerability curve (VC) is a plot of percent loss hydraulic conductivity (PLC) versus the P_x required to cause the PLC by cavitation events. The results of this experiment are illustrated in Figure 13b (see also Jarbeau et al. 1995).

Willow stem segments with leaves were enclosed in a pressure chamber with cut ends protruding into the open air. Water was passed continually through the xylem under positive pressure. While stem conductance was being monitored, the gas pressure, P_a, in the pressure chamber was gradually increased. Initially, the hydraulic conductance of the stem segment did not decrease until a critical pressure of 1 MPa was applied. (Each solid circle in Figure 13b represents the application of pressure for 30–40 minutes.) When P_a was gradually increased beyond the critical value, the stem conductance began to fall (increased PLC). When P_a was gradually decreased, the PLC stopped decreasing. The VC from this experiment was identical to that found for similar branches dehydrated in the air.

D. How Plants Deal with Embolisms

Embolisms may be dissolved in plants if P_x in the xylem become positive or close to positive for adequate time periods (Tyree and Yang 1992, Lewis et al. 1994). Embolisms disappear by dissolution of air into the water surrounding the air bubble. The solubility of air in water is proportional to the pressure of air adjacent to the water (Henry's law). Water in plants tends to be saturated with air at a concentration determined by the average atmospheric pressure of gas surrounding plants. Thus, for air to dissolve from a bubble into water, the air in the bubble has to be at a pressure in excess of atmospheric pressure. If the pressure of water (P_x) surrounding a bubble is equal to atmospheric pressure (P_a), bubbles will naturally dissolve because surface tension (τ) of water raises the pressure of air in the bubble (P_b) above P_a. In general, $P_b = 2\tau/r + P_x$, where r is the radius of the bubble. According to the cohesion theory of sap ascent, P_x is drawn below P_a during transpiration. Since $2\tau/r$ of a dissolving bubble in a vessel is usually less than 0.03 MPa, and since P_x is in the range of -0.1 to -10 MPa during transpiration, P_b is usually $\leq P_a$ and hence bubbles, once formed in vessels, rarely dissolve. Repair ($=$ dissolution) occurs only when P_x grows large via root pressure. One notable exception to this generality has recently been found in *Laurus nobilis* shrubs that appear to be able to refill embolized vessels even while P_x is at -1 MPa (Salleo et al. 1996); the mechanism involved has evaded explanation.

Not all plants deal with embolisms by dissolving them. Embolism dissolu-

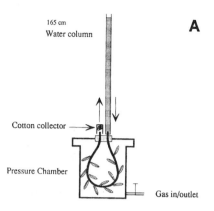

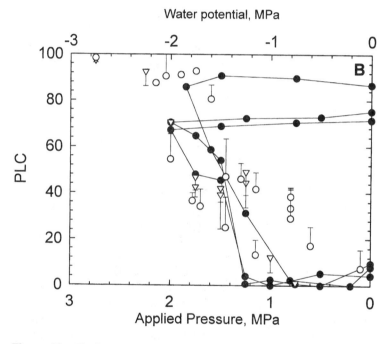

Figure 13 The first experimental test of the air-seeding hypothesis. (A) The experimental apparatus. A willow branch is bent around in a large pressure bomb so that both cut surfaces are outside the bomb. Water continually passes through the stem segment under positive pressure from a water column to a cotton collector. Flow rate is estimated by measuring the weight change of the cotton collector over known time intervals. (B) The results are shown as a vulnerability curve where the y-axis is the percent loss of hydraulic conductivity (PLC) and the x-axis is the negative P_x, or the air pressure in the bomb needed to cause the plotted PLC. Closed circles indicate PLC measured as in Figure 13A. Open circles indicate PLC induced by negative P_x in bench-dehydrated shoots. Open triangles indicate PLC induced by shoot dehydration in a pressure bomb but conductivity measured outside the pressure bomb on subsamples (stem segments excised from dehydrated shoot).

tion is probably most important in monocots, because they lack the ability to grow new vessels by secondary growth. In dicots, however, stems can grow bigger in diameter by secondary growth of the cambium. Cambium divides the phloem (bark in woody plants) from the xylem (wood in woody plants). Cell division on the inner surface of the cambium ultimately results in the growth and differentiation of new, water-filled xylem conduits. So some species simply grow new conduits to replace conduits made dysfunctional by embolism.

VII. FACTORS CONTROLLING THE RATE OF WATER UPTAKE AND MOVEMENT

Environmental conditions control the rate of water movement in plants. The dominant environmental factor is net solar radiation, as mentioned in Section III. Water movement can be explained by the solar energy budget of plants. We will now take a more quantitative look at the solar energy budget of plants at both the leaf and stand level.

A. Leaf Level Energy Budgets

Objects deep in space far away from solar radiation tend to be rather cold (approximately $5°K$), whereas objects on earth tend to be rather warm ($270–310°K$). The reason for the elevated temperature on earth is the warming effects of the sun's solar radiation. The net radiation absorbed by a leaf (R_{NL}, W m^{-2}) can be arbitrarily divided into shortwave and longwave radiation with a demarcation wavelength of 1 μm. Most shortwave radiation comes from the sun and most longwave radiation comes from the earth. As a leaf absorbs R_{NL}, the leaf temperature (T_L) rises, which increases the loss of energy from the leaf by three different mechanisms: black body radiation (B), sensible heat flux (H), and latent heat of vaporization of water (λE, where E is the evaporative flux density of water due to transpiration [mol of water s^{-1} m^{-2}] and λ is the heat required to evaporate a mole of water [J mol^{-1}]). R_{NL} can also be converted to chemical energy by the process of photosynthesis and to energy storage, but these two factors are generally small for leaves and are ignored in the following equation:

$$R_{NL} = B + H + \lambda E \tag{25}$$

The B in Eq. 25 is the black body radiation. All objects emit radiation. Very hot objects such as the sun emit mostly shortwave radiation, whereas cooler objects on earth emit mostly longwave radiation. The amount of radiation emitted increases with the Kelvin temperature according to the Stefan-Boltzmann law: B $= e \, \sigma_{sb} \, T_L^4$, where e is the emisivity (≥ 0.95 for leaves) and σ_{sb} is the Stefan-Boltzmann constant equal to 5.67×10^{-8} W m^{-2} K^{-4}.

H in Eq. 25 is sensible heat flux density, i.e., the heat transfer by heat diffusion between objects of different temperature. For leaves, the heat transfer is between the leaf and the surrounding air. The rate of heat transfer is proportional to the difference in temperature between the leaf and the air, so $H = k (T_L - T_a)$, where T_a is the air temperature and k is the heat transfer coefficient. Heat transfer is more rapid and hence k is larger under windy conditions than in still air. But the value of k is also determined by leaf size, shape, and orientation with respect to the wind direction. Readers interested in more detail should consult reports by Slatyer (1967) or Nobel (1991).

The factors determining E have already been given in Eq. 8. If we combine Eq. 8 with the other equations for B and H, we get:

$$R_{NL} = e \, \sigma_{sb} \, T_L^4 + k \, (T_L - T_a) + \lambda g_L \, (X_i[T_L] - X_o) \tag{26}$$

Every term on the right side of Eq. 25 is a function of leaf temperature. This gives the clue about how the balance is achieved in the solar energy balance equation. At any given R_{NL}, the value of T_L will rise or fall until the sum of B, H, and λE equals R_{NL}. As R_{NL} increases or decreases, the value of T_L increases or decreases to reestablish equality. In practice, this equality is achieved with a leaf temperate near T_a; T_L is rarely less than 5°K below T_a or more than 15°K above T_a. Some examples of leaf energy budgets are shown in Figure 14. Once we know the value of T_L and solve the equation, we can calculate the value of E. If we sum the values of E for every leaf in a stand of plants, then we can calculate stand-level water flow.

Equation 25 provides a good qualitative understanding of the dynamics of energy balance for a community of plants (= the stand level) because the evapora-

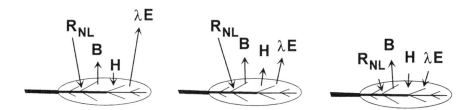

Figure 14 Solar energy budgets of leaves under different conditions. R_{NL}, net adsorbed shortwave and longwave radiation; B, black body radiation from leaves; H, sensible heat flux; λE, latent heat flux. Direction of arrow indicates direction of flux and length of arrow indicates relative magnitude. Specific conditions: (left) high transpiration rate so that leaf temperature is below air temperature because $B + \lambda E > R_{NL}$; (middle) intermediate transpiration rates so that leaf temperature is above air temperature; (right) during dew fall. Because $B > R_{NL}$ at night, leaf temperature is below the dew point temperature of the air.

tion rate from a stand is the sum of the evaporation from all the leaves in the stand. However, Eq. 25 is of little practical value because it is not possible to obtain values of R_{NL}, T_L, k, and g_L for every leaf in a stand to compute the required sum.

B. Stand Level Energy Budgets

Another approach to energy budgets is to measure energy and matter flux in a region of air above an entire stand of plants (Figure 15). A net radiometer can be used to measure net radiation (R_N) above the stand. R_N is the total radiation balance (incoming longwave and shortwave radiation minus outgoing longwave and shortwave radiation, so at the leaf level $R_N = R_{NL} - B$); the energy balance equation for this situation is:

$$R_N = G + H + \lambda E \tag{27}$$

The rate of heat storage in the soil, G (W m^{-2}), can be significant because soil temperature can change a few degrees Kelvin on a daily basis in the upper few centimeters of soil. Heat storage rate is usually measured with soil-heat-flux sen-

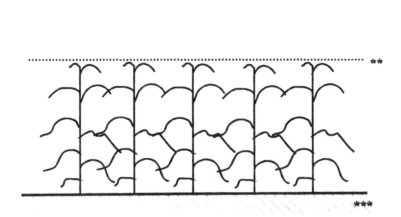

Figure 15 Points at which measurements are made for the solar energy budget of a uniform stand of plants. Net radiation (R_N) is measured at location marked *, air temperature and relative humidity are measured at locations marked * and **, and rate of heat storage in soil is measured at ***.

sor. The H and λE terms are similar to those in Eq. 25, but the equations define the flux densities between the heights marked one and two asterisks in Figure 15. The flux of heat and water vapor across the distance ΔZ is controlled by vertical air convection (eddy) and the defining equations are:

$$H = -C_p\rho_a K_e \Delta T/\Delta Z \tag{28}$$

and

$$\lambda E = \lambda K_e \Delta[H_2O]/\Delta Z \tag{29}$$

where K_e is the eddy transfer coefficient, C_p is the heat capacity of air, ρ_a is the air density, ΔT is the difference in air temperature measured at the two levels in Figure 15 separated by a height difference of ΔZ, and $\Delta[H_2O]$ is the difference in water vapor concentration at the two levels in Figure 15 separated by a height difference of ΔZ.

It is easy to measure ΔT and $\Delta[H_2O]$, but difficult to assign a value for K_e, which changes dynamically with changes in wind velocity. The usual practice is to measure the ratio of $H/\lambda E = \beta$, which is called the Bowen ratio. From the Eq. 28 and 29 it can be seen that

$$\beta = \frac{C_p\rho_a}{\lambda} \frac{\Delta T}{\Delta[H_2O]} \tag{30}$$

Estimates of stand water use are obtained by solving Eq. 29 and 30 for E:

$$E = \frac{R_N - G}{\lambda\,[\beta + 1]} \tag{31}$$

An example of the total energy budget measured over a pasture is reproduced in Figure 16. The data in Figure 16 show again that the main factor determining E is the amount of solar radiation, R_N, as in Figure 4. However, the leaves in a stand of plants do have some control over the value of E. As soils dry and leaf water potential falls, stomatal conductance falls. This causes a reduction in E, an increase in leaf temperature, and thus an increase in H.

Very intensive monitoring of climatic data is needed to obtain data for a solution to Eq. 31. Temperature, R_N, and relative humidity has to be measured every second at several locations. Ecologists prefer to estimate E with a less complete data set. Fortunately, the Penman-Monteith formula (Monteith 1964) permits a relatively accurate estimate of E under some restricted circumstances. The Penman-Monteith formula is derived from energy budget equations together with a number of approximations to turn nonlinear functions into linear relations. After many obtuse steps in the derivation (Campbell 1981), a formula of the following form results:

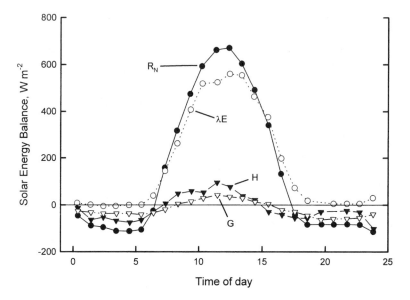

Figure 16 Solar energy budget values measured in a meadow. R_N, net solar radiation of the meadow; λE, latent heat flux; H, sensible heat flux; and G, rate of heat storage in soil. (*Source*: Adapted from Slatyer 1967.)

$$E = \frac{e'(R_N - G) + \rho_a C_p V_{pd} g_a}{\lambda \left[e' + \gamma \left(1 + \dfrac{g_a}{g_c} \right) \right]} \tag{32}$$

where e' is the rate of change of saturation vapor pressure with temperature at the current air temperature, V_{pd} is the difference between the vapor pressure of air at saturation and the current vapor pressure, γ is the psychometric constant, g_a is the aerodynamic conductance, and g_c is the canopy conductance of the stand. The problem with Eq. 32 is that g_a and g_c are both difficult to estimate and are not constant. The value of g_a depends on wind speed and roughness parameters that describe the unevenness at the boundary between the canopy and the bulk air. Surface roughness affects air turbulence and hence the rate of energy transfer at any given wind speed. Surface roughness changes as stands grow and is difficult to estimate in terrain with hills or mountains. The value of g_c is under biological control and difficult to estimate from climatic data. In one study on an oak forest in France, E was estimated independently by Granier sap flow sensors in individual oak trees. This permitted the calculation of g_c versus season, morphological state of the forest, and climate (Granier and Bréda 1996). The value of g_c was found to be a function of global radiation, V_{pd}, leaf area per unit ground area, which changes with season, and relative extractable water, which is a measure of soil dryness. But once all of these factors were taken into account, Eq. 32 provided a reasonable estimate of half-hourly estimates of E over the entire summer.

VIII. PLANTS UNDER STRESS

A. Wilting

Wilting denotes the limp, flaccid, or drooping state of plants during drought. Wilting is most evident in leaves that depend on cell turgor pressure to maintain their shape; hence, wilting occurs when turgor pressure falls to zero. Many plants maintain leaf shape through rigid leaf fiber cells. Wilting in these species is considered to commence at the turgor loss point. Wilted plants generally have low E because stomata are closed and g_s is very small when leaf water potential, Ψ_L, falls during drought. Continued dehydration beyond the wilt point usually causes permanent loss of hydraulic conductance due to cavitation in the xylem. Complete loss of hydraulic conductance usually causes plant death, but there is a lot of species diversity in the water potentials causing loss of hydraulic conductance (Figure 11). Some plants in arid environments avoid drought either by having short reproductive cycles confined to brief wet periods or by having deep roots that can access deep sources of soil water (Kramer 1983).

As plants approach the wilt point there is a gradual loss in stomatal conductance and, hence, a reduction in E and photosynthetic rate (Schulze and Hall 1982). Some species are much more sensitive than others (Figure 17B). The short-term effects of decreased Ψ_L on transpiration are less dramatic than long-term effects (Figure 17A). Long-term effects of drought are mediated by hormone signals from roots that cause a medium-term decline in g_s and by changes in root morphology, e.g., loss of fine roots, suberization of root surfaces, and formation of corky layers (Ginter-Whitehouse et al. 1983). The morphological changes to roots cause a decrease in whole-plant hydraulic conductance (K_p); therefore, Ψ_L becomes more negative at lower values of E because $\Psi_L = \Psi_{soil} - E/K_p$. Very severe drought can further lower K_p due to cavitation of xylem vessels.

B. Waterlogging

Waterlogging denotes an environmental condition of soil water saturation or ponding of water that can last for a just a few hours or for many months. Plants absent from flood-prone sites are damaged easily by waterlogging. On the other

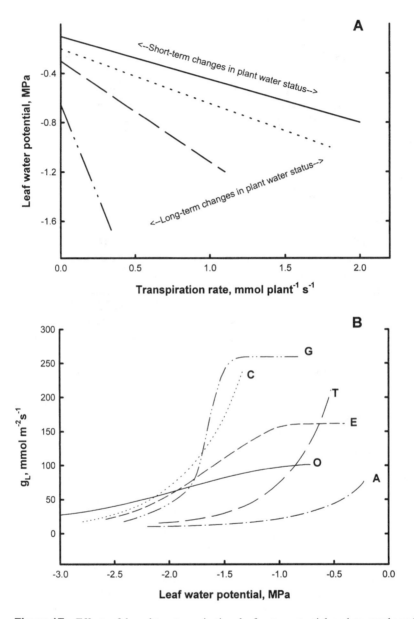

Figure 17 Effects of drought on transpiration, leaf water potential, and stomatal conductance. (A) Short- and long-term effects of drought on leaf water potential and transpiration. The short-term effects are dynamic changes in leaf water potential that might occur in the course of 1 day. The long-term effects are associated with slow drying of soil over many days. The different lines show the short-term relation between leaf water potential versus transpiration rate measured in well-watered plants (—) or after several days without irrigation, e.g., 5 days (· · ·), 10 days (– – –) or 20 days (—··—). (B) Short-term effects of leaf water potential on stomatal conductance. Species represented are: A, *Acer saccharum*; C, *Corylus avellana*; E, *Eucalyptus socialis*; G, *Glycine max*; and T, *Triticum aestivum*. (*Source*: Adapted from Schulze and Hall 1982.)

hand, plants that inhabit flood-prone sites include species that can grow actively in flooded soils or species that survive flooding in a quiescent or dormant state. Paradoxically, the most common sign of tobacco roots with an excess of water is the development of a water deficit in the leaves (Kramer 1983, Kramer and Jackson 1954), but flood-tolerant species are not so easily affected.

Flooding often affects root morphology and physiology. The wilting and defoliation that is found on flooding can be traced to an increased resistance to water flux in the roots (Mees and Weatherley 1957). In most flood-tolerant species, flooding induces morphological changes in the roots. These modifications usually involve root thickening with an increase in porosity. The increase in porosity increases the rate of oxygen diffusion to root tips and thus permits continued aerobic metabolism in the inundated roots. In flood-sensitive species, root and shoot growth are rapidly reduced on flooding, and root tips may be damaged. Growth of roots can be renewed only from regions proximal to the stem. The physiological responses and adaptations to waterlogging are numerous and beyond the scope of this chapter, but interested readers may consult a report by Crawford (1982).

REFERENCES

Bidwell, R.G.S. 1979. Plant Physiology. New York: Macmillan.

Campbell, G.S. 1981. Fundamentals of radiation and temperature relations. In: Lange, O.L., P.S. Nobel, C.B. Osmond, H. Ziegler, eds. Encyclopedia of Plant Physiology New Series, vol 12A. New York: Springer-Verlag, pp. 11–40.

Chang, J-H. 1968. Climate and Agriculture. Chicago: Aldine Publishing.

Cochard, H., P. Cruiziat, and M.T. Tyree. 1992. Use of positive pressures to establish vulnerability curves: further support for the air-seeding hypothesis and possible problems for pressure-volume analysis. Plant Physiology 100: 205–209.

Crawford, R.M.M. 1982. Physiological responses to flooding. In: Lange, O.L., P.S. Nobel, C.B. Osmond, H. Ziegler, eds. Encyclopedia of Plant Physiology New Series, vol 12B. New York: Springer-Verlag, pp 453–477.

Davies, W.J., and J. Zhang. 1991. Root signals and the regulation of growth and development of plants in drying soil. Annual Review of Plant Physiology and Plant Molecular Biology 42: 55–76.

Dawson, T.E. 1993. Hydraulic lift and water parasitism by plants: implications for water balance, performance, and plant-plant interactions. Oecologia 95: 565–574.

Dixon, H.H., and J. Joly. 1894. On the ascent of sap. Philosophical Transactions of the Royal Society London, Series B. 186: 563–576.

Ewers, F.W., J.B. Fisher, and K. Fichtner. 1991. Water flux and xylem structure in vines. In: Putz, F.E. and H.A. Mooney, eds. The Biology of Vines. Cambridge: Cambridge University, pp 127–160.

Ewers, F.W., and M.H. Zimmermann. 1984a. The hydraulic architecture of balsam fir (*Abies balsamea*). Physiologia Plantarum 60: 453–458.

Ewers, F.W., and M.H. Zimmermann. 1984b. The hydraulic architecture of eastern hemlock (*Tsuga canadiensis*). Canadian Journal of Botany 62: 940–946.

Frensch, J., and E. Steudle. 1989. Axial and radial hydraulic resistance to roots of maize (*Zea mays* L.) Plant Physiology 91: 719–726.

Ginter-Whitehouse, D.L., T.M. Hinckley, and S.G. Pallardy. 1983. Spatial and temporal aspects of water relations of three tree species with different vascular anatomy. Forest Science 29: 317–329.

Granier, A., and N. Bréda. 1996. Modeling canopy conductance and stand transpiration of an oak forest from sap flow measurements. Annales des Sciences Forestières 53: 537–546.

Green, P.B., R.O. Erickson, and J. Buggy. 1971. Metabolic and physical control of cell elongation rate—in vivo studies of *Nitella*. Plant Physiology 47: 423–430.

Jarbeau J.A., F.W. Ewers, and S.D. Davis. 1995. The mechanism of water stress-induced embolism in two species of chaparral shrubs. Plant Cell and Environment 126: 695–705.

Kramer, P.J. 1983. Water Relations of Plants. New York: Academic Press.

Kramer, P.J., and W.T. Jackson. 1954. Causes of injury to flooded tobacco plants. Plant Physiology 29: 241–245.

Lewis, A.M., V.D. Harnden, and M.T. Tyree. 1994. Collapse of water-stress emboli in the tracheids of *Thuja occidentalis* L. Plant Physiology 106: 1639–1646.

Machado, J-L., and M.T. Tyree. 1994. Patterns of hydraulic architecture and water relations of two tropical canopy trees with contrasting leaf phenologies: *Ochroma pyramidale* and *Pseudobombax septenatum*. Tree Physiology 14: 219–240.

Mees, G.C., and P.E. Weatherley. 1957. The mechanism of water absorption by roots. II. The role of hydrostatic pressure gradients. Proceedings of the Royal Society London, Series B. 147: 381–391.

Monteith, J.L. 1964. Evaporation and environment. Symposium of the Society for Experimental Biology 19: 205–234.

Nobel, P.S. 1991. Physiochemical and Environmental Plant Physiology. New York: Academic Press.

North, G.B., F.W. Ewers, and P.S. Nobel. 1992. Main root–lateral root junctions of two desert succulents: changes in axial and radial components of hydraulic conductivity during drying. American Journal of Botany 79: 1039–1050.

Patiño, S., M.T. Tyree, and E.A. Herre. 1995. Comparison of hydraulic architecture of woody plants of differing phylogeny and growth form with special reference to free-standing and hemi-epiphytic *Ficus* species from Panama. New Phytologist 129: 125–134.

Pickard, W.F. 1981. The ascent of sap in plants. Progress in Biophysics and Molecular Biology 37: 181–229.

Ren, Z., and E. Sucoff. 1995. Water movement through *Quercus rubra* L. Leaf water potential and conductance during polycyclic growth. Plant Cell and Environment 18: 447–453.

Richards, J.R. and M.M. Caldwell. 1987. Hydraulic lift: substantial nocturnal water transport between soil layers by *Artemisia tridentata* roots. Oecologia 74: 486–489.

Salleo, S., M.A. LoGullo, D. De Paoli, and M. Zippo. 1996. Xylem recovery from cavitation-induced embolism in young plants of *Laurus nobilis*: a possible mechanism. New Phytologist 132: 47–56.

Scholander, P.F., H.T. Hammel, E.D. Bradstreet, and E.A. Hemmingsen. 1965. Sap pressures in vascular plants. Science 148: 339–346.

Schulze, E.D., and A.E. Hall. 1982. Stomatal responses, water loss and CO_2 assimilation rates of plants in contrasting environments. In: Lange, O.L., P.S. Nobel, C.B. Osmond, H. Ziegler, eds. Encyclopedia of Plant Physiology New Series, vol. 12B. New York: Springer-Verlag, pp 181–230.

Slatyer, R.O. 1967. Plant Water Relationships. New York: Academic Press.

Sperry, J.S., and M.T. Tyree. 1988. Mechanism of water stress-induced xylem embolism. Plant Physiology 88: 581–587.

Sperry, J.S., and M.T. Tyree. 1990. Water-stress-induced xylem embolism in three species of conifers. Plant Cell and Environment 13: 427–436.

Steudle, E. 1992. The biophysics of plant water: compartmentation, coupling with metabolic processes, and flow of water in plant roots. In: Somero, G.N., C.B. Osmond, and C.L. Bolis, eds. Water and Life: Comparative Analysis of Water Relationships at the Organismic, Cellular, and Molecular Levels. Heidelberg: Springer-Verlag, pp 173–204.

Steward, F.C. 1964. Plants at Work. Reading, MA: Addison-Wesley.

Turner, N.C., and M.M. Jones. 1980. Turgor maintenance by osmotic adjustment: a review and evaluation. In: Turner N.C., and P.J. Kramer, eds. Adaptation of Plants to Water and High Temperature Stress. New York: Wiley, pp 87–104.

Tyree, M.T. 1976. Physical parameters of the soil-plant-atmosphere system: breeding for drought characteristics that might improve wood yield. In: Cannell, M.G.R., and F.T. Last, eds. Tree Physiology and Yield Improvement. New York: Academic Press, pp 329–348.

Tyree, M.T. 1997. The Cohesion-Tension theory of sap ascent: current controversies. Journal of Experimental Botany 48: 1753–1765.

Tyree, M.T., and J.D. Alexander. 1993. Hydraulic conductivity of branch junctions in three temperate tree species. Trees 7: 156–159.

Tyree, M.T., S.D. Davis, and H. Cochard. 1994a. Biophysical perspectives of xylem evolution: is there a tradeoff of hydraulic efficiency for vulnerability to dysfunction? Journal of the International Association of Wood Anatomists 15: 335–360.

Tyree, M.T., and F.W. Ewers. 1991. The hydraulic architecture of trees and other woody plants. New Phytologist 119: 345–360.

Tyree, M.T., and F.W. Ewers. 1996. Hydraulic architecture of woody tropical plants. In: Smith, A., K. Winter, and S. Mulkey, eds. Tropical Tree Physiology. New York: Chapman and Hall, pp 217–243.

Tyree, M.T., and H.T. Hammel. 1972. The measurement of the turgor pressure and the water relations of plants by the pressure-bomb technique. Journal of Experimental Botany 23: 267–282.

Tyree, M.T., and P.G. Jarvis. 1982. Water in tissues and cells. In: Lange, O.L., P.S. Nobel, C.B. Osmond, and H. Ziegler, eds. Physiological Plant Ecology II. Water Relations and Carbon Assimilation. Berlin: Springer-Verlag, pp 35–77.

Tyree, M.T., S. Patiño, J. Bennink, and J. Alexander. 1995. Dynamic measurements of root hydraulic conductance using a high-pressure flowmeter in the laboratory and field. Journal of Experimental Botany 46: 83–94.

Tyree, M.T., and J.S. Sperry. 1989. The vulnerability of xylem to cavitation and embolism. Annual Review of Plant Physiology and Molecular Biology 40: 19–38.

Tyree, M.T., and S. Yang. 1992. Hydraulic conductivity recovery versus water pressure in xylem of *Acer saccharum*. Plant Physiology 100: 669–676.

Tyree, M.T., S. Yang, P. Cruiziat, and B. Sinclair. 1994b. Novel methods of measuring hydraulic conductivity of tree root systems and interpretation using AMAIZED: a maize-root dynamic model for water and solute transport. Plant Physiology 104: 189–199.

van den Honert, T.H. 1948. Water transport in plants as a caternary process. Discussions of the Faraday Society 3: 146–153.

Yang, S., X. Liu, and M.T. Tyree. 1998. A model of stomatal conductance in sugar maple (*Acer saccharum* Marsh). Journal of Theoretical Biology 191: 197–211.

Yang, Y., and M.T. Tyree. 1993. Hydraulic resistance in the shoots of *Acer saccharum* and its influence on leaf water potential and transpiration. Tree Physiology 12: 231–242.

Yang, Y., and M.T. Tyree. 1994. Hydraulic architecture of *Acer saccharum* and *A. rubrum*: comparison of branches to whole trees and the contribution of leaves to hydraulic resistance. Journal of Experimental Botany 45: 179–186.

Yount, D.E. 1989. Growth of bubbles from nuclei. In: Brubakk, A.O., B.B. Hemmingsen, and G. Sundnes, eds. Supersaturation and Bubble Formation in Fluids and Organisms. Trondheim: Tapir Publishers, pp 131–163.

Zimmermann, M.H. 1983. Xylem Structure and the Ascent of Sap. Berlin: Springer-Verlag.

Zotz, G., M.T. Tyree, S. Patiño, and M.R. Carlton. 1998. Hydraulic architecture and water use of selected species from a lower montane forest in Panama. Trees 12: 302–309.

7
Responses of Plants to Heterogeneous Light Environments

Robert W. Pearcy
University of California, Davis, California

I. INTRODUCTION

Of all the environmental factors affecting plants, light is perhaps the most spatially and temporally heterogeneous. This heterogeneity takes on special importance in tropical forests in particular, because here light is considered to be the single most limiting resource for plant growth and reproduction. Accordingly, the life cycle and physiological responses of many tree and understory species have been shown to be closely keyed to changes in light availability (Bazzaz and Pickett 1980, Denslow 1980, Denslow 1987). Although less studied, heterogeneity in light environments of herbaceous and shrub communities has also been shown to be important in growth, development, and competitive interactions among species and individuals (Ryel et al. 1994, 1996, Tang et al. 1989, Washitani and Tang 1991). Much of the emphasis in ecology has been on the community and ecosystem consequences of environmental heterogeneity. For light in particular, the role of gaps in forest regeneration and diversity has received much attention. Several current models of forest dynamics have as their foundation differential species responses to heterogeneous light environments (Acevedo et al. 1996, Friend et al. 1993, Pacala et al. 1993, Shugart and Smith 1996). Considerable attention has also been given to the responses of individuals in terms of their ability to capture and utilize light as a spatially and temporally heterogeneous resource (Caldwell et al. 1986, Chazdon 1992, Newell et al. 1993, Pearcy et al. 1994). At the individual level, the concepts of heterogeneity and plasticity are intertwined. Phenotypic plasticity has a central explanatory role in how individuals across populations differ in form and function in response to spatial heterogeneity. At the same time, physiological constraints on organism responses to temporal heterogeneity determine to a large extent the nature of the adaptive responses to it.

Several different axes and scales of heterogeneity are of importance in understanding how plants respond to their light environment. The vertical axis is caused by the well-known exponential decrease in photon flux density (PFD) due to the absorption and scattering by leaves in the canopy. In closed forest canopies, most of the PFD is absorbed in the top 1 to 2 m, where much of the leaf area is located. The strong extinction of light results in development of sun and shade leaves within this layer differing in their morphology and physiological capabilities. The PFD ultimately reaching the forest floor may be only 1–2% of that received at the canopy top. Thus, plants in the understory must be able to maximize their light capture and photosynthesis in the extremely dim light of this environment. However, since the leaf area itself is heterogeneously distributed in the canopy, the resulting gaps allow penetration of sunflecks of much higher PFD to lower canopy layers. Sunflecks often contribute a substantial fraction of the total available PFD in the understory (Chazdon 1988, Pearcy 1983). Consequently, the ability to utilize sunflecks is also important for growth and survival.

In this case, the constraints imposed by and the ability to respond to the temporal heterogeneity on fast time scales of seconds to minutes are critical to sunfleck utilization. Species differences among trees in crown leaf area density create differences in the amount of PFD transmitted to the understory and the proportion transmitted as direct (sunfleck) and diffuse light, creating a lateral heterogeneity. This lateral heterogeneity causes differences in performance among individuals in a population. Adding further to the vertical and lateral heterogeneity are larger canopy gaps caused by spacing between tree crowns and by branch and tree falls that allow for much higher PFDs in the areas influenced by them. Temporal responses to gaps involve acclimation and avoidance of photoinhibition operating on a different time scale than the mechanisms involved in sunfleck utilization. In addition, the normal diurnal change in solar radiation and seasonal differences in day length and cloudiness are important causes of temporal heterogeneity that influence light capture and photosynthesis.

II. HETEROGENEITY AND PLASTICITY: SPATIAL AND TEMPORAL ASPECTS

Although it is often convenient to separate spatial and temporal heterogeneity, as is the case in this chapter, the separation is most often rather artificial. Sunflecks cause the smallest scale of spatial heterogeneity and also the most rapid scale of temporal heterogeneity. Moving up the scales through gaps, the spatial and temporal scales are highly correlated. Indeed, in terms of analysis, the spatial and temporal scales are frequently interchangeable. Wavelet analysis or autocorrelation (Baldocchi and Collineau 1994), two commonly used techniques for analysis of heterogeneity, can be applied equally well to either spatial or temporal scales.

However, differences become apparent when the nature of the responses is considered. For temporal heterogeneity, the dynamics of the system as determined by the time constants of the underlying processes must be considered. A dynamic system is one in which the current rate is dependent in some way not only on current conditions, but also on the rate at some previous time. A system is at steady state when it depends only on the current input parameters. Studies of spatial heterogeneity almost invariably treat systems as steady state even though the spatial pattern observed may be a consequence of an underlying dynamics. Of course, essentially all physiological and ecological systems are dynamic if observed on the appropriate time scale. Whether it is necessary to consider the dynamics depends on whether the system is adequately described by a series of steady states occurring in response to the changing temporal heterogeneity. If its dynamic properties are important, they will cause a significant deviation from the steady-state predictions.

The concept in spatial heterogeneity most similar to dynamics in temporal heterogeneity is grain size (Levins 1968). Spatial heterogeneity is considered to be fine grained if the patch size is such that the individual plant may be influenced by more than one patch. It is coarse grained if the plant is wholly within one patch and not influenced significantly by adjacent patches. The gap understory dichotomy is an example of a course-grained patchiness, whereas the understory itself with sunflecks and shade is an example of fine-grained patchiness. Plants within a coarse-grained patch generate a fine-grained heterogeneity within their canopy because of self-shading. Of course, grain size applies equally well to temporal heterogeneity, and, indeed, as illustrated by the sunfleck example, spatial grain size often has a temporal origin. Therefore, the comparison of the effects of grain size to dynamics rests on the question of whether nearby patches influence the response to the environment in each other. Although this question has not been explored extensively, it seems clear that hormonal and hydraulic signals provide a mechanism for such interactions.

It is also necessary to consider how heterogeneity is measured and how the plant perceives and responds to it. Measurements of the environment almost always involve linear scales, whereas the plant may respond in a distinctly nonlinear fashion. The response of photosynthesis to PFD is a particularly relevant example because it strongly increases as PFD increases at low values, but then exhibits saturation at higher PFDs. Depending on the range of PFDs encountered, the apparent heterogeneity in photosynthetic rates may be greater or lesser than the apparent heterogeneity in the PFD. Because photoinhibition occurs in response to excess PFD, the apparent heterogeneity in it would be still different. Therefore, appropriate measures of heterogeneity, preferably based on the biological responses, are important.

At the individual level, there can be several types of responses to environmental heterogeneity, depending on the scale. If the supply of a limiting resource is changed, its acquisition will be changed simply because of changes in diffusion rates and the kinetics of the enzymes involved in uptake. There can also be regulatory responses that may further increase or decrease the capacity for resource acquisition. An example would be a change in stomatal conductance, which influences the diffusion of CO_2 into the leaf. Finally, there can be some change in the properties of the organism itself, such as an increase in the concentration of enzymes in the leaf that influences the capacity for CO_2 assimilation. The latter falls clearly within the realm of a plastic response to the environment, and if it improves the performance of the plant in heterogeneous environments, it is often referred to as an acclimation response. The first would usually not fall within a modern concept of plasticity, which includes changes in an organism that enhance performance in a given environment rather than, as originally envisioned by Bradshaw (1965), as all of phenotypic changes observed when the environment is changed. Thus, a phenotypic change due to some stress injury would not be included in the modern concept of plasticity, but some adjustment that minimizes the stress injury would. Regulatory responses tend to fine tune the capacity of component steps, such as stomatal conductance and the biochemical capacity for CO_2 fixation. The regulated variations in a character such as stomatal conductance in response to changing environments are not an expression of plasticity. But differences in the maximum stomatal conductance and in the time constants expressed as a result of growth in different environments that influence the response to changing environments are rightfully expressions of phenotypic plasticity.

The difference between regulatory responses and acclimation has its greatest consequence for the time constants of the response to environmental change. A change in concentrations of enzymes takes time for synthesis or degradation, whereas regulation of the activity of already-present enzymes can occur on a much faster time scale. Moreover, much of the plasticity expressed in response to changing light environments involves development of new leaves with sun and shade characteristics. This requires even more time. Acclimation is too slow to respond to or provide any significant advantage in sunflecks, but is the dominant mechanism for phenotypic adaptation to gap formation and closure. Regulatory events, although slower than most sunflecks, can enhance their utilization in some circumstances; however, in many circumstances they may also act as a constraint. The primary mechanism allowing utilization of sunflecks is simply the capacity to use the higher resource flux while minimizing the regulatory constraints.

The evolutionary and ecological role of plasticity has received increasing attention over the past decade or so (Sultan 1992, Schlichting 1986). Historically, much of the focus in evolutionary theory has been on adaptive, habitat-based specialization, with plasticity being viewed as an alternative (the specialist/generalist argument). Most of the focus has been on specialization. In fact, in heterogeneous environments, plasticity is not an alternative because it may be required for survival, or at least for maximizing performance. Moreover, plasticity itself may not be an alternative, but a form of specialization (Lortie and Aarssen 1996). Although a characteristic arising as a result of plastic phenotypic variation cannot be the subject of selective forces, the capacity to express plastic variations in phenotype may be under genetic control and therefore subject to selection. This is the basis for the notion that species adapted to certain environments may have a greater capacity for acclimation than others from other environments. The role of plasticity and especially acclimation has been most commonly examined in terms of temporal heterogeneity. Paradoxically, they have been studied most commonly as a static process with plants grown for long term in different light environments rather than as a dynamic process responding to temporal changes in the environment. Functionally, the mechanisms involved in temporal and spatial scales are similar, but a full understanding of the temporal scale requires understanding of the dynamics of the mechanisms and their interaction with development, not just their static end result.

Understanding the adaptive significance of the plasticity of a character requires knowledge of how it scales up to whole-plant performance. If a plastic response in a character functions to improve performance, most often the result at the next level up is a greater homeostasis. For example, sun shade acclimation at the leaf level does not compensate for the large difference in resource supply (light) between sun and shade environments, but it does result in relatively smaller differences in growth between the sun and shade than would be expected in the absence of acclimation. The scaling needs to consider not only the benefits provided by the character itself, but also the costs that may influence other characters and therefore impact the overall performance. The costs are often more difficult to assess and include both the direct (construction and maintenance) and opportunity costs.

III. SPATIAL HETEROGENEITY IN LIGHT ENVIRONMENTS

A. Causes and Scales of Spatial Heterogeneity

Variation in overstory canopy structure is the primary factor causing the great spatial heterogeneity evident in forest understory light environments (Figure 1). Forest overstory canopies never close completely, in part because foliage tends to be clumped at the ends of branches and the statistical probability is that even at the maximum leaf area index, small gaps will still exist. These small gaps create sunflecks, a major source of heterogeneity in understory light environments. Differences in architecture among canopy species influencing light transmission may be important in heterogeneity on a scale equivalent to a tree crown (Canham et al. 1994). Understory trees and shrubs then create further spatial heterogeneity. Crowns of adjacent trees rarely grow into one another, so between them gaps may be more prevalent. Wind-induced abrasion by adjacent crowns may be important in creating gaps between tree crowns, especially in areas prone to high winds. At the next scale up, branch falls create slightly larger canopy gaps. Individual tree falls create gaps 50–600 m^{-2} depending on the size of the individual and the collateral damage. In most lowland tropical forests, 3–15% of the land area is in gaps at various stages of recovery (Brokaw 1985). Estimated turnover rates (mean time between gaps at any one point) range from 60 to 450 years with a mean of approximately 100 years. Because the effect of a gap on the light environment extends into the adjacent understory for some distance, it is likely that every plant in the forest is influenced by a gap at some time in its life cycle, and many may be influenced multiple times.

At the largest scale, hurricanes, fire, landslides, and human activities such as agriculture and logging create spatial heterogeneity. The frequency of these events varies greatly, but in Puerto Rico, for example, recurrent damage to forests from hurricanes is estimated to occur once every 50 years (Scatena and Larsen

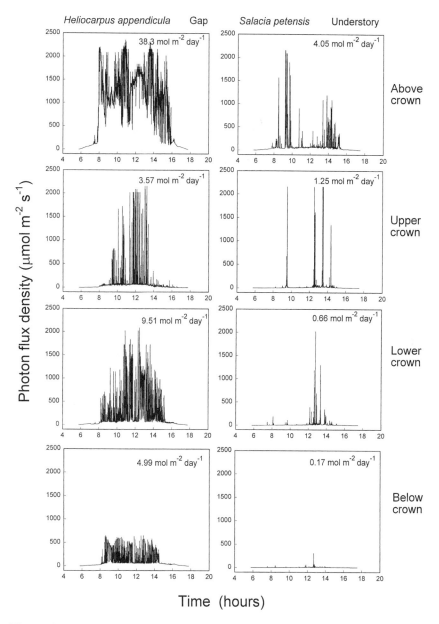

Figure 1 Patterns of spatial and temporal heterogeneity in the light environment as illustrated by the diurnal courses of PFD measured with sensors located at 4 positions in the crowns of a pioneer species, *Heliocarpus appendicula*, growing in a 900-m² gap and understory tree *Salicaia petensis* occuring under a closed canopy. The daily integrated PFD at each sensor is given in each box. (*Source*: Adapted from Kuppers et al. 1996.)

1991). Large-scale disturbances such as fire and hurricanes contain a smaller-scale heterogeneity because the damage they cause is highly heterogeneous.

Within each of these scales of spatial heterogeneity, there exists a further fine-scale heterogeneity. Sunflecks exhibit strong gradients of PFD across them due to penumbral effects arising because of the 0.5° arc diameter of the solar disk. Points from which part of the solar disk is occluded will be in partial shadow or penumbra. Only when the full solar disk is visible (i.e., the gap is large enough and the solar disk is fully within it) will an area of full, direct-beam PFD be present. Most sunflecks are entirely penumbral light with maximum PFDs of 0.1–0.5 of the full, direct-beam PFD. The steepness of the penumbral gradient from the brightest central portion to the edges, where it merges into the background diffuse light, depends on the height to the gap edges. Penumbral effects do not change the total photon flux transmitted through the gap, but do cause the combined size of the umbra and penumbra to be larger than the gap and therefore to potentially result in more carbon gain in the understory. The greater area translates into a greater probability that an understory plant will intercept a sunfleck for a longer time, and because the penumbral light is at a lower PFD than the full direct-beam radiation, it is utilized more efficiently. The spatial scale of sunfleck size has been shown to vary from 0.1 to 1 m, still typically smaller than most plant crowns in the understory, so that often only part of the crown will be influenced by a given sunfleck (Baldocchi and Collineau 1994). Variability within the sunfleck and the self-shading within the crown create an extremely heterogeneous light environment. Spatial autocorrelation analysis revealed that the correlation between light readings from array of photosensors decreased to 0.4 within 0.2 m and to 0.03 by 0.5 m (Chazdon et al. 1988). Cloud cover greatly reduced the spatial heterogeneity both in the short and long term, confirming that most of the spatial heterogeneity was due to sunflecks. Because seasonal changes in solar angle change the location of sunflecks, the large spatial heterogeneity observed for a day could be considerably reduced if monthly or annual means are considered. Analysis of hemispherical photographs taken along transects in a tropical forest reveal only weak spatial autoccorrelation at a 2.5-m scale and none at a 5-m scale (Becker and Smith 1990).

Although the central portion of larger tree fall gaps may receive a relatively uniform light environment, pronounced gradients occur around the edges because of the shading by adjacent canopy trees. In the northern hemisphere, the southern edge of a gap will be shaded even at midday, whereas the northern edge will receive the full direct-beam solar radiation. Similarly, the west side will be full sunlight in the morning, whereas the east side will receive it in the afternoon. The west side of the gap should therefore be more favorable for photosynthesis because of the lower temperatures and higher humidities in the morning. Wayne and Bazzaz (1993a, 1993b) found differences in the diurnal course of photosynthesis on the east and west sides, and small differences in the daily carbon gain, but no differences in growth. Because of variations in solar elevation angle, the effects of gap geometry on heterogeneity of light regimes in them depends on season and latitude. Gap size and geometry influence the total daily PFD available for photosynthesis because of the effect on hours of direct PFD and the transmission of diffuse PFD. Plants on the gap edges are usually in strongly directional light environments because of shading by adjacent plants on one side.

Within-plant heterogeneity has been shown to be as large or larger that the heterogeneity created by the environment. Gradients of PFD occur within canopies because of self-shading, which depends on crown density and the spatial pattern of leaves and stems. The transmission of PFD to a leaf depends on the gap fraction above it, the intersection of gaps with the solar track, and reflected and transmitted PFD from other leaves. Most of the foliage in tree crowns is concentrated in an outer 0.5- to 1-m thick shell, which may intercept 90% of the incident radiation. Therefore, gradients of available PFD in these crowns are quite steep (Figure 1).

B. Responses to Spatial Heterogeneity

1. Photosynthesis and Growth

At the most basic level, spatial heterogeneity in light creates variation in the resources available and hence in photosynthesis and growth. This has been demonstrated for sapling growth rates of a range of light-demanding and shade-tolerant species in a Panamanian forest (King 1994). At intermediate light levels, growth of the light-demanding species was faster than growth of the shade-tolerant species, but the latter had a lower whole-plant light compensation point, where there is just enough carbon gain to replace the current leaf area. The faster leaf turnover in the light-demanding species contributed to their higher light compensation point. Growth of understory saplings of two shade-tolerant Hawaiian tree species was highly correlated with spatial variation in the mean annual potential minutes of sunflecks received as estimated from fisheye photographs (Pearcy 1983). Similar results have been obtained with understory saplings in Costa Rica (Oberbauer et al. 1988) and oak seedlings in Japan (Washitani and Tang 1991). However, spatial heterogeneity in growth has not always been found to be correlated with estimates of sunfleck availability, and in some cases diffuse light availability has been a better predictor of growth (Tang et al. 1992). Pfitsch and Pearcy (1992) found no correlation between estimates of sunfleck light availability and growth of the redwood forest understory herb, *Adenocaulon bicolor*. However, variation in daily carbon gain and in the proportion of the annual carbon gain contributed by photosynthetic utilization of sunflecks was highly correlated with the variation in sunfleck availability (Pearcy and Pfitsch 1991). Moreover, "removal" of sunflecks with shadow bands designed to block sunfleck transmission along the solar track, but pass most of the diffuse PFD, significantly reduced growth and reproductive output of adjacent plants (Pfitsch and Pearcy 1992). Thus, there is no doubt that sunfleck utilization was important to these plants, but other limitations may have prevented translation of the spatial variation in sunfleck PFD into spatial variation in growth. Because there was no evidence for significant loss of carbon gain due to photoinhibition in sunflecks, other factors such as drought stress or increased competition in the brighter microsites need to be considered.

2. Plasticity of the Photosynthetic Apparatus in Response to Spatial Heterogeneity

One of the most universal responses to increased light availability is a plastic response of leaf photosynthetic properties that results in development of "sun" leaves having higher photosynthetic capacities per unit area (A_{max}), greater leaf thickness, and greater leaf mass per unit area (LMA). The plasticity expressed by different species in different growth situations varies widely, ranging from 1.25-to a 3- to 4-fold difference in A_{max} between sun and shade environments (Bjorkman 1981, Chazdon et al. 1996, Pearcy and Sims 1994, Strauss-Debenedetti and Bazzaz 1996). The plasticity of A_{max} can be examined either in terms of the change in photosynthetic rates achieved under light-saturating conditions or in terms of the range of light environments over which this plasticity is expressed. Most studies have focused on the former, generally comparing plants grown in two environments, one with high and the other with low PFDs. When compared in this way, fast-growing species adapted to high-light environments seem to exhibit the greatest plasticity of A_{max}, with more slowly growing, shade-tolerant shrub and tree species having a lower plasticity (Strauss-Debenedetti and Bazzaz 1996). Fast-growing species are also characterized by a higher potential A_{max}, so perhaps in these species there is a greater scope for change. However, there is also a lower PFD limit, below which the response may be more a failure of the leaves to develop normally, even though the result was a reduced A_{max}. Thus, some care must be exercised in interpreting when a change is indeed an adaptive plastic response and when it is perhaps a failure to adapt. Similarly, when the PFD is too high for a species, photoinhibition might occur, eventually reducing the photosynthetic capacity. This could be interpreted as a reduced plasticity of A_{max} when, in fact, it is due more to the range of light environments over which the plasticity can be expressed. Comparisons of plasticity require that more than two and preferably a range of light environments be employed.

The range of light environments over which plasticity is expressed can also differ between species, and may be as ecologically important as the actual change in A_{max} itself. Sims and Pearcy (1989) compared the shade-tolerant, rain-forest herb, *Alocasia macrorrhiza*, to the crop species, *Colocasia esculenta*, in five different light environments ranging from deep shade to 50% of full sun. Both exhibited a 2.2-fold difference in A_{max}, but the range of light environments over which this was expressed was higher in *Colocasia* than in *Alocasia*. Comparisons of 16 gap-dependent and shade-tolerant *Psychotria* species across three light environments from deep shade to 25% of full sun reveals the diversity of responses that can occur (Figure 2). Overall, the plasticity of A_{max} differed little among the species, but the responses to the changes in light environments contrasted markedly. Most species exhibited the largest increase in A_{max} between the low and

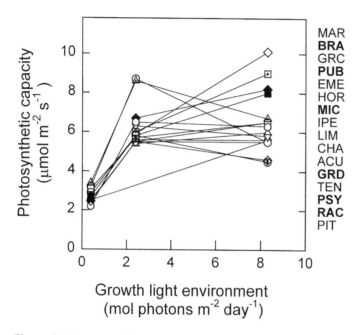

Figure 2 Response of photosynthetic capacity (A_{max}) of 16 *Psychotria* species to growth in three different light environments simulating understory shade to gap conditions. The abbreviations on the right side are for the different species and are ordered identically to the symbols shown for the highest light environment on the right side. The abbreviations are: MAR, *P. marginata*; BRA, *P. brachiata*; GRC, *P. graciflora*; PUB, *P. pubescens*; EME, *P. emetica*; HOR, *P. horizontalis*; MIC, *P. micrantha*; IPE, *P. ipecacuanha*; LIM, *P. limonensis*; CHA, *P. chagrensis*; ACU, *P. acuminata*; GRD, *P. granidensis*; TEN, *P. tenuifolia*; PSY, *P. psychotrifolia*; RAC, *P. racemosa*; PIT. *P. pittieri*. The abbreviations in boldface are for gap-dependent species and show that there is no relationship between gap versus understory species and plasticity of photosynthetic capacity. Each data point is the mean of at least 3 determinations on different plants. (*Source*: Adapted from Valladares et al 1998.)

intermediate light environments. Some species exhibited a moderate increase between the intermediate-and high-light environment, others no increase, and still others exhibited a decline in A_{max} from the intermediate-to the high-light environment. These responses of A_{max} were not correlated with habitat requirement of the species, although an overall measure of plasticity integrating both physiological and morphological characters showed that gap-dependent species had greater plasticity. Of course, A_{max} is only one of a suite of characters, and changes in leaf angle, etc., may function to reduce the actual range of incident PFD experienced by the plant to avoid conditions that may cause photoinhibition. Indeed, the PFD on the leaf surfaces is often close to the PFD that is just saturating for photosynthesis, even though the incident PFD on a horizontal surface is much higher. It would be unrealistic to expect that increasing the PFD would result in a further increase in A_{max}. The more common result may be photoinhibition.

Only a few studies have examined the plasticity of the photosynthetic apparatus to the spatial heterogeneity of light environments in the field. Chazdon (1992) found that A_{max} of an early successionial shrub, *Piper sancti felicis*, was closely related to light availability along transects across natural gaps. A later-successional congener, *P. arieianum*, exhibited a weaker association of A_{max} with light availability, possibly because of a greater susceptibility to photoinhibition in the gap center (Figure 3). Chazdon and Field (1987) found no correlation between light availability and A_{max} in response to microsite variation in the understory, but did find increasing A_{max} with increasing light availability in gap environments. These results suggest a lower threshold PFD below which no further adjustments in photosynthetic capacity occur. Most studies are consistent with the notion that the integrated daily PFD rather than the maximum PFD is the primary signal to which adjustments in A_{max} respond. Sunflecks are the primary source of heterogeneity in daily PFD in the understory, but these are certainly too unpredictable for any advantage in daily carbon gain to accrue from adjustment of A_{max}. Sims and Pearcy (1993) found no differences in A_{max} among *Alocasia macrorhiza* plants grown under either short or long sunflecks or constant PFD environments, where each environment had identical daily PFDs. Pearcy and Pfitsch (1991) found that A_{max}, LMA, and leaf N per unit area of the *Adenocaulon bicolor* plants grown under shadow bands were all positively correlated with diffuse PFD levels. However, no correlation between these measures and either total PFD of sunfleck PFD was found for the adjacent "control" plants that received sunflecks. A filtering mechanism that ignored the unpredictable component of variation in daily PFD associated with sunflecks, but allowed response to the much less variable but more predictable component, would explain these results.

Maximizing Photosynthesis in High Light. Maximizing photosynthetic performance in high light requires investment in the carboxylation and electron

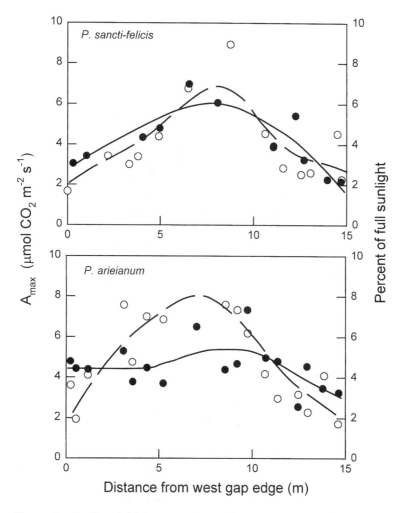

Figure 3 Gradients of light (open circles) and A_{max} (closed circles) of seedling of the gap species, *Piper sancti-felicis* (top), and the understory species, *Piper arieanum* (bottom) across a small gap. Photosynthetic capacity tracks the light environment much better in the gap-dependent species than in the understory species. (*Source*: Adapted from Chazdon 1992.)

transport capacity necessary to sustain high photosynthetic rates. Growth in high versus low light has been shown to result in increases in activities (concentrations) of the primary carboxylating enzyme ribulose-1,5 bisphosphate carboxylase/oxygenase (Rubisco), PSII electron transport capacity, the electron carrier, cytochrome f, and chloroplast coupling factor (CF1) that is involved in adenosine triphosphate (ATP) synthesis (Anderson and Osmond 1987, Bjorkman 1981, Pearcy and Sims 1994). It is likely that nearly all components in the photosynthetic carbon reduction cycle (PCRC) and electron transport must increase in a coordinated manner to maintain a balance between capacities of each component step. In addition, stomatal conductance must also increase to maintain a balance between the capacity for supply and utilization of CO_2.

Insight into the causes of the plasticity in photosynthetic capacity can be gained by comparing rates calculated on a per unit leaf area basis to those calculated on a per unit leaf mass basis. One of the most striking features of sun shade acclimation is the increase in leaf mass per unit area that is due to the development of thicker leaves having, in some cases, multiple layers of palisade cells (Bjorkman 1981, Chabot and Chabot 1977). In contrast, leaves developing in the shade are much thinner with a loosely organized and relatively undifferentiated mesophyll. The large differences in photosynthetic capacities per unit area of sun and shade leaves are much reduced or absent when photosynthetic capacities per unit leaf mass are compared. This is consistent with a constant amount of photosynthetic machinery per unit mass and with a greater LMA associated with a greater concentration of photosynthetic machinery per unit area. Similarly, significant relationships between A_{max} per unit area and N per unit area have been commonly found in comparisons of plants grown in the sun and shade, whereas there is no significant correlation between assimilation rates and N on a mass basis for most species. Indeed, for many species, increased LMA may be the only mechanism involved in the observed plasticity of A_{max}, since photosynthetic rates per unit mass remain nearly constant across light environments (Ellsworth and Reich 1992, Sims and Pearcy 1992). In some others, however, there appears to be a limited capacity to alter concentrations of enzymes and photosynthetic machinery per cell and therefore alter A_{max} per unit biomass in response to changes in the light environment (Chazdon and Kaufmann 1993). Across species and especially across life-forms, however, much of difference in photosynthetic capacity is clearly due to genetically determined differences in concentrations per unit mass of photosynthetic machinery, since LMA is often inversely correlated with A_{max} (Reich et al. 1997).

Since excess light is potentially damaging, resulting in photoinhibition, it is not surprising that sun leaves have a greater capacity for photoprotection than shade leaves. This photoprotection has both morphological and biochemical components. The orientation of leaves toward more vertical surfaces, which is common in sun plants, helps to minimize the interception of excess radiation (Ehleringer and Forseth 1980, Gamon and Pearcy 1989, Ludlow and Bjorkman 1984).

At the biochemical level, down-regulation of PSII through operation of the xanthophyll cycle provides for photoprotection through dissipation of excess energy as heat (Demmig-Adams and Adams 1992b, 1996). Sun leaves possess greater total pool sizes of xanthophyll-cycle pigments and a greater capacity for rapid conversion of violoxanthin to antheroxanthin and zeaxanthin that invokes this photoprotection (Demmig-Adams and Adams 1992a, 1994, Thayer and Bjorkman 1990). In addition, sun leaves have a greater capacity for scavenging of active oxygen-free radicals that can cause photodamage (Grace and Logan 1996). When shade leaves are exposed to PFD in excess of the capacity of these mechanisms, photodamage to PSII may occur, requiring operation of a PSII repair cycle for recovery (Anderson et al. 1995, Geiken et al. 1992, Oquist et al. 1992).

Maximizing Photosynthesis in Shade. Maximizing leaf photosynthetic performance at the low PFDs inherent in shaded, diffuse-light conditions requires: (1) maximizing light absorptance, (2) maximizing the quantum yield with which this light is utilized for CO_2 assimilation, and (3) minimizing respiratory losses. Careful comparisons under conditions where photorespiration is suppressed (low O_2 or high CO_2 partial pressures) have unequivocally established that there are no intrinsic differences in quantum yields between species adapted to sun and shade habitats or, provided that no stress effects interfere, among leaves on plants grown in high and low PFDs (Bjorkman and Demmig 1987, Long et al. 1993). The observed quantum yields are consistent with those derived from stoichiometric requirements for production of the ATP and NADPH necessary for carbon metabolism, small inefficiencies in energy transfer inherent within the chlorophyll antennae and reaction centers, and energy use for other processes such as nitrate reduction. Thus, there is essentially no scope for further improvements in shade compared with sun leaves. Photorespiratory losses under normal atmospheric concentrations reduce the achieved quantum yields under natural conditions, but this is not very different for sun and shade plants. Quantum yields are frequently reduced because of photoinhibition, especially when high light is coupled with high leaf temperatures, water stress, or low nitrogen nutrition (Bjorkman et al. 1981, Ferrar and Osmond 1986, Mulkey and Pearcy 1992). Shade leaves are more susceptible to photoinhibition than sun leaves in this regard, and observed differences in quantum yields may often be due to the occurrence of these other environmental factors and their interaction with high PFD.

Comparisons across many species also reveal no consistent differences in leaf absorptance or chlorophyll (Chl) concentration per unit leaf area between sun and shade leaves (Bjorkman 1981). Chl concentrations per unit area are sufficient to absorb 80–85% of the incident photosynthetically active radiation, with the remainder lost to either reflection or transmission. The lower LMA of shade leaves means that to maintain Chl per unit area constant, there has to be an increased investment in Chl per unit mass. Although shade leaves in particular could benefit in terms of increased photosynthesis from increased Chl per unit area and hence leaf absorptance, the strongly diminishing returns of further investment in Chl–protein complexes required to affect a significant change in light capture may preclude it. A doubling of the Chl concentrations is required to increase leaf absorptance by just 5–6% from the levels previously given. Although there are only 4 mmol N in a mol of Chl, the associated proteins in the Chl–protein complexes contain 21–79 mmol N mol^{-1} Chl (Evans 1986). The lower Chl a/b commonly observed in shade leaves has no significant consequences for light capture or quantum yield in shaded environments, but it does engender a significant savings in the nitrogen investment required for this light capture. Chl b is located in the light harvesting chlorophyll–protein complex II (LHCII), which contains only 25 mmol N mol^{-1} Chl. The lower Chl a/b ratio results from increased LHCII per unit leaf area coupled with decreased concentrations of the photosystem II (PSII) core complex, which contains only Chl a (Evans 1986). Since PSII complexes contain 83 mmol N mol^{-1} Chl, the shift to more LHCII while maintaining total Chl per unit leaf area constant results in a significantly greater light capture per unit N invested. This savings is possible in shade, but not sun leaves, because high capacities of PSII electron transport (ET) are not needed in the shade conditions.

To maintain a positive carbon return, assimilation rates must clearly exceed respiration rates. It is now widely documented that one of the most important differences between sun and shade leaves is the much lower respiration rates of the latter (Bjorkman 1981, Grime 1966). Both comparative measurements and simple carbon balance models using sun and shade leaf characteristics will typically show negative balances for sun leaves in the shade, whereas balances are positive for shade leaves. However, such comparisons may be somewhat misleading. Respiration occurs in response to growth and maintenance processes, and since light severely limits carbon gain and growth in the shade, respiration rates are correspondingly low. Therefore, the observed respiration rates reflect mostly the resource supply, but whether the capacity for respiration also adjusts is less certain. Transfer of sun leaves to the shade resulted in a rapid decrease in respiration rates over the next 4–5 days as carbohydrate supplies were depleted, with little decline in photosynthetic capacity over the same time period (Sims and Pearcy 1991). After 6 days the respiration rates were 1.25% of photosynthetic capacity, which matched the value found in the shade leaves. Although the difference was difficult to quantify directly because of the low rates, the constant ratio respiration to photosynthetic capacity suggests that the maintenance respiration rates are indeed higher in sun than shade leaves, but that the maintenance costs of extra photosynthetic capacity are also rather small. Nevertheless, the lower maintenance costs associated with lower photosynthetic capacity do enhance carbon gain in the shade.

Of more importance may be the construction costs, which occur principally early in the leaf's development but which must be paid back over its lifetime.

Sun leaves are potentially more expensive to construct because of their higher concentrations of photosynthetic enzymes. On the other hand, shade leaves may require more protection from herbivores since, given the very low carbon gain rates, leaf longevity must be greater for a leaf to ultimately turn a net profit. Sun leaves can recover construction costs within a few days, whereas shade leaves may require 60–150 or more days, depending on the light environment (Chabot and Hicks 1982, Jurik 1983, Sims and Pearcy 1992). Although there are clearly pressures for increasing construction costs in both the sun and shade, the available data show that within species, sun leaves cost more per unit area to construct, largely because of their greater leaf thickness and mass per unit area. On a per unit mass basis, shade leaves of *Alocasia macrorrhiza* were found to be slightly more expensive than sun leaves because of lower investments in relatively inexpensive cell wall material compared with protein and membrane components in shade leaves (Sims and Pearcy 1992). In a comparative study of rainforest *Piper* species, Williams et al. (1989) found higher construction costs per unit mass in gap compared with understory shade species, and because the shade species had significantly longer leaf life spans, there was an inverse relationship between leaf life span and construction cost. Among broader surveys, however, there is hardly any evidence for systematic differences in leaf construction costs per unit mass among different functional groups, despite wide differences in leaf morphology (Poorter 1994). The lack of any consistent pattern most likely reflects the diversity of trade-offs between investments in structural, biochemical, and protective compounds that can occur in any given habitat, let alone across habitats.

C. Plasticity of Whole-Plant Responses

Integration of the role of photosynthetic plasticity in enhancing plant function in heterogeneous light regimes requires an understanding of how leaf photosynthetic rates interact with and are determined by allocation patterns at the whole-plant level. Since the carbon content of biomass is nearly constant, growth and whole-plant photosynthesis are closely linked by the conversion efficiency of fixed carbon into biomass (Dutton et al. 1988, Sims et al. 1994). Whole-plant photosynthesis is simply the photosynthetic rate of the leaves integrated over the leaf area. It is necessary to take into account that leaves are in different light microenvironments within the crown and, because of acclimation and aging, phenomena may have different capacities to utilize this light. Moreover, the carbon losses via respiration at the whole-plant level must be subtracted to arrive at a whole-plant carbon balance.

Enhanced whole-plant photosynthesis can therefore be obtained either by increasing the photosynthetic rate per unit area under the prevailing conditions, or by increasing the leaf area per plant. An important linkage in this process is the LMA, since this determines how much leaf area will be produced for a given investment of biomass in leaves. Most studies have found that the leaf weight ratio (LWR: mass of leaves per unit mass of plant) tends to be higher in sun- compared with shade-grown plants, but this difference is, with few exceptions, rather small (see review by Bjorkman 1981). There appears to be few differences in allocation to supporting biomass between sun and shade plants, but shade plants typically invest less in roots, perhaps because of the lower transpirational and nutritional demands in this environment. However, shade-grown plants typi- cally have a much larger leaf area ratio (LAR: area of leaves per unit mass of plant) than sun plants because of the large differences in LMA. Since, as previ- ously discussed, LMA is also a determinant of photosynthetic capacity, it clearly mediates a trade-off between lower photosynthetic capacity but greater leaf area for light capture in the shade, versus higher photosynthetic capacity but less leaf area in the sun. The result in the shade is a large advantage in whole-plant carbon gain for the shade plant (Figure 4). Perhaps counterintuitively, experiments and simulations have shown no carbon gain advantage per unit biomass invested for the sun plant phenotype in high light since the reduced leaf area per plant more than offsets the advantage accrued due to the higher A_{max} per unit leaf area (Sims et al. 1994, Sims and Pearcy 1994). Shade acclimated sunflower plants have been shown to have higher relative growth rates immediately after transfer to high light than plants grown continuously in the high-light environment (Hiroi and Monsi 1963). Subsequent acclimation of the shade plant to the sun environment therefore actually causes the relative growth rate (RGR) to decrease. This does not mean that sun leaves provide no advantage in high light, especially over the long term, since they may well contribute to a greater stress resistance and greater water and nitrogen use efficiency.

D. Responses to Within-Crown Spatial Heterogeneity

Growth itself generates a heterogeneous light environment as upper, younger branches and leaves shade lower branches and older leaves. Senescence of older and more shaded leaves and reallocation of resources to the newly developing

---

Figure 4 Response of instantaneous assimilation rates to PFD for leaves (A,C) and daily assimilation rates of whole plants (B,D) of *Alocasia macrorrhiza* acclimated to sun (open circles) and shade (closed circles) conditions. The curves for whole plants were obtained by growing individual plants in sun or shade conditions and then transferring to a whole-plant chamber for gas exchange conditions. Different daily PFDs were obtained with neutral-density screen and cloth filters under natural sunlight. (*Source*: Adapted from Sims and Pearcy 1994.)

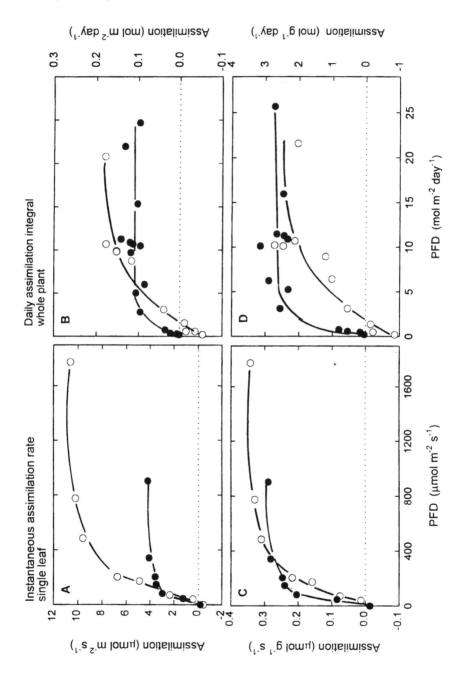

leaves in the higher-light environments is an important mechanism for optimizing the use of resources, particularly nitrogen (N), within the canopy. Optimal N use occurs when canopy photosynthesis is maximized for a given whole-canopy N content (Field 1983, Hirose and Werger 1987a). This demands that more N be partitioned to the upper leaves located in high light, resulting in a higher A_{max} of these leaves compared with lower leaves in more shaded microenvironments. Because most N in leaves is in the photosynthetic apparatus, the result is a gradient of decreasing photosynthetic capacity from the upper to lower leaves. The optimal N gradient is the one that results in a gradient of photosynthetic capacity through the canopy such that light is equally limiting for leaves at all levels (Field 1983). Decreasing photosynthetic capacity causes a decreasing light saturation point, potentially satisfying this optimality requirement. Measurements show that real canopies have nitrogen and photosynthetic-capacity gradients approaching but not equal to an optimal gradient (Hirose and Werger 1987b, Hirose et al. 1988). The discrepancy may be due to the reallocation of N lagging behind canopy development or to constraints on how much N can be efficiently reallocated. In comparison to uniform gradients, the optimal and observed gradients result in substantial increases in canopy photosynthesis. The increase ranges from 1–4% in open-canopied shrubs and trees with relatively low total canopy N concentrations (Field 1983, Leuning et al. 1991) to as much as 20–40% in dense-canopied stands with high canopy N contents (Evans 1993, Hirose 1988, Hirose and Werger 1987a). Both leaf senescence and acclimation to shade contribute to the pool of N that can be reallocated.

Evans (1993) compared the consequences of nitrogen redistribution within a leaf due to acclimation (investment in chlorophyll–protein complexes versus carboxylation capacity) with that due to nitrogen redistribution within a canopy. Leaf photosynthetic acclimation alone (optimal partitioning between photosynthetic components within a leaf at a constant N content) was shown to potentially increase canopy photosynthesis by approximately 4%. On the other hand, redistribution of N among leaves potentially increased canopy photosynthesis by 20% for the same canopy. Thus, in terms of maximizing photosynthesis in heterogeneous light environments within canopies, nitrogen redistribution among leaves is more important than nitrogen redistribution within leaves.

E. Responses to Directional Light Gradients

Plants near the edges of gaps or in competing plant populations will often have much more of their light on one side because of shading by nearby plants on the other. Phototropic reorientation of leaves and stems, which increases light capture from the prevailing direction, can often be observed in these plants. In addition, preferential growth of branches may occur to fill available space on the open side. Since branches are often thought to exhibit semiautonomy, the simplest explanation could be differential carbon gain on the two sides. Indeed, a model of growth in crowded populations based on this principle adequately predicted the general patterns of branch growth and thinning (Takenaka 1994). However, hormonal signals may also be involved, and branches are certainly limited in their extension, perhaps by hydraulic constraints. Gilbert et al. (1995) showed that reflected far-red (FR) light acting through phytochrome inhibited branch growth on the side nearest to a neighboring plant. Novoplansky (1990) showed that *Portulaca oleracea* preferentially branched in a direction away from a reflected or transmitted far-red light source. Additionally, for early-successional herbaceous plants, reflected far-red radiation can be a signal of potential competitors, leading to internode elongation and accelerated height growth at the expense of tillering or branch growth (Ballare 1994, Ballare et al. 1987). This response occurs before any direct shading and therefore allows the plant to effectively anticipate a potential competitive interaction.

F. Differential Growth of Clonal Plants

Clonal plants can exhibit varying degrees of physiological integration, which may be important in their exploitation of patchy environments (Hutchings and de Kroon 1994, Slade and Hutchings 1987). A type of foraging behavior manifested as a greater concentration of ramets and leaves in favorable, high-light patches compared with unfavorable, shaded patches can result from morphological plasticity. In particular, shorter petioles coupled with increased bud activation may be elicited when a clone reaches a favorable patch, whereas in unfavorable patches, longer petioles may help ramets and leaves escape (de Kroon and Hutchings 1995).

Resource sharing via physiological integration of ramets in favorable versus unfavorable microenvironments may modify the responses of ramets. For example, Dong (1995) found that integration in the stoloniferous understory herb *Laminastrum galeobdolon* significantly evened out the morphological differences across high- and low-light patches. In contrast, there was no such dampening in the morphologically similar fenland species, *Hydrocotyle vulgaris*. Since a major source of heterogeneity in the understory is spatially unpredictable sunflecks, the evening out of responses across ramets may be beneficial in terms of whole-clone performance. Patches of light are spatially more predictable in the fenland, and therefore independence in morphological adaptation of ramets may be beneficial.

There is also evidence that clonal integration may enhance utilization of heterogeneous resources via spatial division of labor. In experiments where heterogeneous light and soil water environments were created with strictly negative covariance (high light always associated with low water and vice versa), interconnected ramets of *Trifolium repens* exhibited specialization for uptake of the locally most abundant resource, with resource sharing across ramets (Stuefer et al.

1996). Moreover, total clonal growth was significantly enhanced relative to that of clones subjected to homogeneous environments that were either well watered but shaded or receiving high light but water limited. Inherently negative relationships between patchiness in resources such as light and water may be fairly common in natural environments. However, even in the absence of patchiness of one resource, spatial integration may prove beneficial. For example, in experiments with uniform soil moisture, integration enhanced the growth of ramets of the herbs *Potentilla replans* and *P. anserina* in high light that were interconnected to ramets in low light, apparently because water uptake by the low-light ramets supported, in part, the high transpiration rates of the high-light ramets (Stuefer et al. 1997). Severing the connections reduced the growth of the low-light ramets because of the restriction in carbon supply and also reduced the growth of the high-light ramets because of a reduction in water supply.

IV. TEMPORAL HETEROGENEITY IN LIGHT ENVIRONMENTS

A. Time Scales of Temporal Changes and Responses

Temporal changes in light of interest in ecophysiology involve time scales differing by several orders of magnitude. At one end of the spectrum, the formation and closure of canopy gaps due to tree falls takes place over several to many years. Annual changes occur because of day length, and solar elevation angle changes, as well as seasonal variation in cloudiness. Diurnal changes occur because of the earth's rotation and the resulting path of the sun across the sky. Finally, at the other end of the spectrum, changes on the order of seconds to minutes occur in sunflecks under plant canopies and because of intermittent cloudiness.

The responses to these temporal changes can be conveniently divided into two categories, one involving the acclimation and developmental plasticity occurring in response to long-term changes in the light environment, and the other involving regulatory responses to short-term changes such as those that occur during the normal diurnal course of solar radiation or during more rapidly changing irradiance such as that due to sunflecks in an understory. The time difference for temporal changes for which acclimation and developmental plasticity are of paramount importance, and those for which regulatory mechanisms are prominent appears to be on the order of 1 day. Acclimation to a change in light requires 4–5 days for leaves of fast responding species such as peas (Chow and Anderson 1987a, Chow and Anderson 1987b) and up to 45 days in some slow-responding species, such as the tropical tree *Bischofia javanica* (Kamaluddin and Grace 1992). These types of responses involve changes in enzyme concentrations within leaves that function to alter photosynthetic capacity. Both acclimation and developmental plasticity are sensitive to the level of other resources, especially nitrogen supply (Ferrar and Osmond 1986, Osmond 1983), and set in particular the maximum photosynthetic capacity that a leaf can achieve. Mature leaves typically exhibit through acclimation only part of the possible physiological plasticity that can be expressed by a genotype and in some cases may show little acclimation. Further change affecting whole-plant performance involves development of new leaves having different structural properties such as leaf mass per unit area or leaf size. Development of new leaves, and changes in branching architecture, etc., then allow the plant to express its maximum plasticity with respect to longterm light changes. These types of changes obviously require even more time (weeks to months), with the maximal change in whole-plant physiological properties not occurring until all leaves had turned over. As discussed previously in the context of spatial heterogeneity, growth itself also generates a temporally heterogeneous light environment as newer branches and leaves on top shade lower branches and older leaves.

Although there are enzymes that show marked diurnal variation in concentration (Huffaker and Peterson 1976), photosynthetic enzymes involved in carbon metabolism and electron transport, and the light-harvesting chlorophyll–protein complexes in leaves, are typically already present in high concentrations. Affecting a significant relative change in concentrations due to either synthesis of degradation may require longer than one day and would be energetically costly. Therefore, changes in photosynthetic enzyme concentrations may be evolutionarily precluded as too slow and too costly to be a mechanism for responding to diurnal changes in solar radiation or to sunflecks. Instead, responses to these faster changes involve modulation of enzyme activity by small regulatory molecules such as thioredoxin (Buchanan 1980), regulatory enzymes such as Rubisco activase (Portis 1992), binding of substrates to regulatory sites, chloroplast stromal pH, or energization of the thylakoid membranes (Foyer et al. 1990, Harbinson et al. 1990a, 1990b). In addition, the regulatory mechanisms governing stomatal opening and closure are also important. These regulatory mechanisms appear to match the capacity of the component steps to each other and to the supply of light energy. When the PFD is low, down-regulation of several enzymes in the photosynthetic carbon reduction cycle (PCRC) occurs so that the capacity for carbon metabolism matches the supply of light energy. When the PFD is in excess, regulatory mechanisms involving the xanthophyll cycle in the pigment beds cause the excess energy to be dissipated as heat, thereby affording protection from photoinhibition (Demmig-Adams and Adams 1992b). Stomatal conductance also adjusts to maintain a balance between the supply of CO_2 and the capacity to utilize it, as well as to affect the compromise between water loss and carbon gain (Cowan and Farquhar 1977). The time constants for these regulatory changes typically range from 0.5 to 15 minutes, depending on the particular process. Little is known about the quantitative energetic costs of these regulatory mechanisms, but they are undoubtedly much less than those involved in acclimation and developmental plasticity.

B. Responses to Sunflecks

1. Temporal Nature of Sunfleck Light Regimes

Light environments in canopies and understories present a particularly complex case of temporal heterogeneity. Sunflecks in these environments occur because of the juxtaposition of the solar track on a particular day with a gap in the canopy. Since the sun moves along the track at 15° per hour, sunflecks move across the forest floor at a rate determined by the height to the canopy gap. Penumbral effects cause the PFD in a sunfleck to increase and decrease gradually as the solar disk moves across the gap. However, the canopy is rarely static, and movement due to wind causes the edges of a canopy gap to continually change and even for gaps to open and close. Thus, what may be one gradually changing long sunfleck on a completely still day is more often broken up into many rapidly changing short sunflecks. These then appear as a cluster of brief, continually changing sunflecks superimposed on a background of diffuse shade light. Since the solar track shifts with time of the year, the juxtaposition of the solar disk, a gap, and a point on the surface is very transitory. A sunfleck occurring at a particular time and place may do so for only a few days to weeks, and then only when cloud cover does not intervene. Moreover, seasonal changes in the solar track usually result in changes in sunfleck contributions because lower solar angles cause the solar beam to traverse a longer segment of canopy, decreasing the probability of encountering a gap.

The complexity of sunfleck light regimes makes the task of defining a sunfleck particularly difficult (Chazdon 1988, Smith et al. 1989). The rapid changes in PFD in sunflecks mean that sampling rates must be high to adequately describe the record. Sampling theorem states that the sampling frequency must be double the rate of the most frequent event of interest if the objective is to accurately reconstruct the record. In understories, this dictates sampling of PFDs at intervals of 1–10 seconds, while in canopies where leaf flutter creates highly dynamic light environments, sampling frequencies of 0.1 second may be in order (Pearcy et al. 1990, Roden and Pearcy 1993a). Once an adequate record is obtained, one approach is to define a sunfleck as a transient excursion of the PFD above a threshold level that is just above the background diffuse PFD, or it can be some value that is physiologically meaningful for the studied plant species. This approach has been criticized because the threshold is arbitrary, and different thresholds have to be used in different canopies and with different species (Baldocchi and Collineau 1994). Wavelet analysis provides an objective method for detecting sunfleck events in a data record and does not require a threshold (Baldocchi and

Collineau 1994). Once sunflecks are detected, they can be quantified as to their duration, PFD, etc. However, wavelet analysis depends on selection of the appropriate wavelet transform function. Finding one that works well is difficult given the highly variable and complex nature of sunflecks. Therefore, it is not yet clear whether application of wavelet analysis provides a significant advantage over analyses with detection of sunflecks using the threshold approach. Spectral analysis, in which periodicity in the light environment represented by characteristic frequencies is detected in a record using fast Fourier transform techniques (Desjardins 1967, Desjardins et al. 1973), has not been particularly useful in analyzing temporal sunfleck records since there are usually no characteristic frequencies, or perhaps a smear of overlapping characteristic frequencies. Since leaves often flutter at characteristic frequencies, the effects of this fluttering on the light environment can be detected where it is a significant component of the overall variation, such as in poplar or quaking aspen canopies (Roden and Pearcy 1993a).

Measurements of sunfleck light regimes in a wide variety of forest understories show that on clear days, 10–80% of the daily PFD at any given location may be due to sunflecks (Chazdon 1988, Pearcy et al. 1987). Thus, sunflecks contribute a large fraction of the light potentially available for photosynthesis, and their utilization could be expected to have a large effect on carbon balance of understory plants. The characteristics of sunflecks depend on attributes such as canopy height and flexibility, as well as weather conditions such as wind and cloudiness. Figure 5 shows histograms of sunfleck characteristics for a tropical forest on Barro Colorado Island, Panama, and for sensors mounted within a poplar canopy. Most sunflecks in forest understories are of short duration and low maximum PFD, but the long-duration sunflecks contribute much more of the daily PFD. By contrast, short-duration sunflecks are a much more important contributor of PFD in aspen canopies. In the tropical forest understory on Barro Colorado Island, sunflecks tended to be shorter in the dry compared with the wet season because of the more windy conditions in the former. In the wet season, most sunflecks occurred in the morning hours because clouds typically built up and rain fell in the afternoons. Most days in the dry season were clear in both the morning and afternoon. Overall, the additional cloudiness in the wet compared with the dry season reduced the PFD contributed by sunflecks by approximately 50%. This is consistent with the reductions in PFD above the canopy in the wet versus the dry season.

2. Dynamic Responses of Photosynthesis to Sunflecks

When light is increased, photosynthesis will initially accelerate rapidly up to a level set by the activities of the component enzymes in the pathway. The concentrations of the component enzymes and their level of activation determine these

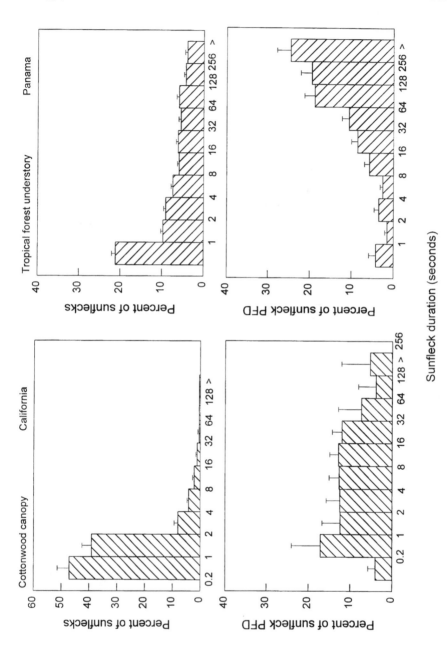

Figure 5 Histograms showing variation in sunfleck characteristics between a *Populus fremontii* canopy in California (left side) and a tropical forest understory on Barro Colorado Island, Panama (right side).

activities. In addition, the stomatal conductance is important in setting the upper limit since it sets the supply of CO_2 available to the carboxylation reaction in the chloroplasts. The initial rapid acceleration is due to the buildup of metabolites (especially RuBP) which are initially limiting to the rates of flux of carbon within the pathway (Sassenrath-Cole and Pearcy 1992). Since the buildup of these metabolites is rapid, the rapid acceleration phase requires only a few seconds to complete. If a leaf has been in the shade for a long period, the initial rapid increase can be small or almost missing because the stomatal conductances are too low and because PCRC enzymes are inactivated. Stomatal opening and activation of these enzymes is a relatively slow process, leading to the well-known induction requirement of photosynthesis (Figure 6) in which the assimilation rate requires 10–30 minutes to reach its maximum value (Pearcy 1988, Walker 1981). If a fully induced leaf is shaded for 1–2 minutes and then re-exposed to saturating PFD, only a few seconds are required for the assimilation rate to recover to its maximum value. The response of a fully induced leaf to a brief shading period

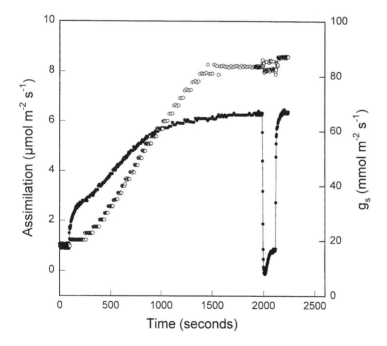

Figure 6 The time course of assimilation (filled circles) and stomatal conductance (open circles) during photosynthetic induction for an *Alocasia macrorrhiza* leaf. The PFD was increased from 12 to 525 μmol photons m^{-2} s^{-1} at about 80 seconds and then decreased back to 12 μmol photons m^{-2} s^{-1} at 2000 seconds for a 120-second period.

such as this one shows that the induction requirement has been eliminated. However, longer shading periods result in the gradual development of the induction requirement. Continuous high PFD is not required for induction since it will occur during a sequence of sunflecks leading to high assimilation rates in later compared with earlier sunflecks (Chazdon and Pearcy 1986a). Thus, in effect, one sunfleck primes the leaf so that it is better able to utilize subsequent sunflecks.

Mechanistically, the induction requirement is known to be due to three factors: (1) light activation of enzymes in RuBP regeneration, (2) light activation of Rubisco, and (3) stomatal opening. In terms of assimilation rates, the key in this process is Rubisco and the limitations on its activity, since assimilation rates are largely a mirror of its kinetics (Woodrow and Berry 1988). Activation of the light-activated enzymes in RuBP regeneration (fructose-1,6-bisphosphatase, seduheptulose 1,7 bisphosphatase, and ribulose-5-P kinase) occurs within 1–2 minutes after a light increase (Pearcy et al. 1996b, Sassenrath-Cole et al. 1994, Woodrow and Walker 1980), thus allowing a sufficient capacity for RuBP regeneration so that RuBP concentrations are not rate-limiting for Rubisco. Light activation of Rubisco itself is a slower process requiring 5–10 minutes for completion (Seemann et al. 1988, Woodrow and Mott 1988). The increase in stomatal conductance that allows for sufficient supply of CO_2 to Rubisco is even slower, requiring 10–30 minutes for completion. The relative limitations imposed by these three factors at any given time during induction determine the in vivo rate of Rubisco and hence the rate of CO_2 assimilation. For example, RuBP regeneration limitations are only important as long as the Rubisco activation and stomatal limitations are small. Thus, after a leaf has been in the shade for a long period so that stomatal conductance (g_s) and the Rubisco activation state is low, the enzymes in RuBP regeneration impose almost no limitation early in induction. However, RuBP regeneration limitations can be significant in leaves that have been shaded for 5–10 minutes since the RuBP-regenerating enzymes are deactivated much more rapidly than Rubisco is deactivated or g_s decreases (Sassenrath-Cole et al. 1994). This can lead to a rather prominent "fast induction" phase during which the photosynthetic rate increases over the first 1–2 minutes before a transition occurs to a slower increase that is due to stomatal opening and Rubisco activation (Kirschbaum and Pearcy 1988a, Tinoco-Ojanguren and Pearcy 1993b).

The relative roles of stomatal and biochemical limitations during induction have been the subject of some controversy. Early experiments were consistent with a greater role for biochemical compared with stomatal limitations (Chazdon and Pearcy 1986a, Usuda and Edwards 1984, Walker 1981). However, separation of the limitations using classical gas exchange approaches in which intercellular CO_2 pressure (c_i) is calculated is subject to a number of sources of error. When stomatal conductances are low, as they can be early in induction, it is necessary to consider the cuticular pathway for water loss, because ignoring it can lead to a considerable overestimation in c_i (Kirschbaum and Pearcy 1988b). Moreover, considerable heterogeneity (patchiness) of stomatal conductance may be present during induction or even before it in the shade, also leading to an overestimation of c_i (Bro et al. 1996, Cardon et al. 1994, Eckstein et al. 1996). Since it is the relationship between assimilation rate and c_i during induction compared with the final steady state that is used to determine the relative roles of stomata versus biochemistry, the above errors lead to an overestimation of the role of biochemistry and an underestimation of the role of stomata. When the g_s in the shade is high, as may often be the case in forest understories in the morning, then much of the limitation to induction may be biochemical. However, lower g_s and slower induction in the afternoon compared with the morning has been observed in tropical forest understory shrubs and redwood forest herbs (Allen and Pearcy, in preparation; Pfitsch and Pearcy 1989). Although classical gas exchange analysis would indicate that changes in biochemical limitations are involved, stomatal heterogeneity may actually be the factor accounting for the morning-to-afternoon differences in induction responses.

Determination of the role of the dynamic responses in sunfleck utilization requires a comparison between actual measurements or simulations in which the dynamic limitations are present to a condition where they are absent. Pfitsch and Pearcy (1989) compared the measured diurnal carbon gain during sunflecks to a steady-state simulation model of the photosynthetic response to PFD. In a steady-state model, it is assumed that photosynthesis responds instantaneously to a change in PFD, and therefore the limitations due to induction or the postlightfleck CO_2 assimilation are not present. This model predicted approximately 20% more carbon gain than measured, indicating that induction limitations predominated and significantly lowered carbon gain in the understory. A semimechanistic, dynamic model that includes light activation of enzymes, stomatal responses, and the buildup and utilization of metabolite pools important in postlightfleck CO_2 assimilation has also been developed and compared with measured values as well as steady-state predictions (Gross et al. 1991, Pearcy et al. 1996a). When applied to assimilation measured under simulated sunflecks for *Alocasia macrorrhiza*, the model and measurements agreed to within 3%, whereas a steady-state model version predicted 50% more carbon gain (Pearcy et al. 1996a). The 50% greater limitation for the dynamic model and the measurements compared with the steady-state version is consistent with a strong dynamic induction limitation. The limitation in this case was for a 3-hour period of sunfleck activity after a long shade period. Comparison of the dynamic and steady-state model predictions for carbon gain with light measurements in different microsites and days showed that the dynamic model gave a 1.5–25.5% lower carbon gain (Pearcy et al. 1994). The lowest values were for microsites and days with little or no sunfleck activity. The comparison of dynamic and steady-state responses reveals that induction is a strong limitation to sunfleck use. Since carbon gain is already extremely limited and maintenance costs must be covered before any growth can occur, reductions in potential carbon gain of this magnitude are likely to be significant.

The dynamic model has also been used to assess the role of specific dynamic elements by selectively modifying the input parameters so that one element becomes static while the others remain dynamic (Pearcy et al. 1994). Removal of the stomatal responses so that they became static caused assimilation rate during sunfleck light regimes to increase by 10%. Changing the time constants for Rubisco so that it activated/deactivated essentially instantaneously resulted in only a small (1%) increase. Having Rubisco respond instantaneously while stomata responded slowly caused c_i to decrease, limiting carbon gain. A coordination between Rubisco activation and stomatal opening appears to necessary to maximize carbon gain. Removing the dynamics of RuBP regeneration, which was simulated as a light regulation of fructose-1,6-bisphosphatase, had no effect. Finally, removing all dynamic limitations except for the metabolite pools responsible for postlightfleck CO_2 assimilation gave an identical carbon gain to the steady-state version of the model. Postlightfleck CO_2 assimilation made no significant contribution to the simulated carbon gain, primarily because, as previously discussed, short lightflecks themselves made only a very small contribution to the available PFD. Simulations for crop and quaking aspen canopies show that postlightfleck CO_2 assimilation can contribute to as much as a 6% enhancement of carbon gain under windy conditions when short sunflecks make a much more significant contribution (He and Pearcy, in preparation). In the understory, induction limitations predominate and significantly constrain carbon gain during sunflecks. Of these, dynamic stomatal limitations and the relationships between stomatal opening and Rubisco activation may be especially important.

Both species and environmentally induced differences in response to dynamic light regimes and lightfleck utilization efficiency have been identified. A higher photosynthetic capacity itself will act to enhance sunfleck utilization (Lei and Lechowicz 1997, Valladares et al. 1997), but the extent to which this benefit is realized will depend on induction limitations. Initial rates of induction are typically fairly rapid in shade-tolerant understory species compared with shade-intolerant species (Kuppers and Schneider 1993, Kuppers et al. 1996, Poorter and Oberbauer 1993). Moreover, induction is lost more slowly in understory compared with high-light-adapted species (Valladares et al. 1997). A slower induction loss allows for a greater carryover of induction and therefore a greater capacity to utilize subsequent lightflecks. Growth conditions can also strongly influence induction and lightfleck utilization. Shade-acclimated plants have been shown in a number of studies to have higher lightfleck utilization efficiencies than sun-acclimated plants (Chazdon and Pearcy 1986b, Kuppers and Schneider 1993, Ogren and Sundlin 1996, Tang et al. 1994). Higher ratios of electron transport to carboxylation capacity have been suggested as one possible explanation (Ogren and Sundlin 1996), but another contributing factor may be a greater induction limitation in the sun-acclimated species. Pons and Pearcy (1992) found no difference in the capacity to utilize very brief lightflecks (0.25 seconds) between sun- and shade-grown soybeans, but longer lightflecks were utilized less efficiently by the sun- compared with the shade-grown plants. Induction should have less influence on utilization of very brief lightflecks, but would impact utilization of longer lightflecks.

Differences in stomatal behavior that influence induction gain and loss in fluctuating light environments appear to have an important role in species differences in the capacity to utilize lightflecks. The capacity for lightfleck utilization was much greater in the shade-tolerant species *Piper aequale* than in a pioneer species *Piper auritum* when both were grown in the shade (Tinoco-Ojanguren and Pearcy 1992). These differences appeared to be related to stomatal behavior since stomatal conductance responded strongly to lightflecks in *P. aequale*, whereas almost no response occurred in *P. auritum*. However, when grown in the sun, the capacity for lightfleck utilization was much enhanced in *P. auritum*, but was suppressed in *P. aequale*. It is likely that low-light stress inhibited the capacity for sunfleck utilization in *P. auritum*, while high-light stress inhibited it in *P. aequale*. Humidity also impacted the stomatal responses and influenced lightfleck utilization, with higher humidities favoring slower closing and hence more carryover of induction from one lightfleck to the next (Tinoco-Ojanguren and Pearcy 1993a). Knapp and Smith (1987, 1990a, 1990b) identified species differences in the tracking response of stomatal conductance and hence photosynthesis to sun–shade transitions. In some species, rapid stomatal opening and closure occurred in response to sun–shade transitions that limited the increase in photosynthesis during the light increase, but overall enhanced water use efficiency. Other species exhibited reduced stomatal opening and closing responses and consequently were able to utilize the light increases more efficiently, but at an overall lower water use efficiency. Although herbs, which tended to exhibit rather large water potential changes compared with the woody species, were initially identified as having the more conservative response (Knapp and Smith 1989), this may not always be the case (Knapp 1992). High stomatal conductances even in the shade in poplar species may enhance the capability of these species to utilize the highly dynamic light environments found in their crowns (Roden and Pearcy 1993b). Because these species typically occur in habitats with abundant soil moisture, the high conductances and resulting low water use efficiency may cost relatively little compared with the enhanced carbon gain.

C. Responses to Gap Dynamics

1. The Temporal Nature of Gap Light Regimes

The most studied case of the responses to long-term temporal heterogeneity in light environments concerns tree-fall gaps in forests (Bazzaz 1991, Bazzaz and Carlson 1982, Canham 1988, Canham et al. 1990, Sipe and Bazzaz 1994). When a tree falls there is an essentially instantaneous increase in PFD, followed by a gradual decrease as the gap fills with new growth. The actual occurrence of higher direct PFD, which may cause photoinhibition, depends on the absence of cloudiness. Therefore, the dynamic response of an understory plant to the formation of a gap may depend on whether it occurs during a period of cloudy or clear weather.

Response to the increase in PFD after gap formation has received much more attention than the response to the much more gradual decrease that occurs as a gap fills. Gap filling occurs both from the bottom up as plants established in the gap gain height, and from the sides as branches extend out into it. Although significant shading may develop within a year or two close to the ground, shading at higher levels may require many years. This gradual refilling stage may be important since many understory shrubs appear to establish most successfully in gaps but then persist and reproduce in the understory long after that gap may have closed. Most trees in tropical forests may first establish as seedlings in the understory and depend on multiple gap events to make the transition from understory seedling to sapling to ultimately a canopy individual. Thus, just how the photosynthetic apparatus and allocation patterns adjust to the gradual decrease in PFD as a gap closes could be important for persistence.

2. The Dynamic Plasticity of Leaves and Plants to Long-Term Temporal Heterogeneity

The transition between shaded understory and gap affects both the pre-existing leaves and the rate of development of new leaves fully acclimated to the new light environment. In addition, changes in resource allocation within the plant are required to adjust for imbalances between root and shoot caused by the change in environment. Most studies of plasticity have examined different plants grown in different environments, and only a few have examined the response of individual plants to a change in environment. Strauss-Debenetti and Bazzaz (1991) distinguished the response to the latter as a measure of acclimation and the response of the former as a measure of plasticity. In many circumstances, the two may be equivalent, but the distinction does point out that constraints may be present, especially when plants are transferred from the shade to bright sun. A good example is that photoinhibition may occur, constraining acclimation at least initially. Similarly, imbalances in the root-to-shoot ratio for the new environment could constrain the response. Later, as imbalances are themselves adjusted and as new leaves develop with better-developed photoprotective mechanisms, photosynthetic plasticity and photosynthetic acclimation (using the distinction of Strauss-Debenetti and Bazzaz) may mostly converge at least at the leaf level. However, with larger plants, whole-plant level constraints on acclimation due to the developmental history may become increasingly important with age or size, and these may influence acclimation at the leaf level. Along a very similar line, it is important to distinguish when acclimation of mature leaves of plants subjected to a light change are studied versus when comparisons of leaves that had fully developed in the respective sun and shade environments are made. Most research has focused on the latter, and unfortunately, many reports are vague about whether the same leaves were or were not monitored.

The capacity for acclimation of A_{max} of already fully expanded leaves after transfer to a new light environment is generally much less than that observed when leaves are allowed to fully develop in the respective light environments. Some species have little or no acclimation of pre-existing leaves after a transfer to a new light environment. Sims and Pearcy (1991) could find little capacity for acclimation in mature leaves of the understory herb *Alocasia macrorrhiza*, even though there was a 2.2-fold difference in A_{max} for leaves that had developed in high- compared with low-light environments. Similarly, Newell et al (1993) found no apparent acclimation of three *Miconia* species 1 week after creation of an artificial tree-fall gap, whereas 4 months later when new leaves had been produced, all three species exhibited an approximate doubling of A_{max} of these leaves compared with those present at gap formation. Significant levels of photosynthetic acclimation were found in mature leaves of three Australian tropical tree species (Turnbull et al. 1993) and in a shade-tolerant tropical tree species from Borneo (Kamaluddin and Grace 1992), but in each case it was less than that expressed by leaves that had developed in the respective light environments. Because many anatomical properties of sun and shade leaves are more or less fixed at the time of completion of leaf expansion, acclimation may be constrained, especially in species such as *Alocasia macrorrhiza*, where changes in A_{max} depend on leaf thickness changes (Sims and Pearcy 1991). The palisade mesophyll in mature *Bischofia javinensis* leaves were shown to increase in thickness by 132% when transferred from shade to high light, but the leaves were still considerably thinner than those developing completely in high light (Kamaluddin and Grace 1992). In this species, A_{max} per unit area increased by a factor of 1.8 in leaves transferred from shade to high light, while A_{max} per unit mass actually declined. Expansion of the palisade mesophyll cells apparently was coupled with an increase in the concentrations of photosynthetic enzymes per unit area, even though there was a decrease per unit mass. Herbaceous high-light-adapted plants such as peas may possess a substantial capacity for acclimation of mature leaves (Chow and Anderson 1987b), but the extent to which this depends on concomitant adjustments to leaf anatomy versus increases in concentrations of photosynthetic enzymes and electron carriers per unit mass is not known.

In many respects, the most revealing and ecologically relevant experiments are those that follow the dynamics of acclimation in response to light increases (Ferrar and Osmond 1986, Mulkey and Pearcy 1992, Turnbull et al. 1993). These experiments typically reveal an initial photoinhibition of photosynthesis as revealed by reductions in quantum yields and the fluorescence ratio F_v/F_m. Provided that conditions are not too severe, recovery from photoinhibition occurs, which is then often followed by an increase in photosynthetic rates in the extant leaves. New leaves then develop with a higher photosynthetic capacity. Ultimately, acclimation at the whole-plant level is complete when there has been a complete turnover of the leaves and adjustments in LAR and root/shoot ratios have occurred. Depending on the species involved and its leaf longevity and production rates, full acclimation could take weeks to years.

After a transfer to high light, there are wide differences among species in the extent and duration of the initial photoinhibition and in the subsequent rates of acclimation. In peas, only a small, 1-day reduction in A_{max} occurred before it increased rapidly within 4 days to a new, 3-fold higher steady-state value (Anderson and Osmond 1987). In the evergreen tropical tree species *Bischofia javanensis*, A_{max} and the chlorophyll fluorescence ratio, Fv/Fm, remained depressed for nearly 8 days, and 30 days were required to reach the new steady-state A_{max} (Kamaluddin and Grace 1992). In the temperate vine, *Hedera helix*, 45 days were required to complete the acclimation process (Bauer and Thoni 1988). Low nitrogen status exacerbates the initial photoinhibitory reduction in A_{max} and slows or prevents full recovery when shade-grown plants are transferred to high light (Ferrar and Osmond 1986). High leaf temperature also strongly interacts with the PFD level, causing a greater inhibition and slower recovery (Mulkey and Pearcy 1992).

A key to the initial response to gap formation is the dynamic interaction between acclimation and photoinhibition. As shown in Figure 7, exposure of shade-grown *Alocasia macrorrhiza* plants to 2 hours of direct sunlight superimposed on a shade-light background resulted in a strong initial reduction in Fv/Fm and A_{max}, followed by an almost complete recovery of Fv/Fm but only a partial recovery of A_{max} in 5–10 days (Mulkey and Pearcy 1992). The recovery during successive days with the same light treatment suggests development of an increased capacity for photoprotection. Leaves that developed under this gap simulation were completely resistant to photoinhibition and had a higher A_{max}. Thus, complete acclimation required replacement of all the leaves, requiring in this species 60–80 days. Although the pre-existing leaves had a reduced photosynthetic capacity compared with later-developing leaves, they still provided substantial carbon gain because of the high light. Thus, protecting them through rapid acclimation of photoprotective mechanisms may be important for maintaining a carbon supply for growth of the new leaves. In this experiment, a strong interaction between high light and high leaf temperatures was found that greatly increased photoinhbition and delayed the recovery. This interaction is undoubtedly important in natural gap formation since leaf temperatures in gaps with bright sunlight can be 10–15° above air temperatures and reach values that would be

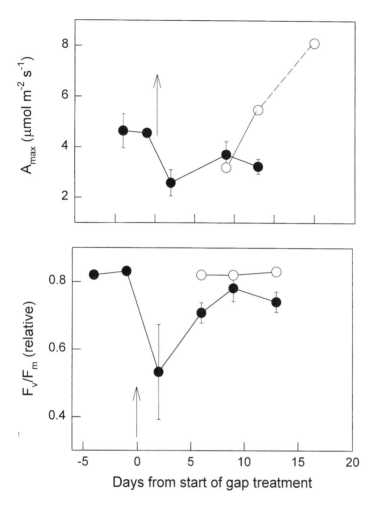

Figure 7 Dynamics of photosynthetic acclimation and photoinhibition as indicated by reduced F_v/F_m ratios following transfer of *Alocasia macrorrhiza* plants from a shade environment simulating the understory to a gap environment in which 2 hours of direct sunlight were received each day. The transfer occurred at the arrow. The filled circles show the response of leaves that had fully developed in the shade before the gap treatment. The open circles show the response of leaves that developed after the gap treatment commenced. The data point connected by the dashed line shows the maximum A_{max} achieved by leaves in the gap environment. Note the strong photoinhibitory decline and recovery (acclimation) in the shade-developed leaves but complete absence of photoinhibition in the gap-developed leaves. (*Source*: Adapted from Mulkey and Pearcy 1992.)

likely to strongly exacerbate photoinhibition. Therefore, different responses might be expected when gaps form during cloudy versus clear periods.

V. CONCLUDING REMARKS

The response of individual plants to spatial and temporal heterogeneity depends on the scale of the heterogeneity. The temporal scale is paramount because it determines whether adjustments in physiological properties rely primarily on relatively rapid and low-cost regulatory mechanisms such as regulation of enzyme catalytic capacity and stomatal opening, or whether plasticity involving regulation of the amounts of enzymes and plant morphological changes will provide a benefit. Responses to temporal changes on a time scale of a day or less primarily involve regulation of activities and stomatal behavior, whereas responses to longer-term changes in the light environment involve acclimation and morphological plasticity (Figure 8). The available evidence suggests that acclimation of photosynthetic capacity depends on the mean daily integral of PFD rather than

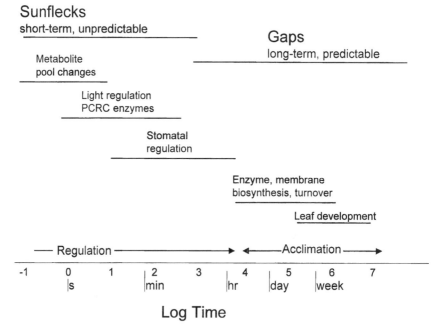

Figure 8 Ranges of time constants for various processes in relation to the time scales of sunflecks and gap dynamics.

the peak PFDs. This provides a mechanism that minimizes potentially costly responses to unpredictable variations such as sunflecks. However, the mechanism by which plants sense the daily integral of PFD and then make adjustments leading to the proper acclimation state are not understood.

The response to the spatial scale of heterogeneity appears to be linked principally to the temporal scale. Sunflecks and light gradients within canopies have similar spatial scales, but since the light gradient is more predictable, acclimation is the principal mechanism for adaptation to this heterogeneity, whereas for sunflecks regulatory mechanisms are overriding. There is evidence that interplant communication may play an important role in acclimation to spatial heterogeneity. The evidence is strongest for clonal plants, but patterns of branch growth and suppression in response to spatial light heterogeneity may also involve some branch-to-branch communication. The signals underlying these communications are poorly understood, but could involve hormones or resource supplies such as carbohydrates. More research is needed to understand the nature and role of these mechanisms in determining how plants respond to environmental heterogeneity. Understanding physiological responses to heterogeneity requires understanding how its consequences scale up. Acclimation of leaf gas exchange properties obviously influences carbon uptake per unit leaf area, but because it also involves changes in LMA, it has consequences for carbon partitioning. Indeed, these latter effects may be more significant to whole-plant carbon gain than the direct effects of photosynthetic acclimation. A consequence of the scaling is that plasticity in photosynthetic properties such as A_{max} may lead to greater homeostasis in growth and resource use efficiency. This plasticity is never enough to offset the wide variation in light levels inherent in natural communities, but nevertheless is important in survival spatially and temporally heterogeneous environments.

REFERENCES

Acevedo, M.F., D.L. Urban, and H.H. Shugart. 1996. Models of forest dynamics based on roles of tree species. Ecological Modelling 87: 267–284.

Anderson, J.M., W.S. Chow, and Y.I. Park. 1995. The grand design of photosynthesis: acclimation of the photosynthetic apparatus to environmental cues. Photosynthesis Research 46: 129–139.

Anderson, J.M., and C.B. Osmond. 1987. Shade-sun responses: compromises between acclimation and photoinhibition. In: Kyle, D.J., C.B. Osmond, and C.J. Arntzen, eds. Photoinhibition. Amsterdam: Elsevier, pp 1–38.

Baldocchi, D., and S. Collineau. 1994. The physical nature of solar radiation in heterogeneous canopies: spatial and temporal attributes. In: Caldwell M.M., and R.W. Pearcy, eds. Exploitation of Environmental Heterogeneity by Plants: Ecophysiological Processes Above-and Belowground. San Diego: Academic Press, pp 21–72.

Ballare, C.L. 1994. Light gaps: sensing the light opportunities in highly dynamic canopy environments. In: Caldwell, M.M., and R.W. Pearcy, eds. Exploitation of Environmental Heterogeneity by Plants: Ecophysiological Processes Above-and Belowground. San Diego: Academic Press, pp 73–110.

Ballare, C.L., R.A. Sanchez, A.L. Scopel, J.J. Casal, and C.M. Ghersa. 1987. Early detection of neighbor plants by phytochrome perception of spectral changes in reflected sunlight. Plant, Cell and Environment 10: 551–557.

Bauer, H., and W. Thoni. 1988. Photosynthetic light acclimation in fully developed leaves of the juvenile and adult life cycle phases of *Hedera helix*. Physiolgia Plantarum 73: 31–37.

Bazzaz, F.A. 1991. Regeneration of tropical forests—physiological responses of pioneer and secondary species. Rain Forest Regeneration and Management 6: 91–118.

Bazzaz, F.A., and R.W. Carlson. 1982. Photosynthetic acclimation to variability in the light environment of early and late successional plants. Oecologia 54: 313–316.

Bazzaz, F.A., and S.T.A. Pickett. 1980. Physiological ecology of tropical succession: a comparative review. Annual Review of Ecology and Systematics 11: 287–310.

Becker, P., and A.P. Smith. 1990. Spatial autocorrelation of solar radiation in a tropical moist forest understory. Agricultural and Forest Meteorology 52: 373–379.

Bjorkman, O. 1981. Responses to different quantum flux densities. In: Lange, O.L., P.S. Nobel, C.B. Osmond, and H. Ziegler, eds. Physiological Plant Ecology I. Physical Environment. Berlin: Springer-Verlag, pp 57–107.

Bjorkman, O., and B. Demmig. 1987. Photon yield of O_2 evolution and chlorophyll flourescence characteristics at 77K among vascular plants of diverse origins. Planta 170: 489–504.

Bjorkman, O., S.B. Powles, D.C. Fork, and G. Oquist. 1981. Interaction between high irradiance and water stress on photosynthetic reactions in *Nerium oleander*. Carnegie Instution of Washington Yearbook 80: 57–59.

Bradshaw, A.D. 1965. Evolutionary significance of phenotypic plasticity in plants. Advances in Genetics 13: 115–155.

Bro, E., S. Meyer, and B. Genty. 1996. Heterogeneity of leaf CO_2 assimilation during photosynthetic induction. Plant Cell and Environment 19: 1349–1358.

Brokaw, N.V.L. 1985. Treefalls, regrowth, and community structure in tropical forests. In: Pickett, S.T.A., and P.S. White, eds. The Ecology of Natural Disturbance and Patch Dynamics. San Diego: Academic Press, pp 53–68.

Buchanan, B.B. 1980. Role of light in the regulation of chloroplast enzymes. Annual Review of Plant Physiology 31: 341–374.

Caldwell, M.M., H.-P. Meister, J.D. Tehunen, and O.L. Lange. 1986. Canopy structure, light microclimate and leaf gas exchange of *Quercus coccifera* L. in a Portugese macchia: measurements in different canopy layers and simulations with a canopy model. Trees 1: 25–41.

Canham, C.D. 1988. Growth and canopy architecture of shade-tolerant trees: Response to canopy gaps. Ecology 69: 786–795.

Canham, C.D., J.S. Denslow, W.J. Platt, J.R. Runkle, T.A. Spies, and P.S. White. 1990. Light regimes beneath closed canopies and tree-fall gaps in temperate and tropical forests. Canadian Journal of Forest Research 20: 620–631.

Canham, C.D., A.C. Finzi, S.W. Pacala, and D.H. Burbank. 1994. Causes and consequences of resource heterogeneity in forests: interspecific variation in light transmission by canopy trees. Canadian Journal of Forest Research 24: 337–349.

Cardon, Z.G., K.A. Mott, and J.A. Berry. 1994. Dynamics of patchy stomatal movements, and their contribution to steady-state and oscillating stomatal conductance calculated using gas-exchange techniques. Plant Cell and Environment 17: 995–1007.

Chabot, B.F., and J.F. Chabot. 1977. Effects of light and temperature on leaf anatomy and photosynthesis in *Frageria vesca*. Oecologia 26: 363–377.

Chabot, B.F., and D.J. Hicks. 1982. The ecology of leaf life spans. Annual Review of Ecology and Systematics 13: 229–259.

Chazdon, R.L. 1988. Sunflecks and their importance to forest understory plants. Advances in Ecological Research 18: 1–63.

Chazdon, R.L. 1992. Photosynthetic plasticity of two rain forest shrubs across natural gap transects. Oecologia 92: 586–595.

Chazdon, R.L., and C.B. Field. 1987. Determinants of photosynthetic capacity in six rainforest *Piper* species. Oecologia 73: 222–230.

Chazdon, R.L., and S. Kaufmann. 1993. Plasticity of leaf anatomy of two rain forest shrubs in relation to photosynthetic light acclimation. Functional Ecology 7: 385–394.

Chazdon, R.L., and R.W. Pearcy. 1986a. Photosynthetic responses to light variation in rain forest species. I. induction under constant and fluctuating light conditions. Oecologia 69: 517–523.

Chazdon, R.L., and R.W. Pearcy. 1986b. Photosynthetic responses to light variation in rain forest species. II. Carbon gain and light utilization during lightflecks. Oecologia 69: 524–531.

Chazdon, R.L., R.W. Pearcy, D.W. Lee, and N. Fetcher. 1996. Photosynthetic responses of tropical forest plants to contrasting light environments. In: Mulkey, S.S., R.L. Chazdon, and A.P. Smith, eds. Tropical Forest Plant Ecophysiology. New York: Chapman and Hall, pp 5–55.

Chazdon, R.L., K. Williams, and C.B. Field. 1988. Interactions between crown structure and light environment in five rain forest *Piper* species. American Journal of Botany 75: 1459–1471.

Chow, W.S., and J.M. Anderson. 1987a. Photosynthetic responses of *Pisum sativum* to an increase in irradiance during growth II. Thylakoid membrane components. Australian Journal of Plant Physiology 14: 9–19.

Chow, W.S., and J.M. Anderson. 1987b. Photosynthetic responses of *Pisum sativum* to an increase in irradiance during growth. I. Photosynthetic activities. Australian Journal of Plant Physiology 14: 1–8.

Cowan, I.R., and G.D. Farquhar. 1977. Stomatal function in relation to leaf metabolism and environment. Symposium of the Society for Experimental Biology 31: 471–505.

de Kroon, H., and M.J. Hutchings. 1995. Morphological plasticity in clonal plants: the foraging concept reconsidered. Journal of Ecology 83: 143–152.

Demmig-Adams, B., and W.W. Adams. 1992a. Carotenoid composition in sun and shade leaves of plants with different life forms. Plant, Cell and Environment 15: 411–419.

Demmig-Adams, B., and W.W. Adams. 1992b. Photoprotection and other responses of plants to high light stress. Annual Review of Plant Physiology and Plant Molecular Biology 43: 599–626.

Demmig-Adams, B., and W.W. Adams. 1994. Capacity for energy dissipation in the pigment bed in leaves with different xanthophyll cycle pools. Australian Journal of Plant Physiology 21: 575–588.

Demmig-Adams, B., and W.W. Adams. 1996. The role of xanthophyll cycle carotenoids in the protection of photosynthesis. Trends in Plant Science 1: 21–26.

Denslow, J.S. 1980. Gap partitioning among tropical forest trees. Biotropica 12 (suppl): 47–55.

Denslow, J.S. 1987. Tropical rainforest gaps and tree species diversity. Annual Review of Ecology and Systematics 18: 431–451.

Desjardins, R.L. 1967. Time series analysis in agrometerological problems with emphasis on spectrum analysis. Canadian Journal of Plant Physiology 47: 477–491.

Desjardins, R.L., T.R. Sinclair, and E.R. Lemon. 1973. Light fluctuations in corn. Agronomy Journal 65: 904–907.

Dong, M. 1995. Morphological responses to local light conditions in clonal herbs from contrasting habitats, and their modification due to physiological integration. Oecologia 101: 282–288.

Dutton, R.G., J. Jiao, M.J. Tsujta, and B. Grodzinski. 1988. Whole plant CO_2 exchange measurements for nondestructive estimation of growth. Plant Physiology 86: 355–358.

Eckstein, J., W. Beyschlag, K.A. Mott, and R.J. Ryel. 1996. Changes in photon flux can induce stomatal patchiness. Plant Cell and Environment 19: 1066–1074.

Ehleringer, J., and I. Forseth. 1980. Solar tracking by plants. Science 210: 1094–1098.

Ellsworth, D.S., and P.B. Reich. 1992. Leaf mass per area, nitrogen content and photosynthetic carbon gain in *Acer saccharum* seedlings in contrasting forest light environments. Functional Ecology 6: 423–435.

Evans, J.R. 1986. A quantitative analysis of light distribution between the two photosystems, considering variation in both the relative amounts of the chlorophyll-protein complexes and the spectral quality of light. Photochemistry and Photobiophysics 10: 135–147.

Evans, J.R. 1993. Photosynthetic acclimation and nitrogen partitioning within a lucerne canopy II. stability through time and comparison with a theoretical optimum. Australian Journal of Plant Physiology 20: 69–82.

Ferrar, P.J., and C.B. Osmond. 1986. Nitrogen supply as a factor influencing photoinhibition and photosynthetic acclimation after transfer of shade-grown *Solanum dulcamara* to bright light. Planta 168: 563–570.

Field, C. 1983. Allocating leaf nitrogen for the maximization of carbon gain: leaf age as a control on the allocation program. Oecologia 56: 341–347.

Foyer, C., R. Furbank, J. Harbinson, and P. Horton. 1990. The mechanisms contributing to photosynthetic control of electron transport by carbon assimilation in leaves. Photosynthesis Research 25: 83–100.

Friend, A.D., H.H. Schugart, and S.W. Running. 1993. A physiology-based gap model of forest dynamics. Ecology 74: 792–797.

Gamon, J.A., and R.W. Pearcy. 1989. Leaf movement, stress avoidance and photosynthesis in *Vitus californica*. Oecologia 79: 475–481.

Geiken, B., C. Critchley, and G. Renger. 1992. The turnover of photosystem II reaction centre proteins as a function of light. In: Murata, N., ed. Research in Photosynthesis: Dordrecht: Kluwer Academic, pp 634–646.

Gilbert, I.R., G.P. Seavers, P.G. Jarvis, and H. Smith. 1995. Photomorphogenesis and canopy dynamics—phytochrome mediated proximity perception accounts for the growth dynamics of canopies of *Populus Trichocarpa* × *Deltoides* Beaupre. Plant Cell and Environment 18: 475–497.

Grace, S.C., and B.A. Logan. 1996. Acclimation of foliar antioxidant systems to growth irradiance in three broad-leaved evergreen species. Plant Physiology 112: 1631–1640.

Grime, J.P. 1966. Shade avoidance and shade tolerance in flowering plants. In: Evans, G.C., R. Bainbridge, and O. Rackham, eds. Light as Ecological Factor. London: Blackwell, pp 187–207.

Gross, L.J., M.U.F. Kirschbaum, and R.W. Pearcy. 1991. A dynamic model of photosynthesis in varying light taking account of stomatal conductance, C_3-cycle intermediates, photorespiration and RuBisCO activation. Plant Cell and Environment 14: 881–893.

Harbinson, J., B. Genty, and N.R. Baker. 1990a. The relationship between CO_2 assimilation and electron transport in leaves. Photosynthesis Research 25: 213–224.

Harbinson, J., B. Genty, and C.H. Foyer. 1990b. Relationship between photosynthetic electron transport and stromal enzyme activity in pea leaves—toward an understanding of the nature of photosynthetic control. Plant Physiology 94: 545–553.

Hiroi, T., and M. Monsi. 1963. Physiological and ecological analysis of shade tolerance of plants 3. Effect of shading on growth attributes of *Helianthus annuus*. Botanical Magazine Tokyo 77: 121–129.

Hirose, T. 1988. Modelling the relative growth rate as a function of plant nitrogen concentration. Physiolgia Plantarum 72: 185–189.

Hirose, T., and M.A. Werger. 1987a. Maximizing daily canopy photosynthesis with respect to the leaf nitrogen pattern in the canopy. Oecologia 72: 520–526.

Hirose, T., and M.A.J. Werger. 1987b. Nitrogen use efficiency in instantaneous and daily photosynthesis leaves in the canopy of a *Solidago altissima* stand. Physiologia Plantarum 70: 520–526.

Hirose, T., M.J.A. Werger, T.L. Pons, and J.W.A. van Rheenen. 1988. Canopy structure and leaf nitrogen distribution in a stand of *Lysimachia vulgaris* L. as influenced by stand destiny. Oecologia 77: 145–150.

Huffaker, R.C., and L.W. Peterson. 1976. Protein turnover in plants and possible means of its regulation. Annual Review of Plant Physiology 25: 363–392.

Hutchings, M.J., and H. de Kroon. 1994. Foraging in plants: the role of morphological plasticity in resource acquisition. Advances in Ecological Research 25: 159–238.

Jurik, T.W. 1983. Reproductive effort and CO_2dynamics of wild strawberry populations. Ecology 64: 1329–1342.

Kamaluddin, M., and J. Grace. 1992. Photoinhibition and light acclimation in seedlings of *Bischofia javanica*, a tropical forest tree from Asia. Annals of Botany 69: 47–52.

King, D.A. 1994. Influence of light level on the growth and morphology of saplings in a Panamanian forest. American Journal of Botany 81: 948–957.

Kirschbaum, M.U.F., and R.W. Pearcy. 1988a. Gas exchange analysis of the fast phase of photosynthetic induction in *Alocasia macrorrhiza*. Plant Physiology 87: 818–821.

Kirschbaum, M.U.F., and R.W. Pearcy. 1988b. Gas exchange analysis of the relative importance of stomatal and biochemical factors in photosynthetic induction in *Alocasia macrorrhiza*. Plant Physiology 86: 782–785.

Knapp, A.K. 1992. Leaf gas exchange in *Quercus macrocarpa* (Fagaceae): rapid stomatal responses to variability in sunlight in a tree growth form. American Journal of Botany 79: 599–604.

Knapp, A.K., and W.K. Smith. 1987. Stomatal and photosynthetic responses during sun/shade transitions in subalpine plants: influence on water use efficiency. Oecologia 74: 62–67.

Knapp, A.K., and W.K. Smith. 1989. Influence of growth form on ecophysiological responses to variable sunlight in subalpine plants. Ecology 70: 1069–1082.

Knapp, A.K., and W.K. Smith. 1990a. Contrasting stomatal responses to variable sunlight in 2 subalpine herbs. American Journal of Botany 77: 226–231.

Knapp, A.K., and W.K. Smith. 1990b. Stomatal and photosynthetic responses to variable sunlight. Physiologia Plantarum 78: 160–165.

Kuppers, M., and H. Schneider. 1993. Leaf gas exchange of beech (*Fagus sylvatica* L.) seedlings in lightflecks: effects of fleck length and leaf temperature in leaves grown in deep and partial shade. Trees 7: 160–168.

Kuppers, M., H. Timm, F. Orth, J. Stegemann, R. Stober, K. Paliwal, K.S.T. Karunaichamy, and R. Ortez. 1996. Effects of light environment and successional status on lightfleck use by understory trees of temperate and tropical forests. Tree Physiology 16: 69–80.

Lei, T.T., and M.J. Lechowicz. 1997. The photosynthetic response of 8 species of *Acer* to simulated light regimes from the centre and edges of gaps. Functional Ecology 11: 16–23.

Leuning, R., Y.P. Wang, and R.N. Cromer. 1991. Model simulations of spatial distributions and daily totals of photosynthesis in *Eucalyptus grandis* canopies. Oecologia 88: 494–503.

Levins, R. 1968. Evolution in Changing Environments. Princeton: Princeton University Press.

Long, S.P., W.F. Postl, and H.R. Bolhar-Nordenkampf. 1993. Quantum yields for uptake of carbon dioxide in C_3 vascular plants of contrasting habitats and taxonomic groupings. Planta 189: 226–234.

Lortie, C.J., and L.W. Aarssen. 1996. The specialization hypothesis for phenotypic plasticity in plants. International Journal of Plant Sciences 157: 484–487.

Ludlow, M.M., and O. Bjorkman. 1984. Paraheliotropic leaf movement in Siratro as a protective mechanism against drought-induced damage to primary photosynthetic reactions: damage by excessive light and heat. Planta 161: 505–518.

Mulkey, S.S., and R.W. Pearcy. 1992. Interactions between acclimation and photoinhibition of photosynthesis of a tropical forest understorey herb, *Alocasia macrorrhiza*, during simulated canopy gap formation. Functional Ecology 6: 719–729.

Newell, E.A., E.P. McDonald, B.R. Strain, and J.S. Denslow. 1993. Photosynthetic responses of *Miconia* species to canopy openings in a lowland tropical rainforest. Oecologia 94: 49–56.

Novoplansky, A., D. Cohen, and T. Sachs. 1990. How *Portulaca* seedlings avoid their neighbors. Oecologia 82: 490–493.

Oberbauer, S.F., D.B. Clark, D.A. Clark, and M.A. Quesada. 1988. Crown light environments of saplings of two species of rain forest emergent trees. Oecologia 75: 207–212.

Ogren, E., and U. Sundlin. 1996. Photosynthetic respnse to dynamic light: a comparison of species from contrasting habitats. Oecologia 106: 18–27.

Oquist, G., J.M. Anderson, S. Mccaffery, and W.S. Chow. 1992. Mechanistic differences in photoinhibition of sun and shade plants. Planta 188: 422–431.

Osmond, C.B. 1983. Interactions between irradiance, nitrogen nutrition, and water stress in the sun-shade responses of *Solanum dulcamara*. Oecologia 57: 316–321.

Pacala, S.W., C.D. Canham, and J.A. Silander. 1993. Forest models defined by field measurements I. The design of a northeastern forest simulator. Canadian Journal of Forest Research 23: 1980–1988.

Pearcy, R.W. 1983. The light environment and growth of C_3 and C_4 tree species in the understory of a Hawaiian forest. Oecologia 58: 19–25.

Pearcy, R.W. 1988. Photosynthetic utilization of lightflecks by understory plants. Australian Journal of Plant Physiology 15: 223–238.

Pearcy, R.W., R.L. Chazdon, L.J. Gross, and K.A. Mott. 1994. Photosynthetic utilization of sunflecks, a temporally patchy resource on a time scale of seconds to minutes. In: Caldwell, M.M., and R.W. Pearcy, eds. Exploitation of Environmental Heterogeneity by Plants: Ecophysiological Processes Above and Belowground. San Diego: Academic Press, pp 175–208.

Pearcy, R.W., R.L. Chazdon, and M.U.F. Kirschbaun. 1987. Photosynthetic utilization of lightflecks by tropical forest plants. Progress in Photosynthesis Research 4: 257–260.

Pearcy, R.W., L.J. Gross, and D.X. He. 1997. An improved dynamic model of photosynthesis for estimation of carbon gain in variable light regimes. Plant Cell and Environment 20: 411–424.

Pearcy, R.W., and W.A. Pfitsch. 1991. Influence of sunflecks on the $\delta^{13}C$ of *Adenocaulon bicolor* plants occurring in contrasting forest understory microsites. Oecologia 86: 457–462.

Pearcy, R.W., J.S. Roden, and J.A. Gamon. 1990. Sunfleck dynamics in relation to canopy structure in a soybean (*Glycine max* (L.) Merr.) canopy. Agricultural and Forest Meteorology 52: 359–372.

Pearcy, R.W., G.F. Sassenrath-Cole, and J.P. Krall. 1996. Photosynthesis in fluctuating light environments In: Baker, N.R., ed. Environmental Stress and Photosynthesis. The Hague: Kluwer Academic, pp 321–346.

Pearcy, R.W., and D.A. Sims. 1994. Photosynthetic acclimation to changing light environments: scaling from the leaf to the whole plant. In: Caldwell, M.M., and R.W. Pearcy, eds. Exploitation of Environmental Heterogeniety by Plants: Ecophysiological Processes Above and Belowground. San Diego: Academic Press, pp 223–234.

Pfitsch, W.A., and R.W. Pearcy. 1989. Daily carbon gain by *Adenocaulon bicolor*, a redwood forest understory herb, in relation to its light environment. Oecologia 80: 465–470.

Pfitsch, W.A., and R.W. Pearcy. 1992. Growth and reproductive allocation of *Adenocaulon bicolor* in response to experimental removal of sunflecks. Ecology 73: 2109–2117.

Pons, T.L., and R.W. Pearcy. 1992. Photosynthesis in flashing light of soybean leaves grown in different conditions. II. Lightfleck utilization efficiency. Plant Cell and Environment 15: 577–584.

Poorter, H. 1994. Construction costs and payback time of biomass: a whole plant perspective. In: A Whole-Plant Perspective on Carbon-Nitrogen Interactions. The Hague: SPB Academic Publishing, pp 111–127.

Poorter, L., and S.F. Oberbauer. 1993. Photosynthetic induction responses of two rainforest tree species in relation to light environment. Oecologia 96: 193–199.

Portis, A.R. 1992. Regulation of ribulose 1,5-bisphosphate carboxylase oxygenase activity. Annual Review of Plant Physiology and Plant Molecular Biology 43: 415–437.

Reich, P.B., M.B. Walters, and D.S. Ellsworth. 1997. From tropics to tundra: global convergence in plant functioning. Proceedings of the National Academy of Sciences 94: 13730–13734.

Roden, J.S., and R.W. Pearcy. 1993a. Effect of leaf flutter on the light environment of poplars. Oecologia 93: 201–207.

Roden, J.S., and R.W. Pearcy. 1993b. Photosynthetic gas exchange response of poplars to steady-state and dynamic light environments. Oecologia 93: 208–214.

Ryel, R.J., W. Beyschlag, and M.M. Caldwell. 1994. Light field heterogeneity among tussock grasses—theoretical considerations of light harvesting and seedling establishment in tussocks and uniform tiller distributions. Oecologia 98: 241–246.

Ryel, R.J., W. Beyschlag, B. Heindl, and I. Ullmann. 1996. Experimental studies on the competitive balance between two central European roadside grasses with different growth forms .1. Field experiments on the effects of mowing and maximum leaf temperatures on competitive ability. Botanica Acta 109: 441–448.

Sassenrath-Cole, G.F., and R.W. Pearcy. 1992. The role of ribulose-1,5-bisphosphate regeneration in the induction requirement of photosynthetic CO_2 exchange under transient light conditions. Plant Physiology 99: 227–234.

Sassenrath-Cole, G.F., R.W. Pearcy, and S. Steinmaus. 1994. The role of enzyme activation state in limiting carbon assimilation under variable light conditions. Photosynthesis Research 41: 295–302.

Scatena, F.N., and M.C. Larsen. 1991. Physical aspects of Hurricane Hugo in Puerto Rico. Biotropica 23: 317–323.

Schlichting, C.D. 1986. The evolution of phenotypic plasticity in plants. Annual Review of Ecology and Systematics 17: 667–693.

Seemann, J.R., M.U.F. Kirschbaum, T.D. Sharkey, and R.W. Pearcy. 1988. Regulation of ribulose 1,5-bisphosphate carboxylase activity in *Alocasia macrorrhiza* in response to step changes in irradiance. Plant Physiology 88: 148–152.

Shugart, H.H., and T.M. Smith. 1996. A review of forest patch models and their application to global change research. Climatic Change 34: 131–153.

Sims, D.A., R. Gebauer, and R.W. Pearcy. 1994. Scaling sun and shade photosynthetic acclimation to whole plant performance. II. Simulation of carbon balance and growth at different daily photon flux densities. Plant Cell and Environment 17: 889–900.

Sims, D.A., and R.W. Pearcy. 1989. Photosynthetic characteristics of a tropical forest understory herb, *Alocasia macrorrhiza*, and a related crop species, *Colocasia esculenta* grown in contrasting light environments. Oecologia 79: 53–59.

Sims, D.A., and R.W. Pearcy. 1991. Photosynthesis and respiration in *Alocasia macrorrhiza* following transfers to high and low light. Oecologia 86: 447–453.

Sims, D.A., and R.W. Pearcy. 1992. Response of leaf anatomy and photosynthetic capacity in *Alocasia macrorrhiza* (Araceae) to a transfer from low to high light. American Journal of Botany 79: 449–455.

Sims, D.A., and R.W. Pearcy. 1993. Sunfleck frequency and duration affects growth rate of the understory plant, *Alocasia macrorrhiza* (L.) G. Don. Functional Ecology 7: 683–689.

Sims, D.A., and R.W. Pearcy. 1994. Scaling sun and shade photosynthetic acclimation to whole plant performance. I. Carbon balance and allocation at different daily photon flux densities. Plant, Cell and Environment 17: 881–887.

Sipe, T.W., and F.A. Bazzaz. 1994. Gap partitioning among maples (Acer) in central New England—shoot architecture and photosynthesis. Ecology 75: 2318–2332.

Slade, A.J., and M.J. Hutchings. 1987. An analysis of the costs and benefits of physiological integration between ramets in the clonal perennial herb *Glechoma hederacea*. Oecologia 73: 425–432.

Smith, W.K., A.K. Knapp, and W.A. Reiners. 1989. Penumbral effects on sunlight penetration in plant communities. Ecology 70: 1603–1609.

Strauss-Debendetti, S., and F.A. Bazzaz. 1991. Plasticity and acclimation to light in tropical Moracae of different successional positions. Oecologia 87: 377–387.

Strauss-Debenedetti, S., and F. Bazzaz. 1996. Photosynthetic characteristics of tropical trees along successional gradients. In: Mulkey, S.S., R.L. Chazdon, and A.P. Smith, eds. Tropical Forest Plant Ecophysiology. New York: Chapman and Hall, pp 162–186.

Stuefer, J.F., H. De Kroon, and H.J. During. 1996. Exploitation of environmental heterogeniety by spatial division of labor in a clonal plant. Functional Ecology 10: 328–334.

Stuefer, J.F., H.J. During, and H. de Kroon. 1997. High benefits of clonal integration in two stoloniferous species in response to heterogenious light environments. Journal of Ecology 82: 511–518.

Sultan, S.E. 1992. What has survived of Darwin's theory? Phenotypic plasticity and the Neo-Darwinian legacy. Evolutionary Trends in Plants 6: 61–71.

Takenaka, A. 1994. A simulation model of tree architecture development based on growth response to local light environment. Journal of Plant Research 107: 321–330.

Tang, Y., K. Hiroshi, S. Mitsumasa, and W. Izumi. 1994. Characteristics of transient photosynthesis in *Quercus serrata* seedlings grown under lightfleck and constant light regimes. Oecologia 100: 463–469.

Tang, Y., I. Washitani, and H. Iwaki. 1992. Effects of microsite light availability on the survival and growth of oak seedlings within a grassland. The Botanical Magazine, Tokyo 105: 281–288.

Tang, Y.-H., I. Washitani, T. Tsuchiya, and H. Iwaki. 1989. Spatial heterogenity of photosynthetic photon flux density in the canopy of *Miscanthus sinensis*. Ecological Research 4: 339–349.

Thayer, S.S., and O. Bjorkman. 1990. Leaf xanthophyll content and composition in sun and shade determined by HPLC. Photosynthesis Research 23: 331–343.

Tinoco-Ojanguren, C., and R.W. Pearcy. 1992. Dynamic stomatal behavior and its role in carbon gain during lightflecks of a gap phase and an understory *Piper* species acclimated to high and low light. Oecologia 92: 223–234.

Tinoco-Ojanguren, C., and R.W. Pearcy. 1993a. Stomatal dynamics and its importance to carbon gain in two rainforest *Piper* species. I. VPD effects on the transient stomatal response to lightflecks. Oecologia 94: 388–394.

Tinoco-Ojanguren, C., and R.W. Pearcy. 1993b. Stomatal dynamics and its importance to carbon gain in two rainforest *Piper* species. II. Stomatal versus biochemical limitations during photosynthetic induction. Oecologia 94: 395–402.

Turnbull, M.H., D. Doley, and D.J. Yates. 1993. The dynamics of photosynthetic acclimation to changes in light quantity and quality in three Australian rainforest tree species. Oecologia 94: 218–228.

Usuda, H., and G.E. Edwards. 1984. Is photosynthesis during the induction period in maize limited by the availability of intercellular carbon dioxide. Plant Science Letters 37: 41–45.

Valladares, F., M.T. Allen, and R.W. Pearcy. 1997. Photosynthetic response to dynamic light under field conditions in six tropical rainforest shurbs occuring along a light gradient. Oecologia 111: 505–514.

Valladares, F., S.J. Wright, E. Lasso, K. Kitajima, and R.W. Pearcy. 1998. Plastic phenotypic response to light of 16 rainforest shrubs (*Psychotria*) differing in shade tolerance and leaf turnover. Ecology, submitted.

Walker, D.A. 1981. Photosynthetic induction. In: Akoyonoglou, G., ed. Proceedings of the 5th International Congress on Photosynthesis, Vol IV. Philadelphia: Balaban International, pp 189–202.

Washitani, I., and Y. Tang. 1991. Microsite variation in light availability and seedling growth of *Quercus serrata* in a temperate pine forest. Ecological Research 6: 305–316.

Wayne, P.M., and F.A. Bazzaz. 1993a. Birch seedling responses to daily time courses of light in experimental forest gaps and shadehouses. Ecology 74: 1500–1515.

Wayne, P.M., and F.A. Bazzaz. 1993b. Morning vs afternoon sun patches in experimental forest gaps: consequences of temporal incongruency of resources to birch regeneration. Oecologia 94: 235–243.

Williams, K., C.B. Field, and H.A. Mooney. 1989. Relationships among leaf construction costs, leaf longevity, and light environment in rainforest species of the genus *Piper*. American Naturalist 133: 198–211.

Woodrow, I.E., and J.A. Berry. 1988. Enzymatic regulation of photosynthetic CO_2 fixation in C_3 plants. Annual Review of Plant Physiology Plant Molecular Biology 39: 533–594.

Woodrow, I.E., and K.A. Mott. 1988. Quantitative assessment of the degree to which ribulosebisphosphate carboxylase/oxygenase determines the steady-state rate of photosynthesis during sun-shade acclimation in *Helianthus annus* L. Australian. Journal of Plant Physiology 15: 253–262.

Woodrow, I.E., and D.A. Walker. 1980. Light-mediated activation of stromal sedoheptulose bisphosphatase. Biochemistry Journal 191: 845–849.

8

Acquisition, Use, and Loss of Nutrients

Frank Berendse, Hans de Kroon, and Wim G. Braakhekke
Wageningen Agricultural University, Wageningen, The Netherlands

I. INTRODUCTION

In many natural environments, nutrient supply is among the most important factors that affect the productivity and the species composition of plant communities (Kruijne et al. 1967, Elberse et al. 1983, Pastor et al. 1984, Tilman 1984). In many grassland, heathland, wetland, and forest communities, increased fertilizer gifts and increased nitrogen inputs through atmospheric deposition have caused not only dramatic changes in species composition, but also important losses of plant species diversity (Aerts and Berendse 1988, Berendse et al. 1992). To understand the changes in plant communities that occur after an increase in nutrient supply, it is essential to understand how plant species are adapted to environments with different nutrient availabilities. The relation between nutrient supply and long-term success of an individual plant in a natural ecosystem is determined by three important components of plant functioning: (1) the acquisition of nutrients in soils that are always more or less heterogeneous; (2) the use of absorbed nutrients for carbon assimilation and other plant functions; and (3) the loss of nutrients determining the length of the time period that nutrients can be used.

This chapter considers these three aspects and attempts to integrate them to conclude how plant species cope with nutrient-poor and nutrient-rich environments. The focus is on plants growing in their natural habitat. Such plant individuals experience a heterogeneous substrate; they have to compete for soil resources and light with other plants and they frequently lose large quantities of nutrients through abscission, disturbances, and herbivory.

II. NUTRIENT UPTAKE KINETICS: BASIC PRINCIPLES

Nutrient uptake is determined by both supply and demand at the root surface. Nutrients arrive at the root surface by the mass flow of water toward the root, which is driven by transpiration. Plants deplete the soil solution near the roots when the nutrient uptake rate exceeds the rate at which nutrients arrive. By doing so, they create concentration gradients around the roots that trigger diffusion of nutrients toward the root surface, which adds to the supply by mass flow. When the depletion at the root surface proceeds, uptake must come in pace with the supply rate. When the supply by mass flow exceeds the demand, nutrients (and other solutes) can either be excluded, accumulating near the root, or enter the root and accumulate in the plant to concentrations that may eventually become deleterious (Marschner 1995, Fitter and Hay 1987).

Depletion and accumulation at the root surface can occur simultaneously for different elements. Table 1 gives an indication of the relation between the demand of the major nutrients and their supply by mass flow. The listed concentrations in plant biomass are considered to be sufficient for adequate growth (Epstein 1965). There is a close relation between biomass production, nutrient demand, and water uptake. Based on the amount of water transpired during the production of a unit biomass (transpiration coefficient), we can calculate the nutrient concentration in the soil solution that would satisfy the nutrient demands as listed in the first column in Table 1 by means of mass flow alone. Actual concentrations in the soil solution of an average agricultural soil illustrate that mass

Table 1 Average Nutrient Concentrations in Plant Biomass (Epstein 1965), Concentrations in Soil Solution Required To Satisfy the Demand by Mass Flow,[a] and Actual Concentrations in the Soil Solution in an Arable Field

Element	Concentration in plant biomass (mmol/kg dry weight)	Sufficient concentration in mass flow (mM)	Actual concentration in bulk soil solution (mM)
N	1000	3.33	3.1
K	250	0.83	0.5
Ca	125	0.42	1.7
Mg	80	0.27	0.5
P	60	0.20	0.002
S	30	0.10	0.6

[a] Assuming the transpiration coefficient is 3 dm^3/g dry weight.
Source: Marschner 1995.

flow rates of S, Mg, and Ca amply exceed the demand, whereas mass flow of P falls entirely short of the demanded rate of supply. In agricultural soils, mass flow rates of N and K are usually sufficient, but in most natural soils, concentrations of N, P, and K are much lower, so that the supply by mass flow alone is insufficient to satisfy the demand. Consequently, diffusion must play an important role in the supply of these nutrients to plants growing on natural soils. This calls for a root system that has the ability to take up nutrients selectively against a concentration gradient.

Selective uptake and transport through cell membranes is an energy-demanding process. Passive, nonselective uptake without energy expenses is only possible when nutrients do not have to pass a cell membrane on their way to the vascular cylinder of the root. Passive uptake can revert into efflux when the concentration in the soil solution falls below the concentration inside the root. Passive cation uptake through cell membranes can also proceed against a concentration gradient, because cells can create an electrochemical gradient by actively pumping out protons across the membrane. Anions, on the other hand, must be transported actively through cell membranes by means of carrier enzymes.

If it were not for the closed structure of the root, passive uptake without energy expenses would be sufficient for nutrients that are required in low quantities and occur in relatively high concentrations in the soil solution. Because passive uptake is not selective and cannot be regulated, plants that rely too much on passive uptake can easily be overloaded with nutrients and toxic ions when concentrations in the soil solution are high. This makes it understandable that plant roots possess an endodermis that prevents passive nutrient transport (see Chapter 5). Solutes can enter the root via the apoplastic pathway between the cortex cells, but no further than the endodermis with its bands of Caspari. Solutes that are transported by mass flow to and into the root at a higher rate than the active uptake rate by rhizodermal, cortex, or endodermis cells will accumulate between the cortex cells and at the root surface. This leads to diffusion in the direction opposite to the water flow, away from the root, and back into the soil. Most nutrients enter the root across a cell membrane somewhere in the cortex by means of active transport and continue their way inside along the symplastic pathway, from one cell to another via intercellular cytoplasmic connections (plasmodesmata), to pass through the endodermis and finally enter the vascular cylinder (Marschner 1995). A small fraction of the nutrients can circumvent the endodermis at the root tip where the bands of Caspari are not yet formed, and at places where the endodermis is pierced by lateral roots, torn or damaged otherwise.

Active uptake against a concentration gradient, by means of an energy-demanding process, is the predominant uptake process for the major nutrients N, P, and K. The uptake rate depends on the nutrient concentration at the root surface. Usually, an asymptotic relation is found between the concentration in a nutrient solution and the uptake rate when measured in short-term experiments with excised roots from plants that have been deprived of nutrients for a few weeks. The uptake as a function of the concentration in the surrounding solution is generally described by:

$$V = V_{max}C_1/(C_1 + K_m) \tag{1}$$

where V is the gross uptake rate (μmol g^{-1} fresh root weight [FW] h^{-1}), V_{max} is the maximum uptake rate, C_1 is the nutrient concentration in the soil solution at the root surface (mM), and K_m is the Michaelis constant, which is the value of C_1 where V is half V_{max}. The Michaelis-Menten equation is typical for the kinetics of enzymatic processes and reflects the fact that the carrier enzymes in the cell membranes become saturated with increasing nutrient concentration. K_m^{-1} expresses the affinity of the carriers for the substrate ion (i.e., the slope of the curve in the origin, which is measured by V_{max}/K_m). Although different nutrient ions are transported by different carriers, the selectivity of the carriers is not perfect. Different nutrients may compete for the same carriers, so that K_m may be increased by the presence of other ions with the same electrical charge.

The maximum uptake rate (V_{max}) is realized when the concentration (C_1) is so high that all carrier enzymes are continuously occupied with a substrate ion. V_{max} depends on the density and activity of carriers in the cell membranes. It depends also on the internal nutrient status of the plant, because the activity of the carriers can be suppressed by high nutrient concentrations in the root (compare V_{max} under deprived and well-fed conditions; Figure 1). This negative feedback control operates when plants are growing under nutrient-rich conditions. It can reduce V_{max} to less than 20% of its value in a deprived plant (Loneragan and Asher 1967) and prevents too-high nutrient concentrations from accumulating in

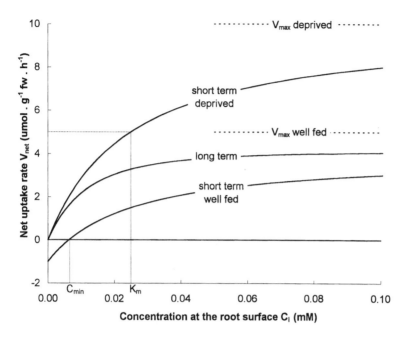

Figure 1 Relation between external nutrient concentration at the root surface (C_l) and net nutrient uptake rate ($V_{net} = V - E$). The upper curve represents the short-term uptake by excised roots from previously deprived plants ($V_{max} = 10$; $K_m = 0.025$; efflux = 0). The lower curve represents short-term uptake by excised roots of previously well-fed plants ($V_{max} = 5$; $K_m = 0.025$; efflux = 1). C_{min} is the concentration where $V_{net} = 0$. The curve marked "long term" represents nutrient uptake by whole plants grown for several weeks on nutrient solutions with constant concentration, so that V_{max} and efflux are in steady state with the internal nutrient concentration and C_l.

the root. On the other hand, it has been found that plants growing under nutrient-poor conditions can temporarily increase V_{max} when the concentration in the soil is increased (Lefebvre and Glass 1982, Jackson et al. 1990).

Consequently, roots of a well-fed plant operate far below their maximum uptake capacity. When the soil becomes depleted and the nutrient concentration in the plant starts to drop, the negative feedback control on V_{max} is relaxed so that V_{max} increases, which compensates for the decrease in uptake rate. The responses of V_{max} to changes in internal and external nutrient concentrations allow a plant to regulate its nutrient uptake and rapidly use temporarily high nutrient concentrations that may occur locally in a predominantly poor soil. Under rich conditions, it allows a plant to maintain its overall uptake rate, even when a large part of the root system is removed.

In addition to uptake, efflux of nutrients may occur. When active uptake takes place, nutrient concentrations inside the root are usually higher than outside the root. Because roots are not perfectly closed, leakage of nutrients can reduce the net uptake rate (V_{net}). When nutrient influx and efflux occur simultaneously, the plant can only decrease the nutrient concentration in the soil solution until it reaches a minimum concentration (C_{min}) at which influx and efflux are equal. At values of C_1 lower than C_{min}, the efflux is larger than the gross influx, so that the net influx is negative and the roots will lose nutrients to the solution until C_1 equals C_{min} (Figure 1).

Like V_{max}, C_{min} is not a constant. Under nutrient-poor conditions, when nutrient concentrations inside the root are low, the efflux will also be low. Together with the release of the feedback control on V_{max}, this results in very low values of C_{min} under nutrient-poor conditions (Figure 1). It is unclear at present whether the K_m value is also able to respond to changes in internal or ambient nutrient concentrations (Marschner 1995). The reported changes in uptake kinetics of roots in response to localized nutrients (Drew and Saker 1975, Jackson et al. 1990) may be due to changes in V_{max} or efflux rate alone. In most plant species, the value of K_m for N, P, and K is so low that the concentration of these nutrients at the root surface can become virtually zero, e.g., 0.35 μM for NO_3^- (Freijsen et al. 1989), 1 μM for K^+ (Drew et al. 1984), and less than 0.01 μM for $H_2PO_4^-$ (Breeze et al. 1984).

III. NUTRIENT ACQUISITION IN SOILS

The aforementioned description of active nutrient uptake applies mainly to uptake by single roots in a well-mixed nutrient solution. However, the relation between uptake and concentration in the soil solution is of little consequence for the overall nutrient uptake by a whole plant growing in poor soil. In soils, uptake of N, P, and K is almost always limited by the rate of transport toward the root surface and not by the capacity of the uptake mechanism. Concentrations of these nutrients in the soil solution are often so low, and the uptake so efficient, that all available nutrients near the root surface can be taken up within a few minutes. When nutrient uptake is not immediately compensated by nutrient transport from the bulk soil toward the root surface, the nutrient concentrations at the root surface will fall and the uptake rate will decrease. Even in nutrient solutions in which transport rates are high, depletion at the root surface may reduce nutrient uptake rates, as appears from the stimulating effect of stirring (Freijsen et al. 1989). The importance of nutrient transport to the root in nutrient-poor soils is illustrated by the following analysis of the balance between nutrient supply and uptake at the root surface (Nye and Tinker 1977).

As previously explained, nutrient uptake is determined by the concentration at the root surface (C_1), which, in turn, is the resultant of nutrient transport to the root and the net nutrient uptake rate (V_{net}). The transport rate (TR) is the sum of the mass flow rate and the diffusion rate. The mass flow rate is the product of water flow (V_w) and nutrient concentration in the bulk of the soil solution (C_b). The diffusion rate is the product of the concentration gradient toward the root surface (dC/dx) and the effective diffusion coefficient (D_e). The effective diffusion constant, in turn, depends on the moisture content of the soil, the tortuosity of the diffusion pathway, and the buffer power of the soil, which accounts for the degree to which nutrient transport is impeded by interaction with soil particles. Phosphate is strongly adsorbed to soil particles, which leads to low values of C_b, resulting in low rates of mass flow and diffusion. At the other end of the spectrum, nitrate is much more mobile because adsorption is negligible. Potassium takes an intermediate position.

The net uptake rate (V_{net}) is calculated as the gross uptake rate (V) minus the efflux rate (E). This leads to the following equations:

$$TR = V_w C_b + D_e dC/dx \tag{2}$$

$$V_{net} = [V_{max} C_1/(C_1 + K_m)] - E \tag{3}$$

When the transport rate equals the net uptake rate, a dynamic equilibrium develops, with equilibrium concentration and uptake rate C_1^* and V_{net}^*. When C_1 is lower than C_1^*, the transport rate is higher than the net uptake rate, so that C_1 will increase until it reaches C_1^* (Figure 2). When C_1 is higher than C_1^*, the transport rate is lower than the net uptake rate, so that C_1 will decrease until it reaches C_1^*. At a short term (hours), the equilibrium concentration (C_1^*) and the corresponding equilibrium uptake rate (V_{net}^*) are stable. When uptake proceeds for a few days, the equilibrium shifts gradually to lower C_1 values, because the depletion zone grows, the diffusion gradient becomes less steep, and the transport rate decreases. Depletion proceeds more slowly around thin roots than around thick roots, due to the radial geometry of roots. The amount of soil per unit root surface present within the same distance from the root surface is larger (and contains more nutrients) when the root diameter is smaller.

The depletion zone around the root will continue to grow until the transport rate becomes so low that it equals the rate at which nutrients are released from the soil within the depletion zone by means of dissolution, desorption, or mineralization. Immobile nutrients such as phosphate are released slowly and have narrow depletion zones, low uptake rates, and low C_1^*. Eventually, when the nutrients adsorbed to soil particles are depleted or when mineralization is interrupted due to low temperatures, the release rate will fall and uptake will stop.

To maximize uptake, plants can reduce the distance over which nutrients

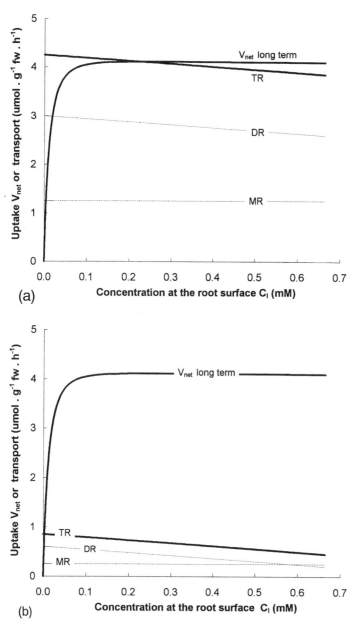

Figure 2 Net uptake rate (V_{net}) and nutrient transport rate (TR) to the root surface as a function of the nutrient concentration at the root surface (C_l). The diffusion rate (DR), the mass flow rate (MR), and their sum (TR) at (a) low ($C_b = 0.3$) and (b) high ($C_b = 5$) concentrations in the bulk soil. In either case, the equilibrium concentration (C_l^*) and equilibrium uptake rate (V_{net}^*) are found at the intersection of the lines V_{net} and TR. The parameter values used are hypothetical, keeping midway between NO_3^- and K^+. V_{net} is identical to the long-term uptake in Figure 1.

are transported through the soil by increasing the density of their root system. Plants can realize higher root densities by increased allocation of carbon to their rooting system, but also by reduced root diameters, which leads to an increased root length and root surface per unit of root biomass. Root hairs are important in this respect because they are thin and require little investment of biomass per unit of soil being explored. However, their vulnerability and short life span make them less profitable when the bulk of the soil is already depleted, so that a plant must "sit and wait" for nutrients that are released from the solid phase. In such nutrient-poor situations, many plant species are living in symbiosis with fungi that form mycorrhizas. Such associations strongly increase the total surface area by which nutrients can be taken up. Fungal hyphae have much smaller diameters than roots (see chapter 5).

Silberbush and Barber (1984) studied the effect of changes in plant and soil characteristics on the equilibrium uptake rate of plants growing in soil. Figure 2 illustrates their results. The equilibrium uptake rate (V_{net}^*) and nutrient concentration (C_i^*) at the root surface are given by the intersection of the lines TR and V_{net}. When nutrient concentrations in the soil are high (Figure 2a, with $C_b = 5$), C_i^* is situated at the horizontal part of the uptake curve. In this case, the equilibrium uptake rate V_{net}^* is determined largely by the maximum uptake capacity of the roots (V_{max}) and not by the affinity of the uptake mechanism (K_m), nor by the mass flow or diffusion rate. Consequently, we may expect that natural selection on rich soils will favor plants with a high maximum uptake capacity (V_{max}).

When nutrient concentrations in the bulk soil are low (Figure 2b, with $C_b = 0.3$), C_i^* is situated at the ascending part of the uptake curve. Here, V_{net}^* is mainly determined by the effective diffusion coefficient D_e and by C_b, which determine the slope of the line that represents the diffusion rate (DR) and the intercept with the horizontal axis, respectively. In this case, the value of V_{net}^* is relatively insensitive to changes in the kinetic parameters that rule the uptake process (K_m and V_{max}). Consequently, we may expect that natural selection of plants in poor soils will not lead to increased affinity or capacity of the nutrient uptake mechanism, but to properties that reduce the transport limitation, bringing the root surface closer to the nutrients, (i.e., by fine and dense root systems and mycorrhizal associations). By increasing the root surface per unit plant biomass, a plant can sustain adequate growth rates with lower nutrient uptake rates per unit root surface and thus with lower nutrient concentrations at the root surface than competitors with a smaller root system.

The nutrient concentration in the bulk of the soil solution (C_b) can differ dramatically from the concentration at the root surface (C_1), implying that C_b is not a good indicator of nutrient availability. When the nutrient pool in the bulk of the soil solution is depleted, the nutrient supply to the root will depend on the rate at which available forms of the nutrient are released from the organic and the mineral substrate. Plants can increase the release rate of nutrients by lowering the concentration in the soil solution or by affecting the chemical conditions or the microbial activity in the rhizosphere. Some species can use chemical forms or physical states (solid, dissolved, adsorbed, or occluded) of a nutrient that other species cannot use. Nitrogen, for example, can be taken up by most species only as NO_3^- and NH_4^+, but some species can also take up amino acids and other dissolved organic molecules that contain nitrogen (Kielland 1994, Northup et al. 1995, Schimel and Chapin 1996, Leadley et al. 1997). Some species can mobilize iron in calcareous soils by lowering the pH or by exudating chelating or reducing substances (Römheld and Marscher 1986). Phosphorus can be taken up by most species only as $H_2PO_4^-$, but some species are able to mobilize solid calcium-phosphate by changing the chemical conditions in the rhizosphere, e.g., by lowering pH, exudation of organic acids, or lowering the Ca concentration in the soil solution (Hoffland 1992). The ability to mobilize insoluble nutrients by altering the chemical conditions in the rhizosphere can be enhanced by forming dense clusters of lateral roots that intensify the rhizosphere effects. These morphological adaptations are called proteoid roots, after the Proteaceae family, but they occur also in species of other taxa (e.g., *Lupine*). An excellent review of the ability of plant species to use alternative nutrient sources is given by Marschner (1995).

Many plant species from nutrient-poor soils have special adaptations that enable them to use nutrients from other sources than the soil solution. The most widespread of these are associations with mycorrhizal fungi and symbiosis with N-fixing bacteria such as *Rhizobium* and *Frankia*. Mycorrhizal fungi are able to decompose dead organic material and to transport the mineralized nutrients to the plant root (see Chapter 5). Recently, Jongmans et al. (1997) suggested that mycorrhizal fungi are also able to penetrate rocky materials, absorb P, Mg, Ca, and K from minerals by the excretion of organic acids, and transport these nutrients to connected roots. More peculiar adaptations that occur mainly in extremely nutrient-poor ecosystems are parasitism on other plants (e.g., *Rhinanthus, Pedicularis*) and carnivory (e.g., *Drosera, Pinguicula, Utricularia*).

IV. ROOT FORAGING IN HETEROGENEOUS ENVIRONMENTS

Thus far we have examined nutrient uptake and transport in a homogenous substrate. However, nutrient availability in soils may vary dramatically beyond the zone of influence of the roots themselves. Soil patches of different quality are created at various scales by abiotic factors (soil type differences, soil depth, microtopography), as well as by biotic factors such as treefalls and stemflow in forests (Gibson 1988a, 1988b, Hook et al. 1991, Lechowicz and Bell 1991). In arid environments, organic matter accumulates in the vicinity of isolated trees, shrubs, and persistent turf grasses, creating "islands of fertility" in a nutrient-deprived matrix (Jackson and Caldwell 1993, Schlesinger et al. 1996, Alpert and Mooney 1996, Ryel et al. 1996). Consequently, from the point of view of the plant individual, in many habitats the spatial distribution of water and nutrients is profoundly heterogeneous from scales as small as a few centimeters to tens of meters and more. How effectively can plants capture the resources in such a heterogeneous world? What fraction of the growth achieved at a homogeneous supply of soil resources can be realized when similar amounts of resources are patchily distributed? Do species from habitats of different resource status have different foraging abilities?

Especially for the less mobile ions such as phosphate, pockets of nutrients may only be captured by the plant if roots expand their surface area into the richer patch (Robinson 1994, 1996, Hutchings and de Kroon 1994). This foraging behavior may be very effective, as is perhaps best illustrated with the classical study by Drew et al. with barley (*Hordeum vulgare*). Single root axes of barley were grown into three compartments in which the concentration of nutrients could be controlled separately. High nitrate concentration in a given compartment promoted the formation of more first- and second-order laterals per unit of primary root length within that compartment, and greater lateral root extension (Drew et al. 1973). When one third of the entire root system received a nutrient-rich solution, total lateral root length per unit of length of the primary axis was ten times higher, and the total root biomass was six times higher, in the high-nutrient compartment than in the low-nutrient compartments. Later in the experiment, when the lateral roots had grown out, whole-plant relative growth rate (RGR) under localized supply of nutrients approached the RGR of control plants growing under homogeneous nutrient supply (Drew and Saker 1975). When phosphate was supplied to 2 cm of the main root axis—a fraction amounting to only a few percent of its total length—whole-plant RGR was more than 80% of its value in control plants, in which the whole root system received phosphate. When applied to 4 cm of the main root axis, the RGR was similar to that of controls. The higher local nutrient uptake from small pockets of nutrients to which part of the root system was exposed was not only due to an enlargement of the local root surface area. Phosphate absorption rates per unit of root length also increased in the enriched compartment compared with other parts of the root system in treated plants and the root system of control plants (Drew and Saker 1975).

In his review of studies on root plasticity, Robinson (1994) found that in 70% of the experiments, a similar root proliferation was observed as in Drew's experiment, although the magnitude of the response was usually smaller. Little or no root proliferation was observed for 30% of the species. Species may differ markedly both in the degree to which root growth is stimulated by local nutrient supply and in the time between nutrient application and response. For example, when exposed to nutrient enrichment, roots of the cold desert species *Agropyron desertorum* showed a fourfold increase in the RGR of root length within 1 day, whereas *Artemisia tridentata* and especially *Pseudoroegneria spicata* responded less vigorously (Jackson and Caldwell 1989). In the latter species, extension growth was not affected until several weeks after nutrient application.

The data collected thus far do not unambiguously support the premise that species from habitats of different nutrient availability possess markedly different patterns of morphological and physiological plasticity (Grime et al. 1986, Grime 1994, Hutchings and de Kroon 1994). Species differ in the extent to which they are morphologically and physiologically plastic, but correlations with habitat type are at best incomplete.

Local conditions determine where lateral root growth and uptake is promoted (Drew et al. 1973, Drew and Saker 1975), but the magnitude of the local response depends on the conditions experienced by the rest of the root system and the entire plant. An experiment by Drew (1975) illustrates this well. He subjected roots of barley plants to either a uniform or localized nutrient supply. Part of the root system given a high phosphate supply produced more and longer lateral roots, while the rest of the root system was receiving low phosphate rather than high phosphate (Figure 3). This suggests that the local morphological response is stronger when phosphate is more limiting to the plant. Broadly similar effects were produced when the nitrate and ammonium supply to different sections of the root system were varied (Drew 1975). However, effects were less clear for nitrate (Drew et al. 1973; Figure 3). In most studies in which nutrients were supplied heterogeneously, root growth was suppressed in the part of the root volume that experiences low nutrient supply (Robinson 1994).

The merits of the ability to forage for patchily distributed nutrients can perhaps best be illustrated by comparing the biomass production of plants grown on homogeneous and heterogeneous substrates, each with the same overall nutrient availability. Fransen et al. (1998) created such treatments by mixing poor riverine sand with black humus-rich soil either homogeneously or by concentrating most of the black soil in a small column within the pot. Five grass species were grown individually in each of these treatments. Their roots readily reached and penetrated the enriched column, but the responses were significant only for the three species characteristic of relatively nutrient-rich habitats. Combined for all species, whole-plant nutrient accumulation and biomass at the end of the experiment were significantly higher in the heterogeneous than in the homogeneous treatment. These results indicate that plant species may profit and grow faster at a heterogeneous distribution of soil nutrients. In this experiment with bunchgrasses, the growth stimulus in the heterogeneous compared with the homogeneous treatment was small. Clonal species that spread horizontally have the ability to take up nutrients locally and produce most of the biomass beyond the nutrient-rich patch. For such species, a severalfold increase in biomass production may occur if the distribution of nutrients is not homogeneous, but concentrated in

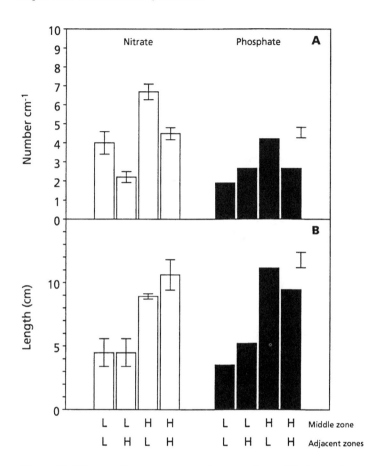

Figure 3 Effects of nitrate and phosphate supply on (A) the numbers of lateral roots per centimeter of main root axis and (B) the lengths of individual lateral roots in barley. Main root axes were divided into three zones and nutrients were supplied independently to each of these zones. Data given are those for the first-order laterals that developed in the middle zone. This zone experiences either a low (L) or a high (H) concentration of nitrate or phosphate. Adjacent rooting zones also grew in either a high or low nutrient solution. In the nitrate experiment, plants were grown hydroponically; in the phosphate experiment they were grown in sand. Nitrate data are given as means ± SE, phosphate as means with separate bars showing the least significant difference at the 5% level. (*Source*: Data from Drew et al. 1973 and Drew 1975. Reprinted from Hutchings and de Kroon 1994, by permission of Academic Press.)

small "hot spots" (Birch and Hutchings 1994, Wijesinghe and Hutchings 1997, Hutchings and Wijesinghe 1997).

Little work has been conducted on whether exploitation of enriched microsites by rapid root formation compromises the ability of the root system to explore the soil volume. Drew and Saker's studies on barley revealed that, in contrast to the growth of the lateral roots, the rate of extension growth of the main root axis was not affected by the nutrient availability in the compartment in which it was growing. Such main roots enable the root system to search the soil volume continuously. It is not clear whether this result more generally applies to other species. Although barley combines both local exploitation and distant exploration, it is likely that local root proliferation reduces the growth of the main root axis, decreasing the chances that distant resource patches are encountered. Vice versa, there are opportunity costs associated with the growth of the main axis. We define opportunity costs as the reduction in lateral root proliferation in the short term as a consequence of the growth of the main axis (Ballaré et al. 1991). The resources invested into main axis growth could have been used for local root proliferation, which would accelerate the acquisition of the locally available resources and hence increase long-term resource gain and plant growth. The height of the opportunity costs depends on the patch structure of the soil, including the size; the number and the distribution of favorable microsites, which determines the chance that exploring roots may encounter them; and the speed with which patches can be depleted (de Kroon and Schieving 1990, Stuefer 1996). To our knowledge, to date, foraging models that calculate the optimal behavior in soils with different patch structure and that take into account the opportunity costs of root extension are lacking (but see Gleeson and Fry 1997 for a start).

Optimal root deployment should maximize the efficiency of nutrient uptake per unit of assimilated carbon as calculated by Eissenstat and Yanai (1997). Note that root efficiency defined in this way does not correspond with foraging efficiency. Foraging theory predicts that it is the strategy with the highest resource gain per unit of time that is most advantageous in a given environmental setting, not the strategy with the highest resource gain per unit of invested resource (Stephens and Krebs 1986). Following Hunt et al. (1990), the efficiency of foraging may be expressed as the amount of resources captured as a proportion of the amount of resources that the environment supplies (Hutchings and de Kroon 1994).

Incomplete information on the costs and benefits allows us only to speculate on the long-term advantages of different foraging behaviors. The data available to date suggest that species from resource-poor versus resource-rich habitats differ only little in root foraging abilities. We suggest that this may be so because the nutrient availability of all habitats is essentially patchy, and all species will have developed a behavior with which these heterogeneous resources may be exploited. Both morphological and physiological plasticity will be important.

However, in extremely nutrient-poor habitats such as tundra, the ability of roots to survive periods of resource depletion seems to be of greater significance than high levels of plasticity. The maintenance of a large viable root mass, despite long periods of low nutrient availability, and the ability to commence absorption of nutrients rapidly when conditions permit, enable species to acquire nutrient pulses of short duration (Crick and Grime 1987, Campbell and Grime 1989, Kachi and Rorison 1990). The high carbon costs of maintaining viable roots (Eissenstat and Yanai 1997) may not be a great problem in these habitats because carbon is not the limiting resource. In very productive environments, however, carbon costs of root maintenance may be significant, and roots generally have a shorter life span than species from less productive habitats (see Section VI). Rapid growth, high nutrient uptake rates, and a high turnover of roots may result in a more fugitive root behavior in these habitats. Enriched microsites will be rapidly exploited, after which the root system shifts its investments toward more profitable parts of the soil volume.

V. ALLOCATION AND USE OF ABSORBED NUTRIENTS

The acquisition of nutrients, their transport within the plant from the roots to the other organs, and their subsequent incorporation into organic compounds require a major carbon expense of the plant (Chapin et al. 1987). Vice versa, the assimilation of carbon requires nutrients, but especially N, in significant quantities. C_3 plants invest approximately 75% of their N in chloroplasts, of which a major part is used in photosynthesis. Approximately one third of this chloroplast N is built into rubisco, the primary CO_2-fixing enzyme (Chapin et al. 1987). As a result, the photosynthetic capacity (the maximum rate of carbon assimilation) is highly positively correlated with leaf nitrogen concentration (Field and Mooney 1986, Evans 1989). The photosynthetic rate per unit of leaf nitrogen is referred to as the photosynthetic nitrogen use efficiency (PNUE) (Lambers and Poorter 1992, Fitter 1997). Beyond a critical level, photosynthesis does not increase further with increasing nitrogen concentration, and may even decline. In such situations, other resources such as light and water may limit photosynthesis.

Although an important part of the assimilated N is allocated to the photosynthetic system, the plant also requires N for a whole variety of other plant functions (Lambers and Poorter 1992). The relationship between nitrogen concentration in the whole plant and relative growth rate may be different for different plant species depending on, among other factors, the fraction of nitrogen that is allocated to the photosynthetic machinery. Such differences may be caused by variation in allocation to plant organs such as leaves, roots, and stems, but also by differences in the allocation to the various organelles and compounds within the leaf. Some rapidly growing species such as *Lolium perenne* allocate an extremely large part of the leaf nitrogen to rubisco, whereas in other species, part of the leaf nitrogen is used for the synthesis of defensive compounds or incorporated in supporting tissues. Ingestad (1979) characterized the relationship between relative growth rate and whole-plant nitrogen concentration by the nitrogen productivity (A), defined as the rate of dry matter production per unit of nitrogen in the plant (g dry weight g^{-1} N day^{-1}). Figure 4 gives the relationships between RGR and nitrogen concentration in the plant for three tree species, which appear to be linear over a broad range of nitrogen concentrations. The slopes of the regression lines represent the nitrogen productivities of each species. All three species increase their growth at higher internal nitrogen concentrations, but the faster-growing species make a more efficient use of the nitrogen that is present in the plant. Figure 4 also shows that the faster-growing species not only has a

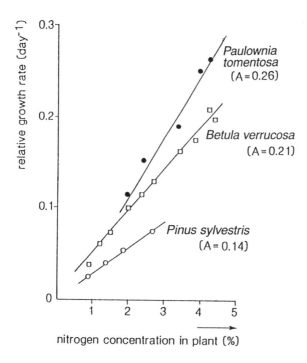

Figure 4 The relative growth rate of seedlings of three tree species versus nitrogen concentration in the total plant. The values of the nitrogen productivity A (g dry weight g^{-1} N h^{-1}) are given by the regression coefficients of the presented lines. (*Source*: Data from Ingestad 1979, Hui-jun and Ingestad 1984, and Ingestad and Kähr 1985. Reprinted from Berendse and Elberse 1990.)

higher growth rate than the slower-growing species at higher nitrogen concentrations, but also at lower concentrations. The difference in nitrogen productivity between the three species is probably caused by differences in allocation to the photosynthetic process, but may be explained as well by differences in costs of biosynthesis of plant tissues.

Whole-plant growth is maximized if all resources are equally limiting (Bloom et al. 1985). As a rule, new biomass is allocated to the plant organs that acquire the most strongly limiting resource. If nutrients are in short supply, there are several ways in which nutrient shortage in the plant may be avoided. More carbon may be invested in root biomass so that a larger soil volume can be explored and the competitive ability for soil nutrients is increased. Tilman (1988) suggested that increased allocation to root biomass would be one of the most important adaptations of plants to nutrient-poor soils. It has been known for a long time that the phenotypic response of all plant species to reduced nitrogen or water supply is an increased carbon and nitrogen allocation to roots (Brouwer 1962), but comparisons of species adapted to nutrient-rich and nutrient-poor sites do not confirm Tilman's hypothesis. Grass species adapted to nutrient-poor soils generally invest less or equal amounts of biomass in belowground parts than species characteristic of more fertile sites (Elberse and Berendse 1993). In a recent review of studies on plasticity in root weight ratio, Reynolds and D'Antonio (1996) showed that species from nutrient-poor and nutrient-rich habitats exhibit a similar increased root allocation in response to nitrogen shortage. The most important difference between species of nutrient-poor and nutrient-rich sites is that the roots of the former species seem to have smaller diameters, leading to an increased root length per unit root weight (Elberse and Berendse 1993, Fitter 1997).

The efficiency of nutrient utilization for growth also depends on other functions to which nutrients are allocated by the plant, such as support, defense, reproduction, and storage (Chapin et al. 1990). Allocation to supporting structures (such as woody tissue), chemical compounds for defense, or reproductive organs may curtail the growth rate of plants (Bazzaz et al. 1987). Plant species with a particularly high allocation to one or more of these functions, or plants in their reproductive phase, will have low growth rates and a low nitrogen productivity. Growth is also curtailed if a significant proportion of the resources is allocated to storage, i.e., reserve formation that involves the metabolically regulated compartmentation or synthesis of storage compounds (Chapin et al. 1990; see also Chapter 5). While reserve formation directly competes for resources with growth, resources may accumulate because resource supply exceeds the demands for growth and other functions during a certain period. This accumulation is commonly referred to as luxury consumption (Chapin 1980) and should be distinguished from reserve formation (Chapin et al. 1990). Luxury consumption allows the slower-growing species to absorb nutrients in excess of immediate growth requirements during nutrient flushes. The reserves built up in this way may be used to support growth in periods of nutrient depletion.

Classical plant ecophysiology often depicts the growth of plants in natural environments simply as resulting from soil nutrient uptake and carbon assimilation. However, in many perennial plant species, growth strongly depends on amounts of nutrients and carbon that have been stored during preceding growing periods (de Kroon and Bobbink 1997). In the alpine forb species *Bistorta bistortoides*, stored N reserves in the rhizomes accounted for 60% of the N allocation to the shoot during the growing season (Jaeger and Monson 1992). In this species, N storage was largely accommodated by increased concentrations of amino acids (Lipson et al. 1996). Resources stored in perennial plant organs may support the aboveground biomass production of plants to a considerable degree, as illustrated in the study by Jonasson and Chapin (1985) with the sedge *Eriophorum vaginatum*. In extremely nutrient-poor tundra, they compared the growth of tillers (with attached belowground stems and roots) in bags without access to soil nutrients with the growth of unbagged tillers. They found that the bagged tillers accumulated as much leaf biomass during one growing season as the unbagged plants. Nutrients were transported from the belowground stems to the leaves during the first 2 months after snow melt. After senescence at the end of the growing season, nutrient contents in the belowground stems of the bagged tillers were only slightly lower than those in the unbagged ones.

VI. LOSSES OF NUTRIENTS THROUGH ABSCISSION AND HERBIVORY

It is clear that the growth of a perennial plant individual is not only determined by the amount of nutrients that it acquires, but also by the amounts of stored nutrients that can be reused. In environments in which nutrients limit plant growth, the long-term dynamics of perennial plant populations is largely determined by the balance between the uptake and the loss of nutrients. Losses of nutrients may occur in various ways, e.g., abscission of leaves and flowers, root death, mortality due to disturbance, nutrient capture by herbivores, leaching from leaves, seed or pollen production, and exudation from roots.

One of the most important pathways by which plants lose nutrients is the seasonal abscission of leaves, roots, and other organs. Several studies show that there is a huge variation in life spans of leaves among vascular plant species. Escudero et al. (1992) found that the life spans of leaves of tree and shrub species in the Pyrenees varied by an order of magnitude from a few months to more than 4 years. A similar large variation (from a few months to 10 years) was reported in a survey of several studies of leaf life spans (Reich et al. 1992). Nutrient losses due to leaf abscission are quantitatively significant but are reduced by active retranslocation of nutrients in the period preceding abscission. Measurements of nutrient withdrawal should take into account not only that the nutrient content is reduced, but also that the dry weight per leaf can decline due to respiration or retranslocation of carbohydrates, implying that nutrient withdrawal should be measured on a whole-leaf basis. In arctic ecosystems, 20–80% of N and 20–90% of P in leaves was withdrawn before abscission, whereas there did not seem to be important differences in percentage withdrawal between graminoids, forbs, and deciduous and evergreen dwarf shurbs (Chapin et al. 1975, Chapin 1989, Chapin and Shaver 1989, Jonasson 1983). Morton (1977) studied the decline of nutrient concentrations in leaves of the deciduous grass *Molinia caerulea* during abscission at the end of the growing season. He compared open plots with plots that were covered with a transparent roof during fall and winter, preventing leaching of nutrients by rain. The reduction in N and P concentrations in dying leaves was measured to be approximately 75% and occurred both in the open and the covered plots, but the decline in concentrations (approximately 90%) of K, Ca, and Mg took place only in the plots without cover. He concluded that the reduction in the N and P content of leaves occurred through active withdrawal from senescing leaves, but that the reduction in K, Ca, and Mg took place through leaching from the leaves to the soil.

Much less data are available about root life spans. Eissenstat and Yanai (1997) showed that the time periods after which 50% mortality has occurred vary between 14 and 340 days, which corresponds with life spans of 20–490 days (assuming a negative exponential decline in the number of living roots). Between-species comparisons are difficult because the life spans vary strongly among root cohorts produced in different seasons, and most studies do not compare species at the same site. In a garden experiment, we followed individual roots of 14 grassland species "from birth to death" using minirhizotrons. Average root life spans varied between 41 days in *Rumex obtusifolius*, which occurs in very fertile habitats, to 381 days in *Succisa pratensis*, which is characteristic of nutrient-poor sites (F. Berendse, unpublished results, 1997). It is not yet clear whether nutrient losses due to root death can be significantly reduced through nutrient withdrawal preceding abscission. A few studies showed that nutrient resorption from dying roots is minimal (Nambiar 1987, Dubach and Russelle 1994), so that nutrient losses by root turnover might be very significant.

In addition to seasonal abscission, nutrient loss occurs due to herbivory by a broad variety of organisms such as grazing mammals, phytophagous insects, parasitic fungi, and root nematodes. The quantities of nutrients that the plant loses due to the activities of herbivores aboveground and belowground have rarely been measured, but may be rather important. We simulated grazing through mammals by clipping plants with 8-week intervals at 5 cm above soil surface. We measured that at low levels of soil fertility the tall grass species *Arrhenatherum elatius* lost 57% of the total amount of nitrogen taken up, whereas the short grass *Festuca rubra* lost 24% (Berendse et al. 1992). These losses increased to more than 90% at higher soil nutrient levels.

The data presented strongly suggest that there is a wide variation in biomass and nutrient turnover among plant species depending on the life spans of plant organs, but they do not supply information about the quantitative significance of whole-plant nutrient loss rates compared with nutrient supply rates. For plants growing in their natural environment, such data are especially extremely difficult to collect. In the 1980s, we conducted a comparative field study in which we attempted to quantify whole-plant nutrient losses and nutrient uptake in populations of the ericaceous dwarfshrub *Erica tetralix* and the perennial, deciduous grass *Molinia caerulea*. In recent decades in many wet heathlands in Europe, *E. tetralix* has been replaced by *M. caerulea*. In competition experiments in containers (Berendse and Aerts 1984) and in field fertilization experiments (Aerts and Berendse 1988, Aerts et al. 1990) we measured that at increased levels of nutrient supply, *M. caerula* is able to outcompete *E. tetralix* whereas *E. tetralix* remains the dominant species under nutrient-poor conditions.

We measured nitrogen losses from populations of *Molinia* and *Erica* plants in adjacent sites for 2 years. Total losses of nitrogen from *Molinia* plants varied between 60% and 100% per year of the total amount of nitrogen present in the plants at the end of the growing season. We calculated the lower turnover rate assuming that 50% of the nitrogen in roots was withdrawn preceding abscission, whereas the higher figure was calculated assuming that no retranslocation took place. It is clear that losses of up to 100% will have important consequences for the success of a population in an environment in which nitrogen limits plant growth. Nitrogen losses from *Erica* were much smaller (approximately 30%). This seems to be an important adaptation to the nutrient-poor habitats that are dominated by this species. In Table 2 we compare the total N losses during 1982 with the N mineralization measured in the upper 10 cm of the soil during this year. The measured N mineralization is approximately equal to the total N supply rate. Almost all organic nitrogen is present in the upper 10 cm of the soil, and the N input through atmospheric deposition is almost completely immobilized by the nutrient-poor litter layer, but is remineralized during the later phases of decomposition. In *Molinia*, the losses of N from the plant appear to be of the same order of magnitude as the rate of N supply to the plant, but in *Erica*, N losses are less than 50% of the N that can be taken up.

If a plant loses a large part of the nutrients in its biomass annually, it must absorb more nutrients to maintain its biomass than a plant that is more economical with its acquired nutrients. To measure the nutrient uptake that plant species need

Table 2 Relative Nitrogen Requirement, Total (Aboveground and Belowground) Biomass at the End of the Growing Season, and Annual Nitrogen Loss from the Plant in Adjacent Populations of *Erica* and *Molinia* and the Annual N Mineralization on These Sites During 1982

	Erica	*Molinia*
Relative nitrogen requirement (mg N g^{-1} biomass yr^{-1})	2.3–3.4	7.4–11.7
Total biomass (g biomass/m^2)	1270	919
Total N-losses (g N m^{-2} yr^{-1})	2.9–4.3	6.8–10.8
N mineralization (g N m^{-2} yr^{-1})	11.5	10.1

in their natural environment, the concept of the relative nutrient requirement was introduced (Berendse et al. 1987). The relative nutrient requirement (L) is defined as the amount of nutrients that a plant population loses per unit of time and biomass. This amount of nutrients should be taken up again to maintain or replace each unit biomass during a given time period (mg N g^{-1} biomass yr^{-1}). In the study referred to earlier, we measured that the relative nitrogen requirement was 2.3–3.4 mg N g^{-1} biomass yr^{-1} in *Erica* compared with 7.4–11.7 mg N g^{-1} biomass yr^{-1} in *Molinia*, depending on the assumption about N withdrawal from dying roots (Table 1). Apparently, *Erica* required much less nitrogen to be taken up to maintain its biomass than did *Molinia*. *Erica* plants appeared to require much less nitrogen because of the longer life spans of their leaves, stems, and roots, and not because of a higher retranslocation efficiency. The withdrawal of nitrogen from dying leaves in *Molinia* is even higher than in *Erica*.

The differences between these two species seem to reflect a general pattern. Escudero et al. (1992) found that the life span of leaves of tree and shrub species in the Pyrenees was strongly correlated with the variation in soil fertility. Plant species dominant on infertile soils had leaves that lived longer than species that were abundant on more fertile soils. This negative correlation was not found between soil fertility and the fraction of N and P that was withdrawn from dying leaves. Recently, we conducted an experiment in which 14 plant species of Dutch grassland and heathland communities were grown in monocultures in experimental plots (F. Berendse, unpublished results, 1997). The direct effects of different soil characteristics were excluded. We found a significant inverse relationship between average leaf life span, as measured in these plots, and the nutrient index of each species, which ranks the average soil fertility of the habitat in which the species involved is most frequently found (Figure 5).

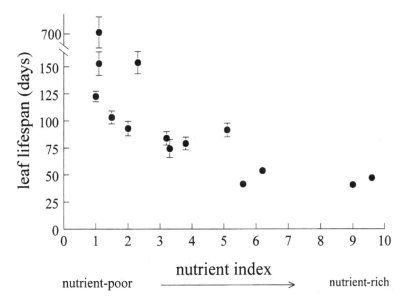

Figure 5 The average leaf life spans of 14 grassland and heathland species versus their nutrient index. Life spans were measured in plants growing in experimental plots under identical conditions. Measurements were taken in 10 plants of each species by monitoring marked leaves with 2-week intervals. The nutrient index is a descriptive parameter that ranks the average soil fertility of the habitats in which the species involved is most frequently found. Bars give standard errors of the mean (Berendse, unpublished results, 1997).

VII. ADAPTATION OF PLANTS TO NUTRIENT-POOR AND NUTRIENT-RICH ENVIRONMENTS

Plant species can increase their success in nutrient-poor habitats along three different lines. First, they can maximize the acquisition of nutrients by increasing their competitive ability for soil nutrients or by exploring nutrient sources that are not available to competing plant populations. The affinity of the uptake system of most plants is sufficiently high to decrease the nutrient concentration at the root surface to practically zero. Further improvement of the uptake capacity is of little use. Uptake kinetics do not differ systematically between species from rich and poor soils. The competitive ability for soil nutrients can be increased by investing more carbon in fine-root biomass or by changes in root morphology that increase root length or root surface area per unit biomass (by reduced root diameter or increased root hair density), so that a greater fraction of the available nutrients can be absorbed relative to competing plant species. In Sections IV and

V, we showed that, in general, plant species of nutrient-poor habitats do not invest more biomass in roots, but that some species of nutrient-poor habitats realize an increased absorbing root surface by producing thinner roots. In addition, species of nutrient-poor habitats do not seem to forage more effectively for patchily distributed nutrients compared with species from more productive habitats. However, many plant species of poor soils can explore additional organic nutrient sources by intimate associations with mycorrhizal fungi, and some genera (e.g., *Drosera, Pinguicula,* and *Utricularia*) can even acquire and use living animal proteins. Deep-rooting species (such as forbs or trees) may absorb nutrients from deeper soil layers that are not available to competing species with a shallow rooting pattern.

The second line along which plant species may be adapted to nutrient-poor sites is by changes in the efficiency with which the nutrients that are present in the plant are used for carbon assimilation and subsequent growth. Different nutrients can be used for different plant functions affecting growth (e.g., N is mainly invested in rubisco, whereas K is required for stomata functioning). The parameter that measures this efficiency is the nutrient productivity, A, as introduced in Section V. A strong selection in favor of an increased nutrient productivity in species adapted to nutrient-poor soils would be expected, but generally, species of nutrient-poor habitats have lower nutrient productivities than species adapted to more fertile sites (Hui-jun and Ingestad 1984, Ingestad and Kähr 1985).

The third line of adaptation is increasing the length of the time period during which nutrients can be used. The length of this time period can be expanded by increased life spans of leaves, roots, and other organs. Life spans can be increased by investing in supporting tissues and in defensive compounds that reduce the risks of herbivory. The residence time of nutrients in the plant can also be increased by retranslocation of a large part of the nutrients in dying plant parts, but we showed earlier that the fraction of nutrients retranslocated from leaves before abscission is not clearly correlated with the soil fertility of the habitat in which the species occurs most frequently (Escudero et al. 1992). The earlier introduced relative nutrient requirement or relative nutrient loss rate measures nutrient losses per unit of biomass, but can also be expressed per unit of nutrient in the plant (Ln; $g \, N \, g^{-1} \, N \, yr^{-1}$). Under steady-state conditions in which nutrient losses are equal to nutrient uptake, the inverse of this parameter (Ln^{-1}) measures the mean residence time of nutrients in the plant.

For a further analysis of the adaptation of plants to nutrient-poor substrates, it is helpful to combine the instantaneous efficiency of nutrient utilization or nutrient productivity (A) with the mean residence time (Ln^{-1}) to the overall nutrient use efficiency (NUE), which measures the amount of biomass that can be produced per unit of nutrient taken up (g biomass produced/g nutrient absorbed):

$$NUE = A/Ln \qquad (4)$$

It would be expected that the NUE would be higher in species of nutrient-poor habitats compared with species from more fertile sites. Berendse and Aerts (1987) calculated this parameter for adjacent field populations of *Erica* and *Molinia* using the amount of biomass and the amount of nutrients in the plant at the end of the growing season (Table 3). It is striking that there is only a relatively small difference in NUE between the two species, but that the NUE values are composed by entirely different combinations of A and Ln^{-1}. *Erica* has a low N loss rate combined with a low N productivity, whereas *Molinia* has a much larger loss rate, but also a higher instantaneous utilization efficiency, A. Apparently, the same overall NUE can be realized by various combinations of plant properties. These data suggest that the components A and Ln^{-1} are especially relevant in the adaptation of plant species to habitats with different nutrient supplies, rather than the NUE itself. Consideration of the differences in growth rate between these two species in relation to their different nutrient loss rates follows.

The dwarfshrub *Erica* is able to maintain itself as the dominant species in nutrient-poor sites because it is much more economical with the nutrients that it has absorbed than the perennial grass species. But this difference does not explain why *Molinia* outcompetes *Erica* after an increase in nutrient supply. In a field experiment with different nutrient supply rates, the potential growth rate of the grass *Molinia* was found to be much higher than that of the dwarf shrubs *Erica* and *Calluna* (Aerts et al. 1990). The higher potential growth rate of *Molinia* enabled this species to increase its biomass much more rapidly than *Erica* after an increase in nutrient supply. In communities where both species are present, an increase or decrease in nutrient supply can result in complete dominance of *Molinia* and extinction of *Erica*, or vice versa. The changes in such communities were calculated using our model for the competition between plant species (Berendse 1994). This model calculates nutrient and light competition and the losses of biomass and nutrients. The model predicts that species 1 outcompetes species 2 under nutrient-rich conditions, whereas species 2 replaces species 1 under nutrient-poor conditions if

$$Gmax_1/Gmax_2 > Ln_1/Ln_2 > 1 \tag{5}$$

Table 3 Nitrogen Productivity, Mean Residence Time, and Nitrogen Use Efficiency as Calculated for *Erica* and *Molinia*

	Erica	*Molinia*
A (g biomass g^{-1} N yr^{-1})	24	94
Ln^{-1} (yr)	4.3	1.4
NUE (g biomass g^{-1} N)	103	132

Source: Berendse and Aerts 1987.

where Gmax$_1$ and Gmax$_2$ (g biomass m^{-2} yr^{-1}) are the potential growth rates in a closed canopy of species 1 and 2, respectively, and Ln$_1$ and Ln$_2$ represent the relative nutrient loss rates of the two species, assuming that all other plant features are equal. This relationship leads us to conclude that interspecific competition is responsible for a strong selection pressure on the potential growth rate and the relative nutrient loss rate of species. Slight changes in these plant characteristics may lead to either complete disappearance or complete dominance. But we can conclude as well that species that combine a low nutrient loss rate with a high potential growth rate will be superior at all nutrient supply rates. The question to be answered is whether plants can easily combine such characteristics.

The difference in potential growth rate (and nitrogen productivity) between the two species can be attributed to differences in allocation of nitrogen to the photosynthetic system and to differences in biosynthesis costs. At the end of the growing season, *Erica* had allocated approximately 12% of total plant nitrogen to its leaves, whereas in *Molinia* plants, 48% of the total plant N was present in leaves and green stems. This difference is not caused by differences in allocation to root biomass, but by the allocation of nitrogen to long-living, woody stems in *Erica*. The two species possibly differ as well in the allocation of nitrogen to the different compounds within the leaf. *Erica* leaves live approximately four times longer than *Molinia* leaves, thanks to their higher lignin content, which results in an increased toughness of the leaf. Lignin is much more expensive to biosynthesize than compounds such as cellulose. In a literature survey, Poorter (1994) did not find any systematic difference in biosynthesis costs of tissues produced by plant species from nutrient-poor and nutrient-rich soils. However, we found that the costs of biosynthesizing *Erica* tissue were higher than those of *Molinia* tissue (1.8 versus 1.4 g glucose/g biomass). We conclude that the adaptation to nutrient-poor environments by minimizing the loss of nutrients has important negative side effects: the allocation of nitrogen to the photosynthetic system is reduced and the biosynthesis costs of tissues are increased, which results in reduced nitrogen productivity and reduced potential growth rate, an important disadvantage when soil fertility increases. Apparently, plant properties that determine nutrient losses and potential growth rates are strongly interconnected. The combinations of low maximum growth rate–low loss rate and high maximum growth–high loss rate strongly correspond with, respectively, the stress-tolerant and competitive strategies that Grime (1979) distinguished much earlier. High maximum growth rates and low biomass losses cannot be easily combined for apparent physiological and morphological reasons.

This leads to the expectation that in plant species adapted to soils with different soil fertilities, biomass turnover rate and maximum growth rate will be correlated. In an experiment in growth chambers, we measured maximum relative growth rates in the 14 grassland species for which we had measured leaf life spans in a garden experiment (Figure 5). We found a significant, negative relationship

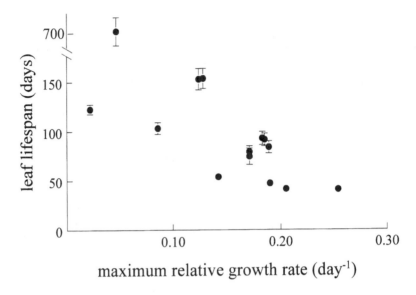

Figure 6 The average leaf life spans of 14 grassland and heathland species versus their maximum relative growth rate. Life spans were measured as given in the legend to Figure 5. Maximum relative growth rates were measured in a growth chamber experiment for seedlings at optimum nutrient supply rates (Berendse and Braakhekke, unpublished results, 1997).

between leaf life span and maximum relative growth rate, which confirms our hypothesis (Figure 6). These results show that the differences between *Erica* and *Molinia* reflect a much more general pattern about the adaptation of wild plant species to soils with low and high nutrient supplies.

REFERENCES

Aerts, R. and F. Berendse. 1988. The effects of increased nutrient availability on vegetation dynamics in wet heathlands. Vegetatio 76: 63–69.

Aerts, R., F. Berendse, H. de Caluwe, and M. Schmitz. 1990. Competition in heathland along an experimental gradient of nutrient availability. Oikos 57: 310–318.

Alpert, P., and H.A. Mooney. 1996. Resource heterogeneity generated by shrubs and topography on coastal sand dunes. Vegetatio 122: 83–93.

Ballaré, C.L., A.L. Scopel, and R.A. Sánchez. 1991. On the opportunity cost of the photosynthate invested in stem elongation reactions mediated by phytochrome. Oecologia 86: 561–567.

Bazzaz, F.A., N.R. Chiariello, P.D. Coley, and L.F. Pitelka. 1987. Allocating resources to reproduction and defense. Bio Science 37: 58–67.

Berendse, F. 1994. Competition between plant populations at low and high nutrient supply. Oikos 71: 253–260.

Berendse, F., and R. Aerts. 1984. Competition between *Erica tetralix* L. and *Molina caerulea* L. Moench as affected by the availability of nutrients. Acta Oecologica 5: 3–14.

Berendse, F., and R. Aerts. 1987. Nitrogen-use-efficiency: a biological meaningful definition? Functional Ecology 1: 293–296.

Berendse, F., and W.Th. Elberse. 1990. Competition and nutrient availability in heathland and grassland ecosystems. In: Grace, J.B., and D. Tilman, eds. Perspectives on Plant Competition. New York: Academic Press, pp 94–116.

Berendse, F., W.Th. Elberse, and R.H.M.E. Geerts. 1992. Competition and nitrogen losses from plants in grassland ecosystems. Ecology 73: 46–53.

Berendse, F., H. Oudhof, and J. Bol. 1987. A comparative study on nutrient cycling in wet heathland ecosystems. I. Litter production and nutrient losses from the plant. Oecologia 74: 174–184.

Birch, C.P.D., and M.J. Hutchings. 1994. Exploitation of patchily distributed soil resources by the clonal herb *Glechoma hederaceae*. Journal of Ecology 82: 653–664.

Bloom, A.J., F.S. Chapin, and H.A. Mooney. 1985. Resource limitation in plants—an economic analogy. Annual Review of Ecology and Systematics 16: 363–392.

Breeze, V.G., A. Wild, M.J. Hopper, and L.H.P. Jones. 1984. The uptake of phosphate by plants from flowing nutrient solution. II. Growth of *Lolium perenne* L. at constant phosphate concentrations. Journal of Experimental Botany 35: 1210–1221.

Brouwer, R. 1962. Nutritive influences on the distribution of dry matter in the plant. Netherlands Journal of Agricultural Science 10: 399–408.

Campbell, B.D., and J.P. Grime. 1989. A comparative study of plant responsiveness to the duration of episodes of mineral nutrient enrichment. New Phytologist 112: 261–267.

Chapin, F.S. 1980. The mineral nutrition of wild plants. Annual Review of Ecology and Systematics 11: 233–260.

Chapin, F.S. 1989. The costs of tundra plant structures: evaluation of concepts and currencies. American Naturalist 133: 1–19.

Chapin, F.S., A.J. Bloom, C.B. Field, and R.H. Waring. 1987. Plant responses to multiple environmental factors. BioScience 37: 49–57.

Chapin, F.S., E.D. Schulze, and H.A. Mooney. 1990. The ecology and economics of storage in plants. Annual Review of Ecology and Systematics 21: 423–447.

Chapin, F.S., and G.R. Shaver. 1989. Differences in growth and nutrient use among arctic plant growth forms. Functional Ecology 3: 73–80.

Chapin, F.S., K. Van Cleve, and L.L. Tieszen. 1975. Seasonal nutrient dynamics of tundra vegetation at Barrow, Alaska. Arctic and Alpine Research 7: 209–226.

Crick, J.C., and J.P. Grime. 1987. Morphological plasticity and mineral nutrient capture in two herbaceous species of contrasted ecology. New Phytologist 107: 403–414.

de Kroon, H. and R. Bobbink. 1997. Clonal plant dominance under elevated nitrogen deposition, with special reference to *Brachypodium pinnatum* in chalk grassland. In: de Kroon, H., and J. van Groenendael, eds. The Ecology and Evolution of Clonal Plants. Leiden: Backhuys Publishers, pp 359–379.

de Kroon, H., and F. Schieving. 1990. Resource partitioning in relation to clonal growth strategy. In: van Groenendael, J., and H. de Kroon, eds. Clonal Growth in Plants— Regulation and Function. The Hague: SPB Academic Publishing, pp 113–130.

Drew, M.C. 1975a. Comparison of the effects of a localized supply of phosphate, nitrate, ammonium and potassium on the growth of the seminal root system, and the shoot, in barley. New Phytologist 75: 479–490.

Drew, M.C., and L.R. Saker. 1975b. Nutrient supply and the growth of the seminal root system in barley. II. Localized, compensatory increases in lateral root growth and rates of nitrate uptake when nitrate supply is restricted to only part of the root system. Journal of Experimental Botany 26: 79–90.

Drew, M.C., L.R. Saker, and T.W. Ashley. 1973. Nutrient supply and the growth of the seminal root system in barley. I. The effect of nitrate concentration on the growth of axies and laterals. Journal of Experimental Botany 24: 1189–1202.

Drew, M.C., L.R. Saker, S.A. Barber, and W. Jenkins. 1984. Changes in the kinetics of phosphate and potassium absorption in nutrient deficient barley roots measured by a solution-depletion technique. Planta 160: 490–499.

Dubach, M., and M.P. Russelle. 1994. Forage legume roots and nodules and their role in nitrogen transfer. Agronomy Journal 86: 259–266.

Eissenstat, D.M., and R.D. Yanai. 1997. The ecology of root lifespan. Advances in Ecological Research 27: 1–60.

Elberse, W.Th., and F. Berendse. 1993. A comparative study of the growth and morphology of eight grass species from habitats with different nutrient availabilities. Functional Ecology 7: 223–229.

Elberse, W.Th., J.P. van den Bergh, and J.G.P. Dirven. 1983. Effects of use and mineral supply on the botanical composition and yield of old grassland on heavy-clay soil. Netherlands Journal of Agricultural Science 31: 63–88.

Epstein, E. 1965. Mineral metabolism. In: Bonner, J., and J.E. Varner, eds. Plant Biochemistry. New York: Wiley, pp 438–466.

Escudero, A., J.M. Del Acro, I.C. Sanz, and J. Ayala. 1992. Effects of leaf longevity and retranslocation efficiency on the retention time of nutrients in the leaf biomass of different woody species. Oecologia 90: 80–87.

Evans, J.R. 1989. Photosynthesis and nitrogen relationships in leaves of C_3 plants. Oecologia 78: 9–19.

Field, C.B., and H.A. Mooney. 1986. The photosynthesis-nitrogen relationship in wild plants. In: Givnish, T.J., ed. On the Economy of Plant Form and Function. Cambridge: Cambridge University Press, pp 25–55.

Fitter, A.H. 1997. Nutrient acquisition. In: Crawley, M.J., ed. Plant Ecology. 2nd edition. Oxford: Blackwell Scientific Publications, pp 51–72.

Fitter, A.H., and R.K.M. Hay. 1987. Environmental Physiology of Plants. 2nd edition. London: Academic Press.

Fransen, B., H. de Kroon, and F. Berendse. 1998. Root morphological plasticity and nutrient acquisition of perennial grass species from habitats of different nutrient availability. Oecologia 115: 351–358.

Freijsen, A.H.J., S.R. Troelstra, H. Otten, and M.A. van der Meulen. 1989. The relationship between the specific absorption rate and extremely low ambient nitrate concentrations under steady-state conditions. Plant and Soil 117: 121–127.

Gibson, D.J. 1988a. The maintenance of plant and soil heterogeneity in dune grassland. Journal of Ecology 76: 497–508.

Gibson, D.J. 1988b. The relationship between sheep grazing and soil heterogeneity to plant spatial patterns in dune grassland. Journal of Ecology 76: 233–252.

Gleeson, S.K., and J.E. Fry. 1997. Root proliferation and marginal patch value. Oikos 79: 387–393.

Grime, J.P. 1979. Plant Strategies and Vegetation Processes. Chichester: Wiley.

Grime, J.P. 1994. The role of plasticity in exploiting environmental heterogeneity. In: Caldwell, M.M., and R.W. Pearcy, eds. Exploitation of Environmental Heterogeneity by Plants. Ecophysiological Processes Above- and Belowground. San Diego: Academic Press, pp 1–19.

Grime, J.P., J.C. Crick, and J.E. Rincon. 1986. The ecological significance of plasticity. In: Jennings, D.H., and A.J. Trewavas, eds. Plasticity in Plants. Cambridge: Biologists Limited, pp 5–29.

Hoffland, E. 1992. Quantitative evaluation of the role of organic acid exudation in the mobilisation of rock phosphate by rape. Plant and Soil 140: 279–289.

Hook, P.B., I.C. Burke, and W.K. Lauenroth. 1991. Heterogeneity of soil and plant N and C associated with individual plants and openings in North American shortgrass priarie. Plant and Soil 138: 247–256.

Hui-jun, J., and T. Ingestad. 1984. Nutrient requirements and stress response of *Populus simonii* and *Plownia tomentosa*. Physiologia Plantarum 45: 149–157.

Hunt, R., J. Warren Wilson, and D.W. Hand. 1990. Integrated analysis of resource capture and utilization. Annals of Botany 65: 643–648.

Hutchings, M.J., and H. de Kroon. 1994. Foraging in plants: the role of morphological plasticity in resource acquisition. Advances in Ecological Research 25: 159–238.

Hutchings, M.J., and D.K. Wijesinghe. 1997. Patchy habitats, division of labour and growth dividends in clonal plants. Trends in Ecology and Evolution 12: 390–394.

Ingestad, T. 1979. Nitrogen stress in Birch seedlings II. N, P, Ca and Mg nutrition. Physiologia Plantarum 45: 149–157.

Ingestad, T., and M. Kähr. 1985. Nutrition and growth of coniferous seedlings at varied relative nitrogen addition rate. Physiologia Plantarum 65: 109–116.

Jackson, R.B., and M.M. Caldwell. 1989. The timing and degree of root proliferation in fertile-soil microsites for three cold-desert perennials. Oecologia 81: 149–153.

Jackson, R.B., and M.M. Caldwell. 1993. Geostatistical patterns of soil heterogeneity around individual perennial plants. Journal of Ecology 81: 683–692.

Jackson, R.B., J.H. Manwaring, and M.M. Caldwell. 1990. Rapid physiological adjustment of roots to localized soil enrichment. Nature 344: 58–60.

Jaeger, C.H., and R.K. Monson. 1992. Adaptive significance of nitrogen storage in *Bistorta bistortoides*, an alpine herb. Oecologia 92: 578–585.

Jonasson, S. 1983. Nutrient content and dynamics in north Swedish tundra areas. Holarctic Ecology 6: 295–304.

Jonasson, S., and F.S. Chapin. 1985. Significance of sequential leaf development for nutrient balance of the cotton sedge, *Eriophorum vaginatum* L. Oecologia 67: 511–518.

Jongmans, A.G., N. van Breemen, U. Lundstrom, P.W. van Hees, R.D. Finlay, M. Srinivasan, T. Unestam, R. Giesler, P.A. Melkerud, and M. Olsson. 1997. Rock-eating fungi. Nature 389: 682–683.

Kachi, N., and I.H. Rorison. 1990. Effects of nutrient depletion on growth of *Holcus lanatus* L. and *Festuca ovina* L. and on the ability of their roots to absorb nitrogen at warm and cool temperatures. New Phytologist 115: 531–537.

Kielland, K. 1994. Amino acid absorption by arctic plants: implications for plant nutrition and nitrogen cycling. Ecology 75: 2373–2383.

Kruijne, A.A., D.M. de Vries, and H. Mooi. 1967. Bijdrage tot de oecologie van de Nederlandse graslandplanten. Agricultural Research Reports 696: 1–65.

Lambers, H., and H. Poorter. 1992. Inherent variation in growth rate between higher plants: a search for physiological causes and ecological consequences. Advances in Ecological Research 23: 187–261.

Leadley, P.W., J.F. Reynolds, and F.S. Chapin. 1997. A model of nitrogen uptake by *Eriophorum vaginatum* roots in the field: ecological implications. Ecological Monographs 67: 1–22.

Lechowicz, M.J., and G. Bell. 1991. The ecology and genetics of fitness in forest plants. II. Microspatial heterogeneity of the edaphic environment. Journal of Ecology 79: 687–696.

Lefebvre, D.D., and A.D.M. Glass. 1982. Regulation of phosphate influx in barley roots: effects of phosphate deprivation and reduction of influx with provision of orthophosphate. Physiologia Plantarum 54: 199–209.

Lipson, D.A., W.D. Bowman, and R.K. Monson. 1996. Luxury uptake and storage of nitrogen in the rhizomatous alpine herb, *Bistorta bistortoides*. Ecology 77: 1277–1285.

Loneragan, J.F., and C.J. Asher. 1967. Response of plants to phosphate concentration in solution culture II. Role of phosphate absorption and its relation to growth. Soil Science 103: 311–318.

Marschner, H. 1995. Mineral Nutrition of Higher Plants. 2nd edition. London: Academic Press.

Morton, A.J. 1977. Mineral nutrient pathways in a Molinietum in autumn and winter. Journal of Ecology 65: 993–999.

Nambiar, E.K.S. 1986. Do nutrients retranslocate from fine roots? Canadian Journal of Forest Research 17: 913–918.

Northup, R.R., Z. Yu, R.A. Dahlgren, and K.A. Vogt. 1995. Polyphenol control of nitrogen release from pine litter. Nature 377: 227–229.

Nye, P.H., and P.B. Tinker. 1977. Solute movements in the root-soil system. Oxford: Blackwell.

Pastor, J., J.D. Aber, and C.A. McClaugherty. 1984. Above ground production and N and P cycling along a nitrogen mineralization gradient on Blackhawk Island, Wisconsin. Ecology 65: 256–268.

Poorter, H. 1994. Construction costs and payback time of biomass: a whole plant perspective. In: Roy, J., and E. Garnier, eds. A whole plant perspective on carbon-nitrogen interactions. the Hague: SPB Academic Publishing, pp 111–127.

Reich, P.B., M.B. Walters, and D.S. Ellsworth. 1992. Leaf life-span in relation to leaf, plant, and stand characteristics among diverse ecosystems. Ecological Monographs 62: 365–392.

Reynolds, H.L., and C. D'Antonio. 1996. The ecological significance of plasticity in root weight ratio in response to nitrogen. Plant and Soil 185: 75–97.

Robinson, D. 1994. The responses of plants to non-uniform supplies of nutrients. New Phytologist 127: 635–674.

Robinson, D. 1996. Resource capture by localized root proliferation: why do plants bother? Annals of Botany 77: 179–185.

Römheld, V., and H. Marschner. 1986. Mobilisation of iron in the rhizosphere of different plant species. In: Tinker, P.B., and A. Läuchli, eds. Advances in Plant Nutrition, volume 2. New York: Praeger Scientific, pp 155–204.

Ryel, R.J., M.M. Caldwell, and J.H. Manwaring. 1996. Temporal dynamics of soil spatial heterogeneity in sagebrush-wheatgrass steppe during a growing season. Plant and Soil 184: 299–309.

Schimel, J.P., and F.S. Chapin. 1996. Tundra plant uptake of amino acid and NH_4^+ nitrogen in situ: plants compete well for amino acid N. Ecology 77: 2142–2147.

Schlesinger, W.H., J.A. Raikes, A.E. Hartley, and A.E. Cross. 1996. On the spatial pattern of soil nutrients in desert ecosystems. Ecology 77: 364–374.

Silberbush, M., and S.A. Barber. 1984. Phosphorus and potassium uptake of field grown soybean cultivars predicted by a simulation model. Soil Science Society of America Journal 48: 592–596.

Stephens, D.W., and J.R. Krebs. 1986. Foraging Theory. Princeton: Princeton University Press.

Stuefer, J.F. 1996. Potential and limitations of current concepts regarding the response of clonal plants to environmental heterogeneity. Vegetatio 127: 55–70.

Tilman, G.D. 1984. Plant dominance along an experimental nutrient gradient. Ecology 65: 1445–1453.

Tilman, D. 1988. Plant strategies and the dynamics and structure of plant communities. Princeton: Princeton University Press.

Wijesinghe, D.K., and M.J. Hutchings. 1997. The effects of spatial scale of environmental heterogeneity on the growth of a clonal plant: an experimental study with *Glechoma hederacea*. Journal of Ecology 85: 17–28.

9
Functional Attributes in Mediterranean-type Ecosystems

Richard Joffre, Serge Rambal, and Claire Damesin*
Center for Functional and Evolutionary Ecology, CNRS, Montpellier, France

Current affiliation: University Paris-Sud, Orsay, France.

I. INTRODUCTION

Since the beginning of the century, numerous studies have highlighted structural
and functional affinities of vegetation communities in Mediterranean-type climate
regions throughout the world, i.e., the Mediterranean basin itself, California, central Chile, the Cape region of South Africa, and parts of southwestern and southern Australia (see Specht 1969a for a summary of early works of Schimper 1903
and Warming 1909). A comprehensive review of the comparative ecological researches since the 1950s that provide data from more than one Mediterranean
region (Hobbs et al. 1995) highlights that the more studied fields are plant species
composition and biological invasions, and among the more neglected, plant–
water relations. Nevertheless, in the 1970s, comparisons of physiological and
structural properties of Mediterranean-type ecosystem (MTE) (Naveh 1967,
Specht 1969a, 1969b, Mooney and Dunn 1970) have shown that, in some cases,
similarities of these plants extend to patterns of growth, morphology, and physiology. In these vegetation types, evergreen species are more abundant than deciduous ones (Mooney and Dunn 1970, Cody and Mooney 1978, di Castri 1981). As
a consequence, ecological studies have largely focused on evergreen and sclerophyllous-leaved species, and less attention was devoted to deciduous broad-
leaved species. Nevertheless, a review of the relative frequencies of growth-form
in the 5 MTEs (Rundel 1995) emphasized that while the woody shrub sclerophyllous growth-form is the dominant element, they never represent the majority
of the total floras.

  In a theoretical point of view, spatial (i.e., habitat-related) and temporal
(i.e., succession-related) characteristics of ecosystems must be considered in any
review of plant functional attributes. Interactions between abiotic factors (e.g.,
climate and lithology) and the frequency and spatial distribution of past and present anthropic perturbations (e.g., fire) will determine the distribution of species
across the landscape. Temporal dynamics of gradients of resources (e.g., water
and light) after perturbations will create a temporary specialized habitat occupied
by opportunistic species (Zedler 1994). As emphasized by Zedler (1994), "these
species need not disperse and are able to persist through multiple generations on
a site because of an ability to tolerate or recover from environmental or other
variation." In contrast, other species could be considered as persistent, with life
spans of many years, and as a consequence must cope with the high temporal

variability of resources and perturbations. These are generally woody species and account for the main part of the community biomass.

The aim of this chapter is to discuss some aspects of the functioning of these species by (1) briefly reviewing the main features of MTE resources; (2) illustrating how individuals and ecosystems cope with variability in water resource and control water loss; (3) presenting the assimilation characteristics in relation to constraints; and (4) evaluating the role of nutrients at the plant and community levels.

II. CHARACTERISTICS OF MTEs

Human occupation, as well as past and present land use, differ strongly between the five MTE regions of the world. This fact has evident and important consequences on biodiversity, landscape structure, and ecosystem functioning, and has been extensively reviewed recently (Hobbs et al 1995, Blondel and Aronson 1995, Davis et al 1996). Fire is considered as a ''natural'' event in California, Australia, and South Africa, but more linked with human pressure and considered as an anthropogenic disturbance in the Mediterranean basin (Moreno and Oechel 1994).

A. Climate Definition and Variability

The most distinctive feature of the Mediterranean climate involves the seasonality in air temperature and precipitation, which leads to a hot drought period in summer and a cool wet period in winter (Köppen 1931, Bagnouls and Gaussen 1953, Emberger 1955, Aschmann 1973). This peculiarity of the Mediterranean climate has important implications for vegetation functioning and limits the most favorable season for growth to brief periods in spring and fall. As quoted by Rundel (1995), ''dry summer conditions limit water availability and thus growth, while cool winter conditions limit growth during the season when water availability is generally highest.'' Based on the total amount of annual precipitation and a combination of averaged maximum temperature of the hottest month and the averaged minimum temperature for the coldest month, the Emberger (1955) pluviometric quotient allows a classification of climates from arid to superhumid and from cold to hot (Daget 1977, Le Houérou 1990).

From an ecological point of view, the variability or unpredictability of precipitation imposes strong constraints on plants that could be extremely important for the survival of individuals. An extreme event is defined here as one in which a climate variable has a low relative frequency of occurrence or is lower (or greater) than a given critical threshold. This concern arises naturally, since the impacts of climate are realized largely through the incidence of variation of normal conditions or extreme events (Wigley 1985). For instance, severe impact would come from droughts, which are evident in the frequency distribution of precipitation rather than its mean alone. Examining annual precipitation at 360 stations around the world, Conrad (1941) found a relationship between interannual variability and mean precipitation. For 73 stations located in a 100 × 100-km area around Montpellier, Rambal and Debussche (1995) found a linear relationship ($r = 0.87$; $P < .01$) between mean and standard deviation. Table 1 presents estimations of the coefficient of variation for a 1000-mm mean. The values range between 0.16 and 0.29. The lowest values (0.16–0.18) are proposed by Le Houérou (1992) for a large set of locations in Africa and in the Near East. Conrad (1941), identifying variability more than the rule or positive anomalies in some dry areas, showed that positive anomalies of 3%, 7%, and 10% were observed in Marseille (southern France), San Diego (California), and Alicante (southern Spain), respectively. These findings agreed with those of Waggoner (1989), who observed that "frequency distributions with much larger variance than expected are in the Mediterranean climates" of California.

B. Substrate

Different parent rocks and geological histories of the five Mediterranean-climate regions have given rise to very distinct soils and fertility levels. Table 2 summarizes the more frequent lithological substratum in each region and the associated nutrient status dependent on leaching and water-driven erosion. We modified the original table first published by Groves et al. (1983) to take into account the presence of siliceous parent rock in large areas of the Mediterranean basin (part of the Iberian Peninsula, Corsica, and part of Provence). South Africa and southern Australia comprise older landscapes, consisting of an inland mass of geologically old material, surrounded by discontinuous strips of more recent marine deposits (Hobbs et al. 1995). Soils of the upland of the Cape mountains as well as south-

Table 1 Coefficient of Variations for Mean Annual Rainfall Amounts

Reference	Source	No. of stations	Coefficient of variation
Conrad (1941)	World	360	0.21
Waggoner (1989)	U.S.	55	0.20
Le Houérou (1992)	North Africa, Near East	407	0.18
	Sahel, Soudan	228	0.17
	East Africa	300	0.16
Rambal and Debussche (1995)	Languedoc (France)	73	0.29

Table 2 Main Lithological Substratum and Associated Nutrient Status from Soils of Mediterranean Climate Regions

Substratum	Nutrient status	South Australia	Southwest Australia	South Africa	California	Chile	Mediterranean basin
Siliceous rocks	Strongly leached nutrient-poor soils	•••	•••	•••	•	—	••
Argillaceous rocks	Strongly leached nutrient-poor soils and moderately-leached nutrient-rich soils	••	Trace	••	•••	•••	Trace
Calcareous rocks	Moderately-leached nutrient-rich soils and shallow, high pH soils	•••	Trace	•	Trace	—	•••
Ultramafic rocks (serpentines)	Mg-rich, Ca-poor soils	—	—	—	•	—	Trace

•••, common; ••, frequent; •, present but not widespread; Trace, found in restricted localities;—, not recorded.
Source: Modified from Groves et al. (1983).

western Australia have been exposed to weathering since the Paleozoic Era. In contrast, soils of the Mediterranean basin, California, and Chile are much younger and reflect Tertiary and Quaternary orogenic events. Analyzing mineral nutrient relations among MTEs, Lamont (1994) concluded that "among Mediterranean regions, the soils in Chile are more fertile than those in California, which in turn are more fertile than soils in south-eastern Australia, which in turn are more fertile than those of southwestern Australia." It has been generally assessed that nutrient availability of soils in the Mediterranean basin and Chile is equivalent to those of South Africa and Australia (Groves et al. 1983).

C. Vegetation, Climate, and Nutrient

The presence of evergreen sclerophyllous-leaved shrubs in MTEs has been discussed in terms of convergent evolution in response to the unique environmental stress associated with Mediterranean-type climate and was a major theme of research in the 1970s (Specht 1969a, 1969b, Mooney and Dunn 1970, Cody and Mooney 1978). Physiologically based models were presented to explain the distribution of broad-leaved evergreen species, and a number of ecological paradigms have been developed to explain the nature of plant adaptation to Mediterranean-type climates (for a comprehensive account see Keeley 1989). Nevertheless, strict climatic control of sclerophyll leaf morphologies is questioned by the presence of sclerophyllous vegetation types in non-Mediterranean environments, e.g., chaparral in Arizona (Rundel and Vankat 1989), heathland communities in eastern Australia (Specht 1979), and fynbos-like vegetation in the Afromontane region of Africa (Killick 1979). Rundel (1995) noted that "these paradigms have developed with the implicit assumption that water availability is the most significant limiting factor whilst in the last decade the complexity of environmental stresses involved has become better understood." The role of nutrients in species and ecosystem convergence was questioned in the 1980s (Kruger 1983) based on observations of the contrasting conditions between the oligotrophic soil conditions of South Africa and western and south Australia, and the relatively richer soil conditions of the other three MTEs. These points are discussed further in Section V.

III. INDIVIDUAL AND ECOSYSTEM RESPONSES TO VARIABILITY IN WATER RESOURCE

Among the numerous mechanisms for drought resistance, Mediterranean plant species have three responses that act together to dampen the effects of variability

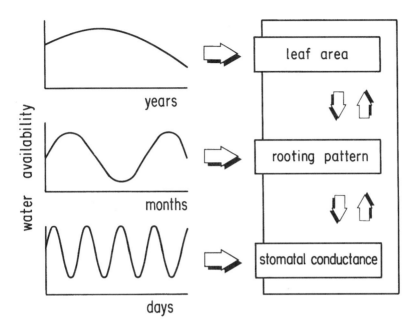

Figure 1   Idealized time-scale-dependent mechanisms of drought resistance for Mediterranean plant species.

in water resource (Figure 1). Change in the leaf area allows the plant to cope with low-frequency oscillations such as a decrease in the annual rainfall amount. The response of the root system dampens the medium-sized oscillations, e.g., changes in seasonal distribution of a given annual rainfall distribution. Finally, the stomatal activity allows quasi-optimization of water use at smaller time scale. Any change in the leaf area induces change in the root system. The new "functional equilibrium" of the plant is controlled by carbohydrates and nitrogen or is hormonally mediated. Any change in the soil water deficit produces a change in the stomatal closure. The plant water status or a hypothetical counterbalancing effect of phytohormones can act as indicators of the root stress. Each mechanism is linked with the one immediately above it, providing an integrated strategy of the plant to improve its water balance under any given set of conditions (Rambal 1993, 1995).

A.   Water Uptake

1.   Rooting Depth

Studies on the root distribution of several Californian chaparral shrub species established that deep root systems are characteristic for the dominant component of the Mediterranean-type vegetation (Hellmers et al. 1955). However, when examined in detail, these species have been shown to exhibit a great diversity in rooting depth (Kummerow et al. 1977) and consequently, in response to drought stress (Poole and Miller 1975). Based on these results, Mediterranean plant species can be roughly divided into two categories: deep-rooted species with root depths greater than 2 m, and shallow-rooted species with root depths less than 2 m. Mediterranean oaks are among the deepest-rooted plant species. For California oaks, Stone and Kalisz (1991) and Canadell et al. (1996) reported roots deeper than 8.5 m for the evergreen *Quercus agrifolia* (10.7 m), *Q. dumosa* (8.5 m), *Q. turbinella* (>9 m), *Q. wislizenii* (24.2 m), and the deciduous *Q. douglasii* (24.2 m). Similar or higher values have been estimated for species of the Chilean matorral (*Lithrea caustica*, 5 m; *Quillaja saponaria*, 8 m) and of the chaparral (*Adenostoma fasciculatum*, 7.6 m). Lower values are found for dominant species of the Mediterranean basin (*Pinus halepensis*, 4.5 m; *Arbutus unedo*, 3.5 m) or of the Australian mallee (*Casuarina* spp., > 2.4 m; *Banksia* spp., 5 m).

2. Soil Water Uptake Patterns

Deep-root systems easily cope with severe water stress. Rambal (1984) distinguished in *Q. coccifera* four patterns of water uptake throughout a drying cycle of 3 months (Figure 2). In late spring, water loss occurred exclusively from the top 0–50 soil layer, which lost approximately 4% of store water per day. The upper meter supplied three quarters of the total. In early summer, root water uptake decreased in the upper layer, which then presented a daily loss of 1.2% of its reserve. Peak water uptake was between 2- and 2.5-m depth. This layer supplied 0.62 of 2.64 mm day^{-1} of transpired water. Late summer was characterized by unevenness of water loss. All of the upper layers were depleted and only the lower layers were able to supply water. During these two late periods, the deepest soil layers were contributing as much water as the top. In early fall at the end of the dry period, all of the layers were depleted. The flat profile did not allow the uptake of water at a rate greater than 0.62 mm day^{-1}. During the first three stages, the daily rates of transpiration were 2.84, 2.64, and 2.35 mm, respectively. Talsma and Gardner (1986) observed the same patterning for various *Eucalyptus* spp.

3. Rooting Patterns

In wet conditions, the major resistance to water uptake appears to be inside the root. Consequently, there is a good correlation between water uptake and rooting density. Hence, the late spring profile of water uptake observed with *Q. coccifera* can be considered as a "picture" of its root density profile. The greatest accumulation of root mass was in the top meter. Below 1 m, root mass decreased gradually with depth. This profile is similar to that of a close species, the shrub live

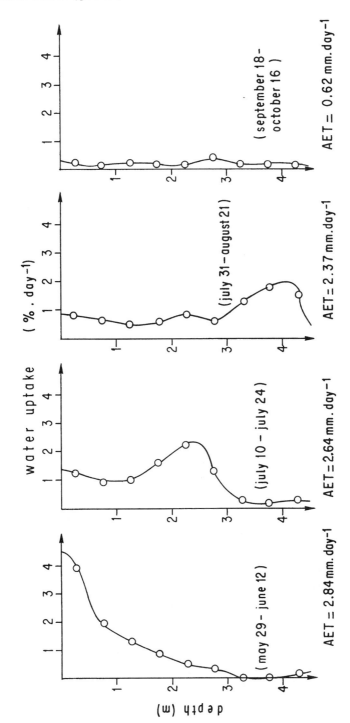

Figure 2 Patterns of water uptake with soil depth for four periods of summer drought. For each layer, the daily water uptake is expressed as a percentage of its total reserve. (*Source*: Rambal 1984.)

oak (*Q. turbinella*) observed by Davis and Pase (1977). The theoretical profile proposed by Jackson et al. (1996) for the sclerophyllous shrub group does not take into account the deep-rooted species identified earlier. This sclerophyllous shrub group included values from some chaparral, fynbos, heath, garrigue, and matorral stands. The root distribution with depth is described by the equation of Gale and Grigal (1987): $y = 1 - \beta^z$, where y is the cumulative root fraction from the soil surface to depth z (cm), and β is an extinction coefficient. For this largely Mediterranean group, $\beta = 0.964$. This means that 67% of the root biomass is in the first 30 cm of soil and 90% in the 0–60-cm layer. These surprisingly high values are far from our estimates for *Q. coccifera* (48% for the 0–50-cm horizon; Rambal 1984) or values for *Q. turbinella* (53% for the 0–60-cm horizon). In the two species, the percentage of roots deeper than 2 m were important (20% and 12%, respectively). Deep roots play a very important role during the summer. As the soil dries, the layers below 2 m contribute an increasing fraction of the total water uptake, reaching between 12% and 23% of the total uptake, depending on the severity of the drought (Rambal 1984).

B. Water Loss

1. Stomatal Regulation

Together with leaf area and rooting pattern, stomatal activity helps regulate water loss. According to their phenology, species avoid or cope with water constraint. For instance, drought-deciduous species of the coastal sage community of California are drought evaders, and chaparral species are drought tolerators. Nevertheless, as stressed by Mooney (1989), ''some coastal sage species that lose most of their leaves during the drought have been shown to tolerate very high water stress, whereas others somehow control stress while maintaining many of their leaves.'' Phenological variations among individuals could also be very high in some species. Ne'eman (1993) found in the case of *Q. ithaburensis*, a deciduous oak of Israel, that some trees were clearly deciduous, whereas others had only a short duration of leaflessness and, as a consequence, could be considered as evergreen. By analogy, there appears to be a continuum of stomatal behavior in response to water stress: no stomata closure, progressive closure of stomata (see Acherar et al. 1991 for Mediterranean oaks, Gollan et al. 1985 for *Nerium oleander*), regulation during the day (Hinckley et al. 1983), and threshold effect leading to a complete closure (*Pinus halepensis*). Stomatal regulation during the day does not characterize Mediterranean species, but could be found in temperate and tropical species (De Lillis and Sun 1990, Mulkey et al. 1996).

Conventionally, stomatal closing is considered to be under the control of leaf turgor pressure or leaf water potential. Recently, however, some investigators showed that leaf conductance is not always closely coupled with leaf water potential or leaf turgor pressure. There is now some evidence that soil water deficit can also induce stomatal closure, even when the leaf water status remains unchanged. These results suggest that leaf conductance is not only affected by leaf water potential, but also, more directly, by soil or root water potential (Turner et al. 1985). One much-discussed possibility is that the counterbalancing effects of two phytohormones, cytokinin and abscisic acid, might provide information on the water status of the roots and induce stomatal closing or opening, as the case may be (Zhang et al. 1987). However, for modeling purposes, the leaf water status still remains the main control variable of the leaf stomatal conductance (Tardieu and Davies 1993).

Daily curves of stomatal conductance in MTE species can be coarsely classified according to three patterns following Hinckley et al. (1978, 1983). Type 1 curves are bell-shaped and represent situations in which soil water potential, leaf water potential, and the vapor pressure deficit do not limit stomatal conductance. Type 2 curves have two maxima, one at the start of the day and the other in the afternoon, both separated by a depression at midday. They correspond to situations in which one or more of the previously mentioned factors limit stomatal conductance. In type 3, the curves have a high point only at the start of the day as the leaf water potential of the plants, according to Hinckley, at or below the turgor loss point for part of the day. The midday stomatal closure was reported by Tenhunen et al. (1987) for many Mediterranean species. This drop in stomatal conductance has been interpreted as a feature that allows Mediterranean species to limit water loss when the atmospheric evaporation is at its maximum. Tenhunen et al. suggested that midday stomatal closure is determined by the leaf to air vapor pressure deficit, whereas in the report by Hinckley et al. (1983), this depended on the interaction among several factors, in particular the instantaneous water potential of the plant.

Cowan and Farquhar (1977) and William (1983) showed how stomatal activity helps optimize water use on a daily time scale in Mediterranean species. With a rapid rate of drying (e.g., species with a shallow-rooted system), an early response of stomata closure was observed in some cases, whereas a delayed response was observed when a slower rate of drying (e.g., species with a deep-rooted system) was imposed. Studies such as those by Poole and Miller (1975, 1978) showed that some chaparral species are more sensitive to water stress than others, as indicated by their water potential after stomata closure. These investigators assume that the degree of drought tolerance is linked to rooting depth. Thus, shallow-rooted species tend to close their stomata at high water potential and to bear tissues with the greatest drought tolerance. Nevertheless, these results were not consistent with observations of Davis and Mooney (1986), who concluded that drought tolerance is not necessarily related to rooting depth.

2. Leaf Area Index

The importance of adjustment in leaf area is emphasized by Passioura (1976): "It is the control of leaf area index and morphology which is often the most powerful means a mesophytic plant has for influencing its fate when subject to long term water stress in the field." Over a large range of climates, changes in leaf area indices have been studied at both the individual and ecosystem scales along gradients from higher to lower rainfall amounts or from moister to dryer habitats in broad- or needle-leaved tree or shrub communities (Ladiges and Ashton 1974, Brown 1981, Gholz 1982, Specht and Specht 1989). In southern Australia, *Eucalyptus viminalis* trees occur over a wide range of rainfall conditions and soil types. Ladiges and Ashton (1974) observed that at moist sites, mature trees are tall and produce large leaves, whereas at drier sites, trees are shorter and tend to produce smaller leaves. Poole and Miller (1981) hold a similar view for Mediterranean shrub species of the California chaparral: "the main response of the shrubs to different precipitation regimes in the chaparral range is to change leaf-area index, not physiological parameters." The importance of this adjustment is largely species-dependent. The ranges of leaf area index of some mature frequent MTE are summarized in Table 3.

C. Water Transfer

1. Soil–Plant Resistance to Water Flow

A simple application of an Ohm's law analogy relating water potential difference from soil to leaf ($\Delta\Psi$) and transpiration rate (T) has been widely used to estimate total flow resistance (R) expressed on a leaf area basis: $\Delta\Psi = TR$, assuming firstly, little capacitance effect in the plant, and secondly, steady-state transpiration conditions. The pathway of water movement in soil and plant can be considered as comprising two main resistances in series: the soil-to-root resistance and the plant resistance. Partitioning soil and plant resistance is difficult. Generally, the plant resistance is assumed constant within a range of Ψ and the relative contribution of the soil resistance is estimated. Gardner (1964) showed that soil resistance is inversely proportional to the hydraulic conductance of the soil. As a consequence, the soil resistance is small at high water content, and any observed difference in resistance should be largely attributable to differences in plant resistance. Under conditions of maximal transpiration (T_{max}) in well-watered conditions: $\Delta\Psi_{max} = T_{max} R_{min}$. Studying a large range of evergreen oak communities, we deduced a hierarchy of soil-to-leaf resistance from the highest R_{min} in xeric sites to the lowest in mesic sites (Rambal 1992). Thus, for *Q. ilex* alone, the ratio of xeric/mesic R_{min} is 1.7. The presence and magnitude of differences in resistance suggest that this attribute could be an important component in drought tolerance. In the same way, Rambal and Leterme (1987) associated a decrease

Table 3 Leaf Area Index Ranges of Mature Stands of Some Frequent Vegetation Types

Vegetation type	Dominant species	LAI range	Reference
Woodland	*Quercus ilex* (evergreen)	2.9–6.0	Damesin et al. (1998)
Woodland	*Quercus pubescens* (deciduous)	2.0–4.2	Damesin et al. (1998)
Shrubland	*Quercus coccifera* (evergreen)	1.5–4.0	Rambal and Leterme (1987)
Chaparral	*Adenostoma fasciculatum*	2.2–3.4	Rambal (1999)
	Ceanothus megacarpus	1.5–1.6	Rambal (1999)
Warm-temperate mallee	*Eucalyptus* spp.	1.5–6.0	Specht and Specht (1989)
Heathland, South Australia		2.5–4.0	Rambal (1999)

LAI, leaf area index.

of leaf area index from 2.5 to 1.5 in the Mediterranean evergreen oak *Q. coc-cifera*, which was growing across a rainfall gradient with changes in canopy structure and plant resistance. The role of the hydraulic resistance in the relative "sensing" of soil water deficit by roots has been emphasized. At a given rate of transpiration and soil water deficit, a plant with high hydraulic resistance will lower its leaf water potential to a greater degree than a plant with low resistance. This plant may be more sensitive to maintain its rate of photosynthesis and growth. On the other hand, with a limited volume of water in the soil, an increase in hydraulic resistance saves water during the wetter periods for use during the drier ones (Turner 1986).

2. Patterns of Changes in $\Delta\Psi$ with Increasing Water Stress

Richter (1976) observed that plant species from sites with pronounced drought periods did not undergo Ψ lower than those of desert plants. His analysis for the former group is largely based on works of Duhme (1974), conducted on 26 species of MTEs. In this study, Duhme measured Ψ of -4.4 MPa for *Q. coccifera*, a similar value to those reported in a synthesis on Mediterranean evergreen oaks (Rambal and Debussche 1995). In this synthesis, whatever the study site and the amount of rainfall during the measurement periods, minimum and predawn leaf water potentials were always higher than -4.4 and -3.8 MPa. Another well-illustrated example of this assumption of lower bound water stress even during every dry year can be found in Griffin (1973). At the end of the driest period, he observed predawn potentials of the evergreen *Q. agrifolia* between -2.5 and -3.1 MPa in the more xeric location. These potentials also remained limited with the deciduous *Q. douglasii* and *Q. lobata*: -3.7 and -2.0 MPa, respectively (see also Damesin and Rambal 1995 for *Q. pubescens* values).

Nevertheless, the trajectories followed by minimum and predawn leaf water potentials to reach their limits were very different according to locations and species. This was particularly true for *Q. ilex* (Figure 3). $\Delta\Psi$ decreased to zero with decreasing soil water availability and predawn potential. As proposed by Ritchie and Hinckley (1975), "it is tempting to compare species based upon these curves. . . . as indicators of species differences." Waring and Cleary (1967) first observed this pattern in the Douglas fir, but it was initially considered as marginal. Indeed, Hickman (1970), from measurements taken from 44 species, concluded that the opposite pattern, in which $\Delta\Psi$ increases with the water stress, is the most common. It corresponds to species characterized as "conformers". Species with the same pattern as our observations were named "regulators." Hickman (1970) suggested "this pattern is probably typical of most plant species in areas with modified (?) Mediterranean climates." It was also described by Oechel et al. (1972) for *Larrea divaricata* and by Aussenac et Valette (1982) for some trees

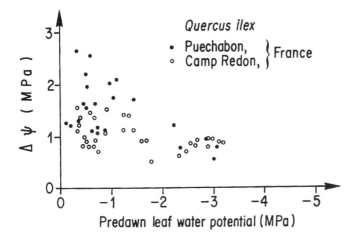

Figure 3 Scatterplot relating predawn leaf water potential to predawn minus minimum leaf water potential for a xeric site (Puéchabon) and a mesic site (Camp Redon) in southern France.

(*Cedrus atlantica, Pinus* sp pl., *Q. pubescens*, and *Q. ilex*) and for the shrub *Buxus sempervirens*.

An important question was asked by Reich and Hinckley (1989): "Does soil-to-leaf hydraulic conductance decrease with decreasing soil moisture due only to increased resistance in the soil, or is there a plant component as well?" Changes in plant resistance under water limitation are attributable to effects on both roots and xylem. As the soil dries, decreased permeability by root suberization and/or increased fine root mortality can reduce the balance between extraction capacity and transpiring leaf area. Xylem loss of vascular transport by cavitation might also cause an increase in plant resistance. There was evidence of embolism formation in Temperate and Mediterranean *Quercus*, one of the most common genera with ring-porous xylem anatomy (Cochard and Tyree 1990). For a 1-year-old twig segment of *Q. ilex*, the loss of conductivity began at -1.8 MPa and linearly increased to reach a total xylem cavitation at -4.35 MPa (Lo Gullo and Salleo 1993). When catastrophic xylem dysfunction occurs, Tyree and Sperry (1988) showed that minor branches begin to die, leading to a loss of leaf area and a reduction in the water flow, which improved water balance of the remaining living stems. Anatomical resistance to the formation and spread of air embolisms in the xylem may be of critical importance. There are a variety of features of xylem anatomy that can increase the "safety" of water conducting systems in Mediterranean species (Carlquist 1989). Vasicentric tracheids adjacent to many vessels act as a subsidiary conducting system and occur in numerous Mediterranean genera such as *Quercus, Arctostaphylos, Phyllirea, Rhus*, and *Banksia*. Vascular tracheids also provide conductive tissues at high stress and are present in drought-deciduous and ericoid evergreen genera such as *Cistus* and *Erica*.

3. Petiole Hydraulic Conductivity

The water relations of the whole tree might be seriously affected by embolism induced by water stress. A large difference in sensitivity has been found between species. Zimmermann (1983) introduced the principle of plant segmentation, stating that embolism should develop first in the terminal part of the trees (i.e., leaves and little branches), thus preserving the other parts of the crown from embolism damage. The risk of xylem dysfunction, especially in the petioles, may determine the ability to resist drought. Are Mediterranean species less vulnerable than other species? There is not much data in the literature on the petioles. Cochard et al. (1992) and Higgs and Woods (1995) compared drought susceptibility by examining hydraulic dysfunction of the xylem vessels in petioles of different oak species. The Mediterranean species (*Q. pubescens* and *Q. cerris*) did not show the lowest vulnerability (or the highest) to embolism formation in comparison to the temperate ones (*Q. petraea, Q. robur*, and *Q. rubra*). Early leaf senescence was also observed near −4 MPa on the Californian deciduous oak *Q. douglasii* (Griffin 1973), whereas all leaves of the deciduous *Q. pubescens* were yellowing at approximately −4.5 MPa in the Languedoc (Damesin and Rambal 1995). There is also some difference between Mediterranean species in their vulnerability to petiole cavitation—partly explained by the distribution of xylem conduit diameter—which can be related to their different distribution within the Mediterranean basin (Salleo and Lo Gullo 1993).

IV. LEAF PERFORMANCES OF SPECIES GROWING IN MTE

A. Assimilation in Relation to Environmental Conditions

1. Net CO_2 Assimilation in Optimal Conditions

The photosynthetic performance of Mediterranean species does not particularly differ from that of species from other biomes (Rambal 1999). For example, for the genus *Quercus*, Damesin (1995) analyzed the literature and found that the Mediterranean species do not differ in their maximum assimilation (mean value of 16.3 μmol m^{-2} s^{-1} for 5 species) from non-Mediterranean species (mean value of 17.2 μmol m^{-2} s^{-1} for 5 species). Numerous studies have been conducted on species of the Californian chaparral and the Chilean matorral (Oechel et al. 1981), on the South African fynbos (Mooney et al. 1983, Van der Heyden and Lewis

1989), and on the shrublands and woodlands around the Mediterranean Sea (Tenhunen et al. 1987, Damesin et al. 1998a). Differences have been demonstrated between growth forms (Oechel et al. 1981) or guilds, and between restoid-ericoid and proteoid species (Van der Heyden and Lewis 1989). A robust relationship between nitrogen content of the leaves and their photosynthetic capacities has often been found in natural vegetation for a wide variety of plants (Field and Mooney 1986). This correlation appears to be a consequence of the limitations on photosynthetic capacity imposed by the levels of the enzyme RuBP carboxylase and of the pigment-protein complexes (Evans 1989, Field and Mooney 1986). Mediterranean species do not deviate from this rule (Field 1991).

2. Response to Water Constraint and Photoprotection

Mediterranean climate leaves must tolerate high irradiance. They must also cope with excess intercepted solar radiation when carbon assimilation is limited either by stomatal closure or a decrease of photosynthetic capacity due to water stress and high temperatures (summer) or low temperatures (winter). Indeed, absorption of light energy may be in excess of that required for carbon fixation and may result in damage to the photosystem. Different physiological regulatory mechanisms have been shown to occur during diurnal cycles to dissipate the excess absorbed energy without any damage to cells: a down-regulation of photosynthesis via a decrease of the photochemical efficiency of PSII (Fv/Fm) (Demmig-Adams et al. 1989, Damesin and Rambal 1995, Faria et al. 1996, Méthy et al. 1996) or via a decrease of chlorophyll content (Kyparissis et al. 1995); a change in the components of the xanthophyll cycle (Demmig et al. 1988); or a high antioxidative potential (Faria et al. 1996). It is difficult to assess if Mediterranean species have more efficient photoprotection mechanisms than temperate (Epron et al. 1992) or tropical species. In particular, it would be interesting to examine if the Mediterranean species' responses to high irradiance are acclimation or adaptation. On that account, a great proportion of the plant species growing in MTEs produce and accumulate aromatic volatile oils that may also serve as an excess energy dissipation system during a period of restricted growth (Ross and Sombrero 1991).

3. Modeling

Functioning at the biochemical level of the photosynthetic system (Farquhar and von Caemmerer 1982) can be summarized by both maximum carboxylation, V_{cmax}, and electron transport rates, J_{max}. Mediterranean species fit in with the general scheme in terms of this functioning, if reference is made to the few Mediterranean species included in the review by Wullschleger (1993), although this investigator did not propose a separate grouping for these species. He observed that they do not deviate significantly from the empirical linear relation between V_{cmax}

and J_{max}. Other data support this trend for Mediterranean oaks and *Arbutus unedo* (Hollinger 1992, Damesin et al. 1998a, Harley and Tenhunen 1991). This leaf photosynthesis model can be next integrated in a canopy level carbon balance model (Hollinger 1992, Sala and Tenhunen 1996). Relations between stomatal conductance (g_s) and assimilation (A) could be simulated following the empirical model proposed by Ball et al. (1987). In this approach, g_F is a dimensionless factor expressing the relation of g_s to A, to relative humidity and to CO_2 partial pressure. The functional dependency of g_F on predawn leaf water potential (Ψ_{pd}) was measured in the field for *Q. ilex* (Sala and Tenhunen 1996) and *Q. pubescens* (Damesin 1996). Both studies showed linear relationships for Ψ_{pd} equal to or lower than -1 MPa during periods of reduced water availability (Figure 4). This relation between g_F and Ψ_{pd} allows us to describe the seasonal patterns of interactions between stomatal conductance and assimilation in the Mediterranean seasonal environment.

B. Assimilation in Relation to Water Loss

One way to estimate water use efficiency (ratio of photosynthesis and transpiration) in C_3 plants is to use leaf carbon isotope composition. Its measurement,

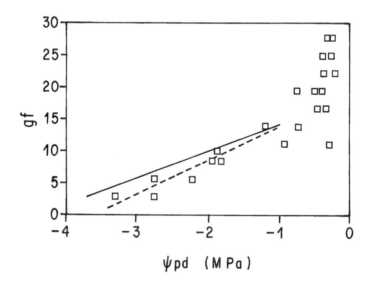

Figure 4   Relationship between g_F factor and predawn leaf water potential of *Quercus pubescens*. Solid line corresponds to the relation obtained by Sala and Tehnunen (1995) for *Q. ilex* in Catalonia (northeast Spain) and dotted line for *Q. pubescens* in the Montpellier region. (*Source*: data from Damesin 1996.)

easier to obtain that those of gaseous exchanges, allows the study of intraspecific and interspecific variability in field conditions (Damesin et al. 1998b).

1. Interspecific Variability

The creation of a superclass for Mediterranean species could be made on the basis of the discrimination (Δ) against $^{13}CO_2$ during the photosynthesis. For the xerophytic woods and scrubs superclass (which includes *Heteromeles arbutifolia*, *Nerium oleander*, and *Eucalyptus socialis*), Lloyd and Farquhar (1994) gave a surprisingly low Δ value of 12.9%. By comparison, this discrimination reaches 18.3% for the superclass cool/cold mixed forest that they proposed, and averages 17.8% for all the C_3 plants. This distinction in term of Δ values implies a segregation of the long-term estimates of the ratio c_i/c_a between the intercellular CO_2 concentration within leaves and the atmospheric CO_2 concentration, and therefore of leaf performance and water use efficiency. This approach has been extended by Beerling and Quick (1995), who used it both for superclasses and at the scale of the individual plant. They estimated V_{cmax} and J_{max} both from Δ and from maximum assimilation. What values of Δ should be adopted for Mediterranean species? From a study conducted at 25 stations in southern France with four species of co-occurring trees or shrubs, we obtained: 18.20 ± 0.65 for *Pinus halepensis*, 18.86 ± 1.53 for *Quercus pubescens*, 19.50 ± 0.52 for *Q. coccifera*, and 19.88 ± 0.68 for *Q. ilex*. Only *P. halepensis* deviated significantly from the three other species. A similar distinction was recorded by Williams and Ehleringer (1996) between *Q. gambelii* and *P. edulis*. Values obtained from Californian oaks (Goulden 1996) and *Q. ilex* at other sites (Fleck et al. 1996) confirm these orders of magnitude. However, all of these values are subject to averaging procedures to overcome the individual variation in Δ (Damesin et al. 1997).

2. Intraspecific Variability

Differential responses of species to local variations of resources can be analyzed through isotopic measurements. For example, Williams and Ehleringer (1996) explained the between-site variability in Δ along a summer monsoon gradient in the southwestern United States using a parameter that integrated both the water balance and the climatic demand throughout the growing season. Similarly, Damesin et al. (1998b) took into account both within- and between-site variability among two deciduous and evergreen Mediterranean oaks using the minimum seasonal leaf predawn potential. This type of response, which tends to optimize water resource utilization, can be extended to entire plant communities growing along a water availability gradient (Stewart et al. 1995).

C. Assimilation in Relation to Leaf Life Span

In Mediterranean communities, evergreen species are more abundant than deciduous ones (Mooney and Dunn 1970, Cody and Mooney 1978, di Castri 1981). As a consequence, ecological studies have largely focused on evergreen and sclerophyllous-leaved species, and less attention was devoted to deciduous species. Nevertheless, Mooney and Dunn (1970) presented a conceptual model to compare the functioning of evergreen chaparral shrubs versus drought-deciduous species of the coastal sage community. The basic assumptions of this model were that the leaves of evergreen sclerophyllous have a relatively low photosynthetic capacity compared with the malacophyllous, drought-deciduous leaves, but can amortize their cost of production over a longer period of time (Mooney and Dunn 1970, Harisson et al. 1971). Although the woody shrub growth-form is an important vegetation element in the MTE, the situation of co-occurrence of winter-deciduous and evergreen trees is not infrequent, at least in California and the Mediterranean basin, where many co-occurring *Quercus* species exhibit the two habits (Hollinger 1992, Damesin et al. 1996, 1998a, 1998b). How do these winter-deciduous species compete with their evergreen neighbors? Research has mainly been oriented toward the mechanisms of seedling installation. However, trees generally have a long life span, and adult survival can thus also determine the abundance of a species in a given region. If the survival of a perennial species depends on its ability to maintain a positive carbon balance over the year (Givnish 1988), the winter-deciduous habit could present two disadvantages compared with the evergreen: (1) it implies a shorter photosynthetically active period, and (2) the active period coincides with the most important constraint imposed by the Mediterranean climate, i.e., the summer drought.

This distinction, recorded by Mooney (1981), is in agreement with a validation of a cost–benefit model at the leaf level (Mooney and Dunn 1970). In this model, deciduous species compensate their lower leaf life span by a higher carbon assimilation by unit of time and a lower leaf production cost. This model was successfully tested for two co-existing Californian oaks (Hollinger 1992), but does not hold for two other oaks co-occurring north of the Mediterranean basin, where *Q. pubescens* has a lower area-based construction cost than *Q. ilex*, but does not have a higher photosynthetic capacity (Damesin et al. 1998a). Despite differences in biochemical composition, size, and mass per unit area, the leaves of the two species respond similarly to water-limited conditions and have similar intrinsic water use efficiencies (Damesin et al 1997, 1998b). These results indicate that key factors distinguishing the functioning of the deciduous species from evergreen ones are more important at a higher level of organization (individual and ecosystem) than at the leaf level.

A high level of secondary metabolites, essential oils, and resins, with their associated volatiles, characterize plants of semiarid regions. Ross and Sombrero

(1991) calculated that 49% of the plant species growing in MTEs produce aromatic volatile oils. These secondary metabolites and volatile organic compounds (VOC) represent an important part of the costs of construction of the leaves. Several studies have stated the role and the evolutionary significance of secondary metabolites (Mabry and Difeo 1973), but not always the emission of VOC (Seufert et al. 1995). Production of VOC shows great individual variability (Bertin and Staudt 1996), and its regulation by environmental conditions must be specified (Loreto et al. 1996, Lerdau et al. 1997).

D. Potential Multieffect of an Increase of CO_2

Anticipating the responses of vegetation to increasing carbon dioxide concentration is critical in attempting to predict the effect of global change on primary production, vegetation spatial distribution, and potential sinks for carbon, which may be provided by terrestrial ecosystems. Results obtained from greenhouse experiments showed that net CO_2 assimilation of *Q. suber* seedlings in conditions of moderate drought was at least twice as great in elevated atmospheric CO_2 conditions, but stomatal conductance was unchanged (Damesin et al. 1996). In addition, shoot and root biomass, stem height, and total leaf area were increased by elevated CO_2, as well as root and stem ramifications. Comparing tree ring chronologies of *Q. ilex* growing in natural CO_2 springs in Italy from those growing in ambient CO_2 control sites, Hättenschwiler et al. (1997) showed that trees grown under high CO_2 for 30 years showed a 12% greater final radial stem width. This stimulation was largely due to responses when trees were young in accordance with the previous cited study.

Increasing CO_2 concentration could affect different leaf functional characteristics. The leaves of *Q. pubescens* grown under elevated CO_2 do not show changes in the number or size of stomata, but do on the size of the guard cells (Miglietta et al. 1995). In contrast, *Q. ilex* leaves (Paoletti and Gellini 1993) and *Pistacia lentiscus* (Peñuelas and Matamala 1990) collected in herbaria showed a reduction of stomatal density with increasing atmospheric CO_2, as shown for temperate species (Woodward 1987).

V. NUTRIENT REGULATION AT INDIVIDUAL AND ECOSYSTEM SCALES

A. Nutrient Uptake and Trophic Types

The five regions supporting MTEs differ largely in their nutrient level (see Section II.B). As a consequence, belowground structures do not show convergence, although the majority of plants form vesicular-arbuscular (VA) mycorrhizae in all regions (Allen et al. 1995). A high proportion of the Proteaceae in South Africa are nonmycotrophic species, and there appear to be no ectomycorrhizal plants. In Australia, nonmycorrhizal species are confined among woody species to root hemiparasites or species bearing proteoid cluster roots (Lamont 1984). The N-fixing plants carry strong root specialization of presumed significance in absorption of P (Pate 1994). The Australian species of Epacridaceae form very peculiar ericoid-type mycorrhizae in extremely nutrient-deficient soils (Read 1992). A high incidence of ectomycorrhizal trees as well as VA mycorrhizal grasses and shrubs characterize the other Mediterranean-type regions. This pattern is probably related to the very old and phosphorous-deficient South African and Australian soils and the relatively high phosphorous Mediterranean, Chilean, and Californian soils (Allen et al. 1995). In Australian nutrient-impoverished heathlands, Pate (1994) emphasized that the mycorrhizal association is just one of many nutrient-acquiring specializations in natural ecosystems, including various forms of parasitism, epiparasitism, and autotrophy with and without mycorrhizal associations.

B. Sclerophylly, Evergreenness, and Nutrient Use Efficiency

The dominance of sclerophyllous—leathery, rigid, and heavily cutinized leaves—evergreen plants in the five MTEs has been interpreted for a long time as a convergent adaptation in response to the unique environmental conditions associated with the Mediterranean climate. Seddon (1974) presented a historical discussion of concepts of sclerophylly and xeromorphy. The possible functional role of sclerophylly has been interpreted in four ways: (1) adaptation to drought was primarily considered; (2) leaf hardness is a epiphenomenon of phosphorous deficiency in soils (Loveless 1961, Monk 1966, Rundel 1988); (3) thick cell walls can better resist negative turgor pressure under water stress (Lo Gullo and Salleo 1988, Salleo et al. 1997); and (4) sclerophylly is an adaptation to herbivory. Correlating scerophyll index (leaf ratio of fiber to nitrogen) with leaf nutrients, Rundel (1988) supported interpretations that leaf nutrition may be a critical factor in the evolution of leaf characteristics. Nevertheless, concerning the regulation of nutrient, it seems that a distinction must be made between sclerophylly and evergreenness or leaf life span. Since the study of Monk (1966), who postulated that the evergreens habit is an adaptation to low nutrient availability, several studies have emphasized that because of their long leaf life spans evergreens have a higher nutrient use efficiency (productivity per unit nutrient uptake) than deciduous species. Recent studies (Pugnaire and Chapin 1993, Aerts 1995 1996, Killingbeck 1996) show that nutrient resorption of nitrogen was significantly lower for evergreens than for deciduous species. In southern France, the proportion of nitrogen in fallen leaves was 78% of that in mature leaves in *Q. ilex* and 44% in *Q. pubescens* (Damesin et al. 1998a). This comparison suggests that the deciduous *Q. pubescens* have a more efficient mechanism for removing nitrogen from senescing leaves. This result is in agreement with that obtained by Del Arco et al. (1991) on five oak species. They found a positive relation between the percentage of nitrogen translocated and the nitrogen concentration of the mature leaves, which was itself negatively related to the life span of the leaves. Leaf longevity appears more important than resorption efficiency as a nutrient-conservation mechanism. Resorption efficiency must be considered as one of the internal mechanisms of nutrient regulation, involving the relative pool sizes of mobile and insoluble nutrients as well as the capacity to store and mobilize energy as carbohydrates and/or lipids. Although substantial differences in seedling storage ability between seeder and resprouter species have been shown in fire-prevalent MTEs of southwestern Australia (Pate et al. 1990), there is actually no strong evidence showing clear differentiation between storage patterns of mature deciduous and evergreen plants in MTEs.

C. Decomposition and Release of Mineral Form

Nitrogen and phosphorous release patterns during leaf litter decomposition varied considerably between species. This section illustrates the importance of growth-form in the nutrient cycling at the ecosystem level, comparing the decomposition patterns of deciduous and evergreen oaks (*Q. pubescens* and *Q. ilex*) in southern France. Gillon et al (1994) showed that nitrogen was released quickly by *Q. ilex*. In this species, the first stage, in which there was a strong decrease in the amount of nitrogen, corresponded to the release of soluble nitrogen (Ibrahima et al. 1995). In contrast, nitrogen was strongly immobilized by *Q. pubescens* litter. Increases in the nitrogen amount immobilized in the litter during decomposition may be partly explained by microbial and fungal incorporation of nitrogen from soil organic matter. This net accumulation of nitrogen in litter during the early stages of decomposition would alter the rates and patterns of nitrogen uptake by trees and may limit tree production. The difference in litter composition is also associated with a difference in the timing of leaf fall. In *Q. ilex* canopies, there are typically two peaks, one in spring and the other in autumn, or sometimes a single peak in spring. In *Q. pubescens* canopies, leaf fall starts in autumn, but since the species is marcescent, there is also some leaf fall in winter. The difference in mineral input to the soil between both species is accentuated by a difference in the intensity of leaching by precipitation. These differences certainly imply difference in the regulation of nitrogen turnover in the forest floor. These results are in agreement with the hypothesis that functional differences between evergreen and deciduous species depend on the management of nutrients, particularly nitrogen (Monk 1966, Moore 1980, Aerts 1995).

VI. SUMMARY

Five regions throughout the world, i.e., the Mediterranean basin itself, California, central Chile, the Cape region of South Africa, and parts of southwestern and southern Australia, are characterized by the same climatic regime, marked by a strong seasonality in temperature and precipitation that leads to a hot drought period in summer and a cool wet period in winter. Their vegetation type presents numerous structural and functional affinities, which has led to the term Mediterranean-type ecosystems. The aim of this chapter was to discuss some aspects of the functioning of the woody Mediterranean species. The main climatic factor controlling the functioning of these ecosystems is water availability, which varies greatly in time and space, imposing strong constraints over the plant. Different mechanisms acting at several scales and levels illustrate how individuals and ecosystems cope with and control water uptake, water loss, and water transfer through the soil–plant–atmosphere continuum.

Carbon assimilation characteristics are presented in relation to environmental conditions. In optimal conditions, the photosynthetic performance of Mediterranean species does not differ particularly from that of species from other biomes. Nevertheless, their leaves must tolerate high irradiance and they must cope with excess intercepted solar radiation when carbon assimilation is limited either by stomatal closure or a decrease of photosynthetic capacity due to water stress and high temperatures (summer) or low temperatures (winter). A great proportion of the plant species growing in MTEs produce and accumulate aromatic volatile oils, which may also serve as an excess energy dissipation system during a period of restricted growth. The functioning of evergreen and deciduous species were presented, and some assumptions based on a cost-benefit model explaining their co-existence were discussed.

The five regions supporting MTEs show very distinct nutrient levels. As a consequence, some differences occur concerning the belowground structures and the role of mycorrhizae. In Australian and South African nutrient-impoverished heathlands, some specific root adaptations were presented. The possible functional role of sclerophylly as a nutrient-conservation mechanism was discussed. Relations between leaf life span, resorption efficiency, and nutrient use efficiency differ between evergreen and deciduous species. Decomposition and release of mineral form have different patterns in these two groups of species. These differences certainly imply difference in the regulation of nitrogen turnover in the forest floor and seem in agreement with the hypothesis that functional differences between evergreen and deciduous species depend on the management of nutrients. In conclusion, the remarkable combination of gradient of natural resources and different anthropic management in the five Mediterranean regions of the world lead to a great diversity of adaptative strategies and functional attributes in the MTEs.

REFERENCES

Acherar, M., S. Rambal, and J. Lepart. 1991. Evolution du potentiel hydrique foliaire et de la conductance stomatique de quatre chênes Méditerranéens lors d'une période de dessèchement. Annales des Sciences Forestières 48: 561–573.

Aerts, R. 1995. The advantages of being evergreen. Trends in Ecology and Evolution 10: 402–407.

Aerts, R. 1996. Nutrient resorption from senescing leaves of perennials: are there general patterns? Journal of Ecology 84: 597–608.

Allen, M.F., S.J. Morris, F. Edwards, and E.B. Allen. 1995. Microbe-plant interactions in Mediterranean-type habitats: shifts in fungal symbiotic and saprophytic functioning in response to global change. In: Moreno, J.M., and W.C. Oechel, eds. Global Change and Mediterranean-type Ecosystems. Ecological Studies 117. Berlin: Springer-Verlag, pp 287–305.

Aschmann, H. 1973. Distribution and peculiarity of Mediterranean ecosystems. In: di Castri, F., and H.A. Mooney, eds. Mediterranean-type Ecosystems: Origin and Structure. Ecological Studies 7. Berlin: Springer-Verlag, pp 11–19.

Aussenac, G., and J.C. Valette. 1982. Comportement hydrique estival de *Cedrus atlantica* Manetti, *Quercus ilex* L. et *Quercus pubescens* Willd. et de divers pins dans le Mont Ventoux. Annales des Sciences Forestières 39: 41–62.

Bagnouls, F., and H. Gaussen. 1953. Saison sèche et indice xérothermique. Bulletin Société d'Histoire Naturelle de Toulouse 88: 192–239.

Ball, J.T., I.E. Woodrow, and J.A. Berry. 1987. A model predicting stomatal conductance and its contribution to the control of photosynthesis under different environmental conditions. In: Biggins, I.J., ed. Progress in Photosynthesis Research, vol IV.5. Dordrecht, The Netherlands: Martinus Nijhoff, pp 221–224.

Beerling, D.J., and W.P. Quick. 1995. A new technique for estimating rates of carboxylation and electron transport in leaves of C_3 plants for use in dynamic global vegetation models. Global Change Biology 1: 289–294.

Bertin, N., and M. Staudt. 1996. Effect of water stress on monoterpene emissions from young potted holm oak (*Quercus ilex* L.) trees. Oecologia (Berlin) 107: 456–462.

Blondel, J., and J. Aronson. 1995. Biodiversity and ecosystem function in the Mediterranean basin: human and non-human determinants. In: Davis, G.W., and D.M. Richardson, eds. Mediterranean-type Ecosystems: The Function of Biodiversity. Ecological Studies 109. Berlin: Springer-Verlag, pp 43–119.

Brown, S. 1981. A comparison of the structure, primary productivity, and transpiration of cypress ecosystems in Florida. Ecological Monographs 51: 403–427.

Canadell, J., R.B. Jackson, J.R. Ehleringer, H.A. Mooney, O.E. Sala, and E-D. Schulze.

1996. Maximum rooting depth of vegetation types at the global scale. Oecologia (Berlin) 108: 583–595.

Carlquist, S. 1989. Adaptative wood anatomy of chaparral shrubs. In: Keeley, S.C., ed. The California Chaparral: Paradigms Reexamined. Los Angeles: Natural History Museum of Los Angeles County, Science Series 34, pp 25–35.

Cochard, H., N. Bréda, A. Granier, and G. Aussenac. 1992. Vulnerability to air embolism of three European oak species (*Quercus petraea* (Matt) Liebl, *Q. pubescens* Willd, *Q. robur* L). Annales des Sciences Forestières 49: 225–233.

Cochard, H., and M.T. Tyree. 1990. Xylem dysfunction in *Quercus*: vessel sizes, tyloses, cavitation and seasonal changes in embolism. Tree Physiology 6: 393–407.

Cody, M.L., and H.A. Mooney. 1978. Convergence versus non convergence in Mediterranean-climate ecosystems. Annual Review of Ecology and Systematics 9: 265–321.

Conrad, V. 1941. The variability of precipitation. Monthly Weather Review 69: 5–11.

Cowan, I.R., and G.D. Farquhar. 1977. Stomatal function in relation to leaf metabolism and environment. Symposium Society of Experimental Biology 31: 471–505.

Daget, P. 1977. Le bioclimat méditerranéen, caractères généraux, modes de caractérisation. Vegetatio 34: 1–20.

Damesin, C. 1996. Relations hydriques, photosynthèse et efficacité d'utilisation de l'eau chez deux chênes méditerranéens caduc et sempervirent cooccurrents. Thesis, Université Paris XI, Orsay.

Damesin, C., and S. Rambal. 1995. Field study of leaf photosynthetic performance by a Mediterranean deciduous oak tree (*Quercus pubescens*) during a severe summer drought. New Phytologist 131: 159–167.

Damesin, C., C. Galera, S. Rambal, and R. Joffre. 1996. Effects of elevated carbon dioxide on leaf gas exchange and growth of cork-oak (*Quercus suber* L.) seedlings. Annales des Sciences Forestières 53: 461–467.

Damesin, C., S. Rambal, and R. Joffre. 1997. Between-tree variations in leaf d13C of *Quercus pubescens* and *Quercus ilex* among Mediterranean habitats with different water availability. Oecologia (Berlin) 111: 26–35.

Damesin, C., S. Rambal, and R. Joffre. 1998a. Cooccurrence of trees with different leaf habit: a functional approach on Mediterranean oaks. Acta Oecologica 19: 195–204.

Damesin, C., S. Rambal, and R. Joffre. 1998b. Seasonal and between-year changes in leaf d13C in two co-occurring Mediterranean oaks: relations to leaf growth and drought progression. Functional Ecology 12: in press.

Davis, E.A., and C.P. Pase. 1977. Root system of shrub live oak: implication for water yield in Arizona chaparral. Journal of Soil and Water Conservation 32: 174–180.

Davis, G.W., D.M. Richardson, J.E. Keeley, and R.J. Hobbs. 1996. Mediterranean-type ecosystems: the influence of biodiversity on their functioning. In: Mooney, H.A., J.H. Cushman, E. Medina, O.E. Sala, and E-D. Schulze, eds. Functional Roles of Biodiversity: A Global Perspective. London: Wiley, pp 151–183.

Davis, S.D., and H.A. Mooney. 1986. Tissue water relations of four co-occurring chaparral shrubs. Oecologia (Berlin) 70: 527–535.

Del Arco, J.M., A. Escudero, and M.M. Vega Garrido. 1991. Effects of site characteristics on nitrogen retranslocation from senescing leaves. Ecology 72: 701–708.

De Lillis, M., and G.C. Sun. 1990. Stomatal response patterns and photosynthesis of dominant trees in a tropical monsoon forest. Acta Oecologica 11: 545–555.

Demmig, B., K. Winter, A. Krüger, and F-C. Czygan. 1988. Zeaxanthin and the heat dissipation of excess light energy in *Nerium oleander* exposed to a combination of high light and water stress. Plant Physiology 87: 17–24.

Demmig-Adams, B., W.W. Adams III, K. Winter, A. Meyer, U. Schreiber, J.S. Pereira, A. Krüger, F-C. Czygan, and O.L. Lange. 1989. Photochemical efficiency of photosystem II, photon yield of O_2 evolution, photosynthetic capacity, and carotenoid composition during the midday depression of net CO_2 uptake in *Arbutus unedo* growing in Portugal. Planta 177: 377–387.

di Castri, F. 1981. Mediterranean-type shrubland of the world In: di Castri, F., D.W. Goodall, and R.L. Specht. Mediterranean-type Shrublands. Ecosystems of the World, vol. 11. Amsterdam, The Netherlands: Elsevier, pp 1–52.

Duhme, F. 1974. Die Kennzeichnung des ökologischen Konstitution von Gehölzen im Hinblick auf den Wasserhaushalt. Dissertationes Botanicae 28: 1–143.

Emberger, L. 1955. Une classification biogéographique des climats. Recueil des Travaux du Laboratoire de Botanique et Géologie. Faculté des Sciences de Montpellier 7: 3–43.

Epron, D., E. Dreyer, and N. Bréda. 1992. Photosynthesis of oak trees (*Quercus petraea* Liebl.) during drought under field conditions: diurnal course of net CO_2 assimilation and photochemical efficiency of photosystem II. Plant, Cell and Environment 15: 809–820.

Escudero, A., J.M. del Arco, I.C. Sanz, and J. Ayala. 1992. Effects of leaf longevity and retranslocation efficiency on the retention time of nutrients in the leaf biomass of different woody species. Oecologia (Berlin) 90: 80–87.

Evans, J.R. 1989. Photosynthesis and nitrogen relationships of C_3 plants: Oecologia (Berlin) 78: 9–19.

Faria, T., J.I. Garcia-Plazaola, A. Abadia, S. Cerasoli, J.S. Pereira, and M.M. Chaves. 1996. Diurnal changes in photoprotective mechanisms in leaves of cork oak (*Quercus suber*) during summer. Tree Physiology 16: 115–123.

Farquhar, G.D., and S. von Caemmerer. 1982. Modeling of photosynthetic response to environmental conditions. In: Lange, O.L., P.S. Nobel, C.B. Osmond, and H. Ziegler, eds. Encyclopedia of Plant Physiology, vol. 12B. Berlin: Springer-Verlag, pp 549–587.

Field, C. 1991. Ecological scaling of carbon gain to stress and resource availability. In: Mooney, H.A., W.E. Winner, and E.J. Pell, eds. Response of Plant to Multiple Stresses. San Diego Academic Press, pp 35–65.

Field, C., and H.A. Mooney. 1986. The photosynthesis-nitrogen relationship in wild plant. In: Givnish, T.J., ed. On the Economy of Plant Form and Function. Cambridge: Cambridge University Press, pp 25–35.

Fleck, I., D. Grau, M. Sanjosé, and D. Vidals. 1996. Carbon isotope discrimination in *Quercus ilex* resprouts after fire and tree-fell. Oecologia (Berlin) 105: 286–292.

Gale, M.R., and D.F. Grigal. 1987. Vertical root distributions of northern tree species in relation to successional plants. Canadian Journal of Forest Research 17: 829–834.

Gardner, W.R. 1964. Relation of root distribution to water uptake and availability. Agronomy Journal 56: 41–45.

Gholz, H.L. 1982. Environmental limits on aboveground net primary production, leaf area, and biomass in vegetation zones of the Pacific Northwest. Ecology 63: 469–481.

Gillon, D., R. Joffre, and A. Ibrahima. 1994. Initial litter properties and decay rate: a microcosm experiment on Mediterranean species. Canadian Journal of Botany 72: 946–954.

Givnish, T.J. 1988. Adaptation to sun and shade: a whole plant perspective. Australian Journal of Plant Physiology 15: 63–92.

Gollan, T., N.C. Turner, and E-D. Schulze. 1985. The responses of stomata and leaf gas exchange to vapour pressure deficits and soil water content. III. In the sclerophyllous woody species *Nerium oleander*. Oecologia (Berlin) 65: 356–362.

Goulden, M.L. 1996. Carbon assimilation and water-use efficiency by neighboring Mediterranean-climate oaks that differ in water access. Tree Physiology 16:417–424.

Griffin, J.R. 1973. Xylem sap tension in three woodland oaks of central California. Ecology 54: 152–159.

Groves R.H., J.S. Beard, H.J. Deacon, J.J.N. Lambrechts, A. Rabinovitch-Vin, R.L. Specht, and W.D. Stock. 1983. Introduction: the origins and characteristics of mediterranean ecosystems. In: Day, J.A., ed. Mineral Nutrients in Mediterranean Ecosystems. South African National Scientific Programmes Report 71. Pretoria, South Africa: Council for Scientific and Industrial Research, pp 1–17.

Harley, P.C., and J.D. Tenhunen. 1991. Modeling the photosynthetic response of C_3 leaves to environmental factors. In: Boote, K.J., and R.S. Loomis, eds. Modeling Crop Photosynthesis from Biochemistry to Canopy, vol. 19. Madison, WI: Crop Science Society of America, pp 17–39.

Hättenschwiler S., F. Miglietta, A. Raschi, and C. Körner. 1997. Thirty years of in situ growth under elevated CO_2: a model for future responses? Global Change Biology 3: 463–471.

Helmers, H., J.S. Horton, G. Juhren, and J. O'Keefe. 1955. Root systems of some chaparral plants in southern California. Ecology 36: 667–678.

Hickman, J.C. 1970. Seasonal course of xylem sap tension. Ecology 51: 1052–1056.

Higgs, K.H., and V. Wood. 1995. Drought susceptibility and xylem dysfunction in seedlings of 4 European oak species. Annales des Sciences Forestières 52: 507–513.

Hinckley, T.M., R.G. Aslin, R.R. Aubuchon, C.L. Metcalf, and J.E. Roberts. 1978. Leaf conductance and photosynthesis in four species of the oak-hickory forest type. Forest Science 24: 73–84.

Hinckley, T.M., F. Duhme, A.R. Hinckley, and H. Richter. 1983. Drought relations of shrub species: assessment of the mecanism of drought resistance. Oecologia (Berlin) 59: 344–350.

Hobbs, R.J., D.M. Richardson, and G.W. Davis. 1995. Mediterranean-type ecosystems: opportunities and constraints for studying the function of biodiversity. In: Davis, G.W., and D.M. Richardson, eds. Mediterranean-type Ecosystems: The Function of Biodiversity. Ecological Studies 109. Berlin: Springer-Verlag, pp 1–32.

Hollinger, D.Y. 1992. Leaf and simulated whole-canopy photosynthesis in two co-occurring tree species. Ecology 73: 1–14.

Ibrahima, A., R. Joffre, and D. Gillon. 1995. Changes in litter during the initial leaching phase: an experiment on the leaf litter of Mediterranean species. Soil Biology Biochemistry 27: 931–939.

Jackson, R.B., J. Canadell, J.R. Ehleringer, H.A. Mooney, O.E. Sala, and E-D. Schulze.

1996. A global analysis of root distributions for terrestrial biomes. Oecologia (Berlin) 108: 389–411.

Keeley, S.C., ed. 1989. The California Chaparral: Paradigms Reexamined Los Angeles: Natural History Museum of Los Angeles County, Science Series 34.

Killick, D.J.B. 1979. African mountain heathlands. In: Specht, R.L., ed. Heathlands and Related Shrubland of the World. Amsterdam, The Netherlands: Elsevier, pp 97–116.

Killingbeck, K.T. 1996. Nutrients in senesced leaves: keys to the search for potential resorption and resorption proficiency. Ecology 77: 1716–1727.

Köppen, W. 1931. Die Klimate der Erde, Grundriss der Klimakunde. 2nd ed. Berlin, Leipzig: De Gruyter.

Kruger, F.J., D.T. Mitchell, and J.U.M. Jarvis, eds. 1983. Mediterranean-type Ecosystems: The Role of Nutrients. Ecological Studies 43. Berlin: Springer-Verlag.

Kummerow, J.D., D. Krause, and W. Jow. 1977. Root systems of chaparral shrubs. Oecologia (Berlin) 29: 163–177.

Kyparissis A, Y. Petropoulou, and Y. Manetas. 1995. Summer survival of leaves in a soft-leaved shrub (*Phlomis fruticosa* L., Labiatae) under Mediterranean field conditions: avoidance of photoinhibitory damage through decreased chlorophyll contents. Journal of Experimental Botany 46: 1825–1831.

Ladiges, P.Y., and D.H. Ashton. 1974. Variation in some central Victorian populations of *Eucalyptus viminalis* Labill. Australian Journal of Botany 22: 81–102.

Lamont, B.B. 1984. Specialized modes of nutrition. In: Pate, J.S., and J.S. Beard, eds. Kwongan—Plant Life of the Sandplain. Nedlands, Australia: University of Western Australia, pp 236–245.

Lamont, B.B. 1994. Mineral nutrient relations in Mediterranean regions of California, Chile, and Australia. In: Arroyo, M.T.K., P.H. Zedler, and M.D. Fox, eds. Ecology and Biogeography of Mediterranean Ecosystems in Chile, California, and Australia. Ecological Studies 108. Berlin: Springer-Verlag, pp 211–238.

Le Houérou, H.N. 1990. Global change: vegetation, ecosystems and land use in the southern Mediterranean basin by the mid twenty-first century. Israel Journal of Botany 39: 481–508.

Le Houérou, H.N. 1992. Relations entre la variabilité des précipitations et celle des productions primaire et secondaire en zone aride. In: Le Floc'h, E., M. Grouzis, and A. Cornet, eds. L'aridité: Une Contrainte au Développement. Paris: ORSTOM, pp 197–220.

Lerdau M., A. Guenther, and R. Monson. 1997. Plant production and emission of volatile organic compounds. Bioscience 47: 373–383.

Lloyd, J., and G.D. Farquhar. 1994. ^{13}C discrimination during CO_2 assimilation by the terrestrial biosphere. Oecologia (Berlin) 99: 201–215.

Lo Gullo, M.A., and S. Salleo. 1988. Different strategies of drought resistance in three Mediterranean sclerophyllous trees growing in the same environmental conditions. New Phytologist 108: 267–276.

Lo Gullo, M.A., and S. Salleo. 1993. Different vulnerabilities of *Quercus ilex* L. to freeze and summer drought-induced xylem embolism: an ecological interpretation. Plant, Cell and Environment 16: 511–519.

Loreto, F., P. Ciccioli, A. Cecinato, E. Brancaleoni, M. Frattoni, and D. Tricoli. 1996.

Influence of environmental factors and air composition on the emission of α-pinene from *Quercus ilex* L. leaves. Plant Physiology 110: 267–275.

Loveless, A.R. 1961. A nutritional interpretation of sclerophylly based on difference in the chemical composition of sclerophyllous and mesophytic leaves. Annals of Botany 25: 168–184.

Mabry, T.J., and D.R. Difeo. 1973. The role of the secondary plant chemistry in the evolution of the Mediterranean scrub vegetation. In: di Castri, F., and H.A. Mooney, eds. Mediterranean Type Ecosystems: Origin and Structure. Ecological Studies 7. Berlin: Springer-Verlag, pp 121–155.

Méthy, M., C. Damesin, and S. Rambal. 1996. Drought and photosystem II activity in two Mediterranean oaks. Annales des Sciences Forestières 53: 255–263.

Miglietta, F., M. Badiani, I. Bettarini, P. van Gardingen, F. Selvi, and A. Raschi. 1995. Preliminary studies of the long-term CO_2 response of Mediterranean vegetaion around natural CO_2 events. In: Moreno, J.M., and W.C. Oechel, eds. Global Change and Mediterranean-type Ecosystems. Ecological Studies 117. Berlin: Springer-Verlag, pp 102–120.

Monk, C.D. 1966. An ecological significance of evergreenness. Ecology 47: 504–505.

Mooney, H.A. 1981. Primary production in Mediterranean-climate regions. In: di Castri, F., D.W. Goodall, and R.L. Specht, eds. Mediterranean-type Shrublands. Ecosystems of the World, vol. 11. Amsterdam, The Netherlands: Elsevier, pp 249–255.

Mooney, H.A. 1989. Chaparral physiological ecology—paradigms revisited. In: Keeley, S.C., ed. The California Chaparral: Paradigms Reexamined. Los Angeles: Natural History Museum of Los Angeles County, Science Series 34, pp 85–90.

Mooney, H.A., and E.L. Dunn. 1970. Photosynthetic systems of Mediterranean-climate shrubs and trees of California and Chile. American Naturalist 104: 447–453.

Mooney, H.A., C. Field, S.L. Gulmon, P. Rundel, and F.J. Kruger. 1983. Photosynthetic characteristics of South African sclerophylls. Oecologia (Berlin) 58:398–401.

Moore, P. 1980. The advantages of being evergreen. Nature 285: 535.

Moreno, J.M., and W.C. Oechel, eds. 1994. The Role of Fire in Mediterranean-type Ecosystems. Ecological Studies 107. Berlin: Springer-Verlag.

Mulkey, S.S., K. Kitajima, S.J. Wright. 1996. Plant physiological ecology of tropical forest canopies. Trends in Ecology and Evolution 11: 408–412.

Naveh, Z. 1967. Mediterranean ecosystems and vegetation types in California and Israel. Ecology 48: 345–359.

Ne'eman, G. 1993. Variation in leaf phenology and habit in *Quercus ithaburensis*, a Mediterranean deciduous tree. Journal of Ecology 81:627–634.

Oechel, W.C., B.R. Strain, and W.R. Odening. 1972. Photosynthesis rates of a desert shrub, *Larrea divaricata* Cav., under field conditions. Photosynthetica 6:183–188.

Oechel, W.C., W. Lawrence, J. Mustafa, and J. Martinez. 1981. Energy and carbon acquisition. In: Miller, P.C., ed. Resource Use by Chaparral and Matorral. A Comparison of Vegetation Function in Two Mediterranean-type Ecosystems. Ecological Studies 39. Berlin: Springer-Verlag, pp 151–182.

Paoletti, E., and R. Gellini. 1993. Stomatal density variation in beech and holm oak leaves collected over the past 200 years. Acta Oecologica 14: 173–178.

Passioura, J.B. 1976. Physiology of grain yield in wheat growing on stored water. Australian Journal of Plant Physiology 3:559–565.

Pate, J.S. 1994. The mycorrhizal association: just one of many nutrient acquiring specializations in natural ecosystems. Plant and Soil 159: 1–10.

Pate, J.S., R.H. Froend, B.J. Bowen, A. Hansen, and J. Kuo. 1990. Seedling growth and storage characteristics of seeder and resprouter species of Mediterranean-type ecosystems of S.W. Australia. Annals of Botany 65: 585–601.

Peñuelas, J., and R. Matamala. 1990. Changes in N and S leaf content, stomatal density and specific leaf area of 14 plant species during the last three centuries of CO_2 increase. Journal of Experimental Botany 41: 1119–1124.

Poole, D.K., and P.C. Miller. 1975. Water relations of selected species of chaparral and coastal sage communities. Ecology 56: 1118–1128.

Poole, D.K., and P.C. Miller. 1981. The distribution of plant water stress and vegetation characteristics in southern California chaparral. American Midland Naturalist 105: 32–43.

Pugnaire, F.I., and F.S. Chapin III. 1993. Controls over nutrient resorption from leaves of evergreen Mediterranean species. Ecology 74: 124–129.

Rambal, S. 1984. Water balance and pattern of root water uptake by a *Quercus coccifera* L. evergreen scrub. Oecologia (Berlin) 62: 18–25.

Rambal, S. 1992. *Quercus ilex* facing water stress: a functional equilibrium hypothesis. Vegetatio 99–100: 147–153.

Rambal, S. 1993. The differential role of mechanisms for drought resistance in a Mediterranean evergreen shrub: a simulation approach. Plant, Cell and Environment 16: 35–44.

Rambal, S. 1995. From daily transpiration to seasonal water balance: an optimal use of water? In: Roy, J., J. Aronson, and F. di Castri, eds. Times Scales of Biological Responses to Water Constraints. Amsterdam, The Netherlands: SPB Academic Publishing, pp 37–51.

Rambal, S. 1999. Spatial and temporal variations in productivity of Mediterranean-type ecosystems: a hierarchical perspective. In: Mooney, H.A., B. Saugier, and J. Roy, eds. Terrestrial Global Productivity: Past, Present, Future. San Diego: Academic Press, in press.

Rambal, S. and J. Leterme. 1987. Changes in aboveground structure and resistances to water uptake in *Quercus coccifera* along a rainfall gradient. In: Tehnunen, J.D., F.M. Catarino, O.L. Lange, and W.D. Oechel, eds. Plant Response to Stress. Functionnal Analysis in Mediterranean Ecosystems. NATO-ASI series Vol. G15. Berlin: Springer-Verlag, pp 191–200.

Rambal, S., and G. Debussche. 1995. Water balance of Mediterranean ecosystems under a changing climate. In: Moreno, J.M., and W.C. Oechel, eds. Global Change and Mediterranean-type Ecosystems. Ecological Studies 117. Berlin: Springer-Verlag, pp 386–407.

Rambal, S., C. Damesin, R. Joffre., M. Methy, and D. Lo Seen. 1996. Optimization of carbon gain in canopies of Mediterranean evergreen oaks. Annales des Sciences Forestières 53: 547–560.

Read, D.J. 1992. The mycorrhizal mycelium. In: Allen, M.F., ed. Mycorrhizal Functioning. London: Chapman and Hall, pp 102–133.

Reich, P.B., and T.M. Hinckley. 1989. Influence of pre-dawn water potential and soil-to-leaf hydraulic conductance on maximum daily leaf diffusive conductance in two oak species. Functional Ecology 3: 719–726.

Richter, H. 1976. The water status in the plant. Experimental evidence. In: Lange, O.L., L. Kappen, and E-D. Schulze, eds. Water in Plant Life. Ecological Studies 19. Berlin: Springer-Verlag, pp 42–58.

Ritchie, G.A., and T.M. Hinckley. 1975. The pressure chamber as an instrument for ecological research. Advances in Ecological Research 9: 165–254.

Ross, J.D., and C. Sombrero. 1991. Environmental control of essential oil production in Mediterranean plants. In: Harborne, J.B., and F.A. Tomas-Barberan, eds. Ecological Chemistry and Biochemistry of Plant Terpenoids. Oxford: Clarendon Press, pp 83–94.

Rundel, P.W. 1988. Leaf structure and nutrition in Mediterranean-climate sclerophylls. In: Specht, R.L., ed. Mediterranean-type Ecosystems: A Data Source Book. Dordrecht, The Netherlands: Kluwer, pp 157–167.

Rundel, P.W. 1995 Adaptative significance of some morphological and physiological characteristics in Mediterranean plants: facts and fallacies. In: Roy, J., J. Aronson, and F. di Castri, eds. Time Scales of Biological Responses to Water Constraints. Amsterdam, The Netherlands: SPB Academic Publishing, pp 119–139.

Rundel, P.W., and J.L. Vankat. 1989. Chaparral communities and ecosystems. In: Keeley, S.C., ed. The California Chaparral: Paradigms Reexamined. Los Angeles: Natural History Museum of Los Angeles County, Science Series 34, pp 127–139.

Sala, A., and J.D. Tenhunen. 1996. Simulations of canopy net photosynthesis and transpiration in *Quercus ilex* L. under the influence of seasonal drought. Agricultural and Forest Meteorology 78: 203–222.

Salleo, S., and M.A. Lo Gullo. 1993. Drought resistance strategies and vulnerability to cavitation of some Mediterranean sclerophyllous trees. In: Borghetti, M., J. Grace, and A. Raschi, eds. Water Transport in Plants Under Climatic Stress. Cambridge: Cambridge University Press., pp 99–113.

Salleo, S., A. Nardini, and M.A. Lo Gullo. 1997. Is sclerophylly of Mediterranean evergreens an adaptation to drought? New Phytologist 135: 603–612.

Schimper, A.F.W. 1903. Plant-Geography upon a Physiological Basis. Oxford: Clarendon Press.

Seddon, G. 1974. Xerophytes, xeromorphs and sclerophylls: the history of some concepts in ecology. Biological Journal of the Linnean Society 6: 65–87.

Seufert, G., D. Kotzias, C. Spartà, and B. Versino. 1995. Volatile organics in Mediterranean shrubs and their potential role in a changing environment. In: Moreno, J.M., and W.C. Oechel, eds. Global Change and Mediterranean-type Ecosystems. Ecological Studies 117. Berlin: Springer-Verlag, pp 343–370.

Specht, R.L. 1969a. A comparison of the sclerophyllous vegetation characteristic of Mediterranean type climates in France, California, and Southern Australia I. Structure, morphology, and succession. Australian Journal of Botany 17: 277–292.

Specht, R.L. 1969b. A comparison of the sclerophyllous vegetation characteristic of Mediterranean type climates in France, California, and southern Australia. II. Dry matter, energy, and nutrient accumulation. Australian Journal of Botany 17: 293–308.

Specht, R.L. 1979 Heathlands and related heathlands of the world. In: Specht, R.L., ed. Heathlands and Related Shrubland of the World. Amsterdam, The Netherlands: Elsevier, pp 1–18.

Specht, R.L., and Specht, A. 1989. Canopy structure in Eucalyptus-dominated communities in Australia along climatic gradients. Acta Oecologica, Oecologia Plantarum 10: 191–213.

Stewart, G.R., M.H. Turnbull, S. Schmidt, and P.D. Erskine. 1995. ^{13}C natural abundance in plant communities along a rainfall gradient: a biological integrator of water availability. Australian Journal of Plant Physiology 22: 51–55.

Stone, E.L., and P.J. Kalisz. 1991. On the maximum extent of tree roots. Forest Ecology and Management 46: 59–102.

Talsma T., and E.A. Gardner. 1986. Soil water extraction by mixed Eucalyptus forest during a drought period. Australian Journal of Soil Research 24: 25–32.

Tardieu F., and W.J. Davies. 1993. Integration of hydraulic and chemical signalling in the control of stomatal regulation and water status of droughted plants. Plant, Cell and Environment 16: 341–349.

Tenhunen J.D., O.L. Lange, J. Gebel, W. Beyschlag, and J.A. Weber. 1984. Changes in photosynthetic capacity, carboxylation efficiency, and CO_2 compensation point associated with midday stomatal closure and midday depression of net CO_2 exchange of leaves of *Quercus suber*. Planta 162: 193–203.

Tenhunen J.D., O.L. Lange, P.C. Harley, W. Beyschalg, and A. Meyer. 1985. Limitations due to water stress on leaf net photosynthesis of *Quercus coccifera* in the Portuguese evergreen scrub. Oecologia (Berlin) 67: 23–30.

Tenhunen J.D., W. Beyschlag, O.L. Lange, and P.C. Harley. 1987. Changes during summer drought in leaf CO_2 uptake rates of macchia shrubs growing in Portugal: limitation due to photosynthetic capacity, carboxylation efficiency, and stomatal conductance. In: Tehnunen, J.D., F.M. Catarino, O.L. Lange, and W.D. Oechel, eds. Plant Response to Stress. Functionnal Analysis in Mediterranean Ecosystems. NATO ASI series, Vol. G15. Berlin: Springer-Verlag, pp 305–327.

Turner, N.C. 1986. Crop water deficits: a decade of progress. Advances in Agronomy 39: 1–51.

Turner, N.C., E-D. Schulze, and T. Gollan. 1985. The responses of stomata and leaf gas exchange to vapour pressure deficits and soil water content II. In the mesophytic herbaceous species *Helianthus annuus*. Oecologia (Berlin) 65: 348–355.

Tyree, M.T., and J.S. Sperry. 1988. Do woody plants operate near the point of catastrophic xylem dysfunction caused by dynamic water stress? Answers from a model. Plant Physiology 88: 574–580.

Van der Heyden, F., and O.A.M. Lewis. 1989. Seasonal variation in photosynthetic capacity with respect to plant water status of five species of the Mediterranean climate region of South Africa. South African Journal of Botany 55: 509–515.

Waggoner, P.E. 1989. Anticipating the frequency distribution of precipitation if the climate change alters its mean. Agricultural and Forest Meteorology 47: 321–337.

Waring, R.H., and B.D. Cleary. 1967. Plant moisture stress: evaluation by pressure bomb. Science 155: 1248, 1953–1954.

Warming, E. 1909. Oecology of Plants. Oxford: Clarendon Press.

Wigley, T.M.L. 1985. Impact of extreme events. Nature 316: 106–107.

William, W.E. 1983. Optimal water-use efficiency in a California shrub. Plant, Cell and Environment 6: 145–151.

Williams, D.G., and J.R. Ehleringer. 1996. Carbon isotope discrimination in three semi-arid woodland species along a monsoon gradient. Oecologia (Berlin) 106: 455–460.

Woodward, F.I. 1987. Stomatal numbers are sensitive to increases in CO_2 from pre-industrial levels. Nature 127:617–618.

Wullschleger, S.D. 1993. Biochemical limitations to carbon assimilation in C3 plants—a retrospective analysis of the A-c_i curves from 109 species. Journal of Experimental Botany 44: 907–920.

Zedler, P.H. 1994. Plant life history and dynamic specialization in the chaparral/coastal sage shrub flora in southern California. In: Arroyo, M.T.K., P.H. Zedler, and M.D. Fox, eds. Ecology and Biogeography of Mediterranean Ecosystems in Chile, California, and Australia. Ecological Studies 108. Berlin: Springer-Verlag, pp 89–115.

Zhang, J., U. Schurr, and W.J. Davies. 1987. Control of stomatal behaviour by abscissic acid which apparently originates in the roots. Journal of Experimental Botany 38: 1174–1181.

Zimmerman, M.H. 1983. Xylem Structure and the Ascent of Sap. Berlin: Springer-Verlag.

10

Plant Survival in Arid Environments

Juan Puigdefábregas and Francisco I. Pugnaire
Arid Zones Experimental Station, Spanish Council for Scientific Research, Almería, Spain

I. INTRODUCTION

In the broadest sense, drylands include areas where rainfall does not meet the atmospheric evaporative demand. Drylands occupy approximately 45% of the terrestrial land surface, contain about 30% of the world's total carbon in above- and belowground biomass (Allen-Diaz et al. 1996), and include grasslands, shrublands, savannas, xerophytic woodlands, and hot and cold deserts. Drylands support approximately 50% of the world's livestock and provide forage for both domestic and wildlife.

In most cases, the evolution of dryland ecosystems has been driven by the episodic depletion of plant biomass, whether caused by drought, fire, or grazing. Therefore, most of these ecosystems are, in a sense, pre-adapted to be exploited, and humans have taken advantage of this feature throughout history (Margalef 1974).

Rainfall variability is a driving force in the dynamics of dryland ecosystems. Small changes in temporal patterns of rainfall interact with fire or grazing factors, and their effects may overrun changes in average values, as incorporated in current climate change scenarios (Allen-Diaz et al. 1996). Dryland ecosystems are probably more resilient to current climate variability than it was thought years ago because of a combination of the opportunistic behavior of their species guilds and the wide range of buffering mechanisms available to them (Puigdefabregas 1998).

This chapter focuses on the adaptations of plants to survive in water-limited environments. The Section II deals with primary adaptations of individual species in terms of the physiology of water relations and canopy architecture. Section III discusses secondary adaptations, those that emerge as side effects of the former, such as the implications of canopy architecture on rainfall partitioning and feedbacks developed by the spatial patterns of plant distribution. Finally, Section IV explores the adaptive value of plant community changes in dryland ecosystems.

II. PRIMARY ADAPTATIONS

Water is the most limiting factor for plant growth in drylands. Plants deal with water stress by several means and have developed tolerance mechanisms that allow them to survive and grow under very harsh circumstances (Lewitt 1980). To gain carbon, plants must necessarily lose water, because the diffusion gradient of water vapor from the leaf to the atmosphere is steeper than the gradient of CO_2 from the atmosphere to the leaf.

Water stress occurs when turgor pressures fall below the maximum potential pressure (Fitter and Hay 1987, Osmond et al. 1987), and the magnitude of such stress is determined by the extent and duration of the deprivation. Therefore, plant responses depend on the length of the water shortage and include physiological responses to short-term changes and adaptations to several levels of water availability. We focus here on long-term adaptations to drought, including physiological mechanisms, biomass allocation, and morphological modifications.

A. Plant Responses to Water Stress: Physiological Mechansisms

Plants may escape drought by either acquiring permanent access to water sources or by restricting their activities to periods of water availability. Large allocations to roots is an important characteristic of plants from dry environments, where root systems typically consist of 60–80% (Caldwell and Richards 1986) and up to 90% of plant biomass (Puigdefábregas et al. 1996). As water availability decreases, root growth may be enhanced at the cost of aboveground productivity (Chapin 1980, Wan et al. 1993), and severe drought may also promote initiation and elongation of lateral roots (Jupp and Newman 1987).

Most dryland species have a dual root system, with a layer of superficial roots exploring surface soil well beyond the projected canopy area as well as a deep tap root. Competition in gaps and open areas restricts root growth, often prevents establishment of seedlings (Aguiar et al. 1992), and causes a regular distribution of mature plants (Haase et al. 1996a). When neighbors are removed, roots usually expand. Lateral shallow roots also compete with annual plants and grasses as they explore the same soil layer (Wan et al. 1994).

The tap roots of plants from arid environments showed the deepest rooting habits in an extensive survey across all terrestrial biomes (Canadell et al. 1996), with roots of species such as *Bossia albitrunca* reaching 68 m in the central Kalahari dessert. Nevertheless, water is available at 2–3-m depth in most habitats, leading to speculation that plants do not fully use all of the water available in the soil (Schulze et al. 1996).

It has been suggested that most water is extracted from the upper soil layers by very dense surface roots (Franco et al. 1994, Wan et al. 1994, Schulze et al. 1996). Subsoil water extraction becomes increasingly important as the surface soil dries out. Thus, plants that switch to more stable water sources at depth during drought are less susceptible than those that rely on surface water (Evans and Ehleringer 1995). The unbranched root systems of plants in dry habitats should tap very deep water sources, although the effectiveness of these roots has been examined only in a few cases, as in the leguminous shrub *Retama sphaerocarpa*, which is functional below 26-m depth (Haase et al. 1996b).

In plants inhabiting desert and steppe environments, there is a shedding of water absorbed in depth from the root into top soil layers. This passive water transport is known as hydraulic lift (Richards and Caldwell 1987) and provides water to annuals and shrubs growing under the canopy (Dawson 1993). It has been shown to occur in only 27 different species, but there is no reason why it should not be more common, and it has important effects on root growth and nutrient cycling (Caldwell et al. 1998).

Most plants adapted to dry environments also have mycorrhizal symbioses, which improve water and nutrient supply to the plant; however, the mycorrhyzae may also be sinks for carbohydrates, consuming 5–10% of total photosynthate.

When water becomes limiting, dryland plants may increase water use efficiency (WUE). WUE expresses biomass production per unit water spent, or the ratio of fixed C to transpiration, and values are highest in species from the driest habitats (Field et al. 1983). However, exceptions to this generalization can be found, e.g., in *Gutierrezia sarothrae*, carbon fixation is maintained at leaf water potentials of -8.2 MPa (Wan et al. 1993). Similar water-spending behavior has been found in other species from dry lands (DeLucia and Heckarthon 1989, Pugnaire et al. 1996). This behavior is the consequence of a high degree of stress tolerance and shows that photosynthesis is favored over water conservation whenever water is available.

Stomatal closure is an effective means of reducing water loss, but can reduce photosynthesis below the compensation point. In dry environments, this may cause heat imbalance because of the reduced transpiration rate, as well as photoinhibition (Powles 1984, Kyle and Osmond 1987). Changes in stomatal sensitivity are variable and can be considered a species-specific response to drought (Fitter and Hay 1987).

Most species are capable of tight stomatal control in response to environmental factors. Leaf conductance of water vapor seems to be a function of leaf-to-air vapor pressure deficit and water status (Franco et al. 1994), although gas exchange may be controlled solely by atmospheric conditions (Terwilliger and Zeroni 1994).

During dry periods, plants may use physiological mechanisms to reduce damage to tissues, but as water deficits continue, dehydration may lead to injury and death. Dehydration tolerance is a species-specific trait, with tissue damage occurring at levels ranging from -1.2 MPa in aquatic plants to -10 MPa or lower in some xerophytes (Kramer 1983). Many species of algae, lichens, and mosses, as well as some 70 higher plant species, can be air dried and later recover (Gaff 1980).

One means of increasing drought tolerance is by decreasing osmotic potential through the accumulation of solutes, so that turgor and turgor-dependent processes may be maintained at a significantly lower water availability. Osmotic adjustment allows cell enlargement and plant growth at decreasing water levels, allowing stomata to be kept open and CO_2 assimilation to occur at otherwise inhibitory levels (Kramer 1983). However, osmotic adjustment can maintain growth only for short periods of time and may not contribute greatly to continued growth in water-stressed plants (DeLucia and Heckarthon 1989). Nevertheless, osmotic adjustment can extend the lifetime of active tissues between ephemeral showers and the period of tissue hardening for drought (McNaughton 1991).

Some studies have indicated that the degree of drought tolerance is associated with the ability to undergo changes in the cell elastic properties. A drought-induced increase in the bulk modulus of elasticity (ε) would allow the maintenance of a large water potential gradient through the soil–plant–atmosphere continuum, with little change in the relative water content (Nobel 1991b). This would increase the ability of the plant to extract soil moisture from progressively drier soil. Although increases in ε have been observed in response to drought stress, studies have shown seasonal patterns to differ among wild plants under the same environmental stress (Davis and Mooney 1986), but observations of cultivated plants were inconclusive (Turner et al. 1987).

Plants with different photosynthetic pathways vary in their metabolic adaptations to water stress. CAM and C_4 photosynthetic pathways are most clearly suited to deal with water shortages. In CAM plants, the daytime closure of stomata combined with dark fixation of CO_2 reduces water loss without limiting photosynthesis. They are mostly desert succulents and show the highest WUE and mostly low growth rates. Nevertheless, the productivity of some CAM species such as *Opuntia ficus-indica* or some *Agave* species may be very high (Nobel 1991a).

The C_4 pathway evolved as a response to a reduction in atmospheric CO_2 levels during the Cretaceous and the Miocene (Ehleringer et al. 1991). Stomata of C_4 species are less sensitive to a desiccating atmosphere than those of C_3 plants, providing them with greater C gains in low-humidity atmospheres (McNaughton 1991). In addition, the C_4 pathway allows a greater WUE than in the C_3 species (Pearcy and Ehleringer 1984). Nevertheless, plant traits other than those related to the photosynthetic pathway may be responsible for the adaptation of some C_4 species to dry habitats (Osmond et al. 1980). When water and N are available, C_4 plants show a high growth rate, and photosynthetic N use efficiency is highest (Field and Mooney 1986). However, when either is limited, C_4 plants have lower productivity than ecologically similar C_3 species (Pearcy and Ehleringer 1984).

Water storage is generally of little importance in dryland species because of high leaf water turnover. Only a few plants, such as the baobab (*Adansonia digitata* L.) and saguaro cactus (*Carnegia gigantea* L.), store water in significant amounts (Kramer 1983). In general, the cost of water storage is high, and most plants have little or nothing in terms of water storing structures (Chapin et al. 1990).

Nutrients can be as limiting for plant growth as water. They are less mobile in dry soils because the pores between soil particles are replaced by air and the pathway from the soil to the root surface is less direct (Nye and Tinker 1977). Since the rate of ion diffusion to the root is very often the step limiting nutrient uptake, a decrease in soil water availability can affect plant growth. Normally, tissue concentrations of growth-limiting nutrients decline during water stress, showing that indirect effects of soil water content on nutrient uptake may be as important as the direct effects of water stress on plant growth (Chapin 1991).

High WUE is intrinsically linked to low nutrient use efficiency (Field et al. 1983, Lajtha and Schlesinger 1989, Nilsen 1992), and net photosynthesis is only correlated with leaf N in the absence of water limitations (Lajtha and Schlesinger 1989).

B. Morphological Adaptations

Leaf and canopy modifications are important adaptations to arid environments. Since the diffusive resistance offered by a leaf to CO_2 uptake is greater than that offered to water loss, any change in the resistance of the common part of the pathway has a greater influence on the transpirational loss of water than on CO_2 intake. Therefore, many species have features that favor photosynthesis over transpiration by increasing the diffusive resistance of stomata. This may take the form of depressions in the epidermis, pores, cutin, or waxes (Fitter and Hay 1987), or mechanisms, such as leaf folding in grasses, that greatly decrease water losses and reduce the surface exposed to light (Pugnaire et al. 1996). By reducing their evaporative surface, plants may reduce water loss, and for this reason leaves tend to be smaller and thicker in dry habitats (Witkowski and Lamont 1991), but maintain a high photosynthetic rate (Field and Mooney 1986). In addition, by reducing leaf size, the convective heat flux to the atmosphere is increased, and by adjustment of leaf angle, the interception of solar radiation can be reduced (Gulmon and Mooney 1977).

Leaf pubescence is a feature that increases light reflectance and decreases leaf temperature (Ehleringer 1980). The presence of hairs permits a higher rate of C fixation under arid conditions because pubescence allows the plant to avoid potentially lethal high leaf temperatures and to lower daily water loss (Sandquist and Ehleringer 1998). Such morphological features, along with steep leaf angles (Pugnaire and Haase 1996, Valladares and Pearcy 1997), curling of resurrection plants during desiccation (Lebkuecher and Eickmeier 1993), leaf folding (Pugnaire et al. 1996), and parahelitropic leaf movements, provide a structural photoprotection that reduces light interception and can be an effective means to reduce photoinhibition, helping to maintain the leaf energy balance and optimize plant growth and functioning (Figure 1).

Drought deciduous shrubs are a characteristic desert group that undergoes morphological changes in foliar biomass. These species typically develop a canopy of mesomorphic leaves when water is available and shed them when it is not. In some species, winter leaves are replaced by smaller, xeromorphic summer leaves as seasonal water stress increases (Rundel 1991). These changes in total

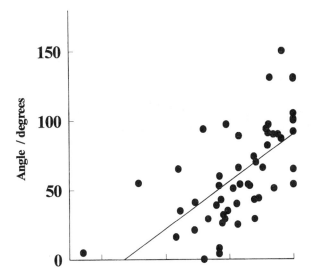

Figure 1   Relationship between the degree of leaf folding (i.e., the angle subtending the two halves of the leaf lamina) and the relative water content (RWC) of *S. tenacissima* leaves. (*Source*: Modified from Pugnaire et al. 1996.)

canopy leaf area sharply reduce the productivity of the xeromorphic phase. However, the increase in WUE, combined with adaptations in tissue water relations, allows photosynthetic activity through all but the most extreme water stress.

The capacity for leaf shedding during drought varies among species (Kozlowsky et al. 1991). Gradual leaf fall seems to be a mechanism to maximize photosynthetic gain and nutrient cycling during periods of water stress (del Arco et al. 1991). Nutrient cycling and nutrient use efficiency are related to leaf fall (Vitousek 1982, Birk and Vitousek 1986), and water stress at the time of leaf shedding may severely affect the plant's nutrient budget by decreasing nutrient resorption from leaves (del Arco et al. 1991, Pugnaire and Chapin 1993).

A permanent state of leaflessness in which plants rely on photosynthetic stem tissues is another feature of dry habitats that allows growth with reduced water loss. Cortical stem tissue is structurally very similar to leaf tissue, but maintains a net positive rate of photosynthesis even in drought-stressed shrubs, and allows a quick recovery from herbivory (Bossard and Rejmanek 1992). Under N-limiting conditions, the proportion of N in photosynthetic stems is higher than in leaves and contributes more to C fixation, thus increasing nutrient use efficiency (Nilsen 1992).

Excess light causes a decline in the efficiency of photosynthesis, and most plants in adverse environments such as arid and semi-arid zones experience some degree of photoinhibition (Powles 1984, Long et al. 1994, Horton et al. 1996). In addition, high light interception may lead to leaf overheating, especially when transpirational leaf cooling is restrained by water deficits (Larcher 1995, Valladares and Pearcy 1997). Two ways to avoid excess light have been observed. The first is physiological photoprotection of the photosynthetic apparatus at high light intensities. Intensive study of this mechanism in the last two decades has shown a remarkable capacity of down-regulation and nonphotochemical dissipation of excitation energy of the photosynthetic apparatus (Demmig-Adams and Adams 1992, Anderson et al. 1993, Long et al. 1994, Horton et al. 1996). The second is structural avoidance of excessive irradiance, which plants achieve by reducing the leaf area directly exposed to the sun.

C. Architecture and Irradiance Interception

Although the importance of structural attributes in plant photoprotection has been frequently noted (Werk and Ehleringer 1984, Ryel and Beyschlag 1995, Valladares and Pearcy 1997), the quantitative measurement of these effects has seldom been reported.

Plants can avoid excessive irradiance by having vertically hanging leaves or a dense crown (Herbert 1996, Valladares and Pearcy 1998). Because dense crowns have high self-shading and poor light penetration to the lower layers of the foliage, they must also have vertically hanging leaves to reach a balance between photoprotection and carbon gain.

Structural avoidance of excessive radiation efficiently prevents the risk of intense irradiance, has no special maintenance costs, and it is biomechanically cheaper than enhanced light harvesting by horizontal canopy elements. For these reasons, structural photoprotection seems to be the most effective strategy to cope with high irradiance stress in semiarid environments (Valladares and Pugnaire 1999).

Crown architecture is thus of great importance in plants living in dry ecosystems, and it affects a number of processes, from irradiance interception to rain partitioning (see Chapter 4 for a functional analysis of aboveground structures).

III. SECONDARY ADAPTATIONS

We have described how leaf morphology and canopy architecture are part of the adaptive syndrome that controls the environment in which leaves perform their work. In addition to this primary role, canopies control the pattern and amount of rainfall that reaches the soil. The resulting spatial heterogeneity in soil moisture represents an important mechanim to cope with water shortage in arid climates.

A. Implications of Canopy Architecture on Rainfall Partitioning

By forming a screen between the sky and the ground, canopies are able to intercept and partially redirect rainfall as throughfall and stemflow. Interception refers to the rainfall water that is retained in the canopies and evaporated back to the atmosphere. Throughfall includes two components: free throughfall and canopy drip. The first is the part of rainfall that reaches the soil surface directly through the holes of the canopy. The second includes the water that drains out from the canopy when it is wetted by rainfall. Stemflow channels water down branches and stems to the root crown and main roots. In many dryland plants, up to 20–40% of rainfall can be directed into stemflow (Hamilton and Rowe 1949, Slatyer 1965, Pressland 1973, Mauchamp and Janeau 1993, Martinez-Meza and Whitford 1996). The sum of throughfall and stemflow is often designated as net rainfall reaching the soil surface.

Interception may reach one third of rainfall (Navar and Bryan 1990, Domingo et al. 1998), which is obviously important in arid climates. Factors determining interception may be grouped in two sets: one depending on atmospheric conditions during the rainfall event and the other depending on properties of the intercepting canopies (Rutter et al. 1971, Gash 1979, Gash et al. 1995).

Atmospheric conditions include the time structure of rainfall and the weather factors controlling the evaporation rate. Interception will be greater in long and low-intensity rainfall events with moderate wind, than in short, intense showers in still atmosphere. Canopy interception factors include the water storage capacity, the canopy drainage curves, the aerodynamic conductance, and the proportion of holes. Canopies with large water storage capacity that drain out slowly and allow free exchanges with the atmosphere are expected to intercept the most rainfall. Recent work has stressed the importance of canopy factors on the interception process, particularly in sparse and dryland vegetation (Teklehaimanot et al. 1991, Teklehaimanot and Jarvis 1991, Domingo et al. 1998).

Net rainfall is further divided into throughfall and stemflow. It is thought that stemflow facilitates the storage of water in deep soil layers, maintaining its availability to plants and reducing direct evaporative loss. Available information and stemflow models are still too crude to allow us to explore the adaptive values of canopy architecture to increase stemflow at the expense of canopy drip. Most approaches consider the canopy as a funnel that conveys water along the branches and stems with a variable efficiency. Vertical projection of the canopy, branch angle, and branch diameter are among the most significant traits (De Ploey 1982,

Van Elewijck 1989), but also important are the shape and junction of leaves, sharp bends of branches, bark surface roughness, etc. (De Ploey 1984).

The role of stemflow as a mechanism of storing water in deep soil layers has been shown both experimentally and by identifying hydromorphic features in the soil next to the shrub or tree bases. Using dye distribution under *Prosopis glandulosa* and *Larrea tridentata* shrubs in the Chihuahuan desert, Martínez-Meza and Whitford (1996) showed that preferential flow paths along the root–soil interface may reach depths of 50 cm. Changes in the soil surrounding the trunk, e.g., more mottled B horizons and thicker elluvial horizons, decreasing pH, decreasing percentage base saturation, and lower free Fe_2O_3 have also been shown (De Ploey 1984).

Managing rainfall partition allows the plant a certain control of water storage and water availability. An example is provided by comparing rainfall partition patterns in two leguminous shrubs, *Anthyllis cytisoides* and *Retama sphaerocarpa*, in southeast Spain. The former is a medium-depth-rooted shrub (approximately 1 m tall) with an outer dense shell of leaves and twigs that leave a very small proportion of holes in the canopy. The latter is a deep-rooted shrub (up to 26 m depth; Haase et al 1996b) with light canopies that allow for a great proportion of free throughfall. With an average annual rainfall of 250 mm and the rainfall partition patterns observed in the field (Table 1), net rainfall estimates per m^2 of canopy under *A. cytisoides* and *R. sphaerocarpa* bushes are 150 mm and 222 mm, with corresponding stemflow amounts of 50 mm and 17 mm, respectively.

Table 1 Empirically Derived Canopy Parameters Controlling Rainfall Interception, and Field Rainfall Partition Percentages from Incoming Rainfall Measured During One Year (1994–1995) in *Anthyllis cytisoides* and *Retama sphaerocarpa* Shrubs in Southeast Spain

	Anthyllis cytisoides	*Retama sphaerocarpa*
Free throughfall coefficient	0.1	0.7
Canopy water storage capacity (mm)	1.80 (0.21)	0.29 (0.02)
ln a (canopy drainage curve, mm)	− 0.62 (0.05)	3.48 (0.36)
b (canopy drainage curve, mm)	4.30 (0.34)	3.12 (0.46)
Canopy conductance per bush, m^3s^{-1} (wind velocity $1–2$ m s^{-1})	0.23 (0.05)	1.21 (0.69)
Interception (% rainfall)	40	21
Throughfall (% rainfall)	40	72
Stemflow (% rainfall)	20	7

Source: Domingo et al. 1998.

Because rainfall occurs mainly during the cold season, *A. cytisoides* can benefit more from stemflow storage than *R. sphaerocarpa*. The latter, because of its deep-water-feeding habit, is less likely to take advantage of stemflow, and it is probably more dependent on larger-scale lateral redistribution of water as a way to replenish deep storage. Available results suggest that the significance of stemflow for increasing the water supply to plants is greater in shallow- or medium-rooted shrubs, with large canopy projected area in relation to their stem base areas.

B. Nucleation and Facilitation

Spatial mosaics of vegetated and bareground patches are common in dryland plant communities. Differences in soil fertility between bare and vegetated patches have been extensively described (see reviews by Schlesinger et al. 1990, Puigdefábregas and Sánchez 1996). Vegetated patches show an increase of soil water storage capacity, higher soil organic carbon and nitrogen, as well as sediment accumulation. They develop as a result of self-reinforcing mechanisms that increase their fertility, and in some cases have been called "resource islands" (Virginia et al 1989). Two reinforcing mechanisms have been identified; one involves redistribution of resources within the landscape (Section III.C.2) and the other is the facilitation between different plant species.

Resource islands themselves are complex systems that support different plant functional types. The permanent structural elements are perennial plants, usually legumes, with long response times that react to low recurrence disturbances. Associated with them are annual species with shorter response times that react to high recurrence fluctuations. It has been shown that between both vegetation types, mutual facilitation mechanisms develop (Pugnaire et al 1996b) that are reinforced with time (Pugnaire et al 1996a) and increase the resilience of the whole assemblage (See Chapter 18 for details).

Resource islands are "hot spots" in the dryland mosaics that support the highest rates of soil and vegetation change. They undergo their own life cycles, with building, mature, and senescent phases, and may be considered as a class of the vegetation in itself (Watt 1947) whose dynamics led to the addition of cyclical or nondirectional change to succession theory (O'Neill et al 1986).

C. Changes of Spatial Structure as a Plastic Response to Disturbances

Important interactions associated with spatial organization arise in response to disturbances in dryland ecosystems. Disturbances reinforce the spatial heterogeneity, and their effects tend to be buffered by a heterogeneous landscape in two ways: by spreading the risks of species disappearance and by contributing to increase local resource retention by nonuniform infiltration and runoff patterns.

1. Spreading the Risks of Species Disappearance Over the Space

The development of spatial heterogeneity of populations living in fluctuating environments may be considered as a source of ecosystem resilience (Bascompte and Sole 1995). When a harmful event occurs, extensive damage may result in some patches, but others may be only slightly affected. In this way, the likelihood of population survival increases overall. However, in event-triggered ecosystems, which are widespread in arid environments, the occurrence of extreme events, such as severe droughts, has a potential of synchronize vegetation stress across large tracts of land. In this way, extreme events would stimulate spatial uniformity rather than differentiation. Plant communities counteract this effect by actively developing spatial heterogeneity as a feedback to stressors using several mechanisms described in the following sections.

Maintaining High Degrees of Genetic Heterogeneity at the Specific and Intraspecific Levels.   Different functional types coexisting in the same landscape are affected in a different manner by rainfall fluctuations (Puigdefábregas et al. 1996). The high level of genetic variation observed in populations of North American deserts have been attributed to the selective pressure of environmental heterogeneity (Shuster et al. 1994). Genetic heterogeneity in fluctuating environments contributes to species coexistence (Westoby 1980) and increases the level of intraspecific diversity in dryland ecosystems (West 1993, Huenneke and Noble 1996). For similar reasons, gene migration via pollen from the center of the distribution area toward the more stressed boundaries has been proposed as a mechanism to increase resilience of these marginal populations to climatic fluctuations (Sage 1996).

Life Span Differences Between the Species that Integrate Plant Communities.   The occurrence of an extreme event may trigger synchronized recruitment of many species, but deaths occur at different times, giving rise to gaps that are available to new guilds (Wiegandet al. 1995).

Spatial Aggregation.   Short dispersal distances and vegetative regeneration may lead to vegetation whose patchiness may be reinforced by nucleation processes (as previously discussed). These processes also contribute to the desynchronization that increases spatial heterogeneity of plant communities. Research on the dynamics of karoo vegetation (Wiegandet al. 1995) using field observations and cellular automata simulations showed that (1) aggregated patterns evolve from randomly distributed populations, and (2) some species can only survive rainfall fluctuations because of their spatial aggregation.

Chaotic Behavior. This mechanism has been theoretically postulated rather than demonstrated experimentally in ecosystems (Bascompte and Sole 1995). Slightly different initial conditions are also a source of asynchrony in chaotic systems. Although the probability of local extintion increases in the chaotic domain, the spatial heterogeneity caused by desynchronization of initially close trajectories lowers the probability of global extinction.

2. Spatial Redistribution of Resources

Canopy density of dryland vegetation tends to minimize water demand (Eagleson 1982, Haase et al. 1998) and is controlled by the evaporative power of the atmosphere and soil water availability (Woodward 1987, Specht and Specht 1989). However, a given canopy density, expressed either as projected canopy area or as leaf area index, in equilibrium with its soil and climatic environment can be distributed in different spatial structures, forming mosaics of clumps and bare areas of different sizes and patterns. Such spatial structures may develop as an adaptation to maximize water harvest by vegetation clumps (Puigdefabregas and Sánchez 1996). Field observations show that clumps harvest water from bare patches through lateral root systems and that dimensions of vegetated and bare phases of the mosaic are not independent from each other. Harvesting runoff from neighboring bare patches has been shown in the perennial tussock grass *Stipa tenacissima* stands (Puigdefábregas and Sánchez 1996), by excluding tussocks from the overland flow generated in the uphill bare patches. Soil moisture and plant growth were significantly lower in runoff-excluded tussocks than in controls (Table 2). The role of tussocks as infiltration sites has also been reported in field studies using rainfall simulation experiments (Bergkamp 1996, Bergkamp et al. 1998).

Table 2 Effect of Excluding Overland Flow from Uphill Bareground Patches on Leaf Growth and Soil Moisture Content of the Uppermost 5-cm Soil Layer of *Stipa tenacissima* Tussocks in Southeast Spain[a]

	Leaf growth (cm/stem/year)	Soil moisture, annual average (0–5 cm, % vol)
Control	23 ± 4	1.6 ± 0.09
Overland flow excluded	10 ± 3	1.3 ± 0.09

[a] Values are mean ± 1 SE. Differences from controls are significant at $P < .01$.
Source: Puigdefábregas and Sánchez 1996.

Because plant clumps and bareground patches are hydrologically linked, we could anticipate that clump sizes adjust to maximize water harvest by vegetation. In *Stipa tenacissima* stands, there is a reasonable agreement between predicted and observed sizes of tussocks and bare patches (Puigdefábregas and Sánchez 1996).

Vegetated patches catch nutrients and sediments from their surroundings as well. The former are carried by overland flow or intercepted by the canopies as dry deposition, and finally leached to the soil. Dry deposition is important in nutrient budgets of semi-arid areas (Domingo et al. 1994). Sediments are also intercepted by canopies, either transported as bulk atmospheric deposition (Soriano 1983, Queralt et al. 1993) or by runoff (Puigdefábregas and Sánchez 1996). Finally, sediments may be trapped underneath canopies by differential splash. Splash can carry sediments into canopy-covered areas, but not away from them, because of the lower rain energy under plant clumps (Parsons et al. 1992). All of these functions are strongly dependent on the canopy structure.

Plant cover works not only as a passive semipermeable umbrella that intercepts rainfall, overland flow, or atmospheric deposition, but also as a living structure, interacting with hillslope fluxes by modifying the shape and spatial pattern of plant clumps. Clump shape is relevant at the patch level, whereas spatial patterns work at the hillslope level. In both cases, implications for plant performance, water supply, and sediment redistribution are significant. In *Stipa tenacissima* grasslands, tussocks intercept sediments that move downhill, and sandy materials are deposited just uphill from the plant clumps, becoming the main infiltration sites (Sánchez and Puigdefábregas 1994, Sánchez 1995, Puigdefábregas and Sánchez 1996). At both sides of the tussock, small swales concentrate the downhill runoff and sediment fluxes (Table 3).

These field observations, incorporated into a cellular automata model (Sánchez and Puigdefábregas 1994), show that plant growth rate, slope, and sediment accumulation influence the tussock shape and stand characteristics (Figure 2). The spatial structure of the vegetation modifies the intensity of the hillslope fluxes, thus affecting the interaction between vegetated patches and sediment or runoff fluxes (Puigdefábregas 1998).

Tussocks are distributed in bands parallel to the slope contour lines to maximize soil and water storage, depending on sediment fluxes. Tussocks are randomly distributed if sediment flow decreases and form rills if flow is very high (Figure 3).

Runoff intensity affects spatial patterns of vegetation in arid climates with low slope gradients and fine textured soils, where vegetation often grows in decametric banded patterns known as ''tiger bush'' in the Sahel (Valentin and d'Herbes 1996) and also in Australia (Tongway and d'Herbes 1990) and in the Chihuahuan desert (Janeauet al 1996). Tiger bush is interpreted as an adaptive

Table 3 Mean Relative Values[a] of Runoff and Sediments
Collected in Different Microsites of Mound and Swale Complexes
in *Stipa tenacissima* Stands in Southeast Spain[b]

	Runoff	Sediment yield
Bare ground	1.21 ± 0.07	1.26 ± 0.10
Tread	0.54 ± 0.08***	0.66 ± 0.11***
Tussock (riser)	0.60 ± 0.11***	0.56 ± 0.14***
Lateral swale	1.67 ± 0.19***	1.65 ± 0.23*

[a] ±1 SE, n = 11 rainfall events.
[b] Every number is a ratio of the amount collected at two positions in the slope.
Significance of the differences against bareground patches: *** < .01;
* < .1.
Source: Puigdefábregas and Sánchez 1996.

structure that evolves in response to the overland flow generated in the bare
stripes, allowing vegetation bands to benefit from this redistribution of water. A
finer pattern emerges within each vegetation band, with an herb layer in the uphill
edge followed downhill by shrubs and small trees. Vegetated bands slowly mi-
grate upward (Valentin and d'Herbes 1996), and the system is highly sensitive
to disturbances that affect or destroy the soil surface structure in the bare stripes.

IV. PLANT COMMUNITY CHANGES AND RAINFALL VARIABILITY

In temperate climates, vegetation states may oscillate after a disturbance pulse.
In the absence of disturbance, vegetation may asymptotically approach an equilibrium condition in which internal regulatory mechanisms (e.g., competition)
occur. In arid zones with unsteady rainfall, this equilibrium behavior rarely occurs
(Ellis et al. 1993). Instead, the occurrence of specific events such as drought,
fire, grazing, or a combination of these may cause different responses, depending
on antecedent local conditions (Westoby et al. 1989, Stafford-Smith and Pickup
1993), with several vegetation states that coexist in a stable fashion, increasing
the spatial heterogeneity of vegetation and buffering the effects of drought events.

The threshold between equilibrium and event-triggered dynamics has been
associated with both increased rainfall variability and decreased annual rainfall
in several African regions (Coppock 1993, Ellis et al. 1993, Wiegand et al. 1995).
Many of the events that maintain dryland ecosystems far from equilibrium have
been incorporated into their evolutionary history and are an integral part of the

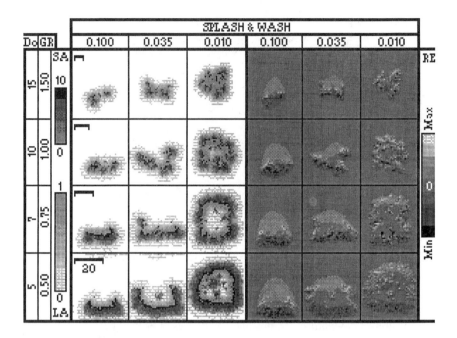

Figure 2 Simulations of tussock (left-hand image) and micro relief (right-hand image) developments in sloping areas using a cellular automata model. The figure shows the outputs of the model using three values of sediment circulation defined by splash and wash parameters that increase to the left, and four values of stem growth rate (GR) that increase upward, and their associated maximum distance of leaf vegetative influence (D_0) on stem bending. The cells in the left column include a scale bar showing the equivalent of 20 length units in each row. In the left-hand image, dark shadings (top) display stem age at each cell, and light shadings (bottom) indicate litter accumulation values. In the right-hand image, grey shadings display relief variations (RE) relative to the original background situation, and light and dark tones indicate increases and decreases of elevation, respectively. (*Source*: Puigdefábregas and Sánchez 1996.)

adaptive dynamics of plant communities. These events do not change the dynamic state of the ecosystem. By contrast, some rare or new events may cause irreversible changes of species composition over a large time scale, and drive a transition toward a different dynamic state of the ecosystem. These two kinds of events have been called integral events and transition triggers, respectively (Wiegand et al. 1995).

In most cases, transition triggers synergetically combine climatic and anthropogenic events. Some of the best-documented examples concern vegetation changes that occurred in central Australia after the European settlement, due to

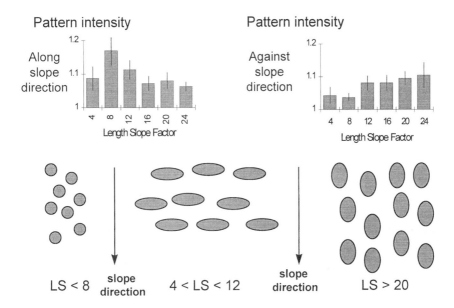

Figure 3 Distribution of mean pattern intensity values among classes of the length slope factor (LS) in *Stipa tenacissima* stands in southeast Spain. The bottom part of the image shows the spatial patterns of tussocks associated with each range of LS. Pattern intensity is calculated as the ratio of the first peak to the first trough of the semivariograms measured in the downhill and contour line directions. Error bars indicate 1 SE. (*Source*: Puigdefábregas and Sánchez 1996.)

the introduction of livestock and the European rabbit and to the modification of fire patterns that became less frequent and more widespread. The impacts of these changes on vegetation were particularly dramatic when associated with climatic fluctuations (Griffin and Friedel 1985).

One of the major changes of dryland ecosystems driven by transition triggers is the replacement of grasslands by shrublands (Puigdefabregas 1998), a change often considered a core process in rangeland degradation and desertification (Schlesinger et al 1990). In most cases, woody components of grasslands first disappear because of overexploitation (Le Houerou 1995) or frequent wildfires and droughts (Griffin and Friedel 1985). Then, the occurrence of further combinations of overgrazing and droughts causes encroachment of different shrub species (Soriano 1983, Schlesinger et al. 1990, Le Houerou 1995). This process involves changes in the spatial structure of grasslands, reinforcing the patchiness and coarsening the grain of spatial heterogeneity. As a result, the contrast between elements of the spatial mosaic becomes sharper, and the spatial redistribution of water, sediments, and nutrients increases by several orders of magnitude (Abrahams et al. 1995, Schlesinger et al. 1990, Bergcamp et al. 1996, Aronson et al. 1993, Pickup et al. 1994, Schlesinger 1996).

Spatial patterns and their associated changes in resource redistribution appear as background processes associated with qualitative changes in dryland vegetation in response to extreme events.

V. CONCLUSIONS

Plant survival in arid environments has to cope with two kinds of stressors: chronic water limitation and rainfall variability, which includes extreme events, including drought, rain spells, or combinations of both. Adaptations have evolved in a wide range of organizational levels, from the biochemical and metabolic to the population and plant community, through morphological and architectural intermediates.

Drought adaptation involves two extreme strategies that are shared by most plants in different degrees: drought tolerance and drought avoidance. Drought tolerance involves maintaining appropriate gradients of water potential in the soil, plant, and atmosphere system, and reducing transpiration while avoiding the harmful side effects of this reduction. The three issues are interconnected and rely on adaptations developed mostly at low organizational levels. The first is achieved through osmotic and cell wall elasticity adjustments. The second is mostly under control of stomata and leaf surface morphology. The third involves adaptations to ensure (1) the CO_2 supply while reducing transpiration, as is the case in CAM metabolism; (2) the nutrient supply through mycorrhizal symbioses; and (3) the avoidance of excessive irradiance and photoinhibition, particularly through morphological and architectural traits in leaves and canopies.

Drought tolerance relies mostly on ensuring access to more reliable sources of water (site selection) or escaping from periods of large atmospheric water deficits (time selection). In both cases, adaptations mostly involve intermediate organizational levels, like large-root systems or phenological adjustments, such as summer leaf fall or whole-branch shedding in dry spells.

Adaptations to rainfall variability concern the improvement of water supply and water storage in the plant environment and the diminution of extinction risks at the population and plant community levels. In the first case, adaptive mechanisms include canopy architectural traits that increase stemflow, as well as building up "fertility islands" and spatial patterns that enhance overland flow and sediment harvesting. In the second case, adaptation mechanisms tend to desynchronize life cycles across space and to enhance the coexistence of multiple vegetation states as a way to avoid large areas being affected by harmful effects of extreme drought events.

REFERENCES

Abrahams, A., A.J. Parsons, and J. Wainwright. 1995. Effects of vegetation change on interrill runoff and erosion, Walnut Gulch, southern Arizona. Geomorphology 13: 37–48.

Aguiar, M.R., A. Soriano, and O.E. Sala. 1992. Competition and facilitation in the recruitment of seedlings in Patagonian steppe. Functional Ecology 6: 66–70.

Allen-Diaz, B., F.S. Chapin, S. Diaz, M. Howden, J. Puigdefábregas, and M. Stafford-Smith. 1996. Rangelands in a changing climate: impacts, adaptations and mitigation. In: Watson, R.T., M.C. Zinyowera, R.H. Moss, and D.J. Dokken, eds. Contribution of Working Group II to the Second Assessment Report of the Intergovernmental Panel on Climate Change. Cambridge: Cambridge University Press, pp 131–158.

Anderson, J.M., W.S. Chow, and G. Öquist. 1993. Dynamics of photosystem II: photoinhibition as a protective acclimation strategy. In: Yamamoto, H.Y., and C.M. Smith, eds. Photosynthetic Responses to the Environment. New York: American Society of Plant Physiologists, pp 14–26.

Archer, S., D.S. Schimel, and E.A. Holland. 1995. Mechanisms of shrubland expansion: land use, climate or CO_2? Climatic Change 9: 1–9.

Aronson, J., C. Floret, E. Le Floch, C. Ovalle, and R. Pointanier. 1993. Restoration and rehabilitation of degraded ecosystems in arid and semi-arid lands. I. A view from the South. Restoration Ecology 1: 8–17.

Bascompte, J., and R.V. Sole. 1995. Rethinking complexity: modelling spatio-temporal dynamics in ecology. Trends in Ecology and Evolution 9: 361–366.

Bergkamp, G. 1996. Mediterranean geoecosystems: Hierarchical organisation and degradation. Ph.D. Thesis, University of Amsterdam, Amsterdam, The Netherlands.

Bergkamp, G., L. Cammeraat, and L. Martínez Fernández. 1996. Water movement and vegetation patterns on shrubland and abandoned fields in two desertification-threatened areas in Spain. Earth Surface Processes and Landforms 21: 1073–1090.

Birk, E.J., and P. Vitousek. 1986. Nitrogen availability and nitrogen use efficiency in loblopolly pine stands. Ecology 67: 69–79.

Bossard, C.C., and R. Rejmanek. 1992. Why have green stems? Functional Ecology 6: 197–205.

Caldwell, M.M., T.E. Dawson, and J.H. Richards. 1998. Hydraulic lift: consequences of water efflux from the roots of plants. Oecologia 113: 151–161.

Caldwell, M.M., and J.H. Richards. 1986. Competing root systems: morphology and models of absorption. In: Givnish, T.J., ed. On the Economy of Plant Form and Function. London: Cambridge University Press, pp 251–273.

Canadell, J., R.B. Jackson, J.R. Ehleringer, H.A. Mooney, O.E. Sala, and E.D. Schulze.

1996. Maximum rooting depth of vegetatin types at the global scale. Oecologia 108: 583–595.

Chapin, F.S. III. 1980. The mineral nutrition of wild plants. Annual Review of Ecology and Systematics 11: 233–260.

Chapin, F.S. III. 1991. Integrated responses of plants to stress. BioScience 41: 29–36.

Chapin, F.S. III E.D. Schulze, and H.A. Mooney. 1990. The ecology and economics of storage in plants. Annual Review of Ecology and Systematics 21: 423–447.

Coppock, D.L. 1993. Vegetation and pastoral dynamics in the southern Ethiopian range-lands: implications for theory and management. In: Behnke, R.H., I. Scoones, and C. Kerven, eds. Range Ecology at Disequilibrium. London: Overseas Development Institute, pp 42–61.

Davis, S.D., and H.A. Mooney. 1986. Tissue water relations of four co-occurring chaparral shrubs. Oecologia 70: 527–535.

Dawson, T.E. 1993. Hydraulic lift and water use by plants: implications for water balance, performance and plant-plant interactions. Oecologia 95: 565–574.

del Arco, J.M., A. Escudero, and M. Vega Garrido. 1991. Effects of site characteristics on nitrogen retranslocation from senescing leaves. Ecology 72: 701–708.

DeLucia, E.H., and S.A. Heckathorn. 1989. The effect of soil drought on water-use efficiency in a contrasting Great Basin desert an Sierran montane species. Plant, Cell and Environment 12: 935–940.

Demmig-Adams, B., and W.W. Adams. 1992. Photoprotection and other responses of plants to high light stress. Annual Review of Plant Physiology and Plant Molecular Biology 43: 599–626.

De Ploey, J. 1982. A stemflow equation for grasses and similar vegetation. Catena 1: 39–52.

De Ploey, J. 1984. Stemflow and colluviation: modelling and implications. Pedologie 34: 135–46.

Domingo, F., G. Sánchez, M.J. Moro, A. Brenner, and J. Puigdefábregas. 1998. Measurement and modelling of rainfall interception by three semi-arid canopies. Agricultural and Forest Meteorology 91: 275–292.

Domingo, F., J. Puigdefábregas, M.J. Moro, and J. Bellot. 1994. Role of vegetation cover in the biogeochemical balances of a small afforested catchment in southeastern Spain. Journal of Hydrology 159: 275–289.

Eagleson, P.S. 1982. Ecological optimality in water limited natural soil-vegetation systems. I: Theory and hypothesis. Water Resources Research 18: 325–342.

Ehleringer, J. 1980. Leaf morphology and reflectance in relation to water and temperature stress. In: Turner, N.C., and P.J. Kramer, eds. Adaptation of Plants to Water and High Temperature Stress. New York: Wiley, pp 295–308.

Ehleringer, J., R.F. Sage, L.B. Flanagan, and R.W. Pearcy. 1991. Climate change and the evolution Of C_4 photosynthesis. Trends in Ecology and Evolution 6: 95–99.

Ellis, J.E., M.B. Coughenour, and D.M. Swift. 1993. Climate variability, ecosystem stability and the implications for range and livestock development. In: Behnke, R.H., I. Scoones, and C. Kerven, eds. Range Ecology at Disequilibrium. London: Overseas Development Institute, pp 31–41.

Evans, R.D., and J.R. Ehleringer. 1994. Water and nitrogen dynamics in an arid woodland. Oecologia 99: 233–242.

Field, C.B., and H.A. Mooney. 1986. The photosynthesis-nitrogen relationship in wild plants. In: Givnish, T.J., ed. On the Economy of Plant Form and Function. Cambridge: Cambridge University Press, p 25.

Field, C.B., J. Merino, and H.A. Mooney. 1983. Compromises between water-use efficiency and nitrogen-use efficiency in five species of California evergreens. Oecologia 60: 384–389.

Fitter, A.H., and R.K.M. Hay. 1987. Environmental Physiology of Plants. London: Academic Press.

Franco, A.C., A.G. de Soyza, R.A. Virginia, J.F. Reynolds, and W.G. Withford. 1994. Effect of plant size and water relations on gas exchange and growth of the desert shrub *Larrea tridentata*. Oecologia 97: 171–178.

Gaff, D.F. 1980. Protoplasmic tolerance of extreme water stress. In: N.C. Turner, N.C., and P.J. Kramer, eds. Adaptations of Plants to Water and High Temperature Stress. New York: Wiley, p 207.

Gash, J.H. 1979. An analytical model of rainfall interception by forests. Quarterly Journal of the Royal Meteorological Society 105: 43–55.

Gash, J.H.C., C.R. Lloyd, and G. Lachaud. 1995. Estimating sparse forest rainfall interception with an analytical model. Journal of Hydrology 170: 79–86.

Griffin, G.F., and M.H. Friedel. 1985. Discontinuous change in central Australia: some implications of major ecological events for land management. Journal of Arid Environments 9: 63–80.

Gulmon, S.L., and H.A. Mooney. 1977. Spatial and temporal relationships between two desert shrubs *Atriplex hymenelytra* and *Tidestromia oblongifolia*. Journal of Ecology 65: 831–838.

Haase, P., F.I. Pugnaire, M. Fernández, J. Puigdefábregas, S.C. Clark, and L.D. Incoll. 1996a. Investigation of rooting depth in the semi-arid shrub *Retama sphaerocarpa* (L.) Boiss. by labelling of ground water with a chemical tracer. Journal of Hydrology 177: 23–31.

Haase, P., F.I. Pugnaire, S.C. Clark, and L.D. Incoll. 1996b. Spatial patterns in a two-tiered semi-arid shrubland in southeastern Spain. Journal of Vegetation Science 5: 27–34.

Haase, P., F.I. Pugnaire, S.C. Clark, and L.D. Incoll. 1999. Photosynthetic rate in relation to canopy demography in the evergreen tussock grass *Stipa tenacissima*. Functional Ecology, in press.

Hamilton, E.L., and P.B. Rowe. 1949. Rainfall interception by chaparral in California. Sacramento, CA: Publications of the California Range and Forest Experimental Station.

Herbert, T. 1996. On the relationship of plant geometry to photosynthetic response. In: Mulkey, S.S., R.L. Chazdon, and A.P. Smith, eds. Tropical Forest Plant Ecophysiology. New York: Chapman and Hall, pp 139–161.

Horton, P., A.V. Ruban, and R.G. Walters. 1996. Regulation of light harvesting in green plants. Annual Review of Plant Physiology and Plant Molecular Biology 47: 655–684.

Huenneke, L.F., and I. Noble. 1996. Ecosystem function of biodiversity in arid ecosystems. In: Mooney, H.A., J.H. Cushman, E. Medina, O.E. Sala, and E.D. Schulze, eds. Functional Roles of Biodiversity: A Global Perspective. London: Wiley, pp 99–128.

Jupp, A.P., and E.L. Newman. 1987. Morphological and anatomical effects of severe drought on the roots of *Lolium perenne* L. New Phytologist 105: 393–402.

Kozlowsky, T.T., P.J. Kramer, and S.G. Pallardy. 1991. The Physiological Ecology of Woody Plants. San Diego: Academic Press.

Kramer, P.J. 1983. Water Relations of Plants. Orlando: Academic Press.

Kyle, D.J., C.B. Osmond, and C.J. Amtzen, eds. 1987. Photoinhibition. Amsterdam: Elsevier.

Lajtha, K., and W.G. Withford. 1989. The effect of water and nitrogen amendments on photosynthesis, leaf demography, and resource-use efficiency in *Larrea tridentata*, a desert evergreen shrub. Oecologia 80: 341–348.

Larcher, W. 1995. Physiological plant ecology. Ecophysiology and Stress Physiology of Functional Groups. 3rd Ed. Berlin-Heidelberg: Springer-Verlag.

Lebkuecher, J.G., and W.G. Eikmeier. 1993. Physiological benefits of stem curling for resurrection plants in the field. Ecology 74: 1073–1080.

Le Houerou, H.N. 1995. Bioclimatologie et biogeographie des steppes arides du Nord de l'Afrique. Diversiteé biologique, développement durable et desertisation. Montpellier: CIHEAM/ACCT.

Levitt, J. 1980. Responses of Plants to Environmental Stress. 2nd Ed. New York: Academic Press.

Long, S.P.,S. Humphries, and P.G. Falkowski. 1994. Photoinhibition of photosynthesis in nature. Annual Review of Plant Physiology and Plant Molecular Biology 45: 633–662.

Margalef, R. 1974. Ecologia. Barcelona: Omega, S.A.

Martinez-Meza, E., and W. Whitford. 1996. Stemflow, throughfall and channelization of stemflow by roots in three Chihuahuan desert shrubs. Journal of Arid Environments 32: 271–287.

Mauchamp, A., and J.L. Janeau. 1993. Water funnelling by the crown of *Flourensia cernua*, a Chihuahuan desert shrub. Journal of Arid Environments 25: 299–306.

McNaughton, S.J. 1991. Dryland herbaceous perennials. In: Mooney, H.A., W.E. Winner, and E.J. Pell, eds. Response of Plants to Multiple Stresses. San Diego: Academic Press, p 307.

Navar, J., and R. Bryan. 1990. Interception loss and rainfall redistribution by three semi-arid growing shrubs in northeastern Mexico. Journal of Hydrology 115: 51–63.

Nilsen, E.T. 1992. Partitioning growth and photosynthesis between leaves and stems during nitrogen limitation in *Spartium junceum*. American Journal of Botany 79: 1217–1223.

Nobel, P.S. 1991a. Achievable productivities of certain CAM plants: basis for high values compared with C_3 and C_4 plants. New Phytologist 119: 183–205.

Nobel, P.S. 1991b. Physicochemical and Environmental Plant Physiology. San Diego: Academic Press.

Nye, P.H., and P.B. Tinker. 1977. Solute Movement in the Soil-Root System. Berkeley, CA: University of California Press.

O'Neill, R.V., D.L. De Angelis, J.B. Waide, and T.F.H. Allen. 1986. A Hierarchual Concept of Ecosystems. Princeton, NJ: Princeton University Press.

Osmond, C.B., K. Winter, and S.B. Powles. 1980. Adaptative significance of carbon dioxide cycling during photosynthesis in water stressed plants. In: Turner, N.C., and

P.J. Kramer, eds. Adaptation of Plants to Water and High Temperature Stress. New York: Wiley, p 139.

Osmond, C.B., M.P. Austin, J.A. Billings, J.S. Boyer, W.H. Dacey, P.S. Nobel, S.D. Smith, and W.E. Winner. 1987. Stress physiology and the distribution of plants. Bioscience 37: 38–48.

Parsons, A.J., A.D. Abrahams, and J.R. Simanton. 1992. Microtopography and soil-surface materials on semi-arid piedmont hillslopes, southern Arizona. Journal of Arid Environments 10: 7–15.

Pearcy, R.W., and J. Ehleringer. 1984. Comparative ecophysiology of C$_3$ and C$_4$ plants. Plant, Cell and Environment 7: 1–13.

Pickup, G., G.N. Bastin, and V.H. Chewings. 1994. Remote-sensing-based condition assessment for nonequilibrium rangelands under large-scale commercial grazing. Ecological Applications 3: 497–517.

Powles, S.B. 1984. Photoinhibition of photosynthesis induced by visible light. Annual Review of Plant Physiology 35: 15–44.

Pressland, A.J. 1973. Rainfall partitioning by an arid woodland (*Acacia aneura* F. Muell) in south-western Queensland. Australian Journal of Botany 2: 35–45.

Pugnaire, F.I., and E. Esteban. 1991. Nutritional adaptations of caper shrub (*Capparis ovata* Desf.) to environmental stress. Journal of Plant Nutrition 14: 151–161.

Pugnaire, F.I., and F.S. Chapin III. 1992. Environmental and physiological factors governing nutrient resorption efficiency in barley. Oecologia 90: 120–126.

Pugnaire, F.I., and F.S. Chapin III. 1993. Controls over nutrient resorption from leaves of evergreen mediterranean species. Ecology 74: 124–129.

Pugnaire, F.I., and P. Haase. 1996. Comparative physiology and growth of two perennial tussock grass species in a semi-arid environment. Annals of Botany 77: 81–86.

Pugnaire, F.I., P. Haase, L.D. Incoll, and S.C. Clark. 1996. Response of the tussock grass *Stipa tenacissima* to watering in a semi-arid environment. Functional Ecology 10: 265–274.

Pugnaire, F.I., P. Haase, and J. Puigdefábregas. 1996. Facilitation between higher plant species in a semiarid environment. Ecology 77: 1420–1426.

Puigdefábregas, J. 1998. Ecological impacts of global change on drylands and their implications on decertification. Land Degradation and Rehabilitation, in press.

Puigdefábregas, J., and G. Sánchez. 1996. Geomorphological implications of vegetation patchiness in semi-arid slopes. In: Anderson, M., and S. Brooks, eds. Advances in Hillslope Processes 2. London: Wiley, pp 1027–1060.

Puigdefábregas, J., C. Aguilera, A. Brenner, S. Clark, M. Cueto, L. Delgado, F. Domingo, L. Gutiérrez, L. Incoll, R. Lazaro, J.M. Nicolau, G. Sánchez, A. Solé, and S. Vidal. 1996. The Rambla Honda Field Site. Interactions of soil and vegetation along a catena in semi-arid SE Spain. In: Brandt, J., and J.B. Thornes, eds. Mediterranean Desertification and Land Use. London: Wiley, pp 137–168.

Queralt, I., F. Domingo, and A. Sole-Benet. 1993. The influence of local sources in the mineral content of bulk deposition over an attitudinal gradient in the Filabres Range (SE Spain). Journal of Geophysical Research 98: 16761–16768.

Richards, J.H., and M.M. Caldwell. 1987. Hydraulic lift: substantial nocturnal water transport between soil layers by *Artemisia tridentata* roots. Oecologia 73: 486–489.

Rundel, P.W. 1991. Shrub life-forms. In: Mooney, H.A., W.E. Winner, and E.J. Pell, eds. Response of Plants to Multiple Stresses. San Diego: Academic Press, p 345.

Rutter, A.J., K.A. Kershaw, P.C. Robins, and A.J. Morton. 1971. A predictive model of rainfall interception in forests. I: Derivation of the model from observations in a plantation of Corsican pine. Agricultural Meteorology 3: 67–84.

Ryel, R.J., and W. Beyschlag. 1995. Benefits associated with steep foliage orientation in two tussock grasses of the American intermountain west. A look at water-use-efficiency and photoinhibition. Flora 190: 251–260.

Sage, R.F. 1996. Atmospheric modification and vegetation responses to environmental stress. Global Change Biology 2: 79–83.

Sánchez, G. 1995. Architecture and dynamics of *Stipa tenacissima* L. tussocks, effects on the soil and interactions with erosion. Ph.D. dissertation, Universidad Autónoma de Madrid, Spain.

Sánchez, G., and J. Puigdefábregas. 1994. Interaction between plant growth and sediment movement in semi-arid slopes. Geomorphology 2: 43–60.

Sandquist, D.R., and J.R. Ehleringer. 1998. Intraspecific variation of drought adaptation in brittlebush: leaf pubescence and timing of leaf loss vary with rainfall. Oecologia 113: 162–169.

Schlesinger, W.H. 1996. On the spatial pattern of soil nutrients in desert ecosystems. Ecology 77: 364–374.

Schlesinger, W.H., J.F. Reynolds, G.L. Cunningham, L.F. Huenneke, W.M. Jarrell, R.A. Virginia, and W.G. Whitford. 1990. Biological feedbacks in global decertification. Science 247: 1043–1048.

Schulze, E.D., H.A. Mooney, O.E. Sala, E. Jobbagy, N. Buchman, G. Bauer, J. Canadell, R.B. Jackson, J. Loreti, M. Oesterheld, and J.R. Ehleringer. 1996. Rooting depth, water availability, and vegetation cover along an aridity gradient in Patagonia. Oecologia 108: 503–511.

Shuster, W.S.F., D.R. Sandquist, S.L. Phillips, and J.R. Ehleringer. 1994. High levels of genetic variation in populations of four dominant arid land plant species in Arizona. Journal of Arid Environments 27: 159–167.

Slatyer, R.O. 1965. Measurements of precipitation interception by an arid zone plant community (*Acacia aneura*, F. Muell). Arid Zone Research 25: 181–192.

Soriano, A. 1983. Desert and semi-desert of Patagonia. In: West, N.E., ed. Temperate Deserts and Semi-Deserts. Amsterdam: Elsevier, pp 181–192.

Specht, R.L., and A. Specht. 1989. Canopy structure in Eucalyptus-dominated communities in Australia along climatic gradients. Acta Oecologica/Oecologia Plantarum 10: 191–213.

Stafford-Smith, M., and G. Pickup. 1993. Out of Africa, looking in: understanding vegetation change. In: Behnke, R.H., I. Scoones, and C. Kerven, eds. Range Ecology at Disequilibrium. London: Overseas Development Institute, pp 196–244.

Teklehaimanot, Z., and P.G. Jarvis. 1991. Modelling of rainfall interception loss in agroforestry systems. Agroforestry Systems 14: 76–80.

Teklehaimanot, Z., P.G. Jarvis, and D.C. Ledger. 1991. Rainfall interception and boundary layer conductance in relation to tree spacing. Journal of Hydrology 123: 261–278.

Terwilliger, V.J., and M. Zeroni. 1994. Gas exchange of a desert shrub (*Zygophyllum dumosum* Boiss.) under different soil moisture regimes during summer drought. Vegetatio 115: 133–144.

Thiery, J.M., J.M. d'Herbes, and C. Valentin. 1995. A model simulating the genesis of banded vegetation patterns in Niger. Journal of Ecology 83: 497–507.

Tongway, C., and J.M. d'Herbes. 1990. Vegetation and soil patterning in semi-arid mulga lands of eastern Australia. Australian Journal of Ecology 15: 23–34.

Turner, N.C., W.R. Stern, and P. Evans. 1987. Water realtions and osmotic adjustment of leaves and roots of lupins in response to water deficits. Crop Science 27: 977–983.

Valladares, F., and R.W. Pearcy. 1997. Interactions between water stress, sun-shade acclimation, heat tolerance and photoinhibition in the sclerophyll *Heteromeles arbutifolia*. Plant, Cell and Environment 20: 25–36.

Valladares, F., and R.W. Pearcy. 1998. The functional ecology of shoot architecture in sun and shade plants of *Heteromeles arbutifolia* M. Roem., a Californian chaparral shrub. Oecologia 114: 1–10.

Valladares, F., and F.I. Pugnaire. 1999. Cost-benefit analysis of irradiance avoidance in two plants of contrasting architecture co-occurring in a semiarid environment. Annals of Botany, submitted.

Van Elewijck, L. 1989. Influence of leaf and branch slope on stemflow amount. Catena 5: 25–33.

Virginia, R.A., and W.M. Jarrell. 1983. Soil properties in a mesquite-dominated Sonoran desert ecosystem. Soil Science Society of America J. 47: 138–144.

Vitousek, P. 1982. Nutrient cycling and nutrient use efficiency. American Naturalist 119: 553–572.

Wan, C., R.E. Sosibee, and B.L. McMichel. 1993. Broom snakeweed responses to drought: II. Root growth, carbon allocation, and mortality. Journal of Range Management 46: 360–363.

Wan, C., R.E. Sosibee, and B.L. McMichel. 1994. Hydraulic properties of shaloow vs. deep lateral roots in a semiarid shrub, *Gutierrezia sarothrae*. American Midland Naturalist 131: 120–127.

Watt, A.S. 1947. Pattern and process in the plant community. Journal of Ecology 35: 1–22.

Werk, K.S., and J.R. Ehleringer 1984. Non-random leaf orientation in *Lactuca serriola* L. Plant, Cell and Environment 7: 81–87.

West, N.E. 1993. Biodiversity of rangelands. Journal of Range Management 46: 2–13.

Westoby, M. 1980. Elements of a theory of vegetation dynamics in arid rangelands. Israel Journal of Botany 28: 169–194.

Westoby, M., B.H. Walker, and I. Noy-Meir. 1989. Opportunistic management for rangelands not at equilibrium. Journal of Range Management 42: 266–274.

White, L.P. 1970. 'Brousse Tiree' patterns in southern Niger. Journal of Ecology 58: 549–553.

Wiegand T., S. Milton, and C. Wissel. 1995. A simulation model for a shrub ecosystem in the semi-arid karoo, South Africa. Ecology 76: 2205–2221.

Witkowski, E.T.F., and B.B. Lamont. 1991. Leaf specific mass confounds leaf density and thickness. Oecologia 88: 486–493.

Woodward, F.I. 1987. Climate and Plant Distribution. Cambridge: Cambridge University Press.

11
Tropical Forests: Diversity and Function of Dominant Life-Forms

Ernesto Medina
Venezuelan Institute for Scientific Investigations, Caracas, Venezuela

I. INTRODUCTION

Tropical forests are geographically limited to the areas located between the tropics of Cancer and Capricorn. They harbor the largest diversity of species and ecosystems types on earth. Their importance for the maintenance of stable trace gas concentrations in the atmosphere is increasingly recognized. As a result, large international projects are currently developed to understand the global interactions derived from biogeochemical cycling in tropical forests and the consequences of large-scale deforestation for loss of biological diversity and stability of the world climate. Numerous fundamental contributions on the ecology of tropical forests have been produced in this century, describing composition, structure, and functioning of these communities (Richards 1952, 1996, Walter 1973, Vareschi 1980, Whitmore 1984, 1990, Jordan 1985, Kellman and Tackaberry 1997, Odum and Pigeon 1970, Unesco 1978, Leigh et al. 1982, Sutton et al. 1983, Jordan et al. 1989, Proctor 1989, Gómez-Pompa et al. 1991, Mulkey et al. 1996). These references served as a conceptual basis for this chapter, which is aimed at highlighting the functional characteristics of tropical-forest plant components. This chapter gives a brief outline of the distribution and edaphoclimatic characteristics of tropical forests as a basis for the more detailed discussion of diversity and function of main life-forms. The relationship between nutrient availability and photosynthetic performance of tropical trees as a basis for considering the tropical forests as natural sinks for atmospheric carbon and nitrogen is discussed.

A. Climate

Climatic delimitation of tropical regions has been discussed by numerous investigators (Richards 1952, 1996, Schultz 1995, Walter 1973, Whitmore 1984). Climatic characterization emphasizes the lack of seasonal variations of temperature and daylight duration. It can still be found in the literature that tropical regions are hot, humid environments, thereby excluding areas located in medium to high tropical mountains or in dry tropical semi-arid lands from the analyses of tropical ecosystems. Using temperature criteria, Walter (1973), described the tropics as areas of the world with a diurnal (''Tageszeiten'') climate in opposition to the marked seasonal (''Jahreszeiten'') climate of extratropical regions. This implies that diurnal temperature fluctuations are ecologically more significant than the difference in average temperatures between the hottest and the coldest month. This characterization applies to the whole range of tropical forests, from the lowlands to the high-altitude formations well represented in all tropical regions of the world. Other treatments separate tropical from nontropical systems using the occurrence of freezing temperatures at altitudes below 100 m.

On the other hand, tropical ecosystems are mostly affected by seasonality of rainfall distribution, a climatic trait associated with the displacement of the intertropical convergence zone. Tropical regions also include large areas with relatively low rainfall levels, which tend to be very dry, particularly in the lowlands, because of large evapotranspiration demands that remain almost constant throughout the year. Most ecological descriptions of tropical ecosystems distinguish humid from dry tropics. The distinction is based essentially on the ratio between precipitation and potential evapotranspiration (P/E ratio). On an annual basis, the humid tropics are characterized by P/E ratios equal to or greater than one, whereas in the dry tropics, P/E ratios are less than one.

B. Ecoclimatic Classifications

Rainfall and temperature account for the large majority of changes in forest structure and species composition in the tropics. Within a certain temperature-rainfall regime, soil structure and fertility regulate the differentiation of forest communities (Hall 1996, Medina and Cuevas 1993, Schulz 1995).

Temperature and rainfall interact to determine water availability at a given site. Temperature is directly correlated with evaporative demand of the atmosphere, whereas rainfall provides water supply through the roots and modifies atmospheric evaporative demand through changes of the atmospheric water content. Humidity indices are frequently calculated using temperature and rainfall data, mainly because of the lack of direct measurements of evaporation (discussions on climate classification including tropical areas can be found in Cramer and Leemans 1993, Lal 1987, Schulz 1995, Walter 1973).

A relatively simple approach to separate large groups of climates in the tropics uses average values of rainfall (as a measure of water availability) and temperature (as a measure of evaporative demand of the atmosphere, but also as a regulator of plant activity). With mean annual data series, climate types can be separated as being associated with increasing water availability (moisture provinces in Figure 1) and with altitudinal belts related to reductions in temperature (altitudinal belts in Figure 1). One widely used system to classify climatic units in connection with forest ecosystems is the Life Zone System of Holdridge (1967). Such a comprehensive summary of climatic data is useful to evaluate patterns of forest distribution on a large scale. For specific studies, the analysis of seasonal variations of conditions for plant growth is essential to understand the ecological process, because in the tropics seasonal variations in rainfall are far more important than variations in temperature. The seasonality of a certain climate is better observed in synthetic displays of the climatological variables (climate diagrams; see Walter 1973 for details). Figure 2 shows the climatic characterization of a number of meteorological stations in northern South America using the criteria of Bailey (1979) for separating humid and dry months. Tempera-

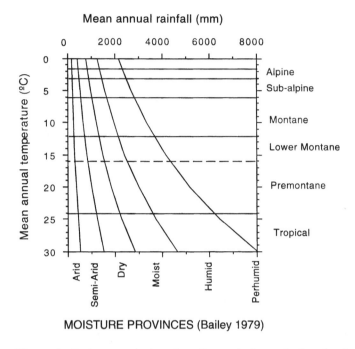

Figure 1 Basic categorization of ecoclimates in the tropics based on humidity provinces after Bailey (1979) and altitudinal belts after Holdridge (1967). Humidity provinces are separated according to rainfall and evaporation ratios, while the altitudinal belts are separated according to annual temperature averages. The dotted horizontal line indicates the altitudinal belt at which frosts are expected to occur.

ture decreases with altitude, and so does the amount of monthly rainfall required to compensate for atmospheric evaporative demand.

C. Soil Fertility

The separation of humid and dry tropics is frequently paralleled by clear-cut changes in edaphic characteristics. Soils of the humid tropics tend to be more leached by heavy rainfall and show slight to strong acid pH and a tendency to high Al mobility. These characteristics represent a condition in which one or more nutrients may limit plant growth and productivity. In these areas, lower geomorphological settings lead to water logging, compromising the nutrient availability by low oxygen supply to the roots during periods of variable duration. In humid tropical forests, water and temperature are not limiting, and nutrients acquire the leading role in differentiating tropical forest types.

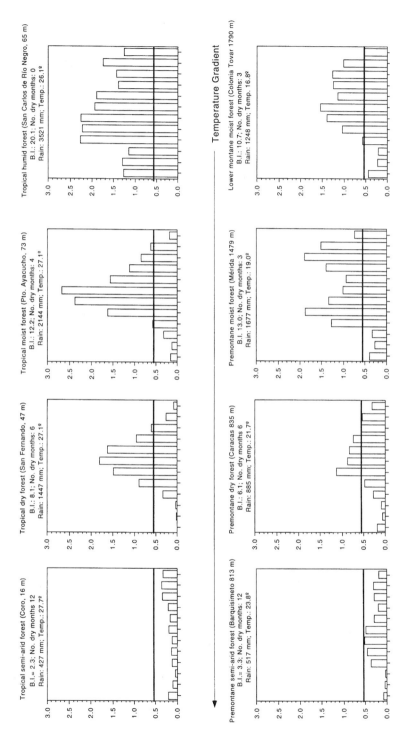

Figure 2 Examples of seasonal variations in water availability along rainfall and temperature gradients in tropical forest sites in northern South America. The Y-axes give the monthly S index of Bailey (1979). The horizontal thick line gives the value of the S index separating the humid from the dry realms.

In the dry tropics, low water availability during annual periods ranging from 2–7 months constitutes the major limiting factor, with soils generally near neutral pH, and a cation saturation of the soil exchange complex generally above 50%.

Large-scale gradients of rainfall and soil fertility put forward the covariance of these two ecological factors in the tropics. Fertile soils, measured with a combination of total sum of bases, percentage of base saturation, phosphorus availability, and pH, are more frequently found in areas with rainfall less than 1500 mm. The large-scale studies conducted in Ghana by Swaine and others (Hall and Swaine 1976, Swaine 1996) showed a good separation of the influence of rainfall (water availability) and soil fertility (nutrient availability). In humid areas (>1500 mm rainfall), soil fertility appeared as a strong determinant of vegetation, whereas in drier areas, other factors seemed to gain more relevance. Investigators were able to separate the group of 48 tree species surveyed throughout the forest zone of Ghana in four distribution groups established by a χ^2 technique: (1) biased to dry-fertile sites; (2) biased to wet-infertile sites; (3) biased to wet-fertile sites; and (4) nonbiased to any specific site. The large majority of the species listed were within the wet-infertile and the nonbiased groups. Similar analyses conducted in drier areas in the tropics will certainly render much insight into the ecological factors determining ecophysiological demands for seedling establishment in these areas (see contributions in Bullock et al. 1995) This differentiation may prove to be very useful in designing experiments to establish nutritional and hydric dependence of seedling establishment and development under natural conditions. Application of this approach to large continuous expanses of tropical forests, as those in the Amazon basin, will certainly uncover a rich set of ecophysiological species–soil–climate groups.

D. Distribution and Extension

The accurate estimation of distribution and extension of tropical forests is essential to: (1) determine the potential carbon fixation capacity of tropical areas of the earth; and (2) assess the potential effects of increasing concentration of atmospheric CO_2 on global climate change. The literature before 1975 has been analyzed and used by Lieth and Whittaker (1975) to calculate the productivity of organic matter by terrestrial ecosystems. For the estimation of productivity of terrestrial ecosystems as a whole, the point values of productivity at specific sites were expanded to an area basis using the extension of each ecosystem unit (Lieth 1975). More recently, estimations using both climatic and geographic criteria based on satellite estimation of Normalized Vegetation Indices have been published (see summarizing reports including distribution and areal extension: Cramer and Leemans 1993, Melillo et al. 1993, Schultz 1995).

Tropical evergreen forests occupy approximately 17 million km², or nearly

Table 1 Area and Net Primary Production of Organic Matter Estimated by Direct Measurements and Using a Process-Based Ecosystem Simulation Model

Vegetation units	Area ($\times 10^6$ km^2)	%	Net primary production (g C m^{-2} yr^{-1})	Total net primary production (10^{15}g C yr^{-1})	%
World total	127.3			53.2	
Tropical evergreen forest	17.4	13.7	1098	19.1	35.9
Tropical deciduous forest	4.6	3.6	871	4.0	7.5
Tropical savanna	13.7	10.8	393	5.4	10.2
Xeromorphic forests	6.8	5.3	461	3.1	5.8
Total tropical	42.5	33.0		31.6	59.4

Source: Adapted from Melillo et al. (1993).

14% of all land surface of the planet, but account for approximately 36% of total net primary production of terrestrial ecosystems (Table 1). As a whole, tropical regions of the earth occupy 33% of the total surface, but their total net primary production accounts for up to 59% of the total primary production of terrestrial ecosystems. This bigger contribution to total organic matter production is largely the result of longer growing periods.

II. SPECIES DIVERSITY

A. Patterns of Species Diversity

Tropical forests contain the largest set of higher plants species known. Gentry (1988) reported values as high as 250 tree species per 0.1 ha in several wet tropical forests in South America. Phillips et al. (1994) analyzed tree species surveys in tropical rain forests throughout the world, and reported species richness of trees $\geq$ 10 cm diameter at breast height (d.b.h.) surpassing 200 species/ha in Malaysia, Sarawak, Ecuador, and Peru, while plots in Uganda, Peru, Venezuela, and Ghana were relatively poor, with approximately 60–90 species/ha. Most sites showed values between 100 and 200 species/ha. The highest species-richness values reached 235 (Lambir, Sarawak) and 267 (Yanamomo, Peru) trees $\geq$ 10 cm d.b.h. per 500 stems, while the lowest values measured in Kibale, Uganda, and Tambopata, Peru, were approximately 50 species/500 stems.

Gentry (1988) found that the number of tree species/0.1 ha increased with rainfall, from approximately 50 at rainfall levels near 1000 mm to 250 species at rainfall levels of $\geq$ 4000 mm. In addition, vascular species diversity decreased strongly with altitude (from 1500–3000 m altitude). No clear relationship was found between soil fertility and diversity, although species composition generally changed strongly in association with soil fertility. Forests without relatively dry months (rainfall ≤ 100 mm) and good soil conditions, such as La Selva and Mersing, had tree-species richness of approximately 100 species/ha, while sites in Lambir, Sarawak, and Mishana, Peru, with no rainfall limitation but very low fertility, had numbers well above 200 species/ha (Phillips et al. 1994; Figure 3). Using data from Holdridge et al. (1971) for Costa Rica, Huston (1980) showed that tree-species richness appeared to be inversely correlated with soil fertility. There were significant negative correlations with available P, exchangeable K

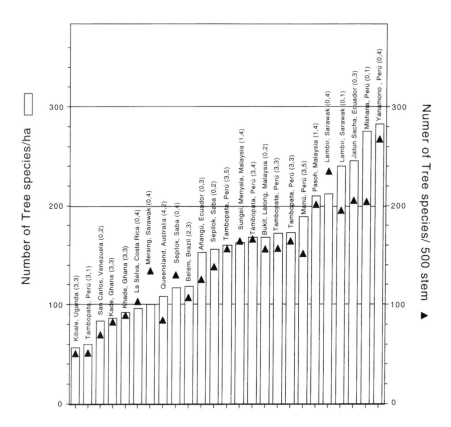

Figure 3 Diversity of tree species in tropical forest sites throughout the world. Sites have been ordered along increasing diversity as number of species per ha or number of species per 500 stems. The first number after the site name indicates the number of dry months (0–4), and the second indicates site fertility increasing from 1 to 5. (*Source*: Phillips et al. 1995.)

and Ca, total bases, base saturation, and cation exchange capacity. These results were interpreted as a support for the hypothesis that higher tree diversity in tropical forests occurs under nutritional conditions that prevent dominance of single, highly competitive species (Huston 1994). The assesment of soil fertility using conventional soil chemical analysis, such as nutrient concentration without considering rooting depth, bulk density, and nutrient availability, is frequently misleading. Therefore, further experimental work is necessary to reach a definitive picture of the role of nutrients in the regulation of diversity patterns in the tropics.

The contribution of nontrees (herbs and epiphytes) to total diversity of vascular plants is substantial, particularly in the neotropics, and increases with rainfall and possibly with soil fertility (Gentry and Dodson 1987a, 1987b, Gentry and Emmons 1987).

There is voluminous literature on the mechanisms explaining the development and maintenance of diversity on local and temporal scales (see review by Barbault and Sastrapadja 1995). Models of particular significance to understand patterns of variations of plant in the tropics are those of Tilman (1988) and Huston (1994). These models open a large framework for experimental testing of prediction based on physiological and behavioral characteristics of plant species.

B. Low-Diversity Tropical Forests

Of considerable theoretical importance is the occurrence of low-diversity tropical forests, because it may provide explanations for the patterns of diversity in the tropics (Newbery et al. 1988, Connell and Lowman 1989, Torti et al. 1997). Connell and Lowman (1989) point out that the majority of those low-diversity forests are constituted by Caesalpinioid legumes, forming ectomycorrhizal symbiosis. Examples of these forests include the following: (1) in Africa: *Gilbertiodendron dewevrei* and *Brachystegia laurentii* (Zaire), *Cynometra alexandri* (Uganda), and *Tetraberlinia tubmaniana* (Liberia); and (2) in tropical South America: *Mora excelsa, M. gonggrijpi*, and *Eperua falcata* (Trinidad and Guyana); and *Pentaclethra macroloba* (Costa Rica).

The Connell-Lowman hypothesis of monodominance is based on the fact that mycorrhizae improve the nutrient and water relations of the host plant (this subject is discussed further in Section III. A). Ectomycorrhizae (EM) tend to be more specific; therefore, their benefits are limited to fewer host species. On the other hand, vesicular-arbuscular mycorrhizae (VAM) are more promiscuous and widely distributed in tropical forests; therefore, their predominance promotes diversity. A relevant ecological situation regarding the argument of mycorrhizal specificity is that of the Dipterocarpaceae, a predominantly ectomycorrhizal family that dominates large tracts of tropical forests in southeast Asia. In this case, ectomycorrhizal fungi are able to form associations with several species. This promiscuity promotes diversity, but only among species of the family that are susceptible to the same fungus. Hart et al. (1989) discussed other possible mechanisms such as reduced fertility, dominance restricted to a certain successional state, gradients of mortality and herbivory based on high recruitment and herbivory pressure on the dominant species, and finally the low disturbance regime that may lead to the dominance of a few more competitive species.

It has recently been shown that in several forests with the tendency to monodominance of caesalpinioid species, the species involved were either exclusively VAM or had a mixture of VAM and EM. Moyersoen (1993) found that *Eperua leucantha*, forming nearly monodominant forests in the upper Río Negro basin, forms only VAM, while *Aldina kundhartiana* (Caesalpiniopid) and species of *Guapira* and *Neea* (Nyctagynaceae) formed both VAM and EM. Torti et al. (1997) showed that two notorious caesalpinioid species forming monodominant forests in Trinidad (*Mora excelsa*) and Panama (*Prioria copaifera*) are exclusively VAM in the areas sampled. Again, these findings open the issue of the fundamental causes for the existence of monodominant forests in the lowland tropics.

In this context, it is worthwhile to mention the contrasting patterns of diversity of mycorrhizal fungi and higher plants (Allen et al. 1995). The coniferous forests in the northern hemisphere have more than 1000 species of ectomycorrhizal fungi, but are dominated by a limited number of higher plants forming EM mycorrhizae. In the tropics, the predominating form of mycorrhiza is endophytic (VAM), but the number of fungi species is very limited despite the high diversity of higher plants. Allen et al. (1995) propose that while EM are taxonomically diverse and tend to be species-specific, VAM are physiologically diverse and tend to be more promiscuous, capable of forming mycorrhizae with a large variety of higher plant species.

III. DIVERSITY AND FUNCTIONAL ECOLOGY OF LIFE-FORMS

Species richness of tropical humid forests is paralleled by the diversity of coexisting life-forms, presumably indicating the variety of ecological niches that can be occupied and exploited by contrasting morphologies and physiologies. Life-forms may be described by a number of morphological features associated with physiological properties of significance for competitive ability and reproductive capacity (Solbrig 1993, Ewel and Bigelow 1996). The association of morphophysiological properties allows a working organization of the species-richness characteristic of tropical forests in a small number of categories that can be associated with ecosystem function (Ewel and Bigelow 1996) and response to environmental change (Denslow 1996).

A number of life-form systems could be used as a classification scheme to differentiate forest types throughout the globe (Box 1981, Schultz 1995). Some of those systems focus mainly on tropical forest communities (Ewel and Bigelow 1996, Vareschi 1980, Halle et al. 1978). The most relevant functional aspects of the larger groups of vascular plants are discussed here, including: (1) trees and shrubs (including palms); (2) vines, lianas, and hemiepiphytes; and (3) epiphytes.

A. Dicotyledonous Woody Plants

Dicot trees constitute the dominant life-form of the majority of tropical forest communities. Predominance of gymnosperms is only observed in some montane forests near the equator (with several species of *Podocarpus*), or toward the limits between the tropical, subtropical, and temperate realms (with species of *Pinus* and *Araucaria*) (Hueck 1966). Palm-dominated forests are restricted to seasonal or permanently flooded sites and will be discussed later.

Forest dynamic is characterized by events of disturbance and regeneration that may be categorized as follows: (1) gap phase: the opening of gaps within the forest matrix caused by natural disturbances; (2) recovery phase: seedling establishment, tree resprouting, sapling and pole development; and (3) mature phase: canopy closure and slow disappearance of pioneer species established during the gap and recovery phases (Whitmore 1991). This successional process is complex in nature due to the space-temporal variability of the interactions between physicochemical factors, the availability of seed sources, and biotic constraints such as seed predation, herbivory, and disease (Bazzaz 1984, Whitmore 1991).

Trees follow several stages during forest regeneration: (1) seedling, (2) sapling, (3) pole, and (4) mature (Oldeman and van Dijk 1991). The first and last phase are better known ecophysiologically and both extremes will be discussed to highlight the present knowledge of environmental constrains and ecophysiological adaptations of tropical forest trees.

Studies of population dynamics under natural conditions have established that tropical rain forest trees present a large range of shade tolerance. The extremes of this continuum are early successional or pioneer trees that grow rapidly in gaps under high light intensity, and late successional (mature forest trees) that are found in the forest understory and persist for prolonged periods, showing slow or nil growth rates (Swaine and Whitmore 1988). Those differences are also associated to population properties such as reproductive effort and frequency, intrinsic growth rate, and sensitivity to nutrient and water availability.

The universe of morphological characteristics, growth patterns, and light requirements may be used for a broad characterization of tree types in tropical forests. They have been described as temperaments that encompass the range of behavior observed under natural conditions and may serve as a guide for the study of forest conservation and regeneration. The main types as identified by

Oldeman and van Dijk (1991) include in one extreme species that reproduce frequently, producing large numbers of seedlings, with little tolerance to low light intensity, and rapid growth rates. Those are the light demander pioneers (Swaine and Whitmore 1988), early successional species, or gamblers in the nomenclature of Bazzaz and Picket (1980). At the other extreme are the species that reproduce less frequently, producing a small number of shade-tolerant seedlings, capable of enduring prolonged periods under the forest canopy. Those are the climax species of Swaine and Whitmore (1988), late successional species, or the strugglers of Bazzaz and Picket (1980). There is a full range of intermediate types, and it is not simple to characterize fully the behavior of a certain species, because frequently their characteristics may also change with age and successional status (Oldeman and van Dijk 1991) (Table 2).

1. Establishment and Development

Forest regeneration requires the establishment of seedlings and saplings within the same (or similar) environment in which the parent trees grow. However, in rain forests, environmental conditions determining performance of adult trees contrast with those under which their seeds germinate and develop. Adult trees occupy a volume of the forest canopy, with levels of light availability at least one order of magnitude higher than those prevalent in the forest understory (Chazdon and Fetcher 1984). Photosynthetically active radiation is the driving force in the production of organic matter; therefore, it has been assumed that openings of the canopy (gaps) allowing enough light to reach the understory are required for the regeneration of forest trees (for a review see Strauss-Debenedetti and Bazzaz 1996).

Light intensity has been identified as the main limitation for seedling establishment in tropical rain forests. Nutrient availability may be also critical as a regulator of seedling growth rate. Root competition and mutualistic symbiosis also affect seedling establishment, at least in forests growing on acidic, highly leached soils. Their influence is mainly related to the regulation of nutrient availability. Genetic factors regulating growth rate determine the demand for environmental resources, and therefore may influence the carbon balance in the shady understory of tropical forests. Herbivory is frequently critical for seedling survival, but it is not within the scope of the present analysis.

2. Plasticity and Acclimation of Photosynthesis

The bulk of literature on the effect of light intensity on growth of rainforest tree seedlings has been thoroughly reviewed elsewhere (Strauss-Debenedetti and Bazzaz 1996, Kitajima 1994). A summary from those reviews indicates the following: (1) independently of successional status, survival and growth rates of all

Table 2 Main Morphological and Ecophysiological Characteristics Separating Types of Tree "Temperaments"

Temperament classification	Leaf arrangement	Successional status	Light requirement	Examples
Hard gambler	Monolayer, spherical to hemispherical, phyllomorphic	Early successional	Light demanding, shade intolerant	*Cecropia, Jacaranda, Didymopanax,* fan palms, *Annona, Apeiba*
Gambler	Multilayer	Later successional	Light demanding, some tolerance	*Trema, Macaranga, Goupia, Ceiba*
Gambling strugglers, struggling gamblers	Umbrella, pagoda, paucilayer	Later to late successional	Shade tolerant to intolerant, favored by light	Feather palms, *Terminalia, Manilkara, Pouruma, Iryanthera, Aspidosperma*
Strugglers	Elongate, diffuse	Late successional	Shade tolerant, favored by moderate light levels	*Anaxagorea, Duguetia, Hirtella, Perebea*
Hard strugglers	Monolayer	Late successional	Very shade tolerant, favored by moderate light levels	*Pavonia, Guarea, Casearia*

Source: Adapted from Oldeman and van Dijk (1991).

species are lower under light regimes similar to those prevalent in the understory of tropical rain forests; (2) photosynthetic traits of species classified as late successional are generally less plastic than those of the species classified as early successional or pioneers; and (3) there is a continuum in the tolerance to shade among tropical forest trees.

Growth under the low-light regimes of the forest understory represents a significant stress jeopardizing seedling establishment and regeneration of the tree species that dominate those forests. The conclusion is that some kind of forest disturbance that results in gap formation of varying sizes might be indispensable for forest regeneration.

Among the multiple responses of the photosynthetic machinery of different species to contrasting light regimes, the interactions between regulation of maximum photosynthetic rate (A_{max}) and the allocation of biomass to photosynthetic structures or roots still must be analyzed and explained in detail. Several studies showed that the responses of seedlings from trees of different successional status may be related to both increased photosynthesis and/or increased allocation into leaf biomass when the plants are grown under relatively high intensity (Tinoco-Oranjuren and Pearcy 1995, Ramos and Grace 1990).

Plants developed under a certain light regime adapt their morphology and biochemical characteristics so that carbon gain is maximized. The degree of change in those properties for a species cultivated under contrasting light regimes represents its photosynthetic plasticity (Fetcher et al. 1983, Oberbauer and Strain 1985, Ramos and Grace 1990, Riddoch et al. 1991a, 1991b, Strauss-Debenedetti and Bazzaz 1991 Turnbull 1991, Turnbull et al. 1993, Ashton and Berlyn 1992, Kammaludin and Grace 1992a, 1992b, Tinoco-Ojanguren and Pearcy 1995). Changes in light conditions during growth, from low to high or in the reverse direction, generate strong physiological changes that can be measured as responses in growth and leaf photosynthetic rates. The speed and efficiency with which those changes take place constitute the acclimation capacity of the species considered (Bazzaz and Carlson 1982). Photosynthesis acclimation can take place within the same leaves developed in a different light regime, or in new modules that develop after the shift of light conditions (Kamaluddin and Grace 1992a, 1992b, Strauss-Debenedetti and Bazzaz 1996). In general, acclimation of leaves formed under a certain light regime is not as complete as the one observed in the newly formed leaves. These results support the view of Bazzaz et al., that light-demanding species are more plastic and acclimate faster to contrasting light regimes than late successional species.

Both A_{max} and dark respiration (R_{dark}) are reduced when plants are transferred from high to low light intensity, regardless of the successional status of the species considered (Turnbull et al. 1993). Transfers from low to high light result in larger increases in A_{max} and R_{dark} by pioneer and early successional species. Pioneer species can modify their allocation of nitrogen to photosynthetic proteins according to levels of light energy available during growth; therefore, they are more plastic, as reported by Bazzaz and Pickett (1980). Late successional or shade types cannot take full advantage of the higher light availability because their photosynthetic protein content does not increase as much as in the sun types (Gauhl 1976, Björkman 1981). Regulation of respiration of leaves, and possibly of all tissues of seedlings developed under low light, may have a more important role in shade tolerance than has been appreciated to date.

Most studies on the photosynthetic plasticity and acclimation capacity of tree seedlings have been conducted at the leaf level only. Conclusions regarding light requirements for survival based on this type of analysis may be erroneous because leaf area development has a number of costs associated with its own construction and maintenance and with support and transport of water and nutrients (Givnish 1988). Ecological significant light compensation points must be calculated on a daily basis, taking into account the photosynthetic surplus of the leaves and the consumption of assimilates by leaf night respiration, and amortization of construction costs. When whole-plant gas exchange is considered, the light requirements for a positive carbon gain increase rapidly. This type of analysis is required for a more fundamental understanding of seedling survival in the understory of tropical forests.

For many tropical rainforests, the light environment in the understory is very dynamic because of the occurrence of sunflecks. Pearcy et al. showed that seedlings and saplings growing in the understory of natural forests respond significantly to the frequency and intensity of sunflecks. These short-term pulses of light contribute to improve the carbon balance of plants in the shade (Chazdon and Pearcy 1986, Pearcy 1988). Photosynthetic capacity of shade plants is induced by sunflecks of several minutes' duration (light activation of ribulose biphosphate [RuBP]-carboxylase). Afterward, photosynthetic responses to subsequent sunflecks are rapid. Fully induced leaves show higher carbon gain than expected with steady-state assimilation rates because of the postillumination CO_2 fixation (Pearcy 1988). Models showed that frequency of light flecks is critical to attain higher carbon balances than those expected under steady state of photosynthesis.

In the understory of rain forests, CO_2 concentrations are frequently above the average concentrations in the atmospheric air above the forest (Medina et al. 1986, Wofsy et al. 1988). The higher CO_2 concentration in the forest understory is a result of the CO_2 production by soil respiration (root respiration + respiration of decomposing organisms in the soil). Higher CO_2 concentration may improve the carbon gain of light-limited leaves by increasing the CO_2 gradient between air and carboxylating sites in the chloroplasts. Plants in the understory of tropical forests have a lower abundance of carbon 13 in their tissues as measured by the $\delta^{13}C$ value (Medina et al. 1986). This results from the contribution of carbon 13–depleted CO_2 from organic matter decomposition and tree respiration to the photosynthesis of the shade flora (Medina et al. 1986, van der Merwe and Medina

1989), and to a reduction in the ratio of internal to external CO_2 concentration (C_i/C_a) in the leaves (Farquhar et al. 1989). I am not aware of any experimental evidence indicating that the increased availability of CO_2 in the understory can compensate for the constraints to photosynthesis imposed by low light intensity.

3. Radiation Load and Photoprotection

Light energy utilization by the photosynthetic apparatus involves light absorption, electron transport against electrochemical gradients leading to energy accumulation in phosphorylated compounds, and synthesis of reduced compounds (photochemical component). This "reducing" power (adenosine triphosphate [ATP] + reduced nicotinamide adenine dinucleotide phosphate [NADPH]) is used to incorporate carbon into energy-rich organic compounds (enzymatic component) and in photorespiration. The use efficiency of absorbed light energy depends on the electron transport capacity of the photochemical component, and the consumption of the ATP and NADPH molecules generated in the process by the enzymatic component. Under natural conditions, when light absorption surpasses the capacity of the photosynthetic machinery to process it, the phenomenon of photoinhibition is observed. It is defined as a reduction in the quantum yield of photosystem II that may lead to a decrease in the CO_2 uptake rate (Long et al. 1994, Osmond and Grace 1995). Higher plants differ in these capacities according to genetic constraints and ecological conditions such as light climate during growth and drought stress. The study of photoinhibition in nature is a rapidly expanding field, resulting from theory development and the availability of portable instrumentation for assessment of leaf fluorescence under natural illumination fields (Schreiber et al. 1995).

Plants occupying different strata in an undisturbed forest are submitted to contrasting light conditions during the day. There is a vertical, logarithmic reduction in integrated light intensity that may vary at least on one order of magnitude (Fetcher et al. 1983, Chazdon 1986). The photochemical efficiency of photosystem II is affected by changes in the light climate in a very dynamic fashion. In leaves of fully sun-exposed plants, intrinsic quantum yield of photosystem II (measured as reduction in the quotient of variable fluorescence to maximum fluorescence in dark adapted leaves, F_v/F_m) decreases as the light energy absorbed by the leaf increases. The reduction in quantum yield of photosystem II can be considered as a protective mechanism of the photosynthetic machinery, and it is accompanied by a highly effective mechanism that dissipate excess light energy as heat, the xantophyll cycle (Björkman and Demming-Adams 1995). Photoinhibition processes under natural conditions are, in general, readily reversed under low light or darkness in the order of minutes (dynamic photoinhibition) or hours (chronic photoinhibition) (Osmond and Grace 1995).

Sensitivity toward photoinhibiton in tropical forests may be assessed by measuring the amount of pigments of the xantophyll cycle (Königer et al. 1995).

Canopy species showed higher maximum photosynthetic rates and had a xanthophyll cycle pool (violaxanthin + anteraxanthin + zeaxanthin) averaging 87 mmol mol^{-1} chlorophyll, leaves of the gap plants had intermediate photosynthetic rates and a xanthophyll cycle pool of 35 mmol mol^{-1} chlorophyll, whereas the understory plants had both the lowest photosynthetic rates and the lowest concentration of the xanthophyll cycle pool, 22 mmol mol^{-1} chlorophyll.

In fully exposed leaves, either in the canopy or in large gaps, the efficiency of photosystem II measured as the ratio F_v/F_m is inversely related to the light intensity previous to the time of measurement; however, this reduction is generally recovered overnight (Königer et al. 1995). The reversion of fluorescence quenching produced by photoinhibition during the day recovers in the shade within 1–2 hours in several gap species in Panama (Krause and Winter 1996). Young leaves appear to be more sensitive to photoinhibition, but they have a similar recovery capacity to that of adult leaves (Krause et al. 1995).

When plants belonging to different successional groups are experimentally grown under low light intensity and then transferred to full sunlight, the result is generally an abrupt reduction in F_v/F_m. This reduction is more pronounced and takes longer to recover in species usually found in understory environments (Lovelock et al. 1994). Other studies did not find differences among species of different successional status regarding chronic photoinhibition when cultivated without water or nutrient stress (Castro et al. 1995).

The intensity of light impinging upon the leaf surface is a function of the cosine of the incidence angle. Therefore, the simplest mechanism to regulate excess of light energy is the change in the angle of leaf inclination. Pronounced leaf angles are frequent in canopy trees in the humid tropics, and had been shown to be critical for leaf energy balance under eventual water stress (Medina et al. 1978). Lovelock et al. (1994) showed that for a set of species growing under varying light climates, degree of leaf inclination is frequently large enough to compensate for differences in leaf exposure to sunlight, resulting in similar F_v/F_m ratios in sun and shade plants.

4. Nutrient Availability and the Role of Mycorrhizal Symbiosis

Soon after exhaustion of seed reserves, seedling growth is strongly dependent on the availability of nutrients from the forest soils (Kitajima 1996). In most tropical forests, P is the most common limiting nutrient (Grubb 1977, Vitousek 1984, Vitousek and Sanford 1986, Medina and Cuevas 1994), but seedling growth and survival may be affected by simultaneous limitations of several nutrients.

In *Melastoma malabathricum* mixtures of P plus Ca, micronutrients Mg, K, and N stimulated growth far beyond that provoked by P alone (Burslem et al. 1994, 1995). Particularly important are the light intensity–nutrient availability interactions. Increasing light intensity elevates nutrient demands, most notably in the case of N (Medina 1971). Increased nutrient supply leads to higher maximum photosynthetic rates in seedlings of plants of different successional status, irrespective of the light intensity of cultivation (Riddoch et al. 1991b, Thompson et al. 1988, 1992). High light intensity may even negatively affect photosynthetic performance under conditions of low nutrient availability (Thompson et al. 1988).

Mycorrhizal symbiosis is widespread in tropical humid forests (Janos 1983, St John and Uhl 1983, Hopkins et al. 1996). Most species develop VAM, but an important group constituted by the Dipterocarpaceae and legumes of the subtribes Amherstieae and Detarieae (Caesalpinioid), and such important genera as *Aldina* and *Swartzia* in the Papilionoid, are ectomycorrhizal and can reach dominance over large areas in the humid tropics (Janos 1983; Alexander and Högberg 1986; Alexander 1989). The occurrence of mycorrhiza is of paramount importance to understand the nutrient balance of humid tropical forests. The widespread limitation of P availability over vast tropical areas emphasizes the importance of this biological interaction, which is considered to increase the capability of water and nutrient uptake, particularly P, by higher plants. Went and Stark (1968) proposed the occurrence in tropical forests of a "closed" nutrient cycle mediated by mycorrhiza preventing or reducing nutrient leaching. It is now generally accepted that predominance of mycorrhizal symbiosis (both ecto-and endomycorrhizae) in the majority of humid tropical forests is certainly associated to low P availability in the soil (Newbery et al. 1988). The frequency and percentage of mycorrhizal infection is inversely related to soil pH and available P (van Noordwijk and Hairiah 1986). In tropical forests with a thick root mat, phosphate solutions sprayed on the soil surface can be efficiently taken up by VAM, thereby preventing nutrient leaching in these forests (Jordan et al. 1979).

The improvement in nutrient supply brought about by mycorrhizal symbioses is related to the increase of surface for nutrient absorption, by penetrating the soil beyond the zone of nutrient depletion around the fine roots. In addition, some mycorrhizal fungi are capable of using organic P sources directly or through previous digestion by extracellular phosphatases (Alexander 1989).

Nutrient availability in a given soil may be markedly affected by root competition. Particularly in forests growing on nutrient poor soils, fine roots tend to accumulate near and above the soil surface, developing a root mat that can exert a strong competitive pressure for water and nutrients (Stark and Jordan 1977, Jordan et al. 1979, Cuevas and Medina 1988). Root trenching experiments in a tropical rain forest of the upper Orinoco basin resulted in increased concentration of N and P in the trenched saplings, and also a marked increase in leaf area development and branching. These experiments show that root competition reduces sapling growth and perhaps survival in the understory due to the competition for nutrients exerted by root biomass of established trees (Coomes 1995).

5. Efficiency of Nutrient Use in Carbon Uptake and Organic Matter Production

Productivity of tropical humid forests is frequently limited by the nutrient supply from the soil substrate. Main limitations have been described for N, P, K, and Ca (Vitousek 1984, Medina and Cuevas 1994). Nutrient limitation is reflected in the photosynthetic capacity of tropical trees, N being the most common limiting nutrient, followed by P, K, and less frequently Ca. The impact of nutrient limitation on the development and performance of the photosynthetic machinery of tropical trees is complex. Nutrient supply affects leaf expansion, leaf thickness, leaf weight/area ratios, and intrinsic capacity for photosynthesis (Medina 1984, Reich et al. 1991, 1995). Efficiency of nutrient use in photosynthesis may be conveniently expressed measuring the relationship between the specific nutrient concentration and the maximum rate of photosynthesis under natural or laboratory conditions. This is particularly significant for the case of N and P, nutrients normally incorporated into organic molecules directly associated with the photosynthetic machinery. Studies in tropical humid forests on sandy soils with strong nutrient limitations confirmed previous findings, showing that photosynthetic rate (μmol CO_2 per unit area or weight) and leaf N concentration are often linearly related (Reich et al. 1991, Raaimakers et al. 1995). These studies indicate that within a certain habitat, the efficiencies of N and P use in photosynthesis (μmol CO_2 mol^{-1} nutrient s^{-1}) are highly correlated (Figure 4). In this data set, photosynthetic N/P efficiencies ratio varies from approximately 50 to 70. However, in communities apparently not limited by P, with leaf P contents much higher than those of the Amazonian trees previously discussed, the P use efficiencies are much lower, and the N/P ratios vary around 20 (Tuohy et al. 1995) (Figure 4).

Both P and N use efficiencies in photosynthesis (and overall plant growth) are significantly higher in species belonging to the pioneer and early succession categories. These species are characterized by larger growth rates, shorter life spans, and higher leaf nutrient concentrations. One significant generalization on the relationships between photosynthetic capacity and nutrient concentration has been achieved combining measurements of photosynthetic performance with demographic studies aimed to measure leaf life spans (Reich et al. 1991). Leaf life span increased along the successional status of the species considered from pioneer, early, mid-, and late successional. Along this gradient, photosynthetic capacity decreased, together with leaf structural properties such as leaf area/weight ratios and N and P concentrations (Reich et al. 1995).

The relationships between structural, nutritional, and functional properties of leaves described for tropical humid forests apply for a wide variety of forest communities across large climatic gradients (Reich et al. 1997).

For the forest as a whole, the efficiency of nutrient use for organic matter productions increases in nutrient-limited environments. This has been shown us-

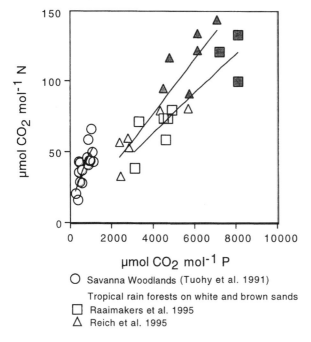

Figure 4 Instantaneous N and P use efficiencies in photosynthesis measured in leaves of native trees under natural conditions. Shaded symbols indicate pioneer and early succession species. (*Source*: data from Raaimaker et al. 1995, Reich et al. 1995, Tuohy et al. 1991.)

ing data on fine litter fall (mainly leaves) and litter nutrient content. The inverse of the nutrient concentration (mass per unit nutrient content) is higher for N, P, and Ca in forests with a limited supply of these nutrients (Vitousek 1984, Medina and Cuevas 1994). As in the case of the nutrient use efficiency for photosynthesis, P shows much higher values than N, as a consequence of the differences in physiologically active concentration of these nutrients. The P/N efficiency ratios measured in large litter fall data sets varies from 59 (Vitousek 1984) to 72 (Proctor 1984), numbers remarkably similar to the ratios reported for instantaneous use efficiency in photosynthesis (Figure 4).

B. Palms

Palms are a ubiquitous component of tropical humid forests that can reach dominance in areas with a tendency to water logging. They are botanically well understood and have more than 200 well-separated genera recognized (Tomlinson

1979). In spite of a relative restricted growth habit, palms have species capable of occupying every habitat in periodically or permanently flooded soils of swamp equatorial forests (from canopy to understory), or constituting nearly monospecific stands in permanently water-logged soils (Kahn and De Granville 1992).

1. Architecture and Growth Patterns

Palms appear to have a precise growth programming characterized by an essentially continuous vegetative growth, lacking dormancy mechanisms that may restrict them to climatically predictable environments such as those predominant in tropical and subtropical environments (Tomlinson 1979).

Their architecture is restricted to two essential types (Halle et al. 1978, Tomlinson 1979):

1. Monoaxial trees, characterized by a single vegetative shoot meristem, without reiteration. This type includes hapaxanthic (monocarpic) species with terminal inflorescences (Holtum's model), such as *Caryota urens, Corypha umbraculifera, Metroxylon salomonense, Raphia regalis*; and species that grow from a single aerial meristem producing one unbranched axis with lateral inflorescences, being therefore policarpic (Corner's model). This second group includes the majority of single-stemmed palms; examples include *Areca catechu, Borassus aethiopium, Cocos nucifera, Mauritia flexuosa, Oenocarpus distichus, Phytelephas macrocarpa, Roystonea oleracea*, and *Socratea exorrhiza*.

2. Polyaxial trees, characterized by repeated development of equivalent orthotropic modules in the form of basal branches initially restricted to the epicotiledonary region of the seedling axis and the basal nodes in the subsequent axes (Tomlinson's model), or those that grow from meristems producing orthotropic or plagiotropic trunks forking at regular but distant intervals by equal dichotomy, with lateral inflorescences but without vegetative lateral branches (Shoute's model). The former group includes almost all multistemmed palms, with species such as *Bactris gasipaes, Euterpe oleracea, Hyphaene guineensis, Oncosperma tigilaria, Phoenix dactilifera*, and *Raphia gigantea*. The second group of this type is less frequent and includes species such as *Allagoptera arenaria, Chamaedorea cataractarum, Hyphaene thebaica, Nannorrhops ritchiana, Nypa fruticans*, and *Vonitra utilis*.

2. Light and Water Relations

Palms are a taxonomically uniform group constituting the only monocots that contribute to the canopy, subcanopy, and forest floor layers of most tropical humid forests. Their ecophysiological differentiation is considerable and is excluded only from the driest tropical forest communities. In the Amazon region they occupy the whole range of available habitats to such an extent that a key for the forest formations in this large expanse of tropical forest communities has been developed on the basis of the palm species (Kahn and de Granville 1992; Table 3). The range of habitats of ecophysiological interest vary from permanently water-logged forests (*Mauritia flexuosa* or *minor*), and seasonally flooded palm savannas with sandy or clay soils (*Sabal mauritiaeformis, Orbignya martiana*), to hyperseasonal sites in savannas with 3–5-month dry periods and water logging during the rainy season (*Copernicia tectorum, C. cerifera*, and *Cocos schizophylla*) (Kahn and de Granville 1992, Hueck 1966, Kalliola et al. 1993, Walter 1973).

The diffuse distribution of the vascular tissue within the stem renders these plants resistant to burning, which is critical for survival of palm species characteristic of some savannas in Africa (*Borassus*) and South America (*Copernicia, Acrocomia, Sabal*). Their tolerance to flooding and comparatively long periods of drought is also related to their anatomy. Holbrook and Sinclair (1992) showed in *Sabal palmetto* that the amount of water accumulated in the stem per unit of living leaf area increases linearly with plant height. In addition, leaf epidermal conductance was in the range reported for xerophytes. Tolerance to prolonged drought is reduced by the fact that palms do not have a resting phase during their life cycle.

Tropical palms frequently constitute a dense understory layer in tropical moist forests (Gentry and Emmons 1986, Kahn and de Granville 1992, Kalliola et al. 1993). In these environments, light is the main factor influencing plant growth rate and time of reproduction. Photosynthesis studies on understory palms showed positive carbon balances at the leaf level when exposed to light intensities as low as 25 μmol m^{-2} s^{-1}. However, their growth under natural conditions was stimulated significantly when growing in gap edges compared with plants under full shade in the forest understory (Chazdon 1986). Computer simulations of carbon gain of full-shade–grown seedlings indicate that 24-hour positive carbon balance is reached whenever daily irradiation is greater than 0.2 mol m^{-2}.

The long-lasting leaf scars on the palm stem have been used as a marker to age palm populations and to detect changes in growth conditions (Piñero et al. 1984). A gap formed by the fall of one or a few trees not only opens the forest canopy, allowing more light to reach the forest floor, but turns down small undercanopy palms without uprooting them. The tilted stems of *Astrocaryum mexicanum* in this study continued to grow vertically, forming a sharp bend that indicates the occurrence of a disturbance. The number of leaf scars after the bending of the stems multiplied by the average leaf life span was found to be a neat measure of the time elapsed since the disturbance. However, although growth is continuous, the rates of height growth and leaf production may change significantly with ecological conditions. In populations of *Prestoea montana* in

Table 3 Forest Types and Palm Genus Diversity in the Amazon Basin

Forest type	Arborescent: single or multistemmed large (leaf > 4 m) or slender (leaf < 4 m)	Medium sized: large or slender, single or multistemmed, erect to creeping	Small: erect, prostrate, or climbing; single, multistemmed or acaulescent	Subterranean stemmed: large leaved
Terra firme forests	*Astrocaryum, Dyctiocaryum, Iriartea, Jessenia, Maximiliana, Oenocarpus, Orbygnia, Socratea*	*Astrocaryum, Oenocarpus, Socratea, Syagrus, Wettinia*	*Aiphanes, Astrocaryum, Bactris, Chamaedorea, Chelyocarpus, Desmoncus, Geonoma, Hyospathe, Iriartella, Pholidostachys*	*Astrocaryum, Orbignya Scheelea*
Forests on periodically flooded alluvial soils	*Astrocaryum, Attalea, Iriartea, Euterpe, Socratea, Oenocarpus, Orbignya, Scheelea*	*Astrocaryum, Chelyocarpus Itaya, Phytelephas*	*Bactris, Geonoma, Desmoncus*	
Forests periodically flooded by Blackwater	*Astrocaryum*	*Leopoldinia*	*Bactris, Leopoldinia, Desmoncus*	
Swamp forests on organic, permanently flooded soils	*Mauritia, Euterpe Socratea*	*Oenocarpus*	*Bactris, Desmoncus*	
Seasonal swamp forest on water-logged, irregularly flooded soils	*Euterpe, Jessenia, Mauritia, Mauritiella Socratea*	*Astrocaryum, Eleaeis, Manicaria, Oenocarpus, Wettinia Mauritiella*	*Asterogyne, Bactris Catoblastus, Desmoncus, Geonoma, Hyospathe Bactris, Desmoncus*	
Forests on dry, white-sandy soils				
Forests on water-logged, white-sandy soils	*Jessenia, Mauritia, Mauritiella*	*Elaeis, Euterpe*	*Bactris, Desmoncus Lepidocaryum, Pholidostachys*	
Submontane and montane forests	*Dictyocaryum, Euterpe Iriartea, Prestoea*	*Geonoma, Oenocarpus Wettinia*	*Aiphanes, Chamaedorea Geonoma*	
Savannahs	*Acrocomia, Astrocaryum Copernicia, Mauritia, Syagrus*	*Mauritiella, Syagrus*	*Bactris, Desmoncus*	*Astrocaryum, Orbignya Scheelea*

Source: Kahn and Granville (1992).

a lower montane forest in Puerto Rico, average leaf production rate over a 36-year observation period was 4 leaves/year, but the production rate changed with the age of the population sampled and was also lower in the suppressed individuals compared with those that reached the canopy (Lugo and Rivera-Battle 1987).

C. Climbing Plants and Hemiepiphytes

A characteristic structural feature of tropical forests is the occurrence of woody plants that use other trees as a support for expanding their photosynthetic area under full sun. These species either climb from the forest floor to the upper canopy (climbers) or descend with their roots from the upper canopy to the forest floor (hemiepiphytes) (Richards 1952, 1996, Walter 1973, Whitmore 1990). Climbing plants may be herbaceous (mostly called vines) or woody (frequently called lianas), although, there may be changes in the structural characteristics of the stem during the lifetime of certain species. They are conveniently distinguished by the method of climbing: (1) scandent climbers (with thorns or hooks); (2) tendrillar climbers (with tendrils); (3) root or bole climbers (with adhesive organs); and (4) twiners (twisting apical organs around the host) (Hegarty 1989).

Biomass allocation in these types of plants is characterized by a larger investment in photosynthetic surface at the expense of the production of self-supporting tissues, at least during a significant portion of their life cycle. The consequences of this pattern of biomass allocation for the ecophysiology of these plants varies in lianas and hemiepiphytes. The former are connected to the forest floor throughout their life cycle, no more limited in their water and nutrient supply than the host trees, whereas the hemiepiphytes have to cope with water and nutrient stress during the establishment phase, until they are able to establish contact with the forest floor through adventituous roots.

1. Lianas

Woody vines (lianas) contribute largely to the total leaf area of lowland forests (Gentry 1983, Putz 1983), and their number and diversity in certain forest types may be comparable to those of tree species (Rollet 1969). Lianas species are capable of rapid growth and thereby affect the development of leaf area and branching of the host trees and may increase mortality (Putz 1984). In fact, the presence of lianas does not greatly affect the total leaf area of the forest, the development of liana leaf area being compensated by a reduction in leaf area production by the tree host (Hegarty 1989). However, the larger ratio of leaf litter production to wood production observed in tropical forests compared with temperate forests may simply be the result of the frequency of lianas in the tropics (Gentry 1983).

The life cycle of a liana begins by germinating in the forest floor, either in a process of natural regeneration or in explosive growth events after natural or man-made disturbances in forest gaps (Rollet 1969). Light requirements of lianas appear to be similar in variety as those reported for rain forest trees. There are very shade-tolerant species with weak regeneration but long survival, intermediate-illumination species with abundant regeneration but small survivorship, and the shade-intolerant species, completely absent from natural regeneration plots (Rollet 1969). During the initial phases lianas are self supporting, but after a few years they begin to climb on neighboring trees in a process that is tightly coupled with the gap dynamics of the forest. Gaps promote development of lianas, which take advantage of their rapid growth in length to gain height in the forest using neighboring trees as a support (Putz 1984).

Lianas are characterized by a larger leaf biomass/stem biomass ratio than self-supporting trees (Putz 1983), and their xylem vessels are large, offering little resistance to water transport (Ewers 1985). In fact, lianas and vines from evergreen and deciduous tropical forests are characterized by higher specific conductivity of the stem than those of the trees of the same communities, at least when stems of similar diameter are compared (Ewers 1985, Gartner et al. 1990). Putz and Windsor (1987) hypothesized that the large lumen of the xylem vessels was more prone to cavitation; therefore, lianas should be very sensitive to drought, shedding their leaves at the beginning of the dry season. However, lianas were found to be evergreen, with a leaf-producing period longer than that of the tree species used for comparison. Therefore, water availability during the dry season in Barro, Colorado, was not as severely curtailed as the investigators expected, or there were other mechanisms preventing cavitation in these plants.

2. Hemiepiphytes

Hemiepiphytes are plants that germinate on branches or stems of trees and that grow and develop leaves and roots using nutrient and water resources that accumulate within the crevices of the stem bark (Richards 1952, 1996, Walter 1973). Their roots eventually reach the soil and develop there in a similar fashion as roots of their host trees. In this process, the plants pass from a truly epiphytic stage, characterized mainly by frequent drought stress, to a self-supporting stage in which the plant functions as a normal tree. A large group of the hemiepiphytic species behave as ''stranglers'' because their developing roots grow around the stems of the host tree, and as they increase in diameter, they compress the stem underneath. This process eventually continues until the host tree stem cannot properly transport water and nutrient upward and therefore dies, leaving the strangler as a self-supporting independent tree. The group of strangler species is rather limited and is frequently found among the genus *Ficus*, while permanent hemiepiphytes remain host-dependent for support despite extensive root systems in contact with the soil (Todzia 1986).

Epiphytic and tree forms of hemiepiphytes differ in their growth habit and water relations. Holbrook and Putz (1996) found that the epiphytic form of hemiepiphytic *Ficus* strangler species produced thinner leaves with larger specific leaf areas and water contents (%), but lower stomatal densities than the leaves produced at the tree stage. In correspondence with their higher water content, osmotic potential at full turgor was lower in the epiphytic forms, but the rate of cuticular water loss was smaller than that measured in excised tree leaves. In this group of hemiepiphytes, the functional properties of the epiphytic stage seems to be directed to conservative water use according to the uncertainty of water supply in the epiphytic habitat, particularly during the dry season. Although nutrient supply is linked to water availability, it did not appear to be a limiting factor for the development of plants at this stage.

One outstanding group among the hemiepiphytes in the neotropics is constituted by the species of the genus *Clusia* that perform a combination of C3 and crassulacean acid metabolism (CAM) photosynthesis. This process was originally described in *Clusia lundelli* (Clusiaceae) (Tinoco-Oranjuren and Vásquez-Yanes 1983), a true strangler, beginning its life cycle as an epiphyte and then developing roots that eventually anchor in the soil. The same photosynthetic metabolism has been thoroughly documented in the large tree strangler *Clusia rosea* in the U.S. Virgin Islands (Ting et al. 1985, Ball et al. 1991) and confirmed in several species of this genus (Ting et al. 1987, Sternberg et al. 1987, Popp et al. 1987, Borland et al. 1992, Winter et al. 1992).

Gas exchange of *Clusia* CAM species has been shown to be highly flexible, going from typical CAM to typical C3, including continuous net fixation of CO_2 during a 24-hour cycle, depending on light intensity during growth, day–night temperature cycles, drought, and even nutrition (Schmitt et al. 1988, Lee et al. 1989, Zotz and Winter 1993). Drought induced contrasting effects in *Clusia* species that fix substantial amounts of CO_2 during the night. In *C. uvitana*, daytime uptake decreased rapidly without irrigation, whereas nocturnal fixation increased throughout the treatment (Winter et al. 1992).

D. Vascular Epiphytes

Epiphytes are plants that can complete their life cycle living on the stem, branches, or even leaves of other plants. They do not need soil contact to acquire water and nutrients. This life-form has been successful from the evolutionary point of view. Within the higher vascular plants (Magnoliophyta), approximately 6% of the genera and 9% of the species have been recorded as epiphytes (Benzing 1990, Kress 1986, Madison 1977). Epiphytic genera and species are more frequent among monocots (Liliopsida) than among dicots (Magnoliopsida). Benzing (1990) registered 520 genera and 16,608 species of epiphytic monocots compared with 262 genera and 4,521 species of epiphytic dicots. The Orchidaceae account for the predominance of epiphytes within the monocots with 67% of the vascular epiphytes.

Epiphytic habitats restrict plant growth mainly through drought stress. Nutrient supply is also restricted because accumulation of organic matter and flow of nutrients in rain water and dust is scarce and erratic. Many epiphytes therefore depend on their capacity to accumulate organic residues produced by the supporting trees, or by animals associated with them (ants, termites), and by their capacity to accumulate water internally via a specialized morphology.

Water and nutrient availability vary with the location of the epiphyte within the forest. Those located in the topmost branches depend much on the interception of dust particles and the absorption of dissolved nutrients in rain water, and are exposed to nearly full sunlight. Exposed epiphytic habitats are characterized by a high evapotranspiration demand. Drought spells may be of short duration, but frequent. The duration of the dry spells increases from wet rain forests to dry seasonal forests.

In the middle to low canopy, evapotranspiration demands are lower, and nutrient availability increases because of tree leaf litter accumulation and enrichment of rain water through nutrient leaching. On the other hand, light availability may be limiting for photosynthesis.

Main adaptations of vascular plants to the epiphytic habitat are related to their water and nutrient economy. They may be summarized as follows:

1. Development of foliose structures capable of retaining rain water and accumulating organic detritus, the so-called "tanks" and "nests." Many bromeliads are tank formers and develop short roots that penetrate into accumulated organic matter and absorb water and nutrients (Pittendrigh 1948). Nests are less efficient than tanks at retaining rain water, but litter accumulation provides a substrate rich in nutrients that are absorbed by the roots of the nest-forming plants. This structure is found among ferns, orchids, and aroids.

2. Development of water-accumulating tissues (succulence). Many epiphytes possess succulent roots, stems, and/or leaves. These organs accumulate amounts of water and are protected against evaporation by epidermises and highly impermeable cuticles. Water accumulated in these tissues maintains turgor of photosynthetic tissues during prolonged dry spells. Succulent leaves are rather frequent in epiphytic species belonging to the families Piperaceae, Gesneriaceae, Orchidaceae (developing also succulents pseudobulbs), and Bromeliaceae.

3. Sclerophyllous photosynthetic organs resistant to desiccation and with high nutrient use efficiency. These characteristics are found within the epiphytic Ericaceae and Rubiaceae that form fiber-rich leaves with highly impermeable cuticles. These leaves are long-lived, have low nutrient content per unit dry weight, and are desiccation resistant.

4. Specialized mechanisms for water and nutrient uptake. The bromeliad leaves, particularly within the subfamily Tillandsioideae, possess specialized trichomes for the uptake of water and nutrients. Orchid roots, on the other hand, are covered by a hygroscopic tissue operative in the absorption of water and nutrients, the velamen. According to Benzing et al. (1983) this structure may be important in explaining the predominance of orchids in epiphytic tropical habitats.

5. Mutualistic symbiosis. Mycorrhizae are a common feature of vascular plants in the tropics, in particular the vesicular-arbuscular type (Janos 1983). Among the richest group of epiphytic plants, the orchids, a highly specialized mycorrhiza dominates. Orchid seeds are not nutritionally independent and associate with fungi to complete their germination process until they reach photosynthetic independence. Later, they develop permanent mycorrhizal associations.

6. High water use efficiency and nocturnal CO_2 fixation. A large number of epiphytes among the Orchidaceae, Bromeliaceae, and Cactaceae, particularly those of strongly seasonal forests, are capable of nocturnal CO_2 fixation via CAM. CO_2 taken up during the night is accumulated in the vacuoles as malic acid. This acid is remobilized during the following day, decarboxylated in the cytoplasm, and the evolving CO_2 is taken up via the photosynthetic reduction cycle in the chloroplasts. In this process, stomata open during the night and remain closed during most of the day, depending on the water status of the photosynthetic tissue and the actual amount of malic acid accumulated. The consequence of this inverted stomatal rhythm is a high water use efficiency (amount of water transpired per unit carbon taken up).

1. Ecophysiology of CAM Epiphytes

CAM plants may be constitutive or facultative (Winter 1985, Medina 1990). The former show CAM characteristics under a wide range of environmental conditions, whereas the latter only show CAM under conditions of water or salt stress. Most CAM plants in the tropics are constitutive, as the number of facultative species described is very limited. There are two variations in the nocturnal organic acid accumulation process that are considered to be related to CAM and are frequently found in epiphytes of humid forests: CAM-idling and CAM-cycling (Kluge and Ting 1978, Benzing 1990). In both processes, nocturnal increases of acid concentration are observed without net CO_2 uptake from the atmosphere. CAM-idling occurs in water-stressed constitutive CAM plants and results from the refixation of respiratory CO_2 (Szarek et al. 1973, Ting 1985), whereas CAM-cycling occurs in tissues without water stress (Ting 1985). CAM-cycling has been hypothesized to be a precursor of full-CAM plants (Monson 1989, Martin 1994), but it is still not clear why the reduction in internal CO_2 partial pressure resulting from fixation of respiratory CO_2 during the night does not influence stomatal behavior.

The following ecophysiological features are essential to understand the abundance and ecological significance of CAM in humid tropical forests:

1. High water use efficiency. Constitutive CAM plants take up CO_2 mainly during the night, when the leaf-air water vapor saturation deficit is lower. High water use efficiency associated with the succulent morphology of photosynthetic organs allows the maintenance of a positive carbon balance in drought-prone habitats, and is therefore of paramount importance for the occupation of exposed epiphytic habitats in rainforests.

2. Respiratory CO_2 recycling. In many constitutive CAM plants, simultaneous measurements of CO_2 exchange and organic acid accumulation during the night frequently reveal that the increase in acid concentration is larger than that expected from the amount of CO_2 taken up by PEP-carboxylase. Dark CO_2 fixation should yield 2 equivalents of acidity (malic acid) per CO_2 molecule taken up. However, this is not the case with several species of the Bromeliaceae and Orchidaceae, of the genus *Clusia*, and in CAM ferns. This difference is apparently due to refixation of respiratory CO_2 (Griffiths 1988). Recycling of respiratory CO_2 may be a crucial carbon conservation mechanism under stressful environmental conditions.

3. Adaptation to shady environments. The numerous taxa with CAM in humid tropical forests that can grow and compete with C3 plants in shady environments indicate that CAM by itself does not appear to be a liability in low-light environments. In humid forest environments, shade-adapted CAM plants have growth rates and net carbon gains similar to those of C3 plants that are limited by low light (Martin et al. 1986, Medina 1990). It seems that CO_2 fixation during the night by CAM plants might be favored by the high CO_2 concentrations of the air under the canopy of humid tropical forests (Martin 1994).

2. Occurrence of CAM in Specific Groups of Epiphytes

The number of epiphytic genera with reported CAM species is substantial (Table 4) and is clearly dominated by orchids.

Pteridophytes.   Only species of the genera *Pyrrosia* (Hew and Wong 1974, Ong et al. 1986, Winter et al. 1986, Kluge et al. 1989b) and *Microsorium* (Earnshaw et al. 1987) have been reported as CAM plants among ferns. *Pyrrosia longifolia* can be found growing under fully exposed or shaded conditions in humid tropical forests, performing CAM in both conditions. In both populations, the proportion of total carbon gain derived from night CO_2 fixation is similar as indicated by their $\delta^{13}C$ values (Winter et al. 1986).

The question of how CAM may have evolved within the Polypodiaceae has been discussed by Kluge et al. (1989a) and Benzing (1990). All relatives of the genus *Pyrrosia* grow in humid, shady environments; therefore, it seems improbable that the CAM species derived from xerophytic ancestors. In this particular case, it appears that high light and frequent low humidity, characteristic of epiphytic environments, may have been the selective factors leading to the development of CAM in this genus. In addition, ferns are a group of plants highly dependent on continuous water availability during their gametophytic stage. This makes it more feasible that CAM ferns are derived from shade forms in humid forests, and that CAM evolved secondarily in the epiphytic environment (Kluge et al. 1989).

Gesneriaceae and Piperaceae.   The Gesneriaceae and Piperaceae have a large number of epiphytic species, many of which are succulent. In a few species (*Codonanthe crassifolia* in the Gesneriaceae and approximately 15 species within the genus *Peperomia*), CAM activity has been detected only as acid accumulation during the night period (Ting et al. 1985, Guralnick et al. 1986). Their isotopic composition indicates that all carbon used for growth derives from daytime CO_2 uptake, but their δD values are more similar to those of CAM plants than to those of C3 or C4 plants (Ting et al. 1985).

Orchidaceae and Bromeliaceae.   Orchids and bromeliads contain the largest number of epiphytic species with constitutive CAM metabolism, and many grow under partial- or heavy-shade conditions. The orchids occupy pantropical epiphytic habitats, whereas the bromeliads are restricted to the neotropics. Both families contain epiphytic and terrestrial species with typical CAM, expressed in gas exchange patterns, nocturnal malic acid accumulation, and reduction of nonstructural carbohydrate content during the night (Coutinho 1969, McWilliams 1970, Medina 1974, Neales and Hew 1975, Goh et al. 1977, Griffiths and Smith 1983, Lüttge et al. 1985, Goh and Kluge 1989). There are several recent accounts on the morphology, physiology, and ecology of epiphytes in these two families (Benzing 1990, Craig 1990, Lüttge 1991, Medina 1990).

Table 4 Families and Genera in Which Epiphytic CAM Species of Humid Tropical Forests Have Been Reported[a]

Family	Genera with CAM species		
Polypodiaceae	*Pyrrosia*		
	Microsorium		
Bromeliaceae	*Acantostachys*	*Canistrum*	*Tillandsia*
	Aechmea	*Hohenbergia*	*Portea*
	Araeococcus	*Neoregelia*	*Quesnelia*
	Billbergia	*Nidularium*	*Wittrockia*
	Bromelia	*Streptocalyx*	
Orchidaceae (~500 genera and more than 20,000 epiphytic species)	*Bulbophyllum*	*Luisia*	*Robiquetia*
	Cadetia	*Micropera*	*Saccolabiopsis*
	Campylocentrum	*Mobilabium*	*Saccolabium*
	Cattleya	*Oberonia*	*Sarchochilus*
	Chilochista	*Oncidium*	*Schoenorchis*
	Cymbidium	*Phalenopsis*	*Taeniophyllum*
	Dendrobium	*Pholidota*	*Thrixspermum*
	Epidendrum	*Plectorrhiza*	*Trachoma*
	Eria	*Pomatocalpa*	*Trichoglottis*
	Flickingeria	*Rhinerrhiza*	*Vanda*
Asclepiadaceae	*Hoya*		
	Dischidia		
Cactaceae (25 epiphytic genera, most probably all CAM)	*Epiphyllum*		
	Hylocereus		
	Rhipsalis		
	Strophocactus		
	Zygocactus		
Clusiaceae	*Clusia*		
	Oedematopus		
Crassulaceae	*Echeveria*		
	Sedum		
Gesneriaceae	*Codonanthe*		
Piperaceae	*Peperomia*		
Rubiaceae	*Myrmecodia*		
	Hydnophytum		

[a] Data from Coutinho 1963, 1969, McWilliams 1970, Medina 1974, 1977, 1989, Griffiths and Smith 1983, Ting et al. 1985, Winter et al. 1983, 1985, Earnshaw et al. 1987, Griffiths 1989, Benzing 1990, Lüttge 1991, Martin 1994, Franco et al. 1994.

Distribution of epiphytic species of orchids and bromeliads is correlated with atmospheric humidity. In montane forests, the proportion of CAM epiphytes decreases with altitude in direct correlation with air humidity (Smith 1989, Earnshaw et al. 1987).

The larger genera of bromeliads and orchids contain both C3 and CAM species. The genus *Tillandsia* in the Bromeliaceae, for instance, has species ranging from dry to humid forests. CAM species predominate in the former, while C3 species predominate in the latter. Another bromeliad genus, *Aechmea*, contains only CAM species, some of which are restricted to the floor of humid forests. In the Orchidaceae, similar photosynthetic diversity can be observed in the genera *Bulbophyllum* (1000 epiphytic species) and *Dendrobium* (900 epiphytic species). In these genera, there is an almost continuous variation in δ ^{13}C values, from less than $-30‰$ to $-12‰$. CAM activity in these orchid genera is correlated with the development of succulence. This morphological property is associated both with the presence of cells with big vacuoles appropriate for accumulation of large quantities of malic acid during the night, and with the development of water-storage parenchyma, often photosynthetically inactive. For example, the leaf thickness of high-altitude orchids in upper montane forests of Papua, New Guinea, is due to the development of a chlorophyll-free hypodermis (Earnshaw et al. 1987), very similar to that described for *Peperomia* spp. (Ting et al. 1985) and *Codonanthe* spp. (Guralnick et al. 1986, Medina et al. 1989).

Roots of many orchid species contain chloroplasts that are photosynthetically active (Benzing et al. 1983). Chloroplast-bearing cells, specialized in photosynthesis, and mycorrhizal cells, specialized in nutrient uptake, coexist in the roots of some leafless species. These roots are capable of performing CAM (Winter et al. 1985).

The process of shade adaptation of CAM plants in view of the relatively higher energetic requirement merits more research at the biochemical level. A number of epiphytic CAM bromeliads grow in a wide range of light conditions, although their abundance tends to increase under full sun exposure (Pittendrigh 1948, Smith et al. 1986).

Cactaceae. Differentiation between terrestrial and epiphytic species in the Cactaceae is more pronounced. There is no genus with both terrestrial and epiphytic species; however, several climbing and decumbent species are found within the genera *Acanthocereus* and *Heliocereus*. The epiphytic cacti are confined within the tribes Rhipsalidanae (8 genera), Epiphyllanae (9 genera), and Hylocereanae (9 genera) within the subfamily Cereoideae (Britton and Rose 1963). The photosynthetic metabolism of epiphytic cacti has been studied in only a few genera (Table 4), but there is little doubt that all the species are most likely constitutive CAM plants. Species growing in deep shade, such as *Strophocactus witti* (Medina et al. 1989), are damaged when exposed to high light intensity, yet they show typical CAM gas exchange and acid accumulation.

IV. CONCLUSIONS

The analysis of the functional properties of tropical forests components, and of the forest as a whole, is developing fast as a result of the international efforts to study the forest–atmosphere interactions within the context of global climate and land-use change. These efforts are directed to develop operative models capable of quantifying the role of tropical forests in the dynamics of trace greenhouse gases in the atmosphere. There is an intense international argument, the subject of which is the impact of increasing atmospheric CO_2 concentrations on the productive performance of tropical forests. A number of groups maintain the notion that elevated CO_2 concentrations in the atmosphere may be stimulating the photosynthetic capacity of tropical forests resulting from a "fertilization" effect. However, others maintain that these effects may be counteracted by water limitations (in water-limited systems) or by nutrient limitations as most humid tropical forests grow on substrates of limited fertility. Some recent developments show that tropical humid forests in the Amazon basin may be acting as a sink for CO_2 (Grace et al. 1995). In addition, there are already some promising modeling efforts integrating the physiological and ecological knowledge on the effects of CO_2 on photosynthesis and the capability for nutrient uptake under natural conditions (Lloyd and Farquhar 1996). We can expect dramatic developments in the understanding of the physiological ecology of the production processes in tropical forests during the first decade of the next century. This will be of the utmost importance for future restoration efforts considering the fast pace of land use change in the tropics.

REFERENCES

Alexander, I. 1989. Mycorrhizas in tropical forests. In: Proctor, J., ed. Mineral Nutrients in Tropical Forest and Savanna Ecosystems. Oxford: Blackwell Scientific, pp 169–188.

Alexander, I.J., and P. Högberg. 1986. Ectomycorrhizas of tropical angiospermous trees. New Phytologist 102: 541–549.

Allen E.B., M.F. Allen, D.J. Helm, J.M. Trappe, R. Molina, and E. Rincón. 1995. Patterns and regulation of mycorrhizal plant and fungal diversity. Plant and Soil 170: 47–62.

Ashton, P.M.S., and G.P. Berlyn. 1992. Leaf adaptations of some *Shorea* species to sun and shade. New Phytologist 121: 587–596.

Bailey, H.P. 1979. Semi-arid climates: their definition and distribution. In: Hall, A.E., G.H. Cannell, and H.W. Lawton, eds. Agriculture in Semi-arid Environments. Berlin: Springer-Verlag, pp 73–97.

Barbault, R., and S. Sastrapradja. 1995. Generation, maintenance and loss of biodiversity. Heywood, V.H., ed. Global Biodiversity Assessment. Cambridge: Cambridge University Press, pp 192–274.

Bazzaz, F.A. 1984. Dynamics of wet tropical forests and their species strategies. In: Medina, E., H.A. Mooney, and C. Vázquez-Yanes, eds. Physiological Ecology of Plants of the Wet Tropics. The Hague: Junk. pp 233–243.

Bazzaz, F.A., and R.W. Carlson. 1982. Photosynthetic acclimation to variability in the light environment of early and late successional plants. Oecologia 54: 313–316.

Bazzaz, F.A., and S.T.A. Pickett. 1980. Physiological ecology of tropical succession: a comparative review. Annual Review of Ecology and Systematics 11: 287–310.

Benzing, D.H. 1990. Vascular Epiphytes. Cambridge: Cambridge University Press.

Benzing, D., W.E. Friedman, G. Peterson, and A. Renfrow. 1983. Shootlessness, velamentous roots and the pre-eminence of Orchidaceae in the epiphytic biotope. American Journal of Botany 70: 121–133.

Björkman, O. 1981. Responses of different quantum flux densities. In: Lange, O.L., P.S. Nobel, C.B. Osmond, and H. Ziegler, eds. Encyclopedia of Plant Physiology. New Series Vol 12 A. Physiological Plant Ecology I. Berlin: Springer-Verlag, pp 57–107.

Björkman, O., and B. Demming-Adams. 1995. Regulation of photosynthetic light energy capture, conversion and dissipation in leaves of higher plants. In: Schulze, E.D., and M.M. Caldwell, eds. Ecophysiology of Photosynthesis. Berlin: Springer-Verlag, pp 17–47.

Bullock, S.H., H.A. Mooney, and E. Medina, eds. 1995. Seasonally Dry Tropical Forests. Cambridge: Cambridge University Press.

Burslem, D.F.R.P., I.M. Turner, and P.J. Grubb. 1994. Mineral nutrient status of coastal hill dipterocarp forest and adinandra belukar in Singapore: bioassays of nutrient limitation. Journal of Tropical Ecology 10: 579–599.

Burslem, D.F.R.P., P.J. Grubb, and M. Turner. 1995. Responses to nutrient addition among shade-tolerant tree seedlings of lowland rain forest in Singapore. Journal of Ecology 83: 113–122.

Castro Y., N. Fetcher, and D.S. Fernández. 1995. Chronic photoinhibition in seedlings of tropical trees. Physiologia Plantarum 94: 560–565.

Chazdon, R.L. 1986. Light variation and carbon gain in rainforest understory palms. Journal of Ecology 74: 995–1012.

Chazdon, R.L., and N. Fetcher. 1984. Photosynthetic light environments in a lowland tropical rainforest in Costa Rica. Journal of Ecology 72: 553–564.

Chazdon, R.L., and R.W. Pearcy. 1986. Photosynthetic responses to light variation in rainforest species. II. Carboxylation gain and photosynthetic efficiency during lightflecks. Oecologia 69: 524–531.

Connell, J.H., and M.D. Lowman. 1989. Low-diversity tropical rain forests: some possible mechanisms for their existence. American Naturalist 134: 88–119.

Coomes, D.A. 1995. The effect of root competition on saplings and seedling in Amazonian Caatinga forests in southern Venezuela. Ph.D. Dissertation, University of Cambridge, Cambridge, UK.

Coutinho, L.M. 1969. Novas observações sôbre a ocorrência do "Efeito de DeSaussure" e suas relações com a suculencia, a temperatura folhear e os movimentos estomáticos. Boletin Faculdáde de Filosofia, Ciencias eletras, Universidade de São Paulo, No. 331: 79–102.

Cramer, W.P., and R. Leemans. 1993. Assessing impacts of climate change on vegetation using climate classification systems. In: Solomon, A.M., and H.H. Shugart, eds. Vegetation Dynamics and Global Change. New York: Chapman and Hall, pp 190–217.

Cuevas, E., and E. Medina. 1988. Nutrient dynamics within Amazonian forest ecosystems. II. Root growth, organic matter decomposition and nutrient release. Oecologia (Berlin) 76: 222–235.

Denslow, J. 1996. Functional group diversity and responses to disturbance. In: Orians, G.H., R. Dirzo, and J.H. Cushman, eds. Biodiversity and Ecosystem Processes in Tropical Forests. Ecological Studies Vol. 122. Berlin: Springer-Verlag, pp 127–151.

Earnshaw, M.J., K. Winter, H. Ziegler, W. Stichler, N.E.G. Cruttwell, K. Kerenga, P.J. Cribb, J. Wood, J.R. Croft, K.A. Carver, and T.C. Gunn. 1987. Altitudinal changes in the incidence of crassulacean acid metabolism in vascular epiphytes and related life forms in Papua New Guinea. Oecologia 73: 566–572.

Ewel, J.J., and S.W. Bigelow. 1996. Plant life-forms and tropical ecosystem functioning. In: Orians, G.H., R. Dirzo, and J.H. Cushman, eds. Biodiversity and Ecosystem Processes in Tropical Forests. Ecological Studies Vol. 122. Berlin: Springer-Verlag, pp 101–126.

Ewers, F.W. 1985. Xylem structure and water conduction in conifer trees, dicot trees, and lianas. IAWA Bulletin (New Series) 6: 309–317.

Fetcher, N., B.R. Strain, and S.F. Oberbauer. 1983. Effects of light regime on the physiology and growth of seedlings of tropical trees: a comparison of gap and pioneer species. Oecologia 58: 314–319.

Gartner, B.L., S.H. Bullock, H.A. Mooney, V. B. Brown, and J.L. Whitbeck. 1990. Water transport properties of vine and tree stems in a tropical deciduous forest. American Journal of Botany 77: 742–749.

Gauhl, E. 1976. Photosynthetic responses to varying light intensity in ecotypes of *Solanum dulcamara* L. from shaded and exposed environments. Oecologia 22: 275–286.

Gentry, A.H. 1988. Changes in plant commmunity diversity and floristic composition on environmental and geographic gradients. Annals of Missouri Botanical Garden 75: 1–34.

Gentry, A.H. and C. Dodson. 1987a. Contribution of nontrees to species richness of a tropical rainforest. Biotropica 19: 149–156.

Gentry, A.H., and C. Dodson. 1987b. Diversity and biogeography of neotropical vascular epiphytes. Annals of Missouri Botanical Garden 74: 205–233.

Gentry, A.H., and L.H. Emmons. 1987. Geographical variation in fertility, phenology and composition of the understory of neotropical forests. Biotropica 19: 216–227.

Gentry, A.H. 1983. Lianas and the ''paradox'' of contrasting latitudinal gradients in wood and litter production. Tropical Ecology 24: 63–67.

Givnish, T.J. 1988. Adaptation to sun and shade: a whole plant perspective. Australian Journal of Plant Physiology 15: 63–92.

Gómez-Pompa, A., T.C. Whitmore, and M. Hadley, eds. 1991. Rain Forest Regeneration and Management. Man and the Biosphere Series, vol. 6. Paris: Unesco and The Parthenon.

Grace, J., J. Lloyd, J.A. Macintyre, A.C. Miranda, P. Meir, H.S. Miranda, C.A. Nobre,

J. Moncrief, J. Massheder, I.R. Wright, and J.H.C. Gash. 1995. Carbon dioxide uptake by undisturbed tropical forest. Science 270: 778–780.

Griffiths, H. 1988. Crassulacean acid metabolism: a re-appraisal of physiological plasticity in form and function. Advances in Botany Research 15: 43–92.

Griffiths, H., and J.A.C. Smith. 1983. Photosynthetic pathways in the Bromeliaceae of Trinidad: relations between life-forms, habitat preference and the occurrence of CAM. Oecologia 60: 176–184.

Grubb, P.J. 1977. Control of forest growth and distribution on wet tropical mountains. Annual Review of Ecological Systems 8: 83–107.

Guralnick, L.J., I.P. Ting, and E.M. Lord. 1986. Crassulacean acid metabolism in the *Gesneriaceae.* American Journal of Botany 73: 336–345.

Hall, J.B. and M.D. Swaine. 1976. Classification and ecology of closed-canopy forest in Ghana. Journal of Ecology 64: 913–951.

Halle, F., R.A.A. Oldeman, and P.B. Tomlinson. 1978. Tropical Trees and Forests: An Architectural Analysis. Berlin: Springer-Verlag.

Hart, T.B., J.A. Hart, and P.G. Murphy. 1989. Monodominant and species-rich forests of the humid tropics: causes for their co-occurrence. American Naturalist 133: 613–633.

Hegarty, E.E. 1989. The climbers-Lianes and vines. In: Lieth, H., and M.J.A. Werger (eds). Ecosystems of the World, vol. 14B. New York: Elsevier, pp 339–353.

Holbrook, N.M., and F.E. Putz. 1996. From epiphyte to tree: differences in leaf structure and leaf water relations associated with the transition in growth form in eight species of hemiepiphytes. Plant, Cell and Environment 19: 631–642.

Holbrook, N.M., and T.R. Sinclair. 1992. Water balance in the arborescent palm *Sabal palmetto.* I. Stem structure, tissue water release properties and leaf epidermal conductance. Plant, Cell and Environment 15: 393–399.

Holdridge, L.R. 1967. Life Zone Ecology. San José, Costa Rica: Tropical Science Center.

Holdridge, L.R., W.C. Grenke, W.H. Hatheway, T. Liang, and J.A. Tosi. 1971. Forest Environments in Tropical Life Zones: A Pilot Study. New York; Pergamon Press.

Hopkins, M.S., P. Reddell, R.K. Hewett, and A.W. Graham. 1996. Comparison of root and mycorrhizal characteristics of primary and secondary rainforest on metamorphic soil in North Queensland, Australia. Journal of Tropical Ecology 12: 871–885.

Hueck, K. 1966. Die Wälder Südamerikas. Stuttgart: Gustav Fischer Verlag.

Huston, M. 1980. Soil nutrients and tree species richness in Costa Rican forests. Journal of Biogeography 7: 147–157.

Huston, M. 1994. Biological Diversity. Cambridge: Cambridge University Press.

Janos, D.P. 1980. Vesicular-arbuscular mycorrhizae affect lowland tropical rain forest plant growth. Ecology 61: 151–162.

Janos, D.P. 1983. Tropical mycorrhizas, nutrient cycles and plant growth. In: Sutton, S.L., T.C. Whitmore, and A.C. Chadwick, eds. Tropical Rain Forest: Ecology and Management. Oxford: Blackwell Scientific, pp 327–345.

Jordan, C.F. 1985. Nutrient Cycling in Tropical Forest Ecosystems. Chichester: Wiley.

Jordan, C.F., ed. 1989. An Amazonian Rain Forest: The Structure and Function of a Nutrient Stressed Ecosystem and the Impact of Slash-and-Burn Agriculture. Man and the Biosphere Series, Vol. 2. London: UNESCO and The Partenon.

Jordan, C.F., R.L. Todd, and G. Escalante. 1979. Nitrogen conservation in a tropical rain forest. Oecologia 39: 123–128.

Kahn, F., and J.-J. de Granville. 1992. Palms in Forest Ecosystems of Amazonia. Ecological Studies Vol. 95. Berlin: Springer-Verlag.

Kalliola, R., M. Puhakka, and W. Danjoy, eds. 1993. Amazonia Peruana: Vegetación húmeda tropical en el llano subandino. PAUT (Finland) and ONERN (Peru). Finland: Jyväskylä.

Kamaluddin, M., and J. Grace. 1992a. Photoinhibition and light acclimation in seedlings of *Bischofia javanica*, a tropical forest tree from Asia. Annals of Botany 69: 47–52.

Kamaluddin, M., and J. Grace. 1992b. Acclimation in seedlings of a tropical tree, *Bischofia javanica*, following a stepwise reduction in light. Annals of Botany 69: 557–562.

Kellman, M., and R. Tackaberry. 1997. Tropical Environments. London: Routledge.

Kitajima, K. 1994. Relative importance of photosynthetic traits and allocation patterns as correlates of seedling shade tolerance of 13 tropical trees. Oecologia 98: 419–428.

Kitajima K. 1996. Ecophysiology of tropical tree seedlings, In: Mulkey, S.S., R.L. Chazdon, and A.P. Smith, eds. Tropical Forest Plant Ecophysiology. New York: Chapman and Hall, pp 559–596.

Königer M., G.C. Harris, A. Virgo, and K. Winter. 1995. Xanthophyll-cycle pigments and photosynthetic capacity in tropical forest species: a comparative field study in canopy, gap and understory plants. Oecologia 104: 280–290.

Krause, G.H., A. Virgo, and K. Winter. 1995. High susceptibility to photoinhibition of young leaves of tropical forest trees. Planta 197: 583–591.

Krause, G.H., and K. Winter. 1996. Photoinhibition of photosynthesis in plants growing in natural tropical forests gaps. A chlorophyll fluorescence study. Botanica Acta 109: 456–462.

Kress, W.J. 1986. The systematic distribution of vascular epiphytes: an update. Selbyana 9: 2–22.

Lal, R. 1987. Tropical Ecology and Physical Edaphology. Chichester: Wiley.

Leigh Jr., E.G., A.S. Rand, and D.M. Windsor, eds. 1982. The Ecology of a Tropical Rain Forest: Seasonal Rhythms and Long-Term Changes. Washington, D.C.: Smithsonian Institution Press.

Lieth, H. 1975. Historic survey of primary productivity research. In Lieth, H., and R. Whittaker, eds. Primary Productivity of the Biosphere. Ecological Studies 14. Berlin: Springer-Verlag, pp 7–16.

Lieth, H. and R. Whittaker, eds. 1975. Primary Productivity of the Biosphere. Ecological Studies 14. Berlin: Springer-Verlag.

Lloyd, J., and G.D. Farquhar. 1996. The CO_2 dependence of photosynthesis, plant growth responses to elevated atmospheric CO_2 concentrations and their interaction with soil nutrient status. I. General principles and forest ecosystems. Functional Ecology 10: 4–32.

Long, S.P., S. Humphries, and P.G. Falkowski. 1994. Photoinhibition of photosynthesis in nature. Annual Review of Plant Physiology and Molecular Biology 45: 633–662.

Lovelock, C.E., M. Jebb, and C.B. Osmond. 1994. Photoinhibition and recovery in tropical plant species: response to disturbance. Oecologia 97: 297–307.

Lugo, A.E., and C. Rivera Batelle. 1987. Leaf production, growth rate, and age of the palm *Prestoea montana* in the Luquillo Experimental Forest. Journal of Tropical Ecology 3: 151–161.

Lüttge, U., ed. 1991. Vascular Plants as Epiphytes. Ecological Studies Vol. 76. Berlin: Springer-Verlag.

Madison, M. 1977. Vascular epiphytes: their systematic occurrence and salient features. Selbyana 2: 1–13.

Martin, C.E. 1994. Physiological ecology of Bromeliaceae. Botanical Review 60: 1–82.

McWilliams, E.L. 1970. Comparative rates of dark CO_2 uptake and acidification in Bromeliaceae, Orchidaceae and Euphorbiaceae. Botanical Gazette 131: 285–290.

Medina, E. 1984. Nutrient balance and physiological processes at the leaf level. In: Medina, E., H.A. Mooney, and C. Vazquez-Yánes, eds. Physiological Ecology of Plants of the Wet Tropics. The Hague: Junk, pp 139–154.

Medina, E. 1990. Eco-fisiología y evolución de las Bromeliaceae. Boletin de la Academia Nacional de Ciencias, Córdoba 59: 71–100.

Medina, E. 1996. CAM and C4 plants in the humid tropics. In: Mulkey, S.S., R.L. Chazdon, and A.P. Smith, eds. Tropical Forest Plant Eco-physiology. London: Chapman and Hall, pp 56–88.

Medina, E., and E. Cuevas. 1994. Mineral nutrition: tropical forests. Progress in Botany 55: 115–129.

Medina, E., and E. Cuevas. 1989. Patterns of nutrient accumulation and release in Amazonian forests of the upper Río Negro basin. In: Proctor, J., ed. Mineral Nutrients in Tropical Forest and Savanna Ecosystems. Oxford: Blackwell Scientific Publications, pp 217–240.

Medina, E., G. Montes, E. Cuevas, and Z. Rokzandic. 1986. Profiles of CO_2 concentration and $\delta^{13}C$ values in tropical rain forests of the Upper Rio Negro basin, Venezuela. Journal of Tropical Ecology 2: 207–217.

Medina, E., M. Delgado, and V. García. 1989. Cation accumulation and leaf succulence in *Codonanthe macradenia* J.D. Smith (Gesneriaceae) under field conditions. Amazoniana 11: 13–22.

Medina, E., M. Sobrado, and R. Herrera. 1978. Significance of leaf orientation for leaf temperature in an Amazonian sclerophyll vegetation. Radiation and Environmental Biophysics 15: 131–140.

Medina, E., V. García, and E. Cuevas. 1990. Sclerophylly and oligotrophic environments: relationships between leaf structure, mineral nutrient content and drought resistance in tropical rain forests of the upper Río Negro region. Biotropica 22: 51–64.

Melillo, J.M., A.D. McGuire, D.W. Kicklighter, B. Moore III, C.J. Vorosmarty, and A.L. Schloss. 1993. Global climate change and terrestrial net primary production. Nature 363: 234–240.

Moyersoen, B. 1993. Ectomicorrizas y micorrizas vesiculo-arbusculares en Caatinga Amazónica al sur de Venezuela. Scientia Guaianae 3: 82 pp.

Mulkey, S.S., R.L. Chazdon, and A.P. Smith, eds. 1996. Tropical forest plant eco-physiology. New York: Chapman and Hall.

Newbery, D.M., I.J. Alexander, D.W. Thomas, and J.S. Gartlan. 1988. Ectomycorrhizal rain-forest legumes and soil phosphorus in Korup National Park, Cameroon. New Phytologist 109: 433–450.

Oberbauer, S.F., and B.R. Strain. 1985. Effects of light regime on the growth and physiology of *Pentaclethra macroloba* (Mimosaceae) in Costa Rica. Journal of Tropical Ecology 1: 303–320.

Odum, H.T., and R.F. Pigeon, eds. 1970. A Tropical Rain Forest: A Study of Irradiation and Ecology at El Verde, Puerto Rico. Oak Ridge, Tennessee: US Atomic Energy Commission.

Oldeman, R.A.A., and J. van Dijk. 1991. Diagnosis of the temperament of tropical rain forest trees. In: Gómez-Pompa, A., T.C. Whitmore, and M. Hadley, eds. Rain Forest Regeneration and Management. Man and the Biosphere Series, Vol. 6. Paris: UNESCO, pp 21–65.

Osmond, C.B, and S.C. Grace. 1995. Perspectives on photoinhibition and photorespiration in the field: quintessential inefficiencies of the light and dark reactions of photosynthesis? Journal of Experimental Botany 46: 1351–1362.

Pearcy, R.W. 1988 Photosynthetic utilization of sunflecks. Australian Journal of Plant Physiology 15: 223–238.

Phillips, O.L., P. Hall, A.H. Gentry, S.A. Sawyer, and R. Vasquez. 1994. Dynamics and species richness of tropical rain forests. Proceedings of the National Academy of Science USA 91: 2805–2809.

Piñero, D., M. Martínez-Ramos, and J. Sarukhan. 1984. A population model of *Astrocaryum mexicanum* and a sensitivity analysis of its finite rate of increase. Journal of Ecology 72: 977–991.

Pittendrigh, C.S. 1948. The bromeliad-*Anopheles*-malaria complex in Trinidad. I. The bromeliad flora. Evolution 2: 58–89.

Popp, M., D. Kramer, H. Lee, M. Diaz, H. Ziegler, and U. Lüttge. 1987. Crassulacean acid metabolism in tropical dicotyledonous trees of the genus Clusia. Trees 1: 238–247.

Proctor, J. 1984. Tropical forests litterfall. II. The data set. In: Sutton, S.L., and A.C. Chadwick, eds. Tropical Rain Forest: The Leeds Symposium. Leeds, UK: Leeds Phylosophical and Literary Society, pp 83–113.

Proctor, J., ed. 1989. Mineral Nutrients in Tropical Forest and Savanna Ecosystems. Oxford: Blackwell Scientific.

Putz, F.E. 1983. Liana biomass and leaf area of a 'tierra firme' forest in the Río Negro basin, Venezuela. Biotropica 15: 185–189.

Putz, F.E. 1984. The natural history of lianas on Barro Colorado Island, Panamá. Ecology 65: 1713–1724.

Putz, F.E., and D.M. Windsor. 1987. Liana phenology on Barro Colorado Island, Panamá. Biotropica 9: 334–341.

Raaimakers, D., R.G.A. Boot, P. Dijkstra, S. Pot, and T. Pons. 1995. Photosynthetic rates in relation to leaf phosphorus content in pioneer versus climax tropical species. Oecologia 102: 120–125.

Ramos, J., and J. Grace. 1990. The effects of shade on the gas exchange of seedlings of four tropical trees from Mexico. Functional Ecology 4: 667–677.

Reich, P.B., D.S. Elsworth, and C. Uhl. 1995. Leaf carbon and nutrient assimilation and conservation in species differing status in an oligotrophic Amazonian forest. Functional Ecology 9: 65–76.

Reich, P.B., M.B. Walters, and D.S. Ellsworth. 1997. From tropics to tundra: global convergence in plant functioning. Proceedings of the National Academy of Science USA 94: 13730–13734.

Reich, P.B., Uhl, C., M.B. Walters, and D.S. Ellsworth. 1991. Leaf lifespan as a determinant of leaf structure and function among 23 Amazonian tree species. Oecologia 86: 16–24.

Richards, P.W. 1952, 1996. The Tropical Rainforest. Cambridge: Cambridge University Press.

Riddoch I., J. Grace, F.E. Fasehun, B. Riddoch, and O. Ladipo. 1991a. Photosynthesis and successional status of seedlings in a tropical semi-deciduous rain forest in Nigeria. Journal of Ecology 79: 491–503.

Riddoch, I., H. Lehto, and J. Grace. 1991b. Photosynthesis of tropical tree seedlings in relation to light and nutrient supply. New Phytologist 119: 137–147.

Rollet, B. 1969. La regénération naturelle en forêt dense humide sempervirente de plaine de la Guyane Vénézuélienne. Bois et Forêts des Tropiques 124: 19–38.

Schreiber, U., W. Bilger, and C. Neubauer. 1994. Chlorophyll fluorescence as a non intrusive indicator for rapid assessment of in vivo photosynthesis. In: Schulze, E.-D., and M.M. Caldwell, eds. Ecophysiology of Photosynthesis. Ecological Studies 100. Berlin: Springer-Verlag, pp 49–70.

Schultz, J. 1995. The Ecozones of the World: The Ecological Divisions of the Geosphere. Berlin: Springer-Verlag.

Smith, J.A.C. 1989. Epiphytic bromeliads. Vascular plants as epiphytes. In: Lüttge, U., ed. Ecological Studies 76. Berlin: Springer-Verlag, pp 109–138.

Smith, J.A.C., H. Griffiths, and U. Lüttge. 1986. Comparative ecophysiology of CAM and C3 bromeliads I. The ecology of Bromeliaceae in Trinidad. Plant, Cell and Environment 9: 359–376.

Solbrig, O.T. 1993. Plant traits and adaptive strategies: their role in ecosystem function. In: Schulze, E.D., and H.A. Mooney, eds. Biodiversity and ecosystem function. Ecological Studies, Vol. 99. Berlin: Springer-Verlag.

St John, T.V., and C. Uhl. 1983. Mycorrhizae in the rain forest of San Carlos de Río Negro, Venezuela. Acta Cientifica Venezuela 34: 233–237.

Stark, N., and C.F. Jordan. 1977. Nutrient retention in the root mat of an Amazonian rain forest. Ecology 59: 434–437.

Sternberg, L.S.L., I.P. Ting, D. Price, and J. Hann. 1987. Photosynthesis in epiphytic and rooted *Clusia rosea* Jacq. Oecologia 72: 457–460.

Strauss-Debenedetti S., and F.A. Bazzaz. 1991. Plasticity and acclimation to light in tropical Moraceae of different successional positions. Oecologia 87: 377–387.

Strauss-Debenedetti S., and F.A. Bazzaz. 1996. Photosynthetic characteristics of tropical trees along successional gradients. In: Mulkey, S.S., R.L. Chazdon, and A.P. Smith, eds. Tropical Forest Plant Ecophysiology. New York: Chapman and Hall, pp 162–186.

Sutton, S.L., T.C. Whitmore, and A.C. Chadwick, eds. 1983. Tropical Rain Forest: Ecology and Management. Oxford: Blackwell Scientific.

Swaine, M.D., and T.C. Whitmore. 1988. On the definition of ecological species groups in tropical rain forests. Vegetatio 75: 81–86.

Swaine, M.D. 1996. Rainfall and soil fertility factors limiting forest species distribution in Ghana. Journal of Ecology 84: 419–428.

Thompson, W.A., G.C. Stocker, and P.E. Kriedemann. 1988. Growth and photosynthesis response to light and nutrients of *Flindersia brayleyana* F. Muell., a rainforest tree with broad tolerance to sun and shade. Australian Journal of Plant Physiology 15: 299–315.

Thompson, W.A., P.E. Kriedemann, and I.E. Craig. 1992. Photosynthetic response to light and nutrients in sun-tolerant and shade-tolerant rainforest trees II. Leaf gas exchange and component processes of photosynthesis. Australian Journal of Plant Physiology 19: 19–42.

Tilman D. 1988. Plant Strategies and the Dynamics and Structure of Plant Communities. Princeton, N.J.: Princeton University Press.

Ting, I.P., J. Hann, N.M. Holbrook, F.E. Putz, S.L. Sternberg, D. Price, and G. Goldstein. 1987. Photosynthesis in hemiepiphytic species of *Clusia* and *Ficus*. Oecologia 74: 339–346.

Ting, I.P., L. Bates, L.O. Sternberg, and M.J. DeNiro. 1985. Physiological and isotopical aspects of photosynthesis in *Peperomia*. Plant Physiology 78: 246–249.

Tinoco-Ojanguren, C., and R.W. Pearcy. 1995. A comparison of light quality and quantity effects on the growth and steady-state and dynamic photosynthetic characteristics of three tropical tree species. Functional Ecology 9: 222–230.

Tinoco-Ojanguren, C., and C. Vázquez-Yanes. 1983. Especies CAM en la selva tropical húmeda de los Tuxtlas, Veracruz. Boletin de la Sociedad Botanica de Mexico 45: 150–153.

Todzia, C. 1986. Growth habits, host tree species, and density of hemiepiphytes on Barro Colorado island, Panamá. Biotropica 18: 22–27.

Torti, S.D., P.D. Coley, and D.P. Janos. 1997. Vesicular-arbuscular mycorrhizae in two tropical monodominant trees. Journal of Tropical Ecology 13: 623–629.

Tuohy, J.M., J.A.B. Prior, and G.R. Stewart. 1991. Photosynthesis in relation to leaf nitrogen and phosphorus content in Zimbabwean trees. Oecologia 88: 378–382.

Turnbull, M.H. 1991. The effect of light quantity and quality during development on the photosynthetic characteristics of six Australian rainforest species. Oecologia 87: 110–117.

Turnbull, M.H., D. Doley, D.J. Yates. 1993. The dynamics of photosynthetic acclimation to changes in light quantity and quality in three Australian rainforest tree species. Oecologia 94: 218–228.

UNESCO. 1978. Tropical Forest Ecosystems. A state-of-knowledge report prepared by Unesco, UNEP, FAO. Natural Resources Research Series 14. Paris: Unesco.

van der Merwe, N.J., and E. Medina. 1989. Photosynthesis and $^{13}C/^{12}C$ ratio in Amazonian rain forests. Geochimica et Cosmochimica Acta 53: 1091–1094.

van Noordwijk, M., K. Hairiah. 1986. Mycorrhizal infection in relation to soil pH and soil phosphorus content in a rain forest of northern Sumatra. Plant and Soil 96: 299–302.

Vareschi, V. 1980. Vegetationsökologie der Tropen. Stuttgart: Verlag Eugen Ulmer.

Vitousek, PM, and R.L. Sanford. 1986. Nutrient cycling in moist tropical forest. Annual Review of Ecology and Systematics 17: 137–167.

Vitousek, P.M. 1984. Litterfall, nutrient cycling, and nutrient limitation in tropical forests. Ecology 65: 285–298.

Walter, H. 1973. Die Vegetation der Erde. I. Die tropischen und subtropishen Zonen. Jena: VEB Gustav Fischer Verlag.

Went, F.W., and N. Stark. 1968. The biological and mechanical role of soil fungi. Proceedings of the National Academy of Science 60: 497–504.

Whitmore, T.C. 1990. An Introduction to Tropical Rain Forests. Oxford: Oxford University Press.

Whitmore, T.C. 1984. Tropical Rain Forests of the Far East. Oxford: Carendon Press.

Whitmore, T.C. 1991. Tropical rain forest dynamics and its implications for management. In: Gómez-Pompa, A., T.C. Whitmore, and M. Hadley, eds. Rain Forest Regeneration and Management. Man and the Biosphere Series, Vol. 6. Paris: UNESCO, pp 67–89.

Winter, K., E. Medina, V. García, M.L. Mayoral, and R. Muñiz. 1985. Crassulacean acid metabolism in roots of a leafless orchid, *Campylocentrum tyrridion* Garay and Dunsterv. Journal of Plant Physiology 111: 73–78.

Winter, K., G. Zotz, B. Baur, and K.J. Dietz. 1992. Light and dark CO_2 fixation in *Clusia uvitana* and the effects of plant water status and CO_2 availability. Oecologia 91: 47–51.

Wofsy, S.C., R.C. Harriss, and W.A. Kaplan. 1988. Carbon dioxide in the atmosphere over the Amazon basin. Journal of Geophysical Research 93: 1377–1387.

Zotz, G., and K. Winter. 1993. Short-term regulation of crassulacean acid metabolism in a tropical hemiepiphyte, *Clusia uvitana*. Plant Physiology 102: 835–841.

12
Plant Diversity in Tropical Forests

S. Joseph Wright
Smithsonian Tropical Research Institute, Balboa, Ancon, Republic of Panama

I. INTRODUCTION

Plant species diversity is greatest in tropical forests. The highest plant species density yet recorded, 365 species in just 1000 m^2, is for the only complete enumeration of tropical forest plants (Gentry and Dodson 1987). The diversity of lowland tropical forest at a local scale may exceed the diversity of extratropical forests at a continental scale. The combined temperate forests of Europe, North America, and Asia support 1166 tree species in 4.2 $\times$ 10^6 km^2 (Latham and Ricklefs 1993). In contrast, 1175 tree species occur in just 0.5 km^2 of lowland dipterocarp forest in Borneo (LaFrankie 1996). Trees comprise just 25% of the plant species in tropical rain forests (Gentry and Dodson 1987). Sixty-five percent of all flowering plant species, 92% of fern species, and 75% of moss species are tropical (Prance 1977).

These levels of diversity challenge modern ecology (Rosenzweig 1995). Theory and experiment confirm that two species cannot coexist if they use resources in an identical manner (Gause 1934). This competitive exclusion principle has become the touchstone against which theory addresses the coexistence of species. Plants compete for light, water, and perhaps a dozen mineral nutrients. Plants must also disperse their progeny and survive the depredations of their pests. Are interspecific differences in these few parameters sufficient to explain the levels of plant diversity observed in tropical forests? Do other mechanisms contribute to the coexistence of tropical forest plants? Or is the competitive exclusion principle an inappropriate touchstone?

This chapter addresses these questions in three steps. First, five mechanisms are evaluated that are frequently invoked to explain the coexistence of extraordinary numbers of plant species in tropical forests. Second, four Appendices present significant attributes of plant diversity within and among tropical forests. A successful theory must explain these observations. Finally, the origination of tropical forest plant diversity is considered. The extraordinary species densities of tropical forests may have more to do with the evolutionary history of plants, especially angiosperms, than with any unique ecological attribute of tropical forests.

II. THE COEXISTENCE OF TROPICAL FOREST PLANTS

This review focuses on five mechanisms that may contribute to the coexistence of very large numbers of species of plants. The review is not intended to be exhaustive. Many potential mechanisms are omitted (see Palmer 1994). The five mechanisms were selected for two reasons: (1) they dominate the literature for tropical forests, and (2) each mechanism is amenable to experimental and/or comparative tests. The relationship to plant species density is briefly described for each mechanism. Principal tests in tropical forests are then evaluated. Finally, the mechanisms are related to one or more of the patterns of plant diversity observed among tropical forests (Appendices 1–4).

A. Niche Differentiation

Species may exploit limiting resources in ways that are sufficiently distinctive to permit stable coexistence in equilibrial communities (Ashton 1969, MacArthur 1969). Coexistence occurs when resources vary spatially, and each species occurs where it is a superior competitor.

Tree falls create spatial heterogeneity. Light intensities are highest near the center of a treefall and lowest nearby, beneath the intact forest canopy. Tree falls also create soil heterogeneity. Clays from depth are exposed when roots tip up, while rich humic material accumulates where fallen canopies decompose. Tropical forest plants may segregate along both light and soil gradients created by treefalls (Ricklefs 1977, Connell 1978, Denslow 1980, 1987, Orians 1982, Platt and Strong 1989).

This treefall gap hypothesis has been evaluated thoroughly. A few species do segregate treefall gaps by size and light intensity (Brokaw 1987). However, in large mapped plots, the spatial distributions of saplings of most tree species were indifferent to treefall gaps (Hubbell and Foster 1986, Lieberman et al. 1995). Saplings of just 5% of the tree species evaluated were specialized to treefall environments in repeated surveys of the performance of 250,000 trees in Panama (Welden et al. 1991). The great majority of species survived well and grew slowly, both in treefall gaps and in the deeply shaded understory. Interspecific differences in performance in response to gaps are necessary to promote coexistence. Recent experimental studies have failed to identify such differences (Uhl et al. 1988, Denslow et al. 1990, Brown and Whitmore 1992, Osunkjoya et al. 1992). The emerging conclusion is that spatial heterogeneity associated with treefalls maintains a relatively small number of light-demanding, shade-intolerant species in the forest landscape and makes a limited contribution to tropical forest plant diversity.

An intriguing extension of the treefall gap hypothesis relates the frequency of treefalls to forest productivity. Tree mortality increases with rainfall in tropical forests and with soil fertility in temperate-zone forests (Grubb 1986, Phillips et al. 1994). More productive forests with higher tree mortality rates have higher rates of gap formation and are predicted to have higher spatial heterogeneity of forest light environments. To the extent that plants differentiate light environments, this enhanced spatial heterogeneity will contribute to the increase in plant species diversity with rainfall in tropical forests and with soil fertility in temperate forests (Grubb 1986, Phillips et al. 1994). However, we have seen that tropical forest plants show limited niche differentiation with respect to light environments. Still, the hypothesis of Phillips et al. (1994) is one of the few that addresses the ubiquitous increase in tropical-forest plant diversity with rainfall (Appendix 1) and merits further examination.

Variation in soil resources related to topography and underlying geological formations may also introduce spatial variation and promote species coexistence. Many tropical forest plant species have restricted distributions along soil moisture and soil fertility gradients (Hubbell and Foster 1983, Lieberman et al. 1985, Baillie et al. 1987, Tuomista and Ruokolainen 1994, Swaine 1996, Sollins 1998). However, in most of these studies, many more species were indifferent to the edaphic gradient and occurred everywhere. Most tropical plant species are generalists with respect to both edaphic gradients and treefall microhabitats.

A modal relationship has been predicted between plant species density and soil fertility when three conditions are met (Tilman 1982): (1) few species survive the least fertile soils; (2) soil resources vary spatially; and (3) the most fertile soils provide ample resources everywhere, effectively negating spatial heterogeneity. Under these conditions, species diversity will increase with the pool of tolerant species from infertile to intermediate soils and then decline on the most fertile soils as effective spatial heterogeneity declines and competitive dominance develops. The predicted modal relationships between tropical tree species' densities and soil fertility have been reported (Appendix 2). The postulated mechanisms remain untested, however, and the declining portion of the diversity–fertility relationship may be an artifact (Appendix 2).

In conclusion, spatially heterogeneous resources and ecologically segregated plant species are obvious to even casual observers of tropical forests. Niche differentiation undoubtedly contributes to tropical forest plant diversity. However, there are many ecologically similar plant species (Appendix 4). Moreover, there is no indication that levels of ecological differentiation among species distinguish dry from wet tropical forest or tropical forest from other forest biomes. Additional mechanisms must contribute to the diversity of tropical forest plants.

B. Pest Pressure

Microbes, fungi, and animals consume living plant tissue. These pests may contribute to the coexistence of their host plant species if the more abundant plant species or the superior competitors suffer disproportionately high levels of damage (Gillett 1962).

Pest pressure is severe in tropical forests. Insect herbivores alone consume an average of 11% of the leaf area produced in the understories of wet tropical forest (Coley and Barone 1996). Granivorous and browsing mammals reduce seedling recruitment, survivorship, and growth in a wide range of tropical forests (Dirzo and Miranda 1991, Osunkjoya et al. 1992, Terborgh and Wright 1994, Asquith et al. 1997). Pathogens kill entire seedling crops, as well as significant numbers of adult trees (Augspurger 1984, Gilbert et al. 1994). Pathogens that impair but do not kill the host plant have received less attention, but are likely to be very important. Herbivores consume a larger proportion of leaf production in tropical forests than in temperate forests, despite greater plant investment in antiherbivore defenses in tropical forests (Coley and Aide 1991, Coley and Barone 1996). High levels of activity of plant pests may set tropical forests apart from other forested biomes.

Pest pressure may vary on several spatial scales. At the smallest scale, pests may cause disproportionately high levels of damage on the concentrations of conspecific seeds or seedlings found near fruiting trees. Conspecifics may be prevented from recruiting successfully near one another, freeing space for other plant species and potentially raising plant diversity (Janzen 1970, Connell 1971). Seeds, seedlings, and saplings experience high mortality rates near conspecific adults (Clark and Clark 1984, Hubbell et al. 1990, Condit et al. 1992). However, the mechanism is not necessarily pest pressure. The alternative is intraspecific competition, occurring either among juvenile plants or between juveniles and the nearby adult (Clark and Clark 1984).

The relationship between pest pressure and the local population density of host plants becomes critical at larger spatial scales (Appendix 3). Pest pressure is positively density dependent when pest pressure increases more rapidly than host density. Positively density-dependent pests cause the highest levels of damage on the most abundant hosts and potentially contribute to the coexistence of host species. Alternatively, pest pressure is negatively density dependent when pests are unable to keep pace with host densities. High host densities satiate pests (Schupp 1992). Negatively density-dependent pests cause the greatest damage on rarer hosts and potentially reduce host diversity. Negative density dependence occurs when factors other than host density control pest populations. Possibilities include control of pest populations by their own predators and parasites or by social behavior, especially territorial behavior. Thus, at spatial scales over which the densities of reproductive plants vary, pest pressure may be positively or negatively density dependent with the potential to increase or decrease plant diversity, respectively.

Keystone predators control community organization through their impact on prey species (Paine 1966). Terborgh (1992) hypothesized that felids and raptors are keystone predators in tropical forests. The hypothesis has two components. First, felids and raptors collectively limit midsized terrestrial mammals. Second, when their numbers are not checked by predation, these prey species alter forest regeneration. This hypothesis has profound implications. Felids and raptors may indirectly control plant diversity in tropical forests just as starfish control the diversity of sessile organisms in some marine environments.

The second component of this hypothesis has recently been falsified. Fences were used to exclude mammals from seeds and seedlings at Cocha Cashu, Peru, where the biota is intact, and at Barro Colorado Island, Panama, where several large predators are missing (Terborgh and Wright 1994). The exclosures had large and virtually identical effects at both sites, enhancing all indices of seedling performance. Browsing and granivorous mammals clearly alter plant regeneration, but the presence or absence of large felids and raptors had no additional effect.

Pest pressure may contribute to the increase in plant species diversity with rainfall (Appendix 1). Wet forests will have higher and more constant pest pressure if seasonal temperature and rainfall reduce pest populations in drier forests. Plant diversity will, in turn, be enhanced in wet forests if the more abundant plant species or the superior competitors suffer disproportionately high levels of damage. The premise of this hypothesis, higher pest pressure in wetter forests, has yet to be established. Levels of herbivory are actually higher in dry forests than in wet forests in the tropics (Coley and Barone 1996). Realized levels of herbivory result from the interaction between herbivores and plant defenses. In dry deciduous forests, leaves live less than 1 year and are poorly defended. In wet evergreen forests, long-lived leaves are better defended. The full cost of pests includes the diversion of resources to pest defenses as well as direct pest damage. Evaluation of pest pressure along tropical rainfall gradients awaits this information, as well as information on levels of activity of fungal and microbial pests.

In conclusion, pest pressure is greater in tropical forests than in other forest biomes. Positively density-dependent pest pressure will increase plant diversity. At very small spatial scales, pest pressure almost certainly reduces recruitment near conspecific adults (Clark and Clark 1984). For larger spatial scales, there have been almost no tests for density-dependent pest pressure for tropical forests (see Schupp 1992). Pest pressure remains a promising but unproven mechanism of plant species coexistence in tropical forests.

C. Intermediate Disturbance

Windstorms, lightning, fire, landslides, and other disturbances kill trees. Connell (1978) predicted that species diversity will vary during succession after such disturbances. Diversity will be low immediately after a disturbance as just those species with the greatest ability to disperse their seeds arrive. Diversity will reach a maximum at intermediate times after a disturbance as many more species arrive. Diversity will then decline with time as the best competitor or the species most resistant to pests or physical stress comes to dominate the forest. Species diversity over entire landscapes will be low at (1) high disturbance frequencies, because recently disturbed patches and the few species with the greatest dispersal abilities dominate the landscape; and (2) low disturbance frequencies, because old patches and the few species best able to persist dominate the landscape. Diversity will be greatest at intermediate disturbance frequencies because the landscape includes patches of a great variety of ages supporting a wide mix of species.

Connell (1978) applied the intermediate disturbance hypothesis to treefall gaps. He predicted that species best able to persist would perform poorly when transplanted into large gaps. These experiments have since been conducted with generally negative results. Performance improves for all species when seedlings are transplanted into gaps (Denslow et al. 1990, Brown and Whitmore 1992, Osunkjoya et al. 1992). Connell (1978) also failed to recognize that saplings of species capable of persisting in deep shade survive the formation and closure of treefall gaps (Uhl et al. 1988). Treefalls do not wipe the ground clean as envisioned by Connell (1978).

Intermediate disturbance may be important at larger spatial scales. Blowdowns caused by downbursts of wind associated with severe storms are an important disturbance. Blowdowns larger than 30 ha can be identified from satellite images. The frequency of large blowdowns over Amazonian Brazil increases with storm activity and annual rainfall (Nelson et al. 1994). Such large disturbances may contribute to the ubiquitous increase in plant species density with rainfall in tropical forests (Appendix 1).

Huston (1979, 1994) extended the intermediate disturbance hypothesis by incorporating environmental factors postulated to affect rates of competitive exclusion. Huston reasoned that productivity controls the rate of competitive exclusion. Given a constant frequency of disturbances, diversity will be low in the least productive environments where few species can tolerate physical stress, highest in intermediate environments where many species can tolerate the environment but recurrent disturbances prevent competitive exclusion, and low again in the most productive environments where competitive exclusion occurs more rapidly than disturbance. Huston (1980, 1993, 1994) used soil fertility as a proxy for productivity in tropical forests. This proxy is evaluated in Appendix 2. Chesson and Huntly (1997) recently overturned the postulated connection between disturbance and rates of competitive exclusion. They demonstrated theoretically that disturbance will either have no effect on or enhance rates of competitive exclusion. Huston's extension of the intermediate disturbance hypothesis must be reconsidered.

In conclusion, the intermediate disturbance hypothesis remains largely untested for tropical forests. One reason is overlap among hypotheses (see Palmer 1994). For example, the mechanisms postulated by Connell (1978) for treefall gaps incorporate interspecific differences in response to light intensity considered here under niche differentiation (see Section II. A). The intermediate disturbance hypothesis also implicitly assumes a tradeoff between the abilities to disperse and persist so that the species with the greatest dispersal abilities are poor competitors and vice versa (Petraitis et al. 1989; see Section II. D). Finally, the intermediate disturbance hypothesis is most likely to be important for large disturbances that kill all advanced regeneration. These types of disturbances await further study.

D. Life History Tradeoffs

Sessile organisms live in communities with rigid spatial structure. Each individual occupies a space. Direct interactions are limited to colonists of the same space and near neighbors. The ability to colonize empty space, compete within a space, and persist once established jointly determines success. Tradeoffs in these abilities, such that a superior colonist was an inferior competitor, might facilitate species coexistence.

Life history tradeoffs occur among tropical forest plants. The number of seeds produced and the mass of individual seeds are inversely related. Presumably, species producing many small seeds disperse to rare high-resource environments, whereas species producing a few large seeds tolerate frequent low-resource environments (Hammond and Brown 1995). At the leaf level, maximum photosynthetic potentials and leaf longevity are inversely related (Reich et al. 1992). Presumably, species with long-lived leaves tolerate low-light environments, and species with short-lived leaves are superior competitors in high-light environments. At the level of individuals, maximum potential growth rates and survivorship in deep shade are inversely related (Kitajima 1994). Presumably, allocation to defense against pests limits growth rates but enhances shade survivorship. These tradeoffs may facilitate species coexistence.

An infinite number of species can coexist when there is a strictly ordered tradeoff between longevity, dispersal ability, and competitive ability (Tilman 1994). However, a critical assumption of this spatial competition hypothesis is violated in tropical forests. Tilman assumed a strict tradeoff between longevity and competitive ability, such that superior competitors had short longevity and vice versa. Superior competitors in tropical forests are likely to tolerate deep shade as seedlings and cast deep shade as adults. Such species invest heavily in defenses against pests, have long-lived tissues including leaves, and are characterized by low growth and mortality rates (Kitajima 1994). Tilman acknowledged the possibility that long lifetimes and superior competitive ability may co-occur, falsifying the spatial competition hypothesis.

A very large number of species can also coexist if dispersal limits recruitment (Hurtt and Pacala 1995). Hurtt and Pacala modelled spatially heterogeneous environments and environmentally dependent competitive abilities. Sites vacated by the death of an adult were won by the best competitor among the species that reached the site. When the absolutely best competitor for the environment at the site did not arrive, the site was "won by forfeit," and dispersal limitation occurred. Hurtt and Pacala proposed a positive feedback between species diversity and dispersal limitation. Dispersal limitation becomes more important as diversity increases because all species become rarer, produce fewer propagules, and reach fewer sites. However, a negative feedback may disrupt dispersal limitation. If a superior competitor became abundant, its seeds would approach ubiquity and only the rarer, less competitive species would be dispersal limited. The two most common tree species on Barro Colorado Island, Panama, have dispersed seeds to each and every one of 200 tiny (0.5 m²) litter traps randomly located throughout a 50-ha forest plot. Both species are superior competitors whose abundant, widespread seedlings grow and survive well both in the deeply shaded understory and in treefall gaps. Such abundant species have escaped dispersal limitation, even in a species-rich tropical moist forest.

Empirically driven simulation models may elucidate the relationship between species diversity and life history traits. Pacala et al. (1996) have successfully applied this approach to North American forests with nine tree species. Model inputs determined empirically for each species included seed dispersal, light-dependent growth rates, growth-dependent death rates, and light interception. The model tracked the performance of individual trees, where each tree competed for light with its immediate neighbors. Important tradeoffs were evident between maximum potential growth rates and (1) dispersal, (2) light interception, and (3) survivorship under low light. A similar approach for a species-rich tropical forest has the potential to identify life history tradeoffs that facilitate the coexistence of tropical forest plants.

E. Chance

A recurrent hypothesis postulates that many tropical forest plants are ecologically equivalent (Appendix 4; Aubréville 1938, Federov 1966, Hubbell 1979, Hubbell and Foster 1986, Gentry 1989, Wright 1992). Ecologically equivalent species might coexist for a very long time given large population sizes and random birth and death processes (Hubbell 1979, Hubbell and Foster 1986). Ecological equivalence might arise through common descent (Federov 1966), convergent evolution for a generalized ability to tolerate diffuse competition (Hubbell and Foster 1986), or the chance dynamics of life in the deeply shaded understory (Gentry and Emmons 1987, Wright 1992). This latter possibility is developed here.

Competitive exclusion requires that species compete and that the better competitor consistently wins. Consider competition among forest understory plants. There is an assymetry between strata in tall forests. Large trees and lianas form the canopy and dominate understory environments. Trees and lianas intercept up to 99.5% of the photosynthetically active radiation reaching tropical forest canopies (Chazdon and Pearcy 1991), and tree and liana roots dominate the underground environment. Trees and lianas suppress understory plants, and, as a result, direct interactions among understory plants are unlikely. Understory environments may violate the two requisites for competitive exclusion.

First, competition among understory plants is unimportant. Pests maintain low understory stem densities that minimize direct competition. For example, the densities of understory plants increase dramatically after the experimental removal of mammalian herbivores (Dirzo and Miranda 1991, Osunkoya et al. 1992, Terborgh and Wright 1994). Manipulations of understory stem densities provide direct evidence for limited competition among understory plants. The removal of all understory plants (<5 cm diameter at breast height [dbh]) had no effect on seedling recruitment and survival in a wet tropical forest (Marquis et al. 1986). Suppression by canopy plants limits direct competitive interactions in the deeply shaded understory.

Second, chance, not competitive ability, determines which individuals succeed in the understory. Chance is introduced by severe light limitation. Understory irradiance has two components. Dim, diffuse irradiance occurs throughout the understory, while direct solar irradiance occurs in sunflecks. Sunflecks contribute 32–65% of the daily carbon gain of understory plants in closed canopy forests (Chazdon and Pearcy 1991). The occurrence of a sunfleck depends on the juxtaposition of the solar path and a canopy opening, cloud cover that diffuses solar irradiance, and wind that moves the vegetation, altering canopy openings. Most sunflecks are small in tall tropical forests (hundreds to thousands of cm^2). Sunfleck location and intensity vary on a variety of time scales as wind, cloud cover, solar declination, and canopy openings change (Chazdon and Pearcy 1991). Sunflecks introduce chance as a primary determinant of performance in the understory.

Treefalls reinforce the importance of chance. Understory plants are temporarily released from canopy suppression when canopy trees die and open large gaps (Denslow 1987). Canopy gaps occur at random with respect to understory plants. Most importantly, canopy gaps close before competition causes mortality among previously suppressed understory plants. For example, more than 80% of the stems present 4 years after gap formation were present before gaps were formed in an Amazonian rain forest (Uhl et al. 1988). Likewise, more than 80% of tree saplings survived the opening and closure of experimental treefall gaps in central Panama (N. Brokaw and A. P. Smith, unpublished data, 1993). Treefall gaps close before competition reduces species richness. Both sunflecks and treefalls introduce chance as an important determinant of the performance of understory plants.

To summarize, pests maintain low stem densities that prevent competition among understory plants, and chance sunflecks and treefall gaps largely determine which individuals succeed. Differences in competitive abilities are never realized among herbs and shrubs that spend their entire lives in the understory or among the seedlings and saplings that form the advanced regeneration for the canopy. The absence of competition and the important role of chance in the understory will contribute to species coexistence in all forest strata.

An extreme form of the chance hypothesis has been advanced by Hubbell (Hubbell 1979, Hubbell and Foster 1986). Species composition is predicted to fluctuate randomly as identical species experience random births and deaths. Species composition is, in fact, very similar for forests in similar abiotic environments, falsifying this prediction (Leigh et al. 1993, Terborgh et al. 1996). A more sophisticated version of the chance hypothesis must incorporate the ecological differences that so obviously exist among species. Chance enhances the potential for coexistence, but is not the sole explanation of population dynamics. It is clear that chance plays a large role in the deeply shaded understory. A renewed theoretical effort is required to explore the consequences for species coexistence.

III. THE ORIGINS OF TROPICAL PLANT DIVERSITY

Perhaps the most robust rule in ecology is that species richness and area increase together (Rosenzweig 1995). Terborgh (1973) ascribed high tropical species richness to the large areas with tropical climates. Terborgh noted that the area within each degree of latitude is greatest at the equator and decreases poleward, and that mean temperatures are uniformly high within 25° of the equator and then decrease poleward. As a consequence, tropical biomes cover four times more area than do subtropical, temperate, or boreal biomes (Rosenzweig 1992). Considerations of area alone suggest that species richness should be exceptionally high in the tropics.

Tropical biomes have been even more important in the past. In particular, the earth was largely tropical while the angiosperms radiated. Angiosperms appeared approximately 140 million years (Myr) ago in the early Cretaceous. The first angiosperms were tropical and required 20 to 30 Myr to spread beyond the tropics (Friis et al. 1987). The mean duration of angiosperm species in the fossil record exceeds 5 Myr (Niklas et al. 1983), making the last few tens of millions of years most relevant to modern angiosperm diversity. Tropical forests extended to 65° latitude in the Eocene (50 Myr ago), contracted to 15° latitude in the Oligocene (30 Myr), expanded to 35° latitude in the Miocene (20 Myr), and finally contracted to 25° latitude today (Behrensmeyer et al. 1992). The angiosperm radiation had a 20–30-Myr headstart in the tropics and occurred when tropical climates covered up to 80% of all land (Eocene). The great species richness of tropical angiosperms may have ancient origins.

IV. CONCLUSIONS

Tree and epiphyte diversity is greatest in everwet tropical forests (Appendix 1), and, when terrestrial herbs and shrubs are included, it is probable that plant diversity is greatest in everwet tropical forests on fertile soils (Appendix 2). Several factors enhance plant diversity in these forests. First, temperature, moisture, and nutrients do not limit the pool of species able to survive in the most diverse tropical forests. Second, negative density dependence characterizes common species, limiting competitive exclusion (Appendix 3). Both intraspecific competition and pest pressure may contribute to negative density dependence. In particular, warm, moist conditions favor pathogens and small insects, and year-round pest pressure may reinforce negative density dependence in moist tropical forests.

High productivity may also indirectly enhance plant diversity in everwet tropical forests. Wet tropical forests are the most productive tierra firme biome, and fertile soils may further enhance production (Vitousek 1984, Grubb 1986, Silver 1994). High tree turnover rates are associated with high production and year-round growth, and high frequencies of large blowdowns are associated with high rainfall (Nelson et al. 1994, Phillips et al. 1994). The most productive forests may be characterized by a mosaic of successional microhabitats on spatial scales ranging from single treefall gaps to blowdowns covering tens to thousands of hectares. This may permit coexistence through microhabitat specialization and/ or life history tradeoffs wherein inferior competitors are superior colonists and vice versa.

These possibilities are all mutually compatible, and there is ample evidence for each of the postulated mechanisms. Performance is habitat dependent (Section II.A), pests do attack concentrations of conspecific plants (Section II.B), disturbances do kill trees (Section II.C), and life history traits do covary (Section II.D). What remains unclear is how these events affect species coexistence.

Future research in three areas will be particularly valuable. First, are plant pests specific to particular hosts, are pest depredations density dependent, and does pest pressure vary among forest biomes? Second, do life history traits covary among the 90% of tropical forest species that persist in deep shade as seedlings and saplings? And, third, does that 90% include species that require the spatial heterogeneity introduced by treefalls and other disturbances to recruit success-fully? Until these questions are answered, the chance events that determine the success of understory plants and a tacit denial of the competitive exclusion principle will continue to fascinate tropical ecologists.

APPENDIX 1: PLANT SPECIES DENSITY INCREASES WITH RAINFALL

Plant species densities increase with rainfall in tropical forests. The increase is greatest for epiphytes, intermediate for terrestrial herbs and shrubs, and least for lianas and trees (Gentry 1988, Gentry and Dodson 1987, Gentry and Emmons 1987). Even for trees, however, species densities increase severalfold along rain-fall gradients in Ghana, the Neotropics, and Southeast Asia (Hall and Swaine 1976, 1981, Gentry 1988, Whitmore 1975, Clinebell et al. 1995). The gross pri-mary production of tropical forests also increases with rainfall (Brown and Lugo

1982, Jordan 1983, Medina and Klinge 1983). Plant species densities are greatest in the most productive tropical forests.

At least three mechanisms may contribute to the increase in plant species densities with rainfall. First, rates of tree turnover and treefall formation are higher in more productive forests (Grubb 1986, Phillips et al. 1994). Treefalls create a mosaic of forest patches recovering from past treefalls. This may increase diversity by increasing the spatial heterogeneity of forest light environments (see Section II.A), by preventing competitive dominance (see Section II.C), or by introducing spatio-temporal variation (see Section II.D).

Pest pressure provides a second mechanism that may contribute to the increase in plant species densities with rainfall. Many microbes, fungi, and small insects are vulnerable to desiccation. Density-dependent attacks by these pests may be more severe in high-rainfall forests that lack a dry, desiccating season (see Section II.B).

Finally, the most productive forests occur where rainfall and soils are most favorable for plant growth. Here, the physiological requirements for moisture and mineral nutrients are fulfilled for the greatest number of species. Allocation can be shifted from roots to photosynthetic functions permitting the greatest number of species to maintain a positive carbon balance and regenerate from the shaded understory (see Section II.E). Regardless of the mechanism, it is clear that the most productive tropical forests also support the greatest plant diversity.

APPENDIX 2: PLANT SPECIES DENSITY AND SOIL FERTILITY

The relationship between plant species density and soil fertility has variously been reported to be positive, modal, negative, or absent for tropical forests. This diversity of results has at least two causes.

First, soils and climate covary. Plant species densities increase (Appendix 1) and soil nutrient concentrations decrease with rainfall (Clinebell et al. 1995). Apparent negative relationships between plant species density and soil fertility may be caused by parallel variation in rainfall. The negative relationships between tree species density and soil fertility reported by Huston (1980, 1993, 1994) are suspect for this reason. Huston (1980) omitted rainfall from a multiple regression analysis even though rainfall was the single best predictor of tree species density. Huston (1993, 1994) reproduced a relationship between plant species density and the first axis of an ordination performed by Hall and Swaine (1976, 1981). Hall and Swaine (1981, p. 31) state that this axis was "closely correlated with the moisture gradient." Huston (1993, 1994) refers to the same axis as a "composite soil fertility index." In a recent multiple regression analysis, annual rainfall and rainfall seasonality were the most important variables explaining tree species richness; soil nutrient concentrations were negatively correlated with rainfall; and, after rainfall was included, soil attributes explained little additional variation in plant species richness. Clinebell et al. (1995) concluded that "tropical forest species richness is surprisingly independent of soil quality."

A second problem suggests that this conclusion is premature. Only surface soils (0–10-cm depths) and shallow subsurface soils (most often 30–40-cm depths) are considered. Forest trees have roots to 12-m depths in Amazonia, and lianas have roots to at least 5-m depths in Panama (Nepstad et al. 1994; M. Tyree and S.J. Wright, personal observation, 1994). Deep roots reach decomposing rock, a rich and overlooked source of mineral nutrients. Many analyses of the relationship between tropical plant species densities and soils have been limited to the most deeply rooted lifeforms: either trees >10 cm dbh (Holdridge et al. 1971, Hall and Swaine 1976, 1981, Huston 1980) or trees and lianas > 2.5 cm dbh (Clinebell et al. 1995). These analyses omit a potentially large source of nutrients from deeper soils.

Shrubs and herbs have relatively shallow roots (Wright 1992), and surface and shallow subsurface soils are more relevant for shrubs and herbs than for trees and lianas. The species densities of ferns and melastomes (mostly shrubs and small treelets) increase with soil fertility in Amazonian Peru (Tuomista and Poulsen 1995, Tuomista and Ruokolainen 1994). More generally, the species densities of fertile understory plants increase with soil fertility throughout the Neotropics (Gentry and Emmons 1987). The number of rain forest genera also increases with soil phosphorus in Australia (Beadle 1966). Species densities increase with soil fertility when fertility is measured for the soil volume reached by roots (i.e., for ferns, herbs, and shrubs). We await comparable analyses that include nutrients available from decomposing rock in the very deep soil horizons reached by tree and liana roots.

APPENDIX 3: DENSITY DEPENDENCE

Negative density dependence occurs when high local densities of conspecifics impair performance. One possible mechanism, intraspecific allelopathy, is unknown for tropical forests. Pest pressure and the competition for resources implicit to niche differentiation also mediate interspecific coexistence through negative density dependence. Negative density dependence prevents any species from becoming dominant and excluding others. The search for negative density dependence has been intense.

Only weak tests for density dependence are possible for long-lived organisms. Strong tests evaluate population fluctuations over many generations, searching for density-dependent temporal variation. Tests for long-lived organisms are limited to spatial variation, specifically to comparisons of performance among sites that differ in conspecific density. An implicit assumption is that population fluctuations are asynchronous among sites. Spatial tests will fail to detect density dependence if populations vary synchronously among sites. Spatial tests are also compromised if resources vary spatially. Spatially variable resources introduce apparent positive density dependence. High resource sites support high performance and high population densities. Spatially variable resources are important in tropical forests (see Section II.A). Negative density dependence must overcome spatial variation in resources and performance. Therefore, we expect evidence for negative density dependence from spatial tests to be rare.

However, this is not the case. Strong evidence for negative density dependence comes from the 50-ha forest dynamics plot on Barro Colorado Island, Panama. The performance of more than 300,000 stems ≥ 1 cm dbh of 314 species was monitored in 1982, 1985, 1990, and 1995. Negative density dependence is evident at scales ranging up to 100 m for recruitment, growth, and survivorship of several of the more abundant species (Hubbell et al. 1990, Condit et al. 1994). Tests for density dependence in rarer species were usually not significant; however, trends were consistently in the direction of negative density dependence. A very local negative density dependence, reduced performance near a conspecific adult, has also been documented in many tropical forests (see Section II.B). Negative density dependence is a fact for the more abundant trees in tropical forests. It only remains to determine the contribution of negative density dependence to interspecific coexistence (Hubbell et al. 1990, Condit et al. 1994).

Negative density dependence has also been reported for much rarer trees (Connell et al. 1984, Wills et al. 1996). The evidence consists of negative correlations between the number of recruits per conspecific adult (R/A) and the number of conspecific adults (A) or the basal area of conspecifics (BA). These correlations are suspect because the independent variable (A) is also the denominator of the dependent variable (R/A). Note that A and BA are closely related.

Wills et al. (1996) performed Monte Carlo simulations to evaluate the significance of R/A versus BA correlations. The size (dbh) and intercensus performance of randomly chosen pairs of conspecific trees were switched while maintaining the mapped location of each tree. Performance included death versus survival for trees present in the first census and recruitment for trees that appeared only in the second census. Adults were larger than a species-specific threshold dbh. After a large number of random switches, simulated correlations between R/A and BA were calculated. This procedure was repeated 999 times for each of a wide range of quadrat sizes, and the observed correlation coefficient for each quadrat size was compared with the distribution of simulated correlations. A subtle bias compromises these simulations for analyses of recruitment.

Two facts unrelated to density dependence affect the relative values of observed and simulated correlations. First, recruits are always small. Second, recruits are never adults. Consider a quadrat for which R/A is large and BA is small. This quadrat includes a relatively large number of recruits. The net effect of the randomization is likely to switch a recruit for an adult. This decreases R/A and increases BA because the adult gained is larger than the recruit lost. The opposite will tend to be true for quadrats for which R/A is small and BA is large. The randomization tends to move quadrats in R/A-BA space so that simulated correlations between R/A and BA are less negative than the observed correlation. This bias is unrelated to density.

This bias explains the spatial pattern of significant results reported by Wills et al. (1996). Recruitment was negatively density dependent for small quadrats but not for large quadrats. The bias identified above only arises when there is a net loss or gain of recruits. Net changes are likely for small quadrats with small numbers of individuals. The randomization is less likely to cause net changes for large quadrats where larger numbers of individuals are switched back and forth. Artifact explains the spatial scale of significant results for recruitment.

A size-stratified randomization would avoid this artifact. Small recruits would only be switched with small survivors. This would eliminate the correlated change in R/A and BA that occurs whenever a small recruit and large adult are switched. Until a size-stratified randomization is performed, we must conclude that negative density dependence only affects recruitment for the most common trees in tropical forests (Hubbell et al. 1990, Condit et al. 1994).

APPENDIX 4: SYNTOPIC CONGENERS

> As species of the same genus have usually, though by no means invariably, some similarity in habits and constitution, and always in structure, the struggle will generally be more severe between species of the same genus, when they come into competition with each other, than between species of distinct genera (Darwin 1859).

Interactions between syntopic congeners often select for character divergence (Grant 1986). Coexistence is then possible through niche differentiation. Hubbell and Foster (1986) proposed instead that syntopic tropical forest plants are selected to converge on a generalized ability to tolerate diffuse competition. Which outcome is more likely? A full answer requires a phylogenetic analysis because convergence and evolutionary stasis are indistinguishable unless the character state of the common ancestor is known (Harvey and Pagel 1991). Still, we can inquire whether syntopic congeners are now ecologically similar or dissimilar.

Large numbers of congeneric plants coexist in tropical forests. Gentry (1982, 1989) drew attention to this phenomenon, describing "species swarms" in *Piper* (Piperaceae), *Miconia* (Melastomataceae), *Psychotria* (Rubiaceae), and several herbaceous genera in southern Central America and northwestern South

America. Other examples include *Eugenia* (Myrtaceae) with 45 tree species in a 50-ha plot at Pasoh, Malaysia; *Pouteria* (Sapotaceae) with 22 tree species in a single hectare in Amazonian Ecuador; and *Inga* (Mimosoideae) with 22 tree species in a 0.16-ha plot also in Amazonian Ecuador (Manokaran et al. 1992; Valencia et al. 1994; R. Foster, personal communication, 1996). Are these syntopic congeners ecologically distinct or ecologically similar?

Reproductive phenologies illustrate both possibilities. Cross-pollination and competition for pollinators and fruit dispersal agents select for divergent phenologies that minimize temporal overlap. The fruiting phenologies of Trinidadian *Miconia* (Melastomataceae) and the flowering phenologies of Malaysian *Shorea* (Dipterocarpaceae) and Costa Rican *Heliconia* (Musaceae) are segregated in time (Snow 1965, Stiles 1977, Ashton et al. 1988). These congeners have diverged despite recent common descent. The attraction of larger numbers of pollinators or fruit dispersal agents and the satiation of pests select for coincident reproductive displays that maximize temporal overlap. Flowering phenologies overlap completely for different species of Panamanian *Costus* (Zingiberaceae) with morphologically similar flowers that attract the same pollinator species to the same microhabitats (Schemske 1981). Both divergence and convergence/stasis are evident among the reproductive phenologies of syntopic congeners. Which is more likely?

When all genera in a local flora are evaluated, a clear answer emerges. Phenologies are remarkably similar in an overwhelming proportion of genera (Wright and Calderon 1995). Strong demonstrable similarities among species motivate the hypothesis that chance population dynamics contribute to the maintenance of tropical forest plant diversity (see Section II. E).

ACKNOWLEDGMENT

I thank John Barone, Richard Condit, Kyle Harms, Egbert Leigh, and many others for stimulating discussions.

REFERENCES

Ashton, P.S. 1969. Speciation among tropical forest trees: some deductions in the light of recent evidence. In: Lowe-McConnell, R.H., ed. Speciation in Tropical Environments. London: Academic Press, pp 155–196.

Ashton, P.S. 1977. A contribution of rain forest research to evolutionary theory. Annals of the Missouri Botanical Garden 64: 694–705.

Ashton, P.S. 1993. Species richness in plant communities. In: Fiedler, P.L., and S.K. Jain, eds. Conservation Biology. New York: Chapman and Hall, pp 4–22.

Ashton, P.S., T.J. Givnish, and S. Appanah. 1988. Staggered flowering in the Dipterocarpaceae: new insights into floral induction and the evolution of mast fruiting in the aseasonal tropics. American Naturalist 132: 44–66.

Asquith, N.M., S.J. Wright, and M.J. Claus. 1997. Does mammal community composition control seedling recruitment in neotropical forests? Evidence from islands in central Panama. Ecology 78: 941–946.

Aubréville, A. 1938. La foret de la Cote d'Ivoire. Bulletin du Comité d'Études Historiques et Scientifiques de l'Afrique Occidentale Française 15: 205–261.

Augspurger, C.K. 1984. Seedling survival of tropical tree species: interactions of dispersal distance, light-gaps and pathogens. Ecology 65: 1705–1712.

Baillie, I.C., P.S. Ashton, M.N. Court, J.A.R. Anderson, E.A. Fitzpatrick, and J. Tinsley. 1987. Site characteristics and the distribution of tree species in mixed dipterocarp forest on Tertiary sediments in central Sarawak, Malaysia. Journal of Tropical Ecology 3: 201–220.

Beadle, N.C.W. 1966. Soil phosphate and its role in molding segments of the Australian flora and vegetation, with special reference to xeromorphy and sclerophylly. Ecology 47: 992–1007.

Behrensmeyer, A.K., J.D. Damuth, W.A. DiMichele, R. Potts, H. Sues, and S.L. Wing. 1992. Terrestrial Ecosystems Through Time. Chicago: University of Chicago Press.

Brokaw, N. 1987. Gap-phase regeneration of three pioneer tree species in a tropical forest. Journal of Ecology 75: 9–19.

Brown, N.D., and T.C. Whitmore. 1992. Do dipterocarp seedlings really partition tropical rain forest gaps? Philosophical Transactions of the Royal Society, London B 335: 369–378.

Brown, S., and A.E. Lugo. 1982. The storage and production of organic matter in tropical forests and their role in the global carbon cycle. Biotropica 14: 161–187.

Chazdon, R.L., and R.W. Pearcy. 1991. The importance of sunflecks for forest understory plants. Bioscience 41: 760–766.

Chesson, P., and N. Huntly. 1997. The roles of harsh and fluctuating conditions in the dynamics of ecological communities. American Naturalist 150: 519–553.

Clark, D.A., and D.B. Clark. 1984. Spacing dynamics of a tropical rain-forest tree: evaluation of the Janzen-Connell model. American Naturalist 124: 769–788.

Clinebell, R.R., O.L. Phillips, A.H. Gentry, N. Stark, and H. Zuuring. 1995. Prediction of neotropical tree and liana species richness from soil and climatic data. Biodiversity and Conservation 4: 56–60.

Coley, P.D., and T.M. Aide. 1991. A comparison of herbivory and plant defenses in temperate and tropical broad-leaved forests. In: Price, P.W., T.M. Lewinsohn, G.W. Fernandes, and W.W. Benson, eds. Plant-Animal Interactions: Evolutionary Ecology in Tropical and Temperate Regions. New York: Wiley, pp 25–49.

Coley, P.D., and J.A. Barone. 1996. Herbivory and plant defenses in tropical forests. Annual Review of Ecology and Systematics 27: 305–335.

Condit, R., S.P. Hubbell, and R.B. Foster. 1992. Recruitment near conspecific adults and the maintenance of tree and shrub diversity in a neotropical forest. American Naturalist 140: 261–286.

Condit, R., S.P. Hubbell, and R.B. Foster. 1994. Density dependence in two understory tree species in a neotropical forest. Ecology 75: 671–680.

Connell, J.H. 1971. On the role of natural enemies in preventing competitive exclusion in some marine animals and in rain forest trees. In: den Boer, P.J., and G.R. Gradwell, eds. Dynamics of Populations. Wageningen, The Netherlands: Centre for Agricultural Publishing and Documentation, pp 298–313.

Connell, J.H. 1978. Diversity in tropical rain forests and coral reefs. Science 199: 1302–1309.

Connell, J.H., J.G. Tracey, and L.J. Webb. 1984. Compensatory recruitment, growth, and mortality as factors maintaining rain forest tree diversity. Ecological Monographs 54: 141–164.

Darwin, C.R. 1859. On the Origin of Species by Means of Natural Selection. London: John Murray.

Denslow, J.S. 1980. Gap partitioning among tropical rainforest trees. Biotropica 12 (suppl): 47–55.

Denslow, J.S. 1987. Tropical rainforest gaps and tree species diversity. Annual Review of Ecology and Systematics 18: 431–451.

Denslow, J.S., J.C. Schultz, P.M. Vitousek, and B.R. Strain. 1990. Growth responses of tropical shrubs to treefall gap environments. Ecology 71: 165–179.

Dirzo, R., and A. Miranda. 1991. Altered patterns of herbivory and diversity in the forest understory: a case study of the possible consequences of contemporary defaunation. In: Price, P.W., P.W. Lewinsohn, G.W. Fernandes, and W.W. Benson, eds. Plant-Animal Interactions: Evolutionary Ecology in Tropical and Temperate Regions. New York: Wiley, pp 273–287.

Federov, A.A. 1966. The structure of the tropical rain forest and speciation in the humid tropics. Journal of Ecology 54: 1–11.

Friis, E.M., Chaloner, W.G., and P.R. Crane, eds. 1987. The Origins of Angiosperms and Their Biological Consequences. Cambridge: Cambridge University Press.

Gause, G.F. 1934. The Struggle for Existence. New York: Hafner.

Gentry, A.H. 1988. Changes in plant community diversity and floristic composition on environmental and geographical gradients. Annals of the Missouri Botanical Garden 75: 1–34.

Gentry, A.H., and C. Dodson. 1987. Contribution of nontrees to species richness of a tropical rain forest. Biotropica 19: 149–156.

Gentry, A.H., and L.H. Emmons. 1987. Geographical variation in fertility, phenology, and composition of the understory of neotropical forests. Biotropica 19: 216–227.

Gilbert, G.S., S.P. Hubbell, and R.B. Foster. 1994. Density and distance-to-adult effects of a canker disease of trees in a moist tropical forest. Oecologia (Berlin) 98: 100–108.

Gillett, J.B. 1962. Pest pressure, an underestimated factor in evolution. Systematics Association Publication 4: 37–46.

Grant, P.R. 1986. Ecology and Evolution of Darwin's Finches. Princeton, NJ: Princeton University Press.

Grubb, P.J. 1986. Global trends in species-richness in terrestrial vegetation: a view from the northern hemisphere. In: Gee, J.H.R., and P.S. Giller, eds. Organization of Communities Past and Present. London: Blackwell Scientific, pp 99–118.

Hall, J.B., and M.D. Swaine. 1976. Classification and ecology of closed-canopy forest in Ghana. Journal of Ecology 64: 913–951.

Hall, J.B., and M.D. Swaine. 1981. Distribution and Ecology of Vascular Plants in a Tropical Rain Forest. The Hague: Dr. W. Junk Publishers.

Hammond, D.S., and V.K. Brown. 1995. Seed size of woody plants in relation to disturbance, dispersal, soil type in wet neotropical forests. Ecology 76: 2544–2561.

Harvey, P.H., and M.D. Pagel. 1991. The Comparative Method in Evolutionary Biology. Oxford: Oxford University Press.

Holdridge, L.R., W.C. Grenke, W.H. Hatheway, T. Liang, and J.A. Tosi Jr. 1971. Forest Environments in Tropical Life Zones: A Pilot Study. New York: Pergamon Press.

Hubbell, S.P. 1979. Tree dispersion, abundance, and diversity in a tropical dry forest. Science 203: 1299–1309.

Hubbell, S.P., and R.B. Foster. 1983. Diversity of canopy trees in a neotropical forest and implications for conservation. In: Sutton, S.L., T.C. Whitmore, and A.C. Chadwick, eds. Tropical Rain Forest: Ecology and Management. Oxford: Blackwell Scientific, pp 25–41.

Hubbell, S.P., and R.B. Foster. 1986. Biology, chance, and history and the structure of tropical rain forest tree communities. In: Diamond, J., and T.J. Case, eds. Community Ecology. New York: Harper and Row, pp 314–329.

Hubbell, S.P., R. Condit, and R.B. Foster. 1990. Presence and absence of density dependence in a neotropical tree community. Philosophical Transactions of the Royal Society, London (B) 330: 269–281.

Hubbell, S.P., and R.B. Foster. 1990. Structure, dynamics and equilibrium status of old-growth forest on Barro Colorado Island. In: Gentry, A., ed. Four Neotropical Forests. New Haven, CT: Yale University Press, pp 522–541.

Hurtt, G.C., and S.W. Pacala. 1995. The consequences of recruitment limitation: reconciling chance, history and competitive differences between plants. Journal of Theoretical Biology 176: 1–12.

Huston, M. 1979. A general hypothesis of species diversity. American Naturalist 113: 81–101.

Huston, M. 1980. Soil nutrients and tree species richness in Costa Rican forests. Journal of Biogeography 7: 147–157.

Huston, M. 1993. Biological diversity, soils, and economics. Science 262: 1676–1680.

Huston, M.A. 1994. Biological Diversity. Cambridge: Cambridge University Press.

Janzen, D.H. 1970. Herbivores and the number of tree species in tropical forests. American Naturalist 104: 501–528.

Jordan, C.F. 1983. Productivity of tropical rain forest ecosystems and the implications for their use as future wood and energy sources. In: Golley, F.B., ed. Ecosystems of the World. 14A. Tropical Rain Forest Ecosystems. Amsterdam: Elsevier, pp 117–136.

Kitajima, K. 1994. Relative importance of photosynthetic traits and allocation patterns as correlates of seedling shade tolerance of 13 tropical trees. Oecologia 98: 419–428.

LaFrankie, J. 1996. Initial findings from Lambir: trees, soils and community dynamics. Center for Tropical Forest Science 1995: 5.

Latham, R.E., and R.E. Ricklefs. 1993. Continental comparisons of temperate-zone tree species diversity. In: Ricklefs, R.E., and D. Schluter, eds. Species Diversity in Ecological Communities. Chicago: University of Chicago Press, pp 294–314.

Leigh, E.G. Jr., S.J. Wright, and F.E. Putz. 1993. The decline of tree diversity on newly isolated tropical islands: a test of a null hypothesis and some implications. Evolutionary Ecology 7: 76–102.

Lieberman, M., D. Lieberman, R. Peralta, and G.S. Hartshorn. 1995. Canopy closure and the distribution of tropical forest trees at La Selva, Costa Rica. Journal of Tropical Ecology 11: 161–177.

Lieberman, M., D. Lieberman, G.S. Hartshorn, and R. Peralta. 1985. Small-scale altitudinal variation in lowland wet tropical forest vegetation. Journal of Ecology 73: 505–516.

MacArthur, R.H. 1969. Patterns of communities in the tropics. Biological Journal of the Linnean Society 1: 19–30.

Manokaran, N., J.V. La Frankie, K.M. Kochummen, E.S. Quah, J.E. Klahn, P.S. Ashton, and S.P. Hubbell. 1992. Stand table and distribution of species in the 50-ha research plot at Pasoh Forest Reserve. Forestry Research Institute of Malaysia, Research Data 1: 1–454.

Marquis, R.J., H.J. Young, and H.E. Braker. 1986. The influence of understory vegetation cover on germination and seeding establishment in a tropical lowland wet forest. Biotropica 18: 273–278.

Medina, E., and H. Klinge. 1983. Productivity of tropical forests and tropical woodlands. In: Lange, O.L., P.S. Nobel, C.B. Osmond, and H. Ziegler, eds. Physiological Plant Ecology IV. Berlin: Springer-Verlag, pp 281–303.

Nelson, B.W., V. Kapos, J.B. Adams, W.J. Oliveira, O.P.G. Braun, and I.L. do Amaral. 1994. Forest disturbance by large blowdowns in the Brazilian Amazon. Ecology 75: 853–858.

Nepstad, D.C., C.R. de Carvalho, E.A. Davidson, P.H. Jipp, P.A. Lefebvre, G.H. Negreiros, E.D. da Silva, T.A. Stone, S.E. Trumbore, and S. Vieira. 1994. The role of deep roots in the hydrological and carbon cycles of Amazonian forests and pastures. Nature 372: 666–669.

Niklas, K.J., B.H. Tiffney, and A.H. Knoll. 1983. Patterns in vascular land versification. Nature 303: 614–616.

Orians, G.H. 1982. The influence of tree-falls in tropical forests in tree species richness. Tropical Ecology 23: 255–279.

Osunkjoya, O.O., J.E. Ash, M.S. Hopkins, and A.W. Graham. 1992. Factors affecting survival of tree seedlings in North Queensland rainforests. Oecologia (Berlin) 91: 569–578.

Pacala, S.W., C.D. Canham, J. Saponara, J.A. Silander, R.K. Kobe, and E. Ribbens. 1996. Forest models defined by field measurements: estimation, error analysis and dynamics. Ecological Monographs 66: 1–43.

Paine, R.T. 1966. Food web complexity and species diversity. American Naturalist 100: 65–75.

Palmer, M.W. 1994. Variation in species richness: towards a unification of hypotheses. Folia Geobotanica & Phytotaxonomica 29: 511–530.

Petraitis, P.S., R.E. Latham, and R.A. Niesenbaum. 1989. The maintenance of species diversity by disturbance. Quarterly Review of Biology 64: 393–418.

Phillips, O.L., P. Hall, A.H. Gentry, S.A. Sawyer, and R. Váquez. 1994. Dynamics and species richness of tropical rain forests. Proceedings of the National Academy of Science (USA) 91: 2905–2809.

Platt, W.J. and D.R. Strong, eds. 1989. Special feature: gaps in forest ecology. Ecology 70: 535–576.

Prance, G.T. 1977. Floristic inventory of the tropics: where do we stand? Annals of the Missouri Botanical Garden 64: 659–684.

Redford, K.H. 1992. The empty forest. Bioscience 42: 412–422.

Reich, P.B., M.B. Walters, and D.S. Ellsworth. 1992. Leaf life-span in relation to leaf, plant and stand characteristics among diverse ecosystems. Ecological Monographs 62: 365–392.

Ricklefs, R.E. 1977. Environmental heterogeneity and plant species diversity: a hypothesis. American Naturalist 376–381.

Rosenzweig, M.L. 1992. Species diversity gradients: we know more and less than we thought. Journal of Mammalogy 73: 715–730.

Rosenzweig, M.L. 1995. Species Diversity in Space and Time. Cambridge: Cambridge University Press.

Schemske, D.W. 1981. Floral convergence and pollinator sharing in two bee-pollinated tropical herbs. Ecology 62: 946–954.

Schupp, E.W. 1992. The Janzen-Connell model for tropical tree diversity: population implications and the importance of spatial scale. American Naturalist 140: 526–530.

Silver, W.L. 1994. Is nutrient availability related to plant nutrient use in humid tropical forests? Oecologia (Berlin) 98: 336–343.

Snow, D.W. 1965. A possible selective factor in the evolution of fruiting seasons in tropical forest. Oikos 15: 274–281.

Sollins, P. 1998. Factors influencing species composition in tropical lowland rain forest: does soil matter? Ecology 79: 23–30.

Stiles, F.G. 1977. Coadapted pollinators: the flowering seasons of hummingbird-pollinated plants in a tropical forest. Science 198: 1177–1178.

Swaine, M.D. 1996. Rainfall and soil fertility as factors limiting forest species distributions in Ghana. Journal of Ecology 84: 419–428.

Terborgh, J. 1973. On the notion of favorableness in plant ecology. American Naturalist 107: 481–501.

Terborgh, J. 1992. Maintenance of diversity in tropical forests. Biotropica 24: 283–292.

Terborgh, J., R.B. Foster, and P. Nunez V. 1996. Tropical tree communities: a test of the nonequilibrium hypotheis. Ecology 77: 561–567.

Terborgh, J., and S.J. Wright. 1994. Effects of mammalian herbivores on plant recruitment in two neotropical forests. Ecology 75: 1829–1833.

Tilman, D. 1982. Resource Competition and Community Structure. Princeton, NJ: Princeton University Press.

Tilman, D. 1994. Competition and biodiversity in spatially structured habitats. Ecology 75: 2–16.

Tilman, D., and S. Pacala. 1993. The maintenance of species richness in plant communities. In: Ricklefs, R.E., and D. Schluter, eds. Species Diversity in Ecological Communities. Chicago: University of Chicago Press, pp 13–25.

Tuomisto, H., and A.D. Poulsen. 1996. Influence of edaphic specialization on pteridophyte distribution in neotropical rain forests. Journal of Biogeography 23: 283–293.

Tuomisto, H., and K. Ruokolainen. 1994. Distribution of Pteridophyta and Melastomataceae along an edaphic gradient in an Amazonian rain forest. Journal of Vegetation Science 5: 25–34.

Uhl, C., K. Clark, N. Dezzeo, and P. Maquirino. 1988. Vegetation dynamics in Amazonian treefall gaps. Ecology 69: 751–763.

Valencia, R., H. Balslev, and G. Paz y Miño C. 1994. High tree alpha-diversity in Amazonian Ecuador. Biodiversity and Conservation 3: 21–28.

Vitousek, P.M. 1984. Litterfall, nutrient cycling and nutrient limitation in tropical forests. Ecology 65: 285–298.

Welden, C.W., S.W. Hewett, S.P. Hubbell, and R.B. Foster. 1991. Sapling survival, growth, and recruitment: relationship to canopy height in a neotropical forest. Ecology 72: 35–50.

Whitmore, T.C. 1975. Tropical Rain Forest of the Far East. Oxford: Clarendon Press.

Wills, C., R. Condit, R.B. Foster, and S.P. Hubbell. 1997. Strong density- and diversity-related effects help to maintain tree species diversity in a neotropical forest. Proceedings of the National Academy of Science (USA) 94: 1252–1257.

Wright, S.J. 1992. Seasonal drought, soil fertility, and the species diversity of tropical forest plant communities. Trends in Ecology and Evolution 7: 260–263.

Wright, S.J., and O. Calderon. 1995. Phylogenetic constraints on tropical flowering phenologies. Journal of Ecology 83: 937–948.

13
Arctic Ecology

Sarah E. Hobbie
University of Minnesota, St. Paul, Minnesota

I. INTRODUCTION

The Arctic encompasses the region of the globe that lies north of the latitudinal treeline. Its environment is extreme in myriad ways, with low mean annual temperatures and precipitation, short growing seasons, and cold soils that are often underlain by permafrost, thawing only incompletely during the growing season.

Consistent with the extreme environment, productivity is lower in the Arctic than in most other regions of the world (Figure 1).

The Arctic comprises a variety of vegetation types at both small and large scales (Bliss and Matveyeva 1992). The High Arctic, consisting primarily of the islands of the Canadian Arctic Archipelago, is characterized by polar desert and semidesert—large patches of bare, unvegetated ground are interspersed with vegetation dominated by cushion plants, forbs, and dwarf shrubs. In contrast, the Low Arctic has a near-continuous cover of tundra vegetation with varying proportions of dwarf shrubs, sedges, mosses, lichens, and forbs. Within both the High and Low Arctic, large variation in both species composition and plant productiv-

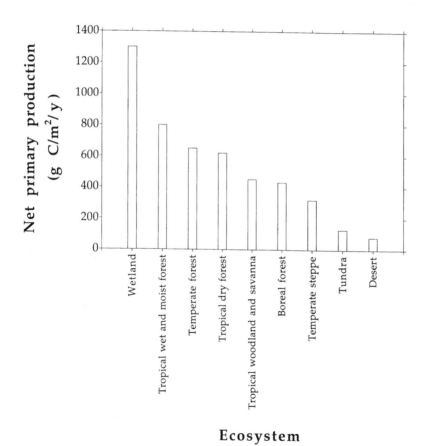

Figure 1 Net primary production in various ecosystems of the world. (*Source*: Houghton and Skole 1990.)

ity occurs at relatively small spatial scales (meters to kilometers) because of variation in topography and parent material, and consequently in drainage, nutrient availability, and snow cover.

Because of its severe environment, the Arctic has long been of interest to plant physiological ecologists (Billings and Mooney 1968). The Arctic has received much less attention from other areas of plant ecology, particularly population and community ecology, for both practical and theoretical reasons. The long-lived nature of most tundra species and their low rates of sexual reproduction (see Section II) make study of plant populations and manipulation of plant densities in situ difficult. Furthermore, some plant ecologists (Savile 1960, Grime 1977) have suggested that competitive interactions are unimportant relative to abiotic factors in structuring communities in high-stress environments, such as the Arctic, perhaps steering population and community ecologists away from working at high latitudes.

This chapter takes a broad view of arctic plant ecology. The unique characteristics of arctic plant populations and the accumulating evidence regarding the importance of interspecific interactions in structuring arctic ecosystems are discussed. The physiological ecology of arctic plants is briefly reviewed, and the reader is referred to numerous reviews and books on this topic (e.g., Billings and Mooney 1968, Bliss et al. 1981, Chapin and Shaver 1985a, Korner and Larcher 1988, Chapin et al. 1992). A major emphasis of this chapter is the synthesis of a large amount of recent experimental work in tundra to evaluate the hypotheses that were formulated during these earlier ecophysiological studies.

II. CHARACTERISTICS OF ARCTIC PLANT POPULATIONS

Relatively little is known about arctic plant populations, perhaps because of the long-lived nature and the infrequent sexual reproduction of arctic plant species. In the closed communities of the Low Arctic, reproduction from seed is relatively rare (Callaghan and Emanuelsson 1985, McGraw and Fetcher 1992), and few annual plant species occur (Hulten 1968). Low rates of sexual reproduction may result from low allocation to total reproductive effort, although allocation to viable seed is not necessarily lower in arctic species than in temperate ones (Chester and Shaver 1982). Low rates of recruitment from seed may also result from the paucity of suitable germination sites in these closed communities and from high mortality of young seedlings (McGraw and Shaver 1982, Callaghan and Emanuelsson 1985). Recruitment from seed increases after natural and human-caused disturbance, particularly for graminoid species, often from seed stored in the seed bank (Chapin and Chapin 1980, McGraw 1980, Freedman et al. 1982, Gartner et al. 1983, 1986, Grulke and Bliss 1988, Ebersole 1989, McGraw and Vavrek 1989). In the more open vegetation of the High Arctic, recruitment from seed is more common (Callaghan and Emanuelsson 1985) and increases with fertilization (Robinson et al. 1998).

In contrast with sexual reproduction, asexual reproduction is nearly ubiquitous in the Arctic. Many arctic plant species possess rhizomes (belowground stems) or stolons, or produce adventitious roots from aboveground stems that are overgrown by mosses and subsequently become "belowground" stems. Ramets can ultimately become independent of one another, making the recognition of individual genets in intact tundra difficult. Although demographic analysis at the level of individual genets is therefore problematic, understanding the demography of plant parts within ramets can enlighten studies of the response of arctic plants to environmental perturbations (McGraw and Fetcher 1992). For example, demographic analysis of shoots or tillers can indicate whether changes in production or biomass at the ecosystem level result from changes in branching, shoot growth, or shoot mortality (McGraw 1985a, Chapin et al. 1995); alternatively, demographic analyses may indicate when turnover of plant parts has changed with no change in total biomass (Fetcher and Shaver 1983).

III. PLANT SPECIES INTERACTIONS IN ARCTIC ECOSYSTEMS

Species interactions have been studied much less in arctic ecosystems than in temperate ecosystems, presumably because of the difficulty of manipulating densities of long-lived, slow-growing, often woody, perennial species that occur only rarely as seedlings. Despite the lack of experimental evidence, many investigators have hypothesized that direct limitation by stressful abiotic factors (e.g., low temperature and nutrient availability) is much more important than species interactions in structuring plant communities in the Arctic (Savile 1960, Warren Wilson 1966, Billings and Mooney 1968, Grime 1977). Other investigators have suggested that competitive interactions are important and help explain community responses to environmental manipulations (Chapin and Shaver 1985b, Chapin et al. 1995).

The few studies that have examined competitive interactions in the Arctic have manipulated densities of mature ramets, primarily by removing the aboveground biomass of one species and observing the response of the remaining species. These studies have found little evidence for strong negative interactions in tundra, demonstrating only a few positive effects of neighbor removal (Fetcher 1985, McGraw 1985a, Jonasson 1992, Hobbie 1995, Shevtsova et al. 1995). For example, in the Scandinavian subarctic, only one of three dwarf shrubs studied (*Vaccinium vitis-idaea*) responded positively to removal of other dwarf shrubs (Shevtsova et al. 1995). In Alaskan tussock tundra, where seven species were removed in separate treatments and numerous species responses were measured, only two pairs of species exhibited negative interactions (Hobbie 1995). In a separate study, *Eriophorum vaginatum* responded to dwarf shrub removal with increased tillering (Fetcher 1985). In the High Arctic, moss removal increased growth of forbs (Sohlberg and Bliss 1986). A study in subarctic Sweden found no positive effects of dwarf shrub removal on any of a number of species in several different tundra types (Jonasson 1992). Similarly, *Dryas octopetala* showed no response to removal of all neighbors after 3 years in Alaskan tundra (McGraw 1985a). The only study to manipulate ramets by outplanting rather than removal demonstrated competitive interactions between two species of *Eriophorum* in Alaska (McGraw and Chapin 1989). None of these studies distinguished competitive interactions from other kinds of negative interactions such as allelopathy or other noncompetitive effects of species on conditions such as soil temperature.

Species removal studies have found as much evidence for positive interactions among plant species in arctic ecosystems as for negative interactions. Several dwarf shrub species responded negatively to removal of other dwarf shrubs in Sweden and Finland (Jonasson 1992, Shevtsova et al. 1995), suggesting that facilitation may be important in determining performance of some species in the Arctic. Such facilitation could explain the clustering of plant species in some tundra environments (Callaghan and Emanuelsson 1985, Carlsson and Callaghan 1991). Mosses in tussock tundra and in boreal Canada also responded negatively to shrub removal, which was attributed to photoinhibition of photosynthesis under high light intensities (Murray et al. 1993) and increased evaporative stress (Busby et al. 1978) when the shrub canopies were removed.

The lack of evidence for strong competitive or other kinds of negative interactions in tundra supports Grime's contention that in high stress environments, plants are directly limited by abiotic factors, and that competitive interactions are relatively unimportant (Grime 1977; see also Chapter 19). However, several other explanations exist for the general failure to demonstrate strong competitive interactions in the Arctic. Arctic species respond individualistically to manipulations of various environmental factors, suggesting that growth of individual species may be limited by several different factors (see Section V). Additionally, arctic plants may partition their use of the nitrogen pool (into inorganic and organic nitrogen; Michelson et al. 1996b, Nadelhoffer et al. 1996). They may also exhibit phenological and spatial differences in nutrient acquisition (Shaver and Billings 1975, Kummerow et al. 1983), thereby minimizing competition for nutrients.

Even if it does occur, competition may be difficult to demonstrate in this system, where annual uptake of nutrients is a relatively small proportion of the annual nutrient requirement of a species. Removing that species may have only a small effect on annual nutrient availability for the remaining species over the time scale of most experiments. Studying competitive interactions by increasing plant densities (e.g., by outplanting or seed sowing) may be a more effective way of demonstrating the importance of competitive interactions than removals; however, such manipulations are problematic with arctic species, since they are generally slow-growing and perennial. The true reason for the lack of strong competitive interactions in tundra deserves more exploration.

IV. ECOPHYSIOLOGY OF ARCTIC PLANTS

Ecophysiological research on arctic plants generally indicates that resource acquisition is not directly limited by cold temperatures. This observation has led to the hypothesis that growth and productivity in arctic ecosystems are limited more by the indirect effects of cold temperatures (i.e., short growing season and low nutrient availability) than by their direct effects (Chapin 1983a). This section reviews the observations that led to these generalizations, followed by an evaluation of how well recent experimental evidence supports them.

A. Carbon

In terms of carbon gain, arctic plants appear limited more by the length of the growing season than by low irradiance or low photosynthetic rates due to cold temperatures per se. Although average irradiance during daytime hours is less in the Arctic than at lower latitudes, the average daily irradiance is comparable because of the 24-hour photoperiod of the northern summer (Chapin and Shaver 1985a). However, many of the long days occur early in the growing season when the ground is still covered with snow or the soil is still frozen. Thus, growing season length, rather than light intensity per se, may limit plant growth.

The photosynthetic temperature optima and carbohydrate status of tundra plants provide additional indirect evidence that arctic species are not carbon limited during the growing season. Although photosynthetic temperature optima are often above ambient air temperatures (Tieszen 1973, Oechel 1976), many tundra species have relatively broad photosynthetic temperature optima, allowing them to achieve near-maximum rates at fairly low temperatures (Tieszen 1973, Johnson and Tieszen 1976, Limbach et al. 1982, Semikhatova et al. 1992). Relatively large pools of total nonstructural carbohydrate in arctic plants also suggest that they are able to acquire carbon in excess of their growth requirements (Chapin and Shaver 1985a).

B. Nutrients

Nutrient uptake in tundra plants is also relatively insensitive to the direct effects of low temperature. Rather, cold temperatures limit nutrient uptake indirectly, by causing low rates of nutrient supply. For example, cold temperatures reduce nutrient inputs from nitrogen fixation (Chapin and Bledsoe 1992) and weathering and recycling of nutrients during decomposition (Nadelhoffer et al. 1992; Figure 2). Although temperatures in arctic soils are colder than optimum temperatures for nutrient uptake (Chapin and Bloom 1976), tundra species have a higher potential for nutrient uptake at low temperatures and low nutrient concentrations than do their temperate counterparts (Chapin 1974, 1983b).

Tundra species use acquired nutrients efficiently, relying heavily on stored nutrients for supporting growth and retranslocating much of the nutrients from senescing tissues. Stored nutrients supply the majority of the nutrients in current growth (Berendse and Jonasson 1992) and may allow arctic plants to grow during times when soil nutrients are frozen and unavailable (Shaver and Kummerow 1992). Many arctic plant species retranslocate a relatively high percentage of nutrients from senescing tissues, although their retranslocation efficiencies are not consistently higher than those of temperate species (Chapin and Shaver 1989, Jonasson 1989, Berendse and Jonasson 1992).

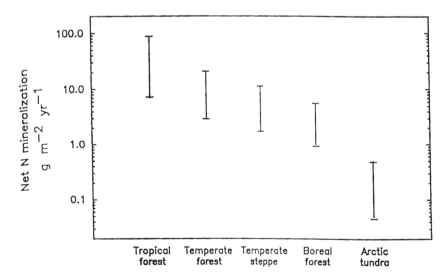

Figure 2 The range of net nitrogen mineralization rates measured in various ecosystems of the world. (*Source*: Nadelhoffer et al. 1992.)

One potential way that arctic plants effectively increase the pool of nutrients available for uptake is by using organic nitrogen directly, rather than having to rely on soil microbes to mineralize nitrogen from organic compounds. There is growing evidence that arctic plants may "short circuit" the nitrogen cycle in this way. Such use of organic nitrogen would help explain the large discrepancy between annual nitrogen uptake and annual net nitrogen mineralization in the Arctic (Giblin et al. 1991, Kielland 1994). Concentrations and turnover of free amino acids are relatively high in tundra soils (Kielland 1995). Both mycorrhizal and nonmycorrhizal species from tussock tundra take up and grow on amino acids in solution culture (Chapin et al. 1993, Kielland 1994), and nonmycorrhizal sedges take up amino acids as readily as ammonium in the field (Schimel and Chapin 1996; Figure 3). Furthermore, both ericoid and ectomycorrhizal species have proteolytic capabilities and can transfer organic nitrogen from soil to the host plant (Read 1991); many of the common arctic species have these types of mycorrhizal associations.

The relative natural abundances of the stable isotopes of nitrogen (^{15}N and ^{14}N) vary widely among tundra plant species (Michelson et al. 1996b, Nadelhoffer et al. 1996; Figure 4). Discrimination against the heavy isotope (^{15}N) during both

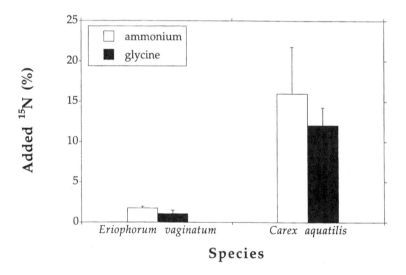

Figure 3 The percentage (+1 SE) of added ^{15}N-labeled ammonium or glycine that was recovered in *Eriophorum vaginatum* and *Carex aquatilis* plant biomass in tussock and wet meadow tundra, respectively, after a 5-day in situ incubation. (*Source*: Schimel and Chapin 1996; used with permission of the Ecological Society of America.)

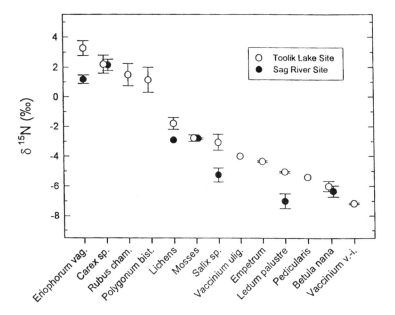

Figure 4 Mean (±1 SE) $\delta^{15}N$ values for plant species in tussock tundra near Toolik Lake (open circles) and the Sagavanirktok River (filled circles), northern Alaska. $\delta^{15}N = [(^{15}N/^{14}N_{sample} \div {}^{15}N/^{14}N_{air}) - 1] \times 1000$. More positive values indicate relatively more ^{15}N. (*Source*: Nadelhoffer et al. 1996; copyright Springer-Verlag.)

biological and physical processes causes soil organic matter to become enriched in ^{15}N relative to plants and fresh litter (Nadelhoffer and Fry 1994). One explanation for the variation in ^{15}N and ^{14}N among arctic species is that different growth forms (e.g., sedges, forbs, mosses, lichens, dwarf evergreen and deciduous shrubs) are accessing a variety of nitrogen forms and sources, including organic nitrogen (Michelson et al. 1996b, Nadelhoffer et al. 1996). Dwarf shrubs (both ericoid and ectomycorrhizal species) are depleted in ^{15}N, suggesting that they are accessing nitrogen from recently fallen litter. Other explanations also exist to explain this pattern. For example, variation in the relative abundances of ^{15}N and ^{14}N may reflect differences in rooting depth (deeper soil organic matter tends to be enriched in ^{15}N) (Nadelhoffer et al. 1996).

By partitioning the nitrogen pool, plants may minimize nitrogen competition in nitrogen-limited tundra ecosystems. Whether this use of organic nitrogen represents a unique characteristic of plants from this extremely low-nitrogen environment is unknown. Further research will reveal whether arctic plants use or-

ganic nitrogen more readily than do temperate species, or whether the availability
of organic nitrogen is higher in tundra than in other ecosystems.

Arctic plant species may also partition their uptake of nutrients in time and
space. The timing of root initiation in the spring differs among species and growth
forms (Shaver and Billings 1977, Kummerow et al. 1983). In particular, evergreens begin root growth earlier in the spring than deciduous species. Deciduous
shrubs begin initiating roots only after leaf expansion has begun. Such interspecific differences in the timing of root growth could have important consequences
for nutrient acquisition in arctic ecosystems that are often characterized by a pulse
of nutrient mineralization in the spring (Kielland 1990, Nadelhoffer et al. 1992).
Different species also show characteristic rooting depths (Shaver and Billings
1975), perhaps partitioning their use of nutrients in space.

C. Growth

Whether cold temperatures directly limit growth of arctic plants is unclear. The
relative growth rates of arctic plants are similar to those of temperate species,
suggesting that length of the growing season, rather than cold temperatures during
the growing season itself, limits biomass accumulation in the Arctic (Chapin and
Shaver 1985b, Semikhatova et al. 1992). On the other hand, ambient temperatures
are often suboptimal for growth (Kummerow and Ellis 1984, Korner and Larcher
1988) and manipulations of temperature in situ sometimes increase plant growth
(see Section V). Thus, growth of some species may be directly limited by cold
temperature.

V. INDIVIDUAL PLANT RESPONSES TO ENVIRONMENTAL
   FACTORS

During the past decade, numerous studies in various arctic habitats have examined the response of plant growth, biomass, reproduction, phenology and/or production to manipulations of environmental factors. These experiments allow us
to evaluate the aforementioned generalizations and hypotheses. Three general
patterns emerge from studies that examined individual plant responses (primarily
growth) to environmental manipulations. First, of the various factors manipulated
(including nutrients, temperature, water, light, and CO_2), plant growth most often
responds to increased nutrient availability. Second, in almost all studies demonstrating a positive response of species growth to nutrient addition, there are exceptions—species that either do not respond or respond negatively to nutrient addition. Third, manipulation of environmental factors besides nutrients has less

predictable results. Each of these points is explored in detail here. For simplicity, a number of different kinds of measurements are subsumed (e.g., current year's shoot length, branching, tillering) by the word "growth." Also included in "growth" are changes in total biomass, although many studies did not determine whether change in biomass resulted from changes in individual shoot growth or from changes in shoot demography (branching or mortality).

Fertilization studies have been conducted in both the Low and High Arctic in North America and in Europe. In subarctic sites in Sweden, addition of nitrogen (N), phosphorus (P), and potassium (K) increased the growth or biomass of numerous species in heath tundra (Havstrom et al. 1993, Parsons et al. 1994, Parsons et al. 1995, Michelson et al. 1996a), in a fellfield (a relatively open, rocky community; Michelson et al. 1996a), and in graminoid and shrub tundra (Jonasson 1992). Fertilization increased growth and/or biomass of nearly all species in upland tussock tundra and wet meadow (sedge-dominated) tundra in Alaska (McKendrick et al. 1978, 1980, Shaver and Chapin 1980, 1986, 1995 Chapin and Shaver 1985b, 1996, Shaver et al. 1986, 1996, 1998) and of some species in heath tundra (Shaver et al. 1996). Although most studies added N, P, and K simultaneously, when nutrients were added separately, this response was usually to N rather than P in upland tundra (Shaver and Chapin 1980, Gebauer et al. 1995; see also McKendrick et al. 1980). Wet meadow tundra, on the other hand, is more often P-limited (Shaver and Chapin 1995, Shaver et al. 1998). In the High Arctic, fertilization also increased growth, biomass, reproduction, and seedling establishment of some species (Henry et al. 1986, Wookey et al. 1994, 1995, Robinson et al. 1998).

Despite numerous examples of positive growth or biomass responses to nutrient addition in a variety of tundra ecosystems, many studies demonstrated exceptions—species that did not respond or responded negatively to nutrient addition (e.g., Shaver and Chapin 1980, 1985b, 1996, Shaver and Chapin 1986, Shaver et al. 1996, 1998, Havstrom et al. 1993, Robinson et al. 1998). The unresponsive species often differed among studies, and intercomparison is further complicated because fertilizer experiments varied in duration, in amount or types of fertilizer applied, in the timing of fertilizer addition relative to plant phenology of nutrient uptake, and in the species whose responses were measured. Several possible reasons for these lack of responses to fertilization exist. Some species in tundra may be limited by factors other than low nutrient availability (Chapin and Shaver 1985b; see later). Nonvascular species (bryophytes and lichens) in particular are highly sensitive to moisture (Murray et al. 1989a, 1989b, Tenhunen et al. 1992), perhaps explaining their lack of (or negative) response to nutrient addition (Jonasson 1992, Chapin et al. 1995). Species may not be able to respond to additional nutrients in the presence of superior competitors, and understory species may be vulnerable to increased shading by overstory species whose biomass increases with nutrient addition (Chapin and Shaver 1985b, 1996, McGraw 1985b, Shaver et al. 1998). However, no studies have combined fertilization with species manipulations to determine whether competitive interactions are indeed responsible for negative responses to nutrient addition.

After nutrients, increased temperature is the environmental factor that most often increases growth or biomass in the Arctic. Warming increased growth of approximately half of the dominant species in tussock tundra (Chapin and Shaver 1985b, 1996, Shaver et al. 1986). Warming also increased growth of some species in Swedish subarctic tundra (Havstrom et al. 1993, Parsons et al. 1994, 1995, Michelson et al. 1996a) and resulted in greater growth and reproduction and earlier phenology in the High Arctic (Svalbard) (Havstrom et al. 1993, Welker et al. 1993, Wookey et al. 1995).

Far fewer manipulations of light, water, and CO_2 have been performed in the Arctic than have manipulations of nutrients or temperature. Shade decreased growth and biomass, particularly of overstory species in both the Low Arctic (Chapin and Shaver 1985b, 1996) and the High Arctic (Havstrom et al. 1993). Water is rarely limiting in tundra (Oberbauer and Dawson 1992, Gold and Bliss 1995), and water addition had few effects on growth or biomass in subarctic Sweden (Parsons et al. 1994) or in the High Arctic (Henry et al. 1986, Wookey et al. 1994, 1995). Addition of flowing water to a tussock tundra slope did increase the growth of some species (Murray et al. 1989b, Oberbauer et al. 1989), but this may be due to greater flow of nutrients rather than a direct water effect (Chapin et al. 1988). In the only study of of its kind in arctic tundra, elevated CO_2 did not alter photosynthetic rate or growth of individual tillers, but increased tillering of *Eriophorum vaginatum* in Alaskan tussock tundra (Tissue and Oechel 1987).

The response of biomass or production at the shoot or whole-plant level to environmental manipulations is not readily predicted from the physiological response of leaves or roots. For example, photosynthetic responses to treatments generally do not translate directly into growth responses (Bigger and Oechel 1982, Wookey et al. 1994, Gebauer et al. 1995, Chapin and Shaver 1996). Similarly, the response of phenology and nutrient uptake provide little indication of how species will respond at the community level (Chapin and Shaver 1996).

In summary, at the individual level, arctic plants are most often nutrient limited; however, some species in some regions may be limited by other environmental factors such as temperature. Some studies have suggested that the indirect effects of temperature (i.e., low nutrient availability) are more likely to limit growth at the southern extent of species' ranges in the Arctic, whereas cold temperatures directly limit growth at the northern extent of species' ranges (Havstrom et al. 1993). Although this idea is compelling, too few species have been studied experimentally over a range of latitudes to determine whether this pattern is general.

VI. ECOSYSTEM RESPONSES TO ENVIRONMENTAL FACTORS

Individual plants respond to both nutrients and temperature, whereas net primary production and total plant biomass respond more consistently to nutrient addition alone than to manipulation of other environmental factors. However, fewer studies have measured total production and biomass than individual plant growth or biomass, and even measurements of "total production" have excluded root production. Vascular aboveground net primary production (ANPP) and total biomass increased with nutrient addition in Alaskan tussock tundra at a number of different sites (Chapin and Shaver 1985b, Shaver and Chapin 1986, Chapin et al. 1995; Figure 5). However, decreased moss biomass compensated for increased vascular plant biomass, resulting in little change in total plant biomass after long-term fertilization (Chapin et al. 1995). Because of higher production:biomass ratios for vascular plants, total aboveground production increased with fertilization. In Alaskan wet sedge tundra, addition of P strongly increased total plant biomass (Shaver et al. 1998). Nutrient addition also increased total biomass in various Swedish graminoid and shrub tundra sites (Jonasson 1992). In the High Arctic, nutrient addition increased total ANPP at two of three sites on Ellesmere Island (Henry et al. 1986) and increased total plant cover on Svalbard (Robinson et al. 1998).

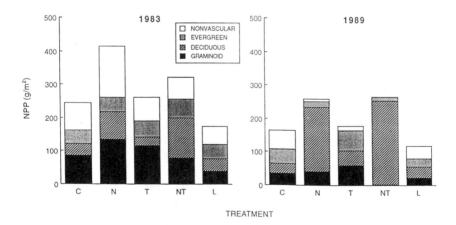

Figure 5 Total aboveground net primary production (NPP) by growth form of tussock tundra in response to environmental manipulations near Toolik Lake, AK, measured 3 (1983) and 9 (1989) years after initiation of treatments. Treatments are control (C), nutrient (N), temperature (T), nutrient–temperature (NT), and light attenuation (L). (*Source*: Chapin et al. 1995; used with permission of the Ecological Society of America.)

Warmer temperatures led to a significant but much smaller increase in ANPP than did nutrients in Alaskan tussock tundra, although this increase was attributed to the indirect effects of temperature on ANPP acting through increased soil nutrient availability (Chapin et al. 1995; Figure 5). By contrast, increased temperature had little effect on total plant biomass and production in Alaskan wet sedge tundra (Shaver et al. 1998). Most studies that increase air temperature also increase soil temperature as well, making it difficult to attribute changes in ANPP to the direct or indirect effects of warmer temperatures. However, one study in Alaskan tussock tundra that manipulated air temperature alone found no response of ANPP (Hobbie and Chapin 1998). In contrast, shading significantly decreased ANPP (Chapin et al. 1995; Figure 5). Elevated CO_2 in Alaskan tussock tundra had little effect on net ecosystem production (Grulke et al. 1990).

In summary, experimental evidence supports the contention that net primary production in the Arctic is limited primarily by the indirect effects of low temperature, namely low soil nutrient availability resulting from slow decomposition in cold soils. Other indirect effects of low temperature, such as short growing season, have not been examined experimentally, but may also limit net primary production. In contrast, tundra shows little capacity to respond to changes in aboveground conditions or resources (i.e., warmer air temperature or elevated CO_2) without an accompanying increase in nutrient availability. Although net primary production is decreased by light attenuation, it is unknown, but unlikely, that increased light availability would stimulate net primary production.

These conclusions based on experimental evidence are generally consistent with patterns of net primary production across Arctic landscapes. Primary production can vary up to 10-fold among different vegetation types within relatively small distances in the Arctic (Shaver and Chapin 1991; Figure 6). Gently sloping mesic areas of tussock tundra have intermediate levels of production (Shaver and Chapin 1991, Shaver et al. 1996). In contrast, the lowest productivity is found on well-drained ridge tops supporting heath tundra and poorly drained low areas supporting wet meadow tundra (Shaver and Chapin 1991). The highest productivity is found in areas of flowing water (riparian areas and water tracks) that support more productive graminoid or shrub tundra (Chapin et al. 1988, Hastings et al. 1989, Shaver and Chapin 1991).

The variation in productivity across arctic landscapes is proximately related to variation in nutrient availability. The highest production is found on deeply thawed soils that offer protection from wind, in sites dominated by N-fixers (Shaver et al. 1996), in sites influenced by animal activity (McKendrick et al. 1980), and in sites influenced by flowing water, which increases bulk flow of nutrients and stimulates net N mineralization (Chapin et al. 1988). In general, vegetation types with the greatest productivity are associated with soils that have

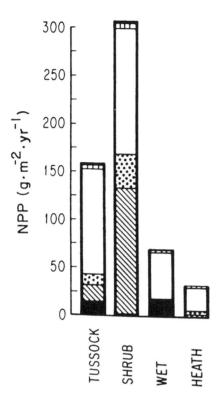

Figure 6 Total net primary production (NPP; excluding roots) of the vascular plants in each of four tundra vegetation types near Toolik Lake, AK. Total NPP is indicated by the height of the bar. Within each bar, inflorescence production is indicated by vertical stripes, leaf production by open space, apical stem growth (current year's twigs) by dots, secondary stem growth by diagonal stripes, and belowground rhizome growth by solid space. (*Source*: Shaver and Chapin 1991; used with permission of the Ecological Society of America.)

relatively high rates of net N mineralization (Chapin et al. 1988, Kielland 1990, Giblin et al. 1991).

ACKNOWLEDGMENTS

Syndonia Bret-Harte, Laura Gough, and Dave Hooper provided helpful comments on the manuscript. Preparation of the manuscript was funded by a National Science Foundation postdoctoral fellowship.

REFERENCES

Berendse, F., and S. Jonasson. 1992. Nutrient use and nutrient cycling in northern ecosystems. In: Chapin, F.S. III, R.L. Jefferies, J.F. Reynolds, G.R. Shaver, and J. Svoboda, eds. Arctic Ecosystems in a Changing Climate. San Diego: Academic Press, pp 337–356.

Bigger, C.M., and W.C. Oechel. 1982. Nutrient effect on maximum photosynthesis in arctic plants. Holarctic Ecology 5:158–163.

Billings, W.D., and H.A. Mooney. 1968. The ecology of arctic and alpine plants. Biological Review 43: 481–529.

Bliss, L.C., O.W. Heal, and J.J. Moore, eds. 1981. Tundra Ecosystems. A Comparative Analysis. Cambridge: Cambridge University Press.

Bliss, L.C., and N.V. Matveyeva. 1992. Circumpolar arctic vegetation. In: Chapin, F.S. III, R.L. Jefferies, J.F. Reynolds, G.R. Shaver, and J. Svoboda, eds. Arctic Ecosystems in a Changing Climate: An Ecophysiological Perspective. San Diego: Academic Press, pp 59–89.

Busby, J.R., L.C. Bliss, and C.D. Hamilton. 1978. Microclimate control of growth rates and habitats of the boreal forest mosses, *Tomenthypnum nitens* and *Hylocomium splendens*. Ecological Monographs 48: 95–110.

Callaghan, T.V., and U. Emanuelsson. 1985. Population structure and processes of tundra plants and vegetation. In: White J., ed. The Population Structure of Vegetation. Dordrecht, The Netherlands: Junk, pp 339–439.

Carlsson, B.A., and T.V. Callaghan. 1991. Positive plant interactions in tundra vegetation and the importance of shelter. Journal of Ecology 79: 973–983.

Chapin, D.M., and C.S. Bledsoe. 1992. Nitrogen fixation in arctic plant communities. In: Chapin, F.S. III, R.L. Jefferies, J.F. Reynolds, G.R. Shaver, and J. Svoboda, eds. Arctic Ecosystems in a Changing Climate: An Ecophysiological Perspective. San Diego: Academic Press, pp 301–320.

Chapin, F.S. III. 1974. Morphological and physiological mechanisms of temperature compensation in phosphate absorption along a latitudinal gradient. Ecology 55: 1180–1198.

Chapin, F.S. III. 1983a. Direct and indirect effects of temperature on arctic plants. Polar Biology 2: 47–52.

Chapin, F.S. III, and W.C. Oechel. 1983b. Photosynthesis, respiration, and phosphate absorption by *Cavex aquatilis* ecotypes along latitudinal and local environmental gradients. Ecology 64: 743–751.

Chapin, F.S. III, and A. Bloom. 1976. Phosphate absorption: adaptation of tundra graminoids to low temperature, low-phosphorus environment. Oikos 26: 111–121.

Chapin, F.S. III, and M.C. Chapin. 1980. Revegetation of an arctic disturbed site by native tundra species. Journal of Applied Ecology 17: 449–456.

Chapin, F.S. III, N. Fetcher, K. Kielland, K.R. Everett, and A.E. Linkins. 1988. Productivity and nutrient cycling of Alaskan tundra: enhancement by flowing soil water. Ecology 69: 693–702.

Chapin, F.S. III, R.L. Jefferies, J.F. Reynolds, G.R. Shaver, and J. Svoboda, eds. 1992. Arctic Ecosystems in a Changing Climate: An Ecophysiological Perspective. San Diego: Academic Press.

Chapin, F.S. III, L. Moilanen, and K. Kielland. 1993. Preferential use of organic nitrogen for growth by a non-mycorrhizal arctic sedge. Nature 361: 150–153.

Chapin, F.S. III, and G.R. Shaver. 1985a. Arctic. In: Chabot, B.F., and H.A. Mooney, eds. Physiological Ecology of North America. New York: Chapman and Hall, pp 16–40.

Chapin, F.S. III, and G.R. Shaver. 1985b. Individualistic growth response of tundra plant species to environmental manipulations in the field. Ecology 66: 564–576.

Chapin, F.S. III, and G.R. Shaver. 1989. Differences in growth and nutrient use among arctic plant growth forms. Functional Ecology 3: 73–80.

Chapin, F.S. III, and G.R. Shaver. 1996. Physiological and growth responses of arctic plants to a field experiment simulating climatic change. Ecology 77: 822–840.

Chapin, F.S. III, G.R. Shaver, A.E. Giblin, K.G. Nadelhoffer, and J.A. Laundre. 1995. Response of arctic tundra to experimental and observed changes in climate. Ecology 76: 694–711.

Chester, A.L., and G.R. Shaver. 1982. Reproductive effort in cottongrass tussock tundra. Holarctic Ecology 5: 200–206.

Ebersole, J.J. 1989. Role of seed bank in providing colonizers on a tundra disturbance in Alaska. Canadian Journal of Botany 67: 466–471.

Fetcher, N. 1985. Effects of removal of neighboring species on growth, nutrients, and microclimate of *Eriophorum vaginatum*. Arctic and Alpine Research 17: 7–17.

Fetcher, N., and G.R. Shaver. 1983. Life histories of tillers of *Eriophorum vaginatum* in relation to tundra disturbance. Journal of Ecology 71: 131–147.

Freedman, B., N. Hill, J. Svoboda, and G. Henry. 1982. Seed banks and seedling occurrence in a high arctic oasis at Alexandra Fjord, Ellesmere Island, Canada. Canadian Journal of Botany 60: 2112–2118.

Gartner, B.L., F.S. Chapin III, and G.R. Shaver. 1983. Demographic patterns of seedling establishment and growth of native graminoids in an Alaskan tundra disturbance. Journal of Applied Ecology 20: 965–980.

Gartner, B.L., F.S. Chapin III, and G.R. Shaver. 1986. Reproduction of *Eriophorum vaginatum* by seed in Alaskan tussock tundra. Journal of Ecology 74: 1–18.

Gebauer, R.L.E., J.F. Reynolds, and J.D. Tenhunen. 1995. Growth and allocation of the arctic sedges *Eriophorum angustifolium* and *E. vaginatum*: effects of variable soil oxygen and nutrient availability. Oecologia 104: 330–339.

Giblin, A.E., K.J. Nadelhoffer, G.R. Shaver, J.A. Laundre, and A.J. McKerrow. 1991. Biogeochemical diversity along a riverside toposequence in arctic Alaska. Ecological Monographs 61: 415–435.

Gold, W.G., and L.C. Bliss. 1995. Water limitations and plant community development in a polar desert. Ecology 76: 1558–1568.

Grime, J.P. 1977. Evidence for the existence of three primary strategies in plants and its relevance to ecological and evolutionary theory. American Naturalist 111: 1169–1194.

Grulke, N.E., and L.C. Bliss. 1988. Comparative life-history characteristics of two high arctic grasses, Northwest Territories. Ecology 69: 484–496.

Grulke, N.E., G.H. Reichers, W.C. Oechel, U. Hjelm, and C. Jaeger. 1990. Carbon balance in tussock tundra under ambient and elevated CO_2. Oecologia 83: 485–494.

Hastings, S.J., S.A. Luchessa, W.C. Oechel, and J.D. Tenhunen. 1989. Standing biomass and production in water drainages of the foothills of the Philip Smith Mountains, Alaska, USA. Holarctic Ecology 12: 304–311.

Havstrom, M., T.V. Callaghan, and S. Jonasson. 1993. Differential growth responses of *Cassiope tetragona*, an arctic dwarf-shrub, to environmental perturbations among three contrasting high- and sub-arctic sites. Oikos 66: 389–402.

Henry, G.H.R., R. Freedman, and J. Svoboda. 1986. Effects of fertilization on three tundra plant communities of a polar desert oasis. Canadian Journal of Botany 64: 2502–2507.

Hobbie, S.E. 1995. The effects of increased temperature on tundra plant community composition and the consequences for ecosystem processes. Ph.D. Thesis, University of California, Berkeley, CA.

Hobbie, S.E., and F.S. Chapin III. 1998. The response of tundra plant biomass, aboveground production, and nitrogen to temperature manipulation. Ecology 79: 1526–1544.

Houghton, R.A., and D.L. Skole. 1990. Carbon. In: Turner, B.L., W.C. Clark, R.W. Kates, J.F. Richards, J.T. Matthews, and W.B. Meyer, eds. The Earth as Transformed by Human Action. Cambridge: Cambridge University Press, pp 393–408.

Hulten, E. 1968. Flora of Alaska and neighboring territories. Stanford, CA: Stanford University Press.

Johnson, D.A., and L.L. Tieszen. 1976. Aboveground biomass allocation, leaf growth, and photosynthesis patterns in tundra plant forms in arctic Alaska. Oecologia 24: 159–173.

Jonasson, S. 1989. Implications of leaf longevity, leaf nutrient reabsorption, and translocation for the resource economy of five evergreen species. Oikos 56: 121–131.

Jonasson, S. 1992. Plant responses to fertilization and species removal in tundra related to community structure and clonality. Oikos 63: 420–429.

Kielland, K. 1990. Processes controlling nitrogen release and turnover in arctic tundra. Ph.D. Thesis, University of Alaska, Fairbanks, AK.

Kielland, K. 1994. Amino acid absorption by arctic plants: implications for plant nutrition and nitrogen cycling. Ecology 75: 2373–2383.

Kielland, K. 1995. Landscape patterns of free amino acids in arctic tundra soils. Biogeochemistry 31: 85–98.

Korner, C., and W. Larcher. 1988. Plant life in cold climates. Symposium of the Society of Experimental Biology 42: 25–57.

Kummerow, J., and B. Ellis. 1984. Temperature effect on biomass production and root/shoot biomass ratios in two arctic sedges under controlled environmental conditions. Canadian Journal of Botany 62: 2150–2153.

Kummerow, J., B.A. Ellis, S. Kummerow, and F.S. Chapin III. 1983. Spring growth of shoots and roots in shrubs of an Alaskan muskeg. American Journal of Botany 70: 1509–1515.

Limbach, W.E., W.C. Oechel, and W. Lowell. 1982. Photosynthetic and respiratory responses to temperature and light of three Alaskan tundra growth forms. Holarctic Ecology 5: 150–157.

McGraw, J.B. 1980. Seed bank size and distribution of seeds in cottongrass tussock tundra. Canadian Journal of Botany 58: 1607–1611.

McGraw, J.B. 1985a. Experimental ecology of *Dryas octopetala* ecotypes. III. Environmental factors and plant growth. Arctic and Alpine Research 17: 229–239.

McGraw, J.B. 1985b. Experimental ecology of *Dryas octopetala* ecotypes: relative response to competitors. New Phytologist 100: 233–241.

McGraw, J.B., and F.S. Chapin III. 1989. Competitive ability and adaptation to fertile and infertile soils in two *Eriophorum* species. Ecology 70: 736–749.

McGraw, J.B., and N. Fetcher. 1992. Response of tundra plant populations to climatic change. In: Chapin, F.S. III, R.L. Jefferies, J.F. Reynolds, G.R. Shaver, and J. Svoboda, eds. Arctic Ecosystems in a Changing Climate: An Ecophysiological Perspective. San Diego: Academic Press, pp 359–376.

McGraw, J.B., and G.R. Shaver. 1982. Seedling density and seedling survival in Alaskan cotton grass tussock tundra. Holarctic Ecology 5: 212–217.

McGraw, J.B., and M.C. Vavrek. 1989. The role of buried viable seeds in arctic and alpine plant communities. In: Leck, M.A., V.T. Parker, and R.L. Simpson, eds. Ecology of Soil Seed Banks. San Diego: Academic Press, pp 91–106.

McKendrick, J.D., G.O. Batzli, K.R. Everett, and J.C. Swanson. 1980. Some effects of mammalian herbivores and fertilization on tundra soils and vegetation. Arctic and Alpine Research 12: 565–578.

McKendrick, J.D., V.J. Ott, and G.A. Mitchell. 1978. Effects of nitrogen and phosphorus fertilization on carbohydrate and nutrient levels in *Dupontia fisheri* and *Arctagrostis latifolia*. In: Tieszen, L.L., ed. Vegetation and Production Ecology of an Alaskan Arctic Tundra. New York: Springer-Verlag, pp 565–578.

Michelson, A., S. Jonasson, D. Sleep, M. Havstrom, and T.V. Callaghan. 1996a. Shoot biomass, $\partial^{13}C$, nitrogen, and chlorophyll responses of two arctic dwarf shrubs to *in situ* shading, nutrient application and warming simulating climatic change. Oecologia 105: 1–12.

Michelson, A., I.K. Schmidt, S. Jonasson, C. Quarmby, and D. Sleep. 1996b. Leaf ^{15}N abundance of subarctic plants provides field evidence that ericoid, ectomycorrhizal and non-and arbuscular mycorrhizal species access different sources of soil nitrogen. Oecologia 105: 53–63.

Murray, K.J., P.C. Harley, J. Beyers, H. Walz, and J.D. Tenhunen. 1989a. Water content effects on photosynthetic response of *Sphagnum* mosses from the foothills of the Philip Smith Mountains, Alaska. Oecologia 79: 244–250.

Murray, K.J., J.D. Tenhunen, and J. Kummerow. 1989b. Limitations on moss growth and net primary production in tussock tundra areas of the foothills of the Philip Smith Mountains, Alaska. Oecologia 80: 256–262.

Murray, K.J., J.D. Tenhunen, and R.S. Nowak. 1993. Photoinhibition as a control on photosynthesis and production of *Sphagnum* mosses. Oecologia 96: 200–207.

Nadelhoffer, K.J., and B. Fry. 1994. Nitrogen isotope studies in forest ecosystems. In: Lajtha, K., and R.H. Michener, eds. Stable Isotopes in Ecology and Environmental Science. Oxford: Blackwell Scientific, pp 22–44.

Nadelhoffer, K.J., A.E. Giblin, G.R. Shaver, and A.E. Linkins. 1992. Microbial processes and plant nutrient availability in arctic soils. In: Chapin, F.S. III, R.L. Jefferies, J.F. Reynolds, G.R. Shaver, and J. Svoboda, eds. Arctic Ecosystems in a Changing Climate: An Ecophysiological Perspective. San Diego: Academic Press, pp 281–300.

Nadelhoffer, K.J., G.R. Shaver, B. Fry, A.E. Giblin, L.C. Johnson, and R.B. McKane. 1996. [15]N natural abundance and N use by plants. Oecologia 107: 386–394.

Oberbauer, S.F., and T.E. Dawson. 1992. Water relations of arctic vascular plants. In: Chapin, F.S. III, R.L. Jefferies, J.F. Reynolds, G.R. Shaver, and J. Svoboda, eds.

Arctic Ecosystems in a Changing Climate: An Ecophysiological Perspective. San Diego: Academic Press, pp 259–279.

Oberbauer, S.F., S.J. Hastings, J.L. Beyers, and W.C. Oechel. 1989. Comparative effects of downslope water and nutrient movement on plant nutrition, photosynthesis, and growth in Alaskan tundra. Holarctic Ecology 12: 324–334.

Oechel, W.C. 1976. Seasonal patterns of temperature response of CO_2 flux and acclimation in arctic mosses growing *in situ*. Photosynthetica 10: 447–456.

Parsons, A.N., J.M. Welker, P.A. Wookey, M.C. Press, T.V. Callaghan, and J.A. Lee. 1994. Growth responses of four sub-arctic dwarf shrubs to simulated environmental change. Journal of Ecology 82: 307–318.

Parsons, A.N., P.A. Wookey, J.M. Welker, M.C. Press, T.V. Callaghan, and J.A. Lee. 1995. Growth and reproductive output of *Calamagrostis lapponica* in response to simulated environmental change in the subarctic. Oikos 72: 61–66.

Read, D.J. 1991. Mycorrhizas in ecosystems. Experientia 47: 376–391.

Robinson, C.H., P.A. Wookey, J.A. Lee, T.V. Callaghan, and M.C. Press. 1998. Plant community responses to simulated environmental change at a high arctic polar semi-desert, Svalbard (79°N). Ecology 79: 856–866.

Savile, D.B.O. 1960. Limitations of the competitive exclusion principle. Science 132: 1761.

Schimel, J.P., and F.S. Chapin III. 1996. Tundra plant uptake of amino acid and NH_4^+ nitrogen in situ: plants compete well for amino acid N. Ecology 77: 2142–2147.

Semikhatova, O.A., T.V. Gerasimenko, and T.I. Ivanova. 1992. Photosynthesis, respiration, and growth of plants in the Soviet Arctic. In: Chapin, F.S. III, R.L. Jefferies, J.F. Reynolds, G.R. Shaver, and J. Svoboda, eds. Arctic Ecosystems in a Changing Climate: An Ecophysiological Perspective. San Diego: Academic Press, pp 169–190.

Shaver, G.R., and D.W. Billings. 1977. Effects of daylength and temperature on root elongation in tundra graminoids. Oecologia 28: 57–65.

Shaver, G.R., and W.D. Billings. 1975. Root production and root turnover in a wet tundra ecosystem, Barrow, Alaska. Ecology 56: 401–410.

Shaver, G.R., and F.S. Chapin III. 1980. Response to fertilization by various plant growth forms in an Alaskan tundra: nutrient accumulation and growth. Ecology 61: 662–675.

Shaver, G.R., and F.S. Chapin III. 1986. Effect of fertilizer on production and biomass of tussock tundra, Alaska, U.S.A. Arctic and Alpine Research 18: 261–268.

Shaver, G.R., and F.S. Chapin III. 1991. Production:biomass relationships and element cycling in contrasting arctic vegetation types. Ecological Monographs 61: 1–31.

Shaver, G.R., and F.S. Chapin III. 1995. Long-term responses to factorial NPK fertilizer treatment by Alaskan wet and moist tundra sedge species. Ecography 18: 259–275.

Shaver, G.R., F.S. Chapin III, and B.L. Gartner. 1986. Factors limiting seasonal growth and peak biomass accumulation in *Eriophorum vaginatum* in Alaskan tussock tundra. Journal of Ecology 74: 257–278.

Shaver, G.R., L.C. Johnson, D.H. Cades, G. Murray, J.A. Laundre, E.B. Rastetter, K.J. Nadelhoffer, and A.E. Giblin. 1998. Biomass and CO_2 flux in wet sedge tundras: responses to nutrients, temperature, and light. Ecological Monographs 68: 75–97.

Shaver, G.R., and J. Kummerow. 1992. Phenology, resource allocation, and growth of arctic vascular plants. In: Chapin, F.S. III, R.L. Jefferies, J.F. Reynolds, G.R. Shaver, and J. Svoboda, eds. Arctic Ecosystems in a Changing Climate: An Ecophysiological Perspective. San Diego: Academic Press, pp 193–211.

Shaver, G.R., J.A. Laundre, A.E. Giblin, and K.J. Nadelhoffer. 1996. Changes in live plant biomass, primary production, and species composition along a riverside toposequence in Arctic Alaska, U.S.A. Arctic and Alpine Research 28: 363–379.

Shevtsova, A., A. Ojala, S. Neuvonen, M. Vieno, and E. Haukioja. 1995. Growth and reproduction of dwarf shrubs in subarctic plant community: annual variation and above-ground interactions with neighbors. Journal of Ecology 83: 263–275.

Sohlberg, E.H., and L.C. Bliss. 1986. Responses of *Ranunculus sabinei* and *Papaver radicatum* to removal of the moss layer in a high-arctic meadow. Canadian Journal of Botany 65: 1224–1228.

Tenhunen, J.D., O.L. Lange, S. Hahn, R. Siegwolf, and S.F. Oberbauer. 1992. The ecosystem role of poikilohydric tundra plants. In: Chapin, F.S. III, R.L. Jefferies, J.F. Reynolds, G.R. Shaver, and J. Svoboda, eds. Arctic Ecosystems in a Changing Climate: An Ecophysiological Perspective. San Diego: Academic Press, pp 213–237.

Tieszen, L.L. 1973. Photosynthesis and respiration in arctic tundra grasses: field light intensity and temperature responses. Arctic and Alpine Research 5: 239–251.

Tissue, D.T., and W.C. Oechel. 1987. Response of *Eriophorum vaginatum* to elevated CO_2 and temperature in the Alaskan tussock tundra. Ecology 68: 401–410.

Warren Wilson, J. 1966. An analysis of plant growth and its control in arctic environments. Annals of Botany 30: 383–482.

Welker, J.M., P.A. Wookey, A.P. Parsons, T.V. Callaghan, M.C. Press, and J.A. Lee. 1993. Comparative responses of subarctic and high arctic ecosystems to simulated climate change. Oikos 67: 490–502.

Wookey, P.A., C.H. Robinson, A.N. Parsons, J.M. Welker, T.V. Callaghan, and J.A. Lee. 1995. Environmental constraints on the growth, photosynthesis and reproductive development of *Dryas octopetala* at a high Arctic polar semi-desert, Svalbard. Oecologia 102: 478–489.

Wookey, P.A., J.M. Welker, A.N. Parsons, M.C. Press, T.V. Callaghan, and J.A. Lee. 1994. Differential growth, allocation and photosynthetic response of *Polygonum viviparum* to simulated environmental change at a high arctic polar semi-desert. Oikos 70: 131–139.

14
Plant Life in Antarctica

T. G. Allan Green
Waikato University, Hamilton, New Zealand

Burkhard Schroeter
University of Kiel, Kiel, Germany

Leopoldo G. Sancho
Complutense University, Madrid, Spain

Antarctica is the coldest, driest, highest and windiest continent, its plants
grow where it is warm, wet, low and calm.
Ecophysiologists should be grateful.

I. INTRODUCTION

Research on terrestrial plants in Antarctica has been most intense over the past
four decades. The International Geophysical Year (IGY, 1958/59) led to the establishment of many national stations in Antarctica and, although research concentrated on the physical sciences, there was always a proportion of natural science. However, progress has certainly been spasmodic, as can be seen in the
summary of the history of terrestrial biota research in the western part of the
Antarctic Peninsula (Smith 1996). Initial studies coincided with a growth in interest in stress survival mechanisms in plants and, as a result, there has been a
considerably higher proportion of plant ecophysiological research in Antarctica
in comparison to work on the same groups elsewhere. To some extent, taxonomy
languished, and only recently have substantial good reviews on some major
groups started to appear, e.g., *Usnea*, (Walker 1985), *Bryum* (Seppelt and Kanda
1986), *Umbilicaria* (Filson 1987), *Stereocaulon* (Smith and Övstedal 1991), Cladoniaceae (Stenroos 1993), and *Caloplaca* (Söchting and Olech 1995). This improvement of knowledge has resulted in a better understanding of the geographical relationships of the flora (Seppelt 1995, Castello and Nimis 1995, 1997).

Despite the mixture of research approaches, we are still far from having a
good understanding of both the distribution and functioning of the terrestrial
plants and animals. This is certainly a reflection of the difficulties of working in

the region, and also of the patchy nature of research in some national programs; this is now being rectified by the development of multidisciplinary, long-term research using standardized methodologies, e.g., long-term ecological research sites in the Taylor Valley, National Science Foundation in the United States, and BIOTAS (Biological Investigations Of Terrestrial Antarctic Systems), initiated by the British Antarctic Survey. This chapter attempts to bring out the major features of the ecophysiology of terrestrial plants in Antarctica with some emphasis on what appears to be controlling the distribution and performance. Here we link information about distribution and abundance with knowledge of the ecophysiological performance, including growth rates, and consider whether the plants show special adaptations for the antarctic environment. This knowledge is of growing importance because of the needs to both conserve and manage the communities, as well as the potential to use the vegetation to detect global change processes such as climate warming (Kennedy 1995). Several excellent review articles exist that can provide more detail about specific plant groups or locations (Ahmadjian 1970, Holdgate 1964, 1977, Longton 1988a, 1988b, Vincent 1988, Kappen 1988, 1993a, Smith 1984, 1996).

A. Definitions

In this chapter, plants are defined as including higher plants (only two angiosperm species occur), all bryophytes (predominantly mosses because liverworts are rare south of approximately 70° latitude), and lichens. The latter, associations between fungi as host and algae or cyanobacteria as symbionts, are today correctly referred to the fungi; however, their prominent role in Antarctica and their phototrophic lifestyle support the tradition of including them as members of the terrestrial vegetation. The algae and cyanobacteria, which are also very common in wet, terrestrial sites (Vincent 1988), are not covered except when in symbiotic association in a lichen.

Antarctica includes the main antarctic continent (continental Antarctica) and the Antarctic Peninsula and its closely associated islands to the west and north (South Shetland Islands, South Orkney Islands); subantarctic islands are not included.

B. Climate Zones

Antarctica spans a large latitudinal range of approximately 27°latitude from the north of the peninsula (63°S) to the pole and an associated large climate range. It is certainly true to say that Antarctica is the coldest, windiest, driest, and highest continent, the latter feature contributing to the very low inland temperatures (Ta-

Table 1   Climatic Information for Various Locations in the Maritime Zone,
Continental Zone, and Polar Plateau in Antarctica

	Mean temperature		
Location	Warmest month	Coldest month	Precipitation (mm rain equivalent)
Maritime			
Signy Island (60°43′ S)	+1.3	−9.0	400
Northwest peninsula (Faraday) (63–69° S)	+0.6	−9.0	380–500
Continent			
Casey (66°17′ S)	+0.2	−15.1	330
Syowa (69°00′ S)	−1.0	−19.4	250
Mawson (67°26′ S)	−0.2	−18.5	100
Davis (68°35′ S)	+1.0	−18.0	350
Cape Hallett (72°19′ S)	−1.4	−26.4	120
Taylor Valley (77°35′ S)	+1.0	−39.3	10 (floor) 50 (mountains)
McMurdo (77°51′ S)	−3.1	−26.1	130
Polar plateau			
Dome Fuji (77°31′ S) (altitude 3810 m)	−26.1	−67.8	100

ble 1). Several investigators have produced subdivisions based on the climate
(summarized in Smith 1996). There is general agreement that two zones can be
separated, reflecting both climate and vegetation.

The first is the maritime Antarctic (Holdgate 1964) or cold-polar zone of
Longton (1988a, 1988b). This comprises the western side of the Antarctic Peninsula north of Alexander Island (about 69° S) and the closely adjacent islands to
the west (Figure 1). The zone is characterized by warmest months with mean
temperatures above freezing point (0–2 °C) and mean winter temperatures rarely
less than −10 °C (see data for Faraday Station in Table 1). Precipitation, which
can come as rain in the summer, is between 350 and 500 mm (rain equivalent),
and the area is essentially a cold semidesert. The climate is controlled by the
strong westerly, maritime influences and is markedly different from the eastern
side of the peninsula. The vegetation includes the only two antarctic angiosperms
(*Colobanthus quitensis* [Kunth.] Bartl. and *Deschampsia antarctica* Desv.) and
substantial numbers of lichens and bryophytes (approximately 200 and 90 species, respectively), including about 10 species of liverworts.

The second zone is the continental Antarctic (Holdgate 1964) or frigid Antarctic of Longton (1988a, 1988b). Geographically, this includes the whole of the

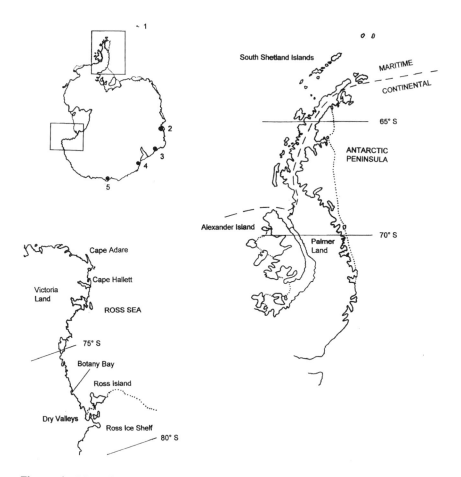

Figure 1 Map of Antarctica with insets of the Antarctic Peninsula and the Ross Sea areas showing the division into maritime and continental zones. The boundary is most clearly seen in the peninsula insert map. The named locations are referred to in the text or in the tables.

main antarctic continent, the east of the Antarctic Peninsula, and the west of the peninsula as far north as, and including, Alexander Island; it is approximately delineated as being south of the antarctic circle (Figure 1). The zone is characterized by the mean temperature of the warmest month being below freezing point, often substantially below (Table 1). The mean winter temperatures are much lower than in the maritime Antarctic, typically around −20°C or below (Table 1). The temperatures, wind, and precipitation are governed by the overall form of the continent and its effect on the atmospheric circulation. The circular shape encourages cyclones to circle the continent rather than moving onto it. Entry of maritime weather systems is also discouraged by the height of the polar plateau, which is predominantly above 2000 m. Precipitation, 300 mm (rain equivalent) as snow, occurs mainly in the coastal belt, and the net result is growing decertification with increase in latitude so that annual precipitation can be very low, around 120 mm (rain equivalent) at McMurdo (77°51′ S) and 50 mm near the pole (Table 1). Precipitation shadows can lead to even lower local values, such as the 120 mm at Cape Hallett (72°19′ S) and the extreme 10–50 mm in the Dry Valley region (around 77°00′–77°50′ S). Cold air descending from the polar plateau causes strong katabatic winds at the continent margins, making some areas exceptionally windy (e.g., Cape Denison, mean wind speed of 24.9 m s^{-1} for July 1913, in Mawson, The Home of the Blizzard). The vegetation in this zone is composed entirely of lichens and mosses, with the rare occurrence of one species of liverwort [*Cephaloziella exiliflora* (Tayl.) Steph.]. The vegetation is confined to ice-free areas, about twenty so-called oases (Pickard 1986), and even there it is very scattered and only locally abundant, being dependent on a moisture supply, warmth, and protection from the wind. Lichens and mosses have been reported from as far south as 86° S and 84° S, respectively (Wise and Gressit 1965, Claridge et al. 1971), but these represent isolated occurrences. Desiccation over winter becomes very important in this zone, and plants are not only frozen but can become completely freeze-dried as temperatures reach −50°C at very low ambient humidities.

II. VEGETATION TRENDS

A. Biodiversity

There is a marked decline in biodiversity with increase in latitude, with a particularly sharp change between the maritime and continental Antarctic zones (Table 2). The two higher plants are confined to the maritime zone and there is, overall, about a 80% loss of species, both bryophytes and lichens, between the peninsula and the main continent. From around 90 species of moss and 200 species of lichen at the northwest of the peninsula, possibly the richest botanical zone on the continent (Lindsey 1940), there is a decrease to only around 7 and 33, respectively, on the continent (e.g., at Birthday Ridge, 70°48′ S [Kappen 1985a, 1985b], and Botany Bay, Granite Harbour, 77°00′ S [unpublished data]). Locally, the decline can be even greater: two mosses and nine lichens are reported for the windy Mawson Rock (67°26′ S; Seppelt and Ashton 1978), and only three mosses and around five lichens survive the extreme climate of the Dry Valleys (77°35′ S; Schwarz et al. 1992, Green et al. 1992, unpublished data). The liverworts, in particular, are only of significance in the maritime zone, and their presence else-

Table 2 Numbers of Species of Flowering Plants, Lichens, Mosses, and Liverworts, Community types, Maximal Biomass, and Presence of Peat at Various Locations in the Maritime and Continental Antarctica[a]

Location	Flowering plants	Lichens	Mosses	Liverworts	Community types[b]	Maximal biomass (g m^{-2})	Peat
Maritime							
Northwest peninsula, 63–69° S	2	> 200	> 90	10	1.1; 1.2; 1.3; 1.4; 1.5; 1.6; 2.1	1000–46,000	1.5 m
Continental							
Southern peninsula, 69–74° S	0	> 100	26	1	1.1; 1.2; 1.3		0
Mawson, 67°26′ S	0	9	2	0	1.1; 1.2; 1.3	< 1100	0
Davis, 68°35′ S	0	34	6	1	1.1; 1.2; 1.3		0
Casey, 66°17′ S	0	35	5	1	1.1; 1.2; 1.3	900	0
Birthday Ridge, 70°48′ S	0	33	5	0	1.1; 1.2; 1.3	50–950	0
Botany Bay, 77°00′ S	0	> 30	6	1	1.1; 1.2; 1.3	> 1000	0
McMurdo, 77°51′ S	0	20 (?)	?	0	1.1; 1.2; 1.3	800	0
Dry Valleys, 77°45′ S	0	6	3	0	1.1; 1.3	800	0

[a] Data from a variety of sources referred to in text.

[b] Communities are according to Smith (1996): 1.1, crustaceous and foliose lichen subformation; 1.2, fruticose and foliose lichen subformation; 1.3, short moss cushion and turf subformation; 1.4, tall moss turf subformation; 1.5, tall moss cushion subformation; 1.6, bryophyte carpet and mat subformation; 2.1, grass and cushion chemophyte subformation.

where must indicate exceptional local conditions such as the occurrence on heated soil on Mt. Melbourne (Broady et al. 1987). Other, less obvious trends exist, such as lichens with cyanobacterial photobionts being confined to the maritime Antarctic (Kappen 1993a, Schroeter et al. 1994) and a general increase in the proportion of crustose lichen species as conditions become more severe (Kappen 1988). Kappen (1988) refers to a group of 13 circumpolar lichen species that tend to occur around the continent, with other species being only of local occurrence.

B. Biomass

Biomass data from a range of antarctic locations are given in Table 2. Within continental Antarctica, there is a remarkable agreement between the maximal biomasses from most sites when expressed as 100% cover. A typical maximal value of around 1000–1500 g m^{-2} is found both for lichens and mosses. Kappen (1993a) gives a data summary showing that such maximal biomasses occur for lichens from Victoria Land (70° S) to King George Island (South Shetland Islands) and Signy Island in the northern maritime Antarctic (61–62° S). This seems to be the maximal biomass for this thin, almost two-dimensional vegetation. However, the situation is different for mosses. In continental Antarctica, similar maximal biomasses are found (Longton 1988a), but in the maritime there can be extensive peat build-up to depths of 50 cm or more, leading to greatly increased values, to 46,000 g m^{-2} in some cases under *Chorisodontium aciphyllum* (Hook f. et Wils.) Broth. and *Polytrichum alpestre* Hoppe. The age of such peat banks is still uncertain, but some have been carbon dated to 5350 years in the South Orkney Islands (Björk et al. 1991). Biomass is lower, about 10,000 g m^{-2}, in the 20-cm-deep layer above the permafrost. A value of 300–1000 g m^{-2} is given by Longton (1988a) for the green, photosynthetic shoots, a number similar to that for maximal biomass without peat production. It is difficult to explain the lack of peat production in the continental Antarctic, where, due to the colder temperatures, decomposition processes would be expected to be even slower (Collins 1976, Davis 1980, 1986).

The data must be treated with considerable caution since they are often adjusted to 100% cover equivalence and, even if this is not done, they only normally apply to where the plants are present and not to the overall ice-free area. Therefore, the data are unrepresentative since vegetation occurrence is very sporadic. Kennedy (1993b) has calculated that, in Ablation Valley on Alexander Island, terrestrial vegetation is confined almost entirely to seven discrete patches, totalling 2300 m^2 in a total ice-free area of 400,000 m^2 (Light and Heywood 1975). Similarly, the rich Canada Glacier flush, totalling some 10,000 m^2, is the only extensive patch of vegetation in an ice-free area of 25,000,000 m^2 (Schwarz et al. 1992). In many ways, the situation is similar to describing the vegetation of the Sahara Desert by extrapolation from surveys at oases. It is unfortunate that the focusing of scientific attention on the plants disguises their relative rarity; if a typical area had to be preserved in these ice-free areas, it would almost entirely be bare ground.

C. Community Structure

There have been several detailed accounts of vegetation and community structure in the Antarctic (Longton 1988a, 1988b, Smith 1996), and only the major trends are considered here. The plant association confined to the maritime is that dominated by higher plants, the grass and cushion chemophyte subformation. Although the two phanerogams can be found occasionally in other formations, their best development occurs on moist to dry soil, especially in sheltered, north-facing coastal habitats where closed swards can sometimes develop. Subdivisions of cryptogamic communities, here from Smith (1996), depend on whether lichens or bryophytes (mosses) are dominant. In the bryophyte carpet and mat subformation (pleurocarpous mosses and liverworts on wet ground), lichens are sparse or absent; in the tall moss cushion subformation (tall moss cushions and some deep carpet forms along melt-stream courses) and in the tall moss turf subformation (predominantly *Polytrichum–Chorisodontium* species), fruticose lichens of the genera *Stereocaulon, Cladonia*, and *Sphaerophorus* are also very common. These formations are only found in the maritime zone. They are much more luxuriant and rich than communities in continental Antarctica, where the predominant association is the short moss cushion and turf subformation (mainly a *Bryum–Ceratodon–Pottia* association occurs). Overall, the bryophytes become shorter, less able to develop continuous stands, and form less extensive patches (Table 2). Location becomes evermore important, and mosses are confined to highly protected sites where melt water can occur, such as cracks in rocks, sites adjacent to permanent snow banks, and areas with running water, usually close to glaciers or in depressions. In some areas, the shoots are restricted to just below or at the surface of the substrate.

Lichen communities show similar trends. In the maritime zone, there are extensive areas where fruticose lichens are dominant, the fruticose and foliose lichen formation (Figure 2), typically on the sides of rocks or boulders, forming marine benches or similar areas (Figure 3). This distribution seems to be determined by improved water relations, and it is easy to see how snow will gather in such areas, how wind will be reduced, and how light will be high but rarely with direct sunlight. Such sites are well illustrated in a report by Kappen (1993a). The genera *Usnea* (Figure 4) and *Umbilicaria* are typically dominant, and plants can reach large sizes, 20+ cm across for *Umbilicaria* specimens. These formations are rarer in the continental zone, but do occur, e.g., Birthday Ridge (Kappen 1985a) and Botany Bay, Granite Harbour (Seppelt, Green, and Schroeter, in preparation). The second lichen formation, crustaceous and foliose lichens, is exten-

Figure 2 Photographs of Botany Bay, Southern Victoria Land, taken looking from the sea and facing south (upper picture) and towards the sea, facing north (lower picture) at approximately the same location. Note how the black *Buellia frigida* forms extensive colonies on the surface of the rock facing away from the sea. The lichens are on the south faces of the rocks, which are in shade for most of the day. The pattern almost certainly reflects the improved moisture regime in this location, with water being predominantly supplied by snowfall. Evaporation would be substantially faster on the north faces.

Figure 3 Photograph of a rich fruticose and foliose lichen subformation (1.2 in Smith 1996) at Livingston Island in the maritime Antarctic. The most visible lichens are specimens of the fruticose genus *Usnea*.

sive in both the maritime and continental Antarctic and is the only association on bare rock faces. It is the predominant formation in the continental zone and can form anywhere where there is protection and a supply of water, either as melt or blown snow. However, there are also considerable limitations to its presence, and the lichens are often confined to a particular rock face or to crevices, depending on wind, light, and snow occurrence (Figure 3). It is a patchy formation, and rarely is 100% cover approached; more typically, the cover is very low. Crustaceous species become almost the only lichens under the more extreme conditions and, under dry conditions but with a water supply and light, the endolithic association appears (Friedmann 1982). In the mountains of the Dry Valley region, this is the most common community, but it is present over almost the entire continent where other growth forms cannot occur. In both the maritime and continental zones, the occurrence of some species is strongly dependent on a rich nutrient supply from perching birds (Olech 1990).

Overall, there is a clear trend for smaller plants, lower biodiversity, and greater confinement to protected sites as latitude increases (Table 2). Where there is a coincidence of shelter, warmth, light, and reliable water supply, rich communities can develop at high latitudes. A good example is the exceptional commu-

Figure 4 Photograph of the fruticose lichen *Usnea aurantiaco-atra* growing abundantly on rocks at Livingston Island, maritime Antarctica. The species forms extensive stands in some locations and also shows a color change from pale in shaded sites to dark in exposed sites.

nity at Botany Bay, Granite Harbour, which, although it is at 77° S, is richer than almost all other continental sites and even has the liverwort *Cephaloziella exiliflora* present (Seppelt and Green 1998). The occurrence of this one species would suggest that the site has summer months with mean temperatures close to or above freezing, although no data are available to test this hypothesis at present.

III. PLANT PERFORMANCE: CO_2 EXCHANGE

Higher plants are considered here only briefly because they are confined to the maritime Antarctic. The majority of the studies have been focused on lichens and bryophytes, which dominate most communities. In the case of the bryophytes, the mosses are clearly the major group; because liverworts are not significant in continental Antarctica and are only locally common in the maritime zone, they have, unfortunately, been little studied. Plant performance has mostly been studied as photosynthetic activity, measured mainly as CO_2 exchange in the field or on samples returned to laboratories. CO_2 exchange is reported as net photosynthesis (NP), dark respiration (DR), photorespiration (PR), and gross photosynthesis (GP, calculated as NP + DR).

A. CO_2 Exchange Response to Major Environmental and Plant Factors

1. Higher Plants

Colobanthus quitensis and *Deschampsia antarctica* studied in the laboratory were found to have optimal temperatures for NP of 13°C and 19°C, respectively. Both species retained NP of about 30% of maximal rate of 0°C, and it is suggested that this ability is the reason for their success in the maritime Antarctic (Edwards and Smith 1988). Light levels for saturation were low, 30 μmol m^{-2} s^{-1} photosynthetic photon flux density (PPFD) at 0°C and 150 μmol m^{-2} s^{-1} PPFD at 10°C, indicating shade-adapted plants. This could also be considered as adaptive for the generally cloudy conditions in this area. Overall, Edwards and Smith (1988) suggested that the plants showed no particular special adaptation for growth in Antarctica, and Convey (1996) suggested that their presence could be a matter of chance. Fowbert and Smith (1994) demonstrated that the populations are increasing and have been doing so since the 1940s, an increase that is suggested to reflect higher summer temperatures in the area (Smith 1994).

2. Bryophytes: Liverworts

The few studies on liverworts have provided some interesting insights into their poor performance in Antarctica. *Marchantia berteroana* Lehm. et Lindenb., studied in Signy Island, was extremely sensitive to both desiccation and low temperatures (Davey 1997a). Photosynthesis was only 16% of original rate at 10°C after 24 hours at −5°C, and there was no activity at all after 24 hours at −25°C. Five freeze/thaw cycles to −5°C caused NP to decrease to approximately 3% of the original value. The plants required a very high water content of 10 gdw^{-1} for maximal GP, and desiccation for 1–6 months led to an almost complete loss of photosynthetic ability. However, in the field the plants were always found to be fully hydrated and never suffered desiccation. Maximal GP and NP were slightly higher than published values for a related species, *Marchantia foliacea* Mitt., in New Zealand (Green and Snelgar 1982). The single study on the performance of *Cephaloziella exiliflora*, the only liverwort to grow in continental Antarctica, revealed that the typical purple color was lost in shaded areas and probably functioned as a form of light protection (Post and Vesk 1992). Green plants, without the pigment, had lower light compensation and saturation values and a greater apparent quantum efficiency. These features, together with the similar Chl *a*/Chl *b* ratios in the two forms, are typical for bryophytes utilizing some form of light filter (Green and Lange 1994). Maximal GP, 100 μmol O$_2$ (mg Chl)$^{-1}$ h^{-1}, was similar to those of other bryophytes. Unfortunately, there were no studies of low-temperature tolerance, but the distribution of the plant suggests that this will not be high.

3. Bryophytes: Mosses

Mosses show a normal saturation response to PPFD, and there is evidence of very low compensation and saturation PPFD (Rastorfer 1970, 1972). Saturation commonly occurs at around 15% of full sunlight (about 300 μmol m^{-2} s^{-1}; Longton 1988a), which would make the mosses more like shade plants. Post (1990) suggested that the ginger and green forms of *Ceratodon purpureus* (Hedw.) Brid. seem to be similar to the sun/shade adaptations of higher plant leaves. NP response to temperature shows an optimum temperature (T_{opt}) between 5°C and 25°C (Longton 1988a). T_{opt} depends strongly on PPFD and is lower at PPFD below saturation (Rastorfer 1970). Below T_{opt}, NP declines due to limitation by low temperature, and excess light energy is handled by increased nonphotosynthetic quench mechanisms (Green et al., unpublished observations). Above T_{opt}, the depression has always been explained as being due to increased DR, since that rises almost exponentially with increase in temperature; however, there is little evidence for this, and the depression could easily represent increased PR as in higher plants. Many mosses show negative NP at higher temperatures, over about 20°C (Convey 1994; Davis and Rothery 1997, Gannutz 1970, Kappen 1989, Longton 1988a, Noekes and Longton 1989, Wilson 1990). At 0°C, NP can still be substantial but is always very low by −5°C and ceases at around −7°C (Kappen et al. 1989). Surprisingly little data exist for the photosynthetic response to water content (WC) for antarctic mosses, which probably reflects the consistently high WC present when the plants are wet. Almost all available data are for species in the northern, maritime Antarctic (Davey 1997b), but little difference was found between these plants and others studied in temperate or north polar areas (Fowbert 1996). It seems unlikely that antarctic species will differ from temperate mosses where ectohydric species have maximal NP at around 300–600% WC. However, there is a strong relationship between moss species distribution and water flow, e.g., *Pottia heimii* typically occupies drier habitats than *Bryum argenteum* Hedw. (Schwarz et al. 1992), and Kappen et al. (1989) found photosynthetic differences between mesic and xeric ecodemes of *Schistidium* (*Grimmia*) *antarctici* Card. in the Windmill Islands; therefore, more detailed investigations are needed. Response to CO_2 concentration has been little investigated for antarctic mosses, but it would be expected to be similar to that found for other species where saturation occurs at levels well above ambient (Adamson et al. 1990). The actual CO_2 concentration around the mosses is a matter of debate, but very high values have been found in *Grimmia antarctici* in continental Antarctica, which is several times the normal ambient levels (Tarnawski et al. 1992). The source of the high CO_2, its seasonal change, and its effect on overall productivity are not yet known.

4. Lichens

Considerably more work has been conducted on antarctic lichens than the mosses (Kappen 1993a). However, being poikilohydric, the lichen and moss photosynthetic responses are similar in form, although differing in detail (see Longton 1988a). PPFD required for saturation can be very high, often around 1000 μmol $m^{-2} s^{-1}$, and light compensation values are dependent on the thallus temperature, e.g., saturation at 1300 μmol $m^{-2} s^{-1}$ PPFD above 1°C and compensation from 5 μmol $m^{-2} s^{-1}$ PPFD at −2°C to 128 μmol $m^{-2} s^{-1}$ PPFD at 20°C for *Leptogium puberulum* Hue (Figure 5; see also Schlensog et al. 1997). Sun and shade forms of *Usnea sphacelata* R. Br. differed little in the form of their photosynthetic responses to PPFD at identical temperatures, although the sun form had considerably larger NP at higher temperatures (Kappen 1983). T_{opt} tends to be lower than for mosses, and values around 5–15°C are common (Longton 1988a, Kappen 1993a), and, as in mosses, declines with fall in PPFD below saturation (Figure 6) (Sancho et al. 1997, Schroeter et al. 1995). In complete contrast to mosses, positive NP has been found to temperatures as low as −17°C for *Umbilicaria*

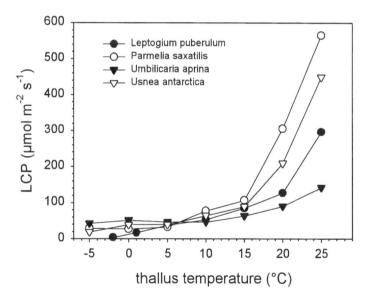

Figure 5 The typical increase in light compensation point (LCP) with temperature for four lichens measured in the field in Antarctica. The most continental species, *Umbilicaria aprina*, shows the least change. (Data from Schlensog et al. 1997 [*Leptogium puberulum*], Sancho et al. 1997 [*Parmelia saxatilis*], Schroeter 1991 [*Usnea antarctica*], and Schroeter 1997 [*Umbilicaria aprina*].)

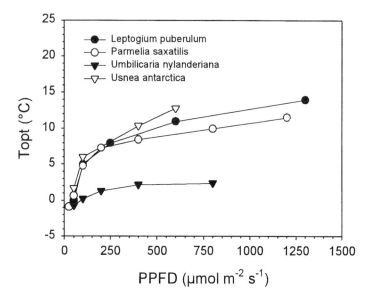

Figure 6   The typical change in T_{opt} (optimal temperature for net photosynthesis) with increase in irradiance (PPFD) for four lichens measured in the field in Antarctica. Sources same as for Figure 5 except *Umbilicaria nylanderiana* (Sancho et al. 1997).

aprina Nyl. measured in the field (Schroeter et al. 1994), although this ability is not confined to antarctic lichens, e.g., $-24°C$ for *Cladonia alcornis* (Lightf.) Rabh. was measured in the laboratory (Lange 1965).

Fewer studies of the effect of WC on photosynthesis exist, but those that do exist show the characteristic low WC range (200–400% maximal WC) compared with mosses, and often with depressed NP at high WC due to increased CO_2 diffusion resistances (Kappen 1985c, Harrisson et al. 1989, Kappen and Breuer 1991, Schroeter 1991). Lichens in the Antarctic normally occupy much drier habitats than mosses and are not found in wet areas unless on the drier tops of mosses. Response to CO_2 concentration seems to have not been measured, but would be expected to be similar to that of temperate lichens, i.e., not saturated at ambient CO_2 levels. The increased ambient CO_2 reported around mosses would not be expected around lichens because of their elevation and lack of substantial organic substrates below them (Tarnawski et al. 1992).

B.   Maximal Rates of Photosynthesis

Longton (1988a) gave an extensive list of maximal NP and T_{opt} for a wide range of polar bryophytes and lichens, although only a few are from the areas covered in this chapter. Kappen (1988, 1993a) gave additional similar data for lichens. The brevity of both lists indicates the lack of knowledge that exists at present. Considerable variation in rates certainly exists for lichens, from 0.08 mg CO_2 g^{-1} h^{-1} for *Rhizoplaca melanophthalma* (Ram.) Leuck and Poelt to 0.8 mg CO_2 g^{-1} h^{-1} for *Umbilicaria aprina*. Although it is occasionally suggested that the abundance of particular lichen species may be related to photosynthetic performance, we do not yet have the data to be certain. Lechowicz (1982) analyzed the relationship between several photosynthetic parameters and latitude for the northern hemisphere. From his analysis, maximal NPs are certainly substantially lower at latitudes greater than 65° N, and these far-northern values are similar to those found in the Antarctic. Kappen (1988) stated that maximal rates of NP are lower for continental antarctic lichens than for those in the maritime zone. No similar analysis exists for mosses, but, in general, the rates found seem to be close to the higher rates reported elsewhere (1–2 mg CO_2 gdw^{-1} h^{-1}; Rastorfer 1972, Longton 1988a, Kappen et al. 1989), and no depression is obvious.

C. Diel and Long-Term Photosynthetic Performance

Continuous measurement of photosynthetic performance over a day or several days has only recently become common in Antarctica, and the few extensive studies have been made in the northern maritime (e.g., Kappen et al. 1991, Schroeter 1991, Schroeter et al. 1991, Sancho et al. 1997 for lichens, Collins 1977 for mosses). A more common approach to obtain an estimate of seasonal production is to log microclimate parameters and to interpolate plant performance from photosynthetic response to PPFD and temperature (e.g., Kappen et al. 1991, Schroeter 1991, Schroeter et al. 1995). The present situation is probably a reflection of the difficulties of working in Antarctica with large and expensive equipment. Kappen et al. (1991) obtained eight diurnal records for *Usnea sphacelata* near Casey Station and recorded positive photosynthesis through the entire light period of the day, even though temperatures were mostly below 0°C. They were able to demonstrate relatively good agreement between their model, based on PPFD and temperature, and in situ rates of photosynthesis. Sancho et al. (1997) obtained diurnal courses for three cosmopolitan lichen species, *Parmelia saxatilis* (L.) Ach., *Pseudophebe pubescens* (L.) M. Choisy, and *Umbilicaria nylanderiana* (Zahlbr.) H. Magn., over 3 months in the summer of 1995, in the maritime Antarctic (Livingston Island) and found activity for only 182 hours. Typical examples of diel patterns for two lichens (*Parmelia saxatilis* and *Umbilicaria aprina*) and one moss (*Bryum argenteum*) are given in Figures 7–9. The two examples from continental Antarctica (Figures 8 and 9) show positive NP 24 hours a day, whereas the maritime example (*P. saxatilis*; Figure 7) had negative NP overnight, and the daily carbon balance was often negative, as was the cumulative balance calculated over 3 months (Sancho et al. 1997).

Parmelia saxatilis

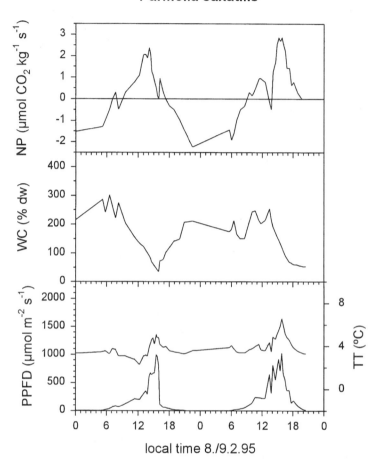

Figure 7   Diel pattern of net photosynthesis (NP, top), water content (WC, middle), and thallus temperature (TT) and irradiance (PPFD, bottom) for *Parmelia saxatilis* on February 8 and 9, 1995, measured at Livingston Island. Since this is a maritime site, there is a brief period of darkness each day leading to a period of negative NP. Measurements were made in the field at a natural growth site using a porometer system, and the lichen was rewetted on February 8 by rainfall. (Data from Sancho et al. 1997.)

Umbilicaria aprina

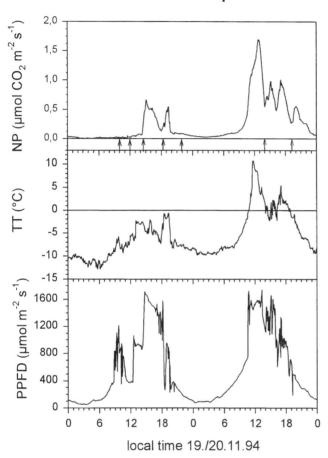

local time 19./20.11.94

Figure 8 Diel pattern of net photosynthesis (NP, top), thallus temperature (TT, middle), and irradiance (PPFD, bottom) for *Umbilicaria aprina* on November 19 and 20, 1994, at Botany Bay, Granite Harbour. The lichen was rewetted at the points marked with arrows in the upper panel. Although this is early in the season at a continental site (77° S), the irradiance reaches high values each day, and there is no period of complete darkness. Therefore, positive NP can occur through the full 24 hours and also on the 19th at subzero thallus temperatures. (Data from Schroeter et al. 1997.)

Bryum argenteum

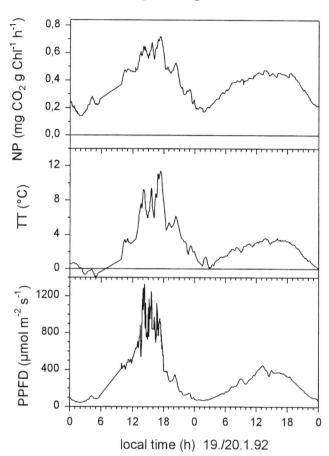

Figure 9 Diel pattern of net photosynthesis (NP, top), thallus temperature (TT, middle), and irradiance (PPFD, bottom) for moist *Bryum argenteum* on January 19 and 20, 1992, at Botany Bay, Granite Harbour just after midsummer. NP is positive through the entire period. Thallus temperatures are almost always positive and are linked to the irradiance. (Data from Schroeter 1997.)

Schroeter et al. (1991) demonstrated that basal chlorophyll *a* fluorescence could be used to monitor photosynthetic activity of *Usnea antarctica* Du Rietz. Unfortunately, the signal did not give a direct estimate of photosynthetic carbon fixation, although it could be estimated using microclimate records of temperature and PPFD (Schroeter 1991, Schroeter et al. 1997, Schroeter et al. 1997). Schroeter et al. (1992) demonstrated that the photosynthetic activity of the crustose lichen *Buellia frigida* (Darb.) Dodge at Granite Harbour could easily be monitored with a chlorophyll *a* fluorescence system. Concurrent measurements of CO_2 exchange indicated a reasonable agreement between relative electron transport rate (ETR) calculated from the fluorescence signal, and photosynthetic rate. Schroeter et al. (1997) measured fluorescence activity of *Buellia frigida* over several days at Granite Harbour. The studies revealed the extremely erratic nature of thallus moistening, which depended on small-scale topography, proximity to snow, degree of snow melt, and snow fall (Figure 10). Although air temperatures remained between $-7.5°C$ and $-12.0°C$, lichen thallus temperatures were much higher, often by approximately 15°C, even when wet. Thus, most of the photosynthetic activity actually occurred at positive temperatures but was detectable down to $-8°C$, and no saturation was evident at 1500 μmol m^{-2} s^{-1} (Figure 11). These results, which were made in November at 77° S, indicate that the growing season could be much longer than previously thought, even at these high latitudes.

Comparative studies on mosses do not seem to have been made. Schroeter (1997) found that *Bryum argenteum*, measured in a climate-controlled gas exchange system that tracked ambient conditions, remained above freezing almost continuously over several days and had positive photosynthesis 24 hours per day because of the continuous daylight at 77° S latitude (Figure 9).

A feature of all the diurnal studies is the elevation of thallus temperatures above ambient air temperatures because of the absorbed irradiation. Kappen (1993a) noted that thallus temperatures of wet lichens were often about 8–10°C at many maritime and continental sites as a result of shelter and sun exposure, although *Buellia frigida* at Granite Harbour could reach 15°C (Schroeter et al. 1997, Kappen et al. 1998b). The net result is that the habitats of the active plants are by no means as extreme as the ambient conditions would suggest and, in fact, may be remarkably constant over large latitudinal ranges; the "relative constancy of habitat conditions" first demonstrated by Poelt (1987), with lichens growing from the Mediterranean to Greenland. If this concept proves correct for Antarctica, then constriction of suitable habitats controls distributions and adaptation may not play as major a role as originally expected. This is certainly an area requiring future investigation.

D. Nutrition Effects

Elemental analyses for mosses (Smith 1996) show considerable variation depending on proximity to birds and species. Total nitrogen ranged from a low

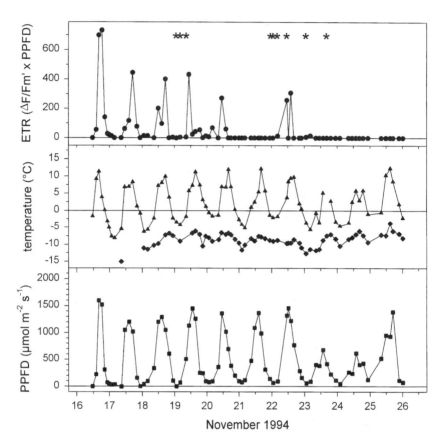

Figure 10 Diel pattern of relative electron transport rate (ETR) through photosystem II (top), thallus and air temperatures (middle), and irradiance (PPFD, bottom) for a *Buellia frigida* thallus over a 10-day period (November 16–25, 1994) at Botany Bay, Granite Harbour. Although air temperatures remained around $-10°C$ through the whole period, the thallus temperature was positive at times of high irradiance and several degrees above air temperature during the "night" because of heat storage in the rock surface. ETR depends on both irradiance and moistening of the thalli. Initially, the lichen received water from melting snow on the rock but, as this retreated from the lichen, it was active only after snowfalls (indicated by asterisks in the top panel). ETR activity is least or zero in the late morning when the lichen has dried out, highest in the afternoon after wetting by melt, and continues overnight at subzero temperatures—the so-called reverse diel pattern (see text). (Data from Schroeter et al. 1997.)

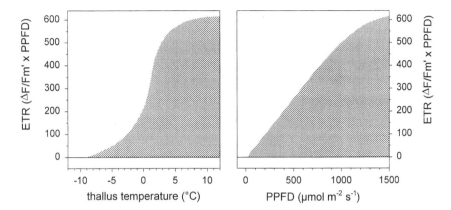

Figure 11 The relationship between relative electron transport rate (ETR) through photosystem II and thallus temperature (left) or incident irradiance (PPFD, right) for *Buellia frigida* thalli measured at approximately 2-hour intervals November 16–25, 1994, at Botany Bay, Granite Harbour. The shaded area covers all of the points, and the left-hand border of the shaded areas represents the response of ETR to temperature or PPFD when other conditions were optimal. Most points lie within the shaded area because thallus water contents (both responses), PPFD (for temperature response), or temperature (for PPFD response) were not optimal. In the left-hand panel, note the substantial activity at temperatures below zero and a clear optimum around 10°C. In the right-hand panel, note the lack of saturation of ETR even at 1500 μmol m^{-2} s^{-1}.

0.55–0.77% for *Polytrichum alpestre* to more than 3% for *Drepanocladus (Sanionia) uncinatus* (Hedw.) Loesk. Rastorfer (1972) found 2.95%, 2.75%, 1.50%, and 2.35% for *Calliergidium austrostramineum, Drepanocladus uncinatus, Polytrichum strictum* Brid. (syn. *P. alpestre*), and *Pohlia nutans* (Hedw.) Lindb., respectively. These values would not seem to be limiting and are not likely to be the major control on production. Lichens, in contrast, can have very low nitrogen contents, e.g., 0.36% and 0.83% for *Usnea sphacelata* and *U. aurantiaco-atra* (Jacq.) Bory, respectively (Kappen 1985c). In essence, the values are similar to those found for temperate lichens and mosses. Lichens with green algal photobionts seem to almost always have less than 1% total nitrogen (Green et al. 1980), and fruticose species are the lowest (Green et al. 1997). However, chlorophyll content, which can be related to nitrogen content (Green et al. 1997), is much lower in antarctic than in temperate species (Sancho et al. 1997).

Very little seems to be known about the effect of nutrients on the performance of mosses and lichens in Antarctica (Crittenden 1996). There is certainly a major effect on community structure with some mosses and, in particular, lichens being characteristic of places locally enriched by perching birds (Olech

1990). Substrate effects have also been found where the concrete of buildings attracts calcicolous species (Kappen 1993a). However, despite this and suggestions that salt can control distributions (Broady 1989), the work still needs to be done.

E. Desiccation, Cold, and Wetting/Drying

All antarctic terrestrial biota face a long period of cold, darkness, and desiccation over the winter. Conditions are particularly extreme in continental Antarctica (Table 1), and there can be little doubt that this limits the distribution of some species of lichens and bryophytes. Both groups are poikilohydric, meaning that their thallus hydration tends to equilibrate with the water status of the environment (Green and Lange 1994). With the exception of the endohydric mosses like *Polytrichum*, water uptake tends to occur over the entire surface. Typically, these groups also show exceptional resistance to desiccation and can survive very low water contents for long periods. However, such resistance is not inevitable, and some rainforest lichens are very sensitive to desiccation (Green et al. 1991), while the antarctic liverwort *Marchantia berteroana* (see Section III.A.2) is clearly also very sensitive (Davey 1997a). In the dry state, poikilohydric organisms tend to be very resistant to environmental extremes. Dry lichens all survived liquid nitrogen temperatures ($-196°C$; Kappen 1973), and moist thalli of antarctic species [*Xanthoria candelaria* (L.) Th. Fr., *Rhizoplaca melanophthalma*] fully tolerated slow or rapid freezing to $-196°C$ (Kappen and Lange 1972). Extended periods of cold and dryness also had little effect. *Alectoria ochroleuca* (Hoffm.) Massal recovered totally after 3.5 years at $-60°C$ (Larson 1978), and the moss *Schistidium antarctici* withstood 18 months at $-18°C$ (Kappen et al. 1989). However, there are several reports of damage through subzero temperatures to young shoot apices in mosses (Longton and Holdgate 1967, Collins 1976), and it does appear that these highly hydrated shoots are at risk.

Despite the impressive figures given, it is equally clear that considerable differences do exist between species, both lichens and mosses, in their abilities to withstand cold. Based on the evidence from *M. berteroana*, liverworts seem to be excluded from continental Antarctica because of their lack of tolerance, as are cyanobacterial lichens (Lange 1965, Schroeter et al. 1994), whereas lichens show a range of abilities to photosynthesise below 0°C (Lange 1965). A more wide-ranging survey of species that have distribution limits within Antarctica might well be very interesting.

Freeze/thaw cycles are extremely common in Antarctica, with up to 110 in 1 year being recorded in the northern maritime (Longton 1988a). The cycles, although rarely falling to temperatures more than a few degrees below freezing point, are thought to provide a severe stress to the plants through intracellular freezing that damages tissue, extracellular freezing that disrupts structure, dehydration by withdrawal of water to external ice, and phase changes in membranes leading to loss of cell contents. Lovelock et al. (1995a, 1995b) studied the effects of freeze/thaw cycles on the photosynthesis of *Grimmia antarctici* as monitored using chlorophyll *a* fluorescence. Every subzero phase caused decreased photosynthetic efficiency, but recovery at low light was also very rapid. The most thorough analysis has been by Kennedy (1993a) for a northern maritime species, *Polytrichum alpestre*. The species proved to be sensitive to free/thaw cycles, with reduction in GP being almost directly proportional to the temperature reached and with no recovery at −5°C or below. Almost all damage occurred on the first freeze cycle, and plants with lower thallus water contents were less sensitive. The mechanism of damage was not clarified but could well have been membrane disruption since increased nitrogen loss from the plants followed the freezing (Greenfield 1988). Kennedy (1993b) was of the opinion that freeze/thaw cycles could easily limit species distribution.

Wetting/drying cycles, also common in the Antarctic, are also thought to be disruptive, and release of carbohydrates has been detected (Davey 1997b, Greenfield 1993, Melick and Seppelt 1992). Friedmann et al. (1993) estimated that examples of the cryptoendolithic community in the Ross Desert would pass through an active/inactive cycle (either wet/dry or freeze/thaw), a minimum of 120–150 days each year with substantial loss of metabolites (Greenfield 1988). The stresses associated with these cycles contributed to the difference between a modelled net production of 106 mg C m^{-2} y^{-1} and actual growth, estimated using a variety of techniques, of 3 mgC m^{-2} y^{-1}.

IV. SPECIAL ANTARCTIC SITUATIONS

A. Lichen Photosynthesis at Subzero Temperatures

Lichens have long been known to carry out photosynthesis at subzero temperatures. Lange (1965) surveyed a wide range of species, including some from Antarctica, and found photosynthesis to −24°C by *Cladonia alcornis*, a temperate lichen. Other species, *Umbilicaria decussata* (Vill.) Zahlbr. and *Physcia caesia* (Hoffm.) Fuernr. (*Parmelia coreyi* in article) from Cape Hallett, had limits of −11°C and −14°C, respectively, values similar to species from hot deserts (Lange and Kappen 1972). Schroeter et al. (1994) have since measured photosynthesis of *Umbilicaria aprina* to −17°C in the field. In their analysis of the response of NP to subzero temperatures, they proposed that tolerance of subzero temperatures had the same physiological basis as tolerance to low water potentials. This impression was gained from the very similar response shown by photosynthesis to low water potentials whether generated by subzero temperatures or by equilibration with low atmospheric humidity (Kappen 1993b). In both cases, water is removed from the cells either to intercellular ice (lichens have external ice-nucleating agents with freezing initiating at about −5°C; Ashworth and Kieft 1992, Schroeter and Scheidegger 1995) or to the atmosphere. Schroeter et al. (1994) suggested that this explained the absence of cyanobacterial lichens in continental Antarctica since they cannot photosynthesize in equilibrium with humid air but need liquid water (Lange et al. 1988, Schroeter 1994). In addition, Schroeter and Scheidegger (1995) demonstrated convincingly that lichen thalli could rehydrate at subzero temperatures only in the presence of ice. Thallus water content depended on the temperature being only 7% of maximal WC at −21°C, 18% at −4.5°C, and nearly 100% at 8°C (Figure 12). Concurrent use of a low-temperature scanning electron microscope showed the progressive refilling of the algal and fungal cells as the temperature was increased with the water coming, via the atmosphere, from extracellular ice (Schroeter and Scheidegger 1995). At the same time, photosynthesis commenced and increased in line with thallus WC (Figure 13).

A common mechanism would also explain the correlation between cold resistance and dry habitats, such as deserts, found for lichens. The concept has even more far-reaching interest when it is realized that the distribution pattern of antarctic plants better reflects their rank order for subzero photosynthesis than

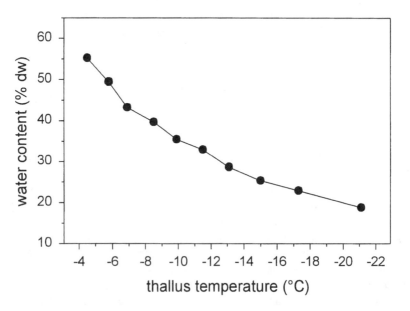

Figure 12 The relationship between thallus water content and thallus temperature for *Umbilicaria aprina* (Schroeter and Scheidegger 1995). Dry lichen thalli were equilibrated in the dark for 24 hours at the selected temperature and in the presence of ice.

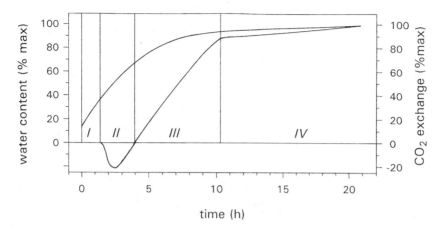

Figure 13 The increase in thallus water content and recovery of photosynthetic activity of a thallus of *Umbilicaria aprina* equilibrated with ice at $-4°C$ (Schroeter et al. 1997). In phase I, the thallus water content immediately starts to increase to a level allowing CO_2 exchange to start, which is initially negative (phase II). In phase III, CO_2 exchange is positive, and both it and thallus water content reach near-maximal values after about 10 hours. Phase IV represents an almost stable period for both CO_2 exchange and water content. During the whole period, the lichen was not in contact with liquid water.

their desiccation or cold tolerance (Table 3). For instance, cyanobacterial lichens have very good resistance to desiccation and to low temperatures, but they have nearly no ability to photosynthesize at subzero temperatures and are excluded from continental Antarctica.

The link between subzero photosynthesis and the ability to utilize humid air seems firm, and the distribution of these major groups within Antarctica thus appears to depend on their underlying physiology.

B. Photosynthesis Under Snow

Kappen (1989) demonstrated that rehydration and reactivation of *Usnea sphacelata* occurred in the field under snow and at subzero temperatures. This was a very important observation since, taken with the later studies of Schroeter and Scheidegger (1995), it is now clear that green algal lichens can reactivate photosynthesis under snow without liquid water being present, i.e., without the need of a thaw cycle. The influence of snow cover has been reviewed by Kappen (1993b). Snow can provide an efficient insulation against wind and extreme temperatures. At East Ongul Island, continental Antarctica, moss temperatures under snow fell to only $-21°C$, while air temperatures reached $-40°C$ (Matsuda 1968).

Table 3 Summary of Physiological Abilities of Cyanobacterial Lichens, Green Algal Lichens, Liverworts, and Mosses with Respect to Photosynthetic Performance at Subzero Temperatures

	Plant type			
	Cynobacterial lichens	Liverworts	Mosses	Green-algal lichens
Distribution	Maritime only	Mainly Maritime	Throughout Antarctica	
Subzero photosynthesis	to −2°C	Little	to −8°C	to −24°C
Positive NP in humid air[a]	None	?	Some ability	to 80% rh
Desiccation/cold tolerance				
Wet	High	Low	Low/medium	High
Dry	Very high	Low	High	Very high

[a] Positive NP in humid air refers to the ability of the group to attain positive net photosynthesis from a dry condition only in the presence of humid (90–95% relative humidity [rh]) air.

Considerable light, equivalent to about 10–30% of incident values, can penetrate to the base of a 15-cm snow pack (Kappen and Breuer 1991), which is more than sufficient to saturate NP at the low temperatures. Although there are no data yet available for the Antarctic, the performances of northern maritime antarctic mosses under snow (Collins and Callaghan 1980) and *Cetraria nivalis* in Sweden (Kappen et al. 1996) were modelled, and considerable photosynthetic production was found. It seems almost certain that substantial photosynthesis can occur long before the temperatures reach above freezing and while the lichens are still covered with snow. Once again, considerable work remains to be done.

C. The Reverse Diel Cycle of Photosynthesis and High Light Stress

Being poikilohydric, lichens and mosses depend on water to rehydrate and become active. Typically, even in Antarctica, this means that liquid water is required (Hovenden and Seppelt 1995); the exception would be the rehydration at subzero temperatures. For mosses, in particular, liquid water is certainly required, and the larger biomasses are found where consistent water flows occur. In temperate zones, water is normally provided by rain, which typically occurs at cloudy times. Mosses and lichens rapidly dry out as soon as brighter light conditions and full sunshine reoccur. Desiccation has often been given as one of the means used by poikilohydric plants to avoid high light and high temperature stress (Kappen 1988).

The situation in the Antarctic tends to be the opposite: melt water only occurs when insolation and temperatures are high so that the lichens and mosses are active when the light is brightest. This can be seen in Figure 10, where the photosynthetic activity of *Buellia frigida* was monitored using chlorophyll *a* fluorescence (Schroeter et al. 1997). Initially, when sunlight reaches the thalli in the early morning, they rapidly dry and become inactive. Then, at about noon, the snowmelt rehydrates the thalli and maximal activity occurs. The thalli then remain photosynthetically active, even when frozen overnight, until drying again occurs the next morning. Mosses also show this reverse pattern since water flow from snow or glacier melt is always at its maximum during the day and when irradiance is greatest.

High light stress is, therefore, of unexpected importance in Antarctica. A combination of cold temperatures and high light has been found to be particularly likely to cause photoinhibition in higher plants, and a similar response might be expected in antarctic mosses and lichens.

In mosses, photoinhibition has, indeed, been reported several times in studies in both the maritime and continental Antarctica. A midday depression in NP was consistently found under high light by Collins (1977) and was built into models of photosynthetic production by Collins and Callaghan (1980) and Davis (1983). Adamson et al. (1988) demonstrated severe photoinhibition in *Grimmia (Schistidium) antarctici*. Depressed photosynthesis occurred after 100 minutes at 500 μmol m^{-2} s^{-1} PPFD, and CO_2 exchange was negative after 150 minutes at 1000–1800 μmol m^{-2} s^{-1} PPFD; there was also a corresponding fall in Fv, the variable fluorescence (Figure 14). Lovelock et al. (1995a, 1995b) extended this work and demonstrated the occurrence of photoinhibition, measured with chlorophyll *a* fluorescence, during freeze/thaw cycles. Plants that were originally under snow showed severe photoinhibition if the snow was removed and they became exposed to full irradiance. Lovelock et al. (1995a, 1995b) showed that full recovery occurred under warmer temperatures and under low light, and they suggested that photoinhibition was more of a protective process than one of damage. Post et al. (1990) showed similar photoinhibition and recovery for *Ceratodon purpureus*. Post (1990) also showed that the ginger pigment found in exposed plants of *Ceratodon purpureus* appeared to serve a protective function against high irradiance. In studies on sun- and shade-adapted *Bryum argenteum*, Green et al. (1999) showed that only the shade-adapted form, which was bright green, was susceptible to high light. It appeared that the silvery shoot points and nonphotochemical quenching within the photosystems could protect the sun form against several hours of full sunlight.

In lichens, the situation is not yet entirely clear. Kappen et al. (1991) re-

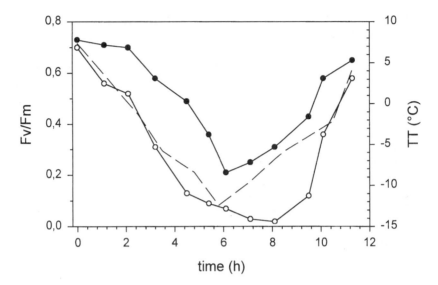

Figure 14 The induction of photoinhibition (Fv/Fm ratio from chlorophyll *a* fluorescence) in *Grimmia antarctici* by subzero temperatures. The photoinhibition is substantially worse in the light (350 µmol m^{-2} s^{-1}) than in the dark, but full recovery rapidly occurs once the temperature rises above freezing. Left-hand axis; Fv/Fm, maximal quantum efficiency of photosystem II from chlorophyll *a* fluorescence; solid circles, Fv/Fm of moss kept in the dark; open circles, Fv/Fm (after predarkening) of moss kept in the light. Right-hand axis and dashed line are thallus temperature (TT). (Adapted from Lovelock et al. 1995.)

ported photoinhibition, indicated by depressed NP, for *Usnea sphacelata* under high light. However, no signs of photoinhibition were found when *Umbilicaria aprina* was monitored for 2 days under full natural sunlight, even though the thalli had been dug from under 70 cm of snow before receiving any sunlight in that summer (Kappen et al. 1998a). Similarly, Schlensog et al. (1997) found no signs of photoinhibition for *Leptogium puberulum* after extensive treatment with high light (Figure 15). It seems that pigments in lichens may act to reduce the light level within the thallus and thus protect the photobionts (Schroeter et al. 1992, Buedel and Lange 1994, Rikkinen 1995, Schlensog et al. 1997).

Much work still remains to be done on the problem of high light and photoinhibition. Since shade-adapted thalli are much more sensitive, future studies must focus on plants in the field with a known light environment history. The role of pigments as protection, such as the ginger color of *C. purpureus* and the blackening of *Usnea* species, also needs further work since this seems to be a major means of protection in Antarctica.

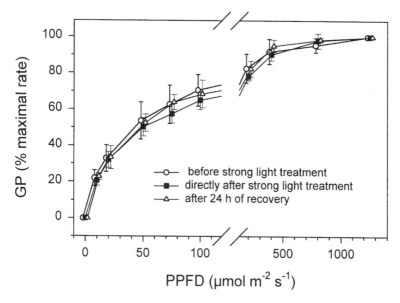

Figure 15 The almost complete lack of photoinhibition in the cyanobacterial lichen *Leptogium puberulum* despite treatment with strong light (3 hours at 1600 μmol m^{-2} s^{-1} PPFD). Immediately after the treatment there was a slight but insignificant decline in GP at intermediate irradiances, but no change in apparent quantum efficiency (initial slope of the response to irradiance). (Modified from Schlensog et al. 1997.)

D. The Endolithic Lichen Community

Under extremely dry conditions in Antarctica, and in other deserts, lichens can adopt an endolithic growth form where they live within the pores of rocks composed of materials such as sandstone and limestone (Friedmann and Galun 1974). The most studied endoliths are the "cryptoendoliths" that penetrate to 2 cm in the Beacon Sandstone of the Dry Valleys, Southern Victoria Land, 77° S, (Friedmann 1982, Kappen 1988). Other species such as *Lecidea phillipsiana* also grow widely in East Antarctica in granites, where they produce a prominent brown color on the surface and cause extensive rock flaking (Hale 1987). The lichens form a layered structure within the rock, with the upper layer containing fungal hyphae and the lichen green algal symbiont *Trebouxia*, which is dark in color, possibly to reduce light intensities. The endoliths grow only on the faces of rocks where higher insolation is received: north-facing or horizontal. On a sunny day, the temperature of the rock can reach +8°C while the air temperature is still lower than 5°C below freezing (Kappen et al. 1981). Humidity within the rock is approximately 80–90%, which is considerably higher than the air, which is normally about 30–40% relative humidity. Because the temperature is so strongly dependent on insolation, it can fluctuate markedly during a day, and freeze/thaw transitions are common. A typical daily pattern of the internal rock environment is shown in Figure 16, and a freeze/thaw or wet/dry transition is expected to occur at least once on all days when metabolic activity occurs, about 120–150 per year. Water is provided to the rock by blowing snow, which then melts into the rock when it is warmed by sunlight. The endoliths are wetted either by equilibration with the high humidity within the rock or by direct moisture uptake after snow melt. The latter method is thought to be the most important in the Dry Valley region (Friedmann 1978).

Carbon metabolism (incorporation of $H^{14}CO_3$) was saturated at 150 μmol $m^{-2} s^{-1}$ at 0°C, and had an optimum at 15°C (Vestal 1988). Measurements of CO_2 exchange showed it to be maximal at 3–6°C and to be still positive at -10°C. In a classic long-term study of the nanoclimate using automated recording systems reporting over satellites, 3 years of continuous data were obtained by Friedmann et al. (1987), allowing modelling of aspects such as the thermal environment (Nienow et al. (1988). This was then connected to the CO_2 exchange studies to produce estimates by modelling of the community productivity throughout the year (Friedmann et al. 1993). The cryptoendoliths showed positive photosynthesis for about 13 hours per day, with annual totals reaching around 1000 hours depending on aspect and slope (Kappen and Friedmann 1983). Net productivity was greatest around -2 to $+2$°C, with significant gains down to -8 to -10°C. Horizontal surfaces proved more productive than north-facing sloping surfaces, although they were colder because of better water relations. One unusual feature of the community is the relative unimportance of respiratory carbon loss during the dark winter; the system is simply too cold. Despite the long productive periods and the low respiration, the estimated carbon gain of 106 mgC $m^{-2} y^{-1}$ only translates into an actual gain of around 3 mgC $m^{-2} y^{-1}$ (estimated from carbon dating techniques and metabolic turnover rates) because of the high stresses, including probable loss of metabolites in the freeze/thaw or wet/dry cycles (Greenfield 1988). Because the endoliths can grow on almost all north-facing rock surfaces (porous rock types) where epilithic species are excluded because of the low temperatures and, in particular, the abrasive wind, they are the most common terrestrial vegetation community in the Dry Valleys (Friedmann 1982). Endolithic communities are one of the better demonstrations of the importance of shelter, aspect, and water supply in continental Antarctica.

V. INTEGRATING PERFORMANCE

A. Annual Productivity

Considerable efforts have been made to obtain values for the productivity, the seasonal net carbon gain, for antarctic plants. Unfortunately, most estimates have

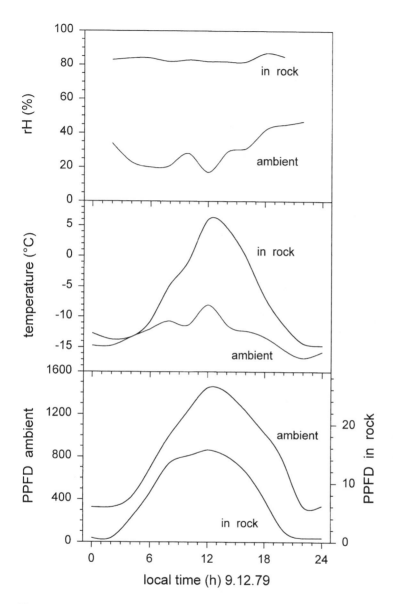

Figure 16 An example of the diel pattern of humidity (rH, top), temperature (middle), and irradiance (PPFD, μmol m^{-2} s^{-1}; bottom) both within the rock and in the surroundings of a north-facing endolithic lichen community on December 9, 1979, at Linnaeus Terrace, Aasgard Range, McMurdo Dry Valleys. In the bottom panel, note the very different scales for the incident PPFD (left-hand axis) and PPFD within the rock (right-hand axis). (Adapted from Kappen et al. 1981.)

been made in the northern maritime area (Signy Island). However, because no species have had their photosynthesis monitored for complete years, the estimates have been produced by the application of models constructed by linking CO_2 exchange and microclimate data sets. The two largest data sets are those for the cryptoendolithic community in the Dry Valleys (see Section IV.D) and for *Usnea aurantiaco-atra* on Livingston Island in the South Shetlands, where microclimate, CO_2 exchange, and activity (from chlorophyll *a* fluorescence) data sets have been linked (Schroeter et al. 1991, Schroeter et al. 1997). In the latter case, productivity gains are predicted throughout the year, with spring and autumn having the higher rates and winter and summer limited by cold and drought. Considerable variation in annual total production was also found from year to year (Schroeter, 1997). When modelled CO_2 gain is plotted against actual PPFD and temperature data, it is clear that the lichen is rarely active under conditions suitable for optimal photosynthesis, but is usually limited by drying out at high PPFD and high temperatures (Figure 17). Most other productivity estimates are from small data sets and with many and varied assumptions. Some estimates of annual production are given in Table 4. There is also little difference between the higher rates for both lichens and mosses, and both show a strong decline from the maritime to the higher latitudes. The extreme climate conditions of the continent depress production, which suggests that there is little adaptation by the plants. It is unfortunate that so many estimates come from the maritime, although, in defense, it is one of the major cryptogam-dominated areas of the world. Overall, however, the data are very thin, and more and better measurements are needed, especially in the continental Antarctic.

B. Growth Rates

It is difficult to be sure just how reliable the productivity data are because of the assumptions involved. One way to check is to compare what is known about growth rates in the same areas. Examples are given in Table 5 from a variety of sources. Once again, the vast majority of estimates are from the northern maritime zone, and there are very few data in total. However, sufficient, data exist to show the same sharp difference between the maritime and continental Antarctica. More data are available for the mosses, and growth rates decline from 10 mm y^{-1} in the northern maritime to around 1 or < 1 mm y^{-1} at the continental margin to <0.2 mm y^{-1} in the Dry Valleys. The data for lichens are much fewer, but a similar trend exists. Growth rates can be measured relatively easily in the maritime but are scarcely detectable in continental Antarctica. Figure 18 shows two photographs of *Buellia frigida* taken 17 years apart in the Asgaard Range, Dry Valleys; growth is close to undetectable and is less than 0.01 mm y^{-1}. It is worth comparing this situation with the photographs in Smith (1990), where increases in area have averaged 41% per year and have occurred for a similar species,

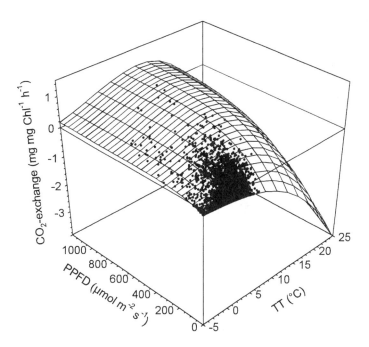

Figure 17 The relationship between thallus temperature (TT), photosynthetic photon flux density (PPFD), and CO_2 exchange for *Usnea aurantiaco-atra* at Livingston Island, maritime Antarctica. Thallus temperature, PPFD, and photosynthetic activity (from chlorophyll *a* fluorescence measurements) were recorded at 10-minute intervals for 1 year. Combinations of temperature and PPFD when the lichen was active were then plotted on a NP response surface to PPFD and temperature generated in the laboratory. The lichen was rarely active under optimal conditions for PPFD and temperature. (Modified from Schroeter 1997.)

Buellia latemarginata Darb, on Signy Island in the maritime Antarctic. While there are photographic records showing colonization of man-made substrates by lichens in the maritime, no such photographs exist for the continent (Kappen 1993a).

VI. ANTARCTIC PLANTS AS INDICATORS OF GLOBAL CLIMATE CHANGE

Plants in Antarctica could provide early indication of global climate change, especially for temperature, where any increase would act to relieve both the geographical isolation and climatic constraints (Kennedy 1995). Several lines of evidence suggest a special sensitivity for antarctic plants and ecosystems.

Table 4 Estimates of Annual Production for Higher Plants, Lichens, and Mosses at Several Locations in Antarctica

Species	Location	Annual production[a]	Comments
Higher plants			
Deschampsia antarctica	Signy Island (60°43′ S)	1700 (2)	Estimate from model based on microclimate and NP data
		390 (2)	Gravimetric (Edwards and Smith 1988, Edwards 1973)
Lichens			
Cladonia rangifera			
Usnea antarctica	Signy Island	80–200 (1)	Gravimetric (Hooker 1980)
U. aurantiaco-atra			
Usnea antarctica	King George Island (62°09′ S)	< 300 (3)	Estimated from model and year-round microclimate studies (Schroeter et al. 1995)
Usnea aurantiaco-atra	Signy Island, Livingston Island (62°40′ S)	250 (2) 85 (3)	(Smith 1984) (Schroeter 1997)
Usnea sphacelata	Casey (66°17′ S)	14 (1)	Estimated from model and assumed season length (Kappen et al. 1991)
Cryptoendolithic lichens	Dry Valleys (77°35′ S)	3 (2)	Estimated from NP data and microclimate (Kappen 1993)
Mosses			
Chorisodontium	Signy Island	315–660 (2)	Harvest plus prediction (Collins 1973, 1977, Longton 1970)
Polytrichum spp.			
Calliergidium and *Drepanocladus* spp.	Signy Island	223–893 (2)	Harvest plus prediction (Davis 1983)
Bryum pseudotriquetrum	Syowa (69°00′ S)	−16 to +4 (2)	Modelling from microclimate (Ino 1983)
Bryum argenteum	Ross Island (77°50′ S)	100 (2)	Modelling from microclimate and NP responses (Longton 1974)

[a] The numbers in parentheses refer to the units for the production: (1) mg gdw^{-1} y^{-1}; (2) g m^{-2} y^{-1}; (3) mg carbon gdw^{-1} y^{-1}.

Table 5 Estimates of Growth Rates for Lichens and Mosses From Various Locations in Antarctica

Species	Location	Growth rate ($mm\ y^{-1}$ unless otherwise stated)	Reference
Lichens			
Crustose spp.	King George Island (on whale bones)	40 mm in 78 y	Kappen 1993
Caloplaca sp.	Singy Island (on concrete)	10 mm in 46 y	Lindsay 1973
Acarospora macrocyclos	Signy Island	1–3	Lindsay 1973
Xanthoria elegans	Signy Island	0.2–0.5	Lindsay 1973
Rhizocarpon geographicum	Livingston Island	0.34	Sancho and Valladares 1993
Buellia frigida	Asgaard Range (Dry Valleys)	< 0.01	Green, unpublished
Mosses			
Polytrichum alpestre	Signy Island	2–5	Longton 1970
Calliergidium austrostramineum	Signy Island	10–32	Collins 1973
Drepanocladus uncinatus	Signy Island	11–16	Collins 1973
Bryum inconnexum	Syowa Coast	<1	Matsuda 1968
Bryum algens	Mawson	1.2	Seppelt and Ashton 1978
Pottia heimii	Taylor Valley	< 0.2	Green, unpublished

Figure 18 Photographs of the same thallus of *Buellia frigida* taken in January 1980 (top) and January 1997 (bottom) at Mt. Falconer summit, Lower Taylor Valley, McMurdo Dry Valley Region. Although there has been very slight growth at some points (less than 0.1 mm over the 17-year period), this cannot be seen at the scale of the photographs. The white bar in the lower right upper panel is 1 cm.

A strong cline in community structure and biodiversity is present in Antarctica. As summarized in Table 2, lichens and mosses decline from a combined total of around 200 and 90 species, in the northern maritime, respectively, to around 6 and 3 species in the McMurdo Dry Valleys respectively. The greatest shift in biodiversity coincides approximately with the 0°C mean temperature of the warmest month, the southern boundary of the maritime Antarctic. At this location, even a slight increase in temperature would be expected to cause a large change, particularly in biodiversity, but also in community structure.

It is also apparent that the lichens and mosses are not performing at their physiological optima under present ambient conditions, so that any increase in temperature would produce a concomitant increase in productivity. However, it is difficult to predict what the balance will be between increased photosynthetic and respiration rates. In particular, thalli that were moist and warm at times of low light could lose substantial production through the extended and increased respiration. There would probably also be a sharp increase in growth rates since these are also much smaller in the continental zone, and growth rate tends to be strongly limited by temperature.

There is growing evidence that a considerable propagule bank exists in antarctic soils, especially in the maritime zone. Use of cloches that can both increase the temperature and water relations has resulted in substantial plant growth where none could previously occur (Smith 1993). Thus, any area influenced by an improvement in temperature or related water supply will almost inevitably be rapidly vegetated.

The three preceding lines of evidence all combine to show how the plants, particularly at some locations, are poised to reveal temperature change.

Global temperature increase is suggested to be greater and more rapid in polar regions. A global increase of 0.3°C per decade (0.2–0.5°C dec^{-1} range) is predicted to rise in Antarctica at 0.5–0.7 times than the global rate in summer and 2–2.4 times the global average in winter (Kennedy 1995). Temperature data from the past 45 years also suggest that an increase has occurred; 0.067°C y^{-1} at Marguarite Bay (67°30′ S), 0.056°C y^{-1} at Faraday Station (65° S), 0.038°C y^{-1} at South Shetland Island (62°30′ S), 0.022°C y^{-1} at Signy Island (63°43′ S), and 0.032°C y^{-1} at Halley Station (76° S). The temperature increases have been greater in the west Antarctic Peninsula region, about double those at other sites, at both higher and lower latitudes. The reasons for this regional difference are not clear, but are suspected to be linked to ocean temperatures or circulation (King 1994). As predicted by global change models, the increase has been particularly large in winter temperatures. Although there is large interannual variability, the trend seems to be confirmed by glacial and ice shelf retreat.

Changes in vegetation have also been reported from the western Antarctic Peninsula. Improved colonization by bryophytes and lichens have been seen particularly in areas of newly exposed ground; however, the most impressive evidence comes from the expansion in abundance of the two phanerogams (Fowbert and Smith 1994, Smith 1994). A study spanning over 27 years has found large increases in populations of *Colobanthus quitensis* and *Deschampsia antarctica*, both in Signy Island (63°43′ S) and the Argentine Islands (65°15′ S). For example, at Skua Island, Argentine Islands, the population of *D. antarctica* was 190 plants in 1964 and 4868 plants in 1990 (Smith 1994). It is suggested that the increase in summer temperatures has led to increased reproduction, germination, and survival.

The overall impression is gained that temperatures have risen in the western Antarctic Peninsula and that the vegetation has rapidly responded. If even the predicted change in mean temperature occurs at other sites, then substantial changes in plant abundance and biodiversity can be anticipated.

VII. SUMMARY

After an initial look at the literature, it might appear that there is very little evidence of special adaptations by the lichens and mosses to antarctic conditions. Lack of special adaptation means that antarctic species show much the same abilities as can be found in temperate species. This point has been made by previous investigators (Longton 1988a, Kappen 1993a). However, the rapid decline in species numbers from the maritime to the continental zone and even more at higher latitudes indicates that even if no new adaptations are occurring, then there is at least strong selection for a group of species that can tolerate antarctic conditions.

Ability to carry out photosynthesis at subzero temperatures and the related tolerance to low water potentials and ability to utilize water vapor from ice or the atmosphere seems to limit the major groups of plants (Schroeter et al. 1994). Liverworts and cyanobacterial lichens seem to be excluded because they lack this ability. It is particularly startling that cyanobacterial lichens are excluded because cyanobacteria are extremely common in Antarctica, producing substantial communities as far south as the Scott Glacier (84° S; Vincent 1988). However, the cyanobacteria are confined to habitats where liquid water occurs, particularly the sides of ponds and streams. Because these habitats cannot be occupied by lichens, the Antarctic is an unusual example of an situation where the formation of a symbiosis (the lichen) does not extend the distribution of the photobiont.

The major decline in biodiversity also suggests that lichens and mosses are differentially sensitive to major limiting factors such as cold and desiccation. Variable resistance to both is known from temperate studies (Green et al. 1991 for desiccation, Lange 1965 for cold), and these factors are also important in controlling distributions. There are very little data for antarctic lichens on cold or desiccation resistance, and considerable work needs to be conducted.

Certain abilities stand out as successful adaptations, perhaps preadaptations might be better, to antarctic conditions. The first is the ability of green algal lichens to photosynthesize at subzero temperatures. The second is the ability of many species studied to tolerate the extremely high light levels that occur when they are active because of the reverse diel pattern (Schroeter et al. 1997). Much more research is needed to discover just how responsive the plants are to their light environment; *Bryum argenteum*, for example, can rapidly change from sun to shade form within days, and there is evidence that it can also reverse this process (Green et al. 1999). More research needs to be conducted on the area of acclimation, and one of the adaptations may turn out to be a substantial ability of species to rapidly acclimate. The third ability is the morphological variability that lichens and mosses show, which is strongly adaptive to the environment. Lichens range from the macrospecies, especially fruticose, with their ability to trap snow, to the extreme endoliths that live entirely within rocks. Even mosses, depending on water and exposure, can change from cushion forms to being almost invisibly buried.

The large decrease in productivity and growth rates between the maritime and the continent in line with the decrease in precipitation, however, indicates that the lichens and mosses are predominantly controlled by the availability of water. The Antarctic is a cold desert with water availability controlled not just by location, but also by being frozen for long periods. Plant growth is confined to locations where precipitation is concentrated and made available by warm conditions. This distribution control by water rather than by temperature has been well explained by Kennedy (1993b). In essence, plants grow wherever liquid water is reliably present as a result of glacier melt stream, by snow melt from stable snow banks, and by trapping of snow (e.g., by warm rock surfaces or fruticose lichens). It is important that water, warmth, and light coincide for plant growth. Kennedy (1993b) pointed out that the Dry Valleys are a relatively warm area, but no mosses or lichens (even cryptoendoliths) can grow in them except where water is consistently present (Green et al. 1992). Antarctica is the coldest, driest, highest, and windiest continent; the lichens and mosses grow where it is warm, wet, low, and protected.

Overall productivity is strongly influenced by the length of period when water is available, and the plants therefore become increasingly confined to areas of exceptionally good microclimate. It is this strong link between microclimate, water availability, and productivity/growth that makes the system so potentially useful for monitoring global climate change, especially temperature increase. The large change in biodiversity between the maritime and the continent closely follows the occurrence of months with mean temperatures above freezing. Even a small increase in temperature will markedly alter the areas over which such months occur and bring with it a marked community shift (Kennedy 1995). Antarctica will therefore remain an important natural laboratory for studying ecosystems under stress and the ecophysiology of cryptogams.

REFERENCES

Adamson, H., M. Wilson, P. Selkirk, and R.D. Seppelt. 1988. Photoinhibition in Antarctic mosses. Polarforschung 58: 103–112.
Adamson, E., A. Post, and H. Adamson. 1990. Photosynthesis in *Grimmia antarctici*,

an endemic antarctic bryophyte, is limited by carbon dioxide. Current Research in Photosynthesis 4: 639–642.

Ahmadjian, V. 1970. Adaptations of Antarctic terrestrial plants. In: Holdgate. M.W., ed. Antarctic Ecology. London: Academic Press, pp 801–811.

Ashworth, E.N., and T.L. Kieft. 1992. Measurement of ice nucleation in lichen using thermal analysis. Cryobiology 29: 400–406.

Björk, S., C. Hjort, O. Ingolfsson, and G. Skog. 1991. Radiocarbon dates from the Antarctic Peninsula region—problems and potential. Quarternary Proceedings 1: 55–65.

Broady, P., D. Given, L. Greenfield, and K. Thompson. 1987. The biota and environment of fumaroles on Mt. Melbourne, northern Victoria Land. Polar Biology 7: 97–113.

Broady, P.A. 1989. Broadscale patterns in the distribution of aquatic and terrestrial vegetation at three ice-free regions on Ross Island, Antarctica. Hydrobiologia 172: 77–95.

Buedel, B., and O.L. Lange. 1994. The role of cortical and eprinacral layers in the lichen genus *Peltula*. Cryptogamic Botany 4: 262–269.

Castello, M., and P.L. Nimis. 1995. A critical revision of antarctic lichens described by C.W. Dodge. Bibliotheca Lichenologica 57: 71–92.

Castello, M., and P.L. Nimis. 1997. Diversity of lichens in Antarctica. In: Battaglia, B., J. Valencia, and D.W.H. Walton, eds. Antarctic Communities, Species, Structure and Survival. Cambridge: Cambridge University Press, pp 15–21.

Claridge, G.G.C., I.B. Campbell, J.D. Stout, M.E. Dutch, and E.A. Flint. 1971. The occurrence of soil organisms in the Scott Glacier region, Queen Maud Range. New Zealand Journal of Science. 14: 306–312.

Collins, N.J. 1973. Productivity of selected bryophytes in the maritime Antarctic. In: Bliss, L.C., and F.E. Wiegolaski, eds. Proceedings of the Conference on Primary Production and Production Processes, Tundra Biome. Edmonton: IBP Tundra Biome Steering Committee, pp 177–183.

Collins, N.J. 1976. Growth and population dynamics of the moss *Polytrichum alpestre* in the maritime Antarctic. Oikos 27: 389–401.

Collins, N.J. 1977. The growth of mosses in two contrasting communities in the maritime Antarctic: measurement and prediction of net annual production. In: Llano, G.A., ed. Adaptations within Antarctic Ecosystems. Washington: Smithsonian Institution, pp 921–933.

Collins, N.J., and T.V. Callaghan. 1980. Predicted patterns of photosynthetic production in maritime antarctic mosses. Annals of Botany 45: 601–620.

Convey, P. 1994. Photosynthesis and dark respiration in antarctic mosses—an initial comparative study. Polar Biology 14: 65–69.

Convey, P. 1996. Reproduction of antarctic flowering plants. Antarctic Science 8: 127–134.

Crittenden, P.D. 1996. The effect of oxygen deprivation on inorganic nitrogen uptake in an antarctic macrolichen. Lichenologist 28: 347–354.

Davey, M.C. 1997a. Effects of physical factors on photosynthesis by the antarctic liverwort *Marchantia berteroana*. Polar Biology 17: 219–227.

Davey, M.C. 1997b. Effects of short-term dehydration and rehydration on photosynthesis and respiration by antarctic bryophytes. Environmental and Experimental Botany 37: 187–198.

Davey, M.C., and P. Rothery. 1997. Seasonal variation in respiratory and photosynthetic parameters in three mosses from the maritime Antarctic. Annals of Botany 78: 719–728.

Davis, R.C. 1980. Peat respiration and decomposition in antarctic terrestrial moss communities. Biological Journal of the Linnean Society 14: 39–49.

Davis, R.C. 1983. Prediction of net primary production in two antarctic mosses by two models of net CO_2 fixation. British Antarctic Survey Bulletin 59: 47–61.

Davis, R.C. 1986. Environmental factors influencing decomposition rates in two antarctic moss communities. Polar Biology 5: 95–103.

Edwards, J.A. 1972. Studies in *Colobanthus quitensis* (Kunth) Bartl. and *Deschampsia antarctica* Desv. V. Distribution, ecology and vegetative performance on Signy Island. British Antarctic Survey Bulletin 28: 11–28.

Edwards, J.A., and R.I.L. Smith. 1988. Photosynthesis and respiration of *Colobanthus quitensis* and *Deschampsia antarctica* from the maritime Antarctic. British Antarctic Survey Bulletin 81: 43–63.

Filson, R.B. 1987. Studies in Antarctic lichens 6: Further notes on *Umbilicaria*. Muelleria 6: 335–347.

Fowbert, J.A., and R.I.L. Smith. 1994. Rapid population increases in native vascular plants in the Argentine Islands, Antarctic Peninsula. Arctic Alpine Research 26: 290–296.

Fowbert, J.A. 1996. An experimental study of growth in relation to morphology and shoot water content in maritime antarctic mosses. New Phytologist 133: 363–373.

Friedmann, E.I. 1978. Melting snow in the dry valleys is a source of water for endolithic microoganisms. Antarctic Journal U.S. 13: 162–163.

Friedmann, E.I. 1982. Endolithic microorganisms in the Antarctic cold desert. Science 215: 1045–1053.

Friedmann, E.I., and M. Galun. 1974. Desert algae, lichens and fungi. In: Brown, G.W., ed. Desert Biology. New York: Academic Press, pp. 166–212.

Friedmann, E.I., L. Kappen, M.A. Meyer, and J.A. Nienow. 1993. Long-term productivity in the cryptoendolithic community of the Ross Desert, Antarctica. Microbial Ecology 25: 25–51.

Friedmann, E.I., C.P. McKay, and J.A. Nienow. 1987. The cryptoendolithic microbial environment in the Ross Desert of Antarctica: satellite-transmitted continuous nanoclimate data, 1984–1986. Polar Biology 7: 273–287.

Gannutz, T.P. 1970. Photosynthesis and respiration of plants in the Antarctic Peninsula Area. Antarctic Journal U.S. 5: 49–52.

Green, T.G.A., B. Buedel, A. Meyer, H. Zellner, and O.L. Lange. 1997. Temperate rainforest lichens in New Zealand: light response of photosynthesis. New Zealand Journal of Botany 35: 493–504.

Green, T.G.A., J. Horstmann, H. Bonnett, A. Wilkins, and W.B. Silvester. 1980. Nitrogen fixation by members of the Stictaceae (Lichenes) of New Zealand. New Phytologist 84: 339–348.

Green, T.G.A., E. Kilian, and O.L. Lange. 1991. *Pseudocyphellaria dissimilis*: a desiccation-sensitive, highly shade-adapted lichen from New Zealand. Oecologia 85: 498–503.

Green, T.G.A., and O.L. Lange. 1994. Photosynthesis in poikilohydric plants: a comparison of lichens and bryophytes. In: Schulze, E.D., and M.M. Caldwell, eds. Ecological Studies, Volume 100, Ecophysiology of Photosynthesis. Berlin: Springer-Verlag, pp. 319–341.

Green, T.G.A., B. Schroeter, and R.D. Seppelt. 1999. Bryophytes and the antarctic environment: effect of temperature and light on the photosynthesis of Bryum argenteum. In: Howard-Williams, C., and W. Davison, eds. Antarctic Ecosystems: Models for Wider Ecological Understanding. Caxton Press, in press.

Green, T.G.A., R.D. Seppelt, and A.-M.J. Schwarz. 1992. Epilithic lichens on the floor of the Taylor Valley, Ross Dependency, Antarctica. Lichenologist 24: 57–61.

Green, T.G.A., and W.P. Snelgar. 1982. A comparison of photosynthesis in two thalloid liverworts. Oecologia 54: 275–280.

Greenfield, L.G. 1988. Forms of nitrogen in Beacon Sandstone rocks containing endolithic microbial communities in southern Victoria Land, Antarctica. Polarforschung 58: 211–218.

Greenfield, L.G. 1993. Decomposition studies on New Zealand and Antarctic lichens. Lichenologist 25: 73–82.

Hale, M.E. 1987. Epilithic lichens in the beacon sandstone formation, Victoria Land, Antarctica. Lichenologist 19: 269–287.

Harrisson, P.M., D.W.H. Walton, and P. Rothery. 1989. The effects of temperature and moisture on CO_2 uptake and total resistance to water loss in the antarctic foliose lichen *Umbilicaria antarctica*. New Phytologist 111: 673–682.

Holdgate, M.W. 1964. Terrestrial ecology in the maritime Antarctic. In: Carrick, R., M.W. Holdgate, and J. Prévost, eds. Biologie Antarctique. Paris: Hermann, pp 181–194.

Holdgate, M.W. 1977. Terrestrial ecosystems in the Antarctic. In: Fuchs, V., and R.M. Laws, eds. Scientific Research in Antarctica. Philosophical Transactions of the Royal Society of London, Series B, Biological Science 279: 5–25.

Hooker, T.N. 1980. Growth and production of *Usnea antarctica* and *U. fasciata* on Signy Island, South Orkney Islands. British Antarctic Survey Bulletin 50: 35–49.

Hovenden, M.J., and R.D. Seppelt. 1995. Uptake of water from the atmosphere by lichens in continental Antarctica. Symbiosis 18: 111–118.

Ino, Y. 1983. Estimation of primary production in moss community on East Ongul Island, Antarctica. Antarctic Record 80: 30–38.

Kappen, L. 1973. Response to extreme environments. In: Ahmadjian V., and M.E. Hale, eds. The Lichens. New York: Academic Press, pp 310–380.

Kappen, L. 1983. Ecology and physiology of the antarctic fruticose lichen *Usnea suphurea* (Koenig) Th. Fries. Polar Biology 1: 249–255.

Kappen, L. 1985a. Vegetation and ecology of ice-free areas of northern Victoria Land, Antarctica. I. The lichen vegetation of Birthday Ridge and an inland mountain. Polar Biology 4: 213–225.

Kappen, L. 1985b. Vegetation and ecology of ice-free areas of northern Victoria Land, Antarctica. II. Ecological conditions in typical microhabitats of lichens at Birthday Ridge. Polar Biology 4: 227–236.

Kappen, L. 1985c. Water relations and net photosynthesis of *Usnea*. A comparison between *Usnea fasciata* (maritime Antarctic) and *Usnea sulphurea* (continental Antarctic). In: Brown, D.H., ed. Lichen Physiology and Cell Biology. New York: Plenum Press, pp 41–56.

Kappen, L. 1988. Ecophysiological relationships in different climatic regions. In: Galun,

M., ed. CRC Handbook of Lichenology. Boca Raton, FL: CRC Press, pp 37–100.

Kappen, L. 1989. Field measurements of carbon dioxide exchange of the antarctic lichen *Usnea sphacelata* in the frozen state. Antarctic Science 1: 31–34.

Kappen, L. 1993a. Lichens in the antarctic region. In: Friedmann, E.I., ed. Antarctic Microbiology. New York: Wiley-Liss, pp 433–490.

Kappen, L. 1993b. Plant activity under snow and ice, with particular reference to lichens. Arctic 46: 297–302.

Kappen, L., M. Bölter, and A. Kühn. 1987. Photosynthetic activity of lichens in natural habitats in the maritime Antarctic. In: Peveling, E., ed. Progress and Problems in Lichenology in the Eighties. Bibliotheca Lichenologica 25 Berlin: J. Cramer, pp 297–312.

Kappen, L., and M. Breuer. 1991. Ecological and physiological investigations in continental Antarctic cryptogams. II. Moisture relations and photosynthesis of lichens near Casey Station, Wilkes Land. Antarctic Science 3: 273–278.

Kappen, L., M. Breuer, and M. Bölter. 1991. Ecological and physiological investigations in continental antarctic cryptogams. 3. Photosynthetic production of *Usnea sphacelata*: diurnal courses, models, and the effect of photoinhibition. Polar Biology 11: 393–401.

Kappen, L., and E.I. Friedmann. 1983. Ecophysiology of lichens of the Dry Valleys of southern Victoria Land, Antarctica. II. CO_2 gas exchange in cryptoendolithic lichens. Polar Biology 1: 227–232.

Kappen, L., E.I. Friedmann, and J. Garty. 1981. Ecophysiology of lichens of the Dry Valleys of southern Victoria Land, Antarctica. I. Microclimate of the cryptoendolithic lichen habitat. Flora 171: 216–235.

Kappen, L., and O.L. Lange. 1972. Die Kälteresistenz einiger Macrolichenen. Flora 161: 1–29.

Kappen, L., R.I.L. Smith, and M. Meyer. 1989. Carbon dioxide exchange of two ecodemes of *Schistidium antarctici* in continental Antarctica. Polar Biology 9: 415–422.

Kappen, L., B. Schroeter, C. Scheidegger, M. Sommerkorn, and G. Hestmark. 1996. Cold resistance and metabolic activity of lichens below 0°C. Advances in Space Research 18: 119–129.

Kappen, L., B. Schroeter, T.G.A. Green, and R.D. Seppelt. 1998a. Chlorophyll *a* fluorescence and CO_2 exchange of *Umbilicaria aprina* under extreme light stress in the cold. Oecologia 113: 325–331.

Kappen, L., B. Schroeter, T.G.A. Green, and R.D. Seppelt. 1998b. Microclimatic conditions, meltwater moistening, and the distributional pattern of *Buellia frigida* on rock in a southern continental antarctic habitat. Polar Biology 19: 101–106.

Kennedy, A.D. 1993a. Photosynthetic response of the antarctic moss *Polytrichum alpestre* Hoppe to low temperatures and freeze-thaw stress. Polar Biology 13: 271–279.

Kennedy, A.D. 1993b. Water as a limiting factor in the antarctic terrestrial environment: a biogeographical synthesis. Arctic Alpine Research 25: 308–315.

Kennedy, A.D. 1995. Antarctic terrestrial ecosystem response to global environmental change. Annual Review of Ecology and Systematics 26: 683–704.

King, J.C. 1994. Recent climate variability in the vicinity of the Antarctic Peninsula. International Journal of Climatology 14: 357–369.

Lange, O.L. 1965. Der CO_2-Gaswechsel von Flechten bei tiefen Temperaturen. Planta 64: 1–19.

Lange, O.L., T.G.A. Green, and H. Ziegler. 1988. Water status related photosynthesis and carbon isotope discrimination in species of the lichen genus *Pseudocyphellaria* with green or blue-green photobionts and in photosymbiodemes. Oecologia 75: 494–501.

Lange, O.L., and L. Kappen. 1972. Photosynthesis of lichens from Antarctica. In: Llano, G.A., ed. Antarctic Terrestrial Biology, Antarctic Research Series 20. Washington, D.C.: American Geophysical Union, pp 83–95.

Larson, D.W. 1978. Patterns of lichen photosynthesis and respiration following prolonged frozen storage. Canadian Journal of Botany 56: 2119–2123.

Lechowicz, M.J. 1982. Ecological trends in lichen photosynthesis. Oecologia 53: 330–336.

Light, J.J., and R.B. Heywood. 1975. Is the vegetation of continental Antarctica predominantly aquatic? Nature 156: 199–200.

Lindsay, D.C. 1973. Estimates of lichen growth rates in the maritime Antarctic. Arctic Alpine Research 5: 341–346.

Lindsey, A.A. 1940. Recent advances in antarctic biogeography. Quarterly Review Biology 15: 456–465.

Longton, R.E. 1970. Growth and productivity of the moss *Polytrichum alpestre* Hoppe in antarctic regions. In: Holdgate, M.W. ed. Antarctic Ecology. London: Academic Press, pp 818–837.

Longton, R.E. 1974. Microclimate and biomass in communities of the *Bryum* association on Ross Island, continental Antarctica. Bryologist 77: 109–127.

Longton, R.E. 1988a. The Biology of Polar Bryophytes and Lichens. Cambridge: Cambridge University Press.

Longton, R.E. 1988b. Adaptations and strategies of polar bryophytes. Botany Journal Linnean Society 98: 253–268.

Longton, R.E., and M.W. Holdgate. 1967. Temperature relations of antarctic vegetation. Philosophical Transactions of the Royal Society of London, Series B 252: 237–250.

Lovelock, C.E., C.B. Osmond, and R.D. Seppelt. 1995a. Photoinhibition in the antarctic moss *Grimmia antarctici* Card. when exposed to cycles of freezing and thawing. Plant Cell Environment 18: 1395–1402.

Lovelock, C.E., A.G. Jackson, D.R. Melick, and R.D. Seppelt. 1995b. Reversible photoinhibition in antarctic moss during freezing and thawing. Plant Physiology 109: 955–961.

Matsuda, T. 1968. Ecological study of the moss community and microorganisms in the vicinity of Syowa Station, Antarctica. Japanese Antarctic Research Expedition Scientific Reports E29: 1–58.

Melick, D.R., and R.D. Seppelt. 1992. Loss of soluble carbohydrates and changes in freezing point of antarctic bryophytes after leaching and repeated freeze-thaw cycles. Antarctic Science 4: 399–404.

Nienow, J.A., C.P. McKay, and E.I. Friedmann. 1988. The cryptoendolithic microbial environment in the Ross Desert of Antarctica. Mathematical models of the thermal regime. Microbial Ecology 16: 253–270.

Noakes, T.D., and R.E. Longton. 1989. Studies on water relations in mosses from the cold-Antarctic. In: Heywood, R.B., ed. University Research in Antarctica. Antarctic Special Topic. Cambridge: British Antarctic Survey, pp. 103–116.

Olech, M. 1990. Preliminary studies on ornithocoprophilous lichens of the arctic and Antarctic regions. NIPR Symposium Polar Biology 3: 218–223.

Pickard, J. 1986. Antarctic oases, Davis station and the Vestfold Hills. In: Pickard, J., ed. Antarctic Oasis: Terrestrial Environments and History of the Vestfold Hills. Sydney: Academic Press Australia, pp. 1–19.

Poelt, J. 1987. Das Gesetz der relativen Standortkonstanz bei den Flechten. Botanische Jahrbuecher der Systematik 108: 363–371.

Post, A. 1990. Photoprotective pigment as an adaptive strategy in the antarctic moss *Ceratodon purpureus*. Polar Biology 10: 241–245.

Post, A., E. Adamson, and H. Adamson. 1990. Photoinhibition and recovery of photosynthesis in antarctic bryophytes under field conditions. Current Research Photosynthesis 4: 635–638.

Post, A., and M. Vesk. 1992. Photosynthesis, pigments, and chloroplast ultrastructure of an antarctic liverwort from sun-exposed and shaded sites. Canadian Journal of Botany 70: 2259–2264.

Rastorfer, J.R. 1970. Effects of light intensity and temperatre on photosynthesis and respiration of two east Antarctic mosses, *Bryum argenteum* and *Bryum antarcticum* Bryologist 73: 544–556.

Rastorfer, J.R. 1972. Comparative physiology of four west antarctic mosses. In: Llano, G.A., ed. Antarctic Terrestrial Biology. Antarctic Research Series, Vol. 20. Washington: American Geophysical Union, pp. 143–161.

Rikkinen, J. 1995. What's behind the pretty colours? A study on the photobiology of lichens. Bryobrothera 4: 1–239.

Sancho, L.G., and F. Valladares. 1993. Lichen colonization of recent moraines on Livingston Island (South Shetland I., Antarctica). Polar Biology 13: 227–233.

Sancho, L.G., A. Pintado, F. Valladares, B. Schroeter, and M. Schlensog. 1997. Photosynthetic performance of cosmopolitan lichens in the maritime Antarctic. Bibliotheca Lichenologica 67: 197–210.

Schlensog, M., B. Schroeter, L.G., Sancho, A. Pintado, and L. Kappen. 1997. Effect of strong irradiance on photosynthetic performance of the melt-water dependent cyanobacterial lichen *Leptogium puberulum* (Collematacege) Hue from the maritime Antarctic. Bibliotheca Lichenologica 67: 235–246.

Schroeter, B. 1991. Untersuchungen zu Primärproduction und Wasshaushalt von Flechten der maritimen Antarktis unter besonderer Berücksichtigung von *Usnea antarctica* Du Rietz. Dissertation, Universitaet Kiel, pp. 1–148.

Schroeter, B. 1994. In situ photosynthetic differentiation of the green algal and the cyanobacterial photobiont in the crustose lichen *Placopsis contortuplicata*. Oecologia 98: 212–220.

Schroeter, B. 1997. Grundlagen der Stoffproduktion von Kryptogamen unter besonderer Berücksichtigung der Flechten - eine Synopse. Habilitationsschrift der mathematisch-naturwissenschaftlichen Fakultät der Christian-Albrechts-Universität zu Kiel, pp. 1–130.

Schroeter, B., T.G.A. Green, R.D. Seppelt, and L. Kappen. 1992. Monitoring photosynthetic activity of crustose lichens using a PAM-2000 fluorescence system. Oecologia 92: 457–462.

Schroeter, B., T.G.A. Green, L. Kappen, and R.D. Seppelt. 1994. Carbon dioxide exchange at subzero temperatures. Field measurements on *Umbilicaria aprina* in Antarctica. Cryptogamic Botany 4: 233–241.

Schroeter, B., L. Kappen, T.G.A. Green, and R.D. Seppelt. 1997. Lichens and the antarctic environment: effects of temperature and water availability on photosynthesis. In: Lyons, W.B., C. Howard-Williams, and I. Hawes, eds. Ecosystem Processes in Antarctic Ice-Free Landscapes. Rotterdam: A. A. Balkema, pp. 103–118.

Schroeter, B., L. Kappen, and C. Moldaenke. 1991. Continuous in situ recording of the photosynthetic activity of antarctic lichens–established methods and a new approach. Lichenologist 23: 253–265.

Schroeter, B., L. Kappen, and F. Schulz. 1997. Long-term measurements of microclimate conditions in the fruticose lichen *Usnea aurantiaco-atra* in the maritime Antarctic. In: Cacho, J., and D. Serrat, eds. Actas del V Simposio de Estudios Antarcticos (1993) CICYT, Madrid, pp 63–69.

Schroeter, B., A. Olech, L. Kappen, and W. Heitland. 1995. Ecophysiological investigations of *Usnea antarctica* in the maritime Antarctic. I. Annual microclimatic conditions and potential primary production. Antarctic Science 7: 251–260.

Schroeter, B., and C. Scheidegger, 1995. Water relations in lichens at subzero temperatures: structural changes and carbon dioxide exchange in the lichen *Umbilicaria aprina* from continental Antarctica. New Phytologist 131: 273–285.

Schroeter, B., F. Schulz, and L. Kappen. 1997. Hydration-related spatial and temporal variation of photosynthetic activity in antarctic lichens. In: Battaglia, B., J. Valencia, and D.W.H. Walton, eds. Antarctic Communities. Species, Structure and Survival. Cambridge: Cambridge University Press, pp 221–225.

Schwarz, A.M.J., T.G.A. Green, and R.D. Seppelt. 1992. Terrestrial vegetation at Canada Glacier, southern Victoria Land, Antarctica. Polar Biology 12: 397–404.

Seppelt, R.D. 1995. Phytogeography of continental antarctic lichens. Lichenologist 27: 417–431.

Seppelt, R.D., and D.H. Ashton. 1978. Studies on the ecology of the vegetation of Mawson Station, Antarctica. Australian Journal of Ecology 3: 373–388.

Seppelt, R.D., and T.G.A. Green. 1998. A bryophyte flora for Southern Victoria Land, Antarctica. New Zealand Journal of Botany, in press.

Seppelt, R.D., and H. Kanda. 1986. Morphological variation and taxonomic interpretation in the moss genus Bryum in Antarctica. Memoirs National Institute Polar Research 37: 27–42.

Smith, R.I.L. 1984. Terrestrial plant biology of the sub-Antarctic and Antarctic. In: Laws, R.M., ed. Antarctic Ecology. London: Academic Press, pp. 61–162.

Smith, R.I.L. 1990. Signy Island as a paradigm of biological and environmental change in antarctic terrestrial ecosystems. In: Kerry, K.R. and G. Hempel, eds. Antarctic Ecosystems, Ecological Change and Conservation. New York: Springer-Verlag, pp. 32–50.

Smith, R.I.L. 1993. The role of bryophyte propagule banks in primary succession: case-study of an antarctic fellfield soil. In: J. Miles, J., and D.W.H. Walton, eds. Primary Succession on Land. Blackwell Scientific, Oxford: pp 55–78.

Smith, R.I.L. 1994. Vascular plants as bioindicators of regional warming in Antarctica. Oecologia 99: 322–328.

Smith, R.I.L. 1996. Terrestrial and freshwater biotic components of the western Antarctic Peninsula. In: Koss R.M., E. Hofmann, and L.B. Quetin, eds. Foundations for Ecological Research West of the Antarctic Peninsula. Antarctic Research Series, Volume 70. Washington: American Geophysical Union, pp 15–59.

Smith, R.I.L., and D.O. Övstedal. 1991. The lichen genus *Stereocaulon* in Antarctica and South Georgia. Polar Biology 11: 91–112.

Söchting, U., and M. Olech. 1995. The lichens genus *Caloplaca* in polar regions. Lichenologist 27: 463–471.

Stenroos, S. 1993. Taxonomy and distribution of the lichen family Cladoniaceae in the Antarctic and peri-Antarctic regions. Cryptogamic Botany 3: 310–344.

Tarnawski, M., D. Melick, D. Roser, E. Adamson, H. Adamson, and R. Seppelt. 1992. In situ carbon dioxide levels in cushion and turf forms of *Grimmia antarctici* at Casey Station, east Antarctica. Journal of Bryology 17: 241–249.

Vestal, J.R. 1988. Primary production of cryptoendolithic microbiota from the Antarctic desert. Polarforschung 58: 193–198.

Vincent, W.F. 1988. Microbial Systems of Antarctica. Cambridge: Cambridge University Press.

Walker, F.J. 1985. The lichen genus *Usnea* subgenus *Neuropogon*. Bulletin British Museum (Natural History) Botany Series 13: 1–130.

Wilson, M.E. 1990. Morphology and photosynthetic physiology of *Grimmia antarctici* from wet and dry habitats. Polar Biology 10: 337–341.

Wise, K.A.J., and J.L. Gressit. 1965. Far southern animals and plants. Nature 207: 101–102.

15

Ecology of Plant Reproduction: Mating Systems and Pollination

Anna V. Traveset
Mediterranean Institute of Advanced Studies, Spanish Council for Scientific Research, Palma de Mallorca, Spain

> . . . That these and other insects, while pursuing their food in the flowers, at the same time fertilize them without intending and knowing it and thereby lay the foundation for their own and their offspring's future preservation, appears to me to be one of the most admirable arrangements of nature.
>
> C.K. Sprengel (1793)

I. INTRODUCTION

One of the main differences between most plants and animals is that the former cannot move in search of a partner to mate and thus need a "messenger," which can be inanimate, such as wind or water, or an animal, vertebrate, or invertebrate, to transport the male gametes (pollen) among flowers. This passivity has caused plants to evolve a great variety of adaptations, either to disperse the pollen—for instance, by attracting animal pollinators with a reward—or to become independent of such messengers (i.e., by asexual reproduction or self-pollination).

The present chapter focuses on the mechanisms by which plants are able to accomplish reproduction. First is a description of how plants reproduce asexually and the advantages that sexual reproduction has, followed by a brief review of the different kinds of plant mating systems and what is known about their evolution, maintenance, and lability. The study of plant breeding systems addresses questions on the genetics of mating patterns, mainly associated with inbreeding depression, and, until the last two decades, it was considered as a separate line of research from that of pollination biology. The two areas have begun to merge into what has been called a "new synthesis" (Lloyd and Barrett 1996) or a "new plant reproductive biology" (Morgan and Schoen 1997) as floral biologists have enlarged their backgrounds with natural history, ecology, genetics, and theoretical approaches. The different systems of self-incompatibility, widespread among flowering plants, will be treated in a short and concise section, and the reader is referred to reports by Nettancourt (1977) and Barrett (1988, 1990, 1992) to explore this topic further. The paternal side of plant reproduction is increasingly receiving more attention in studies of reproductive success, and here existing information on this subject is synthesized, and directions for future research are given. Finally, there is discussion on how pollinators influence the evolution of floral traits and display and on the role they have played in the diversification of angiosperms, with a brief review of the main information obtained to date.

II. ASEXUAL REPRODUCTION

Asexual reproduction is fairly common in plants and allows them to persist in their habitats with complete independence from pollinating vectors. Two types are distinguished, both similar from the genetic viewpoint, although their mechanisms are different: (1) vegetative reproduction, i.e., asexual multiplication of an individual (genet)—that has originally arisen from a zygote into physiologically independent units (ramets) (Harper 1977, Abrahamson 1980); and (2) agamospermy (also termed "apomixis"), the production of fertile seeds without sexual fusion of gametes. Asexual reproduction allows a genet to exploit larger areas and new locations, provided that vegetative propagules are widely dispersed (Janzen 1977, Lovett Doust 1981). It is very often facultative, i.e., most plants can also reproduce sexually depending on soil conditions, light and temperature, level of competition among neighbors, etc. (Abrahamson 1980). In many perennial plants, both asexual and sexual reproduction take place, the latter occurring usually once a growth threshold has been attained (Hirose and Kachi 1982, Weiner 1988, Iwasa and Cohen 1989, Schmid and Weiner 1993, Worley and Harder 1996). A trade-off between asexual and sexual reproduction has been reported in a number of studies (Sohn and Policansky 1977, Law et al. 1983, Westley 1993), although sometimes it is difficult to separate from the effects of plant size (but see Worley and Harder 1996).

A. Vegetative Reproduction

Vegetative reproduction is widespread among angiosperms, especially in herbaceous perennials, but is rare among gymnosperms (possibly due to the predominantly woody habit of this group). Among woody plants, it is much more common in dwarf or creeping shrubs, climbers, and vines than in trees, and is conspicuous in anemophilous monocotyledons, perennial grasses (Gramineae), and sedges (Cyperaceae). Some species, such as those of *Phragmites* or *Ammophila*, often occur in a specialized habitat throughout the world and are among the most widespread species of plants known. Vegetative reproduction is also very successful in hydrophytes, probably because water is an adequate environment for the dispersal of relatively unprotected vegetative propagules. The efficiency of vegetative reproduction in water plants has been demonstrated on different occasions by ecological invaders, which often cause severe environmental and economic problems. A good example is *Caulerpa taxifolia*, a tropical green alga accidentally introduced into the western Mediterranean Sea in 1984, which has rapidly spread over a large area thanks to its efficient reproduction through stolons. This alga not only has the capacity of displacing the native *Posidonia oceanica*, but also produces substances that are toxic to the marine herbivores, with the net result of notably decreasing the benthic species diversity in coastal areas.

The usual organs developed by plants to reproduce asexually are modifications of stems or of axillary buds, which are stem initials. However, underground bulbs and corms are also common and have a protective function, especially during dormancy (hibernation or estivation), besides serving as dispersal units; they can be transported by rainstorms, on the feet of some animals, and even by some rodents that consume a large fraction but carry some for storage, as is the case with *Galanthus nivalis* (Richards 1986).

Vegetative reproduction may be especially disadvantageous in cases where a single clone occupies a large area, as the distance between individuals can be great and genetic variation is much reduced. The whole population may fail to set seed if the species is self-incompatible, as in the case of bamboos (e.g., *Arundinaria* spp.), or is diclinous (the members of a population are not regularly hermaphrodite; all individuals may be entirely female, such as is the case of *Elodea canadensis* in Britain), or due to sterility or hybridity (usually in populations founded by a single genet) (Richards 1986). On the other hand, clonal reproduction may also lose vigor with age, either because of an increased viral load through viral multiplication and re-infection, or because of the accumulation of disadvantageous somatic mutations.

B. Agamospermy

Agamospermy is a phenomenon absent in gymnosperms and limited to a small group (34 families) of angiosperms, occurring mainly (approximately 75% of the agamospermous taxa) in the Compositae, Gramineae, and Rosaceae (Nygren 1967, Richards 1986). It is highly polyphyletic, having arisen on many occasions from sexual taxa, even in closely related genera; every agamospermous genus includes sexual species (e.g., *Taraxacum, Crepis, Hieracium, Sorbus, Crataegus*) (Nygren 1967). There are a few documented cases of evolution of agamospermy from different types of breeding systems such as autogamy (e.g., *Aphanes*), dioecy (e.g., *Antennaria*), or heteromorphy (e.g., *Limonium, Erythroxylum*) (Richards 1986, Berry et al. 1991). Agamospermy can be *sporophytic*, when the whole gametophytic generation is absent, as is the case with *Citrus*, and the sporophyte embryo is budded directly from the old sporophyte ovular tissue, usually the nucellus (adventitious embryony); or more usually *gametophytic*, when a female gametophyte is produced with the sporophytic chromosome number. In the latter, the nonreduction of chromosome number results either from a complete avoidance of female meiosis (apospory and mitotic diplospory) or by a failure in it (meiotic diplospory).

The production of seeds is assured in agamospermous species in the absence of pollination. However, most species with adventitious embryony or with apospory require the stimulus of pollination to fertilize the endosperm nucleus (pseudogamy). The seed habit gives them the advantage of dispersal and the potential for extended dormancy, added to the possibility of fixing a successful genotype through asexual reproduction. Most agamosperms with apospory or adventitious embryony retain good pollen function, which can also be used in sexual reproduction. The reproductive behavior of facultative agamosperms is associated with environmental conditions, as observed in different species of grasses where apospory is related to photoperiod, with short days favoring it (Saran and de Wet 1970).

The main disadvantage of this kind of reproduction is that the agamospermous cell line forms a gigantic linkage in which the advantageous genes cannot "escape" from the accumulated harmful ones. Moreover, such a cell line is unable to recombine novel advantageous mutants and thus cannot adapt to the new conditions after an environmental change, although some genetic variation can exist through somatic recombination (chromosome breakage and fusion), meiotic recombination, chromosome loss and gain, and accumulation of mutants (Richards 1986). That is probably why truly obligate agamospermy, in which all possibility of sexuality has been lost, is rare (Asker 1980) and appears to be limited to a few diplosporous genera in which pollen is absent (unusual, as male-sterile mutants cannot be recombined). Some attributes of flowering plants such as perenniality, hybridity, and polyploidy appear to be closely related to agamospermy and in some way predispose them to this type of reproduction. For some yet unknown reasons, the family Compositae is predisposed to diplospory, whereas the Gramineae and the Rosaceae are predisposed to apospory (Asker 1980, Richards 1986). Common traits among these three families are: (1) nonspecific pollination (by wind or insects); (2) temperate or grassland distributions; and (3) single-seeded fruits. In contrast, adventitious embryony is found mainly in subtropical or tropical areas in woody, diploid species with multiseeded fruits (Richards 1986). The causes of such associations remain obscure.

Although a great deal of information has been accumulated on the origin, distribution, and mechanisms of agamospermy (Darlington 1939, Gustafsson 1946, Asker 1980, Richards 1986, Berry et al. 1991), much research still needs to be conducted to understand the evolution of this phenomenon and for the adequate interpretation of the observed patterns. Relatively few reproductive genes have been identified and mapped thus far, and more information is needed on the role of hormones in female reproductive development. The harnessing of agamospermy could represent, as Vielle-Calzada et al. (1996) have recently claimed, an "asexual revolution" in the next few years, leading to large increases in agricultural production. Therefore, research on this field is certainly worthwhile.

III. ADVANTAGES OF SEXUAL REPRODUCTION

The two most important characteristics of sexuality are: (1) it creates genetic variability through sexual fusion of gametes, chromosome segregation, and allele recombination; and (2) it allows gene migration so that successful mutations can spread between generations and move within and between populations. Moreover, sexuality, and thus meiotic mechanisms, dissipates Müller's Ratchet (accumulation of harmful mutations), breaks up linkage disequilibrium, and also engenders zygotes that are free of virus (Richards 1986). Sexual reproduction is a primitive trait of nearly all eukaryotic organisms that has probably contributed to their success and long-term survival. The genetic variability gives sexual lines evolutionary potential to adapt to new conditions after an environmental change, a feature absent in asexual organisms, as previously mentioned. Sexuality is absent from only a few groups of animals that reproduce parthenogenetically, in agamospermous plants and in sterile (usually hybrid) plant clones. Here, I will refer only to seed plants. The reproductive ecology of algae, bryophytes, and pteridophytes has been reviewed in reports by Lovett Doust and Lovett Doust (1988).

The whole process of embryology in angiosperms (flowering plants) was already described in detail nearly a half century ago in the literature (e.g., Maheshwari 1950). A good introductory chapter to the anatomy and physiology of sexual reproduction in both gymnosperms and angiosperms can be found in Richards (1986), and recent reviews on the origin and evolution of flowers can be found in Friis and Endress (1990) and Doyle (1994). The transition from a free-sporing heterosporous pteridophyte to a plant with gymnospermous reproduction, assessing adaptive explanations for the origin of seeds, is dealt with in Haig and Westoby (1989). According to these two investigators, the first seeds would have originated from heterosporous species, the megaspores of which would have been selected for a larger size; the decisive character in their success would have been related to pollination, by evolving traits to capture microspores before dispersal of the megaspore. In pteridophytes, fertilization always takes after gametes have been dispersed.

The gymnosperms, composed of five polyphyletic groups, are characterized by having the ovule/seed borne externally (gymnosperm actually means *naked seed*), although they are greatly diverse in most reproductive structures. Two general features of their reproduction, relevant to the genetic structure of plant populations, are as follows:

1. There are no hermaphrodite cones. Thus, plants are either monoecious (separate sexes on the same individual plant (e.g., Pinaceae, Taxodiaceae) or dioecious (an individual plant has either all female cones or all male cones, e.g., Cycadaceae, Ginkgoaceae, Taxaceae), although some species have populations with both monoecious and dioecious members (Givnish 1980). And some previously reported monoecious Cupressaceae, such as *Juniperus phoenicea*, have been shown to depart significantly from cosexuality (Jordano 1991 and references therein). If monoecious, there is usually dichogamy (separation of anther dehiscence from stigma receptivity); therefore, outcrossing is always promoted.

2. Pollination is almost always by wind (anemophily). Pollen grains in the Pinaceae even have two lateral air-filled sacs that act as wings and that allow them to fly very long distances. The genus *Ephedra* is an exception because it can be pollinated by insects (entomophily) (Bino and Meeuse 1981; the author has personally observed abundant sirphid flies visiting *E. fragilis* in the Balearic Islands).

The reproductive ecology of gymnosperms has received, in general, less attention than that of angiosperms, and much more information is still needed on the former to infer about the genetic control of mating patterns within or between species. Ellstrand et al. (1990) reviewed the available data for genetic structure of gymnosperms and concluded that they are generally highly diverse but have a low spatial differentiation. Thus far, there seems to be no evidence that nonangiosperms have low genetic diversity or that they are characterized by low gene flow (Midgley and Bond 1991).

Both gymnosperms and angiosperms have two major advantages over pteridophytes: (1) they do not depend on external water for sexual reproduction to take place; and (2) the zygote is protected within a seed, which in turn can be dispersed far from the parent plant. However, only approximately 750 species of gymnosperms survive today, in contrast to 10,000 pteridophytes. The major success has occurred in the angiosperms, a group represented by more than 220,000 species (Cronquist 1981). Such success is due to different factors (see Section IX), perhaps the most important being the wider range of growth forms that allows them to inhabit a wider range of habitats than either pteridophytes or gymnosperms. The latter are mostly trees, which restricts the number of habitats in which they can live, and they have limited breeding systems, limited pollination systems, and unspecialized seed dispersal (Givnish 1980). The development of a gynoecium (pistil) in the angiosperms (term meaning *enclosed seeds*) has probably been one of the most important steps in the evolution of plants. The ovule, which will develop into a seed, is enclosed within a modified megasporophyll, the carpel, which in turn is enclosed in an ovary. This ovary is connected to a structure, the stigma, where pollen reception takes place through the style, down which the pollen tube grows. Such a gynoecium has importantly influenced the reproduction and breeding systems of all angiosperms, from the stage of pollination to seed germination, enabling the evolution of a great diversity of structures and functions of inflorescences and infructescences.

Seed plants have a wide array of reproductive options that have evolved under particular environmental conditions (e.g., scarcity or absence of pollinators) and that are maintained or changed through the process of natural selection. The next section focuses on such reproductive options, on the factors that select for them, and, briefly, on the genetic consequences for the plant population.

IV. SELF-POLLINATION

Most angiosperms bear perfect flowers (containing both anthers and stigmas), and a large fraction of them are self-compatible, and thus potentially selfing, species (Barrett 1988, Bertin and Newman 1993). This breeding system has been studied for only a small fraction of angiosperms, but estimates from some localities suggest that 62–84% of temperate plants (mostly herbs) and 35–78% of tropical plants (including shrubs, trees, vines, and herbs) are at least partially selfers (Arroyo and Uslar 1993). Much theoretical and empirical work has been performed to understand the evolution of selfing, and much more on plants than on animals. Recent comprehensive reviews can be found in Jarne and Charlesworth (1993), Uyenoyama et al. (1993), and Holsinger (1996), and recent models have been developed by Schoen et al. (1996).

Self-pollination can take place within a flower (autogamy) or between flowers of the same genet (geitonogamy). The level of autogamy depends on the degree of separation between anthers and stigma (Table 1). Neither herkogamy nor dichogamy prevents ''geitonogamous'' selfing, although they may reduce it considerably. In the protogynous *Euphorbia dendroides*, for instance, stamen dehiscence is delayed in all inflorescences of some individuals so that it overlaps very little in time with stigma receptivity, clearly decreasing selfing (Traveset

Table 1 Mechanisms of Plants To Avoid Overlapping of Male and Female Functions

Herkogamy: separation in space between anthers and stigma position
Dichogamy: separation in time between stamen dehiscence and stigma receptivity
    I.   Protandry: the male phase is first
    II. Protogyny: the female phase is first

and Sáez 1997). The level of geitonogamous crosses varies greatly both among and within species and depends mainly on (1) pollinator foraging behavior, (2) number of flowers of the same genet simultaneously open, and (3) number (density) of genets with open flowers in the area. The behavior of a pollinator may promote much geitonogamy if it visits many flowers of the same individual successively. This is the case, for instance, in *Podarcis lilfordi*, a lizard that pollinates the flowers of *E. dendroides* in the Balearic Islands but stays on the same plant for a long time (>15 minutes), searching for nectar among many inflorescences in a single visit and going back to the same plant repeatedly, as it is territorial (Traveset and Sáez 1997). The few hermaphrodite species that can avoid geitonogamy must have synchronized dichogamy (unlikely if the plant produces many flowers), or not have more than one flower open at the same time. Geitonogamy can be important as plant size and flower number increase, representing a fitness cost of reduced pollen export ("pollen discounting") and also a cost to female fitness, by interfering with seed set even in self-incompatible species. This has been found, for instance, in *Polemonium viscosum, Ipomopsis aggregata*, and *Malva moschata* (De Jong et al. 1993). The existence of mass blooming plants ("masting" species), especially of those pollinated by animals (mainly trees in the tropics), might be expected to have high fitness costs of selfing (geitonogamy in particular). Geitonogamous pollination has been "the neglected side of selfing" (De Jong et al. 1993) for a long time, and relatively little is known about its ecology (Snow et al. 1996), although it is widespread and can have important implications on sex-allocation theory, the evolution of floral traits that reduce or inhibit selfing, and the evolution of dioecy.

Plants have a variety of mechanisms to promote or prevent selfing. Herkogamy has evolved only in plants that depend on animals for their pollination (LLoyd and Webb 1986), but dichogamy is found both in animal- and wind-pollinated species. An extensive survey on the frequency of dichogamy in angiosperms and on the causes of its evolution has been conducted by Bertin and Newman (1993). These investigators disagree with earlier researchers who believed that selfing avoidance was the main force leading to the evolution of dichogamy, and attribute the prevalence of protandry or protogyny to other factors such as avoidance of interference between pollen export and pollen receipt or imprecise pollen transport. Self-incompatibility is a mechanism that prevents selfing and that is known in approximately 30% of the angiosperm families; it is apparently controlled by one or a few loci and seems to have evolved independently several times (Jarne and Charlesworth 1993 and references therein). Male sterility (gynodioecy) or female sterility (androdioecy) are also mechanisms that reduce selfing, and both might represent early steps in the evolution of dioecy (see Section V). Cleistogamy is a mechanism that promotes selfing, as flowers do not open and can only self-fertilize. All species with cleistogamous flowers also produce hermaphrodite open-pollinated (chasmogamous) flowers (Lord 1981), and they appear to be evolutionary-stable under certain restrictive conditions (Schoen and Lloyd 1984).

Estimates of selfing rates in natural populations, which are needed to understand the evolution of this breeding system, are obtained by: (1) direct observations, tracking pollen flow (using fluorescent dyes); (2) genetic markers (e.g., flower pigments, DNA polymorphisms); (3) electrophoretic markers (allozymes); (4) performing progeny-array analyses, with or without information on maternal genotypes, which provide estimates based on the mixed-mating model (mixed selfing and random outcrossing); and (5) comparing the viabilities between zygotes produced by fertilization under natural conditions and those produced by artificial selfing and outcrossing. All of these methods have advantages and disadvantages and are reviewed by Jarne and Charlesworth (1993).

The immediate genetic consequences of selfing, and especially of obligate selfing, are a decrease in genetic variability commonly associated with high levels of homozygosity and, in the long term, the elimination of unfavorable recessive and partially recessive alleles (''purging''). It also breaks up heterozygous genotypes, whereas homozygous genotypes are reproduced intact. In contrast to outcrossing organisms in which recombination generates variance among progeny, selfing species respond to changes in environments by interline selection (Jarne and Charlesworth 1993). The high levels of homozygosity usually cause a decrease in offspring quality, compared with the progeny of outcrossers. Such decrease is termed inbreeding depression, δ, which was systematically studied by Darwin (1876), and much information has been gathered on its evolutionary consequences (Charlesworth and Charlesworth 1987a, Holsinger 1991, Husband and Schemske 1996). It is considered as the major factor preventing self-fertilization, and is estimated as $1 - w_s/w_x$, where w_s and w_x are the progeny fitnesses of selfers and outcrossers, respectively. An increase in selfing will be selectively favored if the progeny of selfing has a fitness greater than half that of progeny produced by outcrossing. Thus, an allele causing increased selfing will be selected for if $\delta < 0.5$. This condition is considered in models of the evolution of different outbreeding systems, including sporophytic self-incompatibility. In models of gametophytic self-incompatibility, δ must be larger than two thirds (Jarne and Charlesworth 1993 and references therein).

Two hypotheses, non–mutually exclusive, to explain inbreeding depression are: (1) heterozygotes are superior to homozygotes at loci determining fitness; and (2) inbred individuals are homozygous for recessive (or partially recessive) deleterious alleles, which are maintained in the populations through mutation. Theoretical models predict that the magnitude of inbreeding depression will decrease with inbreeding as deleterious recessive alleles are expressed and purged through selection; there is empirical evidence for this prediction (Barrett and Charlesworth 1991, Johnston and Schoen 1996, Husband and Schemske 1996).

The heterozygosis advantage is not easy to test; it requires that the parents to be inbred in the experiment have a zero inbreeding coefficient, which, as pointed out by Jarne and Charlesworth (1993), is not satisfied by most studies. We can experimentally estimate inbreeding depression by promoting selfing and outcrossing in the study population and examine components of fitness, such as fertility values at any stage of the plant's life cycle. It is preferable to obtain these data in the field, as laboratory or greenhouse conditions may underestimate inbreeding depression (for instance, if inbred genotypes are negatively affected by poor environmental conditions such as drought stress, as reported by Hauser and Loeschcke [1996]). Such experiments should thus consider the effects of biotic and abiotic changes in the environment (e.g., competition with other seedlings, nutrient limitation, etc). Although the large majority of studies have focused on just one stage of the plant's life cycle (Jarne and Charlesworth 1993), the ideal is to measure inbreeding depression over the whole life cycle (Ritland 1990, Husband and Schemske 1996).

There are three main factors that promote selfing and that are considered as explanations for the evolution of this breeding system:

1. *Reproductive assurance*. This is the factor that Darwin (1876) believed was the most important for the evolution of selfing. In circumstances in which outcrossing is unlikely, as when pollinators are scarce or absent, selfing can be favored—even if δ is high—to ensure ovule fertilization. A facultative selfing mechanism known as delayed-selfing (Lloyd 1979, Wyatt 1983), which allows outcrossing to occur while also ensuring that seeds are produced in the absence of pollination, is found in plants such as *Kalmia latifolia, Mimulus guttatus, Hibiscus laevis*, and many *Campanula* species (Klips and Snow 1997 and references therein). At least in some species of *Fuchsia*, and in areas where pollinators are rare, such as *F. magellanica* in Tierra del Fuego, the anthers often contact the stigma during late anthesis as filaments keep growing until reaching the style length (Traveset et al. 1998). Depending on the timing of selfing relative to outcrossing, Lloyd and Schoen (1992) classify selfing as: (1) prior, (2) competing, or (3) delayed. The latter is the least likely to displace outcrossing.

2. *Mating costs*. There are two kinds of outcrossing costs: (1) those referring to the transmission of genes, and (2) those referring to the resources needed for copulation and pollination. Due to the parent–offspring relatedness in selfing compared with random mating, selfing alleles have a 50% transmission advantage (the "automatic selection" hypothesis for the evolution of selfing; Jain 1976). This advantage, however, can be reduced by factors such as pollen discounting (when selfing reduces the ability to outcross as a male plant, as is the case in geitonogamous species) (Lloyd 1979). The energetic costs of producing large quantities of pollen, mainly in wind-pollinated plants, plus rewards such as nectar or oil for animal-pollinated ones, are relatively high in most species, and much greater than in highly autogamous plants, in which attractive structures (e.g.,

petals) and male reproductive functions are reduced (Cruden 1977, Charlesworth 1980, Cruden and Lyon 1985, Jarne and Charlesworth 1993, see also Damgaard and Abbott 1995).

   3. *Preservation of successful genotypes.* While environmental conditions are stable, selfing preserves the genotype adapted to those conditions. Evidence for the "local adaptive hypothesis" (which postulates that individuals perform better at their native site, whereas fitness of transplanted individuals declines with increasing distance; Antonovics 1968) has been found, for instance, in *Ipomopsis aggregata* (Waser and Price 1989) or in *Impatiens capensis* (Schmitt and Gamble 1990), although results are inconsistent with other plant species (Jarne and Charlesworth 1993).

   Data available so far on the breeding system of different species suggest that most outbreeders can also self-pollinate, and most selfers can outcross as well, i.e., that mixed mating systems are rather common in nature (Lloyd 1979, Barrett and Eckert 1990). The range of outcrossing rates can vary greatly among populations of the same species, as found, for instance, in *Lupinus, Lycopersicon, Gilia, Avena*, or *Thymus*, due either to genetic or environmental causes such as differences in pollinator attraction, the proximity of stigmas and anthers, the synchrony of stigma and anther maturation, or the level of self-compatibility (Levin 1986). Several models predict stability of intermediate levels of selfing (Uyenoyama 1986, Holsinger 1991, Lloyd 1992, Sakai 1995), although it is not yet clear how often this evolutionary stability occurs in nature (Schemske and Lande 1985, Barrett and Eckert 1990). Intermediate selfing rates are expected to evolve in plants in which selfing reduces either male or female fitness, e.g., when there is pollen discounting or when "competing" selfing reduces the number of fertilized ovules (Lloyd 1979). Mixed mating systems can also be maintained when there is an optimum pollen dispersal distance due to local adaptation (Campbell and Waser 1987) or when inbreeding depression affects dispersed progeny more than nondispersed progeny (Holsinger 1986). Further research is needed to determine correlations between species traits (e.g., flower size, duration of anthesis) and environmental variables comparing both selfing and outbreeding systems, and discerning if a character is a cause or an evolutionary consequence of the breeding system (Jarne and Charlesworth 1993). In addition, more molecular studies (DNA sequence data in particular) will help to assess the consequences of selfing and outcrossing on genetic variability within and between populations.

V. SEXUAL EXPRESSION

There are a number of possibilities with respect to the distribution of male and female organs on a plant that certainly determine the levels of selfing and out-

Table 2 Classification of the Different Possibilities by Which Male and Female Organs are Distributed in a Plant Species

---

Hermaphroditism: all individuals (genets) have "perfect" flowers, all bearing functional stamens and pistils

Monoecy: the two sexes are found on all individuals, but in separate flowers

Andromonoecy: the same individual bears both "perfect" and male flowers

Gynomonoecy: the same individual bears both "perfect" and female flowers

Dioecy[a]: male and female flowers are on separate genets

Androdioecy[a]: male and hermaphrodite flowers are on separate genets

Gynodioecy[a]: female and hermaphrodite flowers are on separate genets

Subdioecy[a]: intermediate stage between monoecy and dieocy in which sex expression of males and females is not constant

Polygamy[a]: different combinations of males, females, and hermaphrodites are possible

---

[a] In these cases, when both male and female functions are not regularly found on the same genet, *dicliny* is said to occur.

crossing (Table 2). For the last two decades, plant biologists have tried to understand this diversity of patterns of sex expression, providing models of sex allocation ("gain curves") that predict how male and female fitness changes with increases in the allocation of limited resources to each sexual function (Lloyd 1975, 1982, Charlesworth and Charlesworth 1978, 1981, 1987b, Bawa 1980, Bawa and Beach 1980, Spalik 1991, Charlesworth and Morgan 1991). Unfortunately, these models have been tested in natural populations only in a very limited number of studies (Emms 1996 and references therein). However, sex allocation theory alone cannot explain many aspects of sex expression, such as the spatio-temporal variation in sex expression found in nondiclinous plants (Thomson and Barrett 1981a, Solomon 1985, Emms 1993), and models of "gamete packaging" have also been developed to explain patterns of sex expression (Lloyd and Yates 1982, Burd 1995).

A. Monoecy

Monoecy is widespread, especially in large wind-pollinated plants such as trees, and sedges, and in water plants; it is more rare in insect-pollinated plants (Richards 1986). At least in some floras (Flores and Schemske 1984), this breeding system is associated with trees and shrubs that produce dry, many-seeded fruits. One of the benefits of having separate sexes on the same individual is that plants have the capacity to invest more on one sex or the other, depending on environmental conditions, to maximize the efficiency of both pollen dispersal and pollen capture. Moreover, monoecious plants benefit from a reduction of inbreeding depression, due to the spatial, and often temporal, segregation of sexes (Freeman et al. 1981). Evolutionary theories based on relative costs and benefits of male and female reproductive structures predict that plants growing under favorable conditions (being larger in size, having a greater resource supply or a greater total reproductive effort) should invest relatively more in female than in male function (the so-called female size advantage hypothesis) (Freeman et al. 1981, Charnov 1982, Lloyd and Bawa 1984, Goldman and Willson 1986). However, wind-pollinated plants do not usually follow those theoretical predictions, increasing relative maleness as patch quality improves (Burd and Allen 1988, Solomon 1989, Traveset 1992, Fox 1993). Large wind-pollinated plants may benefit from a relatively greater male investment if pollen is carried for longer distances and the local mating competition among sib pollen is lowered (male height advantage hypothesis) (Burd and Allen 1988, Ganeshaiah and Shaanker 1991). The little information available on sex allocation related to size in monoecious plants (Burd and Allen 1988, Goldman and Willson 1986, Solomon 1989, Traveset 1992), mostly on herbaceous plants, gives evidence for both hypotheses. More data are needed to test these hypotheses, especially for trees and shrubs, where the potential for fitness differences related to height or size is greater. On the other hand, changes in floral sex ratio may be nonadaptive and result, for instance, from alternations in the architectural arrangement of the inflorescences (Smith 1981, Lloyd and Bawa 1984, Solomon 1989, Traveset 1992); in this case, changes in floral sex ratios may be caused by resource allocation constraints or physiological constraints caused by hormones such as auxins, gibberellins, and cytokinins.

Models of sex allocation predict that the evolution of self-fertilization should result in a reduced allocation to male function and pollinator attraction (Charlesworth and Morgan 1991). In selfing monoecious plants, however, the investment to male function cannot be much reduced compared with hermaphroditic plants, as separate structures (petals, sepals, and pedicels) for male flowers and a higher production of pollen (to be transferred between flowers) are needed. Moreover, evolutionary changes in allocation patterns may be constrained by lack of genetic variation or by genetic correlations among characters (Ross 1990, Mazer 1992, Agren and Schemske 1995). More data on the importance of these genetic constraints, on the genetic and phenotypic correlations between allocations to both sex functions, and on the relationships between sex allocation, mating system, and reproductive success of the two sex functions are needed if we are to understand the evolutionary dynamics of sex allocation. Long-term data on gender variation in natural populations are also necessary in studies of the evolution of sex expression (Primack and McCall 1986, Jordano 1991). There is much individual variation in patterns of sex allocation, and a variety of factors (Goldman and Willson 1986) can cause a lack of consistent results. Variation at a spatial and temporal scale in environmental conditions needs to be considered, as gender expression of a species may vary, for instance, across a climatic gradient (Costich 1995). Documenting such variation at the individual, within-population, and between-population levels in the field is crucial to understand the selective pressures involved in the evolution of gender expression.

B. Andromonoecy

Andromonoecy is a breeding system that has been of particular interest in the study of sex expression patterns. It is uncommon, occurring in less than 2% of plant species (Yampolsky and Yampolsky 1922) and has probably evolved from hermaphrodite ancestors by means of a mutation removing pistils from some perfect flowers and a subsequent regulation of male flower number (Spalik 1991), or by the production of staminate (male) flowers (Anderson and Symon 1989). According to resource allocation models, andromonoecy occurs in species where the cost of maturing a fruit is great and the optimal number of male flowers is greater than the number of flowers that can set fruit (Bertin 1982, Whalen and Costich 1986, Anderson and Symon 1989, Spalik 1991). Pollen from male flowers can be more fertile than pollen from hermaphrodite flowers, as has been found in *Cneorum tricoccon* (Traveset 1995), representing an advantage to andromonoecy, as both male and female fitness may increase through pollen donation by male flowers (Bertin 1982). Another advantage attributed to the production of male flowers in andromonoecious species is the attractiveness to pollinators of a large floral display (Anderson and Symon 1989). Therefore, male flowers can increase both male fitness, through pollen donation, and female fitness, through attraction of pollinators to the plants. Temporal differences in the functioning of male and female organs are a common feature of monoecious and andromonoecious taxa (Thomson and Barrett 1981a, Anderson and Symon 1989, Emms 1993, 1996). In the andromonoecious *Zigadenus paniculatus*, for instance, male flowers are produced at the end of the blooming period, when the returns on female allocation are small or nonexistent (Emms 1996). More data are needed to show that these temporal patterns are adaptive and to answer questions such as: (1) How frequent are the mutations causing pistil loss? (2) Is the production of surplus pistils advantageous (thus selecting against andromonoecy) (Ehrlén 1991)? (3) Does the rechanneling of resources from pistils to other structures (e.g., male flowers) increase fitness in hermaphroditic species? As claimed by Emms (1996), rather than asking why andromonoecy has evolved, it may be more interesting to ask why it is so rare. Moreover, to fully understand this breeding system, the factors that control male fitness also need to be identified. More data are needed, for instance, on variation in pollen production per flower. It is not yet known if in andromonoecious species total pollen output is regulated through an increase in flower number or through the amount of pollen per flower. Data on a few species reveal that pollen grain number does not differ between male and hermaphroditic flowers (*Solanum carolinense*, Solomon 1985; *Cneorum tri-*

coccon, Traveset 1995) or is even lower in males (*Anthriscus sylvestris*, Spalik 1991). Some investigators have suggested that andromonoecy restricts outcrossing (Bertin 1982, Primack and Lloyd 1980, Whalen and Costich 1986), whereas others argue that, depending on the pollinator, it may serve to reduce selfing (Anderson and Symon 1989).

Sex expression in andromonoecious species can be variable among individuals, within and among populations, and through time (Diggle 1993 and references therein, Traveset 1995). Such variation can be genetic or it may be phenotypically plastic, varying with resource availability (e.g., light, water, available nutrients) (Solomon 1985, Diggle 1993). Ecological factors such as the resource status of an individual have often shown to be more important than genetic factors controlling sex ratio (Emms 1993, Diggle 1993, Traveset 1995); however, in general, the relative contribution of each of these sources of variation in sex expression is largely unknown.

C. Gynomonoecy

Gynomonoecy is much more rare than andromonoecy, occurring in only approximately a dozen families, and there seems to be no satisfactory explanation for the difference in frequencies of these two breeding systems. One possible reason may be the more expensive production of fruits compared with flowers (Charlesworth and Morgan 1991).

D. Dioecy

Dioecy is found in a large proportion (approximately 52%; Givnish 1980) of gymnosperms compared with angiosperms (approximately 6%; Renner and Ricklefs 1995), and appears to be strongly associated with woodiness in certain tropical floras (Givnish 1980, Sakai et al. 1995). The incidence of dioecy varies notably among regional floras, ranging from values as low as 2.6% (Balearic Islands) or 2.8% (in California; Fox 1985) to approximately 15% in the Hawaiian flora (Sakai et al. 1995). This mating system has evolved independently many times, as suggested by its scattered systematic distribution (Lewis 1942, Lloyd 1982). The classic hypothesis on the evolution of dioecy states that this breeding system results from selection for outcrossing to avoid the consequences of inbreeding depression (Thomson and Barrett 1981b and references therein). Other hypotheses consider the importance of resource allocation, sexual selection, and ecological factors such as mode of pollination, seed dispersal, or even predation influencing the evolution of dioecy (Bawa 1980, Thomson and Brunet 1990). For the last two decades, much work has been devoted to test these hypotheses, investigating correlations between dioecy and various morphological and ecological traits at different spatial and hierarchical scales. Dioecy has been basically associated with wind pollination, monoecy, woodiness, fleshy fruits, small greenish flowers, water pollination and, "unspecialized" insect pollination (Renner and Ricklefs 1995). However, many of those studies come from local floras and do not control for either correlations among explanatory variables or phylogenetic relationships (but see exceptions in Renner and Ricklefs 1995). Renner and Ricklefs have examined the strength of various of these associations, considering the whole angiosperm flora and taking into account their systematic position, and have found dioecy to be consistently associated with monoecy, wind or water pollination, and climbing growth; animal seed dispersal is not a strong factor at the family level but is overrepresented among dioecious genera. According to Renner and Ricklefs, the association of dioecy with fleshy fruits found in many local floras, although weak in others (Muenchow 1987), is probably spurious, resulting from multiple correlations among variables such as production of fleshy fruits and tree growth form, which in turn predominate in various tropical families. They argue that dioecy has evolved via monoecy most of the time, as already predicted by Yampolsky and Yampolsky (1922) and Lewis (1942). This evolutionary pathway has been described as a gradual divergence in the relative proportions of male and female flowers in the two incipient sexes (Charlesworth and Charlesworth 1978, LLoyd 1982, Ross 1978, 1982) and is presumably easier than the evolution of dioecy via other mating systems as mutations affecting pollen or ovule production have already occurred in the unisexual flowers (Renner and Ricklefs 1995). However, the evolution of dioecy via gynodioecy is not a rare phenomenon (Freeman et al. 1997 and references therein). Dioecy and monoecy are especially abundant in the less advanced dicots, and Renner and Ricklefs attribute this to a relative lack of floral morphological constraints, suggesting that the ontogenic suppression of male or female organs is easier in the less advanced groups.

The association of dioecy (and also monoecy) with abiotic pollination has been frequently documented and interpreted as the result of the imprecise pollen movement by wind or water, promoting selfing in self-compatible hermaphrodites and thus selecting for unisexual flowers. Some studies actually reveal changes in the breeding system associated to changes in the pollination mode (see references in Renner and Ricklefs 1995). The relationship between presence of dioecy and climbing growth, whether a woody liana or a herbaceous vine, is attributed by these two investigators to differential selection for optimal resource allocation to sexual function.

Dioecy has shown to evolve from distyly (see Section VI) in several angiosperm genera. In all of these cases, male plants come from short-style plants, whereas females come from long-style ones. Beach and Bawa (1980) hypothesize that this is caused by a change in the pollination biology (pollinator behavior or composition of pollinator species) of the distylous population, which disrupts the complementary pollen flow between the two morphs and results in directional pollen flow from long stamens to long styles, selecting against short styles and stamens.

Some plants are "cryptically" dioecious, that is, they are morphologically hermaphrodite but functionally dioecious, since either males or females, or both, have sterile opposite-sex structures. This phenomenon has been reviewed by Mayer and Charlesworth (1991) and can reveal, at least in part, the importance of resource allocation to reproduction in the evolution of dioecy. Other species are classified as subdioecious, with populations possessing strictly male or female functions and a variable proportion of hermaphrodites; such proportion may vary depending on how favorable growing conditions are, as Sakai and Weller (1991) found in *Schiedea globosa* (Caryophyllaceae).

Willson (1979) was the first to implicate the role of sexual selection in the evolution of dioecy. Separation of sexes may result in a more efficient use of resources for both male and female functions; for instance, an increase in male reproductive effort may result in a disproportionate increase in pollen dispersal, whereas females, by avoiding the cost of pollen dispersal, can invest more resources in seed production. Moreover, females have the advantage of not getting "contaminated" with their own pollen, as can happen in a hermaphroditic plant. It is unknown at the moment how often dioecy has evolved from sexual selection, but at least in some species, such as in the tree *Myristica insipida* (Armstrong and Irvine 1989), this appears to be the case.

The possibility that differential predation (specifically, seed predation or flower herbivory) could be another force selecting for dioecy was hypothesized a long time ago (Janzen 1971, Bawa 1980), although little evidence for this has been accumulated. Sex-related differences in herbivore damage have been reported for some species (Bawa 1980). In the dioecious *Rhamnus ludovici-salvatoris*, for instance, a cecidomyiid fly induces the production of galls only on male flowers (Traveset, unpublished data, 1996). Dioecy in this plant is not complete, i.e., some female plants bear a few male flowers and some male plants are found to produce a few fruits (always less than 1% of the flower crop). Presumably, the female fly can perfectly discriminite between the two flower sexes as they can detect the few male flowers in female individuals and oviposit on them. Male function is not negatively affected as galls are produced after pollen dispersal. This insect herbivore, also present in other congeneric species such as *R. alaternus* (Traveset, personal observation), may have had an important effect favoring the separation of sexes, although it may well be that the insect has appeared in the scenario well after the evolution of dioecy in this group.

In a recent comprehensive review on the evolution of dioecy, Freeman et al. (1997) argued that the two schools of thought (basically, one believing that dioecy has evolved to avoid inbreeding depression and the other that it has been due to differential selection for male and female aspects of reproduction) are correct, but that both mechanisms act on taxa with different life histories, i.e., in different historical contexts, with particular ecological and pollination conditions. The authors hypothesize that in species in which inbreeding depression is prevented (e.g., self-incompatible species or species with herkogamy, dichogamy, monoecy), dioecy has resulted from selection for sexual specialization, whereas in species lacking such devices, dioecy would have evolved via gynodioecy (mainly in insect-pollinated plants), a route that involves a genetic control of gender. According to Freeman et al., the pathway toward dioecy via monoecy (especially common in wind-pollinated species) might be more controlled by ecological factors, since sex-changing and sexual lability occur mostly among species that arose by this pathway and not via gynodioecy. More research on gynodioecious species and on the diversity of ecological factors that control sex expression in general is crucial to either accept or reject these hypotheses.

E. Gynodioecy

The overall frequency of gynodioecy is generally considered to be low (Yampolsly and Yampolsky 1922, Lloyd 1975), although recent studies have shown that it has been overlooked. For females to be maintained in the population, their lack of male function must be compensated with a higher female fitness (a higher quantity and/or quality of seed production). Sex allocation models predict that female fitness must be at least twofold that of hermaphrodites, if inheritance of male-sterility is governed by nuclear genes, although it can be less than double when both nuclear and cytoplasmic genes control gender (Lewis 1942, Lloyd 1975). This compensation has been found in different gynodioecious species such as *Cucurbita foetidissima* (Kohn 1988), *Geranium maculatum* (Agren and Willson 1991), *Thymus vulgaris* (Atlan et al. 1992), *Phacelia linearis* (Eckhart 1992; Eckhart and Chapin 1997), *Chionographis japonica* (Maki 1993), *Prunus mahaleb* (Jordano 1993), *Sidalcea oregana* (Ashman 1994), *Echium vulgare* (Klinkhamer et al. 1994), *Plantago coronopus* (Koelewijn 1996), and *Plantago lanceolata* (Poot 1997). The evolution of this breeding system, as in the case of dioecy, has been usually interpreted as an escape from inbreeding depression, and this may actually be the main selective factor in species such as *C. japonica* (Maki 1993). However, factors other than inbreeding avoidance select for gynodioecy in other species (e.g., *Ocotea tenera*; Gibson and Wheelwright 1996). A different resource allocation to several floral organs, e.g., corolla size (Eckhart 1992a), anther size, nectar production (Delph and Lively 1992) between females and hermaphrodites, might also be important in the evolution of gynodioecy. Gynodioecy seems to be more common in animal-pollinated species than in those that are abiotically pollinated. We certainly need more studies comparing patterns of pollinator visitation, nectar secretion, and phenology between females (male-steriles) and hermaphrodites to understand the evolution of this breeding system.

F. Androdioecy

Androdioecy has been documented in a very few occasions compared with gynodioecy. It is probably the requirements for its maintenance, predicted by sex allocation theory, that makes it a rare and evolutionarily unstable breeding system. Models predict that the frequency of male plants should be much lower than that of hermaphrodites and, as in the case with gynodioecy, the former should be at least twice as fertile (Charlesworth 1984, Lloyd 1975). One of the first documented cases of functional androdioecy was found in *Datisca glomerata* (Liston et al. 1990), presumably derived from a dioecious ancestor (Rieseberg et al. 1992), against traditional predictions that place androdioecy as an intermediate step toward the evolution of dioecy. Another species in which androdioecy has been found to be functional is *Phillyrea angustifolia*, which is sexually labile: in southern France, Lepart and Dommée (1992) reported functional androdioecy in one population and functional dioecy in another one; in southeastern Spain and in the Balearic Islands, this species is functionally androdioecious, although male individuals appear to be much more fertile than hermaphrodites only in the islands (Traveset 1994). In this case, androdioecy seems more likely to evolve into dioecy than the opposite way, perhaps as a means to avoid inbreeding depression, expected to be greater in an island system than in the continent. Androdioecy has recently been reported in *Neobuxbaumia mezcalaenses* (Valiente-Banuet et al. 1997) and in *Mercurialis annua* (Pannell 1997a, 1997b), and future research may actually find it to be more common than previously thought.

VI. SELF-INCOMPATIBILITY SYSTEMS

Systems of self-incompatibility are widely distributed among flowering taxa, having been recorded from approximately 20 orders and more than 70 families of dicots and monocots, with different life-forms, and from tropical as well as temperate zones (Barrett 1988). Here the major classes of self-incompatibility, their general properties, and the hypotheses to explain the evolution of some of these systems are briefly reviewed.

Self-incompatibility can be (1) gametophytic, determined by the genotype of the pollen grain itself, and where all mating types are morphologically similar (homomorphic) or (2) sporophytic, governed by the genotype of the pollen-producing plant, and where mating types can be either homomorphic or heteromorphic. If heteromorphic, there may be two (distyly) or three (tristyly) mating types or morphs, which differ in style length, anther height, pollen size, pollen production, and incompatibility behavior. However, not all heteromorphic species are self-incompatible (Casper 1985, Barrett et al. 1996). Homomorphic and heteromorphic systems do not seem to co-occur in the same plant families, except the Rubiaceae (Wyatt 1983).

In homomorphic systems, incompatibility is usually controlled by a single locus (S) with multiple alleles, although a polygenic control has been claimed in several cases (Barrett 1988). Heterostyly is governed by a single locus with two alleles in distylous species, or by two loci each with two alleles and epistasis operating between them in tristylous plants. Pollen tube growth is inhibited in the style (in most gametophytic systems) or in the stigmatic surface (in most sporophytic systems), probably depending on the time of the S gene action. Inhibition in the ovary (late-acting incompatibility) is also common, although when the rejection is postzygotic, it is difficult to discern its effect from the effect of inbreeding (Seavey and Bawa 1986). This distinction is also important in species such as *Cheiranthus cheiri* (Bateman 1956, Barrett 1988) and *Dianthus chinensis* (Aizen et al. 1990), where self-incompatibility is cryptic, i.e., where tube growth rate is greater for cross- than for self-pollen.

Some genetic models view gametophytic self-incompatibility as the result of certain allelic combinations between mating partners that affect pollen–pistil interactions (Mulcahy and Mulcahy 1983; see also Lawrence et al. 1985). However, we still need more studies on pollen germination, tube growth, and cross-pollination within and between subpopulations at different spatial scales to gain empirical evidence (Barrett 1988). Data on biochemical and physiological traits of incompatibility systems are relatively scarce, although available information supports the idea of an unitary S gene (Barrett 1988), and recent studies have shown that the protein products of S alleles in the pistil control its ability to recognize and reject self-pollen (Kao and McCubbin 1996).

The main selective force acting on the origin and maintenance of incompatibility systems has been hypothesized to be inbreeding depression, presumably strong in selfed progeny due to the expression of recessive deleterious alleles in homozygotes (Barrett 1988). However, some investigators (e.g., Olmstead 1986) see self-incompatibility independent of the level of inbreeding in the population as a whole, arguing that inbreeding is more influenced by small effective population sizes than by selfing avoidance (the primary outcome of self-incompatibility systems). The monophyletic or polyphyletic origin of self-incompatibility systems is not clear either, and there is controversy on whether gametophytic systems are more primitive than sporophytic ones (Zavada and Taylor 1986).

In particular, heterostyly has a polyphyletic origin, having been reported from approximately 25 families of flowering plants (Barrett 1990). The most visible trait in heterostylous plants is the significant difference between morphs in the height at which stigma and anthers are positioned within the flowers. This polymorphism is usually associated with a sporophytically controlled diallelic self-incompatibility system that prevents self and intramorph fertilizations. The inhibition of pollen tube growth can take place in the stigma, style, or ovary

(Barrett 1990). Herbaceous heterostylous taxa such as *Primula, Oxalis, Linum,* or *Lythrum* have received much attention in genetical studies (see references in Barrett 1990), although mostly in controlled experimental conditions. In the last 15–20 years, a number of studies have been conducted on population biology and on structural, developmental, and physiological aspects of heterostylous species, and much information has been accumulated on the function and evolution of heterostyly (comprehensively reviewed in Barrett 1992).

Different hypotheses have been formulated on the sequence of events occurring in heterostylous plants. The classic model (Charlesworth and Charlesworth 1979) assumes that inbreeding avoidance has selected for a diallelic self-incompatibility, followed by evolution of reciprocal herkogamy and appearance of the different floral polymorphisms to increase the efficiency of pollen transfer between incompatible morphs. This has been challenged by Lloyd and Webb (1992), who believe that reciprocal herkogamy evolved first as a result of selection to increase the efficiency of pollen transfer, and that self-incompatibility appears later as a gradual adjustment of pollen tube growth in the different morphs. Lloyd and Webb's hypothesis, in fact, supports that of Darwin, who viewed the style–stamen polymorphism as promoter of disassortative pollination, and evidence for it is being accumulated in studies of pollen deposition patterns on the stigmas of the different morphs (Lloyd and Webb 1992; Kohn and Barrett 1992).

VII. PATERNAL SUCCESS

For many years, studies of plant reproductive success were strongly biased by examining only the female function (Willson 1994, Schlichting and Delesalle 1997). However, in the last two decades, different aspects of male reproductive success, such as pollen production, pollen removal, and paternity of offspring, have been examined in a number of studies (Bertin 1988, Snow and Lewis 1993). Male fitness is usually expressed in terms of the number of sired offspring surviving to reproductive age. As this is very difficult to measure, or even impossible in many cases, only correlates of fitness—such as pollen germination ability, pollen tube growth rate, ability of pollen to effect fertilization, weight and number of seeds sired, seed germination, and performance of sired seedlings—are usually evaluated. Current research to develop molecular markers for measuring male reproductive success will hopefully help to understand fitness returns from investment in male function (Barrett and Harder 1996).

Resources allocated to male function include those that produce a certain number of pollen grains, of a certain size, and male accessory structures (e.g., petals, sepals, bracts) and substances (e.g., nectar). Such resource allocation may be linked to male fitness, although we still have little experimental evidence supporting this (Bertin 1988, Young and Stanton 1990). In measuring male fitness, it is important to quantify pollen removal but also to monitor the success of removed pollen, as these two variables may not be positively correlated (e.g., Wilson and Thomson 1991; see also Conner et al. 1995); for instance, a bee removing much pollen from a nectar-rich plant may fly short distances or promote much geitonogamy, which may limit potential gains in male fitness.

Pollen longevity is variable both among and within species, is strongly affected by environmental conditions, and does not seem to be relevant for male reproductive success (Bertin 1988). However, male success is influenced by the percentage of pollen grains germinating, by the rapidity of germination on the stigma, and by pollen tube growth rate in the style, all of which in turn are affected by abiotic factors, especially temperature (Bertin 1988, Murcia 1990). The success of a particular pollen grain also depends on the composition and size of the whole pollen load on the stigma. Several studies on pollen tube growth rate have found that the presence of self- or incompatible pollen has a negative effect on tube growth of cross-compatible pollen (Shore and Barrett 1984). The competitive ability of a pollen grain is influenced by its own genotype, which differs among individuals and among pollen grains from the same individual; thus, the genotype of all pollen grains on the stigma influences the success of a particular one (Bookman 1984). Large pollen loads can be advantageous over small loads because the former are more likely to enhance pollen germination as well as tube growth rate (Ter-Avanesian 1978); however, large pollen loads may yield fewer pollen tubes per pollen grain than small loads (Snow 1986) and, thus, the probability that a certain pollen grain is represented in the seed crop can also be lower for large pollen loads. The sequence of pollen deposition on the stigma has also shown to be important in determining the proportion of seeds sired by the different pollen grains (Mulcahy et al. 1983), and it affects the potential for interaction (competition) among grains, which may have been brought by different pollinators (Murcia 1990). In a recent study on *Hibiscus moscheutos*, Snow and Spira (1996) gave strong evidence that pollen-tube competitive ability varies among coexisting plants, arguing that it may be a relevant component of the male fitness in plants. Pollen grains from different donors on the stigma not only race for access to ovules (exploitation competition) but can also interfere with the germination and growth of each other (interference competition), as it has been found in wild radish (Marshall et al. 1996) and in *Palicourea* (Murcia and Feinsinger 1996).

For hermaphroditic plants, the male function has been predicted to be limited by mating opportunities and not by resources, whereas the opposite is expected for the female function. This hypothesis has been termed the "fleurs-du-mâle" hypothesis (Queller 1983), also known as the pollen donation hypothesis (PDH) (Fishbein and Venable 1996a, Broyles and Wyatt 1997). According to the PDH, large floral displays would especially benefit the male function as they would have a greater fraction of their pollen exported. Some investigators even believe that flower number in the angiosperms has been selected by such male function (Sutherland and Delph 1984). Most experimental studies have been performed with milkweeds (*Asclepias* spp.), but there is still controversy on whether the PDH is well supported (Queller 1997) or not (Broyles and Wyatt 1997) in this group. More studies on these and other species, with adequate experimental designs and controlling for variables such as level of resources, are needed to determine whether male function does select for large floral displays. We must also know which are the consequences of self-versus cross-pollination, as large floral displays may be less efficient at exporting pollen if pollinators promote geitonogamy (de Jong et al. 1993). Theoretically, if female fitness (achieved via fruit production) is less affected by geitonogamy than male fitness (achieved via siring of fruits on other plants), we would predict that small plants invest more in male reproduction, whereas large plants emphasize more the female function. Some data seem to support this prediction (de Jong et al. 1993).

By evaluating both female and male reproductive success, it is possible to examine (1) whether they are correlated, (2) the genetic variation for female and male components of fitness, and (3) whether the components of male and female reproductive success are equally affected by environmental factors. Some studies have documented genetic variation in both male and female functions, but there seems to be no consistent pattern for strong positive or negative correlations between them (Schlichting and Devlin 1992, Mutikainen and Delph 1996, Strauss et al. 1996). Likewise, a survey on the consequences of herbivory (and the same is found with other stresses) on male and female functions shows that these are neither equal nor proportional (Mutikainen and Delph 1996), although we still need more data that evaluate the plastic responses of male and female components to different environmental factors. Studies on functional architecture are also necessary to estimate the genetic and phenotypic correlations between both quantitative and qualitative aspects of male and female functions.

VIII.   ROLE OF POLLINATORS IN THE EVOLUTION OF FLORAL TRAITS AND DISPLAY

The evolution of plant mating systems has undoubtedly been linked to the evolution of traits that influence the type of pollination (animal vs. wind pollination) and pollinator's attraction to fertilize flowers (e.g., quantity and quality of floral rewards, petal coloration, flower size, flowering time). By producing large floral displays or great amounts of nectar, for instance, plants can affect the behavior of pollinators, which in turn influences gene flow among plants and, ultimately, plant fitness (see review in Zimmerman 1988). The goal of numerous studies on pollination biology has been to find the adaptedness of floral characters to particular pollinating vectors and to identify pollination syndromes, i.e., suites of structural and functional floral traits that presumably reflect adaptations to different types of pollinating agents (Proctor and Yeo 1973, Faegri and Van der Pijl 1979). The variation in floral characters within a species, and its association with the variation in reproductive success, has received much less attention. Some of these studies have demonstrated phenotypic selection on certain floral traits (Nilsson 1988, Campbell 1989, Galen 1989, Schemske and Horvitz 1989, Robertson and Wyatt 1990, Cresswell and Galen 1991, Herrera 1993, Andersson and Widén 1993, Johnson and Steiner 1997), but we still do not know how frequently this occurs in nature (Herrera 1996). In a survey of 58 insect-pollinated species from a region in southeastern Spain, for instance, this author found that patterns of phenotypic variation in corolla depth were not consistent with the idea of generalized selection by pollinators on this floral trait (Herrera 1996). Likewise, in a study on flower divergence, Wilson and Thomson (1996) found no evidence that floral differences are initiated as adaptations to particular pollinators. Some floral characters may not represent adaptations to current pollinators, but exaptations (Gould and Vrba 1982), having evolved as a consequence of selection by pollinators that are now extinct or not present in the current scenario. Several examples have been reported in the literature. One of them is the case of the liana *Freycinetia arborea* in Hawaii. This species produces flowers with deep corolla tubes that were probably adapted to the pollination by the extinct birds in the Drepanididae family; currently, it is successfully pollinated by *Zosterops japonica*, a bird introduced from Japan (Cox 1983).

A certain floral character is expected to be selected for by a specific pollinator (or by an assemblage of pollinators) if it significantly increases the fitness of the plant possessing such character. To determine the contribution of a pollinator to plant fitness (i.e., the pollinator effectiveness), it is essential to (1) quantify the number of flowers it pollinates (quantitative component of pollination) and (2) evaluate its efficiency as a pollinator (qualitative component). The former depends on the frequency of pollinator visits to a plant and on the flower visitation rate, whereas the latter is a function of the pollen delivered to stigmas, the foraging patterns, and the selection of floral sexual stage by the pollinator (Herrera 1988). The pollinator molding floral traits is expected to be the most effective (Stebbins 1970), especially when pollen is limiting. Correlation between frequency of visitation and efficiency is not usual (Bertin and Willson 1980, Montalvo and Ackerman 1986, Herrera 1988, Ramsey 1988, Schemske and Horvitz 1989, Pettersson 1991, Guitián et al. 1993, García et al. 1995, Pellmyr and Thompson 1996), and effectiveness is determined by the product of the two components.

The strength of selection of floral traits by pollinators and/or the plant's response to such selection may be limited by factors that are either intrinsic (genetic or life history) or extrinsic (environmental) to the plant (Herrera 1996). Among the latter, the spatio-temporal variation in the composition of pollinator assemblages and in their relative abundance is probably the most important factor precluding or strongly reducing selection on floral traits by pollinators. Differences at a spatial and/or at a temporal scale in the assemblage of pollinators have been frequently documented (Herrera 1988, 1995, Sih and Baltus 1987, Wolfe and Barrett 1989, Schemske and Horvitz 1989, Horvitz and Shemske 1990, Eckhart 1992b, Vaughton 1992, Cane and Payne 1993, Conner and Neumeier 1995, Fishbein and Venable 1996b, Traveset and Sáez 1997) and can create a mosaic of selective regimes (Herrera 1988, Thompson 1994). If such mosaic is at a small scale, e.g., within the same geographical area where there is gene flow among plants, selection on floral traits will probably be much weakened. Therefore, to understand the evolution of a particular character, it is important to evaluate such variation in space (embracing the whole species distributional range if possible) and in time (within and between seasons).

Even if phenotypic selection on a floral trait occurs, it may have a small effect on individual variation in maternal fitness relative to that of other factors, such as plant size, herbivory, seed dispersal success, etc. For instance, in both *Calathea ovandensis* (Schemske and Horvitz 1989) and *Viola cazorlensis* (Herrera 1993), individual variation in floral morphology accounted for less than 10% of the variance in fruit production. As mentioned in the previous section, however, selection may occur via the male function, and thus it is necessary to examine both female and male fitness to determine if phenotypic selection is important (Primack and Kang 1989, Devlin and Ellstrand 1990, Dudash 1991, Conner et al. 1996). In species that produce more than one flower, the operational unit of either male or female function, when determining the effect of phenotypic selection of a floral trait on maternal fitness, has to be the whole floral display because, as mentioned previously, the level of geitonogamy determines the incidence of self-pollination and pollen discounting, and ultimately the plant's mating success (Harder and Barrett 1996).

A relevant question that we can ask is *when* is it more likely that pollinators exert phenotypic selection on floral traits and/or display? We may think of at least four necessary conditions: (1) effective pollinators must be few and more or less constant in space and time; (2) life history or genetic constraints need to be negligible; (3) pollination must be more important than other ecological factors, either abiotic (microhabitat, drought, etc.) or biotic (seed predation, pathogens, dispersal success, etc.) determining male and female fitness; and (4) variation in the selected character must be correlated with variation in plant reproductive success.

In species in which geitonogamy or selfing decreases fitness significantly, pollinator behavior can be very important and will probably select for (1) a smaller floral display, (2) a low nectar production per plant, and/or (3) the separation in time or space of sexual functions within an individual plant.

In animal-pollinated plants, the absence or dramatic reduction of pollinators in an area can also influence mating success and select for floral traits that promote selfing. This seems likely to happen in species that have colonized islands or high-mountain habitats where pollinator abundance is low. As mentioned in Section IV, selfing might evolve to assure reproduction, although inbreeding depression could represent an important constraint. If, for some reason, this does not purge the deleterious alleles from the population under certain circumstances (e.g., harsh environmental conditions), selfing would be detrimental (Harder and Barrett 1996). This might explain, for instance, the high incidence of self-incompatible species in high altitudes in Patagonia (Arroyo and Squeo 1990, Jarne and Charlesworth 1993). The reproductive assurance hypothesis, as Harder and Barrett (1996) claim, does need more testing, as only a few studies have examined the correlation between pollinator activity and levels of selfed seeds at the population level (all supporting the hypothesis), and no studies have tested it at the plant or flower level. Even if common, pollinators may also affect the levels of selfing, and thus offspring quality, by being inefficient (Harder and Barrett 1996). For instance, by moving frequently among different plant species, they (1) transport pollen between them, which interferes with fertilization by conspecific pollen (e.g., Thomson et al. 1981, Armbruster and Herzig 1984, Harder et al. 1993) and (2) lose pollen on foreign stigmas (Campbell 1985, Feinsinger and Tiebout 1991).

The view that pollination systems tend toward specialization and that pollinator specialization is critical to plant speciation and evolutionary radiation has been implicit in most pollination biology studies (Grant 1949, Baker 1963, Grant and Grant 1965, Stebbins 1970, Crepet 1983). Recently, however, such tendency has been questioned by Waser et al. (1996), who argue that pollination systems are more generalized and dynamic than previously believed, both in temperate and tropical habitats. Geographical ranges of plants and pollinators are found to rarely match, which indicates that their interactions are not obligate (Thompson 1994). Moreover, pollinators may invade new areas, even replacing the extinct ancestral mutualist, as in the case of *Freycinetia* (Cox 1983). Even the obligate interactions between figs, yuccas, and their pollinators may not be absolutely obligate (Waser et al. 1996, Pellmyr et al. 1996). Herrera's data (1996) also show a relatively loose association of plants to particular insect orders, which supports the idea of a low degree of floral specialization to particular pollinators. The concept of pollination syndrome, in fact, has proven sometimes to be of little use when predicting the pollinators of a certain plant species and when explaining interspecific variation in pollinator composition (Herrera 1996 and references therein). In habitats where pollinators are not common, it is not rare to find plants with both abiotic and biotic pollinating agents (Gómez and Zamora 1996; Traveset, unpublished data, 1995, 1996). Future studies on the pollination systems of different floras, both in the tropics and in the temperate zones, that examine the spatio-temporal variation in pollinator assemblages and in pollen limitation (Dudash and Fenster 1997) are crucial to determine how often and in what conditions plants specialize to particular pollinator agents.

IX.   INFLUENCE OF BIOTIC POLLINATION IN ANGIOSPERM DIVERSIFICATION

A long-standing question in the study of plant evolution is how and to what extent the emergence of animal pollination has driven the great and rapid speciation of flowering plants. Different investigators (Raven 1977, Regal 1977, Burger 1981, Crepet et al. 1991, Eriksson and Bremer 1992) have argued that animal pollinators, referring mostly to the high taxonomic and behavioral diversity of insects, may have importantly influenced the rate of angiosperm diversification by (1) promoting genetic isolation of plant populations through mechanical or ethological mechanisms (see review in Grant 1994), (2) promoting outcrossing, so genetically diverse populations may undergo rapid phyletic evolution, and (3) reducing extinction rates, as they move pollen across long distances among sparse populations (Regal 1977). Similarly, it has been suggested that the biotic dispersal of seeds, mainly referring to dispersal by vertebrates, also has contributed to some extent to angiosperm diversification (Tiffney and Mazer 1995 and references therein).

In contrast to the view that biotic dispersal of pollen and/or seeds has caused or favored the speciation of angiosperms, other investigators (Stebbins 1981, Midgley and Bond 1991, Doyle and Donoghue 1993, Ricklefs and Renner 1994) believe that morphological and physiological characters in flowering plants have played a more important role in their diversification, and that the reason why angiosperms have been more successful than gymnosperms may be more a result of factors such as the greater plasticity in (1) growth forms, (2) type of habitats they can inhabit, (3) ways to exploit the environmental resources, (4) types of reproduction (vegetative reproduction is very rare in gymnosperms), and (5) possibly even types of breeding system, usually less complex in gymnosperms. However, as Ricklefs and Renner (1994) point out, it is important to consider that factors affecting the displacement of gymnosperms by angiosperms may not be the same as those affecting their diversification rate. Angiosperms may be competitively superior for different causes: efficiency of water use in particular dry environments, efficiency of insect pollination in habitats where wind is nearly or totally absent, rapid growth, double fertilization, capacity of vegetative reproduction, etc. However, it is plausible that angiosperm diversification has promoted their proliferation in some habitats and under some circumstances (by being more diverse, angiosperms may be more likely to survive and propagate after an environmental stress such as a period of drought, which may be devastating for a species of gymnosperm).

There is empirical evidence to support both viewpoints, although comprehensive data sets have been used in only a few recent studies (Ricklefs and Renner 1994, Tiffney and Mazer 1995). Previously, Eriksson and Bremer (1992) had evaluated the effect of biotic versus wind pollination on speciation by comparing the rates of diversification in a subgroup of angiosperms (147 families) from their first occurrence in the fossil record and the number of extant species in each family, finding that such rate was significantly higher for insect-pollinated families. However, that work suffers from their assumption that rates of evolution and extinction have been constant through time and from their estimates of first occurrence of families in the fossil record (Crepet et al. 1991, Ricklefs and Renner 1994). Ricklefs and Renner (1994) showed that the earliest fossils provide unreliable estimates of age and that diversification is not absolutely exponential in time. Their data base included all families of flowering plants (n = 365), each with its number of genera and species, geographical distribution, growth form, pollination system, and mode of seed dispersal. Families with animal pollination were found to be richer in species than those with abiotic pollination, although the significance level was much lower than for the effects of geographical distribution and growth form (no effect of biotic seed dispersal was detected). The four variables accounted for only 41% of the variation in species richness, and thus most of the variance was still unexplained. From their results, Ricklefs and Renner concluded that the major factor contributing to speciation is probably the capacity of taxa to exploit a wide range of ecological opportunities by adopting different growth forms and life histories and by differentiating morphologically to be pollinated and dispersed by different vectors (biotic and abiotic). In contrast, the study by Tiffney and Mazer (1995) does show an important effect of biotic dispersal of seeds in angiosperm diversification (unfortunately, they do not include pollination systems in their analysis). The reason for such conflicting results is attributed, by these two authors, to the pooling of angiosperms with different growth forms or other traits, which masks differences among various groups. They perform separate analyses for woody and herbaceous monocots and dicots, finding that dispersal by vertebrates contributes to species richness in woody dicots, and that biotically dispersed families exhibit higher levels of diversification in herbaceous monocots and dicots than vertebrate-dispersed families. Therefore, the possibility exists that the effect of biotic dispersal of both pollen and seeds was underestimated in Ricklefs and Renner's analyses. Other potential problems with this kind of analyses, pointed out by Bawa (1995), are (1) the use of families, rather than genera or species, as independent units; and (2) the broad classification of pollination and seed dispersal into biotic and abiotic categories, as both categories are very heterogeneous. Further analyses that include more variables that might influence diversification (capacity of asexual reproduction, size of flower, fruit and seed, specificity of pollinators, etc.) will certainly reveal new patterns and probably help to explain a larger fraction of the variation in species richness among taxa.

With the information gathered to date, we can be almost sure that insect pollination has relevantly contributed to the diversification of some of the most speciose families (e.g., orchids), but we need much more data to determine its role on the massive mid-Cretaceous angiosperm diversification (Crane et al. 1995). What is certain is that insect pollination was already present when angiosperms originated (Early Cretaceous, approximately 130 and 90 million years ago), as shown by Jurassic fossils of Bennetitales, the closest fossil group to angiosperms, that suggest the presence of a plant–pollinator interaction (Crepet et al. 1991, Crane et al. 1995). The androecium in early angiosperms probably served as the only reward for insects, as it occurred in the Bennetitales, and flowers were presumably small, apetalous, with few structures, either asymmetric or cyclically arranged (Crepet et al. 1991). Such early flowers co-occurred with a greater variety of insects than previously thought. According to Crepet et al., the idea that Coleoptera were the main early pollinators needs to be reviewed, as other insect groups (e.g., pollen-chewing flies and micropterygid moths) were also present at that time. Nectaries appeared later and were present in many of the late Cretaceous rosids, when a rapid radiation of bees took place (Crepet et al. 1991). The radiation of Lepidoptera also coincides with patterns of accelerating radiation in angiosperms, yet Pellmyr (1992) found no evidence that the evolution of this group of insects, for which there exists a reliable phylogeny and information on pollinator function, caused radiation in flowering plants.

Palynological data indicate that the initial increase in angiosperm diversity occurred in low palaeolatitude areas, which were under semiarid or seasonally arid conditions during the Early Cretaceous (Crane et al. 1995). Such conditions might have promoted, according to Crane et al., a "weedy life history with precocious reproduction, . . . and associated effects such as simplication and aggregation of sporophylls to form a flower, enclosure of ovules in a carpel, truncation of the gametophyte phase of the life cycle, and major reorganization of leaf and stem anatomy." Some of these attributes, in turn, together with a more rapid and flexible vegetative growth, might have contributed to the success of angiosperms (perhaps at the base of the angiosperm clade or within a subgroup) and to their rate of speciation. Interestingly, Crane et al. (1995) claim that explanations of angiosperm diversification need to consider the effects of environmental changes during the Early and mid-Cretaceous, such as the high rates of sea-floor spreading, high sea-level stands, and probably high global temperatures, which have been overlooked so far.

X. CONCLUDING REMARKS

If Sprengel were alive today, he would certainly be astonished by the amount of information that has been accumulated on plant reproduction since he conducted his excellent studies on plant natural history. But he would probably be surprised, too, by the fact that almost 200 years of observations, experiments, and modelling have been necessary for plant-reproduction biologists to realize that the study of mating systems and pollination biology cannot be two independent disciplines, that they need much from each other, and that a unified approach is crucial to understand the evolutionary dynamics of plant reproduction. The only explanation for such delay in merging these two fields is probably the inertia from both sides, i.e., students of breeding systems performing genetical studies without observing the pollinators visiting the flowers in nature, and ecologists examining the mutualistic interaction from different viewpoints, but without considering the genetic consequences of pollen transfer, the incompatibility systems, etc. Whatever the reason for the old disconnection between the two areas, the recently emerged new synthesis (Lloyd and Barrett 1996) is expected to solve problems and answer questions. The study of factors that influence pollen transfer (floral morphology, timing of self-vs. outcross pollination, pollinator's effectiveness, etc.) will give valuable information to develop new models that faithfully describe the movement of pollen within and among flowers and reflect the variety of plant–pollinator interactions; these models will certainly help to understand and compare the evolutionary dynamics in the different pollination systems (Holsinger 1996). On the other hand, more studies designed to detect natural and sexual selection on floral traits and display will help to determine the frequency of occurrence of floral adaptations to pollinators (Herrera 1996). More data on whole pollinator assemblages visiting a plant species, examining the spatio-temporal variation in their composition and effectiveness, will also permit evaluation of the degree of plant specialization to pollinators, test whether pollination systems are indeed more generalized than previously thought (Waser et al. 1996), and provide knowledge of the extent to which biotic pollination may influence angiosperm diversification.

If we could come back 200 years from now, we might also be amazed at how much plant reproductive biologists will have learned about (1) asexual reproduction, (2) evolution of selfing and different mating systems, (3) incompatibility systems, (4) the male functions of plants, (5) how, how often, and in what conditions pollinators select for particular floral traits, and (6) what has been the relative importance of biotic pollination on angiosperm speciation, as more reliable phylogenetic trees of plants and pollinators will have been built with both morphological and genetic data, and more information on the other factors affecting diversification will be available.

ACKNOWLEDGMENTS

I especially thank Mary F. Willson for her comments on the manuscript, and Marta Macíes for her efficient help in obtaining references. Rodolfo has been a constant source of encouragement throughout the period I have been writing, and I thank him for his patience.

REFERENCES

Abrahamson, W.G. 1980. Demography and vegetative reproduction. In: O.T. Solbrig, O.T., ed. Demography and Evolution in Plant Populations. London: Blackwell Scientific, pp 89–106.

Agren, J., and D.W. Schemske. 1995. Sex allocation in the monoecious herb *Begonia semiovata*. Evolution 49: 121–130.

Agren, J., and M.F. Willson. 1991. Gender variation and sexual differences in reproductive characters and seed production in gynodioecious *Geranium maculatum*. American Journal of Botany 78: 470–480.

Aizen, M., K.B. Searcy, and D.L. Mulcahy. 1990. Among- and within-flower comparisons of pollen tube growth following self-and cross-pollinations in *Dianthus chinensis* (Caryophyllaceae). American Journal of Botany 77: 671–676.

Anderson, G.J., and D.E. Symon. 1989. Functional dioecy and andromonoecy in *Solanum*. Evolution 43: 204–219.

Andersson, S., and B. Widen. 1993. Pollinator-mediated selection on floral traits in a synthetic population of *Senecio integrifolius* (Asteraceae). Oikos 66: 72–79.

Antonovics, J. 1968. Evolution of closely adjacent plant populations. V. Evolution of self-fertility. Heredity 23: 219–238.

Armbruster, W.S., and A.L. Herzig. 1984. Partitioning and sharing of pollinators by four sympatric species of *Dalechampia* (Euphorbiaceae) in Panama. Annals of the Missouri Botanical Garden 71: 1–16.

Armstrong, J.E., and A.K. Irvine. 1989. Flowering, sex ratios, pollen-ovule ratios, fruit set, and reproductive effort of a dioecious tree, *Myristica insipida* (Myristicaceae), in two different rain forest communities. American Journal of Botany 76: 74–85.

Arroyo, M.T.K., and F. Squeo. 1990. Relationship between plant breeding systems and pollination. In: Kawano, S., ed. Biological Approaches and Evolutionary Trends in Plants. New York: Academic Press, pp 205–227.

Arroyo, M.T.K., and P. Uslar. 1993. Breeding systems in a temperate mediterranean-type climate montane sclerophyllous forest in central Chile. Botanical Journal of the Linnean Society 111: 83–102.

Ashman, T.L. 1994. Reproductive allocation in hermaphrodite and female plants of *Sidalcea oregana ssp. spicata* (Malvaceae) using four currencies. American Journal of Botany 81: 433–438.

Asker, S. 1980. Gametophytic apomixis: elements and genetic regulation. Hereditas 93: 277–293.

Atlan, A., P.H. Gouyon, T. Fournial, D. Pomente, and D. Couvet. 1992. Sex allocation in an hermaphroditic plant: the case of gynodioecy in *Thymus vulgaris* L. Journal of Evolutionary Biology 5: 189–203.

Baker, H.G. 1963. Evolutionary mechanisms in pollination biology. Science 139: 877–883.

Barrett, S.C.H. 1988. The evolution, maintenance, and loss of self-incompatibility systems. In: Lovett Doust, J., and L. Lovett Doust, eds. Plant Reproductive Ecology. Patterns and Strategies. Oxford: Oxford University Press, pp 98–124.

Barrett, S.C.H. 1990. The evolution and adaptive significance of heterostyly. Trends in Ecology and Evolution 5: 144–148.

Barrett, S.C.H. 1992. Evolution and Function of Heterostyly. Berlin: Springer-Verlag.

Barrett, S.C.H., and D. Charlesworth. 1991. Effects of a change in the level of inbreeding on the genetic load. Nature 352: 522–524.

Barrett, S.C.H., and C.G. Eckert. 1990. Variation and evolution of plant mating systems. In: Kawano, S., Biological Approaches and Evolutionary Trends in Plants. New York: Academic Press, pp 229–254.

Barrett, S.C.H., and L.D. Harder. 1996. Ecology and evolution of plant mating. Trends in Ecology and Evolution 11: 73–82.

Barrett, S.C.H., L.D. Harder, and A.C. Worley. 1996. The comparative biology of pollination and mating in flowering plants. Philosophical Transactions of the Royal Society of London (Series B) 351: 1271–1280.

Bateman, A.J. 1956. Cryptic self-incompatibility in the wallflower. I. Theory. Heredity 10: 257–261.

Bawa, K.S. 1980. Evolution of dioecy in flowering plants. Annual Review of Ecology and Systematics 11: 15–39.

Bawa, K.S. 1995. Pollination, seed dispersal and diversification of angiosperms. Trends in Ecology and Evolution 10: 311–312.

Bawa, K.S., and J.H. Beach. 1980. Evolution of dioecy in flowering plants. Annual Review of Ecology and Systematics 11: 15–39.

Beach, J.H., and K.S. Bawa. 1980. Role of pollinators in the evolution of dioecy from distyly. Evolution 34: 1138–1142.

Berry, P.E., H. Tobe, and J.A. Gómez. 1991. Agamospermy and the loss of distyly in *Erythroxylum undulatum* (Erythroxilaceae) from northern Venezuela. American Journal of Botany 78: 595–600.

Bertin, R.I. 1982. The evolution and maintenance of andromonoecy. Evolutionary Theory 6: 25–32.

Bertin, R.I. 1988. Paternity in plants. In: Lovett Doust, J., and L. Lovett Doust, eds. Plant Reproductive Ecology. Patterns and Strategies. Oxford: Oxford University Press, pp 30–59.

Bertin, R.I., and C.M. Newman. 1993. Dichogamy in angiosperms. Botanical Review 59: 112–152.

Bertin, R.I., and M.F. Willson. 1980. Effectiveness of diurnal and nocturnal pollination of two milkweeds. Canadian Journal of Botany 58: 1744–1746.

Bino, R.J., and A.D.J. Meeuse. 1981. Entomophily in dioecious species of *Ephedra*: a preliminary report. Acta Botanica Neerlandica 30: 151–153.

Bookman, S.S. 1984. Evidence for selective fruit production in *Asclepias*. Evolution 38: 72–86.

Broyles, S.B., and R. Wyatt. 1997. The pollen donation hypothesis revisited: a response to Queller. American Naturalist 149: 595–599.

Burd, M. 1995. Ovule packaging in stochastic pollination and fertilization environments. Evolution 49: 100–109.

Burd, M., and T.F.H. Allen. 1988. Sexual allocation strategy in wind-pollinated plants. Evolution 42: 403–407.

Burger, W.C. 1981. Why are there so many kinds of flowering plants? Bioscience 31: 572–581.

Campbell, D.R. 1985. Pollen and gene dispersal: the influences of competition for pollination. Evolution 39: 419–431.

Campbell, D.R. 1989. Measurements of selection in a hermaphroditic plant: variation in male and female pollination success. Evolution 43: 318–334.

Campbell, D.R., and N.M. Waser. 1987. The evolution of plant mating systems: multilocus simulations of pollen dispersal. American Naturalist 129: 593–609.

Cane, J.H., and J.A. Payne. 1993. Regional, annual, and seasonal variation in pollinator guilds: intrinsic traits of bees (Hymenoptera: Apoidea) underlie their patterns of abundance at *Vaccinium ashei* (Ericaceae). Annals of the Entomological Society of America 86: 577–588.

Casper, B.B. 1985. Self-compatibility in distylous *Cryptantha flava* (Boraginaceae). New Phytologist 99: 149–154.

Charlesworth, B. 1980. The cost of sex in relation to mating system. Journal of Theoretical Biology 84: 655–671.

Charlesworth, B., and D. Charlesworth. 1978. A model for the evolution of dioecy and gynodioecy. American Naturalist 112: 975–997.

Charlesworth, B., and D. Charlesworth. 1979. The maintenance and breakdown of heterostyly. American Naturalist 114: 499–513.

Charlesworth, D. 1984. Androdioecy and the evolution of dioecy. Biological Journal of the Linnean Society 23: 333–348.

Charlesworth, D., and B. Charlesworth. 1981. Allocation of resources to male and female functions in hermaphrodites. Biological Journal of the Linnean Society 15: 57–74.

Charlesworth, D., and B. Charlesworth. 1987a. Inbreeding depreesion and its evolutionary consequences. Annual Review of Ecology and Systematics 18: 237–268.

Charlesworth, D., and B. Charlesworth. 1987b. The effect of investment in attractive structures on allocation to male and female functions in plants. Evolution 41: 948–968.

Charlesworth, D., and M.T. Morgan. 1991. Allocation of resources to sex functions in flowering plants. Philosophical Transactions of the Royal Society of London (Series B) 332: 91–102.

Charnov, E.L. 1982. The Theory of Sex Allocation. Princeton, NJ: Princeton University Press.

Conner, J.K., and R. Neumeier. 1995. Effects of black mustard population size on the taxonomic composition of pollinators. Oecologia 104: 218–224.

Conner, J.K., and R. Davis, and S. Rush. 1995. The effect of wild radish floral morphology on pollination efficiency by four taxa of pollinators. Oecologia 104: 234–245.

Conner, J.K., S. Rush, S. Kercher, and P. Jennetten. 1996. Measurements of natural selection on floral traits in wild radish (*Raphanus raphanistrum*). II. Selection through lifetime male and total fitness. Evolution 50: 1137–1146.

Costich, D.E. 1995. Gender specialization across a climatic gradient: experimental comparison of monoecious and dioecious. *Ecballium*. Ecology 76: 1036–1050.

Cox, P.A. 1983. Extinction of the Hawaiian avifauna resulted in a change of pollinators for the ieie, *Freycinetia arborea*. Oikos 41: 195–199.

Crane, P.R., E.M. Friis, and K.R. Pedersen. 1995. The origin and early diversification of angiosperms. Nature 374: 27–33.

Crepet, W.L. 1983. The role of insect pollination in the evolution of the angiosperms. In: Real, L., ed. Pollination and Biology. Orlando, FL: Academic Press, pp 29–50.

Crepet, W.L., E.M. Friis, and K.C. Nixon. 1991. Fossil evidence for the evolution of biotic pollination. Philosophical Transactions of the Royal Society of London (Series B) 333: 187–195.

Cresswell, J.E., and C. Galen. 1991. Frequency-dependent selection and adaptive surfaces for floral character combinations: the pollination of *Polemonium viscosum*. American Naturalist 138: 1342–1353.

Cronquist, A. 1981. An Integrated System of Classification of Flowering Plants. New York: Columbia University Press.

Cruden, R.W. 1977. Pollen-ovule ratios: a conservative indicator of breeding systems in flowering plants. Evolution 31: 32–46.

Cruden, R.W., and D.L. Lyon. 1985. Patterns of biomass allocation to male and female functions in plants with different mating systems. Oecologia 66: 299–306.

Damgaard, C., and R.J. Abbott. 1995. Positive correlations between selfing rate and pollen-ovule ratio within plant populations. Evolution 49: 214–217.

Darlington, D.D. 1939. The Evolution of Genetic Systems. Cambridge: Cambridge University Press.

Darwin, C.R. 1876. The Effects of Cross and Self Fertilization in the Vegetable Kingdom. London: John Murray.

De Jong, T.J., N.M. Waser, and P.G.L. Klinkhamer. 1993. Geitonogamy: the neglected side of selfing. Trends in Ecology and Evolution 8: 321–325.

Delph, L.F., and C.M. Lively. 1992. Pollinator visitation, floral display, and nectar production of the sexual morphs of a gynodioecious shrub. Oikos 63: 161–170.

Devlin, B., and N.C. Ellstrand. 1990. Male and female fertility variation in wild radish, a hermaphrodite. American Naturalist 136: 87–107.

Diggle, P.K. 1993. Developmental plasticity, genetic variation, and the evolution of andromonoecy in *Solanum hirtum* (Solanaceae). American Journal of Botany 80: 967–973.

Doyle, J.A. 1994. Origin of the angiosperm flower: a phylogenetic perspective. Plant Systematics and Evolution Suppl 8: 7–29.

Doyle, J.A., and M.J. Donogue. 1993. Phylogenies and angiosperm diversification. Paleobiology 19: 141–167.

Dudash, M.R. 1991. Plant size effects on female and male function in hermaphroditic *Sabatia angularis* (Gentianaceae). Ecology 72: 1004–1012.

Dudash, M.R., and C.B. Fenster. 1997. Multiyear study of pollen limitation and cost of reproduction in the iteroparous *Silene virginica*. Ecology 78: 484–493.

Eckhart, V.M. 1992a. The genetics of gender and the effects of gender on floral characters in gynodioecious *Phacelia linearis* (Hydrophyllaceae). American Journal of Botany 79: 792–800.

Eckhart, V.M. 1992b. Spatio-temporal variation in abundance and variation in foraging behavior of the pollinators of gynodioecious *Phacelia linearis* (Hydrophyllaceae). Oikos 64: 573–586.

Eckhart, V.M., and F.S. Chapin III. 1997. Nutrient sensitivity of the cost of male function in gynodioecious *Phacelia linearis*. American Journal of Botany 84: 1092–1098.

Ehrlén, J. 1991. Why do plants produce surplus flowers? A reserve-ovary model. American Naturalist 138: 918–933.

Ellstrand, N.C., R. Ornduff, and J.M. Clegg. 1990. Genetic structure of the Australian Cycad, *Macrozamia communis* (Zamiaceae). American Journal of Botany 77: 677–681.

Emms, S.K. 1993. Andromonoecy in *Zigadenus paniculatus* (Liliaceae): spatial and temporal patterns of sex allocation. American Journal of Botany 80: 914–923.

Emms, S.K. 1996. Temporal patterns of seed set and decelerating fitness returns on female allocation in *Zigadenus paniculatus* (Liliaceae), an andromonoecious lily. American Journal of Botany 83: 304–315.

Eriksson, O., and B. Bermer. 1992. Pollination systems, dispersal modes, life forms, and diversification rates in angiosperm families. Evolution 46: 258–266.

Faegri, K., and L. van der Pijl. 1979. The Principles of Pollination Ecology. Third Revised Edition. Oxford: Oxford University Press.

Feinsinger, P. and H.M. Tiebout III. 1991. Competition among plants sharing hummingbird pollinators: laboratory experiments on a mechanism. Ecology 72: 1946–1952.

Fishbein, M., and D.L. Venable. 1996a. Evolution of inflorescence design: theory and data. Evolution 50: 2165–2177.

Fishbein, M., and D.L. Venable. 1996b. Diversity and temporal change in the effective pollinators of *Asclepias tuberosa*. Ecology 77: 1061–1073.

Flores, S., and D.W. Schemske. 1984. Dioecy and monoecy in the flora of Puerto Rico and the Virgin Islands: ecological correlates. Biotropica 16: 132–139.

Fox, J.F. 1985. Incidence of dioecy in relation to growth form, pollination and dispersal. Oecologia 67: 244–249.

Fox, J.F. 1993. Size and sex allocation in monoecious woody plants. Oecologia 94: 110–113.

Freeman, D.C., J.L. Doust, A. El-Keblawy, K.J. Miglia, and E.D. McArthur. 1997. Sexual specialization and inbreeding avoidance in the evolution of dioecy. The Botanical Review 63: 65–94.

Freeman, D.C., E.D. McArthur, K. Harper, and A.C. Blauer. 1981. Influence of environment on the floral sex ratio of monoecious plants. Evolution 35: 194–197.

Friss, E.M., and P.K. Endress. 1990. Origin and evolution of angiosperm flowers. Advances in Botanical Research 17: 99–162.

Galen, C. 1989. Measuring pollinator-mediated selection on morphometric floral traits: bumblebees and the alpine sky pilot, *Polemonium viscosum*. Evolution 43: 882–890.

Ganeshaiah, K.N., and U.R. Shaanker. 1991. Floral sex ratios in monoecious species—why are trees more male-biased than herbs? Current Science 60: 319–321.

García, M.B., R.J. Antor, and X. Espadaler. 1995. Ant pollination of the palaeoendemic dioecious *Borderea pyrenaica* (Dioscoreaceae). Plant Systematics and Evolution 198: 17–27.

Gibson, J.P., and N.T. Wheelwright. 1996. Mating system dynamics of *Ocotea tenera* (Lauraceae), a gynodioecious tropical tree. American Journal of Botany 83: 890–894.

Givnish, T.J. 1980. Ecological constrains on the evolution of breeding systems in seed plants: dioecy and dispersal in gymnosperms. Evolution 34: 959–972.

Goldman, D.A., and M.F. Willson. 1986. Sex allocation in functionally hermaphroditic plants: a review and critique. The Botanical Review 52: 157–194.

Gómez, J.M., and R. Zamora. 1996. Wind pollination in high-mountain populations of *Hormathophylla spinosa* (Cruciferae). American Journal of Botany 83: 580–585.

Gould, S.J., and E.S. Vrba. 1982. Exaptation—a missing term in the science of form. Paleobiology 8: 4–15.

Grant, V. 1949. Pollination systems as isolating mechanisms in angiosperms. Evolution 3: 82–97.

Grant, V. 1994. Modes and origins of mechanical and ethological isolation in angiosperms. Proceedings of the National Academy of Sciences 91: 3–10.

Grant, V., and K.A. Grant. 1965. Flower Pollination in the Phlox Family. New York: Columbia University Press.

Guitián, P., J. Guitián, and L. Navarro. 1993. Pollen transfer and diurnal versus nocturnal pollination in *Lonicera etrusca*. Acta Oecologica 14: 219–227.

Gustafsson, A. 1946. Apomixis in higher plants. I-III. Lunds Univ. Arsskr. 42: 1–67.

Haig, D., and M. Westoby. 1989. Selective forces in the emergence of the seed habit. Biological Journal of the Linnean Society 38: 215–238.

Harder, L.D., and S.C.H. Barrett. 1996. Pollen dispersal and mating patterns in animal-pollinated plants. In: Lloyd, D.G., and S.C.H. Barrett, eds. Floral Biology. Studies on Floral Evolution in Animal-Pollinated Plants. New York: Chapman and Hall, pp 140–190.

Harder, L.D., M.B. Cruzan, and J.D. Thomson. 1993. Unilateral incompatibility and the effects of interspecific pollination for *Erythronium americanum* and *Erythronium albidum* (Liliaceae). Canadian Journal of Botany 71: 353–358.

Harper, J.L. 1977. Population Biology of Plants. London: Academic Press.

Hauser, T.P., and V. Loeschcke. 1996. Drought stress and inbreeding depression in *Lychnis flos-cuculi* (Caryophyllaceae). Evolution 50: 1119–1126.

Herrera, C.M. 1988. Variation in mutualisms: the spatio-temporal mosaic of a pollinator assemblage. Biological Journal of the Linnean Society 35: 95–125.

Herrera, C.M. 1993. Selection on floral morphology and environmental determinants of fecundity in a hawk moth-pollinated violet. Ecological Monographs 63: 251–275.

Herrera, C.M. 1995. Microclimate and individual variation in pollinators: flowering plants are more than their flowers. Ecology 76: 1516–1524.

Herrera, C.M. 1996. Floral traits and plant adaptation to insect pollinators: a devil's advocate approach. In: Lloyd, D.G., and S.C.H. Barrett, eds. Floral Biology. Studies on

Floral Evolution in Animal-Pollinated Plants. New York: Chapman and Hall, pp 65–87.

Hirose, T., and N. Kachi. 1982. Critical plant size for flowering in biennials with special reference to their distribution in a sand dune system. Oecologia 55: 281–284.

Holsinger, K.E. 1986. Dispersal and plant mating systems: the evolution of self-fertilization in subdivided populations. Evolution 40: 405–413.

Holsinger, K.E. 1991. Inbreeding depression and the evolution of plant mating systems. Trends in Ecology and Evolution 6: 307–308.

Holsinger, K.E. 1996. Pollination biology and the evolution of mating systems in flowering plants. Evolutionary Biology 29: 107–149.

Horvitz, C.C., and D.W. Schemske. 1990. Spatiotemporal variation in insect mutualists of a neotropical herb. Ecology 71: 1085–1097.

Husband, B.C., and D.W. Schemske. 1996. Evolution of the magnitude and timing of inbreeding depression in plants. Evolution 50: 54–70.

Iwasa, Y., and D. Cohen. 1989. Optimal growth schedule of a perennial plant. American Naturalist 133: 480–505.

Jain, S.K. 1976. The evolution of inbreeding in plants. Annual Review of Ecology and Systematics 7: 469–495.

Janzen, D.H. 1971. Seed predation by animals. Annual Review of Ecology and Systematics 2: 465–492.

Janzen, D.H. 1977. What are dandelions and aphids? American Naturalist 111: 586–589.

Jarne, P., and D. Charlesworth. 1993. The evolution of the selfing rate in functionally hemaphrodite plants and animals. Annual Review of Ecology and Systematics 24: 441–466.

Johnson, S.D., and K.E. Steiner. 1997. Long-tongued fly pollination and evolution of floral spur length in the *Disa draconis* complex (Orchidaceae). Evolution 51: 45–53.

Johnston, M.O., and D.J. Schoen. 1996. Correlated evolution of self-fertilization and inbreeding depression: an experimental study of nine populations of *Amsinckia* (Boraginaceae). Evolution 50: 1478–1491.

Jordano, P. 1991. Gender variation and expression of monoecy in *Juniperus phoenicea* (L.) (Cupressaceae). Botanical Gazette 152: 476–485.

Jordano, P. 1993. Pollination biology of *Prunus mahaleb* L.: deferred consequences of gender variation for fecundity and seed size. Biological Journal of the Linnean Society 50: 65–84.

Kao, T., and A.G. McCubbin. 1996. How flowering plants discriminate between self and non-self pollen to prevent inbreeding. Proceedings of the National Academy of Sciences 93: 1259–1265.

Klinkhamer, P.G.L., T.J. De Jong, and H.W. Nell. 1994. Limiting factors for seed production and phenotypic gender in the gynodioecious species *Echium vulgare* (Boraginaceae). Oikos 71: 469–478.

Klips, R.A., and A.A. Snow. 1997. Delayed autonomous self-pollination in *Hibiscus laevis* (Malvaceae). American Journal of Botany 84: 48–53.

Koelewijn, H.P. 1996. Sexual differences in reproductive characters in gynodioecious *Plantago coronopus*. Oikos 75: 443–452.

Kohn, J.R. 1988. Why be female? Nature 335: 431–433.

Kohn, J.R., and S.C.H. Barret. 1992. Experimental studies on the functional significance of heterostyly. Evolution 46: 43–55.

Law, R., R.E.D. Cook, and R.J. Manlove. 1983. The ecology of flower and bulbil production in *Polygonum viviparum*. Nordic Journal of Botany 3: 559–565.

Lawrence, M.J., D.F. Marshall, V.E. Curtis, and C.H. Fearon. 1985. Gametophytic self-incompatibility re-examined: a reply. Heredity 54: 131–138.

Lepart, J., and B. Dommée. 1992. Is *Phillyrea angustifolia* L. (Oleaceae) an androdioecious species? Botanical Journal of the Linnean Society 108: 375–387.

Levin, D.A. 1986. Breeding structure and genetic variation. In: Crawley, M.J., ed. Plant Ecology. Oxford: Blackwell Scientific, pp 217–251.

Lewis, D. 1942. The evolution of sex in flowering plants. Biological Review of the Cambridge Philosophical Society 17: 46–67.

Liston, A., L.H. Rieseberg, and T.S. Elias. 1990. Functional androdioecy in the flowering plant *Datisca glomerata*. Nature 343: 641–642.

Lloyd, D.G. 1975. The maintenance of gynodioecy and androdioecy in Angiosperms. Genetica 45: 325–339.

Lloyd, D.G. 1979. Some reproductive factors affecting the selection of self-fertilization in plants. American Naturalist 113: 67–79.

Lloyd, D.G. 1982. Selection of combined versus separate sexes in seed plants. American Naturalist 120: 571–585.

Lloyd, D.G. 1992. Self and cross-fertilization in plants. II. The selection of self-fertilization. International Journal of Plant Sciences 153: 370–380.

Lloyd, D.G., and S.C.H. Barrett. 1996. Floral Biology. Studies on Floral Evolution in Animal-Pollinated Plants. New York: Chapman and Hall.

Lloyd, D.G., and K.S. Bawa. 1984. Modification of the gender in seed plants in varying conditions. Evolutionary Biology 17: 255–338.

Lloyd, D.G., and D.J. Schoen. 1992. Self and cross fertilization in plants. I. Functional dimensions. International Journal of Plant Science 153: 358–369.

Lloyd, D.G., and C.J. Webb. 1986. The avoidance of interference between the presentation of pollen and stigmas in angiosperms. I. Dichogamy. New Zealand Journal of Botany 24: 135–162.

Lloyd, D.G., and C.J. Webb. 1992. The evolution of heterostyly. In: Barrett, S.C.H., ed. Evolution and Function of Heterostyly. Berlin: Springer-Verlag, pp 151–178.

Lloyd, D.G., and J.M.A. Yates. 1982. Intersexual selection and the segregation of pollen and stigmas in hermaphrodite plants, exemplified by *Wahlenbergia albomarginata* (Campanulaceae). Evolution 36: 903–913.

Lord, E.M. 1981. Cleistogamy: a tool for the study of floral morphogenesis, function and evolution. The Botanical Review 47: 421–449.

Lovett Doust, J. 1981. Population dynamics and local specialization in a clonal perennial (*Ranunculus repens*). I. The dynamics of ramets in contrasting habitats. Journal of Ecology 69: 743–755.

Lovett Doust, J., and L. Lovett Doust. 1988. Plant Reproductive Ecology. Patterns and Strategies. New York: Oxford University Press.

Maheshwari, P. 1950. An Introduction to the Embryology of Angiosperms. New York: McGraw-Hill.

Maki, M. 1993. Outcrossing and fecundity advantage of females in gynodioecious *Chiono-*

graphis japonica var. *kurohimensis* (Liliaceae). American Journal of Botany 80: 629–634.

Marshall, D.L., M.W. Folson, C. Hatfield, and T. Bennett. 1996. Does interference competition among pollen grains occur in wild radish? Evolution 50: 1842–1848.

Mayer, S.S., and D. Charlesworth. 1991. Cryptic dioecy in flowering plants. Trends in Ecology and Evolution 6: 320–324.

Mazer, S.J. 1992. Environmental and genetic sources of variation in floral traits and phenotypic gender in wild radish: consequences for natural selection. In: Wyatt, R., ed. Ecology and Evolution of Plant Reproduction. New York: Chapman and Hall, pp 281–325.

Midgley, J.J., and W.J. Bond. 1991. How important is biotic pollination and dispersal to the success of the angiosperms. Philosophical Transactions of the Royal Society of London 333: 209–216.

Montalvo, A.M. and J.D. Ackerman. 1986. Relative pollinator effectiveness and evolution of floral traits in *Spathiphyllum friedrichsthalii* (Araceae). American Journal of Botany 73: 1665–1676.

Morgan, M.T., and D.J. Schoen. 1997. The role of theory in an emerging new plant reproductive biology. Trends in Ecology and Evolution 12: 231–234.

Muenchow, G.E. 1987. Is dioecy associated with fleshy fruit? American Journal of Botany 74: 287–293.

Mulcahy, D.L., and G.B. Mulcahy. 1983. Gametophytic self-incompatibility reexamined. Science 220: 1247–1251.

Mulcahy, D.L., P.S. Curtis, and A.A. Snow. 1983. Pollen competition in a natural population. In: Jones, C.E., and R.J. Little, eds. Handbook of Experimental Pollination and Biology. New York: Van Nostrand, pp 330–337.

Murcia, C. 1990. Effect of floral morphology and temperature on pollen receipt and removal in *Ipomoea trichocarpa*. Ecology 71: 1098–1109.

Murcia, C., and Feinsinger, P. 1996. Interspecific pollen loss by hummingbirds visiting flower mixtures: effects of floral architecture. Ecology 77: 550–560.

Mutikainen, P., and L.F. Delph. 1996. Effects of herbivory on male reproductive success in plants. Oikos 75: 353–358.

Nettancourt, D. 1977. Self-Incompatibility in Angiosperms. Berlin: Springer-Verlag.

Nilsson, L.A. 1988. The evolution of flowers with deep corolla tubes. Nature 334: 147–149.

Nygren, A. 1967. Apomixis in the angiosperms. Handb. der Pflanzenphys. 18:551–596.

Olmstead, R.G. 1986. Self-incompatibility in light of population structure and inbreeding. In: Mulcahy D.L., et al., eds. New York: Springer-Verlag, p 239.

Pannell, J. 1997a. Widespread functional androdioecy in *Mercuarialis annua* L. (Euphorbiaceae). Biological Journal of the Linnean Society 61: 95–116.

Pannell, J. 1997b. The maintenance of gynodioecy and androdioecy in a metapopulation. Evolution 51: 10–20.

Pellmyr, O. 1992. Evolution of insect pollination and angiosperm diversification. Trends in Ecology and Evolution 7: 46–49.

Pellmyr, O., and J.N. Thompson. 1996. Sources of variation in pollinator contribution within a guild. The effects of plant and pollinator factors. Oecologia 107: 595–604.

Pellmyr, O., J.N. Thompson, J.M. Brown, and R.G. Harrison. 1996. Evolution of pollination and mutualism in the yucca moth lineage. American Naturalist 148: 827–847.

Pettersson, M.W. 1991. Pollination by a guild of fluctuating moth populations: option for unspecialization in *Silene vulgaris*. Journal of Ecology 79: 591–604.

Poot, P. 1997. Reproductive allocation and resource compensation in male-sterile and hermaphroditic plants of *Plantago lanceolata* (Plantaginaceae). American Journal of Botany 84: 1256–1265.

Primack, R.B., and H. Kang. 1989. Measuring fitness and natural selection in wild plant populations. Annual Review of Ecology and Systematics 20: 367–396.

Primack, R.B., and D.G. Lloyd. 1980. Andromonoecy in the New Zealand montane shrub Manuke, *Leptospermum scoparium* (Myrtaceae). American Journal of Botany 67: 361–368.

Primack, R.B., and C. McCall. 1986. Gender variation in a red maple population (*Acer rubrum*; Aceraceae): a seven-year study of a "polygamodioecious" species. American Journal of Botany 73: 1239–1248.

Proctor, M., and P. Yeo. 1973. The Pollination of Flowers. London: Collins.

Queller, D.C. 1983. Sexual selection in a hermaphroditic plant. Nature 305: 706–707.

Queller, D.C. 1997. Pollen removal, paternity, and the male function of flowers. American Naturalist 149: 585–594.

Ramsey, M.W. 1988. Differences in pollinator effectiveness of birds and insects visiting *Banksia menziesii* (Proteaceae). Oecologia 76: 119–124.

Raven, P.H. 1977. A suggestion concerning the cretaceous rise to dominance of the angiosperms. Evolution 31: 451–452.

Regal, P.J. 1977. Ecology and evolution of flowering plant dominance. Science 196: 622–629.

Renner, S.S., and R.E. Ricklefs. 1995. Dioecy and its correlates in the flowering plants. American Journal of Botany 82: 596–606.

Richards, A.J. 1986. Plant Breeding Systems. London: George Allen and Unwin.

Ricklefs, R.E., and S.S. Renner. 1994. Species richness within families of flowering plants. Evolution 48: 1619–1636.

Rieseberg, L.H., M.A. Hanson, and C.T. Philbrick. 1992. Androdioecy is derived from dioecy in Datiscaceae. Evidence from restriction site mapping of PCR-amplified chloroplast DNA fragments. Systematic Botany 17: 324–336.

Ritland, K. 1990. Inferences about inbreeding depression based upon changes of the inbreeding coefficient. Evolution 44: 1230–1241.

Robertson, J.L., and R. Wyatt. 1990. Evidence for pollination ecotypes in the yellow-fringed orchid, *Platanthera ciliaris*. Evolution 44: 121–133.

Ross, M.D. 1978. The evolution of gynodioecy and subdioecy. Evolution 32: 174–188.

Ross, M.D. 1982. Five evolutionary pathways to dioecy. American Naturalist 119: 297–318.

Ross, M.D. 1990. Sexual asymmetry in hermaphroditic plants. Trends in Ecology and Evolution 5: 43–47.

Sakai, A.K., and S.G. Weller. 1991. Ecological aspects of sex expression in subdioecious *Schiedea globosa* (Caryophyllaceae). American Journal of Botany 78: 1280–1288.

Sakai, S. 1995. Evolutionarily stable selfing rates of hermaphroditic plants in competing and delayed selfing modes with allocation to attractive structures. Evolution 49: 557–564.

Sakai, A.K., W.L. Wagner, D.M. Ferguson, and D.R. Herbst. 1995. Origins of dioecy in the Hawaiian flora. Ecology 76: 2517–2529.

Saran, S., and J.M.J. de Wet. 1970. The mode of reproduction in *Dicanthium intermedium* (sic) (Gramineae). Bulletin of the Torrey Botanical Club 97: 6–13.

Schemske, D.W., and C.C. Horvitz. 1989. Temporal variation in selection on a floral character. Evolution 43: 461–465.

Schemske, D.W., and R. Lande. 1985. The evolution of self fertilization and inbreeding depression in plants. II. Empirical observations. Evolution 39: 41–52.

Schlichting, C.D., and V.A. Delesalle. 1997. Stressing the differences between male and female functions in hermaphroditic plants. Trends in Ecology and Evolution 12: 51–52.

Schlichting, C.D., and B. Devlin. 1992. Pollen and ovule sources affect seed production of *Lobelia cardinalis* (Lobeliaceae). American Journal of Botany 79: 891–898.

Schmid, B., and J. Weiner. 1993. Plastic relationships between reproductive and vegetative mass in *Solidago altissima*. Evolution 47: 61–74.

Schmitt, J., and S.E. Gamble. 1990. The effect of distance from the parental site on offspring performance in *Impatiens capensis*: a test of the local adaptation hypothesis. Evolution 44: 2022–2030.

Schoen, D.J., and D.G. Lloyd. 1984. The selection of cleistogamy and heteromorphic diaspores. Botanical Journal of the Linnean Society 23: 303–322.

Schoen, D.J., M.T. Morgan, and T. Bataillon. 1996. How does self-pollination evolve? Inferences from floral ecology and molecular genetic variation. Philosophical Transactions of the Royal Society of London (Series B) 351: 1281–1290.

Seavey, S.R., and K.S. Bawa. 1986. Late-acting self-incompatibility in angiosperms. Botanical Review 52: 195–219.

Shore, J.S., and S.C.H. Barrett. 1984. The effect of pollination intensity and incompatible pollen on seed set in *Turnera ulmifolia* (Turneraceae). Canadian Journal of Botany 62: 1298–1303.

Sih, A., and M.S. Baltus. 1987. Patch size, pollinator behavior, and pollinator limitation in catnip. Ecology 68: 1679–1690.

Smith, C.C. 1981. The facultative adjustment of sex ratio in lodgepole pine. American Naturalist 118: 291–305.

Snow, A.A. 1986. Pollination dynamics of *Epilobium canum* (Onagraceae): consequences for gametophytic selection. American Journal of Botany 73: 139–157.

Snow, A.A., and P.O. Lewis. 1993. Reproductive traits and male fertility in plants: empirical approaches. Annual Review of Ecology and Systematics 24: 331–351.

Snow, A.A., and T.P. Spira. 1996. Pollen-tube competition and male fitness in *Hibiscus moscheutos*. Evolution 50: 1866–1870.

Snow, A.A., T.P. Spira, R. Simpson, and R.A. Klips. 1996. The ecology of geitonogamous pollination. In: Lloyd, D.G., and S.C.H. Barrett, eds. Floral Biology. Studies on Floral Evolution in Animal-Pollinated Plants. New York: Chapman and Hall, pp 191–216.

Sohn, J.J., and D. Policansky. 1977. The costs of reproduction in the mayapple, *Podophyllum peltatum* (Berberidaceae). Ecology 58: 1366–1374.

Solomon, B.P. 1985. Environmentally influenced changes in sex expression in an andromonoecious plant. Ecology 66: 1321–1332.

Solomon, B.P. 1989. Size-dependent sex ratios in the monoecious, wind-pollinated annual, *Xanthium strumarium*. American Midland Naturalist 121: 209–218.

Spalik, K. 1991. On evolution of andromonoecy and 'overproduction' of flowers: a resource allocation model. Biological Journal of the Linnean Society 42: 325–336.

Sprengel, C.K. 1793. Discovery of the Secret of Nature in the Structure and Fertilization of Flowers. Berlin: Vieweg.

Stebbins, G.L. 1970. Adaptive radiation of reproductive characteristics in angiosperms, I: pollination mechanisms. Annual Review of Ecology and Systematics 1: 307–326.

Stebbins, G.L. 1981. Why are there so many species of flowering plants? Bioscience 31: 573–577.

Strauss, S.Y., J.K. Conner, and S.L. Rush. 1996. Foliar herbivory affects floral characters and plant attractiveness to pollinators: implications for male and female plant fitness. American Naturalist 147: 1098–1107.

Sutherland, S., and L.F. Delph. 1984. On the importance of male fitness in plants: patterns of fruit-set. Ecology 65: 1093.

Ter-Avanesian, D.V. 1978. The effect of varying the number of pollen grains used in fertilization. Theoretical and Applied Genetics 52: 77–79.

Thompson, J.N. 1994. The Coevolutionary Process. Chicago: University of Chicago Press.

Thomson, J.D., and S.C.H. Barrett. 1981a. Temporal variation of gender in *Aralia hispida* Vent. (Araliaceae). Evolution 35: 1094–1107.

Thomson, J.D., and S.C.H. Barrett. 1981b. Selection for outcrossing, sexual selection, and the evolution of dioecy inplants. American Naturalist 118: 443–449.

Thomson, J.D., B.J. Andrews, and R.C. Plowright. 1981. The effect of a foreign pollen on ovule development in *Diervilla lonicera* (Caprifoliaceae). New Phytologist 90: 777–783.

Thomson, J.D., and J. Brunet. 1990. Hypotheses for the evolution of dioecy in seed plants. Trends in Ecology and Evolution 5: 11–16.

Tiffney, B.H. and S.J. Mazer. 1995. Angiosperm growth habit, dispersal and diversification reconsidered. Evolutionary Ecology 9: 93–117.

Traveset, A. 1992. Sex expression in a natural population of the monoecious annual, *Ambrosia artemisiifolia* (Asteraceae). American Midland Naturalist 127: 309–315.

Traveset, A. 1994. Reproductive biology of *Phillyrea angustifolia* L. (Oleaceae) and effect of galling-insects on its reproductive output. Botanical Journal of the Linnean Society 114: 153–166.

Traveset, A. 1995. Reproductive ecology of *Cneorum tricoccon* L. (Cneoraceae) in the Balearic Islands. Botanical Journal of the Linnean Society 117: 221–232.

Traveset, A., and E. Sáez. 1997. Pollination of *Euphorbia dendroides* by lizards and insects. Spatio-temporal variation in flower visitation patterns. Oecologia 111: 241–248.

Traveset, A., M.F. Willson, and C. Sabag. 1998. Effect of nectar-robbing birds on fruit set of *Fuchsia magellanica* in Tierra del Fuego: a disrupted mutualism. Functional Ecology 12: 459–464.

Uyenoyama, M.K. 1986. Inbreeding and the cost of meiosis: the evolution of selfing in populations practicing biparental inbreeding. Evolution 40: 388–404.

Uyenoyama, M.K., K.E. Holsinger, and D.M. Waller. 1993. Ecological and genetic factors directing the evolution of self-fertilization. Oxford Surveys in Evolutionary Biology 9: 327–381.

Valiente-Banuet, A., A. Rojas-Martinez, M.C. Arizmendi, and P. Vila. 1997. Pollination biology of two columnar cacti (*Neobuxbaumia mezcalaensis* and *Neobuxbaumia macrocephala*) in the Tehuacán Valley, Central Mexico. American Journal of Botany 84: 452–455.

Vaughton, G. 1992. Effectiveness of nectarivorous birds and honeybees as pollinators of *Banksia spinulosa* (Proteaceae). Australian Journal of Ecology 17: 43–50.

Vielle-Calzada, J.P., C.F. Crane, and D.M. Stelly. 1996. Apomixis: the asexual revolution. Science 274: 1322–1323.

Waser, N.M., and M.V. Price. 1989. Optimal outcrossing in *Ipomopsis aggregata*: seed set and offspring fitness. Evolution 43: 1097–1109.

Waser, N.M., L. Chittka, M.V. Price, N.M. Willias, and J. Ollerton. 1996. Generalization in pollination systems, and why it matters. Ecology 77: 1043–1060.

Weiner, J. 1988. The influence of competition on plant reproduction. In: Lovett Doust, J., and L. Lovett Doust, eds. Plant Reproductive Ecology: Patterns and Strategies. New York: Oxford University Press, pp 228–245.

Westley, L.C. 1993. The effect of inflorescence bud removal on tuber production in *Helianthus tuberosus* L. (Asteraceae). Ecology 74: 2136–2144.

Whalen, M.D., and D. Costich. 1986. Andromonoecy in *Solanum*. In: D'Arcy, G., ed. Solanaceae: Biology and Systematics. New York: Columbia University Press.

Willson, M.F. 1979. Sexual selection in plants. American Naturalist 113: 777–790.

Willson, M.F. 1994. An overview of sexual selection in plants and animals. American Naturalist 144 (Suppl): 13–39.

Wilson, P., and J.D. Thomson. 1991. Heterogeneity among floral visitors leads to discordance between removal and deposition of pollen. Ecology 72: 1503–1507.

Wilson, P., and J.D. Thomson. 1996. How do flowers diverge? In: Floral Biology. Studies on Floral Evolution in Animal-Pollinated Plants. Lloyd, D.G., and S.C.H. Barrett, eds. Floral Biology. Studies on Floral Evolution in Animal-Pollinated Plants. New York: Chapman and Hall, pp 88–111.

Wolfe, L.M., and S.C.H. Barrett. 1989. Patterns of pollen removal and deposition in tristylous *Pontederia cordata* L. (Pontederiaceae). Biological Journal of the Linnean Society 36: 317–329.

Worley, A.C., and L.D. Harder. 1996. Size-dependent resource allocation and costs of reproduction in *Pinguicula vulgaris* (Lentibulariaceae). Journal of Ecology 84: 195–206.

Wyatt, R. 1983. Pollinator-plant interactions and the evolution of breeding systems. In: Real, L., ed. Pollination Biology. New York: Academic Press, pp 51–95.

Yampolsky, C., and H. Yampolsky. 1922. Distribution of sex forms in the phanerogamic flora. Bibliotheca Genetica 3: 1–62.

Young, H.J., and M.L. Stanton. 1990. Influences of floral variation on pollen removal and seed production in wild radish. Ecology 71: 536–547.

Zavada, M.S., and T.N. Taylor. 1986. The role of self-incompatibility and sexual selection in the gymnosperm-angiosperm transition: a hypothesis. American Naturalist 128: 538–550.

Zimmerman, M. 1988. Nectar production, flowering phenology, and strategies for pollination. In: Lovett Doust, J., and L. Lovett Doust, eds. Plant Reproductive Ecology. Patterns and Strategies. New York: Oxford University Press, pp 157–178.

16

Seed and Seedling Ecology

Michael Fenner
University of Southampton, Southampton, England

Kaoru Kitajima
University of Florida, Gainesville, Florida

I. INTRODUCTION

In recent decades, ecologists have come to recognize that many features of plant communities can be interpreted in terms of the events surrounding the reproduction of the species present. Characteristics such as the relative abundance of the species, their annual fluctuation in number, their spatial pattern, their response to grazing, and other forms of disturbance are all influenced by differences in the ability of each species to reproduce under the prevailing conditions. The process of succession, in which groups of organisms are sequentially replaced by others, is largely due to differential success in regeneration. The species richness of plant communities may be due to the same cause. Grubb (1977) suggested that the coexistence of so many plant species that appear to have indistinguishable niches as adults, can be explained by their having distinctive requirements in the early stages of their life histories. Differences in the regeneration niche of the species may result in the avoidance of competitive exclusion. Most of these processes involve reproduction by means of seeds. This chapter deals with the ecological aspects of regeneration from seeds in plant communities.

II. SEED SIZE AND COMPOSITION

Each species of plant has seeds whose mean size is characteristic of the species. There is usually some variation in mass between and within populations, and within the progeny of individual plants (and even individual fruits), but the variation is generally much less than that of the vegetative parts. The size adopted by a particular species is partly determined by phylogenetic influences; big (or small) seeds run in families. This is often linked with the size of the parent plants at maturity. Recent studies from a range of floras show strong associations of seed size with plant height and growth form (Leishman and Westoby 1994a, Kelly 1995, Leishman et al. 1995). However, within the constraints of the genetic makeup of the plants, the characteristic seed size of each species is presumed to be the result of natural selection. The selection pressures influencing seed size are likely to have been very numerous, often operating in opposite directions, resulting in a size that may represent the best compromise between opposing forces. For example, Reader (1993) showed that large-seeded species have an advantage when establishing in the presence of ground cover, whereas small-seeded species have an advantage in the presence of seed predators. Under field conditions, where both ground cover and seed predators are present, the selective pressures on seed size are clearly complex and may permit a wide range of equally effective compromise solutions. In another experiment in which a range of seeds were sown into a site subjected to gradients of fertilizer addition and disturbance intensity, Burke and Grime (1996) found that large-seeded species established readily over a wider range of conditions, whereas smaller-seeded species were more dependent on disturbance (Figure 1).

  Seed size should be viewed first in relation to the overall reproductive strategy and life history of the species. For a given allocation of resources to reproduction, the plant can either invest in a small number of large seeds or a large number of small ones, or at some intermediate combination of number and size. The compromise size adopted is conventionally thought to represent the conflicting

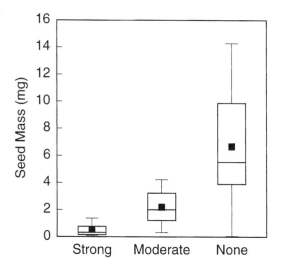

Figure 1 Relationship of seed size to the dependency of non-native species on disturbance for seedling establishment in a temperate limestone grassland. Mixture of seeds of 29 non-native species were sown along an experimental disturbance gradient. Dependency of seedling establishment on disturbance was tested for each species with a χ^2 test (strong, $P < .001$, 19 spp.; moderate, $P < .05$, 5 spp.; none, not significant, 5 spp.). Indicated in the box plots are mean seed mass (square), 90th, 75th, 50th, 25th, and 10th percentiles of seed mass distribution for each category of dependency. (*Source*: adapted from Table 4 of Burke and Grime 1996.)

requirements of dispersal (favoring small seeds) and establishment (favoring large seeds). Several recent studies have interpreted seed size in terms of a trade-off between dispersability and establishment (Ganeshaiah and Uma 1991, Hedge et al. 1991, Allsopp and Stock 1995, Geritz 1995). Short-lived early colonizers of disturbed sites and open ground characteristically produce numerous light, easily dispersed seeds; long-lived species of less disturbed sites tend to have larger, less widely dispersed seeds. There are many exceptions to these generalizations, possibly because of numerous additional modifying factors, such as the means of dispersal adopted, the influence of particular species of animals as dispersers and seed predators, the establishment conditions (degree of shade, drought, nutrient level), the formation of persistent seed banks, cotyledon functional morphologies, and the adoption of a parasitic or hemiparasitic way of life.

For wind-dispersed plants, there is a broad negative relationship between seed weight (or at least diaspore weight) and dispersability (Greene and Johnson 1993). For animal-dispersed species, the position is more complex. Dispersal by mammals (rather than birds) tends to be associated with larger seeds (Kelly 1995).

Within bird-dispersed plants, smaller seeds may be dispersed further than larger ones because they are passed through the gut rather than regurgitated (Hedge et al. 1991). Seeds that are dispersed by ants must clearly have a size that is related to the insects' requirements. In a study in Costa Rica, small seeds were harvested by a wider variety of ants (Kaspari 1996). Small seed size also enhances dispersal in time via formation of soil seed bank (Thompson et al. 1993). In a tropical moist forest in Panama, density of viable seeds in soil is negatively correlated with seed mass (Dalling et al. 1997c).

Ecologists have long identified the advantage of seed mass for successful seedling establishment in shade (Salisbury 1942, Westoby et al. 1992). However, recent studies caution against assuming the generality of association between seed mass and ability to establish in shade independent of phylogeny (Kelly 1995, Leishman et al. 1995). Across and within many taxa, seed size appears to be correlated with ability to survive and establish in shade, although there are exceptions of small-seeded shade-tolerant species and large-seeded light demanders (Augspurger 1984a, Metcalfe and Grubb 1995). Other characteristics of the species associated with seed mass, such as type of dispersers, seed reserve utilization rates, cotyledon functional morphologies, and patterns of seedling growth and development, rather than the amount of seed reserve per se, may be directly responsible for shade tolerance (Leishman and Westoby 1994b, Kitajima 1996a, Saverimuttu and Westoby 1996b, Westoby et al. 1996).

There is also some evidence that plants from dry habitats tend to have larger seeds. Baker (1972) conducted a survey of 2490 species in California and showed a fairly consistent relationship between seed size and dry conditions. It is thought that greater seed reserves might enable the seedling to establish roots quickly, thereby exploiting a greater volume of soil for moisture than would otherwise be possible. However, in a survey of dunes in Indiana, Mazer (1989) was not able to show any significant relationship between seed size and water availability. Jurado and Westoby (1992), in a test involving Australian species, found that seedlings from heavier-seeded species did not (as they hypothesized) allocate a greater proportion of their resources to roots than lighter-seeded species. Glasshouse experiments on seeds of semi-arid species by Leishman and Westoby (1994c) indicated an advantage to larger seeds in dry soil, but field experiments failed to confirm this. Further surveys of the type conducted by Baker, on a range of floras, would help to clarify the relationship between seed size and dry habitats.

Another environmental factor that might be expected to influence the evolution of seed size is the nutrient status of the soil. In poor soils, a large seed nutrient reserve might be an advantage to overcome the deficiencies of the external supply. In experiments in which seeds are deprived of nutrients, large-seeded seedlings generally survive longer than small-seeded ones (Jurado and Westoby 1992). However, the evidence that species of poor soils have larger seeds than species from fertile soils is ambiguous. Lee et al. (1993) found that among species in the grass genus *Chionochloa*, there was a negative correlation between seed size and soil fertility. In contrast, Maranon and Grubb (1993) found that in a selection of 27 Mediterranean annuals, the species with the largest seeds tended to occupy the soils with a richer nutrient supply. For many species, the mineral reserves within the seed may be more important than overall seed size. In a comparison of three tropical woody species in Bignoniaceae, the higher the nitrogen concentration in seeds, the longer the seedlings depend on nitrogen reserves in seeds without a need to use the external nitrogen in soil (Kitajima 1996a). A smaller seed with a high concentration of specific minerals may be more effective than a larger seed with low mineral concentrations. There is a negative correlation between seed size and mineral concentration: small seeds have a higher concentration of mineral nutrients. This was shown, for example, in a comparison of 24 species within the Compositae (Fenner 1983) and a comparison of 70 species within the Proteaceae (Pate et al. 1985).

There is little evidence that plants from poorer soils compensate for this by concentrating higher levels of nutrients in their seeds. Lee and Fenner (1989) sampled seeds from species over a wide range of soil fertilities and found that they all had very similar concentrations of minerals. It is possible that in certain soils with extremely low levels of specific minerals, seed composition may complement the deficiencies in the environment. Stock et al. (1990) investigated early seed mineral use in five species in the family Proteaceae from soils severely deficient in nitrogen and phosphorus. They found these elements to be selectively stored in the seeds, complementing the rich supply of Ca, Mg, and K found in postfire environments where recruitment occurs. It has been suggested that the short-term increases in the availability of nitrogen and phosphorus in the postfire environment may be essential for the establishment of certain species, especially legumes (Hanley and Fenner 1997).

There is no doubt that a myriad of complex natural selective pressures have acted on plants, resulting in the seed sizes observed in contemporary floras. However, natural selection can only operate within the phylogenetic constraints of the organisms involved. Daisies cannot produce seeds the size of coconuts. The fact that seed sizes within one area may range over many orders of magnitude suggests that much seed size variation is not just the result of selection, but is largely determined by plant attributes such as plant size and life-form (Westoby et al. 1992).

III. SEED PREDATION

In general, seed tissue has a much higher concentration of nitrogen, phosphorus, sulphur, and magnesium than other plant tissues, as well as providing a rich source of carbohydrates and, in some cases, oils (Jones and Earle 1966, Vaughan 1970, Barclay and Earl 1974, Penning de Vries and Van Laar 1977, Lee and Fenner 1989). Undispersed seeds represent a particularly concentrated source of potential food of high nutritive quality to any organism able to exploit it. There- fore, it is not surprising to find that in many plant species, a large proportion of seed production is lost to predation. Crawley (1992) provides a useful list of examples from the literature of percentage loss of seeds to predators in different plants. The proportion averages at approximately 45–50%, but often approaches 100%. Two distinct groups of seed eaters exist. Predispersal seed predators are typically highly specialized sedentary larvae of beetles, flies, or wasps that mature within the seed or seedhead. In contrast, postdispersal seed predators are usually vertebrate, more mobile, less specialized feeders, although some tropical insect seed predators attack postdispersally. Whole taxa of granivorous birds and mam- mals have evolved (e.g., finches, rodents) to exploit this rich food source. In the seasonally inundated forests of Amazonia, nearly all the seeds that fall into the water are eaten by fish (Kubitzki and Ziburski 1994). In addition, there are many invertebrates that act as predators of dispersed seeds: various species of ants (Gross et al. 1991), earwigs (Lott et al. 1995), slugs (Godnan 1983), and even crabs (O'Dowd and Lake 1991). There is some evidence that imbibed seeds may be more vulnerable to attack by a wider group of soil organisms (Cardina et al. 1996).

Rather few experimental studies have been conducted to determine the long-term demographic effect of seed predation. In some cases, there is no doubt that seed eaters reduce recruitment. Louda (1982) excluded seed-eating insects from the Californian shrub *Haplopappus squarrosus* by the use of insecticide, and found that the mean number of seedlings established per adult after 1 year was greater in the treated plots by a factor of 23. Further proof that seed predators can reduce subsequent recruitment (and hence lifetime fitness) is provided by a demographic study by Louda and Potvin (1995) in which insecticide was applied to the thistle *Cirsium canescens* (Figure 2). Generally, there are significant in- creases in recruitment when seeds are protected from predators (Molofsky and Fisher 1993, Terborgh and Wright 1994, Asquith et al. 1997). However, the con- sequences of seed predation for a plant population depend on whether regenera- tion is limited by seed numbers or by some other factor such as the availability of safe sites. From a survey of the available evidence in the literature, Crawley (1992) concluded that seed-limited recruitment may be the exception rather than the rule, so that seed predation, even where it seems very severe, often has little effect on plant recruitment. However, predators may influence the genetic makeup of the plant population by differential selection of the seeds. They can also affect the evolution of the structural defenses of the seeds. Benkman (1995) compared the allocation of putative seed defences in limber pine (*Pinus flexilis*) in sites where tree squirrels are present (in the Rocky Mountains) with sites where they are absent (in the Great Basin). He found that allocation of energy to cone,

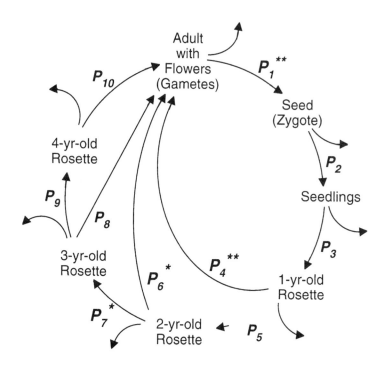

Stage	Transition	Insecticide treatment *P*	Control treatment *P*	N(I,C)
1**	Flower to seed	0.155	0.063	88,92
2	Seed to seedling	0.021	0.020	88,92
3	Seedling to 1-yr old	0.457	0.437	186,71
4**	1-yr old to flowering adult	0.047	0.000	85,31
5	1-yr old to 2-yr old	0.459	0.484	81,31
6*	2-yr old to flowering adult	0.308	0.200	39,5
7*	2-yr old to 3-yr old	0.410	0.667	27,12
8	3-yr old to flowering adult	0.688	0.700	16,8
9	3-yr old to 4-yr old	0.312	0.300	5,3
10	4-yr old to flowering adult	1.000	1.000	5,3

Figure 2 Transition probabilities between two consecutive life stages (P_i) for *Cirsium canescens*. Exclusion of inflorescence feeding insects with an insecticide results in higher transition probability than control at the flower-to-seed stage (P_1). Transition probabilities for subsequent stages are generally not different between treatment and control or significantly higher for progeny from insecticide-treated flowers (**$P < .01$, *$P < .05$). In sum, inflorescence-feeding insects reduced lifetime fitness and altered cohort demography. Exiting arrows indicate death. Paired numbers under N(I,C) indicate sample size for insecticide and control treatments. (*Source*: modified from Figure 5 of Louda and Potvin 1995.)

resin, and seed coat relative to the kernel is greater by a factor of two where the predators are present. This difference in allocation may be a relatively recent evolutionary development since tree squirrels became extinct in the Great Basin only within the last 12,000 years.

Seed predation (mainly by insects, rodents, or birds) is widely thought to provide the selective force that results in the erratic production of seeds (or masting) in many tree species. A common pattern of seed production is the occurrence of bumper crops at irregular intervals with a light seed crop (or total crop failure) in the intervening years. Most of the individuals within a local area tend to mast synchronously. Recently published examples of long-term studies on seed production in trees include rimu (Norton and Kelly 1988), southern beech (Allen and Platt 1990), oak (Crawley and Long 1995), and ash (Tapper 1996). Masting behavior can be interpreted as an evolutionary response to seed predation. It is hypothesized that masting results in the alternate starvation and satiation of the seed predators; in lean years, the predators eat most of the seeds produced, but are overwhelmed by the bounty in bumper years, leaving a surfeit available for regeneration. Although there are alternative explanations for the benefits of masting, such as greater pollination efficiency and the need for large-seeded species to accumulate sufficient reserves for reproduction (Fenner 1991, Sork 1993, Kelly 1994), there is a large body of evidence in support of the general applicability of the predator satiation hypothesis, e.g., species most prone to seed predation show masting behavior most strongly (Silvertown 1980a). Seed predator populations can respond markedly, at least in some cases, to the level of mast and remove all of the crop in most years (Wolff 1996). Seedling establishment can be virtually confined to mast years (Jensen 1985, Forget 1997). Rogue individuals that produce seed in a nonmasting year are targetted by the seed predators, thus selecting for synchronicity. This has been found for a variety of species, including pinyon pine populations (Ligon 1978), the cycad *Encephalartos* (Donaldson 1993), and *Acacia* (Auld 1986). High densities of seeds around and near a parent tree may result in satiation of certain postdispersal seed predators that are territorial (Kitajima and Augspurger 1989, Augspurger and Kitajima 1992, Burkey 1994). All of these observations are at least consistent with the predator satiation hypothesis.

As well as having obvious population effects, seed predation may have important consequences for spatial pattern within a community. Janzen (1970) and Connell (1971) put forward a model indicating that seed predation near trees in tropical rainforests may prevent regeneration of the same species in the immediate vicinity of the parent plant, reducing clumping of species and thereby promoting diversity. Recent tests of the Janzen-Connell model (Augspurger and Kitajima 1992, Condit et al. 1992, Burkey 1994, Houle 1995, Notman et al. 1996) have all yielded results that are somewhat ambiguous in their support of the basic idea. Seed predation appears to affect the spatial pattern of regeneration of each species differently; some comply with Janzen's model, but many do not (Clark and Clark 1984). A high mortality near the parent plant may be due to high offspring density rather than the proximity to parent. A high density of seeds and seedlings often attract predators and promote epidemics of fungal pathogens. Because seed density is almost always confounded with distance from the parent plant, an experimental approach is necessary to determine whether it is density or distance that is responsible for the observed patterns (Augspurger and Kitajima 1992). However, it is important to remember that the overall level of seed mortality is determined by interactions between the seeds and multiple predator species that often exhibit contrasting functional responses to seed density. A given seed density may be high enough to satiate one predator species, but may promote consumption by another.

Seed predation by animals may have had an evolutionary influence on seed size. One means by which a plant could reduce loss to predation would be to reduce seed size (with corresponding increase in number), thereby increasing the foraging cost/benefit ratio of potential predators. Janzen (1969) cited the case of two groups of Central American legumes that adopt contrasting means of coping with predation by beetle larvae. The small-seeded group escape predation by subdivision of their reproductive allocation, whereas the large-seeded group are defended by toxic compounds. A study of predispersal predation of seeds in a number of *Piper* species in Costa Rica found that the large-seeded species lost a much greater proportion of their seeds to insects (Greig 1993). Within a species, the larger seeds may be more vulnerable to attack by predispersal predators. For example, bruchid beetles preferentially oviposit on larger seeds in the sabal palm (Moegenburg 1996). At the same time, extremely large seeds of some tropical trees (seed reserve mass > 50 g) appear to have ample reserves for establishment even after consumption by up to eight bruchid larvae (Dalling et al. 1997). Differential loss is also seen where vertebrate grazers consume seeds as part of their forage. Among legume seeds likely to be eaten by grazing livestock, small seeds may be at an advantage. Tests with sheep have found that small seeds had the highest survival rate after passage through the gut (Russi et al. 1992). Large seeds may therefore need to devote more of their resources to structural defense. Fenner (1983) showed a consistent trend among 24 herbaceous Compositae for relatively greater seed coats in larger seeds. The proportion of seed weight allocated to seed coat varied from 15% in *Erigeron canadense* (seed weight 0.072 mg) to 61% in *Tragopogon pratense* (seed weight 10.3 mg). Thus, defense against seed predation may be another factor in determining the balance between seed size and number.

IV. DISPERSAL

Some degree of spatial distribution of propagules away from the parent plant is generally required for successful regeneration. Dispersal has the advantage of avoiding competition from the parent, escaping the seed predators that the parent may attract, and avoiding pathogens that may become epidemic in a dense seedling population near the parent. The means by which seeds are transported varies from species to species. Many appear at first sight to have no particular adaptation for dispersal. Many of these may be carried in mud on the feet of animals and birds, as was shown in experiments by Darwin (1859), or they may be eaten as part of the forage of grazers and survive passage through the gut and deposition some distance from source (Janzen 1984, Sevilla et al. 1996). The seed itself may be the reward in many scatter-hoarded species, such as oaks and many tropical tree species with fruits and seeds that apparently lack dispersal appendage. Other plant species provide an attractive reward for their dispersers in the form of a fleshy fruit in which the seeds are imbedded. Another large group exploits the wind as means of transport. The latter is characterized by seeds that have structures such as "wings" or "feathers" that decrease the rate of descent, thereby increasing the horizontal distance traveled in a given time (Augspurger 1986). The distance traveled is also a function of the height of release. Recently developed techniques for quantifying the rate of descent of seeds under standardized conditions allow us to make comparisons between species (Askew et al. 1997).

Each species has a characteristic spatial pattern of distribution of its seeds, called its seed shadow. Seed shadows are usually represented as graphs of seed density or seed number in relation to distance from source (Figure 3). The general pattern for wind-dispersed seeds is a more or less exponential decline with distance (Okubo and Levin 1989, Willson 1993), although in practice, seed shadows often peak at a short distance from source before declining (Augspurger 1983, Willson 1992; Figure 3B and C). The shadow is often asymmetrical due to the wind coming from one direction (Augspurger 1984b, Kitajima and Augspurger 1989; Figure 3A) or the land sloping steeply (Lee et al. 1993). Animal-dispersed seeds tend to be more clumped because they are deposited beneath roosting sites (by birds and bats) or in caches (by rodents) (Howe 1989, Forget 1990, Willson 1993).

The range of animals involved in seed dispersal is very wide. The most important groups are birds and mammals, but cases of seed dispersal by other vertebrates are known, e.g., fish (Goulding 1980, Horn 1997), amphibians (Da Silva et al. 1989), and reptiles (Hnatiuk 1978). Seed dispersal by earthworms has also been recorded (McRill and Sagar 1973, Piearce et al. 1994). Some seed may be dispersed twice: first by being deposited in dung, and then being removed by secondary dispersers such as ants (Hughes and Westoby 1992, Levey and Byrne 1993). Survival of seeds may be negligible if they remain in clumps under bat or bird roosts. Their removal by secondary dispersers, either ants (for small-seeded species) or scatter-hoarding rodents (for large-seeded species) (Janzen et al. 1976, Forget and Milleron 1991), greatly increases the likelihood of establishment.

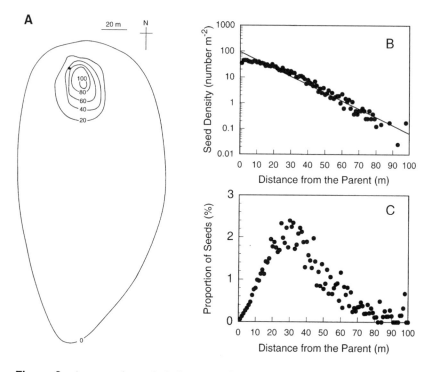

Figure 3 Asymmetric seed shadow around a parent tree of *Tachigalia versicolor*, a wind-dispersed canopy tree. A 1-m wide transect was laid to each of the eight compass directions and seeds in each 1-m interval were counted. (A) Isoclinic lines of density (number per square meter) of dispersed seeds. Black dot represents the position of the parent tree. Line 0 represents the distance of the last seed beyond which no seed was found within 20 m. The seed shadow is elongated to the south due to the strong seasonal wind during the dry season. Seed density was highest at approximately 20 m SE to the parent. The maximum dispersal distance was 225 m to the south. (B) Density of seeds at each distance interval (1 m for 0–80 m, averaged for each 5 m for 80–100 m), averaged for all eight compass directions. The line indicates the exponential fit: $y = 95.0 \, e^{-0.073x}$. (C) Number of seeds at each 1-m interval (estimated for the circular 1-m wide ring area at a given distance), expressed as the proportion of the entire seed shadow. This proportional value also expresses the probability that a seed would disperse to a given distance from the parent. (*Source*: adapted from Kitajima and Augspurger 1989, and unpublished data of C.K. Augspurger.)

Ants are the only invertebrate group that disperses seeds in any appreciable number (Stiles 1992). Dispersal by ants (myrmecochory) is especially prevalent in warm, dry climates and on infertile soils (Westoby et al. 1991). As many as 35% of the species present in a habitat can be involved (Beattie and Culver 1982). The seeds are provided with an oil body (elaiosome) that the ants eat. They retrieve the seed from the ground, carry them off to their nests, remove the elaiosome, and deposit the seed in a refuse heap. The advantages to the plant are thought to be (1) dispersal, although usually only within a few meters of the source; (2) protection from rodents by being buried out of sight; (3) protection from fire; and (4) deposition in a favorable microsite for germination and establishment (Bennet and Krebs 1987). Not all of these features may be equally important in all cases. The importance of the mutualism for the plant can be seen in cases in which native ants have been replaced by less well-adapted invaders, as in the case of fynbos species in South Africa invaded by the Argentine ant (Bond and Slingsby 1984). Not all seeds survive ant transport. In some cases, a proportion of the seeds are eaten as well (Hughes and Westoby 1992, Levey and Byrne 1993).

In general, there is seldom much advantage to a seed being transported very long distances, as it risks being removed from its natural habitat, which may be patchily distributed. Seeds may suffer greater seed predation at the extended tail of a dispersal curve than within a seed shadow where seed predators are satiated (Augspurger and Kitajima 1992). A good example of a situation in which short-distance dispersal may be advantageous is the case of wind-dispersed plants on small, remote islands, where dispersal may result in loss in the sea. Comparisons between related plants on mainlands and islands show that seed dispersabilty is often reduced on islands, presumably due to selective survival of the less mobile seeds. Cody and Overton (1996) show that for certain species of Compositae, reduced dispersability can become established within a remarkably short time. Very remote islands are more likely to be colonized by seed carried by birds than by wind or sea drift. This was well demonstrated by Carlquist's study of the origins of the flora of the Pacific Islands (Carlquist 1965, Fenner 1985). In contrast to the random action of wind and sea, bird movement is from island to island, often on migration routes; therefore, the islands are targetted effectively with seeds deposited in feces and preened from feathers.

V. SEED BANKS

Another strategy for escaping from the parental plant is the formation of long-lived reservoirs of seeds in the soil, undergoing dispersal in time rather than space. The occurrence of persistent seed banks has been regarded as one of a small number of regenerative strategies that have evolved among terrestrial plant species (Grime 1989). Persistent seed banks consist of buried seeds that have the ability to remain dormant but viable for at least several years. They will only germinate if they are brought to the surface by some chance disturbance such as a treefall, an animal digging, or a farmer ploughing. Persistent seed banks are most characteristic of habitats that are prone to frequent but unpredictable disturbance such as cultivation, fire, and floods. Examples of plant communities with large soil seed banks are agricultural fields, heathlands, chaparral, and disturbed wetlands (Thompson et al. 1977). But in many less disturbed communities, species that are characteristic of the early stages of succession and that are habitually the first colonizers of gaps, also form persistent seed banks. Although these species often dominate the seed bank, they usually form only a very small part of the current aboveground vegetation. They represent both the past and potential future species composition of the community (Fenner 1995). Within each species, genetic make-up of a soil seed population will be the result of selection in different years over a period of time, and the appearance of old gene combinations may put a damper on genetic change in the population (Templeton and Levin 1979, Brown and Venable 1986).

The species composition of seedbanks can be altered by management of the vegetation. However, it is not easy to predict how the seed bank will change based on changes in aboveground species composition alone, as different seed-banking species become advantaged when disturbance regimes and vegetation structure change. For example, after 11 years of nitrogen fertilization of the experimental plots in a temperate grassland, species composition of the soil seed bank and of the aboveground vegetation became even less similar to each other than in the unfertilized control plots (Kitajima and Tilman 1996).

The most important requirement for seeds in a persistent seed bank is longevity. To remain viable in the soil requires a high level of dormancy. This dormancy is usually conditional in that it can be readily broken by the application of some specific stimulus, notably light. For such seeds, burial is essential for persistence, and their dormancy prevents them from germinating at depths from which they could not emerge. Many arable weed species can remain viable in soil for up to approximately 10 years, but at least a fraction of the seeds of other species can survive for a century. This was shown in a long-term experiment set up by W. J. Beal in Michigan in 1879 and monitored for 100 years (Kivilaan and Bandurski 1981). Another experiment in which seeds of 20 common weed species were buried under field conditions for 6 years showed that, if left undisturbed, the number of viable seeds remaining declines exponentially with time. Averaged over all 20 species, the rate of decline was 12% per year, giving a half-life of 5.8 years for the population for temperate weeds (Roberts and Feast 1973). Among tropical pioneer tree species, persistence of buried seeds range widely from species dying within a few months to species that do not exhibit any detectable mortality over a few years (Dalling et al. 1997). Many species have recalcitrant seeds, which are notably short-lived and must germinate or die within a short time. The majority of primary tropical forest species fall into this category (Ng 1978, Hopkins and Graham 1987, Garwood and Lighton 1990). The lack of dormancy in these species may be due to their susceptibility to predation. If seeds that do not germinate are eaten within a few days of shedding, there will be no advantage in (and hence no selection for) a high degree of dormancy and potential longevity. Even very small seeds that may remain dormant suffer high mortality from fungal pathogens in tropical forests, especially in the shallow soil layer (Dalling et al. 1998). Loss of viability due to fungal attack deserves greater attention in other types of communities as well (Crist and Friese 1993).

A general trend seen among species that form persistent seed banks is the possession of small, round, smooth seeds. This was first shown by Thompson (1987) for a range of British grasses by comparing species that form persistent seed banks with those that do not. The former were mostly less than 0.3 mg, smooth, and with a low length to width ratio, whereas the latter as a group were bigger, bore appendages, and were more elongated. Bekker et al. (1997) extend these generalizations, indicating that seed size and shape can be used in a predictive way as a guide to probable persistence. The preponderance of small seeds among species with persistent seed banks suggests that selection has favored this feature. At first sight, a large seed containing plentiful provisions of nutrient resources might be thought to be more appropriate to sustain it during years of dormancy. However, metabolic activity during dormancy appears to be negligible, and survival does not appear to be related to use of the stored nutrient material mobilized during germination and establishment. The advantage of small size, coupled with round shape and smooth surface, is probably related to the fact that the strategy depends on the seeds being buried. Survivorship of seeds generally increases with burial depth (Toole 1946, Roberts and Feast 1972, Dalling et al. 1997, 1998). At depth, the seeds are probably protected from inappropriate germination cues and from attack from pathogens that would be more active in shallow well-oxygenated soil. Small, round, smooth seeds can infiltrate more easily to greater depths in the soil by percolating into crevices. In contrast, a large, elongated seed with appendages such as awns or hairs would need an external agent to bury it.

The number of seeds in soil seedbanks has been investigated in a wide range of communities. Estimates are usually made by taking samples (either on a regular grid or at random) using cores of known area and standard depth and extrapolating from these. Most seeds occur in the top few centimeters, with numbers declining very rapidly with depth. Estimates are made either by direct counts of the seeds (often facilitated by sieving or flotation techniques), or by germinating the seeds by incubating the samples on trays under suitable conditions. The latter technique only measures the germinable seeds, but it is probably a good measure for ecological purposes and can be used in comparative studies of different sites. To avoid underestimation in the latter method, soil samples need to be pretreated with likely germination cues to break dormancy and to be spread thinly (<5 mm) during the germination trial (Simpson et al. 1989, Dalling et al. 1995). A major problem in quantifying seedbanks is the difficulty of sampling because of the patchy distribution of the seeds in the soil and large seasonal fluctuations (Thompson and Grime 1979, Thompson 1986, Dessaint et al. 1991, Dalling et al. 1997c). In spite of these difficulties, it is possible to give rough estimates in various habitats: 20,000–40,000 m^{-2} in arable fields, typically less than 1000 m^{-2} in mature tropical forest, down to only only 10–100 m^{-2} in subarctic forests (Leck et al. 1989, Fenner 1995).

VI. DORMANCY AND GERMINATION

Dormancy is a physiological condition in seeds that prevents them from germinating. It plays an important part in preventing seed from germinating at times that would be unfavorable for growth and establishment. Some seeds possess absolute dormancy and will not germinate until certain developmental processes (such as after-ripening) have occurred. However, dormancy can often be a matter of degree. A dormant seed may be induced to germinate, but only under a very restricted set of conditions. The narrower the required conditions, the greater the level of dormancy. This is well illustrated by the cyclical changes in the level of dormancy that occur in the seeds of many annual species. During the summer and autumn months, the seeds of summer annuals in the soil are fully dormant. However, the seeds are gradually released from dormancy by the chilling temperatures experienced during the winter (Washitani and Masuda 1990). This is shown by the fact that if the seeds are taken from the field and tested for germinability in the laboratory, they will germinate over an increasingly wider range of temperatures as spring approaches. As spring advances into summer, the range of permitted germination temperatures narrows, eventually resulting in complete dormancy again. Thus, a window of germinability occurs in spring and germination takes place only when the ambient temperatures overlap with the range of temperatures to which the seed will respond. This set of circumstances occurs only in spring. The mechanism of cyclical dormancy thus ensures that the seeds germinate only at the time that is most favorable for the plants to complete their life cycle. A similar mechanism ensures that winter annuals germinate only in autumn, by having seeds that require high temperatures to release them from dormancy (Vegis 1964, Baskin and Baskin 1980, Bouwmeester 1990). It is important to note that in these examples, there is a clear distinction between the conditions required to overcome dormancy and the conditions needed for germination.

Many of the physiological attributes of seeds can be interpreted as adaptations to ensure that the seeds will only germinate in sites favorable to the long-term success of the plants. For example, if a seed germinates while buried below a given critical depth, it will not be able to emerge. Some seeds are lost in this way (Fenner 1985), but most seeds remain dormant at depth. Exposure of freshly dispersed and imbibed seeds to low red/far red ratio under leaf canopy is important in inducing secondary dormancy to prevent fatal germination after burial (Washitani 1985). Once they are brought to (or near) the surface, usually by some unpredictable disturbance, it is advantageous for them to ensure that their dormancy is not broken unless they are in a suitable gap in the vegetation. A gap may be defined as an area devoid of vegetation suitable for the establishment of a seedling. Some of the responses of seeds to various environmental stimuli may act as gap-detection mechanisms. The requirement for light with a high red/far red ratio means that many seeds will not germinate if shaded by other plants (Gorski et al. 1977, Fenner 1980, Silvertown 1980b). The frequent requirement for fluctuating temperatures (Thompson et al. 1977) could act as both a gap-detecting and a depth-sensing mechanism. Some species in fire-prone communities respond to favorable conditions by requirement of high temperature for scarification (Keeley 1991). The positive response to nitrate seen in many species (Karssen and Hilhorst 1992) could also be related to germinating in gaps, where the disturbed soil will undergo a flush of nitrate availability (Pons 1989). These specific responses may help the seeds to identify favorable sites in which to germinate. Certainly, the seeds of many parasitic species such as *Orobanche, Striga*, etc., can detect the presence of their host plant by a root secretion into the soil (Joel et al. 1995). The concept of gap detection is in principle no different from host detection, although the latter is considerably more specific.

VII. SEEDLING ESTABLISHMENT

After seed production, dispersal, and germination, have been successfully achieved, the final stage in the regeneration process is the establishment of the seedling. Very young seedlings are susceptible to many hazards such as desiccation (Miles 1972, Maruta 1976), pathogens (Augspurger 1983, 1984b), ''winter death'' and grazing (Mack and Pyke 1984), as well as competition from existing vegetation (Aguilera and Lauenroth 1993, Tyler and Dantonio 1995, Kolb and Robberecht 1996). The causes of death may be ascertained and quantified by direct observation, as was attempted for *Bromus tectorum* (Mack and Pyke 1984), but this entails many intractable technical problems (Fenner 1987). Alternatively, the factors limiting recruitment can be investigated experimentally by increasing some resources, e.g., water (Miles 1972, Hoffmann 1997) or nutrients (Grime and Curtis 1976), or by protection from predators (Molofsky and Fisher 1993, Terborgh and Wright 1994, Hanley et al. 1995a, Asquith et al. 1997). These hazard agents may operate sequentially. For example, during the first 2–4 weeks after germination, seedlings of a neotropical tree, *Tachigalia versicolor*, suffer a high mortality rate from mammalian grazers; however, after the fourth week, hypocotyls become woody and mammalian attack ceases, and damping-off pathogens become the main source of mortality (Kitajima and Augspurger 1989).

There are three possible ways in which large seed size translates into an advantage for seedling establishment (Westoby et al. 1992). First, a large seed can create a large seedling. Second, a significant fraction of resources in a large seed may remain in storage instead of being used for immediate seedling development (Garwood 1996). The seedling may augment the limited resources from this storage for a long period, or it may use it for a rapid recovery from loss to herbivory or other physical damage (Dalling et al. 1997a). And third, the advantage of a large seed may be indirect via association of seed size with seedling morphology and development types (Hladik and Miquel 1991, Kitajima 1996a). These three pathways for large-seed advantage are not mutually exclusive. The seedling stage is physiologically dynamic as the seedling switches from complete dependency on seed reserves to complete autotrophy. The relative importance of the first two mechanisms may be evaluated through a careful measurement of seed reserve dependency relative to the development of autotrophy. The third mechanism is a relatively new concept, however, and deserves more attention. In an interspecific comparison of 46 neotropical tree species, the greater the seed mass, the less photosynthetic the cotyledons and the slower the relative growth rates (Kitajima 1992a, 1996a; Figure 4). Phylogeny exerts a strong influence on both seed size and early seedling morphology, and in turn, how advantageous a large seed may be in resource-limited conditions.

All else being equal, the greater the seed reserve mass is, the greater the initial seedling mass. Lipid content in seeds provides a small modifier to this rule; a unit mass of oil-rich seed is converted to a greater mass of seedling than a unit mass of starchy seed (Penning de Vries and Van Laar 1977, Kitajima 1992a). However, interspecific variation in seedling size due to variation in seed lipid content (up to twofold) is completely dwarfed by the variation due to seed mass (up to 10^5-fold). Other factors, such as whether the seedling sets aside a part of seed reserves in storage or what type of seedling tissue is created, appear to be greater modifiers of initial seedling size as a function of seed size. Seed reserves alone are generally sufficient to construct seedling body up to the complete development of the first pair of cotyledons (in the case of species with thin photosynthetic cotyledons) or leaves (in the case of species with storage cotyledons). Certainly, within a species, bigger seeds produce larger seedlings initially (Stanton 1984, Wulff 1986, Gonzalez 1993). However, subsequent rela-

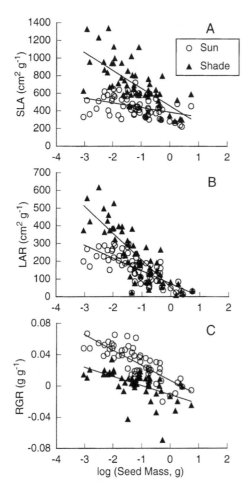

Figure 4 Evolutionary ecological associations of seed mass with morphological traits and relative growth rates of seedlings for neotropical woody species in a seasonal, moist forest in Panama. Seedlings of each species were raised from seeds in controlled sun and shade conditions (23% and 2% of open sky). (A) SLA and (B) LAR were determined at the full expansion of the first true leaves, and (C) subsequent RGR was determined from harvests 10 weeks later. Large-seeded species had significantly smaller SLA, LAR, and RGR in both sun and shade (linear regressions shown; $P < .002$ for SLA in sun, $P < .0001$ for all others; n = 57–61 and 51–58 for sun and shade, respectively). (*Source*: data from Kitajima 1992a.)

tive growth rates are independent of seed size in the absence of competition (Choe et al. 1988) and the differential effect often wears off with maturity (Houssard and Escarré 1991, Bockstaller and Girardin 1994).

In small gaps or shaded environments, the attainment of a minimum size may be necessary to secure an independent existence. Several recent experimental studies showed that bigger seeds have an advantage in shade (Leishman and Westoby 1994b, Osunkoya et al. 1994, Saverimuttu and Westoby 1996b). The increased height obtainable by seedlings would be most useful to seedlings in conditions where there was a steep gradient of light due to shade from surrounding vegetation (Grime and Jeffrey 1965, Leishman and Westoby 1994b), or for seeds germinating below litter (Molofsky and Augspurger 1992). Another way to achieve greater height is elongation of hypocotyl or epicotyl in response to low red/far red ratio of light coming through surrounding vegetation (Ballaré et al. 1988, Leishman and Westoby 1994b). This response seems to be dependent on the ecology and phylogeny of the species (Morgan and Smith 1979, Corré 1983). For example, in a study of 15 tropical tree species, only shade-intolerant species belonging to the Bombacaceae, but not equally shade-intolerant species in other families, responded to a low red/far red ratio with hypocotyl elongation (Kitajima 1994). Stem elongation is perhaps a maladaptive response, as it costs sturdiness without any advantage of escaping shade, unless the height of the surrounding vegetation is low enough, e.g., as in grassland rather than closed forest canopies.

It has long been assumed that large seeds provide a reserve of nutrients on which the seedlings can draw to tide them over in initially inhospitable establishment conditions. Logically, this advantage lasts only as long as the storage of reserves last. For example, Saverimuttu and Westoby (1996a) found that the large-seed advantage of seeding longevity in shade exists only during the cotyledon stage, but not for seedlings transferred to deep shade after full expansion of leaves. However, the existence of attached cotyledons is only a rough evidence of continued dependency on seed reserves. Unfortunately, the actual duration of dependency on seed reserves has rarely been quantified. This is because the transition of resource dependency from internal to external sources occurs gradually and the dependency period may differ for different resources (Fenner 1987, Kitajima 1996a, 1996b, Hanley and Fenner 1997). Initially after radicle emergence, a developing seedling acquires all necessary resources from seed reserves, and its growth rate is independent of external resource availability. Gradually, after the development of organs necessary for autotrophy, such as photosynthetic cotyledons or leaves for acquisition of energy and roots for acquisition of mineral nutrients, a seedling shifts dependency onto external sources. It is possible to apply a functional growth analysis to monitor this change of dependency from heterotrophy to autotrophy for a given resource type, such as carbon or nitrogen (Kitajima 1996b).

There occurs a very general rule that seed size is negatively correlated with seedling relative growth rate between species. This has been found in a wide range of plant families and habitats, e.g., pasture grasses and legumes (Fenner and Lee 1989); species from arid central Australia (Jurado and Westoby 1992); woody plants in various climates (e.g., Figure 4; Kitajima 1992a, 1994, Huante et al. 1995, Cornelissen et al. 1996); Mediterranean annuals (Maranon and Grubb 1993); and a range of Ausralian plants (Swanborough and Westoby 1996). The functional basis of this relationship is via correlation of seed size with biomass allocation patterns, rather than via correlation of seed size with metabolic or photosynthetic rates (Maranon and Grubb 1993, Kitajima 1994, Huante et al. 1995, Cornlissen et al. 1996). First, small-seeded species tend to have thin, epigeal, leaf-like cotyledons. Having photosynthetic cotyledons allows them to start using light as the main source of energy earlier than they could if they had storage-type cotyledons (Kitajima 1992b, 1996a). Second, small-seeded species tend to have a higher specific leaf area (SLA; leaf area divided by leaf mass) and leaf area ratio (LAR; leaf area divided by total plant dry mass) (Figure 4). Several recent studies have identified SLA and LAR as the main source of interspecific variation in potential relative growth rates, rather than photosynthetic rates per unit leaf area or net assimilation rates (Poorter and Remkes 1990, Maranon and Grubb 1993, Kitajima 1994, Huante et al. 1995, Saverimuttu and Westoby 1996a). Interspecific variation in LAR among neotropical trees at the time of full expansion of the first set of true leaves is best explained by the differences in SLA of cotyledons (cotyledon area divided by cotyledon mass). At 10 weeks after full expansion of true leaves, LAR continues to be correlated with cotyledon SLA, but less than with leaf SLA (Kitajima 1996b). Additional correlations of early seedling morphology and seed size, such as the relationship between root architecture and seed size (Huante et al. 1992, Kohyama and Grubb 1994), may also contribute to the relationship between seed mass and relative growth rates. In summary, the diversity of early seedling morphology and biomass allocation patterns exemplifies the natural selection for particular relative growth rates in particular habitats.

Species from resource-rich habitats grow faster than species from resource-poor habitats. The former also tend to exhibit a greater response to experimentally controlled resource availability. However, species ranking according to relative growth rates of seedlings generally does not change in relation to resource availability. Fast-growing species in sun are fast growing in shade (Ellison et al. 1993, Kitajima 1994). Likewise, fast-growing species in nutrient-poor soil are fast growing in rich soil (Huante et al. 1995, Lusk et al. 1997). Although it is tempting to hypothesize that seedlings should allocate preferentially more to leaves for successful recruitment in the shaded habitat, this is not the case. Instead, seedlings of shade-adapted species have lower LAR and lower relative growth rate than light-demanding species when both are grown under a given controlled light regime (Kitajima 1994, Leishman and Westoby 1994b). Thus, the functional basis for shade tolerance cannot be traced to efficient utilization of light, but rather to avoidance and/or tolerance of inevitable tissue loss, such as that caused by grazers and herbivores, in the energy-limited environment. In contrast, high LAR is selected to achieve high potential growth rates in light-demanding species for competitive advantage.

An important determinant of whether a seedling will survive to establish successfully is whether the seed is deposited in a safe site, a place where the seed is provided with (1) the stimuli for breaking dormancy, (2) the conditions and resources required for germination, and (3) the absence of predators, competitors, pathogens, and toxins (Harper 1977). Because different species have different requirements and tolerances as seedlings, a safe site for one species may not be safe for another.

Predation is undoubtedly one of the greatest hazards facing newly germinated seedlings. For example, many tropical tree seedlings are killed by mammalian grazers (Kitajima and Augspurger 1989, Terborgh and Wright 1994). Mollusk grazing of seedlings may be more important in temperate grasslands (Barker 1989, Hulme 1994, Hanley et al. 1996b). Mollusk grazing has been shown to have major effects on numbers of recruits (Hanley et al. 1995b) and on eventual species composition (Hanley et al. 1996a). Very young seedlings appear to be most at risk (Fenner 1987, Hanley et al. 1995a). The likelihood of a seedling being grazed by mollusks is influenced by several factors, including the size of the gap in which it finds itself and the time of year (Hanley et al. 1996b), and even the identity of its nearest neighbor (Fenner 1987). Soil-borne damping-off pathogens also operate as seedling predators, as they often kill seedlings within a few days after infection. Damping-off pathogens operate in a density-dependent manner; crowded seedlings suffer disproportionately greater mortality (Augspurger 1983). The age structure of the seedling population, reflecting its germination chronology, also affects the mortality due to disease (Neher et al. 1992). However, the most important ecological aspect of interaction between seedlings and damping-off pathogens is perhaps that seedlings are more susceptible in shaded environments than in gaps (Augspurger 1983, 1984b). Thus, species differences in survivorship of tree seedlings in the forest understory can be largely attributable to species differences in susceptibility to damping-off pathogens.

A common requirement for many seedlings is the absence of competition from other species within the immediate vicinity. Closed vegetation provides an inhospitable arena for seedling establishment. Breaks in the continuous vegetation cover provide localized, temporary release from competition, and are probably an essential prerequisite for the successful regeneration of many species. Gaps are continually being created in vegetation by both biological and inanimate agents. Examples of gaps caused by animals include molehills, anthills, rabbit scrapes, hoof prints, and wormcasts. Nonbiological sources of gap-forming disturbance are frost-heave, fire, landslip, and flood deposition. A gap may be created by internal processes such as the death of an individual plant or a fallen tree. Each agency creates a gap with characteristic qualities, and many studies indicate that gap type has a large influence on which particular species in a community will regenerate. Clear evidence that regeneration is differentially affected by gap features comes from experiments in which different-sized gaps have been created in vegetation and the fate of seedlings followed (McConnaughay and Bazzaz 1987, Bullock et al. 1995, Gray and Spies 1996). Some species show preferential survival in bigger gaps, others in smaller gaps (Denslow 1980). Although there appears to be a limit to how finely the regeneration niche may be divided along a gap-size gradient in a tropical forest, species do appear to differ in the minimum gap size required for successful regeneration (Brokaw 1987). Due to the large environmental gradient from the center to the edge of a given gap (Brown 1993), a seedling's position within a gap may be more important to its survival than gap size per se (Brown and Whitmore 1992). One species may be favored in the center, whereas others survive better near the margins. Gap shape, which determines the ratio of margin to area, may be important for this reason.

The gap created by a fallen tree provides a particularly complex set of heterogeneous microhabitats favoring distinct groups of species at the root, trunk, and crown (Nuñez-Farfán and Dirzo 1988). It has been shown experimentally that seeds are very sensitive to soil microtopography (Smith and Capelle 1992), with even closely related species showing differential reactions to the precise texture of the surface (Harper et al. 1965, Putz 1983). Germination is partly determined by the degree of contact between the seed and the soil. The coarseness of the topography can also influence the likelihood of entrapment of seeds as well as their germination (Chambers 1995).

The timing at which the disturbance occurs is a crucial determinant of which species will regenerate. Records of similar gaps (e.g., gopher mounds) created at different times of the year show differences in colonization (Hobbs and Mooney 1985). The condition of a given gap changes rapidly and restricts the number of species that can tolerate the increasingly stronger competitive regime (Brokaw 1987). There is clearly also a large stochastic element governing regeneration, influenced by unpredictable factors such as the presence of a seeding parent plant in the vicinity, the absence of grazers, and the occurrence of suitable weather conditions, all coinciding at the right place and time. However, within this haphazard framework, it is still possible to show that certain types of gaps favor the establishment of certain species from seed. A practical example of this is the creation of seedbeds by foresters to encourage natural regeneration. These usually consist of patches of bare ground created in the vegetation on the forest floor, designed to increase the number of safe sites for a particular species (Prevost 1996).

VIII. CONCLUSION

Regeneration from seed is influenced by a wide range of environmental factors, plant characteristics, and stochastic events. The species composition of a plant community is a consequence of the successful regeneration of a selection of the potential species available. The long-term maintenance of each species requires the recurrent creation of suitable regeneration opportunities at appropriate intervals. At any one site, these opportunities are likely not to remain constant with time, due to natural disturbance, human influence, and even climatic change. Knowledge of regeneration requirements of the key species is of great practical importance in vegetation management, either for commercial or conservation purposes. Comparative study of the relationship between seed and seedling traits and regeneration requirements of the species would be particularly useful in this context. The practice of sustainable forestry, the control of invading weed species, and the prevention of soil erosion by revegetation all involve plant regeneration from seed. Detailed research on the regeneration niche of key species in natural communities would enable us to manipulate vegetation much more effectively than at present.

REFERENCES

Aguilera, M.O., and W.K. Lauenroth. 1993. Seedling establishment in adult neighbourhoods intraspecific constraints in the regeneration of the bunchgrass *Bouteloua gracilis*. Journal of Ecology 81: 253–261.

Allen, R.B., and K.H. Platt. 1990. Annual seedfall variation in *Nothofagus solandri* (Fagaceae), Canterbury, New Zealand. Oikos 57: 199–206.

Allsopp, N., and W.D. Stock. 1995. Relationships between seed reserves, seedling growth and mycorrhizal responses in 14 related shrubs (Rosidae) from a low-nutrient environment. Functional Ecology 9: 248–254.

Askew, A.P., D. Corker, D.J. Hodkinson, and K. Thompson. 1997. A new apparatus to measure the rate of fall of seeds. Functional Ecology 11: 121–125.

Asquith, N.M., S.J. Wright, and M.J. Clauss. 1997. Does mammal community composition control recruitment in neotropical forests? Evidence from Panama. Ecology 78: 941–946.

Augspurger, C.K. 1983. Seed dispersal of the tropical tree, *Platypodium elegans*, and the escape of its seedlings from fungal pathogens. Journal of Ecology 71: 759–771.

Augspurger, C.K. 1984a. Light requirements of neotropical tree seedlings: a comparative study of growth and survival. Journal of Ecology 72: 777–795.

Augspurger, C.A. 1984b. Seedling survival of tropical tree species; interactions of dispersal distance, light gaps, and pathogens. Ecology 65: 1705–1712.

Augspurger, C.K. 1986. Morphology and dispersal potential of wind-dispersed diaspores of neotropical trees. American Journal of Botany 73: 353–363.

Augspurger, C.K., and K. Kitajima. 1992. Experimental studies of seedling recruitment from contrasting seed distributions. Ecology 73: 1270–1287.

Auld, T.D. 1986. Variation in predisperal seed predation in several Australian *Acacia* spp. Oikos 47: 319–326.

Baker, H.G. 1972. Seed weight in relation to environmental conditions in California. Ecology 53: 997–1010.

Ballaré, C.L., R.A. Sánchez, A.L. Scopel, and C.M. Ghersa. 1988. Morphological responses of *Datura ferox* L. seedlings to the presence of neighbours. Oecologia 76: 288–293.

Barclay, A.S., and F.R. Earl. 1974. Chemical analyses of seeds III. Oil and protein content of 1253 species. Economic Botany 28: 178–236.

Barker, G.M. 1989. Slug problems in New Zealand pastoral agriculture. In: Henderson, I.F., ed. Slugs and Snails in World Agriculture, BCPC Monograph. Thornton Heath, England: British Crop Protection Council, pp 59–68.

Baskin, J.M., and C.C. Baskin. 1980. Ecophysiology of secondary dormancy in seeds of *Ambrosia artemisiifolia*. Ecology 61: 475–480.

Beattie, A.J., and D.C. Culver. 1982. Inhumation: how ants and other invertebrates help seeds. Nature 297: 627.

Bekker, R.M., J.P. Bakker, U. Grandin, R. Kalamees, P. Milberg, P. Poschlod, K. Thompson, and J.H. Willems. 1997. Seed size, shape and vertical distribution in the soil: indicators of seed longevity. Functional Ecology, in press.

Benkman, C.W. 1995. The impact of tree squirrels (*Tamiasciurus*) on limber pine seed dispersal adaptations. Evolution 49: 585–592.

Bennet, A., and J. Krebs. 1987. Seed dispersal by ants. Trends in Ecology and Evolution 2: 291–292.

Bockstaller, C., and P. Girardin. 1994. Effects of seed size on maize growth from emergence to silking. Maydica 39: 213–218.

Bond, W., and P. Slingsby. 1984. Collapse of an ant-plant mutualism: the Argentine ant (*Iridomyrmex humilis*) and myrmecochorous Proteaceae. Ecology 65: 1031–1037.

Bouwmeester, H.J. 1990. The effect of environmental conditions on the seasonal dormancy pattern and germination of weed seeds. PhD thesis, Agricultural University of Wageningen, Wageningen, The Netherlands.

Brokaw, N.V.L. 1987. Gap-phase regeneration of three pioneer tree species in a tropical forest. Journal of Ecology 75: 9–19.

Brown, J.S., and D.L. Venable. 1986. Evolutionary ecology of seed bank annuals in temporally varying environments. American Naturalist 127: 31–47.

Brown, N.D. 1993. The implications of climate and gap microclimate for seedling growth conditions in a Bornean lowland rain forest. Journal of Tropical Ecology 9: 153–168.

Brown, N.D., and T.C. Whitmore. 1992. Do dipterocarp seedlings really partition tropical rainforest gaps? Philosophical Transactions of Royal Society of London, Series B 335: 369–378.

Bullock, J.M., B.C. Hill, J. Silvertown, and M. Sutton. 1995. Gap colonization as a source of grassland community change: effects of gap size and grazing on the rate and mode of colonization by different species. Oikos 72: 273–282.

Burke, M.J.W., and J.P. Grime. 1996. An experimental study of plant community invasability. Ecology 77: 776–790.

Burkey, T.V. 1994. Tropical tree species diversity: a test of the Janzen-Connell model. Oecologia 97: 533–540.

Cardina, J., H.M. Norquay, B.R. Stinner, and D.A. McCartney. 1996. Postdispersal predation of velvetleaf (*Abutilon theophrasti*) seeds. Weed Science 44: 534–539.

Carlquist, S. 1965. Island Life. New York: Natural History Press.

Chambers, J.C. 1995. Relationships between seed fates and seedling establishment in an alpine ecosystem. Ecology 76: 2124–2133.

Choe, H.S., C. Chu, G. Koch, J. Gorham, and H.A. Mooney. 1988. Seed weight and seed resources in relation to plant growth rate. Oecologia 76: 158–159.

Clark, D.A., and D.B. Clark. 1984. Spacing dynamics of a tropical rain forest tree: evaluation of the Janzen-Connell model. American Naturalist 124: 769–788.

Cody, M.L., and J.M. Overton. 1996. Short-term evolution of reduced dispersal in island plant populations. Journal of Ecology 84: 53–61.

Condit, R., S.P. Hubbell, and R.B. Foster. 1992. Recruitment near conspecific adults and the maintenance of tree and shrub diversity in a neotropical forest. American Naturalist 140: 261–286.

Connell, J.H. 1971. On the role of natural enemies in preventing competitive exclusion in some marine animals and in rain forest trees. In: den Boer, P.J., and G.R. Gradwell, eds. Dynamics of Populations. Proceedings of the Advanced Study Institute on Dynamics of Numbers in Populations, Oosterbeek, 1970. Wageningen, The Netherlands: Centre for Agricultural Publishing and Documentation, pp 298–310.

Cornelissen, J.H.C., P. Castro-Diez, and R. Hunt. 1996. Seedling growth, allocation and leaf attributes in a wide range of woody plant species and types. Journal of Ecology 84: 755–765.

Corré, W.J. 1983. Growth and morphogenesis of sun and shade plants. II. The influence of light quality. Acta Botanica Neerlandica 32: 185–202.

Crawley, M.J. 1992. Seed predators and plant population dynamics. In: Fenner, M., ed. Seeds: The Ecology of Regeneration in Plant Communities. Wallingford: CAB International, pp 157–191.

Crawley, M.J., and C.R. Long. 1995. Alternate bearing, predator satiation and seedling recruitment in *Quercus robur*. Journal of Ecology 83: 683–696.

Crist, T.O., and C.F. Friese. 1993. The impact of fungi on soil seeds: implications for plants and granivores in a semiarid shrub-steppe. Ecology 74: 2231–2239.

Da Silva, H.R., M.C. De Britto-Pereira, and U. Caramaschi. 1989. Frugivory and seed dispersal by *Hyla truncata*; a neotropical tree frog. Copeia 3: 781–783.

Dalling, J.W., K.E. Harms and R. Aizprúa. 1997a. Seed damage tolerance and seedling resprouting ability of *Prioria copaifera* in Panama. Journal of Tropical Ecology 13: 481–490.

Dalling, J.W., M.D. Swaine, and N.C. Garwood. 1995. Effect of soil depth on seedling emergence in tropical soil seed-bank investigations. Functional Ecology 9: 119–121.

Dalling, J.W., M.D. Swaine, and N.C. Garwood. 1997. Soil seed bank community dynamics in seasonally moist lowland tropical forest, Panama. Journal of Tropical Ecology 13: 659–680.

Dalling, J.W., M.D. Swaine, and N.C. Garwood. 1998. Dispersal patterns and seed bank dynamics of pioneer trees in moist tropical forest, Panama. Ecology 79: 564–578.

Darwin, C. 1859. The Origin of Species. London: Murray.

Denslow, J. 1980. Gap partitioning among tropical rainforest trees. Biotropica 12: 47–55.

Dessaint, F., R. Chadoeuf, and G. Barralis. 1991. Spatial pattern analysis of weed seeds in the cultivated soil seed bank. Journal of Applied Ecology 28: 721–730.

Donaldson, J.S. 1993. Mast-seeding in the cycad genus *Encephalartos*: a test of the predator satiation hypothesis. Oecologia 94: 262–271.

Ellison, A.M., J.S. Denslow, B.A. Loiselle, and M.D. Brenes. 1993. Seed and seedling ecology of neotropical Melastomataceae. Ecology 74: 1733–1749.

Fenner, M. 1980. Germination tests on thirty-two East African weed species. Weed Research 20:135–138.

Fenner, M. 1983. Relationships between seed weight, ash content and seedling growth in twenty-four species of Compositae. New Phytologist 95: 697–706.

Fenner, M. 1985. Seed Ecology. London: Chapman and Hall.

Fenner, M. 1987. Seedlings. In: Rorison, I.H., J.P. Grime, R. Hunt, G.A.F. Hendry, and D.H. Lewis, eds. Frontiers of Comparative Plant Ecology. New Phytologist 106(Suppl). London: Academic Press, pp 35–47.

Fenner, M. 1991. Irregular seed crops in forest trees. Quarterly Journal of Forestry 85: 166–172.

Fenner, M. 1995. Ecology of seed banks. In: Kigel, J., and G. Galili, eds. Seed Development and Germination. New York: Marcel Dekker, pp 507–528.

Fenner, M., and W.G. Lee. 1989. Growth of seedlings of pasture grasses and legumes deprived of single mineral nutrients. Journal of Applied Ecology 26: 223–232.

Forget, P.M. 1990. Seed-dispersal of *Vacapoua americana* (Caesalpiniaceae) by caviomorph rodents. Journal of Tropical Ecology 6: 459–468.

Forget, P.M. 1997. Ten-year seedling dynamics in *Vouacapoua americana* in French Guiana: a hypothesis. Biotropica 29: 124–126.

Forget, P.M., and T. Milleron. 1991. Evidence for secondary seed dispersal by rodents in Panama. Oecologia 87: 596–599.

Ganeshaiah, K.N., and S.R. Uma. 1991. Seed size optimization in a wind dispersed tree *Butea monosperma*: a trade-off between seedling establishment and pod dispersal efficiency. Oikos 60: 3–6.

Garwood, N.C. 1996. Functional morphology of tropical tree seedlings. In: Swaine, M.D., ed. Ecology of Tropical Forest Tree Seedlings. Paris: UNESCO, pp 59–129.

Garwood, N.C., and J.R.B. Lighton. 1990. Physiological ecology of seed respiration in some tropical species. New Phytologist 115: 549–558.

Geritz, S.A. 1995. Evolutionarily stable seed polymorphism and small-scale spatial variation in seedling density. American Naturalist 146: 685–707.

Godnan, D. 1983. Pest Slugs and Snails. Berlin: Springer-Verlag.

Gonzalez, E. 1993. Effect of seed size on germination and seedling vigor of *Virola koschny* Warb. Forest Ecology and Management 57: 275–281.

Gorski, T., K. Gorska, and J. Nowicki. 1977. Germination of seeds of various species under leaf canopy. Flora 166: 249–259.

Goulding, M. 1980. The Fishes and the Forest. Berkeley: University of California Press.

Gray, A.N., and T.A. Spies. 1996. Gap size, within-gap position and canopy structure effects on conifer seedling establishment. Journal of Ecology 84: 635–645.

Greene, D.F., and E.A. Johnson. 1993. Seed mass and dispersal capacity in wind-dispersed diaspores. Oikos 67: 69–74.

Greig, N. 1993. Predispersal seed predation on five *Piper* species in tropical rain-forest. Oecologia 93: 412–420.

Grime, J.P. 1989. Seed banks in ecological perspective. In: Leck, M.A., V.T. Parker, and R.L. Simpson, eds. Ecology of Soil Seed Banks. London: Academic Press, pp 15–22.

Grime, J.P., and A.V. Curtis. 1976. The interaction of drought and mineral nutrient stress in calcareous grassland. Journal of Ecology 64: 975–988.

Grime, J.P., and D.W. Jeffrey. 1965. Seedling establishment in vertical gradients of sunlight. Journal of Ecology 53: 621–642.

Gross, C.L., M.A. Whalen, and M.H. Andrew. 1991. Seed selection and removal by ants in a tropical savanna woodland in northern. Australian Journal of Tropical Ecology 7: 99–112.

Grubb, P.J. 1977. The maintenance of species-richness in plant communities: the importance of the regeneration niche. Biological Review 52: 107–145.

Hanley, M.E., and M. Fenner. 1997. Seedling growth of four fire-following Mediteranean plant species deprived of single mineral nutrients. Functional Ecology 11: 398–405.

Hanley, M.E., M. Fenner, and P.J. Edwards. 1995a. An experimental field study of the effects of mollusc grazing on seedling recruitment and survival in grassland. Journal of Ecology 83: 621–627.

Hanley, M.E., M. Fenner, and P.J. Edwards. 1995b. The effect of seedling age on the likelihood of herbivory by the slug *Deroceras reticulatum*. Functional Ecology 9: 754–759.

Hanley, M.E., M. Fenner, and P.J. Edwards. 1996a. The effect of mollusc grazing on seedling recruitment in artificially created grassland gaps. Oecologia 106: 240–246.

Hanley, M.E., M. Fenner, and P.J. Edwards. 1996b. Mollusc grazing and seedling survivorship of four common grassland plant species: the role of gap size, species and season. Acta Oecologica 17: 331–341.

Harper, J.L. 1977. Population Biology of Plants. London: Academic Press.

Harper, J.L., J.T. Williams, and G.R. Sagar. 1965. The behaviour of seeds in soil I. The heterogeneity of soil surfaces and its role in determining the establishment of plants from seed. Journal of Ecology 53: 273–286.

Hedge, S.G., R.U. Shaanker, and K.N. Ganeshaiah. 1991. Evolution of seed size in the bird-dispersed tree *Santalum album* L: a trade off between seedling establishment and dispersal efficiency. Evolutionary Trends in Plants 5: 131–135.

Hladik, A., and S. Miquel. 1991. Seedling types and plant establishment in an African rain forest. In:Bawa, K.S., and M. Hadley, eds. Reproductive Ecology of Tropical Forest Plants. Paris: UNESCO, pp 317–319.

Hnatiuk, S.H. 1978. Plant dispersal by the Aldabran giant tortoise *Geochelone gigantea* (Scheigger). Oecologia 36: 345–350.

Hobbs, R.J., and H.A. Mooney. 1985. Community and population dynamics of serpentine grassland annuals in relation to gopher disturbance. Oecologia 67: 342–351.

Hoffmann, W.A. 1997. The effects of fire and cover on seedling establishment in a neotropical savanna. Journal of Ecology 84: 383–393.

Hopkins, M.S., and A.W. Graham. 1987. The viability of seeds of rainforest species after experimental soil burials under wet lowland forest in northeastern Australia. Australian Journal of Ecology 12: 97–108.

Horn, M.H. 1997. Evidence for dispersal of fig seeds by the fruit-eating characid fish *Brycon guatemalensis* Regan in a Costa Rican tropical rain forest. Oecologia 109: 259–264.

Houle, G. 1995. Seed dispersal and seedling recruitment—the missing link(s). Ecoscience 2: 238–244.

Houssard, C., and J. Escarré. 1991. The effects of seed weight on growth and competitive ability of *Rumex acetosella* from two successional old-fields. Oecologia 86: 236–242.

Howe, H.F. 1989. Scatter and clump-dispersal and seedling demography: hypothesis and implications. Oecologia 79: 417–426.

Huante, P., E. Rincón, and I. Acosta. 1995. Nutrient availability and growth rate of 34 woody species from a tropical deciduous forest in Mexico. Functional Ecology 9: 849–858.

Huante, P., E. Rincón, and M. Gavito. 1992. Root system analysis of seedlings of seven tree species from a tropical dry forest in Mexico. Trees 6: 77–82.

Hughes, L., and M. Westoby. 1992. Fate of seeds adapted for dispersal by ants in Australian sclerophyll vegetation. Ecology 73: 1285–1299.

Hulme, P.E. 1994. Seedling herbivory in grassland: relative impact of vertebrate and invertebrate herbivores. Journal of Ecology 82: 873–880.

Janzen, D.H. 1969. Seed-eaters vs seed size, number, toxicity and dispersal. Evolution 23: 1–27.

Janzen, D.H. 1970. Herbivores and the number of tree species in tropical forests. American Naturalist 104: 501–528.

Janzen, D.H. 1984. Dispersal of small seeds by big herbivores: foliage is the fruit. American Naturalist 123: 338–353.

Janzen, D.H., G.A. Miller, J. Hackforth-Jones, C.M. Pond, K. Hooper, and D.P. Janos. 1976. Two Costa Rican bat-generated seed shadows of *Andira inermis* (Leguminosae). Ecology 57: 1068–1075.

Jensen, T.S. 1985. Seed–seed predator interactions of European beech, *Fagus sylvatica*, and forest rodents, *Clethrionomys glareolus* and *Apodemus flavicollis*. Oikos 44: 149–156.

Joel, D.M., J.C. Steffens, and D.E. Matthews. 1995. Germination of weedy root parasites. In: Kigel, J., and G. Galili, eds. Seed Development and Germination. New York: Marcel Dekker, pp 567–597.

Jones, Q., and F.R. Earle. 1966. Chemical analysis of seeds II: oil and protein content of 759 species. Economic Botany 20: 127–155.

Jurado, E., and M. Westoby. 1992. Seedling growth in relation to seed size among species of arid Australia. Journal of Ecology 80: 407–416.

Karssen, C.M., and H.M.W. Hilhorst. 1992. Effect of chemical environment on seed germination. In: Fenner, M., ed. Seeds: The Ecology of Regeneration in Plant Communities. Wallingford, England: CAB International, pp 327–348.

Kaspari, M. 1996. Worker size and seed size selection by harvester ants in a neotropical forest. Oecologia 105: 397–404.

Keeley, J.E. 1991. Seed germination and life history syndromes in the California chaparral. Botanical Review 57: 81–116.

Kelly, C.K. 1995. Seed size in tropical trees: a comparative study of factors affecting seed size in Peruvian angiosperms. Oecologia 102: 377–388.

Kelly, D. 1994. The evolutionary ecology of mast seeding. Trends in Ecology and Evolution 9: 465–470.

Kitajima, K. 1992a. The importance of cotyledon functional morphology and patterns of seed reserve utilization for the physiological ecology of neotropical tree seedlings. PhD Thesis, University of Illinois, Urbana, IL.

Kitajima, K. 1992b. Relationship between photosynthesis and thickness of cotyledons for tropical tree species. Functional Ecology 6: 582–584.

Kitajima, K. 1994. Relative importance of photosynthetic traits and allocation patterns as correlates of seedling shade tolerance of 13 tropical trees. Oecologia 98: 419–428.

Kitajima, K. 1996a. Cotyledon functional morphology, seed reserve utilization, and regeneration niches of tropical tree seedlings. In: Swaine, M.D., ed. The Ecology of Tropical Forest Seedlings. Paris: UNESCO, pp 193–208.

Kitajima, K. 1996b. Ecophysiology of tropical tree seedlings. In: Mulkey, S.S., R.L. Chazdon, and A.P. Smith, eds. Tropical Forest Plant Ecophysiology. New York: Chapman and Hall, pp 559–596.

Kitajima, K., and C.K. Augspurger. 1989. Seed and seedling ecology of a monocarpic tropical tree, *Tachigalia versicolor*. Ecology 70: 1102–1114.

Kitajima, K., and D. Tilman. 1996. Seed banks and seedling establishment on an experimental productivity gradient. Oikos 76: 381–391.

Kivilaan, A., and R.S. Bandurski. 1981. The one hundred-year period for Dr. Beal's seed viability experiment. American Journal of Botany 68: 1290–1292.

Kohyama, T., and P.J. Grubb. 1994. Below-and above-ground allometries of shade-tolerant seedlings in a Japanese warm-temperate rain forest. Functional Ecology 8: 229–236.

Kolb, P.F., and R. Robberecht. 1996. *Pinus ponderosa* seedling establishment and the influence of competition with the bunchgrass *Agropyron spicatum*. International Journal of Plant Science 157: 509–515.

Kubitzki, K., and A. Ziburski. 1994. Seed dispersal in flood plain forests of Amazonia. Biotropica 26: 30–43.

Leck, M.A., V.T. Parker, and R.L. Simpson, eds. 1989. Ecology of Soil Seed Banks. San Diego, CA: Academic Press.

Lee, W.G., and M. Fenner. 1989. Mineral nutrient allocation in seeds and shoots of twelve *Chionochloa* species in relation to soil fertility. Journal of Ecology 77: 704–716.

Lee, W.G., M. Fenner, and R.P. Duncan. 1993. Pattern of natural regeneration of narrow leaved snow tussock *Chionochloa rigida* ssp. *rigida* in Central Otago, New Zealand. New Zealand Journal of Botany 31: 117–125.

Leishman, M.R., and M. Westoby. 1994a. Hypotheses on seed size: tests using the semi-arid flora of western New South Wales. American Naturalist 143: 890–906.

Leishman, M.R., and M. Westoby. 1994b. The role of large seed size in shaded conditions: experimental evidence. Functional Ecology 8: 205–214.

Leishman, M.R., and M. Westoby. 1994c. The role of seed size in seedling establishment in dry soil conditions: experimental evidence from semi-arid species. Journal of Ecology 82: 249–258.

Leishman, M.R., M. Westoby, and E. Jurado. 1995. Correlates of seed size variation: a comparison among five temperate floras. Journal of Ecology 83: 517–529.

Levey, D.J., and M.M. Byrne. 1993. Complex ant plant interactions: rain-forest ants as secondary dispersers and postdispersal seed predators. Ecology 74: 1802–1812.

Ligon, D.J. 1978. Reproductive interdependence of pinyon jays and pinyon pines. Ecological Monographs 48: 111–126.

Lott, R.H., G.N. Harrington, A.K. Irvine, and S. McIntyre. 1995. Density-dependent seed predation and plant dispersion of the tropical palm *Normanbya normanbyi*. Biotropica 27: 87–95.

Louda, S.M. 1982. Limitations of the recruitment of the shrub *Haplopappus squarrosus* (Asteraceae) by flower-and seed-feeding insects. Journal of Ecology 70: 43–53.

Louda, S.M., and M.A. Potvin. 1995. Effect of inflorescence-feeding insects on the demography and lifetime fitness of a native plant. Ecology 76:229–245.

Lusk, C.H., O. Contreras, and J. Figueroa. 1997. Growth, biomass allocation and plant nitrogen concentration in Chilean temperate rainforest tree seedlings: effects of nutrient availability. Oecologia 109: 49–58.

Mack, R.N., and D.A. Pyke. 1984. The demography of *Bromus tectorum*: the role of microclimate, grazing and disease. Journal of Ecology 72: 731–748.

Maranon, T., and P.J. Grubb. 1993. Physiological basis and ecological significance of the seed size and relative growth rate relationship in Mediterranean annuals. Functional Ecology 7: 591–599.

Maruta, E. 1976. Seedling establishment of *Polygonum cuspidatum* on Mt. Fuji. Japanese Journal of Ecology 26: 101–105.

Mazer, S.J. 1989. Ecological, taxonomic, and life-history correlates of seed mass among Indiana dune angiosperms. Ecological Monographs 59: 153–175.

McConnaughay, K.D.M., and F.A. Bazzaz. 1987. The relationships between gap size and performance of several colonizing annuals. Ecology 68: 411–416.

McRill, M., and G.R. Sagar. 1973. Earthworms and seeds. Nature 243: 482.

Metcalfe, D.J., and P.J. Grubb. 1995. Seed mass and light requirements for regeneration of Southeast Asian rain forest. Canadian Journal of Botany 73: 817–826.

Miles, J. 1972. Early mortality and survival of self-sown seedlings in Glenfeshie, Inverness-shire. Journal of Ecology 61: 93–98.

Moegenburg, S.M. 1996. Sabal palmetto seed size: causes of variation, choices of predators, and consequences for seedlings. Oecologia 106: 539–543.

Molofsky, J., and C.K. Augspurger. 1992. The effect of leaf litter on early seedling establishment in a tropical forest. Ecology 73: 68–77.

Molofsky, J., and B.L. Fisher. 1993. Habitat and predation effects on seedling survival and growth in shade-tolerant tropical trees. Ecology 74: 261–264.

Morgan, C.C.R., and H. Smith. 1979. A systematic relationship beteen phytochrome-controlled development and species habitat for plants grown in simulated natural radiation. Planta 145: 255–258.

Neher, D.A., H.T. Wilkinson, and C.K. Augspurger. 1992. Progression of damping-off epidemics in *Glycine* populations of even-age and mixed-age structure. Canadian Journal of Botany 70: 1032–1038.

Ng, F.S.P. 1978. Strategies of establishment in Malayan forest Trees. In: Tomlinson, P.B., and M.H. Zimmermann, eds. Tropical Trees as Living Systems. Cambridge: Cambridge University Press, pp 129–162.

Norton, D.A., and D. Kelly. 1988. Mast seeding over 33 years by *Dacrydium cupressinum* Lamb. (rimu) (Podocarpaceae) in New Zealand: the importance of economies of scale. Functional Ecology 2: 399–408.

Notman, E., D.L. Gorchov, and F. Cornejo. 1996. Effect of distance, aggregation, and habitat on levels of seed predation for two mammal-dispersed neotropical rain forest tree species. Oecologia 106: 221–227.

Nuñez-Farfán, J., and R. Dirzo. 1988. Within-gap spatial heterogeneity and seedling performance in a Mexican tropical forest. Oikos 51: 274–284.

O'Dowd, D.J., and P.S. Lake. 1991. Red crabs in rain-forest, Christmas Island: removal and fate of fruits and seeds. Journal of Tropical Ecology 7: 113–122.

Okubo, A., and S.A. Levin. 1989. A theoretical framework for data analysis of wind dispersal of seeds and pollen. Ecology 70: 329–338.

Osunkoya, O.O., J.E. Ash, M.S. Hopkins, and A.W. Graham. 1994. Influence of seed size and seedling ecological attributes on shade tolerance of rain forest tree species in northern Queensland. Journal of Ecology 82: 149–163.

Pate, J.S., E. Rasins, J. Rullo, and J. Kuo. 1985. Seed nutrient reserves of Proteaceae with special reference to protein bodies and their inclusions. Annals of Botany 57: 747–770.

Penning de Vries, F.W.T., and H.H. Van Laar. 1977. Substrate utilization in germinating seeds. In: Landsberg, J.J. and C.V. Cutting, ed. Environmental Effects on Crop Physiology. London: Academic Press, pp 217–228.

Piearce, T.G., N. Roggero, and R. Tipping. 1994. Earthworms and seeds. Journal of Biological Education 28: 195–202.

Pons, T.L. 1989. Breaking of seed dormancy by nitrate as a gap detection mechanism. Annals of Botany 63: 139–143.

Poorter, H., and C. Remkes. 1990. Leaf area ratio and net assimilation rate of 24 wild species differing in relative growth rate. Oecologia 83: 553–559.

Prevost, M. 1996. Effects of scarification on soil properties and natural seeding of a black spruce stand in the Quebec boreal forest. Canadian Journal of Forestry Research 26: 72–86.

Putz, R.E. 1983. Treefall pits and mounds, buried seeds, and the importance of soil disturbance to pioneer trees on Barro Colorado Island, Panama. Ecology 64: 1069–1074.

Reader, R.J. 1993. Control of seedling emergence by ground cover and seed predation in relation to seed size for some old-field species. Journal of Ecology 81: 169–175.

Roberts, H.A., and P.M. Feast. 1972. Fate of seed of some annual weeds in different depths of cultivated and undisturbed soil. Weed Research 4: 296–307.

Roberts, H.A., and P.M. Feast. 1973. Emergence and longevity of seeds of annual weeds in cultivated and undisturbed soil. Journal of Applied Ecology 10: 133–143.

Russi, L., P.S. Cocks, and E.H. Roberts. 1992. The fate of legume seeds eaten by sheep from a mediterranean grassland. Journal of Applied Ecology 29: 772–778.

Salisbury, E.J. 1942. The Reproductive Capacity of Plants: Studies in Quantitative Biology. London: Bell.

Saverimuttu, T., and M. Westoby. 1996a. Seedling longevity under deep shade in relation to seed size. Journal of Ecology 84: 681–689.

Saverimuttu, T., and M. Westoby. 1996b. Components of variation in seedling potential relative growth rate—phylogenetically independent contrasts. Oecologia 105: 281–285.

Sevilla, G.H., O.N. Fernandez, D.P. Minon, and L. Montes. 1996. Emergence and seedling survival of *Lotus tenuis* in *Festuca arundinacea* pastures. Journal of Range Management 49: 509–511.

Silvertown, J.W. 1980a. The evolutionary ecology of mast seeding in trees. Biological Journal of Linnean Society 14: 235–250.

Silvertown, J.W. 1980b. Leaf-canopy-induced seed dormancy in a grassland flora. New Phytologist 85: 109–118.

Simpson, R.L., M.A. Leck, and V.T. Parker. 1989. Seed banks: general concepts and methodological issues. In: Leck, M.A., V. Parker, and R.L. Simpson, eds. Ecology of Soil Seed Banks. San Diego, CA: Academic Press, pp 3–8.

Smith, M., and J. Capelle. 1992. Effects of soil surface microtopography and litter cover on germination, growth and biomass production of chicory (*Cichorium intybus* L). American Midland Naturalist 128: 246–253.

Sork, V.L. 1993. Evolutionary ecology of mast seeding in temperate and tropical oaks (*Quercus* spp). Vegetatio 108: 133–147.

Stanton, M. 1984. Seed variation in wild radish: effect of seed size on components of seedling and adult fitness. Ecology 65: 1105–1112.

Stiles, E.W. 1992. Animals as seed dispersers. In: Fenner, M., ed. Seeds: The Ecology of Regeneration in Plant Communities. Wallingford, England: CAB International, pp 87–104.

Stock, W.D., J.S. Pate, and J. Delfs. 1990. Influence of seed size and quality on seedling development under low nutrient conditions in five Australian and South African members of the Proteaceae. Journal of Ecology 78: 1005–1020.

Swanborough, P., and M. Westoby. 1996. Seedling relative growth rate and its components in relation to seed size: phylogenetically independent contrasts. Functional Ecology 10: 176–184.

Tapper, P.G. 1996. Irregular fruiting in *Fraxinus excelsior*. Journal of Vegetation Science 3: 41–46.

Templeton, A.R., and D.A. Levin. 1979. Evolutionary consequences of seed pools. American Naturalist 114: 232–249.

Terborgh, J., and S.J. Wright. 1994. Effects of mammalian herbivores on plant recruitment in two neotropical forests. Ecology 75: 1829–1833.

Thompson, K. 1986. Small-scale heterogeneity in the seed bank of an acidic grassland. Journal of Ecology 74: 733–738.

Thompson, K. 1987. Seeds and seed banks. In: Rorison, I.H., J.P. Grime, R. Hunt, A.F. Hendry, and D.H. Lewis, eds. Frontiers of Comparative Plant Ecology. New Phytologist 106(Suppl). London: Academic Press, pp 23–34.

Thompson, K., S.R. Band, and J.G. Hodgson. 1993. Seed size and shape predict persistence in soil. Functional Ecology 7: 236–241.

Thompson, K., and J.P. Grime. 1979. Seasonal variation in the seed banks of herbaceous species in ten contrasting habitats. Journal of Ecology 67: 893–921.

Thompson, K., J.P. Grime, and G. Mason. 1977. Seed germination in response to diurnal fluctuations of temperature. Nature 267: 147–149.

Toole, E.H. 1946. Final results of the Duval buried seed experiment. Journal of Agricultural Research 72: 201–210.

Tyler, C.M., and C.M. Dantonio. 1995. The effects of neighbours on the growth and survival of shrub seedlings following fire. Oecologia 102: 255–264.

Vaughan, J.G. 1970. The structure and utilization of oil seeds. London: Chapman and Hall.

Vegis, A. 1964. Dormancy in higher plants. Annual Review of Plant Physiology 15: 185–224.

Washitani, I. 1985. Field fate of *Amaranthus patulus* seeds subjected to leaf-canopy inhibition of germination. Oecologia 66: 338–342.

Washitani, I., and M. Masuda. 1990. A comparative study of the germination characteristics of seeds from a moist tall grassland community. Functional Ecology 4: 543–557.

Westoby, M., K. French, L. Hughes, B. Rice, and L. Rogerson. 1991. Why do more plant species use ants for dispersal on infertile compared with fertile soils? Australian Journal of Ecology 16: 445–455.

Westoby, M., E. Jurado, and M. Leishman. 1992. Comparative evolutionary ecology of seed size. Trends in Ecology and Evolution 7: 368–372.

Westoby, M., M. Leishman, and J. Lord. 1996. Comparative ecology of seed size and dispersal. Philosophical Transaction of the Royal Society of London Series B 351: 1309–1317.

Willson, M.F. 1992. The ecology of seed dispersal. In: Fenner M., ed. Seeds: The Ecology of Regeneration in Plant Communities. Wallingford, England: CAB International, pp 61–85.

Willson, M.F. 1993. Dispersal mode, seed shadows and colonization patterns. Vegetatio 107/108: 261–281.

Wolff, J.O. 1996. Population fluctuations of mast-eating rodents are correlated with production of acorns. Journal of Mammalogy 77: 850–856.

Wulff, R.D. 1986. Seed size variation in *Desmodium paniculatum*. II. Effects on seedling growth and physiological performance. Journal of Ecology 74: 99–114.

17

Facilitation in Plant Communities

Ragan M. Callaway
University of Montana, Missoula, Montana

Francisco I. Pugnaire
Arid Zones Experimental Station, Spanish Council for Scientific Research, Almería, Spain

I. INTRODUCTION

Positive relationships among species seem to have been underestimated in community ecology, perhaps unwittingly based on philosophical grounds (Keddy 1990, Barbour 1996). Judging by the number of studies that assume important mutualistic associations at the ecosystem level, but the lack of solid, quantitative data (Cushman and Beattie 1991), we could infer that positive interactions are either difficult to quantify, or that they are unimportant for ecosystem functioning and that ecological interdependence is overestimated in our popular culture (Callaway 1997a), or that the issue has been overlooked. This is particularly so in plant ecology, where mutualistic interactions between higher plant species are virtually lacking.

In plant ecology, this may lead to the question of whether mutualism is absent at all in plant communities, and perhaps the most likely answer is no. Yet few positive relationships between higher plant species have been directly investigated, partly because they may be subtle; partly because where other interactions, such as competition, have been highly emphasized as agents structuring plant communities, a large amount of experimental evidence for interspecific positive interactions has accumulated, suggesting that old conceptual models for the forces that structure communities are inadequate. That different species of plants may benefit each other when growing together is attracting an increasing interest among ecologists, and the number of papers reporting, or somehow involved with positive interactions, has grown considerably in recent years (Callaway 1995). For decades, traditional models have addressed plant interactions by emphasizing competition: plants compete for always-limited resources, such as water, nutrients, light, space, or pollinators, and must try to overthrow each other when in proximity.

Neighboring plant species may compete with one another for resources, but they may also provide benefits for neighbors such as shade, higher nutrient levels, more available moisture, soil oxygenation, protection from herbivores, a more favorable soil microflora, shared resources via mycorrhizae, and increased pollinator visits (Callaway 1995), resulting in an interaction that is positive overall for at least for one of the species involved. Positive interactions, or facilitations, occur when one plant species enhances the survival, growth, or fitness of another, and have been demonstrated in virtually all biome types (DeAngelis et al. 1986, Hunter and Aarssen 1988, Bertness and Callaway 1994, Callaway 1995, Callaway and Walker 1997b), although they have been regarded in the past as interesting anecdotal curiosities rather than principles fundamentally important to communities (Bertness and Callaway 1994).

Facilitation affects plant community structure and diversity in very different ways than competition. Competitive interactions limit coexistence among species, and competition-based theory focuses on how species avoid competitive exclusion (Lotka 1932, Gause 1934, Hardin 1960, Hutchinson 1961). Coexistence in a world dominated by competition has been attributed to "niche partitioning" (Parrish and Bazzaz 1976, Cody 1986), variation in the physical environment and subtle differences in competitive advantages; disturbance that continuously provides competition-free microhabitat and alters competitive hierarchies (Mc-Naughton 1985); heterogeneity in the ratios of limiting resources that alter competitive hierarchies (Tilman 1976, 1985, 1988); the development of local and species-specific resource depletion zones that, under certain conditions, do not strongly affect the resources available to neighbors (Huston and DeAngelis 1994, Grace 1995); and spatial structures that suggest niche facilitation (Van der Maarel et al. 1995). In contrast to the suite of theories that attempt to explain species coexistence despite competition for the few resources that are shared by all plants, positive interactions may directly promote coexistence and increase community diversity.

II. MECHANISMS

A. Shade

Shade provided by the canopies of large plants may protect seedlings and smaller plants from temperature extremes, reduce water loss, and reduce photoinhibition during stomatal closure, but at the cost of reduced energy for photosynthesis. Many species of cacti and other succulents may depend on nurse plants when they are young because seedlings have low surface-to-volume ratios and dissipate heat inefficiently, and exposure of these seedlings to exceptionally hot temperatures on the desert floor may be fatal. For example, saguaro seedlings are sheltered and facilitated by many different species of perennial plants, but predominantly by *Cercidium microphyllum* (Niering et al. 1963, Turner et al. 1966, 1969, Steenberg and Lowe 1969, 1977). Young *Ferrocactus acanthodes* cacti may experience 11°C decreases in maximum stem surface temperatures in the shade of nurse plants (Nobel 1984). Valiente-Banuet and Ezcurra (1991) compared the relative importance of protection from predation and shade in the nurse plant relationship between *Neobuxbaumia tetetzo*, a columnar cactus, and *Mimosa luisana* in the Viscaino Desert and the Gran Desierto de Altar in Mexico. They found that cages improved survival, but that long-term survival was restricted to shade treatments.

In a Chihuahua Desert perennial community, Silvertown and Wilson (1994) found more positive associations between species than expected for a random assemblage, and concluded that community structure relied on species whose establishment was dependent on conditions provided by other species.

Herbaceous species in California oak savannas also show strong preferences to either understory or open microhabitats. Parker and Muller (1982) demonstrated higher shade tolerance of *Bromus diandrus* and *Pholistoma auritum*, two species apparently facilitated by *Quercus agrifolia* (an evergreen oak), than that of *Avena fatua*, a species common in the open. In laboratory experiments, *B. diandrus* had a higher relative growth rate in low light than *A. fatua* (Mahall et al. 1981). Maranon and Bartolome (1993) conducted experiments in which whole soil blocks were transplanted between the understory of coast live oak, adjacent open grassland, and artificial shade treatments. They found that herbaceous species typically found in open habitats in California oak savannas were limited by canopy shade. In some cases, the overstory oaks themselves require shade from nurse plants when they are young. Callaway (1992) tested the facilitative effects of shade and seedling predation on the survival of *Quercus douglasii* (a winter-deciduous oak) seedlings using unshaded and shaded cages. Unshaded cages reduced mortality rates, but eventually all unshaded seedlings died, whereas 35% of shaded and caged seedlings survived for 1 year.

Gradients of both radiation reaching the soil and soil temperature provided by canopies may allow different species to coexist in the more heterogeneous understory environment. Moro et al. (1997) demonstrated how several herbaceous species positioned themselves differently under the canopy of *Retama sphaerocarpa*, a leguminous shrub, in response to several gradients, including irradiance and temperature. Radiation at soil level in a central position under the canopy was 60% of that outside, and temperature differences reached 7°C between both positions.

Shade may also have indirect facilitative effects. Shade provided by salt-tolerant species in salt marshes appears to reduce evaporation from subcanopy soils and consequently maintain lower soil salinities than in soils exposed to direct insolation (Bertness and Hacker 1994, Callaway 1994). Bertness (1991) and Bertness and Shumway (1993) experimentally manipulated soil salinity in the upper zones of a New England salt marsh and found that *Distichlis spicata*, a salt-tolerant colonizer of saline bare patches, facilitated the growth of *Juncus gerardi* when soil salinities were high. When soil salinities were artificially reduced by watering, growth of *Juncus* was not improved by *Distichlis*. Similarly, Bertness and Hacker (1994), Bertness and Yeh (1994), and Hacker and Bertness (1996) found that shade provided by marsh elder shrubs (*Iva frutescens*) decreased soil salinity and facilitated the growth and survival of *Juncus gerardi*.

B. Soil Moisture

The water relations of one species may be enhanced by the presence of another species. Joffre and Rambal (1986) measured a significant delay in soil water loss under *Quercus rotundifolia* and *Q. suber* in Spanish savannas relative to soils in open areas between the trees. They argued that increased soil moisture was accountable for large differences in species composition under trees and in the open. Generally, moisture conditions in the understory are enhanced by overstory plants (Petranka and McPherson 1979, Maranga 1984, Vetaas 1992), but sometimes the reverse is true. Pugnaire et al. (1996a) found that *Retama sphaerocarpa* shrubs in southern Spain had higher water potentials when *Marrubium vulgare* occupied the understory than when the understory was bare ground. This effect was observed in spring but not in summer, suggesting that the presence of a dense understory helped to retain rainfall water in the soil mound that accumulated under the shrub.

In North American shrub steppe, *Artemisia tridentata* (Great Basin sage) transports water from deep, moist soils to dry surface soils during the night via "hydraulic lift" (Richards and Caldwell 1987). Additional experiments using stable isotope analysis indicated that water from hydraulic lift was distributed to neighboring plants, although the magnitude of water transferred was small (Caldwell 1990). However, Dawson (1993) used similar techniques to investigate the magnitude of hydraulic lift conducted by *Acer saccharum* and the effects of the hydraulically lifted water on understory plants. He found hydraulically lifted water in all understory plants he examined, with the proportional use of hydraulically lifted water ranging from 3–60%. Plants that used large proportions of hydraulically lifted water had more favorable water relations and growth than those that did not.

C. Soil Nutrients

Soils directly beneath the canopies of perennials are often richer in nutrients than soils in surrounding open spaces without perennial cover. Subcanopy soil enrichment may occur as a result of "nutrient pumping" as deep-rooted perennials take up nutrients unavailable to shallowly rooted plants and deposit them on the soil surface via litterfall and throughfall. In addition, perennial canopies may trap airborne particles, which eventually are deposited at the base of the plant (Pugnaire et al. 1996b, Whitford et al. 1997). Nutrient enrichment may also occur indirectly via nitrogen fixation.

Enrichment of soil nutrients (and often a corresponding change in species composition and increased productivity) has been reported under perennials in many systems throughout the world, but these effects have been described most often in savannas and other semiarid regions with clearly demarked understory and open microhabitat (Patten 1978, Schmida and Whittaker 1981, Yavitt and Smith 1983, Weltzin and Coughenhour 1990, Callaway et al. 1991, Vetaas 1992, Belsky 1994). Nutrient enrichment has also been reported in more mesic systems, including alder shrublands (Goldman 1961) and Patagonian shrublands (Rostagno et al. 1991), and forest-pasture ecotonehs of central America (Kellman 1989).

Pugnaire et al. (1996a) reported a "facilitative mutualism" between the shrub *Retama sphaerocarpa* and the herb *Marrubium vulgare* based on improvements in soil nutrients and water. *Marrubium* plants under *Retama* had greater leaf specific area, leaf mass, shoot mass, leaf area, flowers, higher leaf nitrogen concentration, and more N per plant than those that were not near a *Retama* shrub. Conversely, *Retama* shrubs with *Marrubium* underneath them had larger cladodes, greater total biomass, nitrogen content, and higher shoot water potentials than shrubs without *Marrubium* in the understory, suggesting that facilitation could be bidirectional rather than unidirectional.

Experimental studies of the relationship between soil enrichment via litterfall and enhanced growth of plants or species shifts are less common. Turner et al. (1966) found that saguaro seedlings survived better on soil collected from under paloverde trees (*Cercidium microphyllum*) than on soils from under either mesquite (*Prosopis juliflora*) or ironwood (*Olneya tesota*). Monk and Gabrielson (1985) found that addition of litter from the floor of deciduous forests in South Carolina generally improved the overall productivity in understory plots, but the effects of litter on individual species ranged from facilitation to interference. Soils under *Quercus douglasii* canopies are much more nutrient rich than surrounding open grassland (Callaway et al. 1991) and improve the growth of *Bromus diandrus*, an annual grass in the understory, in greenhouse tests when compared with open grassland soils. Walker and Chapin (1986) demonstrated facilitative effects of nitrogen-rich soil from under *Alnus tenuifolia* on *Salix alaxensis* and *Populus balsamifera* seedlings. In a second experiment in Glacier Bay, Alaska, they found that *Alnus sinuata* and *Dryas drummondii*, both nitrogen fixers, had strong positive effects on the growth of *Picea sitchensis* (Chapin et al. 1994).

Moro et al. (1997) showed that the interaction between *Retama* and its understory vegetation was strongly affected by gradients of litter accumulation and decomposition, both of which influenced species composition. The abundance of annual herbs produced a more favorable habitat for soil microorganisms, which increased mineralization rate, enhanced litter decomposition, and increased nutrient dynamics under *Retama* shrubs. Litter has both positive and negative effects on plant growth, as has been shown elsewhere (Facelli 1994, Hoffmann 1996).

D. Soil Oxygenation

Wetland emergent plants often passively transport oxygen from leaves to roots through aerenchymous tissues to alleviate belowground oxygen limitation (Armstrong 1979). In some cases, oxygen appears to leak out of submerged roots, oxidize toxic substances and nutrients in the rhizosphere, and oxygenate marsh sediments (Howes et al. 1981, Armstrong et al. 1992). Castellanos et al. (1994) reported that *Spartina maritima* aerates surface sediments in wetlands of southern Spain, creating conditions favorable for the invasion of *Arthrocnemum perenne*.

In eastern saltmarshes in North America, Hacker and Bertness (1996) found that the aerenchymous *Juncus maritimus* increased the redox potential in its rhizosphere, which corresponded with increased growth of *Iva frutescens*, a woody-stemmed perennial, and the extension of *Iva* distribution to lower elevations in the marsh.

Plantago coronopus and *Samolus valerandi* are clumped with tussocks of the aerenchymous *Juncus maritimus* in dune slacks on the coast of Holland, where survival rates appeared to be enhanced by soil oxygenation and oxidation of iron, manganese, and sulfide (Schat and Van Beckhoven 1991). When *P. coronopus* and *Centaurium littorale* were grown with *Juncus maritimus* in the greenhouse, growth and nutrient uptake were improved (Schat 1984).

Substrate oxygenation and facilitation may also occur in freshwater marshes. In greenhouse experiments, undrained pots containing *Typha latifolia* (cattail) had dissolved oxygen contents more than 4 times greater than pots without *Typha*, and other marsh plants grown with *Typha* survived longer and grew larger than in pots without *Typha* when pot substrates were kept between 11 and 12°C (Callaway and King 1996). In the field, *Myosotis laxa* (herbaceous perennials) growing next to transplanted *Typha* were larger and produced more fruit than those isolated from *Typha*.

E. Protection from Herbivores

Positive interactions may be indirectly mediated through herbivores. *Themeda triandra*, an East African savanna grass, is preferred by many herbivores, and suffers approximately 80% mortality from grazers when not associated with less palatable grass species (McNaughton 1978). As codominance with unpalatable species increases, mortality of *Themeda* rapidly decreases. Similar examples of associational defenses have been reported by other investigators (Attsat and O'Dowd 1976).

In the Sonoran desert, paloverde seedlings are protected by various shrub species (McAuliffe 1986). Ninety-two percent of naturally occurring *Cercidium microphyllum* in the open were eaten by herbivores, but only 14% of the seedlings that were touching shrubs were consumed. McAuliffe (1984a) also found that young *Mammillaria microcarpa* and *Echinocereus englemannii* cacti were much more common under live and dead *Opuntia fulgida* where spine-covered stem joints from the nurse plant protected them from herbivores.

In southeast Spain, unpalatable *Artemisia barrelieri* shrubs facilitate seed germination and seedling establishment of more palatable *Anthyllis cytisoides* shrubs in addition to providing shelter from herbivory during early stages of growth, and before being competitively displaced by *Artemisia* (Haase et al. 1997).

In northern Sweden, *Betula pubescens* experiences high herbivory when associated with "plants of higher palatability," *Sorbus aucuparia* or *Populus tremuloides*, but low herbivory when associated with "plants of lower palatability," *Alnus incana* (Hjalten et al. 1993). Hay (1986) found that several species of highly palatable seaweeds were protected from grazers in safe microsites provided by unpalatable seaweed species.

As previously noted, the positive effect of coastal scrub shrubs on *Quercus douglasii* seedling recruitment (Callaway 1992) was caused in part by shade from the shrub canopies; however, analysis of the mortality of individual seedlings showed that the causes and timing of mortality differed significantly under shrubs and in the open grassland. Predation on acorns appeared to be due to rodents and was much higher under shrubs than in the grassland only 1 m away. However, emergent shoots were eaten by deer and experienced much higher predation in the grassland than under shrubs. Consequently, blue oak acorns that survived early rodent predation under shrubs became seedlings that occupied sites relatively free from deer predation on shoots. Similarly, seed and seedling predation appears to be a major factor limiting *Pinus ponderosa* to shrub-free, hydrothermally altered soils in the Great Basin (Callaway et al. 1996).

In most previously mentioned cases, benefactor species appear to physically shelter or hide beneficiary species from herbivores (see Chapter 20). A similar phenomenon, called associational resistance by Root (1972) and associational plant refuges by Pfister and Hay (1988), may occur when some species experience less herbivory as a function of the visual or olfactory complexity of the surrounding vegetation. A number of ecologists have found that community complexity serves as an impediment to search efficiency, in contrast to physical protection from herbivores or close association with unpalatable species (Root 1972, Risch 1981).

F. Pollination

Plants that are highly attractive to pollinators may facilitate their less attractive neighbors by enticing insects into the vicinity. Thomson (1978) found that *Hieracium florentinum* received more visits from pollinators when it was mixed with *H. auranticum* than when alone. In the understory of deciduous forests in Ontario, Laverty and Plowright (1988) recorded higher fruit and seed set in *Podophyllum peltatum* (mayapples) that were associated with *Pedicularis canadensis* (lousewort) than those that were not near *Pedicularis* plants. In later studies, Laverty (1992) found that mayapple, which produces no nectar, depends on infrequent visits from queen bumble bees that accidentally encounter mayapple while collecting nectar from *Pedicularis*, the "magnet species."

Geer et al. (1995) have shown that three co-occurring species of *Astragalus* facilitate each other's visitation rate of pollinators rather than competing for them. There may even be facilitation to attract pollinators between a rust fungus (*Puccinia monoica*) and *Anemone patens* (Ranunculaceae). The presence of rust's pseudoflowers may increase visitation rate to *Anemone*, although the positive effect could be counterbalanced because sticky pseudoflowers remove pollen from visiting insects and fungal spermatia deposited on flower stigmas reduce seed set (Roy 1996).

G. Mycorrhizae and Root Grafts

Woody plants may form intraspecific and interspecific root grafts in which resources and photosynthate move among individuals (Bormann and Graham 1959, Graham 1959, Bormann 1962). Bark girdling of one *Pinus strobus* sapling grafted to a conspecific neighbor resulted in a significant decrease in the diameter growth of the ungirdled member (Bormann 1966). Bormann reported that the intact tree supported root growth in the girdled tree for a period of 3 years. Bark girdling of nongrafted white pines resulted in death within a year, but grafted trees remained alive for 2 or more years after girdling. Bormann argued that the development of naturally occurring white pine stands is shaped by both competition and "a noncompetitive force governed by inter-tree food translocation." However, because both independently acquired and shared resources would require intact transport tissues, it is not clear how grafting overcame the effect of girdling.

Mycorrhizal fungi also facilitate the exchange of carbon and nutrients (Chiarello et al. 1982, Francis and Read 1984, Walter et al. 1996). Grime et al. (1987) found that labeled $^{14}CO_2$ was transferred from *Festuca ovina* to many other plant species in artificial microcosms that shared a common mycorrhizal network, but not to others that did not share the network. The presence of mycorrhizae led to decreased biomass of the dominant species, *Festuca ovina*, and increased biomass of otherwise competitively inferior species, including *Centaurea nigra*. Ultimately, experimental microcosms that were infected with mycorrhizae were more diverse than those that were not infected. Marler et al. (in press) studied the role of mycorrhizae on interactions between *Festuca idahoensis* and *Centaurea maculosa*, a major invasive weed in North America. They found that mycorrhizae mediated strong positive effects of *Festuca* on *Centaurea*. There were no direct effects of the mycorrhizae on the growth of either species, but when *Centaurea* was grown with large *Festuca* in the presence of mycorrhizae, they were 66% larger than in the absence of mycorrhizae. The fact that mycorrhizae mediated much stronger growth responses from large *Festuca* than small *Festuca* suggests that mycorrhizae mediated parasite-like interactions, such as those that occur between myco-heterotrophic parasites (Leake 1994).

III. INTERACTIONS BETWEEN FACILITATION AND INTERFERENCE

In the central Rocky Mountains, Ellison and Houston (1958) showed that herbaceous species in the understory of *Populus tremuloides* (quaking aspen) were stunted unless the tree roots were excluded. However, after trenching, the growth of understory species exceeded that of the surrounding open areas. Facilitative mechanisms of shade or nutrient inputs appeared to have the potential to facilitate understory species, but their effects were outweighed by root interference. More recently, a large number of studies have demonstrated positive and negative interactions operating at the same time. Walker and Chapin (1986) demonstrated that *Alnus tenuifolia* litter had the potential to facilitate *Salix alaxensis* and *Populus balsamifera* in greenhouse experiments and in the field. However, under natural conditions, *S. alaxensis, P. balsamifera*, and *Picea glauca* grew poorer in *A. tenuifolia* stands than in other vegetation. In other experiments, they found that root interference and shading in alder stands was more influential on the other species than nutrient addition via litter, and overrode the effects of facilitation. In a similar study in Glacier Bay, Chapin et al. (1994) found the reverse: that early to mid-successional species, such as *Alnus sinuata*, affected the late-successional *Picea sitchensis* in positive ways (nutrient uptake and growth) and in negative ways (germination and survivorship). *Picea sitchensis* seedlings that were planted in *A. sinuata* stands accumulated more than twice the biomass and acquired significantly higher leaf concentrations of nitrogen and phosphorus than seedlings planted in *P. sitchensis* forests. However, as found at the other site, root trenching in *A. sinuata* stands further increased growth and nutrient acquisition, demonstrating that competitive and facilitative mechanisms were operating simultaneously. In contrast to the interior floodplain, however, the facilitative effects of *Alnus* on it neighbors overrode root interference in natural conditions.

Peterson and Squiers (1995) have also reported a clear facilitation of white pine growth by aspen in northern Michigan, in spite of concurrent signs of strong intraspecific competition among pines in areas where aspen ramets were less dense.

In the Patagonian steppe, Aguiar et al. (1992) found that various shrub species protected grasses from wind and desiccation, but strong facilitative effects were only expressed when root competition was experimentally reduced. In the same system, Aguiar and Sala (1994) found that young shrubs had stronger facilitative effects on grasses, but the positive effects declined as grass densities increased near the shrubs.

During primary succession on volcanic substrates, Morris and Wood (1989) measured both negative and positive effects of *Lupinus lepidus*, the earliest colonist, on species that arrived later than the successional process, but concluded that the net effect of *Lupinus* was facilitative. On the island of Hawaii, the exotic tree *Myrica faya* is a successful invader on volcanic soils where it is replacing the native tree, *Metrosideros polymorpha* (Whiteaker and Gardener, 1985). Walker and Vitousek (1991) found that direct effects of the invading *Myrica* on the native *Metrosideros* were both facilitative and interfering. *Myrica* enriched the nitrogen content of soils, and improved *Metrosideros* growth in greenhouse experiments, and shade from *Myrica* improved *Metrosideros* seedling germination and survival. However, *Metrosideros* germination was sharply decreased by litter from *Myrica*, and the growth of young *Metrosideros* did not improve in the shade of *Myrica* in the field. They concluded that the lack of regeneration of *Metrosideros* under the canopies of *Myrica* in the field was the result of these negative mechanisms dominating the positive mechanisms.

Quercus douglasii deposits large amounts of nutrients to the soil beneath their canopies, and soil and litter bioassays demonstrate strong facilitative effects of these components on the growth of a dominant understory grass, *Bromus diandrus* (Callaway et al. 1991) In the field, however, the expression of this facilitative mechanism is determined by the root architecture of individual trees. *Quercus douglasii*, with low fine root biomass in the upper soil horizons and that appeared to have roots that reached the water table (much higher predawn water potentials), elicited strong positive effects on understory biomass. In contrast, trees with high fine root biomass in the upper soil horizons and that did not appear to root at the water table (much lower water potentials) elicited strong negative effects on understory productivity. In field experiments, soil from beneath all *Q. douglasii*, regardless of root architecture, had strong positive effects, but that root exclosures only improved understory growth under trees with dense shallow roots. Thus, in this ecosystem, as in the Alaskan floodplain studied by Walker and Chapin (1986), root interference, when present, outweighed the positive effects of nutrient addition.

Palatable intertidal seaweed species that depend on less palatable species for protection (Hay 1986, see page 630) are poor competitors with their benefactors. Hay found that, in the absence of herbivores, several highly palatable seaweed species grew 14–19% less when in mixtures with their benefactors than when alone. In the presence of herbivores, however, palatable species survived only when mixed with competitively superior, but unpalatable species. In this system, the effects of competition with neighbors were outweighed by the protection they provided.

The balance of facilitation and interference may change with the lifestages of the interacting plants. Patterns of nurse plant mortality observed in numerous systems indicate that species may begin their lives as the beneficiaries of nurse plants and later become significant competitors with their former benefactors as they mature. McAuliffe (1988) found that young *Larrea tridentata* plants were positively associated with dead *Ambrosia dumosa*, a species critical to the initial establishment of *Larrea*. Similarly, mature saguaros were associated disproportionally with dead paloverde trees, which commonly function as nurse plants to seedling saguaros (McAuliffe 1984b). However, in both of these cases, young *Larrea* may do better in the shade of *Ambrosia* that are already dead because the positive effects come without any competitive cost. In the Tehuacan Valley of Mexico, *Neobuxbaumia tetetzo*, which is nursed by *Mimosa luisana* (Valiente-Banuet et al. 1991b), eventually suppresses the growth and reproduction of its benefactor (Flores-Martinez et al. 1994). The same occurs with *Opuntia rastrera* (Silvertown and Wilson 1994). Archer et al. (1988) found that *Prosopis glandulosa* trees provide a focal point for the regeneration of many other species in southcentral Texas (Figure 1), but *Prosopis* regenerated very poorly in these clusters once they were established.

Other studies have shown that a particular benefactor species may have facilitative effects on some species, but competitive effects on other similar species. *Artemisia tridentata* dominates large regions of the Great Basin of Nevada, but is completely absent from outcrops of phosphorus-poor hydrothermally altered andesite where *Pinus ponderosa* is abundant (DeLucia et al. 1988, 1989). Field experiments demonstrated that *A. tridentata* outcompetes *P. ponderosa* for water off of the altered andesite refuges, and prevents *P. ponderosa* from expanding onto typical soils of the Great Basin. Another pine species, *Pinus monophylla*, is often common in mixtures with *A. tridentata*, but found only as seedlings on altered andesite (Callaway et al. 1996). Experiments indicated that *P. monophylla* is not prevented from maturing on altered andesite by soil conditions,

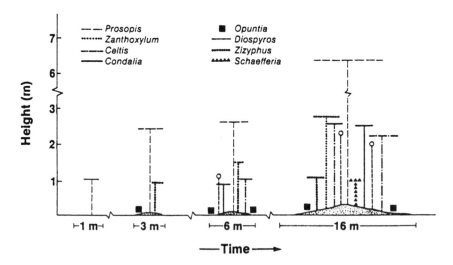

Figure 1 Diagrammatic illustration of the chronosequential development of shrub clusters on dine loamy uplands in the Rio Grande Plains, TX. These landscapes may be moving from the present two-phase configuration—discrete woody clusters scattered throughout the continuous grassland matrix—toward a monophasic woodland, as new clusters are initiated and existing clusters expand and coalesce. (*Source*: Archer et al. 1988.)

but by the absence of *A. tridentata*, which acts as a nurse plant for *P. monophylla* (DeLucia et al. 1989, Callaway 1996). Both species of pines have similar leaf-level physiological characteristics and nutrient requirements.

In the upper zones of southern California salt marshes, the perennial sub-shrub *Arthrocnemum subterminale* exists in a matrix of winter ephemeral species that emerge when soil salinity decreases at the beginning of the rainy season. *Arthrocnemum* has strong facilitative effects on some codominant ephemerals, and they tend to occur in its understory (Callaway 1994). Another, otherwise similar annual is competitively reduced by *Arthrocnemum* and is more common in the open between the shrubs.

Bertness and Callaway (1994) hypothesized that the balance between facilitation and interference is affected by the harshness of physical conditions, and that the importance of facilitation in plant communities increases with increasing abiotic stress or increasing consumer pressure. Alternatively, they hypothesized that the importance of competition in communities would increase when physical stress and consumer pressure were relatively low because neighbors buffer one another from extremes of the abiotic environment (e.g., temperature or salinity) and herbivory.

In support of the abiotic stress hypothesis, Bertness and Yeh (1994) found that the effects of *Iva frutescens* shrubs on conspecific seedlings were positive because soil salinity was moderated by the shade of the benefactors in a New England salt marsh. When patches were watered, however, strong competitive interactions developed between adults and seedlings and among seedlings. Interactions between *I. frutescens* plants were dependent on environmental conditions, competing more intensely under mild abiotic conditions and facilitating each other when abiotic conditions were harsh. Bertness and Shumway (1993) also eliminated the facilitative effects of *Distichlis spicata* and *Spartina patens* on *Juncus gerardi* by watering experimental plots and reducing soil salinity. In the same marsh, the fitness of *Iva* shrubs associated with *Juncus gerardi* was enhanced by neighbors at lower elevations but strongly suppressed by the same species at higher elevations where soil salinity was lower (Bertness and Hacker 1994).

The relationship between abiotic stress and shifts in competition and facilitation has been shown in other systems. In intermountain grasslands of the northern Rockies, *Lesquerella carinata* is positively associated with bunchgrass species in relatively xeric areas, but negatively associated with the same species in relatively mesic locations (Greenlee and Callaway 1996). Greenlee and Callaway studied these spatial pattern experimentally and documented shifts between interference and facilitation between years at a xeric site. Bunchgrasses interfered with *L. carinata* in a wet and cool (low-stress) year, but facilitated *Lesquerella* in a dry and hot year (Figure 2). Other nonexperimental studies indicated that competitive effects are stronger in wet, cool years and facilitative in dry, hot

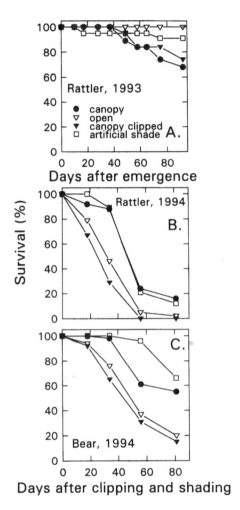

Figure 2 Survival of *Lesquerella* seedlings planted under bunchgrass canopies, in the open, under clipped bunchgrass canopies, and under artificial shade. (A) Rattler Gulch, 1993; χ^2 (for day 92) = 9.61, degrees of freedom (df) = 3, P = .022. (B) Rattler Gulch, 1994; χ^2 (for day 81) = 14.4, df = 3, P = .002. (C) Bear Gulch, 1994; χ^2 (for day 81) = 40.25, df = 3, P < .001. (*Source*: Greenlee and Callaway 1996.)

years (Fuentes et al. 1984, De Jong and Klinkhamer 1988, Frost and McDougal 1989, McClaran and Bartolme 1989, Belsky 1994). In contrast to these studies, Casper (1996) examined survival, growth, and flowering of *Cryptantha flava* in experiments explicitly designed to test for shifts in positive and negative interactions with neighboring shrubs and found no evidence that competition or facilitation changed as soil water varied between years.

Shifts in interspecific interactions have also been found to occur at different temperatures in anaerobic substrates. *Myosotis laxa* appeared to benefit from soil oxygenation when grown with *Typha latifolia* at low soil temperatures (Callaway and King 1996). But at higher soil temperatures, *Typha* had no effect on soil oxygen (presumably because of increased microbial and root respiration), and the interaction between *Typha* and *Myosotis* became competitive.

Changes in interspecific interactions may occur in a different sequence. Haase et al. (1996) have shown that coexisting *Artemisia barrelieri* and *Retama sphaerocarpa* shrubs compete during early stages of succession in abandoned semiarid fields in southeast Spain, but after a time period during which *Artemisia* is competitively displaced, this subshrub is found preferentially under the canopy of larger *Retama* shrubs. The pattern of the interaction suggested that facilitation prevailed over competition because of niche separation that developed over time.

Complex combinations of negative and positive interactions operating simultaneously between species appear to be widespread in nature. Such concomitant interactions indicate that current conceptual models of interplant interactions based on resource competition alone may not accurately depict processes in natural plant communities.

IV. ARE BENEFACTOR SPECIES INTERCHANGEABLE?

The species specificity of positive interactions among plants, or whether benefactor species are interchangeable, is crucial to understanding positive interactions and interdependence in plant communities (Callaway, 1998a). In other words, are the positive effects of plants simply due to the alteration of the biophysical environment, which can be imitated by inanimate objects like rocks, logs, or experimental shade cloth? Or can facilitations depend on the species, with some species eliciting strong positive effects and other, morphologically similar, species producing no effect?

Spatial associations between beneficiaries and benefactors are often disproportional to the abundance of potential beneficiaries. For example, Hutto et al. (1986) reported that saguaros were distributed nonrandomly among potential nurse plants at two locations in Organ Pipe National Monument in the Sonoran Desert. They found that saguaros were proportionally more abundant under many species of shrubs and trees, but significantly more saguaros were associated with

Prosopis juliflora (mesquite) and paloverde trees and less saguaros were associated with *Larrea tridentata* (creosote bush) than expected based on the proportional cover of these species.

Suzán et al. (1996) identified a large number of aborescent, shrub, and cacti species as nurse plants of *Olneya tesota* (ironwood) in the Sonoran Desert of Mexico and the U.S., and argued that, as a habitat modifier, *Olneya* is a "keystone species" for biodiversity (also see Burqúez and Quintana 1994). They described 30 species as shade dependent, with five preferring *Cercidium* species, four preferring *Prosopis* species, and 22 preferring *Olneya tesota*. Franco and Nobel (1989) found that most saguaro seedlings in their Sonoran Desert study sites were associated with *Cercidium microphyllum* and *Ambroisa deltoidea*. In contrast, a second cactus species, the exceptionally heat-tolerant *Ferocactus acanthodes*, was preferentially associated with a bunchgrass, *Hilaria rigida*.

In Californian shrubland and woodland, *Quercus agrifolia* and *Q. douglasii* seedlings are disproportionally associated with shrub species, and experimental manipulations have demonstrated the importance of nurse shrubs for the survival of both species (Callaway and D'Antonio 1991, Callaway 1992). However, not all shrubs had positive effects on *Q. agrifolia*. Forty-three percent of germinating seedlings survived under *Ericameria ericoides*, 34% under *Artemisia californica*, 5% under *Mimulus auranticus*, and 0% under *Lupinus chamissonis* (Callaway and D'Antonio 1991).

In other studies, the species of the benefactor plant does not seem to matter. For example, Steenberg and Lowe (1969) reported that 15 different species can apparently act as nurse plants for saguaro and are found associated with saguaro seedlings in proportion to their frequencies, indicating that no specific biotic factor is involved and that the nurse plant association is only the by-product of microclimatic changes under canopies. Greenlee and Callaway (1996) found that *Lesquerella carinata*, a small perennial herb, occurred commonly under the canopies of bunchgrasses on xeric sites in western Montana. However, *Lesquerella* was distributed among bunchgrass species in proportion to their abundances.

Potential benefactor species (those with similar morphologies) may have specific effects on beneficiaries simply because they alter the physical environment differently. For example, Suzán et al. (1996) attributed the disproportional positive affect of *Olneya tesota* to the fact that it is the only tall evergreen aborescent in the region, and ameliorates climatic conditions throughout the year.

Because facilitation occurs in complex combinations with competition, potential benefactors species may have the same positive effect, but vary in their negative effects. For example, Suzán et al. (1996) suggested that the superior facilitative ability of *Olneya* may be due to its phreatophytic life history and deeply distributed root architecture, thus reducing niche overlap (Cody 1986) and accentuating positive mechanisms.

The consistently poor performance of creosote bush as a nurse plant in the

Sonoran Desert (Hutto et al. 1986, McAuliffe 1988) may be due to the strong negative effects this species has on perennial neighbors (Fonteyn and Mahall 1981). Mahall and Callaway (1991, 1992) found that creosote bush substantially inhibited the root elongation rates of *Ambrosia dumosa*, and that these negative effects were reduced by the addition of small amounts of activated carbon, a strong adsorbent to organic molecules. (Cheremisinoff and Ellerbusch 1978). Thus, creosote bush canopies may have the potential to facilitate shade-requiring plants, but prospective beneficiaries may be eliminated by root allelopathy. In the Chihauhan Desert, however, *Larrea tridentata* has been reported to faciliate *Opuntia leptocaulis* (Yeaton 1978). Muller (1953) documented strong positive associations between *Ambrosia dumosa* and many species of desert annuals; however, *Encelia farinosa* shrubs did not harbor any annual species. He attributed this difference to the allelopathic effect of *Encelia* leaves.

Pugnaire et al. (1996b) showed that edaphic conditions and productivity improved in the understory of *Retama sphaerocarpa* shrubs with age and that there was a replacement of species under *Retama* canopies with a clear successional trend. The species found under youngest shrubs were also found in gaps and, as soil fertility increased and aspects of microclimate (particularly irradiance and temperature) were ameliorated, different species colonized the understory, including agricultural weeds and perennials. Growth conditions improved for *Retama* itself, and there was an increase in cladode mass, photosynthetic area, and fruit production with age, supporting the hypothesis that overstory shrubs benefit from the understory herbs, and that indirect interactions between *Retama sphaerocarpa* and its understory herbs could be considered as a two-way facilitation in which both partners benefit from their association (Pugnaire et al. 1996a).

Species that have similar affects on understory microclimate may vary in their affects on soil nutrients, creating species-specific effects. Turner et al. (1966) found that saguaro seedlings survived better on soil collected from under paloverde trees than on soils from under either mesquite or *Olneya tesota*; however, these differences were confounded by soil albedo and temperature.

Some facilitative mechanisms are produced by unusual plant traits, and thus infer a high degree of species specificity. For example, some of the strongest positive effects are produced by protection from herbivores (Atsatt and O'Dowd 1976, McAuliffe 1984, 1986, Hay 1986). Repelling consumers requires specific morphological traits such as spines or tough tissues, or the possession of chemical defenses. Not all potential benefactors in a community may have these traits.

Many of the traits described here, such as oxygen transport (Callaway and King 1996), shared pollinator attraction (Thomson 1978, Laverty 1992), hydraulic lift (Richards and Caldwell 1987, Dawson 1993), and positive effects mediated through fungal intermediaries (Grime et al. 1987; Marler et al. in press), are not shared by all large numbers of species in a community, suggesting that benefactors are not highly interchangeable.

V. POSITIVE INTERACTIONS AND COMMUNITY THEORY

The consistent observation of independent distributions of plant species along environmental gradients (the individualistic continuum) is a cornerstone of plant community theory (McIntosh 1967, Goodall 1963, Austin 1990, Collins 1993). Plant species are almost never completely associated with another species at all points on environmental gradients; therefore, community ecologists have long assumed that the distributions of plant species are determined by their idiosyncratic responses to the abiotic environment and interspecific competition. However, some evidence suggests that plant species may have some degree of interdependence at some points on gradients, yet interact individualistically at others (Callaway, 1998b).

Pinus albicaulis (whitebark pine) and *Abies lasiocarpa* (subalpine fir) dominate the upper end of many xeric elevation gradients in the Northern Rockies (Daubenmire 1952, Pfister et al. 1977). In many places, *P. albicaulis* is the dominant species at or near timberline, but it also overlaps with *A. lasiocarpa* at lower elevations where the latter is more abundant. Thus, these two species exhibit the classic continuum that is the cornerstone of the individualistic continuum theory. However, a more detailed examination of interactions between these species creates a more complex picture. *Pinus albicaulis* appears to have a cumulative competitive effect on *A. lasiocarpa* at lower elevations, where there are no significant spatial associations and the death of *P. albicaulis* corresponds with higher *A. lasiocarpa* growth rates (Callaway, 1998b). But at timberlines in xeric areas, *A. lasiocarpa* is highly clumped around *P. albicaulis*, and the death of the latter is associated with decreased *A. lasiocarpa* growth rates. Similar processes are also apparent in alpine communities of the central Caucasus Mountains. Kikvidze (1993, 1996) measured numerous significant positive spatial associations and found evidence for facilitation via amelioration of abiotic stress. At high elevations, significant positive spatial associations are four times more common than negative associations. However, at low elevations, gradient positive spatial associations were four times less common than negative associations.

Virtually all important types of interspecific interactions have been shown to vary with changes in abiotic conditions, including predation (Martin, 1998), herbivory (Maschinski and Whitham 1989), parasitism (Gibson and Watkinson 1992, Pennings and Callaway 1996), mutualism (Bronstein 1994), competition (Connell 1983, Kadmon 1995), and allelopathy (Tang et al. 1995).

Shifting positive and negative interactions on environmental gradients indicates that nodes, or fully overlapping discreet groups of species, are not required to demonstrate some level of interdependence among plants in a community. Because plants can have neutral or negative effects on neighbors at one point on an environmental gradient and positive effects at another, and the positive effect of benefactors are often not interchangeable among species, a continuum does not necessarily infer fully individualistic relationships among plant species.

VI. CONCLUSION

Positive interactions are common in nature and are caused by many different mechanisms that are substantially different than those involved in competition. Positive interactions co-occur with competition, and the overall effect of one plant on another is often determined by the balance of several mechanisms in a particular abiotic environment. In many cases, positive interactions are highly species specific and benefactors are not interchangeable, suggesting that plant communities may be more interdependent than has been thought since the widespread acceptance of Gleasonian individualistic communities. Furthermore, plants may have strong competitive effects on a neighbor at one end of its distribution along a physical gradient, but strong positive effects on the same neighbor at the other end. This confounds interpretation of continuous distributions of plants in gradient analyses as evidence for fully individualistic plant communities and supports the concept that plant communities are real entities (Van der Maarel 1996). The growing body of evidence for positive interactions in plant communities, and its theoretical framework, suggests that facilitation plays important roles in determining the structure, diversity, and dynamics of many plant communities.

REFERENCES

Aguiar, M.R., and O.E. Sala. 1994. Competition, facilitation, seed distribution and the origin of patches in a Patagonian steppe. Oikos 70: 26–34.

Aguiar, M.R., A. Soriano, and O.E. Sala. 1992. Competition and facilitation in the recruitment of seedlings in Patagonian steppe. Functional Ecology 6: 66–70.

Archer, S., C. Scifres, C.R. Bassham, and R. Maggio. 1988. Autogenic succession in a subtropical savanna: conversion of grassland to woodland. Ecological Monographs 58: 111–127.

Armstrong, W. 1979. Aeration in higher plants. Advances in Botanical Research 7: 226–332.

Armstrong, J., W. Armstrong, and P.M. Becket. 1992. *Phragmites australis*: venturi- and humidity-induced pressure flows enhance rhizome aeration and rhizosphere oxidation. New Phytologist 120: 197–207.

Atsatt, P.R., and D.J. O'Dowd. 1976. Plant defense guilds. Science 193: 24–29.

Austin, M.P. 1990. Community theory and competition in vegetation. In: Grace, J.B., and D. Tilman, eds. Perspectives on Plant Competition. New York: Academic Press, pp 215–238.

Barbour, M.G. 1996. American ecology and American culture in the 1950s: who led whom? Bulletin Ecological Society America 77: 44–51.

Belsky, A.J. 1994. Influences of trees on savanna productivity: tests of shade, nutrients, and tree-grass competition. Ecology 75: 922–932.

Bertness, M.D. 1991. Interspecific interactions among high marsh perennials in a New England salt marsh. Ecology 72: 125–137.

Bertness, M.D., and R.M. Callaway. 1994. Positive interactions in communities. Trends in Ecology and Evolution 9: 191–193.

Bertness, M.D., and S.D. Hacker. 1994. Physical stress and positive associations among marsh plants. The American Naturalist 144: 363–372.

Bertness, M.D., and S.W. Shumway. 1993. Competition and facilitation in marsh plants. The American Naturalist 142: 718–724.

Bertness, M.D., and S.M. Yeh. 1994. Cooperative and competitive interactions in the recruitment of marsh elders. Ecology 75: 2416–2429.

Bormann, F.H. 1962. Root grafting and noncompetitive relationships between trees. In: Kozlowski, T.T. ed. Tree Growth. New York: Ronald Press, pp 237–245.

Bormann, F.H. 1966. The structure, function, and ecological significance of root grafts in *Pinus strobus* L. Ecological Monographs 36: 1–26.

Bormann, F.H., and B.F. Graham, Jr. 1959. The occurrence of natural root grafting in eastern white pine, *Pinus strobus* L., and its ecological implications. Ecology 40: 677–691.

Bronstein, J. 1994. Conditional outcomes in mutualistic interactions. Trends in Ecology and Evolution 9: 214–217.

Caldwell, M.M. 1990. Water parasitism stemming from hydraulic lift: a quantitative test in the field. Israel Journal of Botany 39: 395–402.

Callaway, R.M. 1992. Effect of shrubs on recruitment of *Quercus douglasii* and *Quercus lobata* in California. Ecology 73: 2118–2128.

Callaway, R.M. 1995. Positive interactions among plants. Botanical Review 61: 306–349.

Callaway, R.M. 1997. Positive interactions and the individualistic-continuum concept. Oecologia 112:143–149.

Callaway, R.M. 1998a. Are positive interactions among plants species-specific? Oikos 82: 202–207.

Callaway, R.M. 1998b. Competition and facilitation on elevational gradients in subalpine forests of the northern Rocky Mts., USA. Oikos 82: 561–563.

Callaway, R.M., E.H. DeLucia, D. Moore, R. Nowak, and W.H. Schlesinger. 1996. Competition and facilitation: contrasting effects of Artemisia tridentata on *Pinus ponderosa* and *P. monophylla*. Ecology 77: 2130–2141.

Callaway, R.M., and L. King. 1996. Oxygenation of the soil rhizosphere by *Typha latifolia* and its facilitative effects on other species. Ecology 77: 1189–1195.

Callaway, R.M., N.M. Nadkarni, and B.E. Mahall. 1991. Facilitation and interference of *Quercus douglasii* on understory productivity in central California. Ecology 72: 1484–1499.

Callaway, R.M., and L.R. Walker. 1997. Competition and facilitation: a synthetic approach to interactions in plant communities. Ecology 1958–1965.

Casper, B. 1996. Demographic consequences of drought in the herbaceous perennial *Cryptantha flava*: effects of density, associations with shrubs, and plant size. Oecologia 106: 144–152.

Castellanos, E.M., M.E. Figueroa, and A.J. Davy. 1994. Nucleation and facilitation in saltmarsh succession: interactions between *Spartina maritima* and *Arthrocnemum perenne*. Journal of Ecology 82: 239–248.

Chapin, F.S. III., L.R. Walker, C.L. Fastie, and L.C. Sharman. 1994. Mechanisms of primary succession following deglaciation at Glacier Bay, Alaska. Ecological Monographs 64: 149–175.

Cheremisinoff, P.N., and F. Ellerbusch. 1978. Carbon Adsorption Handbook. Ann Arbor, MI: Ann Arbor Science Publishers.

Cody, M.L. 1986. Structural niches in plant communities. In: Diamond, J., and T.J. Case, eds. Community Ecology. New York: Harper and Row, pp 381–405.

Collins, S.L., S.M. Glenn, and D.W. Roberts. 1993. The hierarchical continuum concept. Journal Vegetation Science 4: 149–156.

Connell, J.H. 1983. On the prevalence and relative importance of interspecific competition: evidence from field experiments. American Naturalist 122: 661–696.

Cushman, J.H., and Beattie, A.J. 1991. Mutualisms: Assessing the benefits to hosts and visitors. Trends Ecology Evolution 6: 193–195.

Daubenmire, R.F. 1952. Forest vegetation of northern Idaho and adjacent Washington and its bearing on concepts of vegetation classification. Ecological Monographs 22: 301–330.

Dawson, T.E. 1993. Hydraulic lift and water use by plants: implications for water balance, performance and plant-plant interactions. Oecologia 95: 565–574.

DeAngelis, D.L., W.M. Post, and C.C. Travis. 1986. Positive Feedback in Natural Systems. New York: Springer-Verlag.

De Jong, T.J., and P.G.L. Klinkhamer. 1988. Population ecology of the biennials *Cirsium vulgare* and *Cynoglossum officinale* in a coastal sand-dune area. Journal of Ecology 73: 147–167.

DeLucia, E.H., W.H. Schlesinger, and W.D. Billings. 1988. Water relations and the maintenance of Sierran conifers on hydrothermally altered rock. Ecology 69: 303–311.

DeLucia, E.H., W.H. Schlesinger, and W.D. Billings. 1989. Edaphic limitations to growth and photosynthesis in Sierran conifers and Great Basin Vegetation. Oecologia 78: 184–190.

Ellison, L., and W.R. Houston. 1958. Production of herbaceous vegetation in openings and under canopies of western aspen. Ecology 39: 338–345.

Facelli, J.M. 1994. Multiple indirect effects of plant litter affect the establishment of woody seedlings in old fields. Ecology 75: 1727–1735.

Flores-Martinez, A., E. Ezcurra, and S. Sanchez-Colon. 1994. Effect of *Neobuxbaumia tetetzo* on growth and fecundity of its nurse plant *Mimosa luisana*. Journal of Ecology 82: 325–330.

Fonteyn, P.J., and B.E. Mahall. 1981. An experimental analysis of structure in a desert plant community. Journal of Ecology 69: 883–896.

Francis, R., and D.J. Read. 1984. Direct transfer of carbon between plants connected by vesicular-arbuscular mycelium. Nature 307: 53–56.

Franco, A.C., and P.S. Nobel. 1989. Effect of nurse plants on the microhabitat and growth of cacti. Journal of Ecology 77: 870–886.

Frost, W.E., and N.K. McDougald. 1989. Tree canopy effects on herbaceous production of annual rangeland during drought. Journal of Range Management 42: 281–283.

Fuentes, E.R., R.D. Otaiza, M.C. Alliende, A. Hoffman, and A. Poiani. 1984. Shrub clumps of the Chilean matorral vegetation: structure and possible maintenance mechanisms. Oecologia 62: 405–411.

Gause, G.F. 1934. The Struggle for Existence. New York: Hafner.

Geer, S.M., V.J. Tepedino, T.L. Griswold, and V.R. Bowlin. 1995. Pollinator sharing by 3 sympatric milkvetches, including the endangered species *Astragalus montii*. Great Basin Naturalist 55: 19–28.

Gibson, C.C., and A.R. Watkinson. 1992. The role of the hemiparasitic annual *Rhinanthus minor* in determining grassland community structure. Oecologia 89: 62–68.

Goldman, C.R. 1961. The contribution of alder trees (*Alnus tenuifolia*) to the primary productivity of Castle Lake, California. Ecology 42: 282–288.

Goodall, D.W. 1963. The continuum and the individualistic association. Vegetatio 11: 297–316.

Grace, J.B. 1995. In search of the Holy Grail: explanations for the coexistence of plant species. Trends in Ecology and Evolution 10: 263–264.

Graham, B.F. Jr. 1959. Transfer of dye through natural root grafts of *Pinus strobus* L. Ecology 41: 56–64.

Greenlee, J., and R.M. Callaway. 1996. Effects of abiotic stress on the relative importance of interference and facilitation. The American Naturalist 148: 386–396.

Grime, J.P. 1979. Plant Strategies and Vegetation Process. New York: Wiley.

Haase, P., F.I. Pugnaire, S.C. Clark, and L.D. Incoll. 1996. Spatial pattern in a two-tiered semi-arid shrubland in southeastern Spain. Journal of Vegetation Science 7: 527–534.

Haase, P., F.I. Pugnaire, S.C. Clark, and L.D. Incoll. 1997. Spatial pattern in *Anthyllis cytisoides* shrubland on abandoned land in semi-arid south-eastern Spain. Journal of Vegetation Science 8: 627–634.

Hacker, S.D., and M.D. Bertness. 1996. Trophic consequences of a positive plant interaction. American Naturalist 148: 559–575.

Hardin, G. 1990. The competitive exclusion principle. Science 131: 1292–1297.

Hay, M.E. 1986. Associational plant defenses and the maintenance of species diversity: turning competitors into accomplices. American Naturalist 128: 617–641.

Hjalten, J., K. Danell, and P. Lundberg. 1993. Herbivore avoidance by association: vole and hare utilization of woody plants. Oikos 68: 125–131.

Hoffmann, W.A. 1996. The effects of fire and cover on seedling establishment in a neotropical savanna. Journal of Ecology 84: 383–393.

Holmgren, M., M. Scheffer, and M.A. Huston. 1997. The interplay of competition and facilitation in plant communities. Ecology 78: 1966–1975.

Hunter, A.F., and L.W. Aarssen. 1988. Plants helping plants. Bioscience 38: 34–40.

Huston, M.A., and D.L. DeAngelis. 1994. Competition and coexistence: the effects of resource transport and supply rates. American Naturalist 144: 954–977.

Hutchinson, G.E. 1961. The paradox of the plankton. American Naturalist 95: 137–145.

Hutto, R.L., J.R. McAuliffe, and L. Hogan. 1986. Distributional associates of the saguaro (*Carnegia gigantea*). Southwestern Naturalist 31: 469–476.

Joffre, R., and S. Rambal. 1993. How tree cover influences the water balance of Mediterranean rangelands. Ecology 74: 570–582.

Kadmon, R. 1995. Plant competition along soil moisture gradients: a field experiment with the desert annual *Stipa capensis*. Journal of Ecology 83: 253–262.

Keddy, P. 1990. Is mutualism really irrelevant to ecology? Bulletin Ecological Society America 71: 101–102.

Kellman, M. 1985. Forest seedling establishment in Neotropical savannas: transplant experiments with *Xylopia frutescens* and *Calophyllum brasiliense*. Journal of Biogeography 12: 373–379.

Kikvidze, Z. 1993. Plant species associations in alpine-subnival vegetation patches in the Central Caucasus. Journal of Vegetation Science 4: 297–302.

Kikvidze, Z. 1996. Neighbour interactions and stability in subalpine meadow communities. Journal of Vegetation Science 7: 41–44.

Laverty, T.M. 1992. Plant interactions for pollinator visits: A test of the magnet species effect. Oecologia 89: 502–508.

Laverty, T.M., and R.C. Plowright. 1988. Fruit and seed set in Mayapple (*Podophyllum peltatum*): influence of intraspecific factors and local enhancement near *Pedicularis canadensis*. Canadian Journal of Botany 66: 173–178.

Leake, J.R. 1993. Tansley Review Number 69: The biology of myco-heterotrophic (saprophytic) plants. New Phytologist 127: 171–216.

Mahall, B.E., V.T. Parker, and P.J. Fonteyn. 1981. Growth and photosynthetic responses of *Avena fatua* L. and *Bromus diandrus* Roth. and their ecological significance in California savannas. Photosynthetica 15: 5–15.

Mahall, B.E., and R.M. Callaway. 1991. Root communication among desert shrubs. Ecology 88: 874–876.

Mahall, B.E., and R.M. Callaway. 1992. Root communication mechanisms and intracommunity distributions of two Mojave Desert shrubs. Ecology 73: 2145–2151.

Maranga, E.K. 1984. Influence of *Acacia tortilis* trees on the distribution of *Panicum maximum* and *Digitaria macroblephara* in south central Kenya. MS Thesis, Texas A & M University, College Station, Texas.

Marañón, T., and J.W. Bartolome. 1993. Reciprocal transplants of herbaceous communities between *Quercus agrifolia* woodland and adjacent grassland. Journal of Ecology 81: 673–682.

Marler, M., C. Zabinski, and R.M. Callaway. Mycorrhizal mediation of competition between an exotic forb and a native bunchgrass. Ecology, in press.

Martin, T.E. 1998. Are microhabitat preferences of coexisting species under selection and adaptive? Ecology 79: 656–670.

Maschinski, J., and T.G. Whitham. 1989. The continuum of plant responses to herbivory: the influence of plant association, nutrient availability, and timing. American Naturalist 134: 1–19.

McAuliffe, J.R. 1984a. Prey refugia and the distributions of two Sonoran Desert cacti. Oecologia 65: 82–85.

McAuliffe, J.R. 1984b. Saguaro-nurse tree associations in the Sonoran Desert: competitive effects of sahuaros. Oecologia 64: 319–321.

McAuliffe, J.R. 1986. Herbivore-limited establishment of a Sonoran Desert tree: *Cercidium microphyllum*. Ecology 67: 276–280.

McAuliffe, J.R. 1988. Markovian dynamics of simple and complex desert plant communities. American Naturalist 131: 459–490.

McClaran, M.P., and J.P. Bartolome. 1989. Effect of *Quercus douglasii* (Fagaceae) on herbaceous understory along a rainfall gradient. Madrono 36: 141–153.

McIntosh, R.P. 1967. The continuum concept of vegetation. Botanical Review 33: 130–187.

McNaughton, S.J. 1985. Ecology of a grazing ecosystem: the Serengeti. Ecological Monographs 55: 259–294.

McNaughton, S.J. 1978. Serengeti ungulates: feeding selectivity influences the effectiveness of plant defense guilds. Science 199: 806–807.

Monk, C.D., and F.C. Gabrielson. 1985. Effects of shade, litter and root competition on old-field vegetation in South Carolina. Bulletin of the Torrey Botanical Club 112: 383–392.

Moro, M.J., F.I. Pugnaire, P. Haase, and J. Puigdefábregas. 1997. Mechanisms of interaction between *Retama sphaerocarpa* and its understory layer in a semi-arid environment. Ecography 20: 175–184.

Morris, W.F., and D.M. Wood. 1989. The role of lupine in succession on Mount St. Helens: facilitation or inhibition? Ecology 70: 697–703.

Mott, J.J., and A.J. McComb. 1974. Patterns in annual vegetation and soil microrelief in an arid region of western Australia. Journal of Ecology 62: 115–126.

Newman, E.I., and K. Ritz. 1986. Evidence on the pathways of phosphorus transfer between vesicular-arbuscular mycorrhizal plants. New Phytologist 104: 77–87.

Niering, W.A., R.H. Whittaker, and C.H. Lowe. 1963. The saguaro: a population in relation to environment. Science 142: 15–23.

Nobel, P.S. 1984. Extreme temperatures and thermal tolerances for seedlings of desert succulents. Oecologia 62: 310–317.

Parker, V.T., and C.H. Muller. 1982. Vegetational and environmental changes beneath isolated live oak trees (*Quercus agrifolia*) in a California annual grassland. American Midlands Naturalist 107: 69–81.

Parrish, J.A.D., and F.A. Bazzaz. 1976. Underground niche separation in successional communities. Ecology 57: 1281–1288.

Patten, D.T. 1978. Productivity and production efficiency of an upper Sonoran Desert community. American Journal of Botany 65: 891–895.

Pennings, S.C., and R.M. Callaway. 1996. Impact of a parasitic plant on the structure and function of salt marsh vegetation. Ecology 77: 1410–1419.

Peterson C.J., and Squiers E.R. 1995. Competition and succession in an aspen white-pine forest. Journal of Ecology 83: 449–457.

Petranka, J.W., and J.K. McPherson. 1979. The role of *Rhus copallina* in the dynamics of the forest-prairie ecotone in north-central Oklahoma. Ecology 60: 956–965.

Pfister, R.D, B.L. Kovalchik, S.E. Arno, and R. Presby. 1977. Forest habitat types of western Montana. USDA, Forest Service Technical Report, INT-34. Ogden, Utah.

Pfister, C.A., and M.E. Hay. 1988. Associational plant refuges: convergent patterns in marine and terrestrial communities result from differing mechanisms. Oecologia 77: 118–129.

Pugnaire, F.I., P. Hasse, and J. Puigdefabregas. 1996a. Facilitation between higher plant species in a semiarid environment. Ecology 77: 1420–1426.

Pugnaire, F.I., P. Hasse, J. Puigdefábregas, M. Cueto, L.D. Incoll, and S.C. Clark. 1996b. Facilitation and succession under the canopy of the leguminous shrub, *Retama sphaerocarpa*, in a semi-arid environment in south-east Spain. Oikos 76: 455–464.

Richards, J.H., and M.M. Caldwell. 1987. Hydraulic lift: substantial nocturnal water transport between soil layers by *Artemisia tridentata* roots. Oecologia 73: 486–489.

Risch, S.J. 1981. Insect herbivore abundance in tropical monocultures and polycultures: an experimental test of two hypotheses. Ecology 62: 1325–1340.

Root, R.B. 1972. The influence of vegetational diversity on the population ecology of a specialized herbivore, *Phyllotreta cruciferae* (Coleoptera: Chrysomelidae). Oecologia 10: 321–346.

Rostagno, C.M., H.F. del Valle, and L. Videla. 1991. The influence of shrubs on some chemical and physical properties of an aridic soil in north-eastern Patagonia, Argentina. Journal of Arid Environments 20: 179–188.

Roy, B.A. 1996. A plant pathogen influences pollinator behavior and may influence reproduction of nonhosts. Ecology 77: 2445–2457.

Schat, H. 1984. A comparative ecophysiological study on the effects of waterlogging and submergence on dune slack plants: growth, survival and mineral nutrition in sand culture experiments. Oecologia 62: 279–286.

Schmida, A., and R.H. Whittaker. 1981. Pattern and biological microsite effects in two shrub communities in southern California. Ecology 62: 234–251.

Silvertown, J., and J.B. Wilson. 1994. Community structure in a desert perennial community. Ecology 75: 409–417.

Steenberg, W.F., and C.H. Lowe. 1969. Critical factors during the first year of life of the saguaro (*Cereus giganteus*) at Saguaro National Monument. Ecology 50: 825–834.

Steenberg, W.F., and C.H. Lowe. 1977. Ecology of the saguaro. II. Reproduction, germination, establishment, growth, and survival of the young plant. National Park Service Science Monograph Series No. 8. NPS, Washington, D.C.

Suzán, H., G.P. Nablan, and D.T. Patten. 1996. The importance of *Olneya tesota* as a nurse plant in the Sonoran Desert. Journal of Vegetation Science 7: 635–644.

Tang, C., W. Cai, K. Kohl, and R.K. Nishimoto. 1995. Plant stress and allelopathy. In: Inderjit, K.M.M., Dakshini, and F.A. Einhellig, eds. Allelopathy. Washington, D.C.: American Chemical Society, pp 142–157.

Thomson, J.D. 1978. Effects of stand composition on insect visitation in two-species mixtures of *Hieracium*. American Midlands Naturalist 100: 431–440.

Tilman, D. 1976. Ecological competition between algae: experimental confirmation of resource-based competition and predation. Science 192: 463–465.

Tilman, D. 1985. The resource ratio hypothesis of succession. American Naturalist 125: 827–852.

Tilman, D. 1988. Plant Strategies and the Dynamics and Structure of Plant Communities. Princeton Monograph Series. Princeton, NJ: Princeton University Press.

Turner, R.M., S.M. Alcorn, and G. Olin. 1969. Mortality of transplanted saguaro seedlings. Ecology 50: 835–844.

Turner, R.M., S.M. Alcorn, G. Olin, and J.A. Booth. 1966. The influence of shade, soil, and water on saguaro seedling establishment. Botanical Gazette 127: 95–102.

Valiente-Banuet, A., and E. Ezcurra. 1991. Shade as a cause of the association between the cactus *Neobuxbaumia tetetzo* and the nurse plant *Mimosa luisana* in the Tehuacan Valley, Mexico. Journal of Ecology 79: 961–971.

Valiente-Banuet, A., A. Bolongaro, O. Briones, E. Ezcurra, M. Rosas, H. Nunez, G. Barnhard, and E. Vasquez. 1991a. Spatial relationships between cacti and nurse shrubs in a semi-arid environment in central Mexico. Journal of Vegetation Science 2: 15–20.

Valiente-Banuet, A., F. Vite, and J.A. Zavala-Hurtado. 1991b. Interaction between the cactus *Neobuxbaumia tetetzo* and the nurse shrub *Mimosa luisana*. Journal of Vegetation Science 2: 11–14.

Van der Maarel, E. 1996. Pattern and process in the plant community: fifty years after A.S. Watt. Journal of Vegetation Science 7: 19–28.

Van der Maarel, E., V. Noest, and M.W. Palmer. 1995. Variation in species richness on small grassland quadrats—niche structure or small-scale plant mobility. Journal of Vegetation Science 6: 741–752.

Vetaas, O.R. 1992. Micro-site effects of trees and shrubs in dry savannas. Journal of Vegetation Science 3: 337–344.

Walker, L.R., and F.S. Chapin III. 1986. Physiological controls over seedling growth in primary succession on an Alaskan floodplain. Ecology 67: 1508–1523.

Walker, L.R., and P.M. Vitousek. 1991. An invader alters germination and growth of a native dominant tree in Hawaii. Ecology 72: 1449–1455.

Walter, L.E., D.C. Fisher, D. Hartnett, B.A.D. Hetrick, and A.P. Schwab. 1996. Interspecific nutrient transfer in a tallgrass prairie plant community. American Journal of Botany 83: 180–184.

Weltzin, J.F., and M.B. Coughenhour. 1990. Savanna tree influence on understorey vegetation and soil nutrients in northwestern Kenya. Journal of Vegetation Science 1: 325–332.

Whiteaker, L.D., and D.E. Gardner. 1985. The distribution of *Myrica faya* Ait. in the state of Hawaii. Technical Report No. 55. Cooperative National Park Resources Studies Unit, University of Hawaii at Manoa, Honolulu, Hawaii.

Whitford, W.G., J. Anderson, and P.M. Rice. 1997. Stemflow contribution to the "fertile island" effect in creosote bush, *Larrea tridentata*. Journal of Arid Environments 35: 451–457.

Williams, K., M.M. Caldwell, and J.H. Richards. 1993. The influence of shade and clouds on water potential: the buffered behavior of hydraulic lift. Plant and Soil 157:83–95.

Yavitt, J.B., and L. Smith, Jr. 1983. Spatial patterns of mesquite and associated herbaceous species in an Arizona desert upland. American Midlands Naturalist 109: 89–93.

Yeaton, R.I. 1978. A cyclical relationship between *Larrea tridentata* and *Opuntia leptocaulis* in the northern Chihuahuan Desert. Journal of Ecology 66: 651–656.

18

Plant Interactions: Competition

Heather L. Reynolds*

W. K. Kellogg Biological Station, Hickory Corners, Michigan

Not until we reach the extreme confines of life, in the arctic regions or on the borders of an udder desert, will competition cease. The land may be extremely cold or dry, yet there will be competition between some few species, or between the individuals of the same species, for the warmest or dampest spots.

Charles Darwin, 1859

---

* *Current affiliation*: Kalamazoo College, Kalamazoo, Michigan.

I. INTRODUCTION

Organisms are said to compete when shared resource needs have mutual negative effects ($--$) on survival, growth or reproduction. This happens when resources are limiting, i.e., when organismal demand for resources exceeds the supply of resources from the environment. Competition has the potential to profoundly affect the distribution and abundance of species as well as the evolution of their physiology and morphology. However, so do other biotic interactions (e.g., mutualism, predation, herbivory), as well as species' interactions with their abiotic environment.

Controversy has existed over the relative importance of competition versus these other biotic and abiotic interactions to the ecology and evolution of species and the structure of communities. Additional controversy exists about whether the intensity of competition varies with environmental conditions and about what traits are important to competitive ability. In plant ecology, debate about these issues is manifest between the competition–stress–ruderal (C-S-R) model of Grime (1977, 1979, 1988) and the resource competition model of Tilman (1982, 1988, 1990). These controversies are as yet unresolved, in part because of a lack of appropriate experimental tests, in part to disagreement over how competition intensity should be measured, and in part because of failure to distinguish between the concepts of intensity versus importance. Lack of attention to the selection regime commonly experienced by the study species is another issue that contributes to continuing debate about plant competition. This chapter provides an overview of the current state of knowledge about plant competition in the context of these controversies, with focus on the debate about competition intensity and competitive traits, since these issues have received most of the recent attention of plant ecologists.

II. MECHANISMS OF COMPETITION

Competition occurs for resources: quantities of matter or energy necessary for survival, growth, or reproduction that are consumed by organisms. As autotrophs, plants require the inorganic resources carbon dioxide (CO_2), water, light, and mineral nutrients. The major mineral nutrients are nitrogen, phosphorus, potassium, iron, calcium, magnesium, and sulfur; the minor are molybdenum, copper, zinc, manganese, boron, chlorine, sodium (some Chenopodiaceae), aluminum (ferns), cobalt (legumes that symbiotically fix nitrogen), and silicon (diatoms) (Bannister 1976).

There are two mechanisms by which organisms compete for resources: exploitation competition and interference competition (Figure 1). Because plants are sessile and do not exhibit behavior, the main mechanism of plant competition

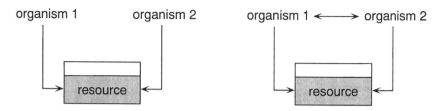

Figure 1 Mechanisms of competition. (Left) In exploitation competition, organisms interact only indirectly, via shared resource use. (Right) In interference competition, organisms interact directly for the shared resource. Interference competition is common in animals (e.g., fighting over mates or food), although plants can sometimes interfere by growing over one another, or via chemical production (allelopathy).

is exploitation; competitors interact solely by consuming (i.e., depleting or preempting) resources (see Weiner and Thomas 1986 and Thomas and Weiner 1989 for discussion of resource depletion versus resource preemption). Conversely, interference competition involves direct interactions between competitors that prevent access to resources, such as fighting between animals. In plants, interference competition can occur by overgrowth of one individual by another. Allelopathy, or production of chemicals that inhibit plant function, may also be (but is not necessarily) an evolved mechanism of interference competition (Williamson 1990). The following sections focus solely on exploitation competition. See Williamson (1990) for a review of the relatively little-studied subject of allelopathy.

III. C-S-R AND RESOURCE COMPETITION MODELS

It has sometimes been asserted that different definitions of competition are at the root of any disagreement between the C-S-R and resource competition models (Thompson 1987, Grace 1991). However, both Grime and Tilman define competition in terms of exploitation (Tilman 1987): "Here, competition is defined as the tendency of neighboring plants to utilize the same quantum of light, ion of a mineral nutrient, molecule of water, or volume of space" (Grime 1973), and "The mechanism of competitive displacement is resource consumption" (Tilman 1988).

Competition can only occur for resources when they are limiting. Some resources required by plants (e.g., CO_2) are generally superabundant, and use by one organism has no affect on other organisms requiring the same resource. Resources become limiting when potential demands on resources by organisms exceed resource supply. The major source of disagreement between the C-S-R and resource competition models is about when resource demand has the potential to exceed resource supply.

A. C-S-R

The C-S-R model is a conceptual model that categorizes vegetation into general strategies based on adaptation to combinations of two types of selection pressures: stress and disturbance (Figure 2). Stress is defined as any environmental condition that limits plant production (e.g., extremes of pH or temperature), and disturbance as any environmental condition that removes plant biomass (e.g., herbivory, pathogens, wind, fire, human activities) (Grime 1977). Low availability of mineral nutrients (low soil fertility) is most often associated with stress in this model. Stress and disturbance are thought to reduce competition by preventing development of a dense vegetation and thereby reducing demand for resources. Competition is hypothesized to be of highest intensity and greatest importance in highly productive vegetation, such as that which develops on fertile soil under relatively undisturbed conditions. Proponents of the C-S-R model (Grime 1977, Campbell et al. 1991) reason that dense, highly productive vegetation will result in increased demand for both light and belowground resources (water, mineral nutrients). Therefore, traits important to the competitive strategy are considered to be those that maximize acquisition of all resources. These traits

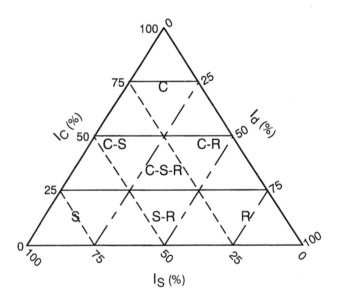

Figure 2 Model showing the various equilibria between competition (C), stress (S), and disturbance (d) in vegetation and the location of primary and secondary strategies. I_C, relative importance of competition (———); I_S, relative importance of stress (----); I_d, relative importance of disturbance (— · —). (*Source*: Grime 1977.)

include high growth rates, large size and height, and high plasticity in roots and shoots to allow "active foraging" for undepleted resource pockets. Investment in resource acquisition comes at the expense of investment in traits important for adapting to stress and disturbance. Such trade-offs in allocation to competition, stress, and disturbance mean that each species cannot be adapted to the extremes of more than one strategy, although adaptation to intermediate intensities of competition, stress, and disturbance may occur (Grime 1977, 1988).

B. Resource Competition

The resource competition model is a mechanistic, mathematical model of plant population dynamics that considers essentially similar selection pressures (or environmental constraints): soil resource supply rates and loss (disturbance, mortality) rates. The model consists of two types of coupled equations, one describing plant dynamics as a function of resource-dependent growth and loss, and the other describing resource dynamics as a function of resource supply and plant consumption (Figure 3). The model is explicit about the fact that plants grow by consuming resources, and exploitative competition is an integral component of the model because the resources consumed (light, soil resources) are depletable (i.e., plant consumption affects availability) and can therefore become limiting.

Competition is not restricted mainly to dense vegetation of high productivity habitats, as in the C-S-R model, but occurs under every combination of re-

rate of biomass change = growth - loss

(a)
$$\frac{dB_i}{B_i dt} = f_i(R) - m_i$$

rate of resource change = supply rate - sum of consumption rates

(b)
$$\frac{dR}{dt} = y(R) - \sum_{i=1}^{n} [Q_i B_i f_i(R)]$$

Figure 3 Simple consumer-resource model of exploitative competition for a limiting soil nutrient. (a) Per-unit biomass rate of change of a population, where B_i is biomass of species i; $f_i(R)$ is a function that describes the dependence of net growth for species i on the resource (R); and m_i is loss rate of species i. (b) Resource dynamics, where $y(R)$ is a function that describes resource supply; Q_i is the nutrient content per unit biomass of species i, and n is the total number of consumer species. This simple model can be extended to include more than one limiting resource. (*Source*: Tilman 1990.)

source supply and loss rate. The reason for this is that the model considers plant demand for resources relative to resource supply. It is the ratio of supply to demand that determines whether resources are limiting (and consequently whether competition will occur) (Taylor et al. 1990; see also a related discussion by Chesson and Huntly 1997). The absolute demand for soil resources may be high in high productivity sites, but soil resource supply rates are also high, and thus the ratio of supply to demand is not necessarily different than in low productivity sites (in which absolute demand is lower, but so is supply).

In fact, the resource competition model predicts that competition is equally important at both low and high soil fertility (low and high plant productivity). Only the nature of the resources for which there is competition is predicted to change along fertility gradients, as the limiting resources change from soil nutrients at low soil resource supply rates (causing strong belowground competition) to light at high soil resource supply rates (causing strong aboveground competition). Loss rates can similarly influence the nature of competition (below- versus aboveground) through their influence on availabilities of soil nutrients and light (Tilman 1988, Smith and Huston 1989).

Traits important to competitive ability are also predicted to vary with resource supply and loss rates. For example, at low soil resource supply rates, allocation to acquisition and retention of soil resources is predicted (e.g., traits such as high root to shoot ratio, long-lived tissues). At high soil resource supply rates, allocation to acquisition of light may become important (e.g., traits such as high stem allocation) (Tilman 1988). Investment in belowground resource acquisition and use necessarily comes at the expense of investment in traits important for acquisition and use of aboveground resources. Such trade-offs impose limits on the ability of any one organism to dominate in all habitat types (Tilman 1988, 1990, Smith and Huston 1989).

By acquiring depletable resources for itself, an individual reduces availability of those resources to other species. The level to which the individuals of a population deplete a given resource at equilibrium is known as R^* (Tilman 1982, 1988, 1990). Tilman (1990) and Reynolds and Pacala (1993) have shown analytically how traits can influence R^*s. Resource reduction is the mechanism of exploitation competition, and thus, the lower the R^* of a species, the more resource it has gained for itself and the better is its competitive ability for that resource. At equilibrium, the species with the lowest R^* (i.e., greatest ability to consume the resource) is predicted to displace all competitors (Tilman 1977, 1982, 1988, 1990).

It is important to note that the resource competition model can also be solved numerically to predict the competitive dominants under nonequilibrium conditions (Tilman 1987, 1988). However, these so-called ''transient dominants'' are not considered to be the true superior competitors for a given set of soil resource supply and loss rates. According to resource competition theory, the true superior competitor for a given set of soil resource supply and loss rates is defined as that species that dominates at equilibrium (Tilman 1987, 1988). For practical reasons, many competition experiments are very short term, and thus the competitive outcomes may sometimes reflect transient dynamics rather than equilibrium conditions. Failure to distinguish transient from equilibrium outcomes has likely contributed to the continuing debate between the C-S-R and resource competition perspectives (Tilman 1987).

IV. EXPERIMENTAL DESIGN AND COMPETITION INDICES

There are four main types of experimental designs for measuring competition between two species: (1) the replacement series (or substitutive design); (2) the additive design; (3) the bivariate factorial design; and (4) the neighborhood design. In a replacement series (de Wit 1960), species are grown in pure stands and in mixtures, with one species at proportion p of its pure stand density and the other species at proportion (1-p) of its pure stand density (Connolly 1986). The pure stand densities of each species need not be equal; however, the standard replacement design consists of one pure stand of each species grown at the same density, and a single 1:1 mixture of the two species, with each at half the pure stand density (Figure 4a). The replacement approach thus attempts to measure interspecific competition relative to intraspecific competition.

The additive design also uses pure stands and mixtures of two species, but the density of each species is the same in pure and mixed stands. The standard additive design consists of pure stands of each species at the same density and a single 1:1 mixture of the two species, with each at its pure stand density (Figure 4b). The additive approach thus attempts to measure the magnitudes of interspecific competition while holding intraspecific competition constant. Knowledge of the magnitudes of both types of competition is necessary for predicting the effects of competition on species distribution and abundance (Roughgarden 1979, Underwood 1986). Additive designs can easily be extended to measure absolute magnitudes of intraspecific competition (Underwood 1986).

For both designs, many indices of competition are based on performance of species in mixture compared with their performances in pure stand (Table 1). Replacement designs have been criticized because this comparison confounds change in intraspecific density with change in interspecific density (Firbank and Watkinson 1985, Underwood 1986, Connolly 1986, Silvertown and Dale 1991, Snaydon 1991). Such confounding causes dependence of the competition indices on species densities in pure stands versus mixtures (Connolly 1986, Snaydon 1991), as well as on species' proportions and patterns of response to density (Snaydon 1991). Inherent differences in species' sizes can also cause dependence of competition indices on species' densities (Connolly 1986, Grace et al. 1992).

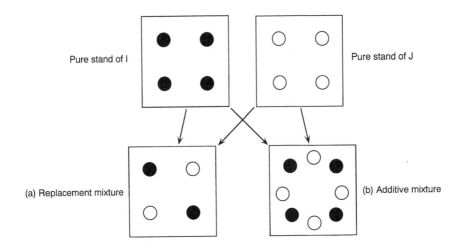

Pure stand of I

Pure stand of J

(a) Replacement mixture

(b) Additive mixture

(c)

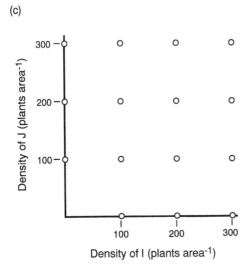

Figure 4 Planting arrangements for pure stands of I (filled circle) and J (open circles) and for (a) replacement mixtures and (b) additive mixtures of I with J. (c) Diagrammatic representation of a bivariate factorial design, in which densities of the two species are varied independently. (*Source*: Snaydon 1991.)

Table 1 Common Indices of Competition for Replacement, Additive, Simplified Neighbor, Neighbor, and Bivariate Factorial Designs

Design[a]	Index	Abbreviation	Formula[b] Interspecific[c]	Formula[b] Intraspecific	Reference
r, a	Relative yield total	RYT	$(Y_{ij}/Y_{ii}) + (Y_{ji}/Y_{jj})$	—	de Wit 1960
r, a	Relative crowding coefficient[d]	RCC	$(Y_{ij}/Y_{ii})/(Y_{ji}/Y_{jj})$	—	Harper 1977
r, a	Aggressivity	A	$1/2[(W_{ij}/W_{ii}) - (W_{ji}/W_{jj})](r)$ $(W_{ij}/W_{ii}) - (W_{ji}/W_{jj})(a)$ or $(Y_{ij}/Y_{ii}) - (Y_{ji}/Y_{jj})$ (r, a)	—	McGilchrist and Trenbath 1971 Snaydon 1991
r, a	Absolute severity of competition	ASC	$\log_{10}(W_{io}/W_{ij})$	$\log_{10}(W_{io}/W_{ii})$	Snaydon 1991
r, a	Relative severity of competition	RSC	$\log_{10}(W_{if}/W_{ij})$		Campbell and Grime 1992
a, sn	Absolute competition intensity	Absolute CI	$Y_{ii} - Y_{ij}$ (a) $Y_{io} - Y_{ij}$ (a, sn)	$Y_{io} - Y_{ii}$ (a, sn)	Wilson and Keddy 1986, Wilson and Tilman 1991
a, sn	Relative competition intensity	Relative CI	$(Y_{ii} - Y_{ij})/Y_{ii}$ (a) $(Y_{io} - Y_{ij})/Y_{io}$ (a, sn)	$(Y_{io} - Y_{ii})/Y_{io}$ (a, sn)	
n	Per-amount competitive effect	X	$W_i = W_{io}/(1 + XB_j)$	$W_i = W_{io}/(1 + XB_i)$	Goldberg 1987
a	Competition coefficient	α	$W_i = W_{io}[1 + a_i(N_i + \alpha_{ij}N_j)]^{-b_i}$	—	Firbank and Watkinson 1990
b	Competition coefficient	c	$W_i = W_{io}/(1 + N_i^{c_{ii}} + N_j^{c_{ij}})$	Same	Law and Watkinson 1987
n	Interference coefficient[e]	c	$W_i = W_{io}/(1 + c_{ii}N_i + c_{ij}N_j)$	Same	Silander and Pacala 1990

[a] r, replacement design; a, additive design; sn, simplified neighbor design; n, neighbor design; b, bivariate factorial design.

[b] Variables: Y, yield per unit area; W, yield or weight per plant; B, total neighbor biomass; N, total number of neighbors; a and b, fitted regression parameters. Subscripts: i, species i; j, species j; i0, species i grown with no competition; ii, species i grown in pure stand; jj, species j grown in pure stand; ij, for Y and W, refers to species i when grown in mixture with species j, whereas for α and c, refers to effect of species j on species i; ji, for Y and W, refers to species j when grown in mixture with species i, whereas for α and c, refers to effect of species i on species j.

[c] Note that interspecific formulas are given in terms of species i when grown with species j.

[d] For 1:1 ratios only (see Willey and Rao 1980, for application to other component ratios). Snaydon (1991) recommends $\log_{10}$ or $\log_e$ transformations to facilitate statistical and biological analysis.

[e] See Pacala and Silander 1985, 1990, and Silander and Pacala 1985, 1990, for methods of determining the best neighborhood size.

Sackville Hamilton (1994) defends replacement designs, but only for a restricted set of conditions (Snaydon 1994).

Because species' densities are the same in pure stands and mixtures in additive designs, competition indices are unaffected by pure stand densities or by species' patterns of response to density (Snaydon 1991). However, additive (and replacement) designs are, of course, sensitive to total density. If total densities are low enough, resources will not be limiting, and species will not compete even if they would experience strong competition at higher densities (Taylor and Aarssen 1989). Additive designs may also be sensitive to species proportions, i.e., it is possible that relative competitive abilities change as species proportions in mixture change. To deal with these issues, additive designs may be repeated over a range of densities or proportions, yielding additive series (Snaydon 1991). When species' densities are varied factorially, producing a complete range of total densities and species' proportions, the resulting design has been termed both an addition series (Firbank and Watkinson 1985, 1990) and a bivariate factorial design (Snaydon 1991). Additive series and bivariate factorial designs can be analyzed by regressing weight per plant in mixture on the density and/or proportion of each component species (Firbank and Watkinson 1985, 1990) or by calculating separate competition indices for each mixture (Snaydon 1991).

A modification of the additive design that has become popular is the neighborhood design, in which the density of a neighbor species around a single individual of a target species is varied (Goldberg and Werner 1983, Goldberg 1987) (Figure 5). Performance (e.g., growth rate, survival) of the target species can be regressed on the amount (e.g., biomass, density) of neighbor species. The slope of this regression measures the per-amount competitive effect of the neighbor species on the target species (Goldberg and Werner 1983), and the coefficient of determination measures the importance of competition (the proportion of variation in performance that is explained by variation in neighbor amount) (Welden and Slauson 1986). The total competitive effect of a neighbor species is given by the product of its per-amount effect and its abundance. Per-biomass effects are indicative of physiological and/or morphological effects, whereas per-individual effects are often related to inherent size (Goldberg 1987). This approach has also been developed for populations of two species whose densities vary randomly, and can also be used to measure intraspecific competition, to compare the effects of different neighbor species on one target species or the responses of different target species to one neighbor species, and to account for the effects of neighbor distance or angular dispersion (Mack and Harper 1977, Weiner 1982, Pacala and Silander 1985, Silander and Pacala 1985, Pacala and Silander 1987). Gaudet and Keddy (1988) have used a simplified neighborhood approach to identify traits important to competitive ability (see Section VI).

All of the aforementioned designs measure competition between only two species at a time, yet natural communities are typically composed of many more than two species. The number of experiments necessary to study all possible pairs

INITIAL FIELD CONDITIONS

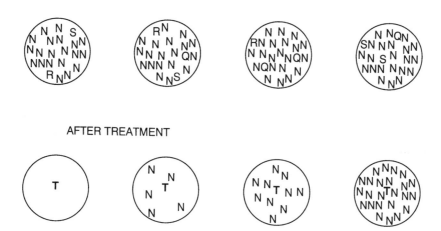

Figure 5 Example of the experimental design for evaluating competitive effects of one neighbor species (N) on a target species (T). R, Q, and S represent individuals not belonging to the neighbor species selected for study. Only four steps of the neighbor density gradient (after treatment) are shown: the experiments must include a much wider range of densities to estimate accurately the slope of the regression of target performance on amount of neighbors. (*Source:* Goldberg and Werner 1983.)

of species in a multispecies assemblage quickly becomes unreasonably large—even without considering intraspecific interactions, different species' densities or proportions, or the effect of environmental heterogeneity on competitive interactions (Keddy 1990, Tilman 1990). Ability to predict the outcome of competition in a community from pairwise competition experiments may also be complicated by indirect effects. Indirect interactions between two species arise via direct interactions with a third species (Connell 1990).

Many studies have used a simplified neighborhood approach to multispecies competition: comparing the performance of target individuals in the presence versus the absence of neighbor vegetation (Reader and Best 1989, Reader 1990, Reader et al. 1994, Wilson 1994) or in the presence versus absence of neighbor roots (Wilson 1993a) or roots and shoots (Aerts et al. 1991, Wilson and Tilman 1991, 1993, 1995, Wilson 1993b, Belcher et al. 1995, Gerry and Wilson 1995). This approach has been useful in examining changes in total, above and belowground competition along environmental gradients (such as soil fertility gradients; see Section V).

Another approach to multispecies competition is to compare the performance of each species when grown in a single, multispecies community to that when grown in pure culture (additive design: Campbell and Grime 1992; replacement series design: Turkington et al. 1993). Alternatively, species can be removed from existing communities, and the performance of the remaining species can be compared with that in no-removal controls (Pinder 1975, Abul-Fatih and Bazzaz 1979, Fowler 1981, Silander and Antonovics 1982, Keddy 1989, Gurevitch and Unnasch 1989). Goldberg et al. (1995) suggested a design called the community density series that additionally varies neighbor abundance. In this design, community density is varied while initial relative abundances of component species are held constant. The slope of a regression of eventual relative abundance on initial community density for each species yields a measure of its community-context competitive ability. Although these multispecies approaches do not separate the effects of different neighbor species, or the effects of total neighbor biomass or density from per-amount effects of neighbors, they provide information about species' competitive performances under conditions more similar to that in the field.

Tilman (1990) suggested that a focus on the mechanism of competition—resource reduction (see Section III)—leads to the simplest and most predictive approach to understanding multispecies competition. Theoretically, the species that reduces the limiting resource (e.g., available nitrogen, water, light) to the lowest level in monoculture is eventually able to displace all competitors (Tilman 1982, 1988). Many fewer experiments are required to determine the R* (or R*s, if more than one resource is limiting) of each species in a community than to run pairwise competition experiments (Tilman 1990). R*s for soil nitrogen were found to predict the outcome of pairwise competition among four perennial grass species (Tilman and Wedin 1991). Similarly, greater rates of water extraction and greater depletion of available water were correlated with the superior competitive ability of one aridland tussock grass compared with another (Eissenstat and Caldwell 1988). This approach has yet to be applied to other terrestrial plant systems or to competition among multispecies assemblages.

V. COMPETITION INTENSITY

In the last decade and a half, experimental work on competition has moved from demonstrating the existence of plant competition in the field (Connell 1983, Schoener 1983) toward a focus on testing the contrasting predictions of the C-S-R and resource competition models (Goldberg and Barton 1992, Gurevitch et al. 1992). Two questions are typically asked: (1) Does competition intensity (competitive ability) change along gradients of soil fertility and/or disturbance? (2) What physiological and morphological traits are important to competitive ability? As yet, no consensus has been reached on either of these questions. This is partly due to the relative novelty of such studies, especially those examining more than one species and more than one fertility and disturbance treatment

(Goldberg and Barton 1992, Goldberg 1994). There are also differences of opinion about the appropriate index of competition intensity to use. Two indices of competition intensity (CI) are commonly in use: absolute reduction in performance (physiological state, growth rate, fecundity, size, fitness) due to the presence of competition and relative reduction in performance (Table 1). Absolute CI will almost inevitably be greater as plant biomass increases (with fertilization or inherent size differences), simply because the maximum possible difference between monoculture and mixture performance increases as biomass increases (Campbell and Grime 1992, Grace 1993). Relative CI avoids this effect because it is standardized to a per-unit biomass basis. Furthermore, Grace (1995) has shown that relative CI accurately reflects the trajectories of species grown in mixtures (i.e., which species "wins"), while absolute CI does not.

Failure to clearly distinguish between the intensity of competition and its importance has also been raised as a complicating issue in the debate over whether competition intensity changes with environmental gradients (Weldon and Slauson 1986, Grace 1991). Intensity (CI, as above) measures the reduction below optimum in some measure of performance due to competition, whereas importance measures the relative degree that competition reduces performance below optimum compared with other processes (e.g., herbivory, disturbance, direct abiotic effects) (Weldon and Slauson 1986). Weldon and Slauson argue that high intensity of competition does not necessarily translate into high importance (or vice versa), as is typically assumed, and that the two concepts are often confused.

Continuing debate about competition intensity and about the traits important to competitive ability is also due to inattention to the selection regime commonly experienced by the study species. Most field experiments examining changes in competition intensity with changes in soil fertility (or disturbance): (1) measure competition between one or more target species (usually transplanted) and surrounding vegetation over a natural fertility or productivity gradient (Reader and Best 1989, Reader et al. 1994, Belcher et al. 1995), sometimes in both the presence and absence of disturbance (Reader 1992, Bonser and Reader 1995); (2) measure competition between target species and surrounding vegetation with and without fertilization (Reader 1990, DiTommaso and Aarssen 1991, Wilson and Tilman 1991, Vila and Terradas 1995) or with application of fertilization and disturbance (Wilson and Shay 1990, Wilson and Tilman 1993, 1995); or (3) measure competition in experimental mixtures of species with and without fertilization (Helgadottir and Snaydon 1985, McGraw and Chapin 1989, Aerts et al. 1990, 1991) or with factorial application of fertilization and disturbance (Campbell and Grime 1992, Turkington et al. 1993).

A common expectation in such experiments, particularly those using target species, is that species will act as "phytometers" of competition intensity that are dependent only on the experimental conditions (e.g., soil fertility, disturbance) and independent of the environmental conditions to which the study species are adapted. This expectation is reasonable only if the C-S-R model holds, in which case species should experience greater competition intensity at high versus low soil fertility regardless of whether they are adapted to low or high fertility soil (Taylor et al. 1990).

However, if the resource competition model holds, a species adapted to a lower fertility soil (i.e., a good competitor for soil resources) should experience greater competition intensity when competing on a higher fertility soil against better light competitors than when competing on its own soil type (Aarssen 1984, Tilman 1988, Smith and Huston 1989). The converse is true for a species adapted to high fertility soil. Similar reasoning can be applied to the relationship between disturbance and competition intensity. This section discusses the experimental evidence from field studies on competition intensity in light of the selection regime-dependent predictions of the resource competition model. The focus is on field studies conducted in natural or partially natural soil because artificial soil mixes are likely to have depleted microbe communities compared with natural soils or natural soil mixes (Diaz et al. 1993). Mycorrhizae (Hartnett et al. 1993), associations with nitrogen-fixing bacteria (Chanway et al. 1991), endophytic fungi (Clay 1990, Marks et al. 1991), and pathogenic microorganisms (Bever 1994) are all known to affect plant nutrition, growth, and/or competitive ability. Results from soil fertility (or productivity) gradients are discussed first, followed by results from disturbance gradients. Some studies calculate both relative and absolute CI, but only results using relative CI are discussed here.

A. Soil Fertility or Productivity Gradients

Several studies using target species most common at (and presumably most adapted to) low soil fertility or low productivity sites found that competition was more intense at high compared with low soil fertility (Reader and Best 1989, Reader 1990, Bonser and Reader 1995). All concluded that these results supported the C-S-R model. However, as previous argued, the resource competition model predicts higher competition intensity at high versus low soil fertility for species adapted to low fertility soil. These results could therefore be consistent with both models. Additional information, such as whether target species adapted to high fertility soil experienced more intense competition at low versus high soil fertility and/or the relative intensities of below- versus aboveground competition, would be required to distinguish which model was supported.

In contrast, other studies using target species found that competition intensity did not vary with soil fertility, regardless of whether the study species were common to low- or high-fertility soil (Wilson and Shay 1990, Wilson and Tilman 1991, 1993, Reader et al. 1994, Belcher et al. 1995). Two studies that also measured the ratio of below- to aboveground competition found that belowground competition was more important at low soil fertility, whereas aboveground competition became important at high soil fertility (Wilson and Tilman 1991, 1993; Belcher et al. 1995 found that aboveground competition was consistently insignificant). These studies can be interpreted as support for the resource competition model, with the caveat that some species can be plastic in their competitive abilities, such that they are able to compete well at both low and high soil fertility. Many of the target species used in these studies occurred over a range of soil fertilities or productivities, and plasticity in competitive ability might be expected in species with such broad ranges. Differences in plasticity among species could explain why some studies find that the relationship between competition intensity and soil fertility varies with target species (DiTommaso and Aarssen 1991, Reader 1992, Wilson and Tilman 1995).

Results from studies examining competition within experimental mixtures of low and high fertility–adapted species also tend to support the resource competition model. A number of these studies found the predicted reversals of competitive ability among low and high fertility–adapted species with soil fertility (low fertility–adapted species favored on low-fertility soil, high fertility–adapted species favored on high-fertility soil) (Helgadóttir and Snaydon 1985, McGraw and Chapin 1989, Aerts et al. 1990; see also Aerts et al. 1991). Two other studies again suggested the existence of plasticity in competitive ability, finding that for most study species, relative (although not absolute) competition intensity was constant over a range of soil fertilities (Campbell and Grime 1992, Turkington et al. 1993).

A cluster of studies using the annual desert grass *Stipa capensis* as a target species found that competition intensity increased along a productivity gradient driven by a topographic soil moisture gradient (Kadmon and Schmida 1990a, 1990b, Kadmon 1995). Because this grass is most common at the high-productivity (low topography) end of the gradient, it cannot be argued that the higher competition intensity observed at high productivity was due to poor adaptation of this grass to higher productivity conditions. Therefore, the results are more in line with the C-S-R than resource competition model, and suggest that for this grass species, the ratio of resource supply to demand is higher at the low compared with the high end of the productivity gradient. A higher resource supply to demand ratio at low productivity is most likely a result of lower plant demand due to lower population density at low productivity rather than to higher resource supply (the effective supply of water is actually lower at the low productivity end [slope areas] due to rapid drainage).

The studies with *S. capensis* are distinct from the studies discussed earlier in this section in that the productivity gradient was largely determined by water availability rather than by soil fertility. It has recently been hypothesized that the C-S-R model is more likely to apply to productivity gradients driven by water than by mineral nutrients (Goldberg and Novoplansky 1997). A key observation behind this "two-phase resource dynamics" hypothesis is that resource supply is pulsed (as is often true for water), such that plants experience pulse periods of high resource availability and interpulse periods of low resource availability (with interpulse periods of longer duration at lower productivity—also true for water). Competition is expected to be less intense during interpulse compared with pulse periods if resource availability during interpulses is determined more by abiotic factors (e.g., drainage, evaporation) than by plant uptake, or more precisely, when the resource supply to resource demand ratio is higher in interpulse periods than in pulse periods. This could occur for species that are better adapted to pulse periods than interpulse periods, such that the onset of an interpulse period causes severe enough mortality or limited enough growth to increase the resource supply to demand ratio relative to pulse periods. To date, little data exist with which to test the two-phase resource dynamics hypothesis, and Goldberg and Novoplansky's (1997) call for more studies that focus on competition under pulsed resource regimes is well justified.

One of the most straightforward ways of addressing whether competition intensity varies with soil fertility is to compare the intensity of above- and belowground competition among low fertility–adapted species growing on low-fertility soil with that among high fertility–adapted species growing on high-fertility soil (Taylor et al. 1990). The same approach could be used to examine competition intensity along disturbance or other environmental gradients. This type of experiment has hardly ever been conducted. A partial example is provided by Wilson (1993), who examined belowground competition intensity in forest (higher fertility) versus prairie (lower fertility). In support of resource competition theory, Wilson found that belowground competition was stronger where soil resources were lowest.

In conducting this type of experiment, it is important to consider the soil heterogeneity of the study sites. For example, a site may have a matrix of low soil fertility but contain patches of higher-nutrient soil (and the converse may be true for a site of mostly high soil fertility). Competitive abilities may shift with such local-scale soil heterogeneity (Reynolds et al. 1997), depending on whether species are adapted to particular soil patches or to a range of soil patches (i.e., plastic species). On the other hand, spatial and temporal heterogeneity may be so fine-grained and the spatial arrangements of species so random that species sharing the same habitat coevolve to be essentially similar in their competitive abilities (Aarssen 1983, 1989, 1992).

B. Disturbance Gradients

Effects of disturbance on competition intensity have been examined using a variety of experimentally imposed treatments (fire, tilling, clipping, trampling) and natural gradients (herbivory). The study species used in such studies vary in the extent to which they are adapted to each disturbance. Annual species are presumably adapted to tilling; perennial prairie grasses to fire, grazing (clipping), and possibly trampling; and species typically found where herbivory is intense are presumably adapted to herbivory. The C-S-R model (see also Taylor et al. 1990) predicts that species will experience lower competition intensity in the presence of disturbance regardless of how well they or their competitors are adapted to the disturbance. In support of this, Reader (1992) found that exposure to herbivory appeared to decrease the intensity of competition (as measured by flower production) experienced by pasture species. In addition, annual tilling reduced the relative intensity of belowground (Wilson and Tilman 1995) and total competition (Wilson and Tilman 1993) for old-field species, regardless of whether they were annuals or perennials.

However, accounting for differences in species' adaptations to disturbance could be important in interpreting experimental results. Disturbance is thought to reduce competition intensity by reducing population density or biomass and so reducing demand for resources. But adaptation to disturbance (e.g., defense against herbivores) may require increased expenditures of energy and resources by organisms, i.e., disturbance may result in fewer organisms that require more resources, so the effective demand for resources may not really be changed at all by disturbance (see Holt 1985 and Chesson and Huntly 1997 for related discussion). Thus, whether disturbance reduces competition intensity may depend on whether the competing species are adapted to the disturbance (equivalently, on how novel the disturbance is). Competition intensity might be lower in the presence versus absence of a novel disturbance if the disturbance knocks species abundances far enough below the resource-supplying power of the environment. Alternatively, competition intensity among species similarly adapted to a disturbance may show no change in the presence versus absence of that disturbance. Competition among species similarly adapted to grazing (Turkington et al. 1993) and fire (Wilson and Shay 1990) may explain why no change in relative competition intensity with disturbance was typically observed in two studies of grassland species.

VI. TRAITS

In most cases, it is not feasible to experimentally manipulate traits within any one species. The influence of traits on competitive ability must therefore be examined by correlating traits (e.g., R^*s, root to shoot ratio, height, seed size) across species with some measure of species performance in competition. Traits are usually measured on species grown in the absence of interspecific competition. If plastic responses to competition occur, values of traits may be different under control (no competition) versus competition treatments. However, as long as relative species differences are constant, ability to predict competitive outcomes from traits measured in either treatment should still be possible (although it is certainly more useful to be able to predict competitive outcome from knowledge about individually grown species). Clearly, studies examining as many traits, species, and environmental conditions (e.g., soil fertility, disturbance gradients) as possible are needed to uncover any general patterns of traits and competitive abilities and to address the contrasting predictions of the C-S-R and resource competition models. Relatively few studies have examined the relationship between traits and competitive ability; therefore, this section includes greenhouse studies and studies conducted in artifical soils as well as field studies.

An approach that allows comparison of a large number of species is the neighborhood design of Gaudet and Keddy (1988; see Section IV). The competitive ability of different neighbor species is measured as relative ability to suppress the growth of a common target species, or "phytometer." Multiple regression is used to determine which morphological and/or physiological traits of each neighbor species correlate with competitive ability. The generality of results can be assessed by testing the neighbor species (or some subset of them) against different phytometer species. Thus far, this approach has been used to compare species of highly productive wetland communities growing in high fertility soil (Gaudet and Keddy 1988). Results showed that competitive ability was most strongly correlated with aboveground biomass, followed by total and belowground biomass. Plant height, canopy diameter and area, and leaf shape were of secondary importance. These results could be interpreted as supporting either the C-S-R or resource competition models. The C-S-R model predicts that large size and height are among traits that allow acquisition of both above- and belowground resources at high soil fertility, whereas the resource competition model emphasizes large aboveground size as important for light acquisition at high soil fertility.

It would be enlightening to use this method to compare traits and competitive abilities among species adapted to lower-fertility soils, as well as between species of low- and high-fertility soils. Do the traits important to competitive ability change with resource supply rates, as resource competition theory predicts? Another good idea would be to focus more explicitly on the mechanism of resource competition, and correlate traits with above-and belowground resource depletion. It is also important to study competitive traits within a multispecies context, since this is typically the relevant context for understanding field distributions. One approach would be to grow different target species within multispecies neighborhoods and correlate target species traits with either resource reduction or target competitive performance. (Note that more than one target individual would likely be required to effectively measure its effects on resource levels.)

To date, relatively few traits and species have been evaluated in most other studies of plant competitive ability. As a group, however, these studies support resource competition theory in suggesting that traits related to acquisition of aboveground resources are important at high soil fertility and traits related to acquisition of belowground resources are important at low soil fertility.

For example, size (biomass, seed size, height) is often a good predictor of competitive ability when light competition becomes important (i.e., conditions of high soil fertility and consequently high productivity) (Black 1958, Schoener 1983, Gross 1984, Dolan 1984, Stanton 1984, Weiner and Thomas 1986, Wilson and Tilman 1991, Goldberg 1987, Miller and Werner 1987, Bazzaz et al. 1989, Reekie and Bazzaz 1989, Houssard and Escarré 1991, Grace et al. 1992). A reason suggested for this is that light is a directional resource, and thus competition for light is asymmetric; large plants are often able to obtain proportionately more light than smaller plants (Weiner and Thomas 1986, Thomas and Weiner 1989). Asymmetric competition may explain the existence of competitive hierarchies and transitivity (i.e., consistent rankings of competitive ability; A > B, B > C, ∴ A > C) in some competition studies (Keddy and Shipley 1989; but see Silvertown and Dale 1991 for a critique of these studies).

However, size is not always a good indicator of ability to preempt light. The degree to which biomass or productivity negatively correlates with light interception may, instead, depend on canopy architecture. For example, Tremmel and Bazzaz (1993) found that neighbor biomass was a poor predictor of the ability of neighbors to surpress targets, but that an index of neighbor light interception was a good predictor of neighbor competitive ability. Also, Wilson (1994) found that size made no difference in seedling competitive ability when seedlings were competing against mature (larger) vegetation.

Conversely, when competition is mostly for soil resources (i.e., conditions of low soil fertility and consequently low productivity), root allocation, specific root length, and ability to retain or to efficiently use nutrients become good predictors of competitive success (Eissenstat and Caldwell 1988, McGraw and Chapin, 1989, Tilman and Wedin 1991, Berendse et al. 1992, Wilson 1993a, 1993b). R^*s for soil resources also correspond with competitive success at low soil fertility (Tilman and Wedin 1991), and Tilman (1990) has shown analytically how R^* summarizes the effects of these other traits on ability to acquire resources.

VII. SUMMARY

Plants compete when demands on resources (water, light, mineral nutrients, CO_2) exceed resource supply. Plant competition is usually exploitative—competitors affect one another soley through resource consumption. Competition has been measured in a variety of habitats, from desert to temperate grassland, yet controversy remains about its significance for the ecology and evolution of species and the structure of communities; about whether the intensity of competition varies with environmental conditions; and about what traits are important to competitive ability.

In the last decade and a half, attention has been directed mostly at the latter two issues in the form of debate between the C-S-R and resource competition models. The crux of the debate between these models lies in when resource demand will exceed resource supply. The C-S-R model focuses on demand and hypothesizes that plant demand for above- and belowground resources will exceed supply only in the dense, highly productive vegetation that develops on fertile soil. In contrast, the resource competition theory considers demand for resources relative to resource supply. Consequently, this theory predicts that competition may occur in any habitat type, regardless of its level of soil fertility or degree of disturbance. It is the nature of competition that is predicted to change along gradients of soil fertility and disturbance. For example, competition for belowground resources is predicted to predominate at low soil fertility, whereas competition for light is predicted to become increasingly important as soil fertility increases (and light availability decreases due to shading).

Field experiments on competition intensity have as yet failed to produce any resolution to the debate between the C-S-R and resource competition models. This is partly due to the relative novelty of such studies, especially those examining more than one species and more than one fertility and disturbance treatment. There is also disagreement over whether to use absolute or relative competition intensity, although the weight of argument favors the relative index in studies of competition intensity. Failure to clearly distinguish between the intensity of competition versus its importance has also contributed to the ongoing controversy.

The lack of consensus may also be due to inattention to the selection regime commonly experienced by the study species, and to what experimental outcome each model should predict. If the C-S-R model is correct, competition intensity will be categorically lower on low versus high soil fertility. In contrast, if the resource competition model is correct, the intensity of competition experienced by a species will depend on the level(s) of soil fertility to which that species and its competitors are adapted. The same reasoning applies to competition along other environmental gradients, such as disturbance gradients. Thus, studies finding that competition intensity is more intense at high versus low soil fertility for species commonly found in low-fertility soil are consistent with both the C-S-R and resource competition models. A number of studies have found that the intensity of competition experienced by individual species does not change with changes in soil fertility, but that belowground competition is often more intense at low versus high soil fertility. These studies are consistent with the resource competition model, but suggest that the study species were plastic in their competitive abilities. Still other studies find the competitive reversals (low fertility–

adapted species favored at low soil fertility, high fertility–adapted species favored at high soil fertility) that the resource competition model predicts. Studies along productivity gradients driven by soil moisture availability rather than soil fertility tend to support the C-S-R model, and whether this is a feature of soil moisture gradients or more generally of gradients characterized by strongly pulsed resource supply, remains to be established.

Studies examining changes in competition intensity with changes in disturbance are much less common. Effects of disturbance on competition intensity have been examined using a variety of experimentally imposed treatments (fire, tilling, clipping, trampling) and natural gradients (herbivory). In support of the C-S-R model, exposure to herbivory and annual tilling have been found to decrease the intensity of competition in pasture and old-field species. However, other studies have found no change in competition intensity with grazing and fire among species similarly adapted to these disturbances, supporting the idea that the relationship between disturbance and competition intensity may depend on the extent to which the competing species are adapted to the disturbance.

Although there are relatively fewer studies about plant traits and competitive ability, they have been more consistent in supporting the resource competition model. As a group, such studies suggest that size (biomass, seed size, height) is often a good predictor of competitive ability when light competition becomes important (i.e., conditions of high soil fertility and consequently high productivity). Conversely, when competition is mostly for soil resources (i.e., conditions of low soil fertility and consequently low productivity), R*s for soil nutrients, root allocation, specific root length, and ability to retain or to efficiently use nutrients become good predictors of competitive success.

Further progress in resolving controversy over the importance of plant competition, its intensity in different habitats, and the traits most important in predicting competitive outcomes will require studies that (1) distinguish between importance and intensity; (2) include as many species and traits as possible (including R*, which is potentially a useful summary trait); (3) include as many environmental constraints (e.g., soil fertility, disturbance) as possible, including pulsed resources; and (4) pay close attention to the extent to which each study species is adapted to the particular environmental constraints used. It may also be useful to distinguish two aspects of the question of whether competition intensity changes along fertility (or disturbance) gradients: (1) Is competition equally intense among low fertility–adapted species growing on low-fertility soil and among high fertility–adapted species growing on high-fertility soil (low disturbance–adapted species growing in undisturbed environments versus high disturbance–adapted species growing in disturbed environments)? and (2) Does competition play a role in maintaining community boundaries, for example, in preventing low fertility–adapted species from invading high-fertility soil and vice versa?

ACKNOWLEDGMENTS

I thank Kay Gross for comments that improved this chapter. Financial support was provided by the National Science Foundation under a postdoctoral fellowship awarded in 1995, and by Michigan State University's Center for Microbial Ecology.

REFERENCES

Aarssen, L.W. 1983. Ecological combining ability and competitive combining ability in plants: toward a general evolutionary theory of coexistence in systems of competition. American Naturalist 122:707–731.

Aarssen, L.W. 1984. On the distinction between niche and competitive ability: implications for coexistence theory. Acta Biotheoretica 33:67–83.

Aarssen, L.W. 1989. Competitive ability and species coexistence: a 'plant's-eye' view. Oikos 56:386–401.

Aarssen, L.W. 1992. Causes and consequences of variation in competitive ability in plant communities. Journal of Vegetation Science 3:165–174.

Abul-Fatih, H.A., and F.A. Bazzaz. 1979. The biology of *Ambrosia trifida* L. I. Influence of species removal on the organization of the plant community. New Phytologist 83:813–816.

Aerts, R., F. Berendse, H. de Caluwe, and M. Schmitz. 1990. Competition in heathland along an experimental gradient of nutrient availability. Oikos 57:310–318.

Aerts, R., R.G.A. Boot, and P.J.M. van der Aart. 1991. The relation between above- and belowground biomass allocation patterns and competitive ability. Oecologia 87: 551–559.

Bannister, P. 1976. Introduction to Physiological Plant Ecology. Oxford: Blackwell Scientific.

Bazzaz, F.A., K. Garbutt, E.G. Reekie, and W.E. Williams. 1989. Using growth analysis to interpret competition between a C_3 and a C_4 annual under ambient and elevated CO_2. Oecologia 79:223–235.

Belcher, J.W., P.A. Keddy, and L. Twolan-Strutt. 1995. Root and shoot competition intensity along a soil depth gradient. Journal of Ecology 83:673–682.

Berendse, F., W.Th. Elberse, and R.H.M.E. Geerts. 1992. Competition and nitrogen loss from plants in grassland ecosystems. Ecology 73:46–53.

Bever, J.D. 1994. Feedback between plants and their soil communities in an old-field community. Ecology 75:1965–1977.

Black, J.N. 1958. Competition between plants of different initial seed sizes in swards of subterranean clover (*Trifolium subterraneum* L.) with particular reference to leaf area and the light microclimate. Australian Journal of Agricultural Research 9:299–318.

Bonser, S.P., and R.J. Reader. 1995. Plant competition and herbivory in relation to vegetation biomass. Ecology 76:2176–2183.

Campbell, B.D., J.P. Grime, and J.M.L. Mackey. 1991. A trade-off between scale and precision in resource foraging. Oecologia 87:532–538.

Campbell, B.D., and J.P. Grime. 1992. An experimental test of plant strategy theory. Ecology 73:15–29.

Campbell, B.D., J.P. Grime, and J.M.L. Mackey. 1992. Shoot thrust and its role in plant competition. Journal of Ecology 80:633–641.

Chanway, C.P., R. Turkington, and F.B. Holl 1991. Ecological implications of specificity between plants and rhizosphere micro-organisms. In: Begon, M., A.H. Fitter, and A. Macfadyen, eds. Advances in Ecological Research, Volume 21. San Diego, CA: Academic Press, pp 121–169.

Chesson, P., and N. Huntly. 1997. The roles of harsh and fluctuating conditions in the dynamics of ecological communities. The American Naturalist 150:519–553.

Clay, K. 1990. The impact of parasitic and mutualistic fungi on competitive interactions among plants. In: Grace, J.B., and D. Tilman, eds. Perspectives on Plant Competition. San Diego, CA: Academic Press, pp 391–412.

Connell, J.H. 1983. On the prevalence and relative importance of interspecific competition: evidence from field experiments. American Naturalist 122:661–696.

Connell, J.H. 1990. Apparent versus "real" competition in plants. In: Grace, J.B., and D. Tilman, eds. Perspectives on Plant Competition. San Diego, CA: Academic Press, pp 9–26.

Connolly, J. 1986. On difficulties with replacement-series methodology in mixture experiments. Journal of Applied Ecology 23:125–137.

Darwin, C. 1859. On the Origin of Species. Cambridge, MA: Harvard University Press.

de Wit, C.T. 1960. On competition. Verslagen van Landouwkundige Onderzoekingen 66: 1–82.

Diaz, S., J.P. Grime, J. Harris, and E. McPherson. 1993. Evidence of a feedback mechanism limiting plant response to elevated carbon dioxide. Nature 364:616–617.

DiTommaso, A., and L.W. Aarssen. 1991. Effect of nutrient level on competition intensity in the field for three coexisting grass species. Journal of Vegetation Science 2:513–522.

Dolan, R.W. 1984. The effect of seed size and maternal source on individual seed size in a population of *Ludwigia leptocarpa* (Onagraceae). American Journal of Botany 71:1302–1307.

Eissenstat, D.M., and M.M. Caldwell. 1988. Competitive ability is linked to rates of water extraction. A field study of two aridland tussock grasses. Oecologia 75:1–7.

Firbank, L.G., and A.R. Watkinson. 1985. On the analysis of competition within two-species mixtures of plants. Journal of Applied Ecology 22:503–517.

Firbank, L.G., and A.R. Watkinson. 1990. On the effects of competition: from monocultures to mixtures. In: Grace, J.B., and D. Tilman, eds. Perspectives on Plant Competition. San Diego, CA: Academic Press, pp 165–192.

Fowler, N. 1981. Competition and coexistence in a North Carolina grassland. II. The effects of the experimental removal of species. Journal of Ecology 69:843–854.

Gaudet, C.L., and P.A. Keddy. 1988. A comparative approach to predicting competitive ability from plant traits. Nature 334:242–243.

Gerry, A.K., and S.D. Wilson. 1995. The influence of initial size on the competitive responses of six plant species. Ecology 76:272–279.

Goldberg, D.E., and P.A. Werner. 1983. Equivalence of competitors in plant communities: a null hypothesis and a field experimental approach. American Journal of Botany 70:1098–1104.

Goldberg, D.E. 1987. Neighborhood competition in an old-field plant community. Ecology 68:1211–1223.

Goldberg, D.E., and A.M. Barton. 1992. Patterns and consequences of interspecific competition in natural communities: a review of field experiments with plants. American Naturalist 139:771–801.

Goldberg, D.E., and A. Novoplansky. 1997. On the relative importance of competition in unproductive environments. Journal of Ecology 85:409–418.

Goldberg, D.E., R. Turkington, and L. Olsvig-Whittaker. 1995. Quantifying the community-level consequences of competition. Folia Geobotanica Phytotaxonomica 30: 231–242.

Grace, J.B. 1990. On the relationship between plant traits and competitive ability. In: Grace, J.B., and D. Tilman, eds. Perspectives on Plant Competition. San Diego, CA: Academic Press, pp 51–65.

Grace, J.B. 1991. A clarification of the debate between Grime and Tilman. Functional Ecology 5:583–587.

Grace, J.B. 1993. The effects of habitat productivity on competition intensity. Tree 8: 229–230.

Grace, J.B. 1995. On the measurement of plant competition intensity. Ecology 76:305–308.

Grace, J.B., J. Keough, and G.R. Guntenspergen. 1992. Size bias in traditional analyses of substitutive experiments. Oecologia 90:429–434.

Grime, J.P. 1973. Reply to Newman. Nature 244:311.

Grime, J.P. 1977. Evidence for the existence of three primary strategies in plants and its relevance to ecological and evolutionary theory. American Naturalist 111:1169–1194.

Grime, J.P. 1979. Plant Strategies and Vegetation Processes. Chichester: Wiley.

Grime, J.P. 1988. The C-S-R model of primary plant strategies—origins, implications and tests. In: Gottlieb, L.D., and S.K. Jain, eds. Plant Evolutionary Biology. London: Chapman and Hall, pp 371–393.

Gross, K.L. 1984. Effects of seed size and growth form on seedling establishment of six monocarpic perennial plants. Journal of Ecology 72:369–387.

Gurevitch, J., and R.S. Unnasch. 1989. Experimental removal of a dominant species at two levels of soil fertility. Canadian Journal of Botany 67:3470–3477.

Gurevitch, J., L.L. Morrow, A. Wallace, and J.S. Walsh. 1992. A meta-analysis of competition in field experiments. American Naturalist 140:539–572.

Harper, J.L. 1977. Population Biology of Plants. London: Academic Press.

Hartnett, D.C., B.A.D. Hetrick, G.W.T. Wilson, and D.J. Gibson. 1993. Mycorrhizal influences on intra- and interspecific neighbour interactions among co-occurring prairie grasses. Journal of Ecology 81:787–795.

Helgadóttir, A., and R.W. Snaydon. 1985. Competitive interactions between populations of *Poa pratensis* and *Agrostis tenuis* from ecologically-contrasting environments. Journal of Applied Ecology 22:525–537.

Holt, R.D. 1985. Density-independent mortality, non-linear competitive interactions, and species coexistence. Journal of Theoretical Biology 116:479–493.

Houssard, C., and J. Escarré. 1991. The effects of seed weight on growth and competitive ability of *Rumex acetosella* from two successional old-fields. Oecologia 86:236–242.

Kadmon, R. 1995. Plant competition along soil moisture gradients: a field experiment with the desert annual *Stipa capensis*. Journal of Ecology 83:253–262.

Kadmon, R. and A. Schmida. 1990a. Competition in a variable environment: an experimental study in a desert annual plant population. Israel Journal of Botany 39:403–412.

Kadmon, R. and A. Schmida. 1990b. Patterns and causes of spatial variation in the reproductive success of a desert annual. Oecologia 83:139–144.

Keddy, P.A. 1989. Effects of competition from shrubs on herbaceous wetland plants: a 4-year field experiment. Canadian Journal of Botany 67:708–716.

Keddy, P.A. 1990. Competitive hierarchies and centrifugal organization in plant communities. In: J.B. Grace and D. Tilman, eds. Perspectives on plant competition. San Diego, CA: Academic Press, pp 265–290.

Keddy, P.A., and B. Shipley. 1989. Competitive hierarchies in herbaceous plant communities. Oikos 54:234–241.

Law, R. and A.R. Watkinson. 1987. Response-surface analysis of two-species competition: An experiment with *Phleum arenarium* and *Vulpia fasciculata*. Journal of Ecology 75:871–886.

Mack, R.N., and J.L. Harper. 1977. Interference in dune annuals: spatial pattern and neighborhood effects. Journal of Ecology 65:345–363.

Mahmoud, A., and J.P. Grime. 1976. An analysis of competitive ability in three perennial grasses. New Phytologist 77:431–435.

Marks, S., K. Clay, and G.P. Cheplick. 1991. Effects of fungal endophytes on interespecific and intraspecific competition in the grasses *Festuca arundinacea* and *Lolium perenne*. Journal of Applied Ecology 28:194–204.

McGilchrist, C.A., and B.R. Trenbath. 1971. A revised analysis of plant competition experiments. Biometrics 27:659–671.

McGraw, J.B., and F.S. Chapin III. 1989. Competitive ability and adaptation to fertile and infertile soils in two *Eriophorum* species. Ecology 70:736–749.

Miller, T.E., and P.A. Werner. 1987. Competitive effects and responses between plant species in a first-year old-field community. Ecology 68:1201–1210.

Newman, E.I. 1973. Competition and diversity in herbaceous vegetation. Nature 244:310.

Pacala, S.W., and J.A. Silander, Jr. 1985. Neighborhood models of plant population dynamics. I. Single-species models of annuals. American Naturalist 125:385–411.

Pacala, S.W., and J.A. Silander, Jr. 1987. Neighborhood interference among velvet leaf, *Abutilon theophrasti*, and pigweed, *Amaranthus retroflexus*. Oikos 48:217–224.

Pinder, J.E. III 1975. Effects of species removal on an old-field plant community. Ecology 56:747–751.

Reader, R.J. 1990. Competition constrained by low nutrient supply: an example involving *Hieracium floribundum* Wimm & Grab. (Compositae). Functional Ecology 4:573–577.

Reader, R.J. 1992. Herbivory, competition, plant mortality and reproduction on a topographic gradient in an abandoned pasture. Oikos 65:414–418.

Reader, R.J., and B.J. Best. 1989. Variation in competition along an environmental gradient: *Hieracium floribundum* in an abandoned pasture. Journal of Ecology 77:673–684.

Reader, R.J., S.D. Wilson, J.W. Belcher, I. Wisheu, P.A. Keddy, D. Tilman, E.C. Morris, J.B. Grace, J.B. McGraw, H. Olff, R. Turkington, E. Klein, Y. Leung, B. Shipley, R. van Hulst, M.E. Johansson, C. Nilsson, J. Gurevitch, K. Grigulis, and B.E. Beisner. 1994. Plant competition in relation to neighbor biomass: an intercontinental study with *Poa pratensis*. Ecology 75:1753–1760.

Reekie, E.G., and F.A. Bazzaz. 1989. Competition and patterns of resource use among seedlings of five tropical trees grown at ambient and elevated CO_2. Oecologia 79: 212–222.

Reynolds, H.L., and S.W. Pacala. 1993. An analytical treatment of plant competition for soil nutrient and light. American Naturalist 141:51–70.

Reynolds, H.L., B.A., Hungate, F.S. Chapin III, and C.M. D'Antonio. 1997. Soil heterogeneity and plant competition in an annual grassland. Ecology 78:2076–2090.

Roughgarden, J. 1979. Theory of Population Genetics and Evolutionary Ecology: An Introduction. New York: Macmillan; London: Collier Macmillan.

Sackville Hamilton, N.R. 1994. Replacement and additive designs for plant competition studies. Journal of Applied Ecology 31:599–603.

Schoener, T.W. 1983. Field experiments on interspecific competition. American Naturalist 122:240–285.

Silander, J.A., and J. Antonovics. 1982. Analysis of interspecific interactions in a coastal plant community—a perturbation approach. Nature 298:557–560.

Silander, J.A. Jr., and S.W. Pacala. 1985. Neighborhood predictors of plant performance. Oecologia 66:256–263.

Silander, J.A. Jr., and S.W. Pacala. 1990. The application of plant population dynamic models to understanding plant competition. In: Grace, J.B. and D. Tilman, eds. Perspectives on Plant Competition. San Diego, CA, USA. Academic Press, pp 67–91.

Silvertown, J., and P. Dale. 1991. Competitive hierarchies and the structure of herbaceous plant communities. Oikos 61:441–444.

Smith, T., and M. Huston. 1989. A theory of the spatial and temporal dynamics of plant communities. Vegetatio 83:49–69.

Snaydon, R.W. 1991. Replacement or additive designs for competition studies? Journal of Applied Ecology 28:930–946.

Snaydon, R.W. 1994 Replacement and additive designs revisited: comments on the review paper by N.R. Sackville Hamilton. Journal of Applied Ecology 31:784–786.

Stanton, M.L. 1984. Seed variation in wild radish: effect of seed size on components of seedling and adult fitness. Ecology 65:1105–1112.

Taylor, D.R., and L.W. Aarssen. 1989. On the density dependence of replacement-series competition experiments. Journal of Ecology 77:975–988.

Taylor, D.R., L.W. Aarssen, and C. Loehle. 1990. On the relationship between r/K selection and environmental carrying capacity: a new habitat templet for plant life history strategies. Oikos 58:239–250.

Thomas, S.C., and J. Weiner. 1989. Including competitive asymmetry in measures of local interference in plant populations. Oecologia 80:349–355.

Thompson, K. 1987. The resource ratio hypothesis and the meaning of competition. Functional Ecology 1:297–315.

Tilman, D. 1977. Resource competition between planktonic algae: an experimental and theoretical approach. Ecology 58:338–348.

Tilman, D. 1982. Resource Competition and Community Structure. Monographs in Population Biology 17. Princeton, NJ: Princeton University Press.

Tilman, D. 1987. On the meaning of competition and the mechanisms of competitive superiority. Functional Ecology 1:304–315.

Tilman, D. 1988. Plant Strategies and the Dynamics and Structure of Plant Communities. Monographs in Population Biology 26. Princeton, NJ: Princeton University Press.

Tilman, D. 1990. Mechanisms of plant competition for nutrients: the elements of a predictive theory of competition. In: Grace, J.B., and D. Tilman, eds. Perspectives on Plant Competition. San Diego, CA: Academic Press, pp 117–141. USA.

Tilman, D., and D. Wedin. 1991. Plant traits and resource reduction for five grasses growing on a nitrogen gradient. Ecology 72:685–700.

Tremmel, D.C., and F.A. Bazzaz. 1993. How neighbor canopy architecture affects target plant performance. Ecology 74:2114–2124.

Turkington, R., E. Klein, and C.P. Chanway, 1993. Interactive effects of nutrients and disturbance: an experimental test of plant strategy theory. Ecology 74:863–878.

Underwood, T. 1986. The analysis of competition by field experiments. In: Kikkawa, J., and D.J. Anderson, eds. Community Ecology: Pattern and Process. Boston: Blackwell Scientific, pp 240–268.

Vilà, M., and J. Terradas. 1995. Effects of nutrient availability and neighbors on shoot growth, resprouting and flowering of *Erica multiflora*. Journal of Vegetation Science 6:411–416.

Weiner, J. 1982. A neighborhood model of annual-plant interference. Ecology 63:1237–1241.

Weiner, J., and S.C. Thomas. 1986. Size variability and competition in plant monocultures. Oikos 47:211–222.

Welden, C.W., and W.L. Slauson. 1986. The intensity of competition versus its importance: an overlooked distinction and some implications. Quarterly Review of Biology 61:23–44.

Williamson, G.B. 1990. Allelopathy, Koch's postulates, and the neck riddle. In: Grace, J.B., and D. Tilman. Perspectives on Plant Competition. San Diego, CA: Academic Press, pp 143–162.

Wilson, S.D. 1993a. Belowground competition in forest and prairie. Oikos 68:146–150.

Wilson, S.D. 1993b. Competition and resource availability in heath and grassland in the Snowy Mountains of Australia. Journal of Ecology 81:445–451.

Wilson, S.D. 1994. Initial size and the competitive responses of two grasses at two levels of soil nitrogen: a field experiment. Canadian Journal of Botany 72:1349–1354.

Wilson, S.D., and P.A. Keddy. 1986. Measuring diffuse competition along an environmental gradient: results from a shoreline plant community. American Naturalist 127:862–869.

Wilson, S.D., and J.M. Shay. 1990. Competition fire, and nutrients in a mixed-grass prairie. Ecology 71:1959–1967.

Wilson, S.D., and D. Tilman. 1991. Components of plant competition along an experimental gradient of nitrogen availability. Ecology 72:1050–1065.

Wilson, S.D., and D. Tilman. 1993. Plant competition and resource availability in response to disturbance and fertilization. Ecology 74:599–611.

Wilson, S.D., and D. Tilman. 1995. Competitive responses of eight old-field plant species in four environments. Ecology 76:1169–1180.

19

Plant–Herbivore Interaction: Beyond a Binary Vision

**Regino Zamora, José A. Hódar, and
José M. Gómez**
University of Granada, Granada, Spain

I. INTRODUCTION

Herbivory is currently defined as the interaction that results when an animal consumes the live tissues of a plant (i.e., a heterotroph preying on an autotroph). This makes herbivory the most basic trophic interaction in the food chains. Herbivory has given rise to the appearance of marvelous phenotypic traits both in plants and animals, has moulded the vegetation as well as entire landscapes of virtually all the ecosystems known on earth, and has determined the success of many species, including our own. Herbivory, therefore, deserves attention from ecologists.

Recently, at least two reviews have been published on the importance of herbivory among other ecological concepts. For example, in a survey among the members of the British Ecological Society, the concept of plant–herbivore interactions was scored 21st in importance among 50 concepts (Cherrett 1989), although the list included broad concepts such as ecosystem and succession, under which more narrow concepts like herbivory are clearly subsumed. By contrast, in a more recent bibliographic review of articles published in three leading journals (*Ecology, Oikos*, and *Oecologia*), Stiling (1994) found that plant–herbivore interactions scored third place in terms of the number of reports dedicated to this subject. In contrast to this large body of current research, Stamp (1996) indicated that ecology text books offer poor coverage of the plant–herbivore topic.

We have conducted a review to identify the features that drive current research in plant–herbivore interactions in terrestrial ecosystems. From 1990 to 1996, we reviewed the journals *American Naturalist, Ecology, Ecological Monographs, Evolution*, and *Oikos*, and from 1994 to 1996, we reviewed *Functional Ecology, Journal of Applied Ecology, Journal of Ecology*, and *Oecologia*. From this sample (6730 articles), we selected all of the articles dealing with herbivory in terrestrial systems in which the effects of the interaction were evaluated with respect to the plant, in line with the present book's scope (although some of the selected articles also analyzed the herbivore's point of view). To determine how this interaction is analyzed, we classified the articles according to several aspects of the study:

1. life phase studied: seed, seedling, sapling, adult, or several at the same time
2. life habit: woody, herbaceous, or other (studies with species having different life habits)
3. traits measured in plant: chemistry, mechanical, growth, reproduction, or several at the same time
4. tissue affected by herbivory: root, leaf, reproductive parts, others (wood, xylem, phloem, bark, cambium, etc.), or several at the same time
5. kind of herbivore: invertebrate (mainly mollusks and nematodes), insects, vertebrates, pathogens, or several at the same time
6. study system: bitrophic, tritrophic, or simulated herbivory
7. study level: plant module, individual, population, community, ecosystem, or several at the same time

We also noted whether plant resources (e.g., N, H_2O, P, light) were controlled. Because not every aspect could be evaluated in all articles, the number of articles for each aspect varied.

The number of articles dealing with herbivory from the plant's point of view was 455, or 6.8% of the total. The proportion of articles selected oscillated between the lowest figures of the more evolutionarily oriented journals, *Evolution* and *American Naturalist*, with 3.2% and 3.4%, respectively, to the maximum 11.2% in *Oecologia*, followed by *Journal of Ecology* with 10.0%. Figure 1 shows the results of our review.

Some 76.2% of reports dealt with adult plants. This is a revealing bias because plants are subjected to herbivory during their entire life cycle. In addition, scarcely 5.2% treated more than one life phase in the same article. Almost the same number of reports concerned woody or herbaceous plants.

The most common traits analyzed were plant chemistry (40.5%) and growth (48.9%). A scant 14.9% quantified the effect on reproduction, which is remarkable since theoretically the ultimate measure of the effect on the plant is fitness. An even lower 6.1% quantified mechanical defensive traits on the plant, which is attributable to the boreal-temperate bias of most field studies. On the other hand, only 21.0% of the reports evaluated more than one type of response.

More than half (58.7%) of the reports quantified herbivory only on leaves, some 12.4% on reproductive organs, and a scant 17.4% quantified herbivory in

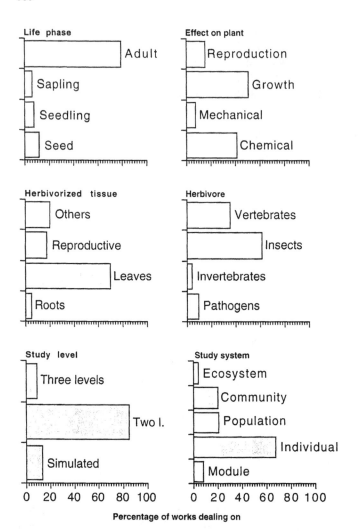

Figure 1 Frequency distribution (in percentage) of works dealing with the different topics analyzed in the literature. Note that some of the categories within an aspect were sometimes equally applicable to the same article, for which total percentages could add up to more than 100.

more than one plant's organs. But most noteworthy is the scarcity of works on root herbivory, despite the fact that roots comprised almost half of the biomass of most plants. Of the herbivores studied, most were insects (59.0%) and vertebrates (32.6%, almost exclusively mammals); other kinds of herbivores were almost incidental. Only 9.9% of the reports combined more than one kind of herbivore.

In general, the systems studied were simple pairs (one plant vs. one herbivore, 68.1%) of interacting elements. The number of experiments using direct manipulation was also notably limited (17.6%). Most of the studies (54.5%) analyzed the plant responses at the individual level, and more rarely at the population (9.6%) or community (13.6%) level. At lower organizational levels, it is also noteworthy that only 2.4% of the reports analyzed differences within the individual, which is rather surprising given the modular structure of plant species. Finally, 20.9% of the studies investigated the effect of resources available to the plant in the interaction.

In summary, research on herbivory typically concerns adult plants (woody or herbaceous) affected by defoliation either by insects or vertebrates, and mainly analyzes plant chemistry or growth in a simple pair of protagonists. In contrast, there is a lack of information on life phases such as seedlings or seeds, on aspects such as reproduction or mechanical defenses, on effect upon roots and tissues other than leaves, on population regeneration, and on communities and ecosystems. Consequently, the vision we have of herbivory may be biased by these gaps, and thus we should become aware of these limitations and fill these gaps.

In conclusion, herbivory has been viewed mainly as a binary interaction. However, to remain at this level is to say little about what herbivory truly represents in nature. The aim of this chapter is to try to transcend this limited view, examining from a phytocentric point of view the effect of herbivory within a broader ecological framework.

The literature on plant–herbivore interactions in terrestrial ecosystems is vast, and we have selectively cited more recent studies. The readers can consult many excellent reviews on more specific aspects of plant–herbivore interactions published in such journals as *Annual Review of Ecology and Systematics, Trends in Ecology and Evolution*, and the minireview section in *Oikos*. Finally, as a starting point for a general overview of plant–herbivore interactions, we suggest books by Crawley (1983), Howe and Westley (1988), and Abrahamson (1989).

II. PLANT TRAITS THAT DETERMINE HERBIVORY

There is compelling evidence that numerous plant traits reduce the feeding, growth, and survival of herbivores. These plant traits can be considered *resistance* when they represent an exclusively ecological trait (regardless of whether they are a response to herbivory), and *defense* when they involve an evolutionary response. In addition, resistance traits can in turn be split into two subgroups: constitutive and inducible. The former maintain a steady level, while the latter are physical, nutritional, and/or allelochemical traits that change in response to herbivory (Karban and Myers 1989). Another way to classify resistance traits is to split them into two main groups: tolerance and avoidance mechanisms. The former refers to the capacity of plants to regrow after an herbivory attack. Avoidance refers to the capacity of plants to deter herbivory; this can in turn be divided into three main mechanisms: probability of being found, physical barriers, and quality of plants as food.

A. Probability of Being Found

As the first step in herbivory, the plant must be found by the herbivore. There are several traits that reduce the plant's probability of being discovered, for example, remaining inconspicuous within a given habitat, in terms of both morphology and abundance, occupying enemy-free sites, imitating another well-defended plant or damaged tissue, completely lacking chemical attraction (being odorless), having a short life cycle, and having asynchronous timing—that is, the plant produces edible tissues in the period least likely to coincide with herbivore presence or feeding. Long-lived plants suffering from herbivores that have a short life cycle (such as insects) can opt to reproduce only at times that do not coincide with the onslaught of the insect population.

Another way to escape detection at the individual level includes synchronizing as closely as possible the production of tissues subject to attack (flowers, leaves, or fruit), both within the same plant and among plants of a population. This strategy, termed mass flowering or fruiting, attempts to overwhelm the herbivore's capacity to consume all the tissues available. Plants with mass flower or fruit production benefit from dense populations. In this way, the effect of time is multiplied by the effect of space (Kelly 1994).

B. Physical Barriers

After finding a host plant, herbivores must overcome other barriers. Plants have a myriad of structures to repel prospective herbivores. These include scales, barbs, thorns, and spines, which are particularly effective against mammals (Cooper and Owen-Smith 1986, Grubb 1992), and trichomes, leaf hairs, and similar structures meant to discourage invertebrate approach (Bernays and Chapman 1994). In the same way, some plants (e.g., Cariophyllaceae) have glands that secrete an adhesive to hamper small pests from crawling freely over the plant; these glands, which trap primarily small arthropods, are thought to have given rise to carnivory in plants. Sclerophyllous leaves are also avoided by herbivores, although sclerophylly may be considered either as mechanical barrier or digestibility reducer (see Section II.C), since it consists mainly of thickened cell walls in the leaf (Turner 1994).

Spines appear more in zones subject to heavy herbivore pressure, and even within an individual plant, parts less accesible to mammal herbivores tend to bear fewer spines than vulnerable parts. Furthermore, it has been demonstrated that spines have a negative effect on herbivore performance, reflected by consumption rate (Cooper and Owen-Smith 1986, Milewsky et al. 1991, Gowda 1996), and that simulated herbivory induces spinescence (Myers 1987, Milewsky et al. 1991, Obeso 1997).

C. Quality of Plants as Food

In most habitats, plants are abundant, and therefore food quantity is, in principle, not a problem to herbivores. The cell wall and cytoplasm of plant cells are both potential food for herbivores, but are different in chemical composition. The cell wall of higher plants consists mainly of cellulose, hemicellulose, and lignin. These three substances comprise the primary component of the biomass of trees (approximately 90%) and grasses (approximately 65%). Most herbivore animals cannot directly use this superabundant food resource because they cannot produce the enzymes necessary for decomposing cellulose. On the other hand, the cell cytoplasm, abundant in reproductive and photosinthetic tissues, is rich in lipids, starches, and proteins, but also in secondary compounds. Animals that feed on plant cytoplasm must solve the problem of plant's chemical defenses (Howe and Westley 1988). Clearly, not everything that is green is edible for a herbivore.

Herbivores are thus limited not by energy available but by the nutritional quality of plant tissues (White 1993). This is because there are strong stoichiometric differences between the composition of plant and animal tissues, i.e., plant matter, compared with animal tissues, is higher in carbon and lower in other essential elements such as N, P, and S (Sterner and Hessen 1994). As a whole, animal herbivores contain nearly 10 times more nitrogen than the plants that they eat. Thus, the plant–animal interface is characterized by a marked disparity in the biochemical make-up of consumer and resource. Thus, herbivores live in a "green desert" and are critically dependent (especially for female reproduction; Moen et al. 1993) on relatively rare, high-quality plants or plant organs. Herbivores must therefore selectively ingest and assimilate essential limiting minerals. For example, McNaughton (1988) found that the heterogeneous distributions of African ungulates correlated significantly between animal density and levels of minerals such as magnesium, sodium, and phosporous in vegetation. Similar cases have been found for many different herbivores (insects: Mattson 1980; sea urchins: Renaud et al. 1990; small mammals: Batzli 1983). Furthermore, not all mineral nutrients present in vegetal tissues are equally available to herbivores. For instance, some nitrogen sources such as proteins can be easily assimilated, whereas several nitrogenous compounds (alkaloids, cyanogenic glycosides) may be poisonous. In this case, the total nitrogen concentration may not reflect the nutritional value of the plant tissue (Bentley and Johnson 1992).

Despite this dietary selectivity, the chemical composition of herbivore diets is generally unbalanced. The result is that herbivores must consume large quantities of carbon to obtain enough nitrogen and other essential nutrients. The resulting nutritional imbalance can decrease growth efficiency for herbivores and could filter down from the top of the food chain, causing reduced production at all trophic levels. For example, assimilation efficiency of herbivores is lower (20–50%) in comparison with that of carnivores (nearly 80%; Begon et al. 1990). This nutritional limitation has forced some herbivore species to develop opportunistic feeding strategies to obtain alternative and/or complementary food nutritionally richer than vegetal tissues (see White 1993 for a thorough review).

Another aspect of the poor quality of plants as a food is the fact that a large fraction of vegetal tissues consists of nondigestible material (digestibility reducers). These are compounds from both primary (e.g., cellulose and hemicellulose) and secondary (lignin, cutins, tannins, and silicate particles; Howe and Westley 1988) metabolism. In general, they are quantitative (or chronic) metabolites that exert their effects according to their concentration within the plant, and do not act instantaneously, but rather gradually depress the growth or fecundity of the herbivore. In addition to reducing digestibility, silicate particles also accelerate tooth wear by the abrasion of mouthparts, contributing to the development of esophageal canker, and may cause fatal urolithiasis due to the formation of calculi in the urinary tracts (McNaughton et al. 1985). As a consequence, herbivores generally reject plant parts containing high concentrations of this compound.

Other compounds that reduce the quality of vegetal food are toxins. These compounds, accumulated from enzyme catalysis in biosynthesis (Harborne 1997), wield their influence by their sheer presence (qualitative secondary metabolites), striking with immediacy and often inflicting death on the herbivore. Normally, these compounds accumulate in tissues unprotected by quantitative substances—new leaves, immature fruits, or flower buds. The principal chemical compounds of this type are alkaloids, glucosinolates, toxic amino acids, terpenoids, and cyanogenic compounds.

D. Effects of Resource Availability on Plant–Herbivore Interaction

Variation in plant resource availability (light, nutrients, water) affects a host of interrelated plant traits, such as plant biomass, morphology, and chemistry, that can condition herbivore selection. These resource variations may alter the quality of plants as food, first by altering the nitrogen (and other mineral elements) concentration in plant tissues, and second, by changing the quantity of defensive structures and compounds. Enhanced nitrogen availability usually improves the quality of host plants as food for herbivores by increasing the nutritional quality and water content of plant tissues (Feller 1995, Hartley et al. 1995, Ruohomäki et al. 1996) as well as plant biomass (Dudt and Shure 1994), which in turn results in enhanced herbivore growth and survival (Kytö et al. 1996). In the same way, shade decreases the photosynthetic rate, which in turn increases the relative quantity of nutrients in tissues (Hartley et al. 1995, Ruohomäki et al. 1996), whereas sunlight availability has a significant positive influence on levels of carbon-based secondary metabolites (Dudt and Shure 1994). Appendix 1 shows the most widely known hypotheses explaining the effects of these environmental factors on secondary metabolism.

Plants stressed by an extreme situation of any environmental factor tend to increase the availability of nutrients for herbivores in tissues (e.g., soluble amino acids in phloem), and herbivores may feed preferentially on stressed plants (plant-stress hypothesis; White 1993). There are many reports of abundance of insects on plants stressed by air pollution, soil acidity, shade, soil nutrient deficiency, or previous herbivory. However, not every kind of stress provokes the same response in plants, and there are also cases in which stressed plants represented suboptimal food with respect to control plants (see, e.g., Mopper and Whitham 1992, and references therein). For instance, water stress reduces the water content in plant tissues, thus diminishing insect performance. According to Mopper and Whitham (1992), sustained environmental stress (poor soils or persistent drought), by causing numerous metabolic changes (e.g., increasing soluble nitrogen availability, reducing secondary compounds), can be benefitial to insects, whereas a brief drought period during insect oviposition may harm herbivore performance. By contrast, Price (1991) stated that some specialized herbivores feed preferentially on vigorous plants or plant modules (plant-vigor hypothesis). It seems that the latter applies especially to insect herbivores most closely associated with plant growth processes, as endophytic gallers and shoot borers (Price 1991).

A recent issue of interest is the effect of increased atmospheric CO_2, as a consequence of the ongoing global change. According to the predictions, increased CO_2 will lead to poorer plant tissues, both because of the increased C/N rate in tissues (due to enhanced CO_2 photosynthetic assimilation by the plant) and because of the increase of carbon-based secondary metabolites. This implies direct negative consequences for herbivores. However, global change scenarios should also consider the effects of increased temperature and changes in rainfall and cloud cover, in combination with increased CO_2, and limited information is available in this sense (Ayres 1993).

III. EFFECT ON PLANT PERFORMANCE AND POPULATIONS

A. Herbivory and Plant Performance

It is a general assumption that herbivory has a negative effect on plant performance, usually measured in terms of survival, growth, and reproduction. Some factors that determine the potential effect of herbivory are discussed in the following sections.

1. The Type of Herbivorized Tissue

Most herbivores process their food by chewing and then ingesting leaves and stems (Weis and Berenbaum 1989). However, other tissues may be also affected, such as flowers and buds. The damage to both buds and leaves is likely to be especially injurious because, not only are photosynthetic organs removed, but also the sites for plant growth. Perhaps the most damaging herbivore for the life of a tree is one that takes cambial tissue and destroys the ability of the plant to maintain a regenerating link between the root and shoot system. Flowers and fruits are highly nutritive because their tissues act as nutrient sinks, and floral and fruit damage implies a greater loss of female reproductive success than does injury to other tissues (Dirzo 1984, Hendrix 1988). This loss may be reflected directly by the diminished number of ovules after an herbivore feeds on flowers (Dirzo 1984).

An alternative to ingesting plant tissue is to suck the liquid contents from the vascular system (phloem and xylem), thus avoiding the indigestible cellulose and lignin. Sap-feeding insects are generally regarded as less damaging than leaf chewers because presumably they can feed on the products of photosynthesis without destroying the photosynthetic machinery. However, many plants support sap-feeding insects, with serious repercussions on plant performance (Meyer 1993).

The impact of root herbivory has been a much neglected facet of plant–herbivore interaction, despite the fact that belowground herbivory can have dramatic consequences for plant performance (Prins et al. 1992, Müller-Schärer and Brown 1995).

2. Plant Age

The impact of herbivory also depends on the age of the plant. Although herbivory on adult woody plants may decrease plant survival (Crawley 1983, Edenius et al. 1995), it is at the seed and seedling stages that herbivores principally influence plant mortality (Harper 1977, Hulme 1996). Herbivory in this case is analogous to predation.

Herbivores represent a major factor in seedling mortality, both directly by consumption and indirectly by trampling. Herbaceous seedlings are most often killed by insects and mollusks, whereas shrub and tree seedlings are killed primarily by rodent and browser mammals. Typical grazers looking for grasses can also kill seedlings of woody species.

3. The Amount, Timing, and Pattern of Herbivorized Tissue

Nutritional quality and defensive chemistry vary within an individual plant, with new leaves having a higher content of water and nitrogen, as well as defensive compounds, than older leaves. In addition, food quality of leaves may change over time. Foliar loss can have a detrimental or stimulatory effect, or no effect at all, on reproductive outcome, depending on the intensity of the loss (Maschinski and Whitham 1989). These effects may depend on the pattern and timing of damage because certain defoliation patterns may upset the relative resource competition between branches (Marquis 1992a, Honkanen and Haukioja 1994). Even within a modular unit (as a leaf), the timing of herbivory can have differential effects. For the plant, herbivory losses are more harmful when leaves are young than when leaves are older, especially toward senescence (Harper 1977).

4. The Frequency of Herbivory

In saplings of some tree species, the repeated loss of the apical meristems forcibly modifies the architecture of the tree, from tall to dwarf individuals, and retards the timing of the first reproductive season. In Sierra de Cazorla (southeast Spain), the sum of wild and domestic ungulates feeding on holm hak *Quercus ilex* (Fagaceae) produces a large number of prostrate, cushion, or cone-shaped forms in the younger trees (Cuartas and García-González 1992). The same occurs in the Sierra Nevada (southeast Spain), where saplings of Scots pine *Pinus sylvestris nevadensis* (Pinaceae), maple *Acer granatense* (Aceraceae), and other tree species develop stunted morphologies because of the browsing by Spanish ibex (*Capra pyrenaica* Bovidae) and domestic goats (Hódar et al. 1998).

5. The Capacity of Compensation

Compensation enables damaged plants to maintain their fitness through extra growth and reproduction, and may involve a variety of physiological and morphological mechanisms, including increased photosynthesis and altered patterns of resource allocation, differential balance in vegetative and reproductive tissues, changes in nutrient uptake, and altered hormonal balance (Rosenthal and Kotanen 1994). Compensation may result from damage due to a variety of factors such as drought, fire, frost, and wind, although the one that has received the most attention recently in the literature is herbivory (Aarssen 1995). Several studies have found that compensation follows herbivore attack (Hendrix and Trapp 1989) or experimental clipping (Ehrlén 1992, Obeso and Grubb 1994), and at low levels of herbivory, plants may even overcompensate for damage (Belsky 1986). Most plants produce more ovules than seeds, so that extra ovules can eventually be developed to replace damaged ones (Ehrlén 1991). Often, when flowers are removed from a plant, assimilates initially targeted for them are subsequently detoured to remaining flowers nearby, thus enhancing the probability of seed production among the surviving flowers (Ho 1992).

Life habit can affect capacity of compensation. In this respect, Obeso (1993) found that the effect of defoliation in seed production was more detrimental in woody plants than in herbaceous ones because the higher capacity of compensation of the latter. Furthermore, many extrinsic factors influence a plant's physiological state and, as a consequence, its ability to compensate for damage (Herms and Mattson 1992). For instance, the availability of resources needed for compensation may be strained by competition by neighboring plants (Coley et al. 1985, Hjältén et al. 1993a). However, most studies on compensation have dealt with grasses losing vegetative tissues while growing in resource-rich habitats (Rosenthal and Kotanen 1994), but have not investigated how compensation varies with environmental conditions. In fact, plant productivity can affect the plant's capacity to respond to herbivory. For example, Danell et al. (1991) found that Scots pine growing in habitats of low and medium productivity suffer more than pines in highly productive habitats, where the possibility for regrowth is better. Slow-growing plants are less able to compensate for biomass losses than fast-growing plants, and are likely to be more susceptible to herbivory if attacked (Coley et al. 1985).

B. Herbivory and Plant Population Regeneration

1. Plant Recruitment and Population Dynamics

Herbivory affects plant recruitment when the number of propagules that are integrated in the population is smaller than number of deaths by herbivory (that is, the recruitment rate is smaller than death rate; Harper 1977). In field conditions, plant recruitment depends on availability of propagules (propagule limitation) or on suitable microsites (microsite limitation; Eriksson and Ehrlén 1992). Works treating the relationship of herbivores to plant regeneration are rare (Hulme 1996). As a consequence, there is scant information about: (1) whether herbivores generally deplete seeds enough to reduce the densities of seedlings; (2) which plant species or life histories might be particularly vulnerable, or how environmental conditions affect the limitation of seedling density; and (3) the extent to which subsequent density-dependent compensation counteracts the effect of seed and seedling reduction caused by herbivores (Louda 1989, Crawley 1992, Louda and Potvin 1995).

The effect of herbivory on plant population regeneration may be disproportionate to the amount of biomass removed, especially if biomass is removed during sensitive stages of the life cycle or if meristems are damaged (Crawley 1983). Although the amount of biomass in seedling consumption might be minor, the reduction in the density of woody seedlings may have direct effects on population regeneration and successional processes (Facelli 1994). A common mistake is to assume that a strong herbivore effect on plant performance implies an equally intense effect on the plant population dynamics (Andersen 1989). For example, herbivores destroy roughly 95% of the total seeds in four species of long-lived perennials; however, there was negligible impact on population recruitment because in most years, recruitment appears to be limited by a rarity of safe sites, and not by seed supply, and the losses did not prevent the establishment of large seed banks potentially capable of exploiting temporary conditions favorable for recruitment (Andersen 1989).

The effect on population dynamics depends on life stage affected by herbivory. When grazing on saplings and adult plants, herbivory affects plant resource partitioning. Attacks on propagules via predispersal or postdispersal predation or seedling herbivory potentially diminish population regeneration (Hulme 1996), although there is no overall agreement about the actual effect of seed predation on plant population dynamics (Crawley 1992, Louda 1994).

For example, the importance of postdispersal seed predation has been considered negligible in perennial grasslands, in which microsite limitation was considered the main limitation, but the few studies that have additionally excluded seed predators suggest that the failure of many species to establish on dense vegetation may be due to high rates of seed predation rather than increased interference with established vegetation (Reader 1993). These studies support the view that the importance of seed limitation may be currently underestimated (Eriksson and Ehrlén 1992).

2. Herbivores as Plant Dispersers

Some herbivores can act as frugivores, dispersing some fleshy (Nogales et al. 1995) and nonfleshy fruits (Miller 1994). In these cases, the plant withstanding the most consumption would also be the most dispersed, and the outcome between antagonism–mutualism may be beneficial for the plant population because the seed dispersion enables the colonization of new areas. Granivorous ants also act as secondary dispersers of many plant species in semiarid and Mediterranean habitats (Aronne and Wilcock 1994). Other examples of herbivorous, seed-predators species acting as dispersers are corvids and rodents, which not only consume but also gather acorns, nuts, and pine nuts of many tree species. This type of interaction has even brought about some adaptations in plants. For instance, stone pines produce heavy, wingless seeds dispersed by corvids (Lanner 1996). These seeds are much stronger than those belonging to wind-dispersed species of pine.

Herbivores can also have positive effects in grasses by providing dispersal for the seeds of attacked plants. Foliage of small-seeded plants may function ecologically as fruit, attracting large herbivores just as arils and fleshy berries attract and reward frugivores (the "foliage is the fruit" hypothesis; Janzen 1984). In fact, there is growing evidence that some herbivorous mammals, such as lagomorphs and ungulates, act as dispersers of grass and forb seeds consumed incidentally while feeding on foliage (Malo and Suárez 1995). They consume grasses from tip to base, so the first thing consumed are reproductive parts. This kind of endozoochorous dispersal of seeds by herbivores has received only scant attention, despite the fact that it has been demonstrated for many different types of habitats and areas, such as Spanish dehesas (Malo and Suárez 1995, 1996), North American grasslands (Quinn et al. 1994), and, to a lesser extent, southern Asian flood plains (Dinerstein 1989), as well as for many different types of herbivores, including rabbits, cattle, deer, and even rhinoceros. Moreover, some grasses increase in germination rate and seedling growth after passage through the gut of herbivores (Quinn et al. 1994, Malo and Suárez 1995).

IV. THE EVOLUTIONARY PLAY

A. Plant Traits as Defenses Against Herbivory

The two main features of plants that are important for herbivores, nutritional quality, and secondary compounds related to defense, may vary among individual plants or genotypes within a population (Karban 1992). In this context, herbivores may result in differential fitness of plant genotypes. Herbivores can exert selective pressure without having a great effect on host–plant population dynamics (Andersen 1989), for which it is important to contrast the evolutionary and the demographic impact of herbivory (Crawley 1992).

A traditional assumption among evolutionary ecologists is that herbivory tends to lead to coevolution (Erhlich and Raven 1964), which implies that there is simultaneous evolution of ecologically interacting populations, consisting of a continuous evolutionary arms race between plants evolving new defenses and herbivores evolving adaptations to these defenses. Contrary to this coevolutionary thinking, the theory of sequential evolution (Jermy 1993) states that plants evolve by selective pressures far more imposing that those exerted by herbivores. Thus, this author contends that plants shape herbivore evolution, not vice versa.

To consider a plant trait to be a defense, we must accept that herbivores have to act as a selective pressure on plants. Four conditions are necessary (but not sufficient) to demonstrate that a herbivore selects plant traits: (1) intraspecific variation in the damage; (2) intraspecific variation in the set of plant traits correlated with the damage; (3) correlation between the trait (and the damage, of course) and plant fitness (Marquis 1992b); and (4) heritability of those putative defensive traits in plants.

The main plant traits studied from an evolutionary perspective are chemical compounds. Most investigators agree that these compounds have appeared as a response against herbivores, and several hypotheses have emerged to account for the evolution of chemical resistance, under the assumption that herbivory has been a major selective pressure. The most influential of these hypotheses are listed in Appendix 2. All of the hypotheses assume that (1) natural selection has lead to a high degree of optimization of defenses; (2) a plant's commitment to defense reflects the selective pressure by herbivores, and that herbivory has acted as a major selective force in plant evolution; and (3) defense traits are costly (Edwards 1989, Simms and Fritz 1990).

Although experiments, works, and theories on chemical defense are flooding the literature (Hartley and Jones 1997), there is an evident lack of information on mechanical barriers such as spines (Grubb 1992; see also Figure 1). Few would doubt the role of mechanical structures as resistance against herbivores, or even the evolution of these structures with this aim (Cooper and Owen-Smith 1986, Grubb 1992), but these assumptions have rarely been tested. The sketchy evidence suggests that nutrient availability does not influence the production of spines (Campbell 1986, Myers 1987). However, Grubb (1992) argued that spinescence is higher both in very poor (dry), and very rich environments. Furthermore, it seems that this type of mechanical defense does not imply an appreciable cost for the plant (Ågren and Schemske 1993) because of its smaller dependence on physiological pathways (Skogsmyr and Fagerström 1992). Spiny plants are widespread around the world—they cover vast surfaces, include a large numbers of species, and defend against a variety of herbivores. We cannot understand plant defense evolution solely by examining the chemical side of plants.

However, to date, insufficient evidence is available about the actual role of herbivores as selective pressure (Marquis 1992a). Some factors synergically appear to reduce the successful systems, showing selective impact of herbivores. First, empirical evidence seems to show that not every defense trait is costly (Bergelson and Purrington 1996). Second, there is no reason to assume that individual variation in damage has a genetic basis (Marquis 1992a), since the relative contribution of plant genotype with respect to environmental variability seems to be low in many of the studied systems (Karban 1992). Third, although most studies quantify phenotypic selection, no consistency has been established between phenotype and natural selection. For example, Núñez-Farfán and Dirzo (1994) detected phenotypic selection for some traits of *Datura stramonium* (Solanaceae), but no selection on their breeding values. Studies such as this suggest that evolutionary inferences based solely on phenotypic analysis must be considered with caution. Fourth, there are few studies showing any evidence of relationship between plant fitness and resistance-trait value. Finally, although paired plant–animal coevolution is theoretically possible, such pairing remains unlikely in nature because most plants interact with an array of herbivorous species, and vice versa. A plant species must often respond to the selective pressures exerted by a multispecific system (Simms and Rausher 1989, Meyer and Root 1993). This view postulates that the evolution of defense to one herbivore species is not independent of the presence of other species. The final result can be a dilution of all selective pressures, because the pressure of one herbivore species on plant traits is often opposed, constrained, or modified by pressures of other herbivore species.

B. An Alternative View

Plant secondary metabolism could have evolved as active defense against herbivores, or their current contribution to resistance could represent a secondary phenomenon that fortuituously benefits the plant (Tuomi et al. 1990). Not every plant trait conferring resistance should be interpreted as a defense against herbivory in evolutionary terms, since some traits provide neutral resistance without being aimed distinctly at animals (Edwards 1989). In fact, several investigators have recently proposed that yield increase after herbivory is an indirect consequence of selection induced by plant competition (Aarssen 1995). Most of these traits currently related to herbivory would have evolved in a world without animals as a result of abiotic selection for vegetative growth and survival. For example, sclerophyllous leaves may be an adaptive mechanism related to water and nutrient conservation (Turner 1994). Spiniscence may enhance resistance to drought as well as represent a barrier to herbivory. Terpenoid compounds act as feeding deterrents against generalist herbivores, but resins may also protect plants from drying in arid environments, and some chemical compounds may act as signals for other plants or animals (Bruin et al. 1995). Even secondary compounds are not always exclusively related to plant defense. For example, tannins may contribute to drought tolerance by increasing the elasticity of the cell wall. Plant exposure to UV-B generally induces production of secondary metabolites such as flavonoids, tannins, and lignin, which absorb UV-B. These increased metabolites could secondarily affect herbivory (Rozena et al. 1997). Furthermore, secondary metabolites may also directly increase competitive ability through allelopathic effects on neighboring plants. For example, Schmitt et al. (1995) showed that the secondary metabolites that defend the brown alga *Dyctiota menstrualis* (Dyctiotales) against herbivores also have the allelopathic function of eliminating competitors; thus, these compounds fulfil simultaneously different ecological functions. Similarly, the ability of plants to produce regrowth shoots from dormant buds is a clear response to abiotic (fire, drought, frost) as well as biotic (herbivore) damage.

In conclusion, many plant traits are part of a broad set of dynamic adaptive features of the flora to a diverse source of abiotic disturbances (Edwards 1989, Perevolotsky 1994), although herbivores may have synergically reacted by producing more intense, or even new, responses.

V. MULTISPECIFIC CONTEXT OF HERBIVORY

The great majority of plant species interact with more than one plant or animal species at the same time. Caught up in competition against other plants for substrate and nutrients, each plant species may be simultaneously eaten by many herbivores and/or pollinized by many species of floral visitors. The importance of the community context becomes apparent with the observation that the strength and even the sign of the interaction between two species may change in the presence of others by the action of the so-called indirect effects (Strauss 1991). Under these circumstances, scenarios in which only two or three species interact generally offer an overly simplistic and even inappropriate view of what in fact occurs in nature.

In this section, we provide examples of multi-species systems in plant-herbivore interaction, which illustrate that further research on this issue is necessary if we wish to gain a full understanding of herbivory and its effect on plant populations. A useful approach to the study of the enormous complexity of ecological communities are the "community modules" proposed by Holt (1997), involving a small number of species (3 to 6) linked in a specified structure of interactions. Most ecological studies on herbivory venture beyond the paired species traditionally analysed in these types of subsystems (see below).

A. Effect of Herbivores on Plant–Plant Interaction

1. Affecting Competition Between Plants

Herbivores can impair the competitive abilities of their host plants (Harper 1977, Edwards 1989; Figure 2, left). The selective consumption of individual plants can result in a hierarchy of sizes within a given plant population, thereby increasing the likelihood of asymmetrical competition between the plants (Weiner 1993). McEvoy et al. (1993) showed that herbivory on ragwort *Senecio jacobaea* (Asteraceae) by the beetle *Longitarsus jacobaeae* (Chrysomelidae) intensifies the competition of the plant with other species, speeding the ragwort's elimination, which would otherwise come about slowly.

Herbivory can also produce apparent competition among plants that share herbivores (Figure 2, center). An increase in density of one plant species results in a decreased density of another, not because they compete for the same resources, but because they are consumed by the same herbivore (Huntly 1991).

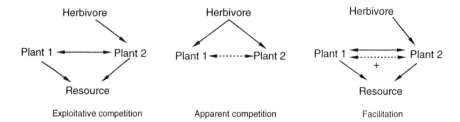

Figure 2 Schematic representation of the modules referring to more than one plant interacting with one herbivore. Solid arrows, direct effects; dotted arrows, indirect effects. All effects are negative except those marked with ''+.''

2. Associations Among Plants Sharing Herbivores

The probability that a plant will be attacked by a herbivore depends not only on the characteristics of the individual plant, but also on the quality and abundance of its neighbors. A plant species may have a positive net effect on another species by deterring the amount of herbivory that would otherwise be inflicted on that other species (Figure 2, right). In this context, antagonistic plant–plant relationships can be turned to positive ones (facilitation) under herbivore presence. This nonconsumer positive interaction can result when palatable plant species grow among unpalatable ones. For instance, Hay (1986) demonstrated a competitive cost for a palatable alga that grows in association with unpalatable algae; however, that cost is lower than the benefit from reduced herbivory under some conditions. As a result, in the presence of dense numbers of herbivorous fish, palatable algae remain more abundant where there is at least a 20% cover of unpalatable algae. In the same way, birch shoots were less consumed by voles and hares when associated with plants of lower palatability (Hjältén et al. 1993b).

Another type of facilitation can appear in plant–parasite interactions. Gómez (1994) has shown that parasitism by dodder *Cuscuta epithymum* (Cuscutaceae) on *Hormathophylla spinosa* (Cruciferae) changes from harmful to neutral or beneficial because parasitized individuals are less herbivorized by ungulates than are healthy plants. As a consequence, healthy and parasitized plants produced roughly equivalent numbers of seeds.

Facilitation can also result from physical protection provided by nurse plants. Plants growing in the immediate vicinity can block mammalian herbivore access to the vulnerable plant by creating a thorny barrier. The advantage of facilitation increases parallel to herbivore pressure (Bertness and Callaway 1994), to the point that in some situations the only seedlings and juvenile trees that survive remain within the islands formed by spiny shrubbery. Beyond these protectorates, natural regeneration can be completely arrested by strong herbivore pressure. For example, *Geranium purpureum* (Geraniaceae) and *Iberis contracta* (Cruciferae) are restricted in the National Park of Doñana (southwestern Spain) almost exclusively to areas beneath the cover of shrubby species, where these flowering plants are free of mammalian herbivory (Herrera 1991, Fernández-Haeger et al. 1996). A similar example is found in overgrazed Mediterranean mountains, where spiny shrub species, such as *Berberis hispanica* (Berberidaceae) and *Prunus ramburii* (Rosaceae), protect and favor the growth and survival of seedlings and saplings of Scots pine, maple *Acer granatense*, yew *Taxus baccata* (Taxaceae), and other canopy species (Hódar et al. 1996).

B. More than One Herbivore

Interspecific relationships between two herbivorous species can range from mutually competitive to mutually beneficial (Crawley 1983, Strauss 1991). When one plant becomes the host of several different herbivore species, it is difficult to understand the result of an interacting pair of species without taking into account the effect of the other herbivores.

1. Exploitative Competition Between Herbivores

With more than one herbivore species living on the same plant, the impact on one herbivore can alter the performance of the other herbivore by changing plant productivity, architecture, phenology, chemistry, etc. Thus, the possibility for exploitative competition arises (Figure 3, left). There is mounting evidence that interspecific competition exists between herbivores, at least between insect herbivores (Denno et al. 1995). Recently, Hudson and Stiling (1997) have shown that folivory on *Baccharis halimifolia* (Asteraceae) by larvae of *Trirhabda bacharidis*

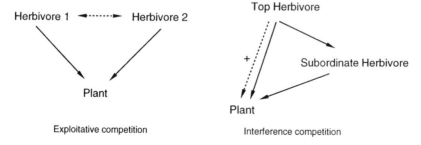

Figure 3 Schematic representation of the modules referring to more than one herbivore interacting with one plant. Solid arrows, direct effects; dotted arrows, indirect effects. All effects are negative except those marked with ''+.''

(Chrysomelidae) depresses densities of the most common herbivores of the host plant (two species of leaf-mining flies and one species of gall-making fly).

Masters and Brown (1997) differentiated three types of indirect interactions between insect herbivores sharing a host plant: (1) spatial interaction, where herbivores live on different parts of the host plant (different feeding niche); (2) temporal interaction, where the insect herbivores use the same niche at different times; and (3) spatio-temporal interactions, where temporally separated herbivores use different feeding niches. Most studies concentrate on temporally separated herbivores. For example, Faeth (1986) showed that early feeders negatively affect later feeders. Moreover, if the effects of the late feeders are carried over to the next growing season, the late feeder may have also a detrimental impact on the next season's early feeders. Competition among spatially separated herbivores is evident when considering interactions among aboveground and belowground insects. According to Masters et al. (1993), the interaction between these two types of herbivores is a plus–minus one, in which root feeders can help foliar feeders by intensifying plant stress, but foliar feeders hinder root feeder by decreasing root (resource) growth.

2. Asymmetrical Competition Between Vertebrate and Invertebrate Herbivores

Many studies indicate that asymmetric competitive interactions are more common among herbivores than symmetrical ones (Baines et al. 1994, Tscharntke 1997). This asymmetry can affect both competitors. For example, small sap-sucking herbivores can depress plant growth, thereby indirectly affecting mammal herbivores. Furthermore, in cases of a huge size difference between herbivores, asymmetry favors an interference mechanism (Baines et al. 1994, Tscharntke 1997; Figure 3, right). For example, Zamora and Gómez (1993) reported interference between a mammalian herbivore (Spanish ibex *Capra pyrenaica*) browsing on plant tissue where tiny herbivores (gall-makers) live. Although both the ibex and the gall-maker are herbivores of *Hormathophylla spinosa*, their interaction is direct—the ibex acts accidentally as a true predator of the gall-maker, apparently unable to distinguish the galls from the vegetal tissues. The spatial pattern of the galls is altered by the Spanish ibex browsing on *H. spinosa*, since gall ingestion depends on gall position on infrutescence (Zamora and Gómez 1993).

3. Interactions Between Herbivores and Pathogens

Interaction between herbivores, (e.g., insects, slugs, snails, birds, or mammals) and pathogens (e.g., bacteria, fungi and viruses) has been studied only recently (Faeth and Wilson 1997). Bowers and Sachi (1991) recorded an increase in disease levels of the rust *Uromyces trifolii* (Pucciniaceae) on clover (*Trifolium pratense* Fabaceae) in fenced exclosure plots compared with control plots. This increase resulted from an increased host plant density in the exclosures. On the other hand, some studies have reported that macroherbivores preferentially attack plants bearing microherbivores. Mollusks graze more heavily on rust-infested plants than on healthy ones (Ramsell and Paul 1990). Similarly, Ericsson and Wensström (1997) analyzed the interaction between the fungus *Urocystis tridentalis* (Ustilaginales), its host plant *Trientalis europaea* (Primulaceae), and two herbivores (scale insects and voles) in a 2-year experiment. The results indicate that both the scale insects and the voles preferred smut-infected shoots to healthy shoots. Fencing out the voles resulted in a significant boost in host density and a significantly higher disease level.

The interaction between herbivores and pathogens is not always antagonistic. For example, herbivores can transmit diseases to the host plant, increasing its harmful effect without substantial direct consumption of plant tissue. For example, European elm (*Ulmus* spp. Ulmaceae) forests have declined in the last 20 years because of a parasitic fungus that produces Dutch elm disease. This fungus is transmitted by some species of herbivorous beetles belonging to the family Scolytidae (Gil 1990). The detrimental effect of the beetle increases elm mortality not by direct consumption of cambium, but by acting as vector of the parasite.

C. Effect of Herbivores on Mutualism Involving Host Plant

Plants are involved in a diverse array of mutualistic interactions, including pollination, seed dispersal, or mycorrhiza symbiosis. Herbivores can influence any of these mutualistic interactions displayed by the host plant.

1. Effect on Pollen- and Seed-Dispersal System

Leaf damage decreases pollen production and performance in *Cucurbita texana* (Cucurbitaceae; Quesada et al. 1995) and produces selective fruit abortion in *Lindera benzoin* (Lauraceae; Niesenbaum 1996). Moreover, flower consumption by herbivores also affects the pollen-dispersal system indirectly by altering visitation rate of pollinators in entomophilous plants (Marquis 1992a).

Seed predators and fruit pests can affect the outcome of mutualistic interactions between plants and vertebrate seed dispersers. In this sense, most studies show that vertebrates reject fruits attacked both by insects (pulp eaters or seed predators) and by fungi (Sallabanks and Courtney 1992). For instance, juniper (*Juniperus communis* Cupressaceae) fruits attacked either by the seed-predator wasp *Megastigmus bipunctatus* (Thorymidae) or by the pulp-sucker scale *Carulaspis juniperi* (Diaspididae) were less consumed than healthy fruits by the avian dispersers *Turdus torquatus* and *Turdus viscivorus* (Turdidae) in the Mediterranean high mountains (García et al., in press). However, there are some cases in which vertebrate frugivores consume preferently insect-attacked fruits, obtaining a significant increase of proteins in the nutritional content of the fruit (Valburg 1992).

2. Effect on Plant–Mychorrhiza Interaction

Herbivore damage in aboveground biomass reduces the carbon source of plants to meet the demands of mycorrhiza fungi, leading to reductions in mycorrhizal colonization and, hence, a reduced nutrient uptake (Gehring and Whitham 1994). In addition, belowground herbivores and mycorrhiza fungi share the same habitat and therefore are likely to interact with each other (Gange and Bowler 1997). Insects such as Collembola graze on mycorrhizae, thereby affecting plant growth either directly or (above all) indirectly by N mobilization (Finlay 1985). On the other hand, the higher nutrient concentrations resulting from mycorrhizal colonization might make plants more attractive to herbivores, although data on this subject remain spotty and inconclusive.

D. Tritrophic Interactions

The basic food chain is composed of a plant, its herbivore, and the predator of the herbivore (Figure 4). The most widely studied tritrophic systems consist of a plant or a seed, a parasitic herbivore (seed predator, gall-maker, or the like), and parasitoids, although interest is growing with respect to the types of tritrophic systems, such as those in which insectivorous vertebrates (birds, reptiles, or mammals) intervene in the relationship between herbivores and plants (Tscharntke 1997). For instance, Marquis and Whelan (1994) found that white oak saplings caged from insectivorous birds suffered twice the leaf area loss by caterpillars

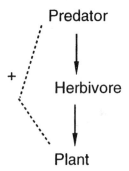

Figure 4 Schematic representation of a basic food chain. Solid arrows, direct effect; dotted line, indirect effect. All effects are negative except those marked with "+."

than uncaged saplings, growing one third less in total aboveground biomass. Thus, they concluded that insectivorous birds have a significant impact on oak growth by consumption of folivorous insects.

Parasitism represents a crucial mortality factor for many species of herbivorous insects. For this reason, the parasitoids can improve plant performance, although this has been reported only infrequently (Tscharntke 1997). Gómez and Zamora (1994) tested the totality of direct and indirect forces in a tritrophic system composed of a guild of three parasitoid species, a single weevil seed predator, and the host plant. When parasitoids were experimentally excluded, the percentage of attacked fruits increased from 20% to 43%, the parasitoids thus enhancing plant reproductive performance.

E. Multispecific Interactive Scenarios

1. The Importance of Guilds

Multispecific interactions can occur throughout guilds as interactive units, when there are functionally equivalent animals or plants (i.e., from the plant's or the herbivore's perspective). Plants may interact with a guild of ecomorphological similar herbivore species rather than with a particular species. The degree of generalization determines the breadth of the filter of the interaction, and the real possibility that the system might be facultative (different species with the same role). For example, Maddox and Root (1990), studying the trophic organization of the herbivorous insect community (more than 100 species distributed among 5 orders) of *Solidago altissima* (Asteraceae), suggested that the functionally similar herbivore groups may constitute selective units more powerful than individual species; this opens the possibility of synergetic responses as opposed to the same blocks of selective pressures (broad-spectrum responses). In this way, Krischik et al. (1991) indicated that nicotine was inhibitory to growth of both herbivores and pathogens, suggesting that certain secondary plant chemicals with high toxicity are of a generalized nature and affect multiple species. Hay et al. (1994) reported on a marine alga that produces secondary broad-spectrum metabolites to oppose herbivores. The combined effect of calcification of the algae and the presence of secondary metabolites act synergetically against a wide variety of herbivorous consumers.

2. Hierarchical Structures: *Hormathophylla spinosa* and its Herbivores

Not all herbivores affect plant performance in the same way. When a plant species is attacked by a multispecific system of herbivores differing strongly in taxonomical and ecomorphological traits (for example, ungulates, insects, and fungi), a hierarchically structured web of interactions might emerge, where one or two species can affect not only the reproductive success of the plant, but also the interactions involving each of the other herbivores (Figure 5). In this case, the top herbivores are capable of competing asymmetrically, and even of interfering with the population dynamics of the other herbivore species that infest the plant (see Section V.B.1). This precludes the possibility of evolutionary interaction between a host plant and certain minor herbivores. For example, *Hormathophylla spinosa* is an abundant stunted shrub of the high mountains of the western Mediterranean. In the Sierra Nevada (Granada, Spain), this species loses more than 55% of its ovules and/or seeds to the combined action of gall makers (presumably *Systasis encyrtoides* Pteromalidae), seed predators (*Ceutorhynchus* sp nov. Curculionidae), floral herbivores (*Baetica ustulata* Tettigonidae, *Oiketicoides febretta* Psychidae, *Pimelia monticola* Tenebrionidae, and *Timarcha lugens* Chrysomelidae), dodder (*Cuscuta epithymum*), granivorous birds (Linnet *Carduelis cannabina* Fringillidae), and ungulates (Spanish ibex and domestic sheep *Ovis aries*, Bovidae). However, the main herbivores are ungulates—domestic sheep in lower populations and the Spanish ibex in upper populations. This relationship between *H. spinosa* and each species of herbivore is not additive and independent of the occurrence of shared herbivores. When analyzing the effect of each species of herbivore on plant performance, we found that for several herbivores (for example, Linnets), the most herbivorized plants also dispersed the most seeds. This was because ungulates browsing on *H. spinosa* mask the consequences of other species of secondary herbivores, limiting the ecological importance of those minor herbivores. When we integrated the activity of all herbivores, we perceived

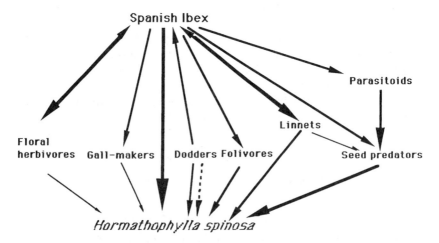

Figure 5 Schematic diagram showing the hierarchical relationships between the herbivores of *Hormathophylla spinosa* (Cruciferae) in the Sierra Nevada mountains.

a hierarchical structure in the various mortality factors that appeared in the multi-specific system (Figure 5).

VI. HERBIVORY AND THE DYNAMICS OF PLANT COMMUNITIES

A. Herbivory and Plant Diversity

Given that herbivores often strongly influence the growth and survival of individual plants, we would expect dramatic changes in plant community composition if herbivores either change the ability of plants to acquire limiting resources or eliminate other plants as competitors (Louda et al. 1990). In this context, differential selectivity among herbivores mediates responses at the population level and subsequently alters plant community composition. In the few cases in which insect and vertebrate herbivores have been excluded singly and in combination, the vertebrate herbivores have consistently proven to have the greater impact on plant dynamics (Crawley 1989). However, these conclusions could be biased because the effects of mammalian herbivory on plant communities are more documented (and even more evident) than those of invertebrate herbivory (Bach 1996).

Analogous to some marine keystone predators in relation to their invertebrate prey, some generalist herbivores have been viewed as keystone species because they appear to regulate plant diversity by selective grazing (Hunter 1992). When herbivorized species are competitively dominant, then herbivory may effectively free subdominant, unpalatable species from competitive suppression and thereby enhance diversity; on the contrary, when palatable species are competitively subdominant, then species diversity may decline as herbivory intensifies. For example, Ostfeld and Canahm (1993) showed that voles act as keystone species in North American old-fields because they delay succession by reducing the number of invading trees, create patchiness by concentrating surviving seedlings within certain microsites, and alter relative abundance of invading species via differences in palatability or apparency. Similarly, in the southeastern United States, Weltzin et al. (1997) illustrated how transitions from grassland to woodland vegetation can be regulated by a rodent herbivore, the prairie dogs (*Cynomys ludovicianus* Sciuridae). Widespread erradication of this abundant rodent has thereby facilited transitions to shrubs and woodland stages.

B. Herbivory and Plant Succession

Herbivory can be considered as a form of disturbance that removes living biomass from a community. In this context, herbivory is one of the many processes that contribute interactively to successional change (Pickett et al. 1987). A typical successional sequence goes from fast-growing, undefended herbaceous plants to slow-growing, well-defended woody plants. Herbivory can alter the trajectory and rate of succession, depending on the successional stage to which the preferentially herbivorized plants species belong (Hixon and Brostoff 1996). First, herbivores may slow down the rate of succession by inhibiting later successional species to the benefit of the earlier species. Alternatively, herbivory may speed up the rate of succession by inhibiting earlier successional species in favor of the invasion of later species. A third possibility is that, irrespective of the succession rate, herbivory forces the community to follow a different trajectory than would be followed in the absence of herbivores. This altered trajectory generally results from intense herbivore pressure that allows only resistant, nonpalatable species to persist.

Vertebrate as well as invertebrate herbivores have been found to feed on early successional species, thus aiding colonization by later species and thereby accelerating succession (Bryant 1987). In other cases, insect and vertebrate herbivores feeding on later successional species have reportedly retarded the succession rate by creating conditions advantageous to early successional species (Davidson 1993). In contrast to herbivory, granivory is found to fall differentially on large-seeded, later succesional species (Davidson 1993). Granivory can also inhibit primary succession in plant communities where plant recruitment is not microsite limited. Brown and Gange (1992) showed that the direction and rate of succession may be modified by the action of aboveground and belowground insect herbivores: secondary terrestrial succession was decelerated by foliar-feeding insects, but accelerated by root-feeding insects. By decelerating succession, herbivores can prolong the high-diversity midsuccesional stage. A case in point was described by Hixon and Brostoff (1996) for coral-reef algae communities, and by Bowers (1993) for a terrestrial grazing system. These results support Connell (1975) in his "intermediate disturbance hypothesis," which states that species richness should peak when disturbances (herbivore pressure) are of intermediate intensity, and should decrease under higher or lower disturbance intensities.

Noy-Meir et al. (1989) provided a thorough picture of the effects of grazing by domestic herbivores on the dynamics of Mediterranean grassland communities: (1) the first effect of grazers on grasslands is the removal of living plant parts, which is selective and hence differential between species; (2) grazing shifts the balance of species abundance; (3) grassland composition closely reflects the established grazing regime; and (4) relative abundance consistently decreases for some plant species in response to intensified grazing (decreasers: mostly taller, erect plants), but consistently increases for others (increasers: more prostrate or rosette plants); some species appear only above a certain threshold of grazing intensity (invaders). As grazing increases from light to medium, the dominant species are grazed down and suppressed. Small-scale disturbance in the horizontal plane is the major mechanism of change at low to medium grazing pressure, whereas differential defoliation along the vertical axis becomes the dominant mechanism at medium to high herbivore pressure. For example, only small and prostrate annual and rossette forms remained abundant under heavy grazing. Noy-Meir et al. (1989) also indicated that plant response to grazing depends not only on grazing intensity, but also on the grazing schedule at the site (year-long, seasonal, intermittent, etc.) and the grazer species. Furthermore, the plant–community response to grazing interact in a complex way with other abiotics factors such as soil nature, fire regime, and rainfall conditions.

C. Herbivory and Primary Production and Nutrient Cycling

The primary productivity of an ecosystem sets a maximum limit on the population density of herbivores that can be supported. Changes in plant species composition due to herbivory increase with rising productivity and with longer, more intense evolutionary histories of grazing (Milchunas and Lauenroth 1993). The proportion of aboveground net primary production consumed by herbivores varies greatly both in space and time, ranging widely from less than 1% to nearly 100% (Cebrián and Duarte 1984). For example, removal of annual primary production by grazing mammals on the Serengeti grasslands reaches 94% (McNaughton 1985). In contrast, herbivore rate was lowest in forests, where herbivores consumed less than 1% of production. As a result, the quantitative significance of detrital chains relative to grazing chains tends to be greater in forests than in grasslands (Begon et al. 1990).

Comparative analyses of the extend of herbivory in different ecosystems have demonstrated a positive linear relationship between primary productivity and herbivory rate, both in terrestrial (McNaughton et al. 1989) and aquatic (Cyr and Pace 1993) ecosystems. These results suggest that herbivores remove the same proportion of annual net primary production in poor as in rich ecosystems. On the other hand, herbivore pressure tends to intensify with increasing plant growth rate (Cebrián and Duarte 1984). Thus, herbivory was more likely to exert a stronger control on the biomass and production of fast-growing, high-nutrient plant communities compared with slow-growing, low-nutrient plant communities where most plant biomass passes directly to the detrital food chain.

In addition to removing plant biomass, herbivores provide litter and detritus. In doing so, they can also determine both the quantity and quality of litter return to the soil, and hence nutrient mineralization. For example, the moose changes the plant communities and the array of litter decomposed from them, thus controlling the productivity of the boreal forest (Pastor et al. 1993). Insect herbivores can also influence plant communities by altering the availability of resources required by plants. Brown (1994) indicated that net N mineralization and nitrification were higher in monocultures grazed by leaf beetles than in ungrazed monocultures.

Depending on the herbivore species and the plants species eaten, herbivores fail to digest 20–80% of the food they ingest and generate larges quantities of feces and urine. Ungulates excrete more than 90% of their ingested phosphorous, nearly 99% of this through feces, and 65–95% of their ingested nitrogen, nearly 85% of this through urine (Ruess 1987). Excreted phosphorous and nitrogen are readily available forms for uptake by plants and soil microbes. Dung also serves as a carbon source for soil microbial decomposition and can thus stimulate microbial nitrogen uptake. Because nutrients are often released more rapidly from the feces of herbivores than from the plants from which they are composed, herbivory increases the rate of cycling of the major nutrients. Consequently, herbivores may influence plant productivity and nutrient availability. For example, in the Serengeti, plant productivity is stimulated by the grazing of large mammals due to accelerated nutrient recycling through the deposition of dung and urine by animals. The positive associations between aboveground production and herbivore consumption, and consumption and dung deposition (i.e., nutrient return) to sites from herbivores suggest that production and grazing are coupled to herbivore-facilitated nutrient cycling (McNaughton 1985).

Herbivores also determine the spatial distribution of nutrients in ecosystems. The spatially heterogeneous removal of vegetation by herbivory generates an irregular pattern of grazed and ungrazed areas or patches (McNaughton 1985). Animals that forage over large areas but defecate over small areas may substantially concentrate and redistribute nutrients, creating patches of differing nutrient levels and thereby altering species composition. Furthermore, dung disperses a high number of viable seeds. Endozoochorous dispersal encourages the spread of species that lack morphological structures for dissemination, and favors the small-scale spatial variability of dunging (Malo and Suárez 1995). On a larger scale, however, endozoochory can play a homogenizing role in the plant community through long-distance seed dissemination (Malo and Suárez 1996).

D. Nontrophic Effects of Herbivores

The impact of herbivores on plants goes beyond direct trophic relationships. Nontrophic animal activity can alter the physicochemical environment and thus the plant community (Huntly 1995). Activities such as disturbing and restructuring soil by trampling and crushing plants, burrowing, forming trails, compacting, or loosening soil, etc., can produce spatial heterogeneity, thereby promoting patchworks of varying amounts of herbivore pressure and, hence, differing successional stages. Furthermore, many herbivore species such as beavers, gophers, rabbits, ants, and termites build large structures above or below ground. These activities affect soil properties, water movement, and nutrient dynamics and can therefore have power implications for the entire ecosystem (Jones et al. 1994).

The strongest nontrophic effects are attributable to large herbivore species in high densities, such as herds of domestic or wild ungulates.

VII. HERBIVORY AND HUMAN ECOLOGY: OVERGRAZING

Humans raise livestock in high numbers, eliminate the large carnivorous verte-brates from the wild, and even introduce exotic herbivorous mammals to agricul-tural and protected areas. The biomass of herbivores supported per unit of primary productivity is approximately one order of magnitude greater in agricultural than in natural ecosystems for a given level of primary productivity (Oesterheld et al. 1992). Consequently, an adequate density of wild ungulates may improve forage resources (Milchunas and Lauenroth 1993), whereas a high density of livestock depletes them. This fact is especially evident in Mediterranean and arid habitats, grazed by domestic livestock over thousands of years (Noy-Meir et al. 1989, Perevolotsky 1994).

Enclosure experiments have shown that overgrazing by grazers alter the ecosystem process by consuming most of the aboveground plant biomass, reduc-ing the cover of herbaceous plants and litter, disturbing and compacting soils, reducing water infiltration, and increasing the proportion of bare ground and soil erosion (Belsky and Blumenthal 1997). For example, Schulz and Leininger (1990) found that grazed areas of a riparian meadow had 400% more bare ground than ungrazed areas. Such reduction in litter may cause severe consequences because litter is a critical source of soil nutrients, promoting water infiltration, and shielding the soil from the erosive force of rain as well as from freezing. Lower soil moisture content in turn reduces plant productivity and vegetative cover, creating negative feedbacks loops. The result depends largely on the pro-ductivity of the habitat (as well as the density and species of herbivore). In high-productive habitats, there is usually a development of grazing lawns. However, in less-productive, drought-stressed habitats, the result is bare ground, with some scattered, heavily browsed, thorny, unpalatable plants. This process leads to strong spatial heterogeneity of soil resources that characterize much of the arid landscape (Figure 6).

Alteration of community structure and processes caused by intense herbiv-ory is also a real concern in protected areas where wild ungulates are overabun-dant (MacCracken 1996). For example, moose browsing prevented saplings of preferred species from growing into the tree canopy, resulting in a forest with fewer canopy trees and a well-developed undestory of shrubs and herbs (McInnes et al. 1992). Moreover, consumption by abundant native large herbivores aver-aged 45% in grazed Yellowstone National Park (Frank and McNaughton 1992), where herbivory was over six times higher than other similar temperate grassland sites. In the Natural Park of Sierra Nevada, a low-productive mountain range in

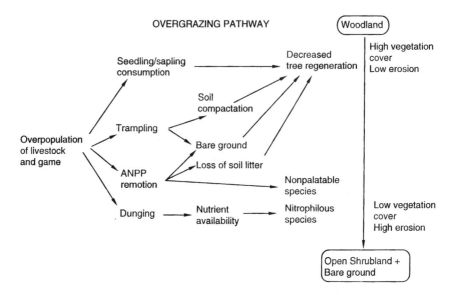

Figure 6 Effect of high density of livestock and game husbandry on woodland degradation. The causal chain shows the pathway from a Mediterranean woodland to an open shrubland with spiny, unpalatable plant species. ANPP, aboveground net primary production.

Southeast Spain, we studied a relict high-mountain forest of Scots pine and found that herbivory by browsing ungulates (Spanish ibex and domestic goats) was intense in pine saplings, representing more than 95% of total damage; the other kinds of damage (insects, freezing, etc.) were very scarce (Figure 7A). This damage was heavier in clearer zones of the woodland (treeline) than in the closer zones, i.e., the scarce woody areas suffer the most herbivore pressure at the borders, a situation that precludes the possibility for expansion (Figure 7B). Furthermore, in wet years, only 20% of the saplings suffer some herbivore attack, probably because ungulates have alternative resources, such as pasture or other more palatable woody species. However, in dry years, with low pasture production, up to 80% of saplings suffer browsing damage (Figure 7B; Hódar et al. 1998). This situation applies to most of the woody species in the area, being particularly critical for the most palatable ones, such as maple *Acer granatense* or rowan *Sorbus aria* (Rosaceae). In these species, the bank of seedlings and saplings does not translate to juvenile ones due to the intense herbivory by browsing ungulates. The result is that forest regeneration is severely constrained, a common problem in most of the Mediterranean basin (Le Houérou 1981).

An ecologically based definition of herbivore overpopulation was provided

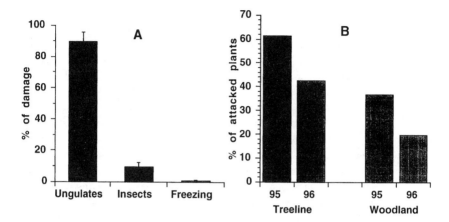

Figure 7 Herbivory on *Pinus sylvestris* saplings in the Sierra Nevada mountains. (A) Relative importance of ungulates with respect to other sources of damage; "% of damage" refers to the proportion of vegetative buds lost per plant. (B) Percentage of saplings herbivorized by ungulates in a dry (95) and a wet (96) year, differentiating between one area with low density (Treeline) and another area with high density (Woodland) of trees.

by Caughley (1981), who recognized wildlife overabundance as a state in which the herbivore–plant relationship loses equilibrium. The overgrazing problems previously described depict the situation in most of the humanized landscapes and protected areas around the world today. While pasturelands have been relegated the main function of providing animal goods to man, protected areas are designated with the aim of preserving a portion of nature to be guarded against human exploitation. However, social and political matters have complicated the labor of park conservation and preservation by forcing them to provide many forms of entertainment to people. The overpopulation of ungulates in protected areas (their populations kept high to make them more visible to hunters and park visitors) induces a direct conflict between natural conservation and human enjoyment. There is not a magic or universal recipe to achieve the appropriate balance between conservation and public entertainment, and this is a major challenge of the future.

APPENDIX 1: HYPOTHESES ON THE PRODUCTION OF SECONDARY METABOLITES

Two conceptually similar hypotheses have been advanced to predict environmental sway over the phenotypic expression of secondary metabolism:

1. The carbon-nitrogen balance (CNB) hypothesis predicts that carbon-based secondary metabolites tend to accumulate when growth is limited by the availability of mineral nutrients but not of carbon (nutrient stress), whereas nitrogen-based metabolites accumulate under conditions of nutrient abundance and carbon deficiency, for example, when photosynthesis is a limiting factor (Bryant et al. 1983).

2. The growth-differentiation balance (GDB) hypothesis assumes that plants allocate more to defense (differentiated structures) when resources are limited and growth is depressed (Herms and Mattson 1992). Growth dominates under favorable conditions, whereas differentiation prevails whenever conditions other than the supply of photosynthates are suboptimal for growth (Tuomi 1992). The fundamental premise of GDB hypothesis is the presumption of a physiological trade-off between growth and differentiation, including secondary metabolism, and any environmental factor that slows growth more than photosynthesis can augment the resource pool available for allocation to secondary metabolism.

Predictions

These two hypotheses are highly similar in their predictions and implications. Both hypotheses predict that the outcome after herbivory, in terms of plant performance and the abundance of C-based metabolites, will depend on nutrient availability and plant productivity. The main difference between these two hypotheses is that CNB emphasizes site differences in resource availability, whereas GDB emphasizes temporal variation in resource availability (Tuomi 1992).

APPENDIX 2: MAIN HYPOTHESES ON PLANT-DEFENSE EVOLUTION

1. The apparency hypothesis (Feeny 1976, Rhoades and Cates 1976) is an optimally animal-centered hypothesis. This hypothesis suggests that plants that have a high risk of being discovered (those dominating the landscape) will bear a large investment in broadly effective defenses (quantitative defenses, especially tannins) that reduce the growth rate and fitness of all enemies. In contrast, less apparent plants, which are more likely to escape herbivory, require a much lower investment in defense, and may employ relatively small concentrations of toxins (qualitative defenses such as alkaloids and glucosinolates), which are

highly effective against nonadapted enemies, although easy, in an evolutionary sense, for attackers to overcome.

2. The resource-availability hypothesis (Coley et al. 1985) is an optimally plant-centered hypothesis. According to this hypothesis, habitat quality constitutes the major influence behind the evolution of both the type and amount of antiherbivore defenses. In habitats or microhabitats with high levels of available resources, rapid growth is possible and can compensate for relatively high levels of herbivory. In these habitats, plants will invest less in defenses, principally those of the mobile sort, such as alkaloids. By contrast, in habitats where growth is restricted by low levels of resources, the consequences of herbivory are likely to be more severe and plants must invest more in defense. Plants that grow in resource-poor habitats and that are less capable of resprouting after foliar loss can be expected to invest more generously in defense, particularly in immobile constitutive defense systems such as tannins and cellulose, which are not continuously formed and permanent throughout the life of the plant.

ACKNOWLEDGMENTS

We thank the Agencia de Medio Ambiente, Junta de Andalucía, for permitting our field work in the Natural Park of Sierra Nevada and the Natural Park of the Sierra de Baza. Conchita Alonso and José R. Obeso gave valuable suggestions to a first draft of the manuscript. The dedication we showed in writing this chapter tested the patience of our colleagues, Daniel García and Jorge Castro. David Nesbitt reviewed the English version of the text. This work was supported by CICYT grant AMB95-0479.

REFERENCES

Aarssen, L.W. 1995. Hypotheses for the evolution of apical dominance in plants: implications for the interpretation of overcompensation. Oikos 74: 149–156.

Abrahamson, W.G. 1989. Plant-Animal Interactions. New York: McGraw-Hill.

Ågren, J., and D.W. Schemske. 1993. The cost of defense against herbivores: an experimental study of trichome production in *Brassica rapa*. American Naturalist 141: 338–350.

Andersen, A.N. 1989. How important is seed predation to recruitment in stable populations of long-lived perennials? Oecologia 81: 310–315.

Aronne, G., and C.C. Wilcock. 1994. First evidence of myrmecochory in fleshy-fruited shrubs of the Mediterranean region. New Phytologist 127: 781–788.

Ayres, M.P. 1993. Plant defense, herbivory, and climate change. In: Kareiva, P.M., J.G. Kingsolver, and R.B. Huey, eds. Biotic Interactions and Global Change. Sunderland, MA: Sinauer Associates, pp 75–94.

Bach, C.E. 1996. Effects of a specialist herbivore (*Altica subplicata*) on *Salix cordata* and sand dune succession. Ecological Monographs 64: 423–445.

Baines, D., R.B. Sage, and M.M. Baines. 1994. The implications of red deer grazing to ground vegetation and invertebrate communities of Scottish native pinewoods. Journal of Applied Ecology 31: 776–783.

Batzli, G.O. 1983. Responses of arctic rodent populations to nutritional factors. Oikos 40: 396–406.

Begon, M., J.L. Harper, and C.R. Townsend. 1990. Ecology: Individuals, Populations and Communities. 2nd ed. Oxford: Blackwell Scientific.

Belsky, A.J. 1986. Does herbivory benefit plants? A review of the evidence. American Naturalist 127: 870–892.

Belsky, A.J., and D.M. Blumenthal. 1997. Effects of livestock grazing on stand dynamics and soils in upland forests of the interior west. Conservation Biology 11: 315–327.

Bentley, B.L., and N.D. Johnson. 1992. Plants as food for herbivores: the roles of nitrogen fixation and carbon dioxide enrichment. In: Price, P.W., T.M. Lewinsohn, G.W. Fernandes, and W.W. Benson, eds. Plant-Animal Interactions: Evolutionary Ecology in Tropical and Temperate Regions. New York: Wiley, pp 257–272.

Bergelson, J., and C.B. Purrington. 1996. Surveying patterns in the cost of resistance in plants. American Naturalist 148: 536–558.

Bernays, E.A., and R.F. Chapman. 1994. Host Plant Selection by Phytophagous Insects. New York: Chapman and Hall.

Bertness, M.D., and R. Callaway. 1994. Positive interactions in communities. Trends in Ecology and Evolution 9: 191–193.

Bowers, M.A. 1993. Influence of herbivorous mammals on an old-field plant community: years 1–4 after disturbance. Oikos 67: 129–141.

Bowers, M.A., and C.F. Sacchi. 1991. Fungal mediation of a plant-herbivore interaction in an early successional plant community. Ecology 72: 1032–1037.

Brown, D.G. 1994. Beetle folivory increases resource availability and alters plant invasion in monocultures of goldenrod. Ecology 75: 1673–1683.

Brown, V.K., and A.C. Gange. 1992. Secondary plant succession–how is it modified by insect herbivory? Vegetatio 101: 3–13.

Bruin, J., M.W. Sabelis, and M. Dicke. 1995. Do plants tap SOS signals from their infested neighbours? Trends in Ecology and Evolution 10: 167–170.

Bryant, J.P. 1987. Fletleaf willow-snowshoe hare interactions: plant carbon/nutrient balance and floodplain succession. Ecology 68: 1319–1327.

Bryant, J.P., F.S. Chapin III, and D.R. Klein. 1983. Carbon/nutrient balance of boreal plants in relation to vertebrate herbivory. Oikos 40: 357–368.

Campbell, B.M. 1986. Plant spinescence and herbivory in a nutrient poor ecosystem. Oikos 47: 168–172.

Caughley, G. 1981. Overpopulation. In: Jewell, P.A., and S. Holt, eds. Problems in Management of Locally Abundant Wild Mammals. New York: Academic Press, pp 7–19.

Cebrián, J., and C.M. Duarte. 1984. The dependence of herbivory on growth rate in natural plant communities. Functional Ecology 8: 518–525.

Cherrett, J.M. 1989. Key concepts: the results of a survey of our members' opinion. In: Cherrett, J.M., ed. Ecological Concepts. Oxford: Blackwell Scientific, pp 1–16.

Coley, P.D., J.P. Bryant, and F.S. Chapin III. 1985. Resource availability and plant antiherbivore defense. Science 230: 895–899.

Connell, J.H. 1975. Some mechanisms producing structure in natural communities: a model and evidence from field experiments. In: Cody, M.L., and J. Diamond, eds. Ecology and Evolution of Communities. Cambridge, MA: Harvard University Press, pp 460–490.

Cooper, S.M., and N. Owen-Smith. 1986. Effects of plant spinescence on large mammalian herbivores. Oecologia 68: 446–455.

Crawley, M.J. 1983. Herbivory: The Dynamics of Animal-Plant Interactions. Oxford: Blackwell Scientific.

Crawley, M.J. 1989. Insect herbivory and plant population dynamics. Annual Review of Entomology 34: 531–564.

Crawley, M.J. 1992. Seed predators and plant population dynamics. In: M. Fenner, M., ed. Seeds: The Ecology of Regeneration in Plant Communities. Wallingford, England: CAB International, pp 159–171.

Crawley, M.J. 1997. Plant-herbivore dynamics. In: Crawley, M.J., ed. Plant Ecology, 2nd edition. Oxford: Blackwell Scientific Publications, pp 401–474.

Cuartas, P., and R. García-González. 1992. *Quercus ilex* browse utilization by Caprini in Sierra de Cazorla and Segura (Spain). Vegetatio 99/100: 317–330.

Cyr, H., and M.L. Pace. 1993. Magnitude and patterns of herbivory in aquatic and terrestrial ecosystems. Nature 361: 148–150.

Danell, K., P. Niemelä, T. Varvikko, and T. Vuorisalo. 1991. Moose browsing on Scots pine along a gradient of plant productivity. Ecology 72: 1624–1633.

Davidson, D.W. 1993. The effects of herbivory and granivory on terrestrial plant succession. Oikos 68: 23–35.

Denno, R.F., M.S. McClure, and R.J. Ott. 1995. Interspecific interactions in phytophagous insects: competition reexamined and resurrected. Annual Review of Entomology 40: 297–331.

Dinerstein, E. 1989. The foliage-as-fruit hypothesis and the feeding behavior of South Asian ungulates. Biotropica 21: 214–218.

Dirzo, R. 1984. Herbivory: a phytocentric overview. In: Dirzo, R., and J. Sarukhán, eds. Perspectives in Plant Population Ecology. Sunderland, MA: Sinauer Associates, pp 141–165.

Dudt, J.F., and D.J. Shure. 1994. The influence of light and nutrients on foliar phenolics and insect herbivory. Ecology 75: 86–98.

Edenius, L., K. Danell, and H. Nyquist. 1995. Effects of simulated moose browsing on growth, mortality, and fecundity in Scots pine—relations to plant productivity. Canadian Journal of Forest Research 25: 529–535.

Edwards, P.J. 1989. Insect herbivory and plant defence theory. In: Grubb, P.J., and J.B. Whittaker, eds. Toward a More Exact Ecology. Oxford: Blackwell Scientific, pp 275–297.

Edwards, P.J., and S.D. Wratten. 1980. Ecology of Insect-Plant Interactions. Studies in Biology 121. London: Edward Arnold.

Ehrlén, J. 1991. Why do plants produce surplus flowers? A reserve-ovary model. American Naturalist 138: 918–933.

Ehrlén, J. 1992. Proximate limits to seed production in a herbaceous perennial legume, *Lathyrus vernus*. Ecology 73: 1820–1831.

Erhlich, P.R., and P.H. Raven. 1964. Butterflies and plants: a study of coevolution. Evolution 18: 586–608.

Ericson, L., and A. Wennström. 1997. The effect of herbivory on the interaction between the clonal plant *Trientalis europaea* and its smut fungus *Urocystis trientalis*. Oikos 80: 107–111.

Eriksson, O., and J. Ehrlén. 1992. Seed and microsite limitation of recruitment in plant populations. Oecologia 91: 360–364.

Facelli, J.M. 1994. Multiple indirect effects of plant litter affect the establishment of woody seedlings in old fields. Ecology 75: 1727–1735.

Faeth, S.H. 1986. Indirect interactions between temporally separated herbivores mediated by the host plant. Ecology 67: 479–494.

Faeth, S.H., and D. Wilson. 1997. Induced responses in trees: mediators of interactions among macro- and micro-herbivores? In: Gange, A.C., and V.K. Brown, eds. Multitrophic Interactions in Terrestrial Ecosystems. Oxford: Blackwell Scientific, pp 201–215.

Feeny, P. 1976. Plant apparency and chemical defense. Recent Advances in Phytochemistry 10: 1–40.

Feller, I.C. 1995. Effects of nutrient enrichment on growth and herbivory of dwarf red mangrove (*Rhyzophora mangle*). Ecological Monographs 65: 477–505.

Fernández-Haeger, J., D. Jordano, E.C. Retamosa, V.E. Caballero, and P. Fernández. 1996. Herbivore-plant interactions as a basis for the management of protected areas. The case of *Iberis contracta*. In: Chacón, J. and J.L. Rosúa, eds. Proceedings of the 1st International Conference on Sierra Nevada: Conservation and Sustainable Management, vol. 2. Granada: University of Granada, pp 541–556 (in Spanish with English summary).

Finlay, R.D. 1985. Interactions between soil, micro-arthropods and endomicorrhizal associations of higher plants. In: Fitter, A.H., D. Atkinson, D.J. Read, and M.B. Usher, eds. Ecological Interactions in Soil. Oxford: Blackwell Scientific, pp 319–331.

Frank, D.A., and S.J. McNaughton. 1992. The ecology of plants, large mammalian herbivores, and drought in Yellowstone National Park. Ecology 73: 2043–2058.

Gange, A.C., and E. Bowler. 1997. Interactions between insects and mycorrhizal fungi. In: Gange, A.C., and V.K. Brown, eds. Multitrophic Interactions in Terrestrial Ecosystems. Oxford: Blackwell Scientific, pp 115–132.

García, D., R. Zamora, J.M. Gómez, and J.A. Hódar. Bird rejection of unhealthy fruits reinforces plant-seed disperser mutualism: the case of *Juniperus communis* L. in the Mediterranean mountains. Oikos, in press.

Gehring, C.A., and T.G. Whitham. 1994. Comparisons of ectomycorrhizae on pinyon pines (*Pinus edulis*, Pinaceae) across extremes of soil type and herbivory. American Journal of Botany 81: 1509–1516.

Gil, L. 1990. The elms and the Dutch elm disease in Spain. Madrid, Spain: Technical Collection of the Ministry of Food, Agriculture, and Fisheries.

Gómez, J.M. 1994. Importance of direct and indirect effects in the interaction between a parasitic plant (*Cuscuta epithymum*) and its host plant (*Hormathophylla spinosa*). Oikos 71: 97–106.

Gómez, J.M., and R. Zamora. 1994. Top-down effects in a tritrophic system: parasitoids enhance plant fitness. Ecology 75: 1023–1030.

Gowda, J.H. 1996. Spines of *Acacia tortilis*: what do they defend and how? Oikos 77: 279–284.

Grubb, P.J. 1992. A positive distrust in simplicity—lessons from plant defences and from competition among plants and among animals. Journal of Ecology 80: 585–610.

Harborne, J.B. 1997. Plant secondary metabolism. In: Crawley, M.J., ed. Plant Ecology, 2nd edition. Oxford: Blackwell Scientific, pp 132–155.

Harper, J.L. 1977. Population Biology of Plants. New York: Academic Press.

Hartley, S.A., K. Nelson, and M. Gorman. 1995. The effect of fertiliser and shading on plant chemical composition and palatability to Orkney voles, *Microtus arvalis orcadensis*. Oikos 72: 79–87.

Hartley, S.E., and C.G. Jones. 1997. Plant chemistry and herbivory, or why the world is green. In: Crawley, M.J., ed. Plant Ecology, 2nd edition. Oxford: Blackwell Scientific, pp 284–324.

Hay, M.E. 1986. Associational plant defenses and the maintenance of species diversity: turning competitors into accomplices. American Naturalist 128: 617–641.

Hay, M.E., Q.E. Kappel, and W. Fennical. 1994. Synergisms in plant defenses against herbivores: interactions of chemistry, calcification, and plant quality. Ecology 75: 1714–1726.

Hendrix, S.D. 1988. Herbivory and its impact on plant reproduction. In: Lovett-Doust, J., and L. Lovett-Doust, eds. Plant Reproductive Ecology: Patterns and Strategies. Oxford: Oxford University Press, pp 246–263.

Hendrix, S.D., and E.J. Trapp. 1989. Floral herbivory in *Pastinaca sativa*: do compensatory responses offset reductions in fitness? Evolution 43: 891–895.

Herms, D.A., and W.J. Mattson. 1992. The dilemma of plants: to grow or defend. The Quarterly Review of Biology 67: 283–335.

Herrera, J. 1991. Herbivory, seed dispersal, and the distribution of a ruderal plant living in a natural habitat. Oikos 62: 209–215.

Hixon, M.A., and W.N. Brostoff. 1996. Succession and herbivory: effects of differential fish grazing on Hawaaian coral-reef algae. Ecological Monographs 66: 67–90.

Hjältén, J., K. Danell, and L. Ericson. 1993a. Effects of simulated herbivory and intraspecific competition on the compensatory ability of birches. Ecology 74: 1136–1142.

Hjältén, J., K. Danell, and P. Lundberg. 1993b. Herbivore avoidance by association: vole and hare utilization of woody plants. Oikos 68: 125–131.

Ho, L.C. 1992. Fruit growth and sink strength. In: Marshall, C., and J. Grace, eds. Fruit and Seed Production, Aspects of Development, Environmental Physiology and Ecology. Cambridge: Cambridge University Press, pp 101–124.

Hódar, J.A., J.M. Gómez, D. García, and R. Zamora. 1996. Effect of ungulate herbivory on the growth of *Pinus sylvestris nevadensis*. In: Chacón, J. and J.L. Rosúa, eds. Proceedings of the 1st International Conference on Sierra Nevada: Conservation and Sustainable Management, vol. 2. Granada: University of Granada, pp 529–539 (in Spanish with English summary).

Hódar, J.A., J. Castro, J.M. Gómez, D. García, and R. Zamora. Effects of herbivory on growth and survival of seedlings and saplings of *Pinus sylvestris nevadensis* in SE Spain. In: Papanastasis, V.P. and D. Peter, eds. Ecological Basis of Livestock Grazing in Mediterranean Ecosystems. Luxembourg: Official Publications of the European Communities, pp 264–267.

Holt, R.D. 1997. Community modules. In: Gange, A.C., and V.K. Brown, eds. Multitrophic Interactions in Terrestrial Ecosystems. Oxford: Blackwell Scientific, pp 333–350.

Honkanen, T., and E. Haukioja. 1994. Why does a branch suffer more after branch-wide than after tree-wide defoliation? Oikos 71: 441–450.

Howe, H.F., and L.C. Westley. 1988. Ecological Relationships of Plants and Animals. New York: Oxford University Press.

Hudson, E.E., and P. Stiling. 1997. Exploitative competition strongly affects the herbivorous insect community on *Baccharis halimifolia*. Oikos 79: 521–528.

Hulme, P.E. 1996. Herbivory, plant regeneration, and species coexistence. Journal of Ecology 84: 609–615.

Hunter, M.D. 1992. Interactions within herbivore communities mediated by the host plant: the keystone herbivore concept. In: Hunter, M.D., T. Ohgushi, and P.W. Price, eds. Effects of Resource Distribution on Plant-Animal Interactions. San Diego:Academic Press, pp 287–325.

Huntly, N. 1991. Herbivores and the dynamics of communities and ecosystems. Annual Review of Ecology and Systematics 22: 477–503.

Huntly, N. 1995. How important are consumer species to ecosystem functioning? In: Jones, C.G., and J.H. Lawton, eds. Linking Species and Ecosystems. New York: Chapman and Hall, pp 72–83.

Janzen, D.H. 1984. Dispersal of small seeds by big herbivores: foliage is the fruit. American Naturalist 123: 338–353.

Jermy, T. 1993. Evolution of insect-plant relationships—a devil's advocate approach. Entomologia Experimentalis et Applicata 66: 3–12.

Jones, C.G., J.H. Lawton, and M. Shachak. 1994. Organisms as ecosystem engineers. Oikos 69: 373–386.

Karban, R. 1992. Plant variation: its effects on populations of herbivorous insects. In: Fritz, R.S., and E.L. Simms, eds. Plant Resistance to Herbivores and Pathogens. Ecology, Evolution, and Genetics. Chicago: Chicago University Press, 195–215.

Karban, R., and J.H. Myers. 1989. Induced plant responses to herbivory. Annual Review of Ecology and Systematics 20:331–348.

Kelly, D. 1994. The evolutionary ecology of mast seeding. Trends in Ecology and Evolution 9:465–470.

Krischik, V.A., R.W. Goth, and P. Barbosa. 1991. Generalized plant defense: effects on multiple species. Oecologia 85: 562–571.

Kytö, M., P. Niemelä, and S. Larsson. 1996. Insects on trees: population and individual response to fertilization. Oikos 75: 148–159.

Lanner, R.M. 1996. Made for Each Other, a Symbiosis of Birds and Pines. New York: Oxford University Press.

Le Houérou, H.N. 1981. Impact of man and his animals on Mediterranean vegetation. In: di Castri, F., D.W. Goodall, and R.L. Spetch, eds. Ecosystems of the World 11: Mediterranean-Type Shrublands. Amsterdam, The Netherlands: Elsevier, pp 479–521.

Louda, S.M. 1989. Differential predation pressure: a general mechanism for structuring plant communities along complex environmental gradients? Trends in Ecology and Evolution 4: 158–159.

Louda, S.M. 1994. Experimental evidence for insect impact on populations of short-lived, perennial plants, and its application in restoration ecology. In: Bowles, M.L., and C.J. Whelan, eds. Restoration of Endangered Species. Cambridge, MA:Cambridge University Press, pp 118–138.

Louda, S.M., and M.A. Potvin. 1995. Effects of inflorescence-feeding insects on the demography and lifetime fitness of a native plant. Ecology 76: 229–245.

Louda, S.M., K.H. Keeler, and R.D. Holt. 1990. Herbivore influence on plant performance and competitive interactions. In: Grace, J.B., and D. Tilman, eds. Perspectives in Plant Competition. New York: Academic Press, pp 413–444.

MacCracken, J.G. 1996. Managing and understanding wild ungulate population dynamics in protected areas. In: Wright, R.G., ed. National Parks and Protected Areas. Cambridge, MA: Blackwell Science, pp 249–276.

Maddox, G.D., and R.B. Root. 1990. Structure of the selective encounter between goldenrod (*Solidago altissima*) and its diverse insect fauna. Ecology 71: 2115–2124.

Malo, J.E., and F. Suárez. 1995. Herbivorous mammals as seed dispersers in a Mediterranean "dehesa." Oecologia 104: 246–255.

Malo, J.E., and F. Suárez. 1996. New insights into pasture diversity: the consequences of seed dispersal in herbivore dung. Biodiversity Letters 3: 54–57.

Marquis, R.J. 1992a. The selective impact of herbivores. In: Fritz, R.S., and E.L. Simms, eds. Plant Resistance to Herbivores and Pathogens. Ecology, Evolution, and Genetics. Chicago: Chicago University Press, pp 301–325.

Marquis, R.J. 1992b. A bite is a bite is a bite? Constraints on response to folivory in *Piper aireianum* (Piperaceae). Ecology 73: 143–152.

Marquis, R.J., and C.J. Whelan. 1994. Insectivorous birds increase growth of white oak through consumption of leaf-chewing insects. Ecology 75: 2007–2014.

Maschinski, J., and T.G. Whitham. 1989. The continuum of plant responses to herbivory: the influence of plant association, nutrient availability, and timing. American Naturalist 134: 1–19.

Masters, G.J., and V.K. Brown. 1997. Host-plant mediated interactions between spatially separated herbivores: effects of community structure. In: Gange, A.C., and V.K. Brown, eds. Multitrophic Interactions in Terrestrial Ecosystems. Oxford:Blackwell Scientific, pp 217–237.

Masters, G.J., V.K. Brown, and A.C. Gange. 1993. Plant mediated interactions between above-and below-ground insect herbivores. Oikos 66: 148–151.

Mattson, W.J. Jr. 1980. Herbivory in relation to plant-nitrogen content. Annual Review of Ecology and Systematics 11: 119–161.

McEvoy, P.B., N.T. Rudd, C.S. Cox, and M. Huso. 1993. Disturbance, competition, and herbivory effects on ragworth *Senecio jacobaea* populations. Ecological Monographs 63: 55–75.

McInnes, P.F., R.J. Naiman, J. Pastor, and Y. Cohen. Effects of moose browsing on vegetation and litter of the boreal forest, Isle Royale, Michigan, USA. Ecology 73: 2059–2075.

McNaughton, S.J. 1985. Ecology of a grazing ecosystem: the Serengeti. Ecological Monographs 55: 259–294.

McNaughton, S.J. 1988. Mineral nutrition and spatial concentrations of African ungulates. Nature 334: 343–345.

McNaughton, S.J., J.L. Tarrants, M.M. McNaughton, and R.H. Davis. 1985. Silica as a defense against herbivory and a growth promotor in African grasses. Ecology 66: 528–535.

McNaughton, S.J., M. Oesterheld, D.A. Frank, and K.J. Williams. 1989. Ecosystem-level patterns of primary productivity and herbivory in terrestrial habitats. Nature 341: 142–144.

Meyer, G.A. 1993. A comparison of the impacts of leaf-and sap-feeding insects on growth and allocation of goldenrod. Ecology 74: 1101–1116.

Meyer, G.A., and R.B. Root. 1993. Effects of herbivorous insects and soil fertility on reproduction of goldenrod. Ecology 74: 1117–1128.

Milchunas, D.G., and W.K. Lauenroth. 1993. Quantitative effects of grazing on vegetation and soils over a global range of environments. Ecological Monographs 63: 327–366.

Milewsky, A.V., T.P. Young, and D. Madden. 1991. Thorns as induced defenses: experimental evidence. Oecologia 86: 70–75.

Miller, M.F. 1994. The costs and benefits of *Acacia* seed consumption by ungulates. Oikos 71: 181–187.

Moen, J.H. Gardfjell, L. Oksanen, L. Ericson, and P. Ekerholm. 1993. Grazing by food-limited microtine rodents on a productive experimental plant community: does the "green desert" exist? Oikos 68: 401–413.

Mopper, S., and T.G. Whitham. 1992. The plant stress paradox: effects on pinyon sawfly sex ratios and fecundity. Ecology 73: 515–525.

Müller-Schärer, H., and V.K. Brown. 1995. Direct and indirect effects of above-and below-ground insect herbivory on plant density and performance of *Tripleurospermum perforatum* during early plant succession. Oikos 72: 36–41.

Myers, J.M. 1987. Nutrient availability and the deployment of mechanical defenses in grazed plants: a new experimental approach to the optimal defense theory. Oikos 49: 350–351.

Niesenbaum, R.A. 1996. Linking herbivory and pollination: defoliation and selective fruit abortion in *Lindera benzoin*. Ecology 77: 2324–2331.

Nogales, M., A. Valido, and F.M. Medina. 1995. Frugivory of *Plocama pendula* (Rubiaceae) in xerophytic zones of Tenerife (Canary Islands). Acta Oecologica 16: 585–591.

Noy-Meir, I.N. Gutman, and Y. Kaplan. 1989. Responses of Mediterranean grassland plants to grazing and protection. Journal of Ecology 77: 290–310.

Núñez-Farfán, J., and R. Dirzo. 1994. Evolutionary ecology of *Datura stramonium* L. in central Mexico: natural selection for resistance to herbivorous insects. Evolution 48: 423–436.

Obeso, J.R. 1993. Does defoliation affect reproductive output in herbaceous perennials and woody plants in different ways? Functional Ecology 7: 150–155.

Obeso, J.R. 1997. The induction of spinescence in European holly leaves by browsing ungulates. Plant Ecology 129: 149–156.

Obeso, J.R., and P.J. Grubb. 1994. Interactive effects of extent and timing of defoliation, and nutrient supply on reproduction in a chemically protected annual *Senecio vulgaris*. Oikos 71: 506–514.

Oesterheld, M., O.E. Sala, and S.J. McNaughton. 1992. Effect of animal husbandry on herbivore-carrying capacity at a regional scale. Nature 356: 234–236.

Ostfeld, R.S., and C.D. Canham. 1993. Effects of meadow vole population density on tree seedling survival in old fields. Ecology 74: 1792–1801.

Pastor, J.B. Dewey, R.J. Naiman, P.F. McInnies, and Y. Cohen. 1993. Moose browsing and soil fertility in the boreal forest of Isle Royale National Park. Ecology 74: 467–480.

Perevolotsky, A. 1994. Tannins in Mediterranean woodland species: lack of response to browsing and thinning. Oikos 71: 333–340.

Pickett, S.T.A., S.L. Collins, and J.J. Armesto. 1987. Models, mechanisms and pathways of succession. Botanical Review 53: 335–371.

Price, P.W. 1991. The plant vigor hypothesis and herbivore attack. Oikos 62: 244–251.

Prins, A.H., H.W. Nell, and G.L. Klinkhammer. 1992. Size-dependent root herbivory on *Cynoglossum officinale*. Oikos 65: 409–413.

Quesada, M.,K. Bollman, and A.G. Stephenson. 1995. Leaf damage decreases pollen production and hinders pollen performance in *Cucurbita texana*. Ecology 76: 437–443.

Quinn, J.A., D.P. Mowrey, S.M. Emanuele, and R.D.B. Whalley. 1994. The "foliage is the fruit" hypothesis: *Buchloe dactyloides* (Poaceae) and the shortgrass prairie of North America. American Journal of Botany 81: 1545–1554.

Ramsell, J., and N.D. Paul. 1990. Preferential grazing by molluscs of plants infected by rust fungi. Oikos 58: 145–150.

Reader, R.J. 1993. Control of seedling emergence by ground cover and seed predation in relation to seed size for some old-field species. Journal of Ecology 81: 169–175.

Renaud, P.E., M.E. Hay, and T.M. Schmitt. 1990. Interactions of plant stress and herbivory: intraspecific variation in the susceptibility of a palatable versus an unpalatable seaweed to sea urchin grazing. Oecologia 82: 217–226.

Rhoades, D.F., and R.G. Cates. 1976. Toward a general theory of plant antiherbivore chemistry. Recent Advances in Phytochemistry 10: 168–213.

Rosenthal, J.P., and P.M. Kotanen. 1994. Terrestrial plant tolerance to herbivory. Trends in Ecology and Evolution 9: 145–148.

Rozena, J., J. van de Staaij, L.O. Björn, and M. Caldwell. 1997. UV-B as an environmental factor in plant life: stress and regulation. Trends in Ecology and Evolution 12: 22–28.

Ruess, R.W. 1987. The role of large herbivores in nutrient cycling of tropical savannas. In: Walker, B.H., ed. Determinants of Tropical Savannas. Oxford: IRL Press, pp 67–91.

Ruohomäki, K., F.S. Chapin III, E. Haukioja, S. Neuvonen, and J. Suomela. 1996. Delayed induced resistance in mountain birch in response to fertilization and shade. Ecology 77: 2302–2311.

Sallabanks, R., and S.P. Courtney 1992. Frugivory, seed predation, and insect-vertebrate interactions. Annual Review of Entomology 37: 377–400.

Schmitt, T.M., M.E. Hay, and N. Lindquist. 1995. Constraints on chemically mediated coevolution: multiple functions for seaweed secondary metabolites. Ecology 76: 107–123.

Schulz, T.T., and V.C. Leininger. 1990. Differences in riparian vegetation structure between grazed areas and exclosures. Journal of Range Management 43: 295–299.

Simms, E.L., and M.D. Rausher. 1989. The evolution of resistance to herbivory in *Ipomoea purpurea*. II: Natural selection by insects and cost of resistance. Evolution 43: 573–585.

Simms, E.L., and R.S. Fritz. 1990. The ecology and evolution of host-plant resistance to insects. Trends in Ecology and Evolution 5: 356–360.

Skogsmyr, I., and T. Fagerström. 1992. The cost of anti-herbivory defence: an evaluation of some ecological and physiological factors. Oikos 64: 451–457.

Stamp, N.E. 1996. Developing a theory of plant-insect herbivore interactions: are we there yet? Bulletin of the Ecological Society of America 80: 51–61.

Sterner, R.W., and D.O. Hessen. 1994. Algal nutrient limitation and the nutrition of aquatic herbivores. Annual Review of Ecology and Systematics 25: 1–29.

Stiling, P. 1994. What do ecologists do? Bulletin of the Ecological Society of America 78: 116–121.

Strauss, S.Y. 1991. Indirect effects in community ecology: their definition, study and importance. Trends in Ecology and Evolution 6: 206–210.

Tscharntke, T. 1997. Vertebrate effects on plant-invertebrate food webs. In: Gange, A.C., and V.K. Brown, eds. Multitrophic Interactions in Terrestrial Ecosystems. Oxford: Blackwell Scientific, pp 277–297.

Tuomi, J. 1992. Toward integration of plant defence theories. Trends in Ecology and Evolution 7: 365–367.

Tuomi, J., P. Niëmëla, and S. Sirén. 1990. The Panglossian paradigm and delayed inducible accumulation of foliar phenolics in mountain birch. Oikos 59: 399–410.

Turner, J.M. 1994. Sclerophylly: primarily protective? Functional Ecology 8: 669–675.

Valburg, L.K. 1998. Eating infested fruits: interactions in a plant-disperser-pest triad. Oikos 65: 25–28.

Weiner, J. 1993. Competition, herbivory and plant size variability: *Hypochaeris radicata* grazed by snails (Helix aspersa). Functional Ecology 7: 47–53.

Weis, A.E., and M.R. Berenbaum. 1989. Herbivorous insects and green plants. In: Abrahamson, W.G., ed. Plant-Animal Interactions. New York: McGraw Hill, pp 123–162.

Weltzin, J.F., S. Archer, and R.K. Heitschmidt. 1997. Small-mammal regulation of vegetation structure in a temperate savanna. Ecology 78: 751–763.

White, T.C.R. 1993. The Inadequate Environment: Nitrogen and The Abundance of Animals. Berlin: Springer-Verlag.

Zamora, R., and J.M. Gómez. 1993. Vertebrate herbivores as predators of insect herbivores: an asymmetrical interaction mediated by size differences. Oikos 66: 223–228.

20
Biological Diversity and Functioning of Ecosystems

Helena Freitas
University of Coimbra, Coimbra, Portugal

I. INTRODUCTION

Biodiversity includes the variety of species of plants, animals, and other organisms, and also the genes they contain and the communities and ecosystems of which they are a part. Genotypic variability is the basis upon which diversity at the population, species assemblage, and ecosystem levels is built. Biodiversity can be considered to include not only the compositional aspects at each hierarchical level, but also structural and functional aspects (Noss 1990).

Biodiversity has recently emerged as an issue of both scientific and political concern, primarily because of an increase in extinction rates caused by human activities. In fact, fragmentation, grazing, forestry, and nutrient deposition are decreasing the biological diversity of many of the world's remaining seminatural ecosystems (Ehrlich and Ehrlich 1981, Wilson 1988, 1992, Ehrlich and Daily 1993, Pimm 1993), causing a dramatic depauperation of the earth's biotic capital.

Recent attention has focused on the potential importance of biodiversity for the adequate functioning of ecosystems (Schulze and Mooney 1993). Functioning of ecosystems represents the assemblage of processes such as primary production, decomposition, and nutrient cycling, and their interactions.

It is known that species perform diverse ecological functions. A species may regulate biogeochemical cycles (Vitousek 1990, Zimov et al. 1995), modify disturbance regimes (Dublin et al. 1990, D'Antonio and Vitousek 1992), or change the physical environment (Jones et al. 1994, Naiman et al. 1994). Other species regulate ecological processes indirectly through trophic interactions such as predation or parasitism (Kitchell and Carpenter 1993, Prins and Jeud 1993), or functional interactions such as seed dispersal (Brown and Heske 1990) and pollination (Fleming and Sosa 1994).

It will be not practicable to develop models for every ecosystem of the globe or represent every species within those ecosystems. The complexity of the models can be reduced by treating a smaller number of functional types, and the essential dynamics of ecosystems can be captured by grouping species into a limited number of functional types (Steffen et al. 1992).

The biodiversity–function relationship, or how biological diversity relates to ecological function, is of considerable importance since, except in the case of natural parks or preserves, decisions on the management of land and seascapes will increasingly be made on a functional basis rather than on the basis of conservation of the numbers or kinds of species per se (Mooney 1997). It will be crucial to know how landscape and ecosystem processes are controlled by species function and interactions.

II. A BRIEF HISTORY OF FUNCTIONAL CLASSIFICATIONS

There has always been a need for global classification of plants in terms of their ecophysiological properties or functional attributes (Westoby and Leishman 1997). The first known evidences for a functional classification based on physiognomy trace back to the ancient Greek civilization, namely in the system developed by Teophrastus around 300 BC (du Rietz 1931), which attempted to define and understand the climate and edaphic controls of plant distribution.

Later, the need was for defining appropriate crop and garden plants for a specific climate and soil. It was then obvious that a functional classification could predict development and yield more accurately than the typical phylogenetically based taxonomic classification. (von Humboldt 1806, du Rietz 1931, Cain 1950).

Functional definitions may cut across the traditional taxonomic classification of plants, often with taxonomically closely related species showing more marked differences in environmental sensitivity than unrelated species with very similar ecology. The recognition of a possible but undefined ecological classification has challenged generations of plant geographers and ecologists to define a workable scheme (Barkman 1988, Westoby and Leishman 1997).

Systematics and population biologists have been the leaders in alerting society to the increase in species extinctions that is occurring worldwide and of the consequent losses of potentially useful food and drug products. On the other hand, it has been the ecological community that has focused on the functional consequences of these losses. This is a fundamental divergence in viewpoint between these two groups. "It can be stated somewhat simplistically that systematics and population biologists in general focus on biotic differences and ecologists on the biotic similarities" (Mooney 1997).

The history of ecology illustrates why ecologists have concentrated on the functional similarities among species rather than on their differences. Warming (1909) lumped the biota into such groups as xerophytes, mesophytes, and hidrophytes. These groups are easily recognized as functional groups. Large numbers of species of different taxonomic relationships share ecological characteristics, in this case, an evolutionary response to habitat water abundance. Another significant example is the classification of Raunkiaer (1934). He described a scheme in which the functional capabilities of plants were defined in terms of the disposition and seasonality of meristems (buds). This scheme attempted to define the functional characteristics of plants on the basis of easily observed structural properties. Although a useful classification (it remains in wide use today), it does not help to differentiate between plants with the same structural properties that occur in very different ecological situations.

The functional grouping of organisms has continued to modern times. Odum (1963) lumped all organisms into a few functional categories: producers, consumers, and decomposers, in a functional configuration of ecosystems. Mooney and Dunn (1970) and Mooney (1977) developed the concept of convergent evolution, confining the number of "strategy types" to a given region. More recently, those concerned with the metabolism of the biosphere have divided the vegetation of the world into a very few large groupings that encompass information on the roughness characteristics of the environment and very simple physiological attributes, such as abundances of C_3 versus C_4 (Sellers et al. 1992).

III. HOW TO DEFINE FUNCTIONAL TYPES

Functional types have often been described as biotic components of ecosystems that perform the same function or set of functions within the ecosystems. Friedel et al. (1988) defined them as groups that respond similarly to the same perturbation. Noble (1989) discussed a classification based on a set of physiological, reproductive, and life history characters, where variation in each character has specific ecologically predictive (rather then descriptive) value. Barbault et al. (1991) argued that functional types should be defined in terms of morphology and physiology, particularly as these properties relate to resources and species instruction and thus represent feeding guilds or plant growth forms. Keddy (1992) stated that species can be aggregated into functional groups sharing similar traits.

Functional types are thus groups of species with analogous functional traits in a ecosystem. This is a nonphylogenetic classification leading to a grouping of organisms that respond in a similar way to a syndrome of environmental factors (a combination of biotic and abiotic processes that change as a result of a perturbation to the system). Species can be grouped on the basis of different aspects of ecosystem function and/or adaptive responses to environmental variables, for example (Hobbs 1997):

> Resource use
> Ecosystem function: production, consumption, decomposition, N fixation
> Response to disturbance
> Response to environment: mesophytes, hidrophytes, xerophytes, etc.
> Reproductive strategy, pollination systems
> Tolerance to stress: halophytes, etc.
> Physiognomic types: Raunkiaer, etc.
> Physiological types: C_3, C_4, etc.
> Phenology

Walker (1992) argued that the most practical approach may be to group species in relation to limiting or dominant ecosystem processes. For example, the

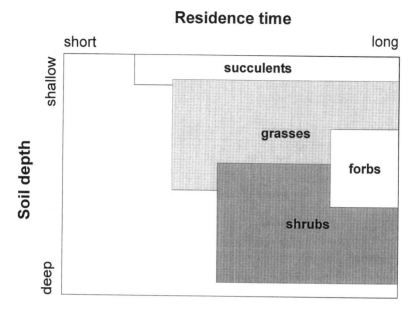

Figure 1 Conceptual model of the distribution of water resources among the four functional types of the Patagonian steppe: grasses, shrubs, succulents, and forbs. Soil depth represents the depth from which each functional type is able to absorb water, and the residence time of water is the period of time during which water remains within the range of water potential available for plants. (*Source*: Sala et al. 1997.)

four plant functional types defined for the Patagonian steppe—grasses, shrubs, succulents and forbs (Sala et al. 1997)—were primarily defined based on morphological and phenological characteristics. The conceptual model established for this arid region that relates these four functional types focuses on water relations because water is the critical resource explaining both the structure and dynamics of these arid ecosystems (Figure 1).

If these functional types or groups could be identified, they would tell us how much redundancy there is in an ecosystem, and this information could be very useful for the management of the system. Of course, to have significant applicability and utility in models, the number of functional types should be smaller than the number of plant species.

IV. MODELS OF DIVERSITY–FUNCTION

Ecosystem functioning refers to the capacity of an ecosystem to perform the primary ecosystem processes of capturing, storing, and transfering energy, carbon dioxide, nutrients, and water (Woodward 1993, 1997, Shugart 1997). The functioning of ecosystems may depend on their biological diversity.

Questions involving the effects of eliminating species on ecosystem behavior often had at their root an underlying assumption that, if species function contributed to the dynamic behavior of an ecosystem and if enough species were eliminated, then the functioning of the ecosystem would be disrupted to some degree. In some explanations, sets of species were thought of as having parallel functions. Thus, the presence of several species provided a redundancy that might allow ecosystems to maintain their function even in the face of species extinction.

Many models attempted to describe the relationship between biodiversity and ecosystem functioning, ranging from one that states that each species plays a unique role in the functioning of ecosystems and that therefore deletion of any species results in a change in ecosystem functioning, to those that consider that most species are redundant and that changes in ecological complexity should not result in changes in functioning (Vitousek and Hooper 1993; Figure 2).

Essentially, we can consider four models or hypotheses: (1) rivet (Ehrlich and Ehrlich 1981); (2) redundant (Lawton and Brown 1993); (3) idiosyncratic (Lawton 1994); and (4) null (Vitousek and Hooper 1993, Lawton 1994).

A. Rivet

The rivet hypothesis suggested by Ehrlich and Ehrlich (1981) likens the ecological function of species to the rivets that attach a wing to a plane. A certain number

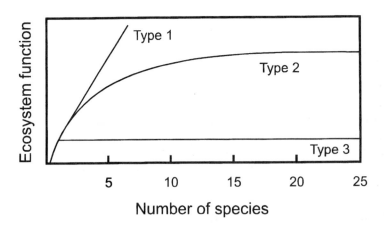

Figure 2 Possible functional relationships between biological diversity and ecosystem functioning. Type 1, a linear effect of diversity; type 2, an asymptotic relationship; and type 3, no effect of diversity. (Source: Vitousek and Hooper 1993.)

of rivets can be lost before the wing falls off. This model proposes that the ecological functions of species overlap, so that even if a species is removed, ecological function may persist due to the compensation of other species with similar functions. In the rivet model, an ecological function will not disappear until all the species performing that function are removed from an ecosystem. Overlap of ecological function allows an ecosystem to persist.

B. Idiosyncratic

In this model proposed by Lawton (1994), ecological function varies idiosyncratically as species richness increases. This model argues that the contribution of each species to ecological function is strongly influenced by interactions among species. Therefore, the effects of the introduction or removal of species to an ecosystem can be either insignificant or major, depending on the nature of the species introduced or removed, and the nature of the species with which it interacts. Essentially, it suggests that ecosystem processes change when diversity changes, but the response is unpredictable.

C. Null

The null hypothesis suggests that ecosystem processes are insensitive to species deletions or additions in the short term (Vitousek and Hooper 1993, Lawton 1994).

D. Redundant

All of the other hypothesis deal with the relationship between diversity and ecosystem processes in a constant environment, whereas the insurance hypothesis (Lawton and Brown 1993) suggests that diversity may serve to maintain ecosystem processes in the face of environmental perturbations, when certain species may be important in the face of environment variation, e.g., drought, fire, or other disturbances. It suggests that there is a minimal species diversity necessary for ecosystem functioning, but beyond that, most species are redundant in their roles.

   Several experiments provide the evidence to support or reject the different diversity-function models. Primary production and ecological stability and their relationship to plant species diversity are two of the best-studied relationships.

V. PRODUCTIVITY VERSUS DIVERSITY

There is considerable information on the role of diversity in controlling the rates of production of communities. A number of observations and experiments support the statement that "productivity is independent of species numbers, at least for a given growth form" (Mooney and Gulmon 1983). The addition of species may increase productivity, but only to a limited extent. For example, a new growth form can be able to use unutilized resources of the habitat or enhance ecosystem properties related to nutrient cycling.

The diversity–productivity hypothesis is based on the assumption that interspecific differences in the use of resources by plants allow more diverse plant communities to use more fully limiting resources, therefore attaining greater productivity (McNaughton 1993, Naeem et al. 1994, Naeem et al. 1995). A related hypothesis is that nutrient leaching losses from ecosystems should be a decreasing function of plant diversity because of greater nutrient capture and/or immobilization in more diverse ecosystems.

In the Serengeti grasslands, removal of grasses with different contributions to total productivity shows the limits of ecosystems to compensate for the deletion of different species (McNaughton 1985). Removal of rare plants resulted in full compensation of production by remaining species, removal of species of intermediate abundance resulted in partial compensation, and removal of dominant species resulted in a significant reduction in production.

VI. STABILITY VERSUS DIVERSITY

The variety of functions that a species can perform are limited, and consequently ecologists frequently have proposed that an increase in species richness also increases functional diversity, producing an increase in ecological stability (Tilman et al. 1996).

Many experimental studies showed that increasing species number increases the stability of ecosystem function (Schindler 1990, Naeem et al. 1994, Frost et al. 1995, Tilman 1996), but apparently no investigations on the relationship between species richness and stability have indicated that additional species continue to increase stability at a constant rate, indicating that the species–diversity model is excessively simplistic (Peterson et al. 1997).

Elton (1958) suggested that decreased diversity would lead to decreased ecological stability and functioning. However, there has been continuing debate about the diversity–stability hypothesis (May 1973, McNaughton 1977, 1985, Pimm 1979, 1984, Schultze and Mooney 1993, Tilman and Downing 1994, Tilman 1996).

May (1973) showed that population dynamics were progressively less stable as the number of competing species increased, and concluded that there need not be any relationship between diversity and stability. In contrast, McNaughton (1977) presented data on plant productivity in the Serengeti that supported Elton. King and Pimm (1983) modeled systems like McNaughton's and found that higher plant diversity generally led to greater biomass stability in response to changes in herbivory, just as McNaughton had observed.

Ecological resilience can be defined as the rate at which a system returns to a single steady or cyclic state after a perturbation (Pimm 1984). McNaughton (1993) showed how greater diversity can lead to greater resilience to grazing in an African grassland (Figure 3), due to the presence of a pool of species capable of regrowth during early showers after the drought. Thus, in this case, the increased resilience with greater diversity was due to the presence of a specialized "strategy," or functional group. McNaughton suggests that the question should be "How and why does diversity affect ecosystem function?" rather than "Does biodiversity affect ecosystem function?".

Biodiversity may affect not only average ecosystem functioning, but also the system response to extreme conditions. The diversity–stability hypothesis suggests that more diverse systems will be less affected by environmental perturbations. Tilman and Downing (1994) analyzed the effects of a severe drought along a diversity gradient created as a result of an experimental fertilization where diversity was maximum in the native system and decreased as fertility increased.

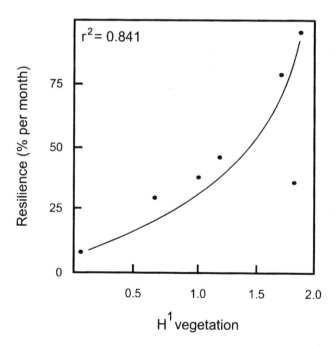

Figure 3 Resilience of grassland communities to grazing as related to community diversity. (*Source*: McNaughton 1993.)

In both cases, the effect of perturbation on production was maximum in simple systems and minimum in the most diverse systems. Some of the conclusions drawn from these experiments have been criticized because fertilization simultaneously modified diversity and selected for species with a lower root–shoot ratio, higher leaf conductance, and greater photosynthetic capacity, which are characteristics that result in lower drought resistance (Givnish 1994). Consequently, the lower productivity during the drought year in low-diversity plots could have been the result of those plots being dominated by drought-sensitive plants.

VII. THE ISSUE OF REDUNDANCY

There is no doubt that the functional-type approach can be very useful in reducing the biotic complexity of a given system to answer specific questions about ecosystem organization and function. However, it does not totally help in answering the species redundancy question because of the multiple-function possibilities of species and the possibilities that these functions change with changing environmental conditions and biotic assemblages.

Apparent species redundancy is observable in most ecosystems, at least in regard to the major ecosystem processes (Walker 1992). Examples of two or more plant species performing the same role in one process (such as primary productivity) but different roles in another (such as herbivory) are common. The special significance of some species may only become apparent under particular circumstances (exceptional droughts, intense fire, etc.). The species that constitute a single functional type with respect to water-use efficiency may respond differently to an increase in minimum temperature.

Although it appears that there may be considerable redudancy in ecosystems in relation to ecosystem function, this does not mean that all species are equal and substitutable. There is increasing evidence, some experimental, that some species count to an inordinate degree in controlling community structure. These species were termed *keystone species* by Paine (1966), and they control the stability of communities. Their removal can result in a large alteration of community structure and function.

VIII. THE CONCEPT OF KEYSTONE SPECIES

The keystone species concept (Paine 1966, Power et al. 1996) offers a very important guideline for targeting species with particularly important roles in the functioning of ecosystems. Their removal from the ecosystem can result in a large alteration of community structure and function. Bond (1993) noted that keystone species include predators (otters), herbivores (rabbits), competitors (dominant trees), mutualists (certain pollinators), earth movers (pocket gophers), and certain species that play a role in ecosystem processes, such as nitrogen fixers.

The keystone concept is an important addition to our basic thinking about species and function. Not only may a few species perform the bulk of the gross ecosystem functions, but it also appears that not all species are equal in the role that they play in ecosystem integration (Mooney 1997).

IX. CONCLUSIONS

Both empirical and theoretical research in ecology suggests that the processes that maintain ecosystem functioning could be adversely affected by the loss of diversity (Chapin et al. 1997), but debate continues as to whether this has been adequately demonstrated (André et al. 1994, Givnish 1994, Huston 1997).

Several experimental results have shown that the establishment and functioning of grassland ecosystems depend on their species richness, with more diverse communities being more productive and having lower nutrient losses than less diverse ecosystems (Naeem et al. 1994, 1995, Tilman et al. 1996). The rapid loss of species on earth and management practices that decrease local biodiversity threaten ecosystem productivity, sustainability, and stability.

To achieve a useful classification of plant functional types, there is a need for better definition and understanding of the critical ecosystem processes and environmental factors that determine the structure and function of an ecosystem. In the context of global change, there is a particular distinction between functional types relating to the role plants perform in ecosystem processes and their functionally different responses to environmental shifts. The special significance of some species may only become apparent under particular circumstances (exceptional droughts, intense fire, temperature increase).

It will not be possible to model every species or every ecosystem. The grouping of species sharing common traits and a common behavior in the ecosystem are the functional types. Species within each functional type should have a similar function in the ecosystem, and they should have a common ecological role. Functional groupings of species will be crucial to develop global change models (Cramer 1997) and to predict how ecosystems will respond to global change.

REFERENCES

André, M., F. Bréchignac, and P. Thibault. 1994. Biodiversity in model ecosystems. Nature 371:565.

Barbault, R., R.K. Colwell, B. Dias, D.L. Hawksworth, M. Huston, P. Laserre, D. Stone, and T. Younes. 1991. Conceptual framework and research issues for species diversity at the community level. In: Solbrig, O.T., ed. From genes to Ecosystems: A Research Agenda for Biodiversity. Cambridge, MA: IUBS, pp 37–71.

Barkman, J.J. 1988. New systems of plant growth forms and phenological plant types. In: Werger, M.J.A., ed. Plant Forms and Vegetation Structure. The Hague, The Netherlands: SPB Academic, pp 9–44.

Bond, W.J. 1993. Keystone species. In: Schulze, E.D., and H.A. Mooney, eds. Biodiversity and Ecosystem Function. Berlin: Springer-Verlag, pp 237–253.

Brown, J.H., and E.J. Heske. 1990. Control of a desert-grassland by a keystone rodent guild. Science 250: 1705–1707.

Cain, S.A. 1950. Lifeforms and phytoclimate. Botanical Review 16: 1–32.

Chapin, F.S. III, H.L. Reynolds, C.M. D'Antonio, and V.M. Eckhart. 1997. The functional role of species in terrestrial ecosystems. In: Walker, B., and W. Steffen, eds. Global Change and Terrestrial Ecosystems. IGBP Book Series 2. Cambridge: Cambridge University Press, pp 403–428.

Cramer, W. 1997. Using plant functional types in a global vegetation model. In: Smith, T.M., H.H. Shugart, and F.I. Woodward, eds. Plant Functional Types. Their Relevance to Ecosystem Properties and Global Change. Cambridge: Cambridge University Press, pp 271–288.

D'Antonio, C.M., and P.M. Vitousek. 1992. Biological invasions by exotic grasses, the grass/fire cycle and global change. Annual Review of Ecology and Systematics 23: 63–87.

Dublin, H.T., A.R.E. Sinclair, and J. McGlade. 1990. Elephants and fire as causes of multiple stable states in the Serengeti-Mara woodlands. Journal of Animal Ecology 59: 1147–1164.

du Rietz, G.E. 1931. Lifeforms of terrestrial flowering plants. Acta Phytogeographica Suecica 3: 1–95.

Ehrlich, P.R., and G.C. Daily. 1993. Population extinction and saving biodiversity. Ambio 22: 64–68.

Ehrlich, P.R., and A.H. Ehrlich. 1981. Extinction. The Causes and Consequences of the Disappearance of Species. New York: Random House.

Elton, C.S. 1958. The Ecology of Invasions by Animals and Plants. London: Methuen.

Fleming, T.H., and V.J. Sosa. 1994. Effects of nectarivorous and frugivorous mammals on reproductive success of plants. Journal of Mammalogy 75: 845–851.

Friedel, M.H., G.N. Bastin, and G.F. Griffin. 1988. Range assessment and monitoring of arid lands: the derivation of functional groups to simplify vegetation data. Journal of Environmental Management 27: 85–97.

Frost, T.M., S.R. Carpenter, A.R. Ives, and T.K. Kratz. 1995. Species compensation and complementarity in ecosystem function. In: Jones, C.G., and J.H. Lawton, eds. Linking Species and Ecosystems. New York: Chapman and Hall, pp 224–239.

Givnish, T.J. 1994. Does biodiversity beget stability? Nature 371: 113–114.

Hobbs, R.J. 1997. Can we use plant functional types to describe and predict responses to environmental change? In: Smith, T.M., H.H. Shugart, and F.I. Woodward, eds. Plant Functional Types. Their Relevance to Ecosystem Properties and Global Change. Cambridge: Cambridge University Press, pp 66–90.

Humboldt, A. von. 1806. Ideen zu einer Physiognomik der Gewachse. Stuttgart, Germany: Cotta.

Huston, M.A. 1997. Hidden treatments in ecological experiments: re-evaluating the eco-system function of biodiversity. Oecologia 110: 449–460.

Jones, C.G., J.H. Lawton, and M. Shachak. 1994. Organisms as ecosystem engineers. Oikos 69: 373–386.

Keddy, P.A. 1992. Assembly and response rules: two goals for predictive community ecology. Journal of Vegetation Science 3: 157–164.

King, A.W., and S.L. Pimm. 1983. Complexity, diversity, and stability: a reconciliation of theoretical and empirical results. American Naturalist 122: 229–239.

Kitchell, J.F., and S.R. Carpenter. 1993. Synthesis and new directions. In: Carpenter, S.R., and J.F. Kitchell, eds. The Trophic Cascade in Lakes. New York: Cambridge University Press, pp 332–350.

Lawton, J.H., and V.K. Brown. 1993. Redundancy in ecosystems. In: Schulze, E.D., and H.A. Mooney, eds. Biodiversity and Ecosystem Function. Berlin: Springer-Verlag, pp 255–270.

Lawton, J.H. 1994. What do species do in ecosystems? Oikos 71: 367–374.

MacArthur, R.H. 1955. Fluctuations of animal populations and a measure of community stability. Ecology 36: 533–536.

May, R.M. 1973. Stability and Complexity in Model Ecosystems. Princeton, NJ: Princeton University Press.

McNaughton, S.J. 1977. Diversity and stability of ecological communities: a comment on the role of empiricism in ecology. American Naturalist 111: 515–525.

McNaughton, S.J. 1985. Ecology of a grazing ecosystem: the Serengeti. Ecological Mono-graphs 55: 259–294.

McNaughton, S.J. 1993. Biodiversity and function of grazing ecosystems. In: Schulze, E.D., and H.A. Mooney, eds. Biodiversity and Ecosystem Function. Berlin: Springer-Verlag, pp 361–384.

Mooney, H.A., and E.L. Dunn. 1970. Convergent evolution of Mediterranean-climate ev-ergreen sclerophyll shrubs. Evolution 24: 292–303.

Mooney, H.A. 1977. Convergent Evolution of Chile and California-Mediterranean Climate Ecosystems. Strondsburg, PA: Dowden, Hutchinson and Ross.

Mooney, H.A., and S.L. Gulmon. 1983. The determinants of plant productivity-natural versus man-modified communities. In: Mooney, H.A., and M. Godron, eds. Distur-bance and Ecosystems. Berlin: Springer-Verlag, pp 146–158.

Mooney, H.A. 1997. Ecosystem function of biodiversity: the basis of the viewpoint. In: Smith, T.M., H.H. Shugart, and F.I. Woodward, eds. Plant Functional Types. Their Relevance to Ecosystem Properties and Global Change. Cambridge: Cambridge University Press, pp 341–354.

Naeem, S., L.J. Thompson, S.P. Lawler, J.H. Lawton, and R.M. Woodfin. 1994. Declining biodiversity can alter the performance of ecosystems. Nature 368: 734–737.

Naeem, S., L.J. Thompson, S.P. Lawler, J.H. Lawton, and R.M. Woodfin. 1995. Empirical evidence that declining species diversity may alter the performance of terrestrial ecosystems. Proceedings of the Royal Society of London, Series B 347: 249–262.

Naiman, R.J., G. Pinay, C.A. Johnston, and J. Pastor. 1994. Beaver influences on the long-term biogeochemical characteristics of boreal forest drainage networks. Ecology 75: 905–921.

Noble, I.R. 1989. Attributes of invaders and the invading process: terrestrial and vascular plants. In: Drake, J.A., H.A. Mooney, F. di Castri, R.H. Groves, F. J. Kruger, M. Rejmánek, and M. Williamson, eds. Biological Invasions: A Global Perspective. Scientific Commitee on Problems of the Environment. Chichester: Wiley, pp 301–311.

Noss, R.F. 1990. Indicators for monitoring biodiversity: a hierarchical approach. Conservation Biology 4: 355–364.

Odum, E.P. 1963. Ecology. New York: Holt, Rinehart and Winston.

Paine, R.T. 1966. Food web complexity and species diversity. American Naturalist 100: 65–75.

Peterson, G., C.R. Allen, and C.S. Holling. 1997. Ecological resilience, biodiversity and scale. Biodiversity 1: 1–15.

Pimm, S.L. 1979. Complexity and stability: another look at MacArthur's original hypothesis. Oikos 33: 351–357.

Pimm, S.L. 1984. The complexity and stability of ecosystems. Nature 307: 321–326.

Pimm, S.L. 1993. Biodiversity and the balance of nature. In: Schulze, E.D., and H.A. Mooney, eds. Biodiversity and Ecosystem Function. Berlin: Springer-Verlag, pp 347–359.

Power, M.E., D. Tilman, J.A. Estes, B.A. Menge, W.J. Bond, L.S. Mills, G. Daily, J.C. Castilla, J. Lubchenco, and R. Paine. 1996. Challenges in the quest for keystones. Bioscience 46: 609–620.

Prins, H.H.T., and H.P.V.D. Jeud. 1993. Herbivore population crashes and woodland structure in East Africa. Journal of Ecology 81: 305–314.

Raunkiaer, C. 1934. The Life Forms of Plants and Statistical Plant Geography. Oxford: Clarendon Press.

Sala, O.E., W.K. Lauenroth, and R.A. Golluscio. 1997. Plant functional types in temperate semi-arid regions. In: Smith, T.M., H.H. Shugart, and F.I. Woodward, eds. Plant Functional Types. Their Relevance to Ecosystem Properties and Global Change. Cambridge: Cambridge University Press, pp 217–233.

Schindler, D.W. 1990. Experimental perturbations of whole lakes as tests of hypothesis concerning ecosystem structure and function. Oikos 57: 25–41.

Schultze, E.D., and H.A. Mooney. 1993. Biodiversity and Ecosystem Function. Berlin: Springer-Verlag.

Sellers, P.J., M.D. Heiser, and F.G. Hall. 1992. Relations between surface conductance and spectral vegetation indices at intermediate (100 m^2 to 15 Km^2) length scales. Journal of Geographical Research 97: 19033–19059.

Shugart, H.H. 1997. Plant and ecosystem functional types. In: Smith, T.M., H.H. Shugart, and F.I. Woodward, eds. Plant Functional Types. Their Relevance to Ecosystem Properties and Global Change. Cambridge: Cambridge University Press, pp 20–43.

Steffen, W.L., B.H. Walker, J.S.I. Ingram, and G.W. Koch, eds. 1992. Global Change and Terrestrial Ecosystems: The Operational Plan. Stockholm: IGBP and ICSU.

Tilman, D., and J.A. Downing. 1994. Biodiversity and stability in grasslands. Nature 367: 363–365.

Tilman, D. 1996. Biodiversity: Population versus ecosystem stability. Ecology 77: 350–363.

Tilman, D., D. Wedin, and J. Knops. 1996. Productivity and sustainability influenced by biodiversity in grassland ecosystems. Nature 379: 718–720.

Vitousek, P.M. 1990. Biological invasions and ecosystems processes: towards an integration of population biology and ecosystem studies. Oikos 57:7–13.

Vitousek, P.M., and D.U. Hooper. 1993. Biological diversity and terrestrial ecosystem biogeochemistry. In: Schulze, E.D. and H.A. Mooney, eds. Biodiversity and Ecosystem Function. Berlin: Springer-Verlag, pp 3–14.

Walker, B. 1992. Biological diversity and ecological redundancy. Conservation Biology 6: 18–23.

Warming, J.E.B. 1909. Ecology of Plants. London: Oxford University Press.

Westoby, M., and M. Leishman. 1997. Categorizing plant species into functional types. In: Smith, T.M., H.H. Shugart, and F.I. Woodward, eds. Plant Functional Types. Their Relevance to Ecosystem Properties and Global Change. Cambridge: Cambridge University Press.

Wilson, E.O. 1988. Biodiversity. Washington D.C.: National Academy Press.

Wilson, E.O. 1992. The Diversity of Life. Cambridge, MA: Belknap Press of Harvard University Press.

Woodward, F.I. 1993. How many species are required for a functional ecosystem? In: Schulze, E.D., and H.A. Mooney, eds. Biodiversity and Ecosystem Function. Berlin: Springer-Verlag, pp 271–291.

Woodward, F.I., T.M. Smith, and H.H. Shugart. 1997. Defining plant functional types: the end view. In: Smith, T.M., H.H. Shugart, and F.I. Woodward, eds. Plant Functional Types. Their Relevance to Ecosystem Properties and Global Change. Cambridge: Cambridge University Press, pp 355–359.

Zimov, S.A., V.I. Chuprynin, A.P. Oreshko, F.S. Chapin, J.F. Reynolds, and M.C. Chapin. 1995. Steppe-tundra transition: a herbivore-driven biome shift at the end of the Pleistocene. American Naturalist 146: 765–794.

21
Resistance to Air Pollutants: From Cell to Community

Jeremy Barnes and Alan Davison
University of Newcastle, Newcastle upon Tyne, England

Luis Balaguer and Esteban Manrique-Reol
Complutense University, Madrid, Spain

The race is not to the swift, nor the battle to the strong. . . . . . .but time and chance happeneth to them all.

Ecclesiastes, 9:11

I. INTRODUCTION

The generation of energy by the burning of fossil fuels, all manner of industrial processes, the biodegradation of wastes, and some farming operations lead to the release of a wide range of contaminants into the air. Most have little or no discernible effect on the environment, because the resulting concentrations in the atmosphere are well below levels known to be toxic or because they are not toxic to biological systems. Others attain levels that are known to threaten human health and to damage both fauna and flora. The situation is not new because there have been air pollution problems of one kind or another since fire was first used and metals were first smelted, but the unbridled expansion of industry in many parts of the world over the past century has resulted in problems on an unprecedented scale, with impacts extending from the local or regional to global level.

    Although air pollution can take various forms (i.e., dusts, smoke, fumes, aerosols, or mist), this chapter focuses on resistance and adaptation to the most common gaseous pollutants. Stringent control measures have resulted in a steady decline in the emissions of several pollutants in developed regions (e.g., sulphur dioxide [SO_2]); however, ground-level concentrations of some of the most potent gases (e.g., ozone [O_3]) continue to increase (Penkett 1988, Stockwell et al. 1997, Boubel et al. 1994). Locally, ground-level concentrations of some pollutants may be high enough to result in severe foliar injury under conditions favoring accumulation in the atmosphere (i.e., periods of high solar radiation, favorable temperatures or temperature inversions), while potentially damaging concentrations of others (e.g., O_3) may be generated at considerable distances from the source (Bell 1984, Boubel et al. 1994, Krupa 1996). Long distance transport is usually favored by the high levels of irradiance and stable atmospheric conditions associated with slow-moving high-pressure systems in the northern hemisphere. Under such conditions, there is poor dispersal of polluted air masses, and pollutant concentrations, although typically lower than those experienced near to point sources, may be high enough to result in subtle changes in plant physiology, growth, and community composition. Such effects are not necessarily associated with the appearance of typical visible symptoms of injury, but are more common and just as debilitating (Davison and Barnes 1998, Wolfenden and Mansfield 1991, Davison et al. 1992).

    It would seem that all plants possess the encoded capability for the perception, signaling, and response to air pollutants, but differential expression under the influence of genetic and/or environmental factors can result in constitutive and inducible differences between the "reaction norms" of plants within and between populations to the same air pollution insult. In this chapter, we discuss aspects related to the ecotoxicology of airborne pollutants. We begin by reviewing what is known about the mechanisms underlying differential resistance to the most common gaseous pollutants, and then attempt to scale-up from responses at the cellular level to those affecting resistance at the plant, population, and community level. A generic model is used to provide a conceptual framework within which to discuss the mechanisms underlying differential resistance. Where appropriate, we have elected to focus on plant responses to ozone (O_3). Not only because the authors are more familiar with the literature relating to this pollutant than any other, but also because O_3 is now recognized to be one of the most potent and widespread toxic agents to which vegetation is exposed in the field (Davison and Barnes 1992, Kärenlampi and Skärby 1996, Fuhrer et al. 1997, Davison and Barnes 1998). Furthermore, increasing concentrations of the pollutant pose a growing threat to vegetation in many regions (Penkett 1988, Stockwell et al. 1997), and exciting advances have recently been made in our understanding of the mechanisms underlying, and the genetic basis of, resistance to O_3 (Kangasjärvi et al. 1994, Schraudner et al. 1994, Alscher et al. 1997, Pell et al. 1997, Schraudner et al. 1997).

II. THE CELLULAR LEVEL

The processes controlling differential resistance to pollutants are considered within the framework of a conceptual model (Figure 1), discussed first by Ariens et al. (1976) and later by Tingey et al. (Tingey and Taylor 1981, Hogsett et al. 1988, Tingey and Andersen 1991), where resistance* is envisaged to be governed by constitutive and inducible differences in a complex sequence of events that either reduce pollutant penetration to the target (i.e., avoidance mechanisms) and/ or enhance the ability of plant tissues to withstand the pollutant and its products once it has penetrated to the target (i.e., tolerance mechanisms). However, various feedbacks can also influence plant response, especially in relation to pollutant detoxification and the repair of injury. These processes are initially dependent on the constitutive resources available, whereas subsequent responses may be governed by the regulation of gene expression, posttranslational modification of enzymes (e.g., phosphorylation/dephosphorylation), and the synthesis of secondary defense-related metabolites. If these responses are not sufficient to prevent damage at the cellular level, then there will be destabilization and injury that will be reflected in downstream consequences at the level of the individual, population, and community. The mechanisms conferring resistance may be independent (i.e., pollutant-specific) or broadly based (i.e., a number of different pollutants trigger the same coordinated defense reaction); broadly based responses can result

---

* Herein defined after Roose et al. (1982) as the "relative ability of a genotype to maintain normal growth and remain free from injury in a polluted environment. A trait that is quantitative, rather than qualitative, as resistance needs not be complete."

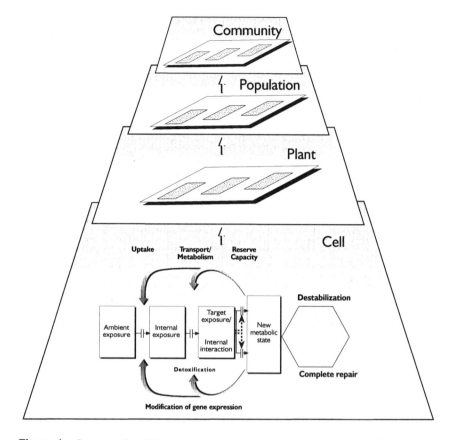

Figure 1 Conceptual model showing the processes that govern the sensitivity of plants to gaseous air pollutants. Pollutant levels that exceed the capacity of avoidance/tolerance mechanisms will result in cellular destabilization—an effect that underpins changes in the performance of the individual and shifts in the genetic composition of both populations and communities.

in cross-resistance to several pollutants (and possibly to a number of other environmental stresses), while there is growing evidence that cross-tolerance may be restricted to pollutants (and/or other stresses) that provoke similar insult on the same target (Schraudner et al. 1997, Barnes and Wellburn 1998).

A. Uptake

The dose of a pollutant absorbed by plant tissues plays a key role in determining effects on metabolism and physiology, and in the description and quantification of dose-response relationships (Taylor et al. 1988, Runeckles 1992). Uptake is predominantly controlled by rates of foliar gas exchange, with conventional approaches focusing on the importance of the cuticle and stomata in controlling the rate at which pollutants diffuse into individual leaves (Mansfield and Freer-Smith 1984). However, it is important to recognize that factors operating at different scales of resolution influence the rate of uptake, in addition to those operating at the level of the individual leaf. At a higher scale of resolution (i.e., scaling-up), aerodynamic factors influence uptake at the canopy level, while at a finer scale of resolution (i.e., scaling-down) physicochemical factors determine uptake at the interface between the plant and its external surroundings (Figure 2).

The size, density, and shape of the canopy all have a pronounced effect on the concentration of a pollutant to which individual leaves are exposed. Relatively few studies have addressed this issue, but it has been shown that the concentration of O_3 (and other pollutants) declines as one passes down through the canopy to soil level (Bennett and Hill 1973). As a result, leaves within a dense canopy tend to be exposed to lower concentrations than those on the surface or at the edges. However, because the air movement within a dense canopy is reduced, concentrations within it tend to be less prone to short-term fluctuations (Runeckles 1992). Other features of the leaves (e.g., leaf thickness, Rubisco content, ratio of mesophyll cell surface area to projected surface area, etc.) and of the environment (e.g., reduced levels of irradiance, increased temperature, higher

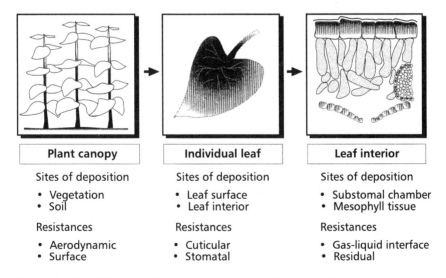

Plant canopy	Individual leaf	Leaf interior
Sites of deposition	Sites of deposition	Sites of deposition
• Vegetation • Soil	• Leaf surface • Leaf interior	• Substomal chamber • Mesophyll tissue
Resistances	Resistances	Resistances
• Aerodynamic • Surface	• Cuticular • Stomatal	• Gas-liquid interface • Residual

Figure 2 The different scales of resolution that need to be considered in determining rates of pollutant uptake. (*Source*: redrawn from Taylor et al. 1988.)

humidity, etc.) also differ within the canopy in relation to those on the outside. This may strongly influence the uptake and effects of pollutants on leaves at different positions within the canopy and is an important consideration when attempting to extrapolate to the field from laboratory-based studies, identify the magnitude of the response of different species in a mixture, and establish the impact of a pollutant at different developmental stages, since certain species or particular growth stages (e.g., seedlings) may be "protected" from exposure to potentially damaging pollutant concentrations by other elements of the canopy.

At the leaf-level, the flux of the pollutant to the leaf interior (J) is a function of the concentration gradient between the atmosphere (i.e., the concentration of the pollutant in the surrounding air [C_a]) and the leaf interior (i.e., the concentration of the pollutant in the intercellular air spaces [C_i]), and the sum of the physical, chemical, and biological resistances (ΣR) to diffusion from "source" to "sink." Mathematically, this is generally expressed in a form analogous with Ohm's law, where:

$$J = (C_a - C_i)/\Sigma R$$

The various impedances (ΣR) are generally visualized as a network of resistances to gas flow (Figure 3), using Gaastra's (Gaastra 1959) fundamental principles governing the tortuous route taken by effluxing water vapor molecules. The most important resistances at the leaf level are recognized to be those governing the movement of the pollutant into the leaf interior (i.e., stomatal resistance, r_1, and cuticular resistance, r_2), the rate of deposition on the hydrated surface of mesophyll cells, and the extent of sorption and reaction on the surface of the cuticle and in the substomatal cavities. These resistances are under genetic and environmental control, as well as being influenced by the physicochemical characteristics of the gas in question.

The rate of diffusion of gaseous pollutants through the cuticular membrane is commonly several orders of magnitude lower than that through the stomata (Lendzian 1984) and may be considered negligible for some reactive gases, such as O_3 (Kersteins and Lendzian 1989). Hence, the dose of the pollutant absorbed at the leaf level is predominantly controlled by factors determining stomatal conductance (i.e., stomatal aperture and frequency). Some plants are known to exhibit intrinsically higher stomatal conductance than others (Körner 1994), and, in general, these tend to be more susceptible to damage (Reich 1987, Becker et al. 1989, Darrall 1989). However, the response of stomata to external stimuli (e.g., irradiance, vapor pressure deficit (VPD), soil moisture content, the presence of pollutants in the atmosphere, atmospheric CO_2 concentrations, etc.) can strongly influence the rate of pollutant uptake through effects on stomatal aperture (Darrall 1989, Wolfenden and Mansfield 1991, Wolfenden et al. 1992, Mansfield and Freer-Smith 1984, Heath and Taylor 1997). Differences in stomatal conductance between sensitive and resistant individuals are rarely sufficient to result in

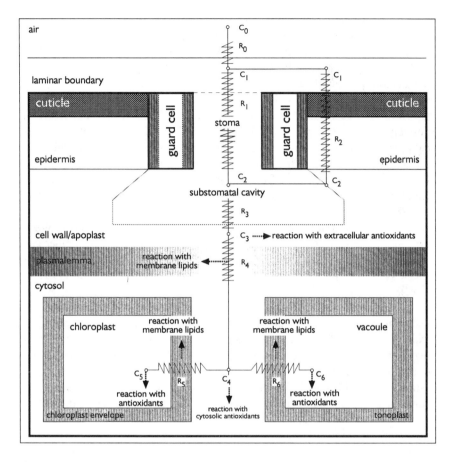

Figure 3 Resistance analog model indicating the network of physical, chemical, and biological impedances influencing the rate of diffusion of gaseous pollutants from the external atmosphere to the target.

complete exclusion of the pollutant from the leaf interior; therefore, it is generally concluded that mechanisms resulting in the avoidance of pollutant uptake are not the only factor determining resistance to airborne pollutants.

Physical leaf characteristics such as leaf thickness, mesophyll cell surface area, internal air space volume, cell wall thickness, and the volume of the aqueous matrix of the cell wall, influence the eventual concentration of the pollutant and its dissolution products in the apoplast. Plants show considerable variation in all of these attributes. Mesophyll cell surface area: projected leaf area, for example, ranges from typical values of between 10 and 40 for mesophytes, but may be as high as 70 for some xerophytes (Nobel and Walker 1985, Pfanz 1987). There may also be systematic differences in anatomy along altitudinal gradients, which can contribute to differential resistance (Körner et al. 1989). In addition, pollutant molecules in the intercellular space or substomatal cavity are partitioned across the gas-to-liquid interface at a rate determined by the solubility of the gas in the extracellular fluid (Nobel 1974) and its chemical reactivity in the liquid phase (Heath 1988, Heath and Taylor 1997), factors influenced by temperature and possibly radiation (Barnes et al. 1996, Cape 1997). Differences in physicochemical properties between the most common gaseous pollutants result in substantial differences in their solubility in water and rates of diffusion in air (Nobel 1974). Factors which, independent of other considerations, result in substantial differences in the rate at which individual gaseous pollutants are taken up (Taylor et al. 1988, Runeckles 1992). Leaf surface characteristics (such as surface wetness, wax composition, micromorphology, etc.) can also influence the extent of sorption onto foliar surfaces (Wellburn et al. 1997), while reactions with other gases in the boundary layer or in the substomatal cavities may constitute a significant "sink" for some pollutants, e.g., O_3 (Hewitt and Terry 1992, Salter and Hewitt 1992).

In most instances, difficult and time-consuming measurements of physical leaf characteristics are not undertaken, so it has become common practice to express rates of pollutant uptake on the basis of the flux to the leaf interior (i.e., that impinging on the mesophyll cell surface). This represents what is often termed the "absorbed" or "effective" dose of the pollutant, and can be readily estimated using Fickian diffusion principles from a knowledge of boundary layer, stomatal and cuticular conductances, correcting for differences in the diffusivities between water vapor and the pollutant of interest. Hence, the flux of O_3 to the leaf interior (J_{O3}) may be described as:

$$J_{O3} = g_b + 0.612 g_{H_2O} (O_a - O_i)$$

where g_b is the turbulent boundary layer conductance, 0.612 is the difference in the binary diffusivities of water vapor and ozone in air (Nobel 1983), g_{H_2O} represents the stomatal conductance to water vapor, and O_a and O_i represent the concentrations of O_3 in the external atmosphere and in the intercellular spaces, respectively. No correction needs to be made for uptake through the cuticle in this case, since the cuticle is considered to represent a virtually impermeable barrier to O_3 (Kersteins and Lendzian 1989), while measurements of the intercellular O_3 concentration (O_i) suggest it is close to zero (Laisk et al. 1989). On a cautionary note, it is important to emphasize two points. First, such calculations represent the gross flux to the leaf interior—for some gases where the plant may act as a source as well as a sink (e.g., H_2S, NH_3, NO), correction for effluxing gas molecules is required to enable estimates of the net flux of the pollutant. Second, fluxes determined in the above manner take no account of differences in physical leaf characteristics or internal resistances influencing the rate at which the pollutant (and/or its products) is delivered to the eventual target. Potentially more informative models are available that enable estimates of the extent of penetration to the plasmalemma and beyond (Chameides 1989, Ramge et al. 1992, Plöchl et al. 1993), but these have rarely been used due to the uncertainties surrounding the complex solution chemistry of certain gases (e.g., O_3) and the relative importance of scavenging/transformation in the cell wall region.

B. Metabolism

Once the pollutant has penetrated as far as the mesophyll cell surface, it may be metabolized, sequestered, or excreted. Recent attention has focused on the rate at which pollutants (and/or their reactive products, including a variety of free radical species) are immobilized and/or detoxified at either the first barrier encountered after entry into the leaf (i.e., in the apoplast) or subsequently, after penetration into the cell proper. In any consideration of the processes underlying pollutant detoxification, it is important to recognize that dissolution in the apoplast can result in different types of stress on cellular constituents. Many gases (e.g., HF, NO_x, NH_3, and SO_2) induce acidification of subcellular compartments, whereas others (e.g., O_3, PAN, NO_x) result in oxidative stress, and some (e.g., SO_2) produce both (Malhotra and Khan 1984, Hippeli and Elstner 1996, Mudd 1996). The extent of damage resulting from the former is related to the ability of cells to buffer the increase in acidity or to excrete protons to the external media (Slovik 1996, Burkhardt and Drechsel 1997), whereas the influence of the latter is dependent on the efficiency of endogenous antioxidant systems that scavenge free radical and active oxygen species before they can react with cellular constituents (Kangasjärvi et al. 1994, Luwe and Heber 1995, Alscher et al. 1997).

Detoxification systems capable of protecting sensitive targets from the oxidative stress imposed by pollutants and their derivatives are common in plants, as in animals, and are subject to strict genetic control. Most attention has focused on those systems located in the various intracellular compartments (in chloroplasts, mitochondria, cytosol, and peroxisomes); however, similar systems are intimately associated with the plasmalemma and cell wall (Figure 4). In recent years, the latter have attracted increased attention since there is growing evidence that some pollutants (e.g., O_3, SO_2, NO_2) and/or their dissolution products may be scavenged and detoxified/transformed at the mesophyll cell surface (i.e., in the apoplast). The aqueous matrix of the cell wall is now recognized to contain significant quantities of ascorbic acid (vitamin C) and polyamines as well as isoforms of Cu/Zn superoxide dismutase (SOD), ascorbate peroxidase (APX), and nonspecific peroxidases (GPODs) (Polle and Rennenberg 1993, Luwe and Heber 1995, Ogawa et al. 1996, Dietz 1997), which are known to function as

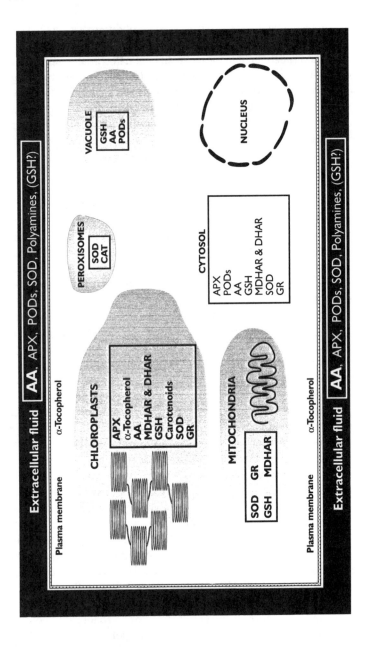

Figure 4 Subcellular localization of the antioxidant systems. AA, ascorbic acid; GSH, glutathione; APX, ascorbate peroxidase EC. 1.11.1.11; MDHAR, monodehydroascorbate radical reductase EC. 1.1.5.4; DHAR, dehydroascorbate radical reductase EC. 1.8.5.1; GR, glutathione reductase EC 1.6.4.2; SOD, superoxide dismutase EC. 1.15.1.1; CAT, catalase EC. 1.11.1.6; PODs, nonspecific peroxidase (sometimes referred to as guaiacol peroxidase) EC. 1.11.1.7.

antioxidants (Polle and Rennenberg 1993, Kangasjärvi et al. 1994, Alscher et al. 1997, Polle 1997.) Research is still at an early stage and many questions remain to be answered, but preliminary model calculations based on the scavenging of O_3 (rather than its dissolution products) by apoplastic ascorbic acid indicate that detoxification processes operating in the apoplast may be sufficient to provide at least limited protection against O_3 (Chameides 1989, Polle and Rennenberg 1993, Lyons et al. 1998a), a finding supported by several independent lines of evidence (Lyons et al. 1998b). Although the relative importance of scavenging in the cell wall region remains to be established, there is a growing opinion that factors such as the extracellular concentration of ascorbic acid may play a central role in determining resistance to O_3 (Conklin et al. 1996, Lyons et al. 1998b), as well as influencing the impacts of SO_2 (Dietz 1997) and NO_2 (Ramge et al. 1992).

Pollutants and/or their reactive products that breach the extracellular defenses, or are produced from the reaction with plasmalemma constituents, must be scavenged by intracellular detoxification systems if damage is to be averted. There is, for example, strong evidence linking SO_2 tolerance with the activity of intracellular enzymes such as catalase (CAT), superoxide dismutase (SOD), glutathione sythetase (GS), glutathione reductase (GR), glutathione peroxidase (GPX), and glutathione transferase (GST) (Tanaka et al. 1988, Ranieri et al. 1992, Lea et al. 1998), as well as with levels of glutathione (GSH) and ascorbic acid (ASC) (Madamanchi and Alscher 1994). Furthermore, recent work has drawn attention to the importance of the subcellular localization of these systems in relation to the protection afforded against different pollutants. For example, SO_2 tolerance in transgenic plants engineered to overexpress GR in different cellular compartments indicates that it is the cytoplasmic activity of this enzyme, rather than that of the plastidic forms, that is important in determining SO_2 tolerance (Aono et al. 1993, Broadbent 1995, Lea et al. 1998). There is also considerable evidence linking components of the cellular antioxidant system (ASC, GSH, polyamines, α-tocopherol, carotenoids, SOD, ascorbate peroxidase [APX], GR, and CAT) with O_3 tolerance (Heath 1988, Kangasjärvi et al. 1994, Hippeli and Elstner 1996, Mudd 1996, Alscher et al. 1997, Pell et al. 1997, Heath and Taylor 1997). However, the degree of protection afforded by these intracellular systems in relation to that achieved by scavenging in the cell wall region (the primary site of O_3 action) remains poorly understood (Lyons et al. 1998b), and it is interesting to note that transgenic plants overproducing particular antioxidant enzymes in different intracellular compartments rarely display enhanced O_3 resistance (Pitcher et al. 1991, Van Camp et al. 1994, Pitcher and Zilinskas 1996, Torsethaugen et al. 1997).

In addition, toxicity thresholds are influenced by the efficiency of metabolic processes resulting in the utilization, sequestration, and/or excretion of pollutants. The plasmalemma, for example, is known to constitute an important obstacle impeding the penetration of pollutants (and/or their products) to their intracellular sites of action (Herschbach et al. 1995). Some pollutants (e.g., O_3) react readily with membrane constituents (Mudd 1996, Heath and Taylor 1997). The uptake of others (i.e., SO_2-derived sulphate [SO_4^{2-}], sulphite [SO_3^{2-}], and bisulphite [HSO_3^-] is limited by the activity of membrane-bound carriers and transformation processes (Pfanz et al. 1990). Indeed, research on the inheritance of SO_2 resistance in *Cucumis sativus* L. suggests that differences in intrinsic membrane properties may contribute to variations in SO_2 resistance (Bressan et al. 1981). Once inside the cell, the products of some (e.g., SO_4^{2-}, NO_3^-, NH_4^+)and NH_4^+) may be metabolized via the usual channels, sequestered for later use, stored indefinitely, or volatilized. For example, SO_2-derived SO_4^{2-} can either be metabolized to yield elevated levels of water-soluble nonprotein thiols (such as cysteine, γ-glutamylcysteine, and glutathione), which can be degraded at a later date to provide reduced S to support new growth (De Kok 1990), or it can be sequestered on a semipermanent basis in the vacuole (presuming there is sufficient available energy and H^+ ions to facilitate its transport across the tonoplast [Cram 1990, Kaiser et al. 1989, Slovik 1996]). In contrast, SO_3^- may be photoreduced and, after volatilization, be re-emitted, mainly as H_2S. This pathway was originally considered to form a possible pathway for the detoxification of SO_2 (Rennenberg 1984), but current opinion suggests that the contribution of such emissions to the detoxification of environmentally relevant SO_2 concentrations may be negligible (Stuhlen and De Kok 1984).

C. Gene Expression

The photoautotrophic habit adopted by plants has resulted in the evolution of a sophisticated battery of mechanisms that renders them, with the exception of all but a few microbes, the most adaptable of all multicellular organisms on the planet (Smith 1990). This flexibility includes the capacity to sense and react to the presence of airborne pollutants, as to other environmental stimuli, in a manner directed at sustaining survival to reproduction. A goal that may or may not be achieved, depending on the extent of metabolic flexibility and the degree of stress imposed at the cellular level.

Exposure to pollutants and/or other oxidative stresses induces changes in the expression of defense-related genes, posttranslational modification of enzymes (e.g., phosphorylation/dephosphorylation), and the synthesis of secondary metabolites—resulting in increases in the threshold for damage, i.e., "acclimation" (Kangasjärvi et al. 1994, Schraudner et al. 1994). Recent work suggests that the pattern of changes induced by some pollutants (e.g., O_3) reflects an orchestrated series of events, triggered by disparate oxidative syndromes, that resembles the "hypersensitive response" provoked by pathogen attack (Alscher et al. 1997, Pell et al. 1997, Schraudner et al. 1997). This raises the question of whether the patterns of defense-related gene expression triggered by one pollutant are similar to those induced by another, whether the same signal transduction pathways are involved, and whether the similarity in response results in enhanced tolerance to a range of pollutants, and possibly other oxidative stresses. Opinions differ. However, based on the fact that particular genotypes, specific transformants, and plants treated with specific protectants (eg., EDU) may be sensitive to one pollutant but not to another (Barnes and Wellburn 1998 and references therein), while the direction and extent of the responses of the same species to different pollutant combinations vary in individual genotypes grown under common conditions (Bender and Weigel 1992), we take the view that despite the common responses observed at the level of gene expression, specific mechanisms must underlie tolerance to different air pollutants (i.e., tolerance to one pollutant is independent of that to another), as the action and subcellular localization of the stresses imposed by pollutants differ. This conclusion is supported by observations that differences in stomatal behavior/conductance (i.e., avoidance mechanisms) commonly govern the similarity in response to different pollutants (Winner et al. 1991). It is also interesting to note that some pollutants (e.g., SO_2 and NO_2) are much less effective than others (e.g., O_3) in eliciting changes in antioxidant gene expression (Schraudner et al. 1994, Schraudner et al. 1997), whereas other stresses (such as wounding, necrotizing pathogens, or elevated levels of UV-B radiation), which are known to elicit strong and rapid changes in defense-related gene expression, have been shown to reduce the extent of O_3 injury (Yalpani et al. 1994, Rao et al. 1996, Övar et al. 1997).

III. THE PLANT LEVEL

Where pollutant uptake exceeds the capacity of the detoxification/repair systems to prevent damage, there may be a host of adverse consequences on plant physiology resulting, ultimately, in the death of plant tissue. The oxidative stress imposed by O_3, for instance, is reflected in a decline in the photosynthetic capacity of individual leaves (Kangasjärvi et al. 1994, Pell et al. 1997), increased rates of maintenance respiration (Darrall 1989, Wellburn et al. 1997), enhanced retention of fixed carbon in leaves (Balaguer et al. 1995, Cooley and Manning 1987), and accelerated rates of leaf senescence (Alscher et al. 1997, Pell et al. 1997)—effects that are reflected in reduced growth and reproductive potential. Under some circumstances, the pollutant can induce localized cell death, resulting in typical visible symptoms of foliar injury. However, what aspect of performance should be used as an indication of resistance in an ecological context? In the case of crops, this is relatively straightforward since the impact of the pollutant on yield and/or marketable product is clearly the important feature. However, it is more difficult for natural vegetation. Because annual or monocarpic species must produce seeds to survive, seed output is an obvious criterion to use. But what should be used to rank the performance of perennial species? Many iteroparous perennials live for decades or even centuries (Harberd 1960, Harper 1977). Yet, more often than not, assessments of resistance have been based on the degree of visible foliar damage or impacts on plant growth rate relative to that of controls, with little consideration or understanding of whether these features are important in an ecological context (Davison and Barnes 1998). It has, for example, become common practice to rank species in terms of susceptibility on the basis of visible symptoms of injury. This tends to lead to the intuitive conclusion that the affected species must suffer from some ecological disadvantage in the field, and conversely that unmarked species do not. This may be a serious misconception. First, the expression of symptoms is affected by many factors, including soil water deficit, vapor pressure deficit, photon flux density, temperature (Balls et al. 1996), and possibly UV-B radiation (Thalmair et al. 1995). Second, there is usually little relation between relative sensitivity in terms of visible symptoms and effects on growth or seed production (Heagle 1979, Fernandez-Bayon et al. 1992, Bergmann et al. 1995); some taxa show highly significant effects of O_3 on growth but no visible symptoms, whereas others exhibit extensive visible symptoms but no effects on growth (Davison and Barnes 1998 and references therein). Third, and possibly most importantly, species that show injury are not necessarily debilitated in competition with other species that do not. This is clearly demonstrated in the work of Chappelka and colleagues (Chappelka et al. 1997, Barbo et al. 1998) on communities containing "sensitive" species such as blackberry (*Rubus cuneifolius*) and tall milkweed (*Asclepias exaltata*), shown to proliferate in field studies, despite the development of extensive visible symptoms of O_3 injury (Figure 5). Observations such as these led Davison and Barnes (1998) to conclude that "visible symptoms are best regarded as evidence of a biochemical response to ozone. They do not necessarily indicate sensitivity in terms of growth reduction and they are not evidence of an ecological impact."

Many other assessments of resistance of wild species have been based on the ratio of harvest weight in treated plants to that of controls, whereas others have used the ratio of mean relative growth rate. Most have used aboveground weight, and few have measured root weight or parameters of ecological importance such as specific root length or root length/leaf area ratio (Pell et al. 1997). This is important because the choice of measure can influence both the apparent magnitude of the ozone response and the relative ranking (Macnair 1991, Ashmore and Davison 1996). The difficulties are apparent in the hypothetical example given by Davison and Barnes (1998), where straightforward comparisons of the effects of O_3 on the growth of a fast-growing ruderal and a slower-growing stress tolerator (Grime 1979) would yield erroneous results. Attempts to find broad relationships between resistance and adaptive or ecological characters have proved disappointing. Preliminary studies conducted by Reiling and Davison (1992) on 32 species revealed a weak negative relationship between the effect of O_3 and plant

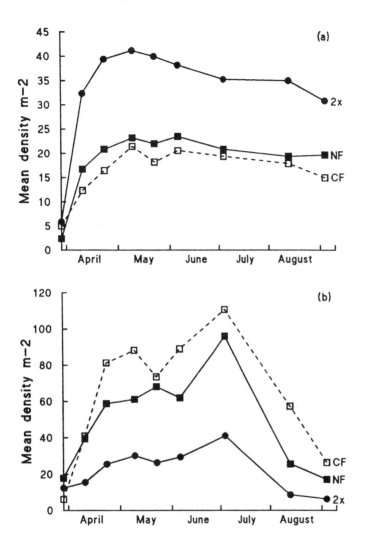

Figure 5 Effects of open-top chamber ozone exposure on two understory species in an early successional forest community; changes in mean density of (a) blackberry (*Rubus cuneifolius*), a species that develops extensive visible symptoms of O_3 injury and has therefore often been classified as sensitive, and (b) bahia grass (*Paspalum notatum*), a species that does not show visible injury and has therefore often been classified as resistant. Treatments: NF, nonfiltered ambient air; CF, charcoal-filtered air; 2×, 2 × ambient O_3. (*Source*: Davison and Barnes 1998, based on original data presented by Barbo et al. 1994.)

relative growth rate (R) in clean air, implying that approximately 30% of the variation between taxa was related to differences in inherent growth rate. However, in a more comprehensive study of 43 species, in which one of the principle aims was to investigate relationships between O_3 sensitivity and CSR strategy (Grime 1979), O_3 resistance was found to be significantly correlated with only one other trait, mycorrhizal status, and no relation existed between O_3 sensitivity and R in clean air or plant growth strategy (Grime et al. 1997). In contrast, SO_2 resistance was significantly correlated with 12 other traits. However, conclusions drawn from such studies are difficult to interpret because the effects on growth may be influenced by many factors, including seed characteristics and seed provenance; some experiments have used seeds collected in the field, whereas others have used commercial seeds or a mixture of sources (Davison and Barnes 1998). Hence, maternal effects caused by the parental environment (Roach and Wulff 1987) may contribute to some of the differences in ranking reported for the same species, e.g., *Phleum alpinum* (Mortensen 1994, Pleijel and Danielsson 1997). There may also be substantial intraspecific variation in the response of different genotypes within a population to the same air pollution insult (see Section IV).

Much attention has focused on the relative partitioning of dry matter between the root and shoot of crop plants, and the observed impacts of specific pollutants (e.g., O_3) are often interpreted as being universal (Övar et al. 1997). However, experiments with a range of wild species show that the situation is much more complicated, and subtle effects on allocation that are probably of greater ecological significance than changes in mass are common. In the legume *Lotus corniculatus*, Warwick and Taylor (1995) found that O_3 had no effect on allometric root/shoot growth, but caused a large reduction in specific root length, and there are other cases in which decreased allocation to the root has been found to be associated with compensatory changes in thickness, so length is unaffected (Taylor and Ferris 1996). Some species (e.g., *Arrhenatherum elatius, Rumex acetosa*) may even show an increased allocation to the root when exposed to ozone (Reiling and Davison 1992); others such as clover show the greatest decrease in allocation not to the roots, but to the storage and overwintering organs, i.e., the stolons (Wilbourn et al. 1995, Fuhrer 1997). One of the most instructive studies of pollutant impacts on resource allocation in wild species was performed by Bergmann et al. (1995, 1996). They exposed 17 herbaceous species from seedling stage to flowering to two O_3 regimes with different dynamics: CF + 70 nl l^{-1} 8 h d^{-1} and CF + 60% ambient + 30 nl l^{-1}. Responses varied with exposure regime, and the weight of some species was reduced to about 60% of controls, but the most striking differences were in resource allocation (Figure 6). Most showed a proportionate change between shoot mass and reproductive effort (bottom left quadrant of Figure 6), but two species (*Chenopodium album* and *Matricaria discoidea*) showed a greater vegetative shoot weight and reduced reproduc-

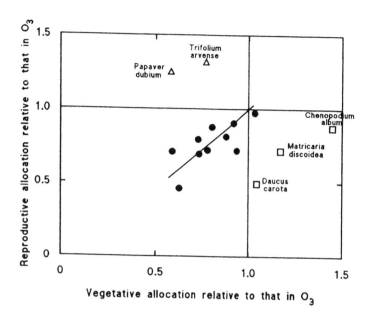

Figure 6 Effects of O_3 on the relative allocation of resources between vegetative and reproductive organs of herbaceous species exposed from seedling to flowering to two O_3 regimes; charcoal-filtered air plus 70 nl l^{-1} O_3 8 h d^{-1} or charcoal-filtered air plus 60% ambient O_3 plus 30 nl l^{-1}. Changes in allocation expressed as the ratio of that in O_3 to that in CF air. (*Source*: Barnes and Davison 1998, based on the data collected by Bergmann et al. 1995, 1996.)

tive allocation. Conversely, *Papaver dubium* and *Trifolium arvense* exhibited reduced shoot mass and increased allocation to seed/flowers. Such shifts in resource allocation may help to explain why O_3 is sometimes found to stimulate growth and highlight the need for greater understanding of the control of resource allocation in species having different reproductive and survival strategies.

Relative rankings of resistance may also be biased by growth stage/developmental status, since plants do not appear to be equally sensitive to pollutants at all stages in their life cycle (Davison and Barnes 1998). Our own work (Lyons and Barnes 1998) on *Plantago major*, for example, shows that seedlings are much more sensitive to O_3 than juvenile or mature plants; O_3-induced declines in accumulated biomass appeared to be almost entirely due to effects on seedling relative growth rate in this species, while seed production is most affected during the early stages of flowering (Figure 7; Davison and Barnes 1998). Compensatory changes in growth and morphology may also limit the impacts of prolonged expo-

(a)

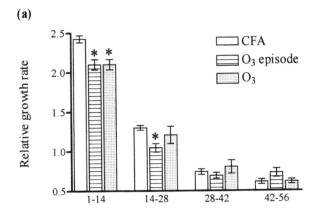

(b)

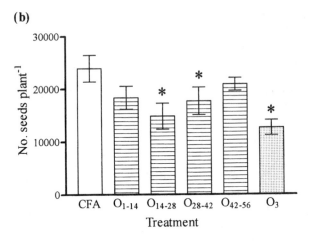

Figure 7 Effects of O_3 on (a) growth and (b) seed production in *Plantago major* Valsain exposed in duplicate controlled environment chambers to charcoal/Purafil-filtered air (CFA), O_3 (CFA plus 70 nl l^{-1} O_3 7 h d^{-1}), or 14-d episodes of O_3 (i.e., windows) administered between 1–14 d (O_{1-14}), 14–28 d (O_{14-28}), 28–42 d (O_{28-42}), or 42–56 d (O_{42-56}). Asterisks indicate significant ($P = 0.05$) differences from plants maintained in CFA. (*Source*: Lyons and Barnes 1998.)

sure to O_3 (Lyons and Barnes 1998 and references therein). In the absence of such effects, it is conceivable that the impacts of O_3 (and other pollutants) would be considerably greater than they are.

The impacts of pollutants in the field may also be modified by a multitude of factors, including management practices, soil water deficit, mineral nutrition, other pollutants, frost, disease, and herbivory (Davison and Barnes 1992, Barnes et al. 1996, Wellburn et al. 1997, Barnes and Wellburn 1998, Davison and Barnes 1998). Work on crops, for example, indicates that the impacts of O_3 are commonly reduced under conditions in which soil water deficit results in a decline in stomatal conductance and hence pollutant uptake (Tingey and Taylor 1981, Darrall 1989, Tingey and Andersen 1991, Wolfenden et al. 1992, Wellburn et al. 1997). However, field observations relating to possible pollutant–water stress interactions in wild species are rare and anecdotal. Showman (1991) reported that visible oxidant injury in wild species in Ohio and Indiana was virtually absent in a year when it was dry and ozone was high, but widespread in a year when it was not as dry and ozone was lower. Similarly, Davison and Barnes (1998) drew attention to the fact that in heavily polluted regions of southern Europe, visible symptoms of oxidant injury are common in irrigated crops, and there are effects on yield; however, in nonirrigated areas subjected to severe summer drought, there are virtually no records of symptoms in wild species.

IV. THE POPULATION LEVEL

There is growing evidence that pollutants, like other novel stresses imposed by human activities, can bring about changes in the genetic composition of populations (Roose et al. 1982, Taylor and Pitelka 1992). The phenomenon was first revealed through the work of Bell et al. (1991) on the evolution of SO_2 resistance in grassland species in industrialized regions of the United Kingdom. However, convincing experimental evidence of evolution of resistance to regional-scale pollutants (such as O_3) has only recently been reported. In a way, this is rather surprising, since the requirements to drive the evolution of O_3 resistance in wild species have been recognized for many years (Lyons et al. 1997, Davison and Barnes 1998). Our own artificial selection studies (Whitfield et al. 1997) on resistant and sensitive populations of *Plantago major* indicate the potential for evolution of resistance/sensitivity to O_3 within a matter of only a few generations in this short-lived species. Based on the effects of a 2-week exposure to 70 nl l^{-1} O_3 for 7 h d^{-1} on rosette diameter, selection from an initially sensitive population led to a line with significantly enhanced O_3 resistance, although it was possible to select a line with greater sensitivity than the original population. Conversely, selection from an initially resistant population led to a line with increased sensitivity, but not to a line with enhanced resistance (Figure 8). Subsequent experi-

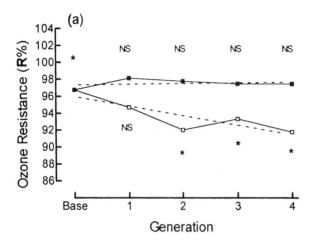

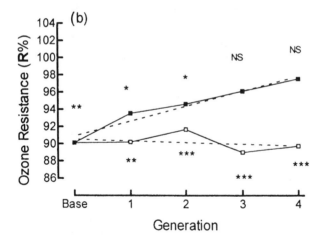

Figure 8 Change in O_3 resistance over four generations in lines selected for sensitivity (□) and resistance (■) from two populations of *Plantago major* based on effects on rosette diameter. Data presented indicate effects on plant relative growth rate (R); R% = [R_{O3}/R_{CF}] × 100. Plants were exposed in controlled-environment chambers to charcoal/Purafil-filtered air or O_3 (CFA plus 70 nl l^{-1} O_3 7 h d^{-1}). Dashed lines represent linear regression fits for selections. Data for each generation in each selected line were subjected to ANOVA. Probabilities (*P < .05; **P < .01; ***P < .001) indicate the significance of O_3 effects on each generation. (*Source*: Whitfield et al. 1997.)

ments have shown that the resistance of the selected lines is maintained, and differences are reflected in contrasting effects on growth and seed production.

The earliest suggestion that ambient levels of O_3 may be high enough to drive the selection of resistant genotypes in the field was provided by Dunn (1959), who worked on *Lupinus bicolor* in the Los Angeles basin. He attributed differences in the performance of populations of this species to oxidant smog and commented that "the stress was so severe that some populations failed to set seed"—an effect expected to contribute to rapid evolution. It is only through recent work (reviewed by Macnair 1993, Davison and Barnes 1998) on *Populus tremuloides, Trifolium repens*, and *Plantago major* that convincing experimental evidence has been provided to support the evolution of resistance to O_3 in the field. Our own studies on *Plantago major* represent the only case in which heritable change in resistance has been shown to occur in the field over a period of time when O_3 levels increased (Reiling and Davison 1992, 1995). However, the crucial question in all of the studies conducted to date is whether the observed differences in resistance between populations have arisen in response to O_3, or to other correlated environmental factors (Bell et al. 1991). It was suggested by Roose et al. (1982) that because traits affecting sensitivity to air pollutants may simultaneously reflect adaptations to other natural stresses and vice versa, resistance might be indirect. Our own work on 41 European populations of *Plantago major* indicates that O_3 resistance is significantly correlated with the O_3 concentration at or near the site of collection (Figure 9), and similar findings have been reported by Berrang et al. for *Populus tremuloides* originating from parts of the U.S. with different O_3 climates (Davison and Barnes 1998). However, in some cases, O_3 resistance has also been shown to correlate with other variables (Reiling and Davison 1992). Although the available evidence is consistent with the evolution of O_3 resistance, correlations do not definitively prove a cause–effect relationship. Therefore, it has been necessary to try to eliminate other possibilities. This has been achieved using partial correlations to remove the effect of climatic differences between collection sites. In some cases, this has reduced the significance of regressions consistent with the evolution of O_3 resistance (Berrang et al. 1991, Reiling and Davison 1992); in other studies it has made little difference to the significance of regressions (Lyons et al. 1991). Although the experimental data are generally consistent with the evolution of resistance to O_3, the work graphically illustrates the difficulties in interpreting the observed spatial variability in pollution resistance.

Based on the assertion of Bell et al. (1991) that "if populations are evolving it should be possible to demonstrate a change in resistance over time, as with evolution to other novel stresses," Davison and Reiling (1995) compared the ozone resistance of *P. major* populations grown from seed collected from the same sites over a 6-year period in which ambient O_3 concentrations increased. They demonstrated that two populations increased in resistance, and Wolff and

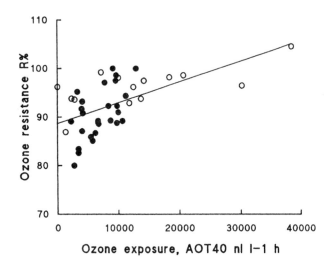

Figure 9 Ozone resistance of seed-grown *Plantago major* populations plotted against the O_3 exposure (based on the accumulated O_3 exposure above a 40 nl l^{-1} threshold, i.e., AOT40) at the collection sites. Resistance determined as the mean relative growth rate in O_3 (70 nl l^{-1} O_3 7 h d^{-1}) as a percentage of that in charcoal/Purafil-filtered air. ●, UK populations; ○, continental European populations. Regression = 88.7 + 0.00043(AOT40), r = 0.538, P < .0001. (*Source*: Davison and Barnes 1998; based on data presented by Reiling and Davison 1992 [●], and Lyons et al 1997 [○].)

Morgan-Richards (unpublished data, 1997) have recently proven (using random amplified polymorphic DNA primers [RAPDs]) that the later populations are subsets of the earlier ones. This is consistent with directional in situ selection rather than a catastrophic loss and replacement of the populations by migration. However, difficulties in interpretation remain, since one of the reasons that O_3 levels increased was because it was sunnier and warmer. This probably led to a greater incidence of soil moisture deficit and photoinhibition, but no records are available for the collection sites. Hence, the possibility that other factors may have contributed to the evolution of O_3 resistance cannot be dismissed. It is also important to recognize that the evolution of resistance to O_3 may be associated with costs or a loss of fitness in unpolluted environments, as with other cases of directional selection motivated by novel stresses (Roose et al. 1982, Macnair 1993, Reiling and Davison 1995), but little is known about the nature of these costs with respect to O_3 resistance (Davison and Barnes 1998). Furthermore, the rate of evolution of resistance would be expected to be influenced by many factors, including the mode of reproduction, the form of sexual reproduction (determining the degree of inbreeding), the dynamics of gene flow within and between populations, generation time, the presence of seed banks, the extent of the loss of fitness induced by the pollutant, and the timing of selection in relation to the plants' life cycle (Roose et al. 1982, Taylor and Pitelka 1992, Macnair 1993, Lyons and Barnes 1998). Thus, even in areas exposed to potentially damaging pollutant concentrations for long periods, it is possible to find species that persistently show typical visible symptoms of injury (Davison and Barnes 1998).

V. THE COMMUNITY LEVEL

Studies on isolated taxa indicate that there is wide variation in resistance between species to gaseous pollutants. This is exemplified in our own screening of the O_3 sensitivity of 30 species using seed collected from central England, using a standard O_3 exposure of 70 nl l^{-1} for 7 h d^{-1} over 2 weeks (Reiling and Davison 1992). The minimum detectable effect of O_3 on growth rate in this test was approximately 5%, and species exhibited a range of sensitivities (Table 1); the most sensitive species affected (in terms of effects on plant relative growth rate) to a similar extent as some of the most sensitive crop species (e.g., tobacco "Bel-W3"). These and other studies (Fuhrer et al. 1997) indicate that herbaceous species exhibit wide-ranging sensitivity to O_3. However, there may be as much variation within species as between species (see Section IV), and there is no guarantee that controlled trials are reminiscent of responses in the field, where a range of additional factors must be considered. Consequently, such studies contribute little other than indicating the range of potential responses to gaseous pollutants.

Where pollutants emanate from a point source, it is possible to determine effects on biodiversity by standard ecological and multivariate methods, especially where records can be repeated over time (Musselman et al. 1992). There is, for example, a wealth of literature documenting the effects of acidifying air pollutants on epiphytic lichen communities (Nimis et al. 1991). For regional pollutants such as O_3 it is more difficult, because the pollutant does not usually show sharp gradients within defined boundaries. Consequently, measurable effects on the structure of plant communities may be restricted to a few special cases. Westman (1979, 1985), for example, showed that percentage cover and species richness were strongly influenced by O_3 (and other oxidants) in Californian coastal sage scrub. Preston (1985) reported similar effects for SO_2. Interestingly, ordination approaches suggested a greater effect at sites with low foliar cover than at high ones, but Preston did not comment on this. The same types of studies are likely to prove unfruitful in less impacted areas, because of the lack of sharp gradients in oxidant concentrations and the difficulty in locating appropriate control sites. Consequently, progress in understanding the impacts of regional pollutants (such as O_3) on plant communities has, and will probably continue, to depend

Table 1 Impact of O_3 on Growth and Root/Shoot Dry Matter Partitioning in 32 Taxa[a]

	R wk^{-1}			K = R_{root}/R_{shoot}		
	Control	+O_3	R%	Control	+O_3	K%
Arrhenatherum elatius	1.96	1.85	−6	0.84	1.01	+20***
Avena fatua	1.97	2.00	+2	0.87	0.80	−8
Brachypodium pinnatum	1.58	1.48	−6	0.75	0.81	+8*
Bromus erectus	1.93	1.92	0	1.05	1.18	−12*
Bromus sterilis	1.77	1.73	−2	1.02	1.08	+6
Cerastium fontanum	2.18	1.96	−10***	0.67	0.65	−3
Chenopodium album	1.99	1.97	−1	0.62	0.53	−14***
Deschampsia flexuosa	1.81	1.73	−4	0.95	0.96	+1
Desmazeria rigida	1.21	1.10	−9	1.23	0.75	−39**
Epilobium hirsutum	1.61	1.58	−2	0.83	0.62	−25**
Festuca ovina	1.35	1.27	−6	0.58	0.60	+4
Holcus lanatus	1.64	1.57	−4*	0.93	0.91	−2
Hordeum murinum	1.84	1.74	−5*	0.83	0.79	−5
Koeleria macrantha	1.73	1.63	−6***	0.77	0.42	−45*
Lolium perenne Talbot	2.03	1.98	−2	0.65	0.88	+35*
Nicotiana tabacum Bel-W3	2.27	1.91	−16***	0.98	0.89	−9
Pisum sativum Conquest	1.95	1.80	−8***	1.07	1.24	+16
Plantago coronopus	2.28	1.98	−13***	0.76	0.79	+4
Plantago lanceolata	2.17	1.98	−9**	1.10	0.72	−34**
Plantago major[b]						
1	2.46	1.88	−24***	0.95	0.86	−9***
2	2.36	1.81	−23**	0.90	0.83	−8**
Plantago major Athens	1.79	1.77	−1	0.90	0.96	+7
Plantago maritima	1.76	1.66	−6	0.87	0.95	+9
Plantago media	1.29	1.33	+3	1.16	1.25	+8
Poa annua	1.56	1.45	−7*	1.03	0.92	−11*
Poa trivialis	1.34	1.29	−4	0.95	0.89	−6
Rumex acetosa	1.72	1.71	−1	0.63	0.85	+35*
Rumex acetosella	1.71	1.54	−10*	1.06	1.08	+2
Rumex obtusifolius	2.16	1.97	−9**	0.87	0.83	−5
Teucrium scorodomia 0221	1.74	1.56	−10**	0.70	0.70	0
Teucrium scorodonia 0223	1.74	1.57	−10**	0.84	0.70	−17**
Urtica dioica	2.59	2.29	−12***	1.54	1.35	−12***

[a] Plants were raised in duplicate controlled environment chambers ventilated with charcoal/Purafil-filtered air (control; <5 nl l^{-1} O_3) or O_3 (70 nl l^{-1} O_3 for 7 h d^{-1} for 2 weeks). R, mean plant relative growth rate; K, allometric root/shoot coefficient; R%, the % change in R; K%, the % change in K. Asterisks denote probability of difference between control and fumigated plants: *0.05; **0.01; ***0.001. With the exception of pea (*Pisum sativum* L cv. Conquest, supplied by Batchelors foods) and tobacco (*Nicotiana tabacum* L. cv. Bel-W3, supplied by IPO, Wageningen), which were included for comparative purposes, all seed was supplied by the UCPE seed bank, Sheffield University, UK.

[b] Because of limited space in the fumigation chambers, only five to six taxa could be grown and tested at a time. To ensure reproducibility, one species (*Plantago major*) was tested twice—once at the beginning (1) and once at the end (2) of the series of experiments.

Source: Reiling and Davison 1992.

on experimental exposures using open-top chambers and relatively simple species mixtures. To date, the majority of research has focused on herbaceous plant communities, especially seminatural grasslands (Ashmore and Davison 1996, Kärenlampi and Skärby 1996, Fuhrer 1997, Fuhrer et al. 1997, Davison and Barnes 1998). Consequently, there is very little known about the responses of the wide range of vegetation types that are found across Europe and the U.S., although there is no reason to believe that these will necessarily respond the same as some of the simple mixtures that have been studied experimentally (Fuhrer et al. 1997).

Investigations on simple grass/clover mixtures indicate that the effects of O_3 depend on the relative sensitivity of the competing species (Fuhrer 1997) and on management practices (Fuhrer et al. 1997, Davison and Barnes 1998). Because red clover (*Trifolium pratense*) and timothy (*Phleum pratense*) are about equally sensitive to ozone, both are equally affected in competition and the ratio in biomass is unaffected. In contrast, because white clover (*Trifolium repens*) is more sensitive than its usual companion grasses, it tends to decline in competition. The primary effect is on the stolons; if O_3 concentrations decrease then plants may recover, but there can be lasting effects on stolon density (Wilbourn 1991). Figure 10 shows the recent collation of data by Fuhrer (1997) from white clover experiments performed in the U.S. and Europe. This reveals two important points: (1) there is good agreement in dose-response relationships between experimental

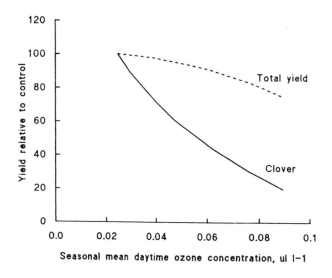

Figure 10 Effects of O_3 (seasonal daytime mean) on relative yield and clover content of managed grassland. (*Source*: redrawn from Fuhrer 1997; based on the regression of four datasets.)

studies, despite differences in varieties, exposure techniques, and climate; and (2) total forage yield tends to be much less affected than that of the sensitive component (in this case, clover).

The few open-top chamber (OTC) studies conducted on communities other than grass–clover mixtures indicate that ambient levels of SO_2 and O_3 may already be high enough to modify species composition in parts of Europe and the U.S. Working on O_3, Ashmore and Ainsworth (1995) found little affect on the total biomass of mixtures of two grasses and two forbs sown as seeds in pots of unamended acid soil, but the forb component declined with increasing exposure. Similar changes in forbs (*Campanula rotundifolia, Leontodon hispidus, Lotus corniculatus, Sanguisorba minor*) were reported by Ashmore et al. (1995) in a simulated calcareous grassland mixture. Bearing in mind the simplified nature of the community and the absence of interacting factors such as water deficit, these data provide a tentative indication that species composition might be affected in parts of central and northern Europe in high-O_3 years. Where pollutant concentrations are variable from year to year, long-term effects would be expected to depend on the magnitude of the changes in years with high pollutant concentrations and the capacity of the community to recover in between. The strongest experimental evidence that ambient levels of O_3 have a significant ecological impact comes from work conducted in the U.S. Duchelle et al. (1983) determined the effects of ambient O_3 on the productivity of natural vegetation in a high meadow in the Shenandoah National Park. Over 3 years, aboveground biomass was increased by filtration and the cumulative dry weights for charcoal-filtered (CF), nonfiltered (NF), and ambient air (AA) treatments were significantly different at 1.38, 1.09, and 0.89 kg m^{-2}, respectively, but data relating to effects on species richness were not collected. Working in the same area, Barbo et al. (1994, 1998) exposed an early successional forest community to AA, CF, NF, and 2 × AA. They found changes in species performance, canopy structure, species richness, and diversity index consistent with the view that oxidants have resulted in a shift in vegetation dominance in some heavily polluted regions. Comparably large effects have not been reported in Europe, but Evans and Ashmore (1992) showed that filtration affected the forb component of a seminatural grassland in a year when O_3 concentrations were relatively high (17% of days with maximum hourly average > 60 ppb). In contrast, there is much evidence that ambient levels of SO_2 are high enough to change community structure in some regions, and several excellent reviews are available on the subject (Kozlowski 1985, Bell et al. 1991, Armentano and Bennett 1992, Taylor and Pitelka 1992).

VI. CONCLUSIONS

In this chapter we have attempted to evaluate some of the features that underlie the observed variation in the resistance of plants to the most common gaseous air pollutants. However, many key questions remain to be answered. Comparative analysis at the molecular level is beginning to reveal evidence of similar defense-related responses to a range of stresses, and the use of mutants and genetic transformation techniques, already of importance in our improved understanding of specific genetic sequences, should eventually allow the dissection of the molecular mechanisms underlying resistance, as well as yielding approaches that may be used to manipulate the resistance of crop plants.

We have chosen to focus on O_3 since this is one of the most potent gases to which plants are regularly exposed in the field, and it is likely to constitute a continuing threat to vegetation for the foreseeable future. Although there is growing literature documenting the effects of this and other gaseous pollutants on wild species, most studies have measured a small number of response variables, commonly under controlled or semicontrolled conditions, with little regard to their ecological significance. This work demonstrates which aspects of growth and reproduction are sensitive to pollutants and the potential range of responses that exist both within and between species, but it is of limited value in assessing how pollutants affect the fitness of plants in communities in the field where a range of additional factors must be taken into consideration. The few studies that have been performed on simple plant communities in the field indicate that there may be "winners" and "losers" at the individual, population, and community levels. These differential responses may result in the most fit genotypes predominating in future generations in polluted regions, and there can be changes in the genetic structure of populations and dominance relationships within individual communities. In many cases, however, it is difficult to ascribe directional selection to individual pollutants, as other factors of the physical and biological environment interact in a variety of ways that may collectively influence the direction and extent of selection. Even in circumstances in which the pollutant may not be the principle factor underlying evolutionary changes in population structure, the existence of resistant populations in polluted regions is of considerable ecological significance. A major challenge will be to devise experiments in the future in which the impacts of gaseous pollutants on natural ecosystems can be assessed; to use the words of Smith (1990): "this will require the marriage—or at least the co-habitation—of specialists in different disciplines (e.g. molecular biologists, biochemists, ecologists & physiologists), at present not the most congenial of bed-fellows."

ACKNOWLEDGMENTS

The authors acknowledge support from the Royal Society, the Spanish Ministry of Science and Culture, and the Spanish Interministerial Commission for Science and Technology, the Natural Environment Research Council, the European Union, the Swales Foundation (administered by Newcastle University), and the

U.K. Overseas Development Agency. The manuscript was written during J.B.'s tenure as a Royal Society Research Fellow.

REFERENCES

Alscher, R.G., J.L. Donahue, and C.L. Cramer. 1997. Reactive oxygen species and antioxidants: relationships in green cells. Physiologia Plantarum 100: 224–233.

Aono, M., A. Kubo, H. Saji, K. Tanaka, and N. Kondo. 1993. Enhanced tolerance to photoxidative stress of transgenic *Nicotiana tabacum* with high chloroplastic glutathione reductase activity. Journal of Plant Cell Physiology 34: 129–135.

Ariens, E.L., A.M. Simonis, and J. Offermerier. 1976. Introduction to General Toxicology. New York: Academic Press.

Armentano, T.V., and J.P. Bennett. 1992. Air pollution effects on the diversity and structure of communities. In: Barker, J., and D.T. Tingey, eds. Air Pollution Effects on Biodiversity. New York: Van Nostrand Reinhold, pp 159–176.

Ashmore, M.R., and A.W. Davison. 1996. Towards a critical level of ozone for natural vegetation. In: Kärenlampi, L., and L. Skärby, eds. Critical Levels for Ozone in Europe: Testing and Finalizing the Concepts. UN-ECE Workshop Report, Kuopio: University of Kuopio, pp 58–71.

Ashmore, M.R., and N. Ainsworth. 1995. The effects of cutting on the species composition of artificial grassland communities. Functional Ecology 9: 708–712.

Ashmore, M.R., R.H. Thwaites, N. Ainsworth, D.A. Cousins, S.A. Power, and A.J. Morton. 1995. Effects of ozone on calcareous grassland communities. Water, Air and Soil Pollution 85: 1527–1532.

Balaguer, L., J.D. Barnes, A. Panicucci, and A.M. Borland. 1995. Production and utilization of assimilates in wheat (*Triticum aestivum* L.) leaves exposed to elevated CO_2 and/or O_3. New Phytologist 129: 557–568.

Balls, G.R., D. Palmer-Brown, and G.E. Sanders. 1996. Investigating microclimatic influences on ozone injury in clover (*Trifolium subterraneum*) using artificial neural networks. New Phytologist 132: 271–280.

Barbo, D.N., A.H. Chappelka, and K.W. Stolte. 1994. Ozone effects on productivity and diversity of an early successional forest community. Proceedings of the Eighth Biennial Southern Silvicultural Research Council, Auburn, Alabama, November) 1–3, pp 291–298.

Barbo, D.N., A.H. Chappelka, G.L. Somers, M.S. Miller-Goodman, and K.W. Stolte. 1998. Diversity of an early successional plant community as influenced by ozone. New Phytologist 138: 653–662.

Barnes, J.D., and A.R. Wellburn. 1998. Air pollutant combinations. In: de Kok, L.J., and I. Stuhlen, eds. Responses of Plant Metabolism to Air Pollution and Global Change. Leiden: Backhuys, pp. 147–164.

Barnes, J.D., M.R. Hull, and A.W. Davison. 1996. Impacts of air pollutants and rising CO_2 in wintertime. In: Yunus, M., M. Iqbal, eds. Plant Response to Air Pollution. London: Wiley, pp 35–166.

Becker, K., M. Saurer, A. Egger, and J. Fuhrer. 1989. Sensitivity of white clover to ambient ozone in Switzerland. New Phytologist 112: 235–243.

Bell, J.N.B. 1984. Air pollution problems in western Europe. In: Koziol, M.J., and F.R. Whatley, eds. Gaseous Air Pollutants and Plant Metabolism. London: Butterworths, pp 3–24.

Bell, J.N.B., M.R. Ashmore, and G.B. Wilson. 1991. Ecological genetics and chemical modification of the atmosphere. In: Taylor, G.E., L.F. Pitelka, and M.T. Clegg, eds. Ecological Genetics and Air Pollution. New York: Springer-Verlag, pp 33–59.

Bender, J., and H-J. Weigel. 1992. Crop responses to mixtures of air pollutants. In: Jäger H.J., M. Unsworth, L. De Temmerman, and P. Mathy, eds. Effects of Air Pollution on Agricultural Crops in Europe. CEC Air Pollution Research Report No. 46. Brussels: CEC, pp 445–454.

Bennett, J.H., and A.C. Hill. 1973. Absorption of gaseous air pollutants by a standardized plant canopy. Journal of Air Pollution Control and Waste Management 23: 203–206.

Bergmann, E., J. Bender, and H. Weigel. 1996. Effects of chronic ozone stress on growth and reproduction capacity of native herbaceous plants. In: Knoflacher, M., R. Schneider, and G. Soja, eds. Exceedance of Critical Loads and Levels. Vienna: Federal Ministry for Environment, Youth and Family, pp 177–185.

Bergmann, E., J. Bender, and H-J. Weigel. 1995. Growth responses and foliar sensitivities of native herbaceous species to ozone exposure. Water, Air and Soil Pollution 85: 1437–1442.

Berrang, P., D.F. Karnosky, and J.P. Bennett. 1991. Natural selection for ozone tolerance in *Populus tremuloides*: An evaluation of nationwide trends. Canadian Journal of Forest Research 21: 1091–1097.

Boubel, R.W., D.L. Fox, D.B. Turner, and A.C. Stern. 1994. Fundamentals of Air Pollution, 3rd ed. London: Academic Press, pp 19–60.

Bressan, R.A., L. LeCurreux, L.G. Wilson, P. Filner, and L.R. Baker. 1981. Inheritance of resistance to sulphur dioxide in cucumber. Horticultural Science 16: 332–333.

Broadbent, P., G.P. Creissen, B. Kular, A.R. Wellburn, and P.M. Mullineaux. 1995. Oxidative stress responses in transgenic tobacco containing altered levels of glutathione reductase activity. Plant Journal 8: 247–255.

Burkhardt, J., and P. Drechsel. 1997. The synergism between SO_2 oxidation and manganese leaching on spruce needles—a chamber experiment. Environmental Pollution 95: 1–11.

Cape, J.N. 1997. Photochemical oxidants—what else is in the atmosphere besides ozone? Phyton 37: 45–58.

Chameides, W.L. 1989. The chemistry of ozone deposition to plant leaves: role of ascorbic acid. Environmental Science and Technology 23: 595–600.

Chappelka, A.H., J. Renfro, G. Somers, B. Nash. 1997. Evaluation of ozone injury on foliage of black cherry (*Prunus serotina*) and tall milkweed (*Asclepias exaltata*) in the Great Smoky Mountains National Park. Environmental Pollution 95: 13–18.

Conklin, P.L., E.H. Williams, and R.L. Last. 1996. Environmental stress sensitivity of an ascorbic acid-deficient *Arabidopsis* mutant. Proceedings of the National Academy of Sciences USA 93: 9970–9974.

Cooley, D.R., and W.J. Manning. 1987. The impact of ozone on assimilate partitioning in plants: a review. Environmental Pollution 47: 95–113.

Cram, W.J. 1990. Uptake and transport of sulphate. In: Rennenberg, H., C. Brunold, L.J. De Kok, and I. Stuhlen, eds. Sulphur Nutrition and Assimilation in Higher Plants. The Hague: SPB Academic Publishing, pp 3–11.

Darrall, N.M. 1989. The effects of air pollutants on physiological processes in plants. Plant, Cell and Environment 12: 1–30.

Davison, A.W., and J.D. Barnes. 1992. Patterns of air pollution: the use of the Critical Loads concept as a basis for abatement strategy. In: Newsom, M.D., ed. Managing the Human Impact on the Natural Environment: Patterns and Processes. New York: Bellhaven, pp 109–129.

Davison, A.W., and J.D. Barnes. 1998. Effects of ozone on wild plants. New Phytologist 139: 135–151.

Davison, A.W., and K. Reiling. 1995. A rapid change in ozone resistance in *Plantago major* after summers with high ozone concentrations. New Phytologist 131: 227–343.

De Kok, L.J. 1990. Sulphur metabolism in plants exposed to atmospheric sulfur. In: Rennenberg, H., C. Brunold, L.J. De Kok, and I. Stuhlen, eds. Sulphur nutrition and assimilation in higher plants. The Hague: SPB Academic Publishing, pp 111–130.

Dietz, K-J. 1997. Functions and responses of the leaf apoplast under stress. In: Progress in Botany, Berlin: Springer-Verlag, pp 221–254.

Duchelle, S.F., J.M. Skelly, T.L. Sharick, B. Chevone, Y.S. Yang, and J.E. Nielessen. 1983. Effects of ozone on the productivity of natural vegetation in a high meadow of the Shenandoah National Park of Virginia. Journal Environmental Management 17: 299–308.

Dunn, D.B. 1959. Some effects of air pollution on *Lupinus* in the Los Angeles area. Ecology 40: 621–625.

Evans, P., and M.R. Ashmore. 1992. The effects of ambient air on a semi-natural grassland community. Agriculture, Ecosystems and Environment 38: 91–97.

Fernandez-Bayon, J.M., J.D. Barnes, J.H. Ollerenshaw, and A.W. Davison. 1992. Physiological effects of ozone on cultivars of watermelon (*Citrullus lanatus*) and muskmelon (*Cucumis melo*) widely grown in Spain. Environmental Pollution 81: 199–206.

Fuhrer, J. 1997. Ozone sensitivity of managed pastures. In: Chereminisoff, P.N., ed. Ecological Advances and Environmental Impact Assessment. Advances in Environmental Control Technology Series. Houston: Gulf Publishing Company, pp 681–706.

Fuhrer, J., L. Skärby, and M.R. Ashmore. 1997. Critical levels for ozone effects on vegetation in Europe. Environmental Pollution 97: 91–106.

Gaastra, P. 1959. Photosynthesis of crop plants. Mededelingen van de Landbouwhogeschool te Wageningen 59: 1–68.

Grime, J.P. 1979. Plant Strategies and Vegetation Processes. London: Wiley.

Grime, J.P., K. Thompson, R. Hunt, J.G. Hodgson, J.H.C. Cornelissen, I.H. Rorison, G.A.F. Hendry, T.W. Ashenden, A.P. Askew, S.R. Band, R.E. Booth, C.C. Bossard, B.D. Campbell, J.E.L. Cooper, A.W. Davison, P.L. Gupta, W. Hall, D.W. Hand, M.A. Hannah, S.H. Hillier, D.J. Hodkinson, A. Jalili, Z. Liu, J.M.L. Mackey, N.

Matthews, M.A. Mowforth, A.M. Neal, R.J. Reader, K. Reiling, W. Ross-Fraser, R.E. Spencer, F. Sutton, D.E. Tasker, P.C. Thorpe, and J. Whitehouse. 1997. Integrated screening validates primary axis of specialisation in plants. Oikos 79: 259–281.

Harberd, D.J. 1960. Observations on population structure and longevity in *Festuca rubra* L. New Phytologist 60: 202–206.

Harper, J.L. 1977. Population Biology of Plants. London: Academic Press.

Heagle, A.S. 1979. Ranking of soybean cultivars for resistance to ozone using different ozone doses and response measures. Environmental Pollution 19: 1–10.

Heath, R.L. 1988. Biochemical mechanisms of pollutant stress. In: Heck, W.W., O.C. Taylor, and D.T. Tingey, eds. Assessment of Crop Loss from Air Pollutants. London: Elsevier, pp 259–286.

Heath, R.L., and G.E. Taylor. 1997. Physiological processes and plant responses to ozone exposure. In: Sandermann, H., A.R. Wellburn, and R.L. Heath, eds. Forest Decline and Ozone: A Comparison of Controlled Chamber and Field Experiments. Berlin: Springer-Verlag, pp 317–368.

Herschbach, C., L.J. De Kok, and H. Rennenberg. 1995. Net uptake of sulfate and its transport to the shoot in spinach plants fumigated with H_2S or SO_2: does atmospheric sulphur affect the ''inter-organ'' regulation of sulfur nutrition? Botanica Acta 108: 41–46.

Hewitt, N., and G. Terry. 1992. Understanding ozone-plant chemistry. Environmental Science and Technology 26: 1890–1891.

Hippeli, S., and E.F. Elstner. 1996. Mechanisms of oxygen activation during plant stress: biochemical effects of air pollutants. Journal of Plant Physiology 148: 249–257.

Hogsett, W.E., D.T. Tingey, and E.H. Lee. 1988. Ozone exposure indices: concepts for development and evaluation of their use. In: Heck, W.W., O.C. Taylor, and D.T. Tingey, eds. Assessment of Crop Loss from Air Pollutants. London: Elsevier, pp 107–138.

Kaiser, G., E. Martinoia, G. Schroppel-Meier, and U. Heber. 1989. Active transport of sulphate into the vacuole of plant cells provides halotolerance and can detoxify SO_2. Journal of Plant Physiology 133: 756–763.

Kangasjärvi, J., J. Talvinen, M. Utriainen, and R. Karjalainen. 1994. Plant defence systems induced by ozone. Plant Cell and Environment 17: 783–794.

Kärenlampi, L., and L. Skärby. 1996. Critical levels for ozone in Europe: testing and finalizing the concepts. UN-ECE Workshop Report, Kuopio: University of Kuopio, pp 1–355.

Kersteins, G., and K.J. Lendzian. 1989. Interactions between ozone and plant cuticles. I. Ozone deposition and permeability. New Phytologist 112: 13–19.

Körner, C. 1994. Leaf diffusive conductance in the major vegetation types of the globe. In: E-D. Schulze, E-D., and M.M. Caldwell, eds. Ecophysiology of Photosynthesis. Berlin: Springer-Verlag, pp 463–490.

Körner, C.H., M. Neumayer, S.P. Menendez-Riedl, and A. Smeets-Schiel. 1989. Functional morphology of mountain plants. Flora 182: 353–383.

Kozlowski, T.T. 1985. SO_2 effects on plant community structure. In: Winner, W.E., H.A. Mooney, and R.A. Goldstein, eds. Sulphur Dioxide and Vegetation. Stanford: Stanford University Press, pp 431–453.

Krupa, S.V. 1996. The role of atmospheric chemistry in the assessment of crop growth and productivity. In: Yunus, M., and M. Iqbal, eds. Plant Response to Air Pollution. London: Wiley, pp 35–73.

Laisk, A., O. Kull, and H. Moldau. 1989. Ozone concentration in leaf intercellular air spaces is close to zero. Journal of Plant Physiology 90: 1163–1167.

Lea, P.J., F.A.M. Wellburn, A.R. Wellburn, G.P. Creissen, and P.M. Mullineaux. 1998. Use of transgenic plants in the assessment of responses to atmospheric pollutants. In: de Kok, L.J., and I. Stuhlen, eds. Responses of Plant Metabolism to Air Pollution and Global Change. Leiden: Backhuys, pp 241–250.

Lendzian, K.J. 1984. Permeability of plant cuticles to gaseous air pollutants. In: Koziol, M.J., and F.R. Whatley, eds. Gaseous Air Pollutants and Plant Metabolism. London: Butterworth, pp 77–81.

Luwe, M., and U. Heber. 1995. Ozone detoxification in the apoplasm and symplasm of spinach, bean and beech leaves at ambient and elevated concentrations of ozone in air. Planta 197: 448–455.

Lyons, T.M., J.H. Ollerenshaw, and J.D. Barnes. 1998a. Impacts of ozones on *Plantago major*: apoplastic and symplastic antioxidant status. New Phytologist, in press.

Lyons, T.M., M. Plöchl, E. Turcsanyi, and J.D. Barnes. 1998b. Extracellular ascorbate: a protective screen against ozone? In: Agrawal, S.B., S. Madhoolika, M. Agrawal, and D.T. Krizek, eds. Environmental Pollution and Plant Responses. New York: CRC Press/Lewis Publ., in press.

Lyons, T.M., and J.D. Barnes. 1998. Influence of plant age on ozone resistance in *Plantago major*. New Phytologist 138: 83–89.

Lyons, T.M., J.D. Barnes, and A.W. Davison. 1997. Relationships between ozone resistance and climate in European populations of *Plantago major*. New Phytologist 136: 503–510.

Macnair, M.R. 1991. Genetics of the resistance of plants to pollutants. In: Taylor, G.E., L.F. Pitelka, and M.T. Clegg, eds. Ecological Genetics and Air Pollution. New York: Springer-Verlag, pp 127–136.

Macnair, M.R. 1993. The genetics of metal tolerance in vascular plants. New Phytologist 124: 541–559.

Madamanchi, N.R., and R.G. Alscher. 1994. Metabolic bases to differences in sensitivity of two pea cultivars to sulphur dioxide. Journal of Plant Physiology 97: 88–93.

Malhotra, S.S., and A.A. Khan. 1984. Biochemical and physiological impact of major pollutants. In: Treshow, M., ed. Air Pollution and Plant Life. London: Wiley, pp 123–146.

Mansfield, T.A., and P.H. Freer-Smith. 1984. The role of stomata in resistance mechanisms. In: Koziol, M.J., and F.R. Whatley, eds. Gaseous Air Pollutants and Plant Metabolism. London: Butterworth, pp 131–146.

Mortensen, L. 1994. Further studies on the effects of ozone concentration on growth of subalpine plant species. Norwegian Journal of Agricultural Science 8: 91–97.

Mudd, J.B. 1996. Biochemical basis for the toxicity of ozone. In: Yunus, M., and M. Iqbal, eds. Plant Response to Air Pollution. London: Wiley, pp 267–283.

Musselman, R.C., D.G. Fox, C.G. Shaw III, and W.H. Moir. 1992. Monitoring atmospheric effects on biological diversity. In: Barker, J., and D.T. Tingey, eds.

Air Pollution Effects on Biodiversity. New York: Van Nostrand Reinhold, pp 52–71.

Nimis, P.L., G. Lazzarin, and D. Gasparo. 1991. Lichens as bioindicators of air pollution by SO_2 in the Veneto Region (NE Italy). Studia Geobotanica 11: 3–76.

Nobel, P.S. 1974. Introduction to Biophysical Plant Physiology, 2nd ed. San Francisco: Freeman.

Nobel, P.S. 1983. Biochemical Plant Physiology and Ecology. New York: Freeman.

Nobel, P.S., and D.B. Walker. 1985. Structure of leaf photosynthetic tissue. In: Barber, J., and N.R. Baker, eds. Photosynthetic Mechanisms and the Environment. New York: Elsevier, pp 501–536.

Ogawa, K., S. Kanematsu, and K. Asada. 1996. Intra- and extra-cellular localization of "cytosolic" CuZn-superoxide dismutase in spinach leaf and hypocotyl. Journal of Plant Cell Physiology 37: 790–799.

Örvar, L.B., J. McPherson, and B.E. Ellis. 1997. Pre-activating wounding response in tobacco prior to high-level ozone exposure prevents necrotic injury. Plant Journal 11: 203–212.

Pell, E.J., C.D. Schlagnhaufer, and R.N. Arteca. 1997. Ozone-induced oxidative stress: mechanisms of action and reaction. Physiologia Plantarum 100: 264–273.

Penkett, S.A. 1988. Indications and causes of ozone increase in the troposphere. In: Rowland, F.S., and I.S.A. Isaksen, eds. The Changing Atmosphere. London: Wiley, pp 91–103.

Pfanz, H., K-J. Dietz, I. Weinerth, and B. Oppmann. 1990. Detoxification of sulphur dioxide by apoplastic peroxidase. In: Brunold, C.H., L.J. De Kok, and I. Stuhlen, eds. Sulphur Nutrition and Assimilation in Higher Plants. The Hague: SPB Academic, pp 229–233.

Pfanz, H. 1987. Uptake and distribution of sulphur dioxide in plant cells and plant organelles. Effects on metabolism. Masters Thesis, Maximillian University, Wurzburg, 136 pp.

Pitcher, L.H., and B.A. Zilinskas. 1996. Overexpression of copper/zinc superoxide dismutase in the cytosol of transgenic tobacco confers partial resistance to ozone-induced foliar necrosis. Journal of Plant Physiology 110: 583–588.

Pitcher, L.H., E. Brennan, A. Hurley, P. Dunsmuir, J.M. Tepperman, and B.A. Zilinskas. 1991. Overproduction of petunia chloroplastic copper/zinc superoxide dismutase does not confer ozone tolerance in transgenic tobacco. Journal of Plant Physiology 97: 452–455.

Pleijel, T.H., and H. Danielsson. 1997. Growth of 27 herbs and grasses in relation to ozone exposure and plant strategy. New Phytologist 135: 361–367.

Plöchl, M., P. Ramge, F-W. Badeck, and G.H. Kohlmaier. 1993. Modeling of deposition and detoxification of ozone in plant leaves: a contribution to subproject BIATEX. In: Borrell, P.M., ed. Proceedings of the EUROTRAC Symposium '92. The Hague: SPB Academic, pp 748–752.

Polle, A. 1997. Defense against photoxidative damage in plants. In: Scandalios, K., ed. Oxidative Stress and the Molecular Biology of Antioxidant Defenses. Cold Spring Harbor: Laboratory Press, pp 623–665.

Polle, A., and H. Rennenberg. 1993. Significance of antioxidants in plant adaptation to environmental stress. In: Fowden, L., T.A. Mansfield, and J. Stoddart, eds.

Plant Adaptation to Environmental Stress. London: Chapman and Hall, pp 263–273.

Preston, K.P. 1985. Effects of sulphur dioxide pollution on a Californian sage scrub community. Environmental Pollution 51: 179–195.

Ramge, P., F-W. Badeck, M. Plöchl, and G.H. Kohlmaier. 1992. Apoplastic antioxidants as decisive elimination factors within the uptake process of nitrogen dioxide into leaf tissues. New Phytologist 125: 771–785.

Ranieri, A., M. Durante, A. Volterrani, G. Lorenzini, and G.F. Soldatini. 1992. Effects of low SO_2 levels on superoxide dismutase and peroxidase isoenzymes in two different wheat cultivars. Biochem Physiol Pflanzen 188: 67–71.

Rao, M.V., G. Paliyath, and D.P. Ormrod. 1996. Ultraviolet-B and ozone-induced biochemical changes in antioxidant enzymes of *Arabidopsis thaliana*. Journal of Plant Physiology 110: 125–136.

Reich, P.B. 1987. Quantifying plant response to ozone: a unifying theory. Tree Physiology 3: 63–91.

Reiling, K., and A.W. Davison. 1992. Spatial variation in ozone resistance of British populations of *Plantago major* L. New Phytologist 122: 699–708.

Reiling, K., and A.W. Davison. 1995. Effects of ozone on stomatal conductance and photosynthesis in populations of *Plantago major* L. New Phytologist 129: 587–594.

Reiling, K., and A.W. Davison. 1992. The response of native, herbaceous species to ozone; growth and fluorescence screening. New Phytologist 120: 29–37.

Rennenberg, H. 1984. The fate of the excess sulphur in higher plants. Annual Review of Plant Physiology 35: 121–153.

Roach, D.A., and R. Wulff. 1987. Maternal effects in plants. Annual Review of Ecological Systems 18: 209–235.

Roose, M.L., A.D. Bradshaw, and T.M. Roberts. 1982. Evolution of resistance to gaseous air pollution. In: Unsworth, M.H., and D.P. Ormrod, eds. Effects of Gaseous Air Pollutants in Agriculture and Horticulture. London: Butterworth, pp 379–409.

Runeckles, V.C. 1992. Uptake of ozone by vegetation. In: Lefohn, A.S., ed. Surface Level Ozone Exposures and their Effects on Vegetation. Chelsea: Lewis, pp. 157–188.

Salter, L., and N. Hewitt. 1992. Ozone-hydrocarbon interactions in plants. Phytochemistry 31: 4045–4050.

Schraudner, M., C. Langebartels, and H. Sandermann. 1997. Changes in the biochemical status of plant cells induced by the environmental pollutant ozone. Physiologia Plantarum 100: 274–280.

Schraudner, M., U. Graf, C. Langebartels, and H. Sandermann. 1994. Ambient ozone can induce plant defence reactions in tobacco. Proccedings of the Royal Society of Edinburgh 102B: 55–61.

Showman, R.E. 1991. A comparison of ozone injury to vegetation during moist and drought years. Journal of Environmental Management Association 41: 63–64.

Slovik, S. 1996. Chronic SO_2- and NO_x- pollution interferes with the K^+ and Mg^{2+} budget of Norway spruce trees. Journal of Plant Physiology 148: 276–286.

Smith, H. 1990. Signal perception, differential expression within multigene families and the molecular basis of phenotypic plasticity. Plant, Cell and Environment 13: 585–594.

Stockwell, W.R., G. Kramm, H-E. Scheel, V.A. Mohnen, and W. Seiler. 1997. In: Ozone formation, destruction, and exposure in Europe and the United States. Sandermann, H., A.R. Wellburn, and R.L. Heath, eds. Forest Decline and Ozone: A Comparison of Controlled Chamber and Field Experiments. Berlin: Springer-Verlag, pp 1–38.

Stuhlen, I., and L.J. De Kok. 1990. Whole plant regulation of sulfur metabolism a theoretical approach and comparison with current ideas on regulation of nitrogen metabolism. In: Rennenberg, H., C. Brunold, L.J. De Kok, and I. Stuhlen, eds. Sulphur Nutrition and Assimilation in Higher Plants. The Hague: SPB Academic Publishing, p 77–91.

Tanaka, K., I. Furusawa, N. Kondo, and K. Tanaka. 1988. SO_2 tolerance of tobacco plants regenerated from paraquat-tolerant callus. Journal of Plant Cell Physiology 29: 743–746.

Taylor, G., and R. Ferris. 1996. Influence of air pollution on root physiology and growth. In: Yunus, M., and M. Iqbal, eds. Plant Response to Air Pollution. London: Wiley, pp 375–394.

Taylor, G.E., and L.F. Pitelka. 1992. Genetic diversity of plant populations and the role of air pollution. In: Barker, J., and D.T. Tingey, eds. Air Pollution Effects on Biodiversity. New York: Van Nostrand Reinhold, pp 111–130.

Taylor, G.E., P.J. Hanson Jr., and D.B. Baldochi. 1988. Pollutant deposition to individual leaves and plant canopies: sites of regulation and relationship to injury. In: Heck, W.W., O.C. Taylor, and D.T. Tingey, eds. Assessment of Crop Loss from Air Pollutants. London: Elsevier, pp 227–257.

Thalmair, M., G. Bauw, S. Thie, T. Dohring, C. Langebartels, and H. Sandermann. 1995. Ozone and ultraviolet-B effects on the defence-related proteins beta-1,3-glucanase and chitinase in tobacco. Journal of Plant Physiology 148: 222–228.

Tingey, D.T., and C.P. Andersen. 1991. The physiological basis of differential plant sensitivity to changes in atmospheric quality. In: Taylor, G.E., L.F. Pitelka, and M.T. Clegg, eds. Ecological Genetics and Air Pollution. Berlin: Springer-Verlag, pp 209–236.

Tingey, D.T., and G.E. Taylor. 1981. Variation in plant response to ozone: a conceptual model of physiological events. In: Unsworth, M., and D.P. Ormrod, eds. Effects of Gaseous Air Pollutants in Agriculture and Horticulture. London: Butterworth, pp 113–138.

Torsethaugen, G., L.H. Pitcher, B.A. Zilinskas, and E.J. Pell. 1997. Overproduction of ascorbate peroxidase in the tobacco chloroplast does not provide protection against ozone. Journal of Plant Physiology 114: 529–537.

Van Camp, W., H. Willekens, C. Bowler, M. Van Montagu, D. Inzé, P. Reupold-Popp, H. Sandermann, and C. Langebartels. 1994. Elevated levels of superoxide dismutase protect transgenic plants against ozone damage. Biotechnology 12: 165–168.

Warwick, K.R., and G. Taylor. 1995. Contrasting effects of tropospheric ozone on five native herbs which coexist in calcareous grassland. Global Change Biology 1: 143–151.

Wellburn, A.R., J.D. Barnes, P.W. Lucas, A.R. McLeod, and T.A. Mansfield. 1997. Controlled O_3 exposures and field observations of O_3 effects in the UK. In: Sandermann, H., A.R. Wellburn, and R.L. Heath, eds. Forest Decline and Ozone: A Comparison of Controlled Chamber and Field Experiments. Berlin: Springer-Verlag, pp 201–248.

Westman, W.E. 1979. Oxidant effects on California coastal sage scrub. Science 205: 1001–1003.

Westman, W.E. 1985. Air pollution injury to coastal sage scrub in the Santa Monica Mountains, Southern California. Water, Air and Soil Pollution 26: 19–41.

Whitfield, C.P., A.W. Davison, and T.W. Ashenden. 1997. Artificial selection and heritability of ozone resistance in two populations of *Plantago major* L. New Phytologist 137: 645–655.

Wilbourn, S. 1991. The effects of ozone on grass-clover mixtures. PhD Thesis, University of Newcastle Upon Tyne, U.K.

Wilbourn, S., A.W. Davison, and J.H. Ollerenshaw. 1995. The use of an unenclosed field fumigation system to determine the effects of ozone on a grass-clover mixture. New Phytologist 129: 23–32.

Winner, W.E., J.S. Coleman, C. Gillespie, H.A. Mooney, and E.J. Pell. 1991. Consequences of evolving resistance to air pollution. In: Taylor, G.E., L.F. Pitelka, and M.T. Clegg, eds. Ecological Genetics and Air Pollution. Berlin: Springer-Verlag, pp 177–202.

Wolfenden, J., and T.A. Mansfield. 1991. Physiological disturbances in plants caused by air pollutants. Proceedings of the Royal Society Edinburgh 97B: 117–138.

Wolfenden, J., P.A. Wookey, P.W. Lucas, and T.A. Mansfield. 1992. Action of pollutants individually and in combination. In: Barker, J., and D.T. Tingey, eds. Air Pollution Effects on Biodiversity. New York: Van Nostrand Reinhold, pp 72–92.

Yalpani, N., A.J. Enyedi, J. León, and I. Raskin. 1994. Ultraviolet light and ozone stimulate accumulation of salicylic acid, pathogenesis-related proteins and virus resistance in tobacco. Planta 193: 372–376.

22

Canopy Photosynthesis Modeling

Wolfram Beyschlag
University of Bielefeld, Bielefeld, Germany

Ronald J. Ryel
Utah State University, Logan, Utah

I. INTRODUCTION

Photosynthesis models are an important development for estimating gas-flux rates of plants. These models have been used to estimate fluxes from the level of the single leaf to entire plant communities. Questions addressed with photosynthesis models involve community responses to climate change and basic ecological concepts concerning resource acquisition, competition for light, and effects of stress.

Primary productivity of a whole plant or canopy is the cumulative carbon gain from all photosynthetically active organs. Because of differences in age, physiology, and exposure to microclimatic conditions, organs are not equally productive, and measurements of individual elements generally will not represent the behavior of the whole plant or canopy. However, by accounting for structural and microclimatic differences between foliage elements, photosynthesis models can provide integrated estimates of photosynthesis and water–vapor exchange.

In this chapter, a class of photosynthesis models is presented that scale up from single-leaf estimates to the whole plants and entire canopies. A general description, perspectives on use, and recent developments are presented, as well as relevant details for model development. Model formulations described here range from simple for homogeneous single-species canopies, to complex for diverse multispecies canopies, and are suitable for addressing a range of ecological questions. This presentation is broken into two parts: Section II is designed to introduce the reader to the general approach used in constructing canopy-photosynthesis models, whereas Section III contains sufficient mathematical detail to aid in model use and development.

II. MODEL OVERVIEW AND PERSPECTIVES

Whole-plant and canopy photosynthesis models contain integrated submodels for single-leaf photosynthesis and the linkage between foliage elements and the physical environment (Figure 1). Whole-plant or canopy photosynthesis is modeled by dividing the canopy structure into subregions of similar foliage characteristics (density, orientation, and physiological properties). Microclimatic conditions affecting foliage elements within these subunits are determined at regular intervals, and photosynthesis rates are then calculated for defined foliage classes within each subunit. Whole-plant or canopy rates are calculated as sums of rates for canopy subunits weighted by foliage density. Canopy photosynthesis models often include three general assumptions: (1) the photosynthetic activity of a plant organ relates to its maximum photosynthetic capacity and reflects age, phenology, acclimation, and physiological condition of the plant; (2) the photosynthetic rate of individual foliage elements depends on interactions with microclimatic conditions (e.g., intercepted radiation, leaf temperature, CO_2 partial pressure within

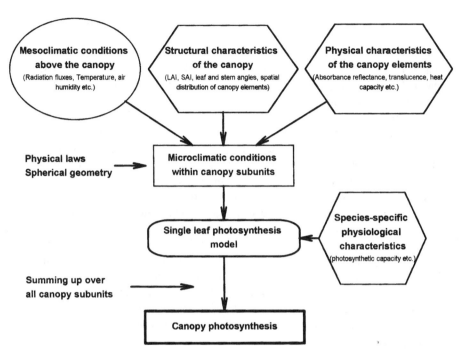

Figure 1 Flow diagram of typical canopy photosynthesis model. Input parameters are in oval and hexagonal boxes, the latter representing parameters provided for each canopy subsection. Variables calculated by the model are shown within rectangular boxes.

the leaf); and (3) microclimatic conditions within plant canopies result from interactions of macroclimate above the canopy, structural and physical properties of canopy elements, and relative position of foliage elements. An additional assumption is often included: (4) the photosynthesis rate of each foliage element is independent of the rates of the other elements.

A. Single-Leaf Photosynthesis and Conductance Models

Single-leaf photosynthesis is the basic unit of canopy photosynthesis models, and carbon assimilation for this basic unit depends on photosynthetic characteristics and surrounding microclimatic conditions. There are two basic types of models: (1) empirical models, where mathematical relationships are formulated between measured variables, and (2) mechanistic models, where mathematical formulations are more closely linked to the physiology of photosynthesis. Because mechanistic models are more closely linked to physiological processes, they are more suitable for addressing how changes in environmental conditions affect photosynthesis (Harley et al. 1986, Tenhunen et al. 1994). Because parameterization is often simpler, however, empirical models are often used when questions concern canopy structure or light competition (Ryel et al. 1990, 1994). Models exist for both C_3 and C_4 metabolic pathways of single-leaf photosynthesis. Development details for mechanistic and empirical models for single-leaf C_3 and C_4 photosynthesis are contained in Section III.

Important companions to models for single-leaf photosynthesis are models for stomatal conductance. Because CO_2 is supplied to the mesophyll through the stomata, and water vapor exits the leaf through these same pores, stomatal diffusive conductance is an essential component in modeling single-leaf photosynthesis. Although many factors affect stomatal conductance (including irradiance, temperature, air humidity, CO_2 partial pressure, plant water status, and endogenic rhythms), controlling mechanisms of stomatal regulation are not fully understood. As a result, truly mechanistic models do not exist for stomatal conductance. The existing empirical models can be divided into uncoupled models where stomatal conductance is calculated as a function of external environmental conditions, and coupled models where the rate of leaf photosynthesis is linked to conductance. Models for stomatal conductance are usually linked with formulations that account for foliage boundary-layer diffusional effects (Gates 1980, Nobel 1983, Schuepp 1993).

B. Whole-Plant/Canopy Models

Whole-plant/canopy photosynthesis is accomplished by simultaneously calculating single-leaf fluxes at multiple locations within the plant canopy. Because microclimatic differences in light intensity, temperature, humidity, and CO_2 concentration result in variable rates of photosynthesis and transpiration throughout the canopy, describing the interaction between canopy structure and its environment is essential to providing realistic predictions. Of the microclimatic variables affecting gas-flux rates, variability in incident light intensity is usually responsible for much of the heterogeneity in rates of net photosynthesis and transpiration within the canopy. This occurs primarily because of the strong light and temperature dependence of photosynthesis and stomatal conductance, but also because of the effect of vapor pressure deficit on transpiration as manifested by radiation-induced increases in leaf temperature. Other factors that add to variability of gas-flux rates within the canopy include photosynthetic characteristics that vary with depth in the canopy (Beyschlag et al. 1990, Niinemets 1997) and turbulence in the canopy, which can significantly alter temperature and humidity gradients.

Foliage intercepts both longwave (> 3000 nm) and shortwave (400–3000 nm) radiation. The portion of the shortwave spectrum where absorption by chlorophyll a and b is high is often referred to as photosynthetically active photon flux (PFD), and may vary from full sunlight at the top of the canopy to less than 1% of full sunlight deep within the canopy (Pearcy and Sims 1994). Shortwave radiation (including PFD) incident on foliage is the sum of three fluxes: direct solar beam, diffuse radiation from the sky, and diffuse radiation reflected and transmitted by other foliage elements (Baldocchi and Collineau 1994). Position of the sun and cloud cover affect fluxes of direct solar beam radiation, and both solar altitude and azimuth are important in relationship to foliage. Solar direct beam flux depends on latitude, date, time of day, and orientation of the foliage elements. Diffuse radiation from the sky emanates from the hemisphere of the sky, and may be relatively constant across the hemisphere with clear or uniformly overcast skies (but also see Spitters et al. 1986). Reflection and transmission of direct beam and sky diffuse radiation within the canopy constitutes leaf diffuse radiation, with flux a function of the proximity and optical properties (transmittance and reflectance) of adjacent foliage.

Absorbed shortwave (I_S) and longwave (I_L) radiation affect the leaf energy balance, and in conjunction with convection, leaf transpiration, and leaf longwave emittance, affect leaf temperature. Longwave radiation emanates to the leaf surface from the sky, soil surface, and surrounding foliage, and fluxes are related to the temperature and emissivity of the radiation surfaces. Convective heat transfer (C_l) between the leaf and the surrounding air varies with air and leaf temperatures and wind speed across the leaf surface. Leaf transpiration rate affects latent heat loss (H_l) from the leaf. Leaf temperature results from a balance of energy gains and losses, which may be written as:

$$I_S a_S + I_L a_L = C_l + H_l + L_l \qquad (1)$$

where a_S and a_L are the fractions of intercepted shortwave and longwave radiation, respectively, and L_l is longwave radiation emittance from the leaf surface. Formulations for convective and latent heat transfer and leaf emittance may be found in reports by Norman (1979) and Gates (1980). Energy balance routines to calculate leaf temperature require iterative calculation procedures when linked to stomatal conductance, and resulting model formulations are generally more complex (Caldwell et al. 1986, Ryel and Beyschlag 1995). However, when leaves are small or narrow in stature, the assumption that leaf and air temperature are identical is often made (Ryel et al. 1990, 1993, Wang and Jarvis 1990).

1. Uniform Monotypic Plant Stands

Single-species plant communities with relatively homogeneous foliage distributions are modeled with the simplest canopy photosynthesis models. Generally, this model structure is limited to grass (e.g., lawns, pastures) or crop canopies, but may also include forest canopies with relatively uniform tree cover. Models for these plant stands divide the canopy into layers of approximately uniform

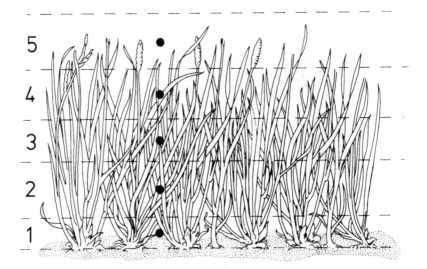

Figure 2 Uniform single-species grass canopy subdivided into five layers. Foliage density and orientation are assumed similar within each layer. Calculations for light interception and photosynthesis are conducted for the points as shown within the layers.

foliage density and orientation (Figure 2), and interception of radiation and photosynthesis is calculated for points located within the center of these layers. These models are considered one dimensional since foliage is assumed to be horizontally uniform, and differences in radiation interception occur only in the vertical dimension. Details for modeling light relations within uniform monotypic plant stands are contained in Section III.

2. Uniform Multispecies Plant Canopies

Canopies with relatively homogeneous mixtures of two or more species (e.g., grasslands and crop/weed mixtures) can be modeled as simple extensions of the model for uniform monotypic plant stands. Relatively uniformly distributed foliage is assumed for all species, but the vertical distribution may vary by species. A simple situation arises when one species overtops another (Tenhunen et al. 1994), but the typical canopy has foliage elements mixed within canopy layers (Figure 3; Beyschlag et al. 1992, Ryel et al. 1990). As with the uniform multispecies plant canopy model, intercepted radiation and photosynthesis is calculated for points located within the center of layers defined by uniformity in foliage

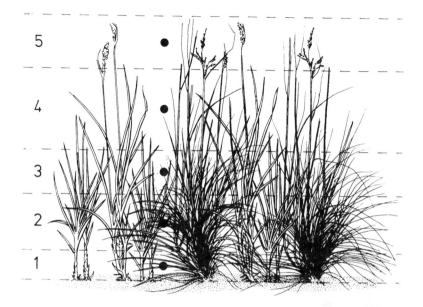

Figure 3 Two-species uniform canopy subdivided into five layers. Foliage density and orientation is assumed similar by species within each layer. Calculations for light interception and photosynthesis are conducted for the points as shown within the layers.

density and orientation for each species. Details for this model type are contained in Section III.

3. Inhomogeneous Canopies

Plant canopies with clumped or discontinuous vegetation cannot be realistically represented with models that only vary foliage density vertically. Within these canopies, foliage interception of light is affected by neighboring vegetation that may not be positioned at uniform distances or compass direction. If gaps occur within the canopy, light may reach plants from the sides, and not simply above the foliage. With this complexity, a three-dimensional light-interception model is necessary to represent these canopies. The common approach to modeling heterogeneous canopies is to fit individual plants or clumps of vegetation with suitable three-dimensional geometric shapes, including cubes (Fukai and Loomis 1976), cones (Oker-Blom and Kellomäki 1982a, 1982b, Kuuluvainen and Pukkala 1987, Oker-Blom et al. 1989), ellipses (Charles-Edwards and Thornley 1973, Mann et al. 1979, Norman and Welles 1983, Wang and Jarvis 1990), and cylinders (Brown and Pandolfo 1969, Ryel et al. 1993). Cescatti (1997) developed a model structure allowing for radial heterogeneity within individual tree crowns. Regions within these shapes are assumed to have relatively similar foliage density and orientation, and light interception and photosynthesis are calculated at points within these subregions (Figure 4).

4. "Big-Leaf" Models

In contrast to models for inhomogeneous canopies, "big-leaf" models simplify rather than increase canopy structural complexity. In these models (Sellers et al. 1992, Amthor, 1994), properties of the whole canopy are reduced to that of a single leaf, and modified equations for single-leaf net photosynthesis and conductance are used for calculating whole-canopy (big-leaf) flux rates. These models have the advantage of having fewer parameters, greatly reduced complexity of development, and substantially less time required for model simulation. Big-leaf

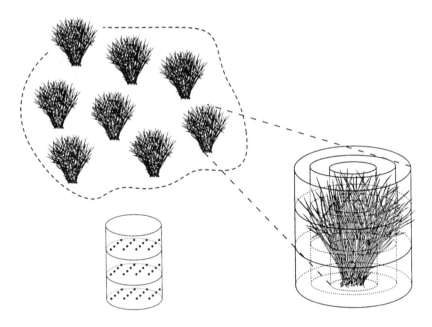

Figure 4 Individual plant represented as a series of concentric cylinders subdivided into layers as used by the model of Ryel et al. (1993). Foliage density and orientation is assumed similar within each layer for an individual plant. Individual plants can be grouped to form a multi-individual canopy with calculations conducted for each member or representative members. Light attenuation for an individual plant would be affected by neighboring plants when such a canopy is defined. The matrix of points indicates locations where light interception and photosynthesis are calculated.

models are often used when flux rates are modeled for several vegetation communities at the landscape scale (Kull and Jarvis 1995).

Although attractive because of their simplicity, big-leaf models have serious limitations. Parameters for big-leaf models cannot be directly measured, and simple arithmetic means of parameters for individual leaves are inadequate because most functions involving light transmission and gas fluxes are nonlinear (Leuning et al. 1995, Jarvis 1995). McNaughton (1994) illustrates this problem by showing that the average canopy conductance preserving whole-canopy transpiration flux differed from the conductance necessary to preserve whole-canopy CO_2 assimilation. Despite these problems, big-leaf models may be suitable if calculated fluxes are calibrated to outputs from more detailed canopy models across an appropriate range of meteorological conditions, or to whole-canopy flux measurements (Fan et al. 1995). In addition, De Pury and Farquhar (1997) developed a modified big-leaf model that separately integrates the sunlit and shaded fractions of the canopy, resulting in calculated fluxes that are similar to that of a multilayer model.

C. Examples

Canopy photosynthesis models have been used to address a wide variety of topics, including basic plant ecophysiology (Beyschlag et al. 1990, Barnes et al. 1990, Ryel et al. 1993), environmental change (Ryel et al. 1990, Reynolds et al. 1992), and crop management (Grace et al. 1987a, 1987b). Model outputs can also provide carbon-gain inputs to allocation and growth models (Johnson and Thornley 1985, Charles-Edwards et al. 1986, Reynolds et al. 1987, Buwalda 1991, Webb 1991). Two examples of model use are briefly discussed in the following sections.

1. Roadside Grasses

The neophytic grass *Puccinellia distans* has recently invaded roadsides in central Europe, which are dominated by the highly competitive grass *Elymus repens*. Beyschlag et al. (1992) showed that *P. distans* could coexist in garden plots with the highly competitive grass *E. repens* when regular mowing reduced the competitive advantage of *E. repens* for light. Ryel et al. (1996) conducted in situ experiments along roadsides where mowed and unmowed portions of the same roadway were compared. A multispecies, homogeneous canopy photosynthesis model was used to estimate reductions in net photosynthesis for *P. distans* due to the presence of *E. repens* within the mowed and unmowed plots. Simulations indicated that little difference in net photosynthesis occurred between mowed and unmowed plots (Figure 5), eliminating mowing as the primary factor contributing to this coexistence. Subsequent experiments indicated that shallow soil depth was the primary factor contributing to coexistence (Beyschlag et al. 1996).

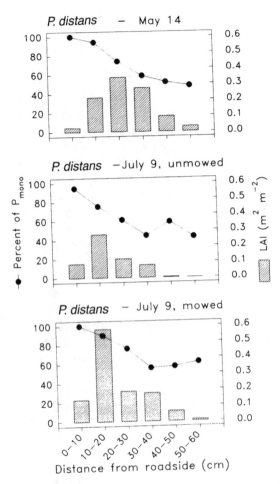

Figure 5 Model calculations of relative photosynthesis rates (lines) for *Puccinellia distans* for roadside canopies at six distances from the road edge at the beginning of the mowing experiment in May (top) and for unmowed (center) and mowed (bottom) plots in July. Relative photosynthesis rates are expressed as a percent of the rate with *Elymus repens* removed from the canopy. Total foliage area of *P. distans* is also shown (bars). (*Source*: Ryel et al. 1996.)

2. Effects of Needle Loss on Spruce Photosynthesis

Needle loss in conifers is a prevalent symptom of forest decline. Beyschlag et al. (1994) assessed the effect of needle loss on whole-plant photosynthesis for forests of young *Picea abies*. A photosynthesis model for inhomogeneous canopies was used to evaluate light interception and net photosynthesis in these canopies. Simulation results indicated that in sparse canopies, needle loss resulted in significant reductions in whole-plant net photosynthesis. However, in more dense canopies, little reduction occurred (Figure 6) as in canopies without needle loss, shaded foliage contributed little to whole-plant net photosynthesis.

D. Future Directions

Canopy photosynthesis models are one method of effectively estimating whole-canopy gas fluxes (Ruimy et al. 1995) and provide a link between measurable

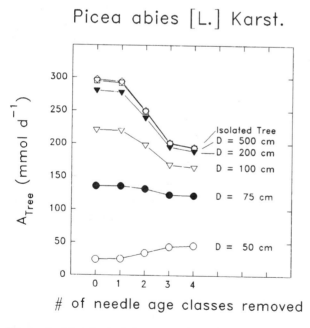

Figure 6 Model calculations of absolute changes in daily overall tree photosynthesis of experimental 7-year-old spruce tree as a function of the number of needle age classes removed and the stand density of a simulated canopy. Simulated stands were created with trees equally spaced (D = distance center to center) as indicated. (*Source*: Beyschlag et al. 1994.)

single-leaf photosynthesis and whole-plant rates. Development of these models is an ongoing endeavor, with both increasing complexity and simplification characteristic of new advances. Research objectives will play an important role in influencing the direction of future model developments.

Model complexity will be increased through the addition of other phenomena and more realistic structural design. Additional phenomena may include stomatal patchiness (Pospíšilová and Šantrůček 1994, Eckstein et al. 1996, Beyschlag and Eckstein 1997), sunflecks (Pearcy and Pfitsch 1994, Küppers et al. 1997), penumbra (Oker-Blom 1985), leaf flutter (Roden and Pearcy 1993), and non–steady state stomatal dynamics (Pearcy et al. 1997). Macroclimate and microclimate linkages with the canopy may also be improved, with better representations of air turbulence and concentration gradients, particularly using large-eddy simulation of airflow within canopies (Dwyer et al. 1997). Plant structural complexity may also increase as indicated by the developments of Cescatti (1997). Modular format of model structure allowing for relatively easy replacement of components with new routines will enhance the increase in model complexity (Reynolds et al. 1987).

Canopy photosynthesis models may also be reduced in complexity, particularly with research directed at landscape level fluxes. Approaches similar to the big-leaf models may be important, but necessitate dealing with problems inherent with model structure and parameterization. Canopy-level models of photosynthesis based on remote sensing data may also see more development in the future (Ustin et al. 1993, Field et al. 1994).

Canopy photosynthesis models will be increasingly linked to growth models (Charles-Edwards 1986, Webb 1991, Hoffmann 1995), as growth models become further developed. Development of growth models will require better knowledge of linkages between carbon gain and structural changes within the canopy, particularly as affected by light climate. Other linkages may occur between canopy photosynthesis models and cellular automaton models (Wolfram 1983) to address plant succession and the formation of stable vegetation patterns (Hogeweg et al. 1985, van Tongeren and Prentice 1986, Czaran and Bartha 1989, Silvertown et al. 1992).

III. MODEL DEVELOPMENT

This section reviews model development for both single-leaf photosynthesis and light attenuation by foliage. Models described in this section have been selected because they have been successfully used to address a broad range of ecological questions, and because they characterize the formulations within this class of models. The level of detail provided is sufficient for the reader to gain a basic understanding of the modeling process and to aid in development of similar models. The section concludes with a brief discussion of model parameterization and validation.

A. C_3 Single-Leaf Photosynthesis

C_3 photosynthesis is the most widespread metabolic pathway for plant carbon assimilation. Because of this, substantially more effort has been focused on developing models for C_3 photosynthesis than for C_4 photosynthesis. Portions of C_3 photosynthesis models are also contained within C_4 photosynthesis models.

1. Mechanistic

The model developed by Farquhar et al. (1980) and Farquhar and von Caemmerer (1982) for C_3 single-leaf photosynthesis is the most commonly used mechanistic model. Their original model assumed limitations in photosynthesis by the activity of the CO_2 binding enzyme RubP-1,5-carboxidismutase (Rubisco) and by the RuBP regeneration capacity of the Calvin cycle as mediated by electron transport. Sharkey (1985) added an inorganic phosphate limitation to photosynthesis that was incorporated into models by Sage (1990) and Harley et al. (1992). This presentation follows the model development of Harley et al. (1992), which is suitable for ecological applications.

With 0.5 mol of CO_2 released in the cycle for photorespiratory carbon oxidation for each mole of O_2 reduced, net photosynthesis (A) may be expressed as:

$$A = V_c - 0.5 \cdot V_o - R_d = V_c \cdot \left(1 - \frac{0.5 \cdot O}{\tau \cdot C_i}\right) - R_d \qquad (2)$$

where V_c and V_o are carboxylation and oxygenation rates at Rubisco, C_i and O are the partial pressures of CO_2 and O_2 in the intercellular air space, R_d is CO_2 evolution rate in the light excluding photorespiration, and τ is the specificity factor for Rubisco (Jordan and Ogren 1984).

As previously discussed, the rate of carboxylation (V_c) is limited by three factors and is set as the minimum of:

$$V_c = \min \{W_c, W_j, W_p\} \qquad (3)$$

where W_c is the carboxylation rate limited by the quantity, activation state, and kinetic properties of Rubisco, W_j is the carboxylation rate limited by the rate of

RuBP regeneration in the Calvin cycle, and W_p is the carboxylation rate limited by available inorganic phosphate. A becomes:

$$A = \left(1 - \frac{0.5 \cdot O}{\tau \cdot C_i}\right) \cdot \min\{W_c, W_j, W_p\} - R_d \tag{4}$$

Michaelis-Menten kinetics are assumed for W_c with competitive inhibition by O_2, and W_c is written:

$$W_c = \frac{V_{c_{max}} \cdot C_i}{C_i + K_c(1 + O/K_o)} \tag{5}$$

where $V_{c_{max}}$ is the maximum carboxylation rate with K_c and K_o the Michaelis-Menten constants for carboxylation and oxygenation, respectively.

W_j is assumed proportional to the electron transport rate (J) and is written:

$$W_j = \frac{J \cdot C_i}{4 \cdot (C_i + O/\tau)} \tag{6}$$

with the additional assumption that sufficient adenosine triphosphate (ATP) and reduced form of nicotinamide adenine dinucleotide phosphate (NADPH) are generated by four electrons to regenerate RubP in the Calvin cycle (Farquhar and von Caemmerer 1982). J is a function of incident PFD and was formulated empirically by Harley et al. (1992) with the equation of Smith (1937) as:

$$J = \frac{\alpha \cdot I}{(1 + \alpha^2 \cdot I^2/J_{max}^2)^{1/2}} \tag{7}$$

where α is quantum efficiency, I is incident PFD, and J_{max} is the light-saturated rate of electron transport.

Carboxylation rate as limited by phosphate (W_p) is expressed as:

$$W_p = 3 \cdot TPU + \frac{V_o}{2} = 3 \cdot TPU + \frac{V_c \cdot 0.5 \cdot O}{C_i \cdot \tau} \tag{8}$$

where TPU is the rate at which phosphate is released during starch and sucrose production by triose-phosphate utilization.

The temperature dependencies of factors K_c, K_o, R_d, and τ were expressed empirically by Harley et al. (1992) as:

$$K_c, K_o, R_d, \text{ and } \tau = \exp[c - \Delta H_a/(R \cdot T_K)] \tag{9}$$

where c is a scaling factor, ΔH_a is the energy of activation, R is the gas constant, and T_K is leaf temperature.

J_{max} and $V_{c_{max}}$ are also temperature dependent, and Harley et al. (1992) used activation and deactivation energies based on Johnson et al. (1942):

$$J_{max} \text{ and } V_{c_{max}} = \frac{\exp[c - \Delta H_a/(R \cdot T_K)]}{1 + \exp[(\Delta S \cdot T_K - \Delta H_d)/(R \cdot T_K)]} \tag{10}$$

where ΔH_d is the deactivation energy and ΔS is an entropy term.

2. Empirical

The model of Thornley and Johnson (1990) has been widely used and gives good fits to measured data. Gross photosynthesis (P) is expressed as:

$$P = \frac{\alpha \cdot I_l \cdot P_m}{\alpha \cdot I_l + P_m} \tag{11}$$

where α is the quantum efficiency, I_l is incident PFD, and P_m is the maximum gross photosynthesis rate at saturating PFD. A factor (θ) for resistance between the CO_2 source and site of photosynthesis, determined from fitting measured data, can be added to obtain:

$$P = \frac{1}{2\theta} \{\alpha \cdot I_l + P_m - [(\alpha \cdot I_l + P_m)^2 - 4\theta \cdot \alpha \cdot I_l \cdot P_m]^{1/2}\} \tag{12}$$

for $0 < \theta < 1$. A report by Johnson et al. (1989) contains a typical application of this model.

Another useful empirical model was developed by Tenhunen et al. (1987), which uses formulations from Smith (1937) for the light and CO_2 dependency of net photosynthesis. Equation 7 is used for the light dependency of A, and a similar formulation is used for the CO_2 dependency. This equation is also used for the light dependency of carboxylation efficiency. Equation 10 is used for the temperature dependence of the maximum capacity of photosynthesis.

B. C$_4$ Single-Leaf Photosynthesis

C_4 photosynthesis involves a CO_2 concentrating mechanism coupled with the C_3 photosynthesis cycle. In the concentrating process, phosphoenol pyruvate (PEP) is carboxylated in the mesophyll cells, transferred to the bundle sheath cells, and decarboxylated before entering the C_3 cycle (Peisker and Henderson 1992). An initial submodel calculates the pool of inorganic carbon in the bundle sheath cells.

1. Mechanistic

A simple C_4 photosynthesis model is that of Collatz et al. (1992), which assumes that the carboxylation catalyzed by PEP carboxylase is linearly related to CO_2 concentration of the internal mesophyll air space. The model does not consider light dependencies of PEP carboxylase activity and other activation processes (Leegood et al. 1989). Chen et al. (1994) proposed a more complex model with the C_4 cycle controlled by PEP carboxylase. The rate of this cycle (V_4) is described by Michaelis-Menten kinetics and related to mesophyll CO_2 concentration by:

$$V_4 = \frac{V_{4m} \cdot C_m}{C_m + k_p} \tag{13}$$

where C_m is mesophyll CO_2 concentration and k_p is a rate constant. The maximum reaction velocity (V_{4m}) is related to incident PFD (I_p) by:

$$V_{4m} = \frac{\alpha_p \cdot I_p}{(1 + \alpha_p^2 \cdot I_p^2/V_{pm}^2)^{1/2}} \tag{14}$$

with α_p a fitted parameter and V_{pm} the potential maximum activity of PEP carboxylase. The C_3 and C_4 cycles are linked as:

$$V_4 = V_b + A_n \tag{15}$$

where V_b is the diffusion flux of CO_2 between the bundle sheath and the mesophyll, and A_n (net photosynthesis) is the net CO_2 exchange rate between the atmosphere and the mesophyll intercellular air space. A_n is calculated using equation 2, with C_i replaced with the CO_2 concentration in the bundle sheath cells. Equations 13 and 15 are solved iteratively to obtain A_n by balancing the CO_2 and O_2 concentrations in the bundle sheath cells (see Chen et al. 1994 for details).

2. Empirical

Two approaches to empirical C_4 photosynthesis model are discussed here. Dougherty et al. (1994) used the minimum of photosynthetic capacities limited by light (A_1) and by intercellular CO_2 (A_2). A_1 is expressed with a nonrectangular hyperbola as:

$$A_1 = \frac{A_m + \alpha I^2 \sqrt{A_m^2 - 2A_m \alpha I(2\beta - 1) + \alpha^2 I^2}}{2\beta} \tag{16}$$

where α is quantum efficiency, β is an empirical shape parameter, I is incident PFD, and A_m is maximum photosynthetic capacity. A_2 is expressed as:

$$A_2 = A_m \frac{c_i}{c_i + 1/E_c} \tag{17}$$

where E_c is an empirical index of leaf CO_2 efficiency and c_i is intracellular CO_2. The temperature dependence of A_m uses Equation 10.

   Thornley and Johnson (1990) described two empirical formulations of C_4 photosynthesis. In one formulation, energy necessary for pumping CO_2 into the bundle sheath is assumed independent of that available to the bundle sheath for

C_3 photosynthesis and photorespiration. In the second model, a common supply of energy to both mesophyll and bundle sheath is assumed.

C. Stomatal Conductance

Models for stomatal diffusive conductance are either coupled or uncoupled with leaf photosynthesis rates, and are coupled with various environmental factors. Presently, all models are empirical in design. Coupled stomatal models are recommended for addressing questions more physiological in nature.

1. Coupled Models

A simple but effective coupled model for stomatal conductance was developed by Ball et al. (1987). Stomatal conductance (g_s) is related linearly to net photosynthesis (A) and relative humidity (h_s), and expressed as:

$$g_s = k \cdot A \cdot (h_s/c_s) \tag{18}$$

where c_s is the mole fraction of CO_2 at the leaf surface. Although g_s does not respond directly to net photosynthesis, relative humidity, or CO_2 at the leaf surface, this model often corresponds well to measured data.

This model has been modified by Leuning (1995) to be consistent with the findings of Mott and Parkhurst (1991), that stomata respond to the rate of transpiration. Interactions between gas molecules leaving and entering stomata (Leuning 1983) are also considered, and g_s is expressed as:

$$g_s = g_0 + a_1 \cdot A/[(c_s - \Gamma) \cdot (1 + D_s/D_0)] \tag{19}$$

where α_1 and D_0 are the empirical coefficients, Γ is the CO_2 compensation point, and g_0 is stomatal conductance when A approaches zero; D_s and c_s are humidity deficit and CO_2 concentration at the leaf surface, respectively.

2. Uncoupled Models

In uncoupled models of stomatal conductance, factors considered to influence g_s include incident PFD, leaf temperature, water vapor mole fraction difference, air humidity, and leaf water potential (Jarvis 1976, Whitehead et al. 1981, Caldwell et al. 1986, Jones and Higgs 1989, Lloyd 1991, Beyschlag et al. 1990, Ryel et al. 1993). Regression equations often relate g_s to one or more of these environmental factors. Uncoupled conductance models can be easier to parameterize than coupled models, and often give good correspondence to measured data.

D. Whole-Plant and Canopy Models

Models for light attenuation through canopies and interception by foliage are linked with single-leaf photosynthesis models to form the basis of whole-plant and canopy photosynthesis models. Although other environmental factors affecting photosynthesis throughout the canopy can be included in model developments, modeling of canopy light relations is the most important aspect of the canopy photosynthesis models. In this section, we illustrate the approach commonly used to model light attenuation through plant canopies. Models for attenuation and interception of direct beam, sky diffuse and leaf reflected, and transmitted diffuse within canopies with varying complexity are presented, with incident PFD calculated for both sunlit and shaded leaves.

1. Uniform Monotypic Plant Canopies

Models for single-species plant communities with relatively homogeneous distribution of foliage are constructed by dividing the canopy into layers of relatively homogeneous foliage density and orientation, and interception of radiation is calculated for points at the center of these layers (Figure 2). A random distribution of foliage elements is typically assumed in these models (but see Caldwell et al. 1986). The thickness of layers is determined by the foliage density distribution in the canopy, and ideally, layers have a leaf-area index less than 0.5 m^2 foliage area per m^2 ground area to facilitate attenuation of leaf-diffuse radiation (Norman 1979). Foliage area of leaves and stems (or branches) are often separated as they typically have different optical properties, photosynthetic rates, and orientations. Leaf inclination, which is the angle of the major axis of the foliage element from horizontal, and azimuth angle, which is the directional alignment, comprise the components of foliage orientation. Leaf inclination is most simply defined as a constant for a layer, but can have a defined distribution (Norman 1979, Campbell 1986). Azimuth orientation of foliage elements may be considered random or nonrandom in distribution (Lemeur 1973, Caldwell et al. 1986). The model development presented here considers constant inclination and random azimuth orientation of foliage in each layer, which is applicable to most uniform canopies.

With randomly distributed foliage within a layer, the penetration of direct-beam PFD declines exponentially with passage through increasing amounts of foliage. The relative area or fraction of sunlit foliage (P_i) for the sample point within in layer i (bottom layer $= 1$, top layer $= n$) is:

$$P_i = \exp\left(\sum_{m=i}^{n} (L_m Kl_m + S_m Ks_m)l_m \right) \tag{20}$$

where L_i and S_i are leaf and stem densities (m^2 m^{-3}) in layer i, respectively, and l_i is the path length (m) of PFD through layer i. Kl_i and Ks_i are the light extinction coefficients for leaves and stems, respectively, and for fixed leaf inclination (α_i) and random azimuth can be calculated (Duncan et al. 1967) for inclination of sun above horizon (β) as:

$$K_i = \cos\alpha_i \sin\beta \tag{21}$$

for α_i less than β, and otherwise as:

$$K_i = \frac{2}{\pi}\sin\alpha_i\cos\beta\sin\theta_i + \left(1 - \frac{\theta_i}{90}\right)\cos\alpha_i\sin\beta \tag{22}$$

where θ_i is the angle (0–90°) that satisfies $\cos\theta_i = \cot\alpha_i\tan\beta$.

The flux of direct-beam PFD does not decline with attenuation through foliage (Figure 7), but the flux incident on foliage is a function of the orientation of the leaf surface relative to the sun. Incident flux of PFD can hit either the

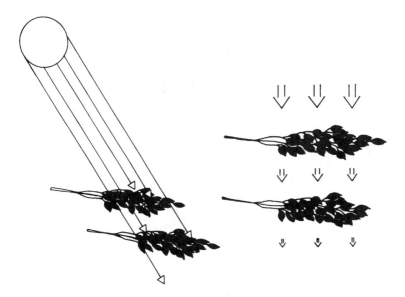

Figure 7 Attenuation of direct-beam (left) and sky-diffuse (right) radiation through foliage. The fraction of foliage illuminated by direct-beam sunlight declines with interception by foliage, but the radiation flux does not change in the sunlit portion. In contrast, radiation flux for sky diffuse (and for scattered diffuse) declines with passage through foliage, but all foliage receives similar flux.

upper or lower surface of the leaf surface, and the flux for the upper leaf surface can be expressed (Burt and Luther 1979) as:

$$Qu_i = B(\cos\alpha_i \sin\beta - \sin\alpha_i\cos\beta\cos\delta_i) \tag{23}$$

and for the lower surface as:

$$Ql_i = B[\cos(180 - \alpha_i)\sin\beta - \sin(180 - \alpha_i)\cos\beta\cos\delta_i] \tag{24}$$

where B is the PFD flux on a surface normal to the solar beam and δ is the azimuth angle of the major axis of the leaf surface relative to the sun.

Attenuation of sky-diffuse PFD through the canopy is considered from the midpoint of concentric bands of equal width or area (skybands) that divide the hemisphere of sky above the canopy. Unlike direct-beam PFD flux, the flux of sky-diffuse PFD declines with passage through increasing amounts of foliage (Figure 7), and can be expressed as:

$$D_{i,w} = D_{sky}A_w\exp\left(\sum_{m=i}^{n} (L_mKl_m + S_mKs_m)l_m\right) \tag{25}$$

where w is the skyband (e.g., $w = 1, 2, \ldots, 9$ for inclination bands $= 0-10°$, $10-20°, \ldots, 80-90°$), D_{sky} is the sky-diffuse PFD flux on a horizontal surface, and A_w (view factor, see Duncan et al. 1967 for formulations) is the fractional portion of the sky hemisphere within band w.

Diffuse radiation reflected and transmitted by foliage is difficult to accurately portray in canopy models. While leaf optical properties, foliage density, and characteristics of adjacent foliage contribute to scattering, the process of scattering is very complex. Simplistic approaches have been proposed, with one approach assuming that radiation striking the upper foliage surface reflects upward or transmits downward, with the opposite occurring for radiation striking the lower surface. Using this approach (Norman 1979, Ryel et al. 1990), downward PFD flux from the midpoint of layer i would be:

$$Td_i = t_iL_i (Qu_iP_i + Du_i) + r_iL_i(Ql_iP_i + Dl_i) \\ + Tu_{i-1}r_iL_i + Td_{i+1}L_i + Td_{i-1}(1 - L_i - S_i) \tag{26}$$

while the upward flux would be:

$$Tu_i = r_iL_i (Qu_iP_i + Du_i) + t_iL_i(Ql_iP_i + Dl_i) \\ + Td_{i+1}r_iL_i + Tu_{i-1}L_i + Tu_{i-1}(1 - L_i - S_i) \tag{27}$$

where r_i and t_i are leaf reflectance and transmittance for PFD, respectively.

Sunlit leaves have incident PFD fluxes from direct beam, sky diffuse, and reflected and transmitted diffuse. The total PFD incident on both sides of a sunlit leaf in layer i is:

$$l_i = Qu_i + Ql_i + Du_i + Dl_i + Tu_{i-1} + Td_{i+1} \tag{28}$$

Incident fluxes for shaded leaves are calculated similarly, but without direct-beam components Qu_i and Ql_i.

2. Uniform Multispecies Plant Canopies

The model structure for uniform multispecies canopies is a generalization of the single-species models. Equations describing interception of PFD can be extended from those for monotypic plant stands. The sunlit fraction of foliage in layer i for species x is simply:

$$P_{x,i} = \exp \left(\sum_{m=i}^{n} \left(\sum_{y=x}^{N} (L_{y,m} Kl_{y,m} + S_{y,m} Ks_{y,m}) \right) l_m \right) \tag{29}$$

where N is the number of species in the canopy, and corresponds to equation 20. Equations 25–27 are similarly extended to apply to multiple species (Ryel et al. 1990).

3. Inhomogeneous Canopies

The cylinder model of Ryel et al. (1993) is used as an example of models for complex canopy structure as it contains the spectrum of complexity found in these models.

Individual plants within the canopy are fitted to concentric cylinders and horizontal layers of relatively uniform foliage density and orientation (Figure 4), a process analogous to the layer divisions in uniform canopies. A three-dimensional array of points (Figure 4) is used to "sample" the plant canopy for calculation of light interception and gas exchange. Consistent calculations of flux rates throughout the course of a day may require 1000 or more points within an individual plant crown (Ryel et al. 1993, Falge et al. 1997). Values for points within the canopy are appropriately weighted, averaged, and summed to generate whole-plant flux rates. Simulations may be conducted for all plants within the canopy (Grace et al. 1987b), for individuals representing plants of similar structure and stature (Falge et al. 1997), or for one individual when all plants are assumed to be similar in structure and relatively uniformly distributed (Beyschlag et al. 1994, Ryel et al. 1994). Whole-canopy rates can be calculated as the sum of individual plants and expressed per ground area if desired for comparisons between vegetation communities or plots.

Many of the model equations are analogous to the homogeneous canopy model and are simply generalizations of those formulations. The fraction of sunlit foliage (analogous to Eq. 20) is calculated for each sample point k in plant x as:

$$P_{x,k} = \exp\left(\sum_{y=1}^{N_P}\left(\sum_{i=1}^{n_{y,l}}\sum_{m=1}^{n_{y,c}}(L_{y,i,m}Kl_{y,i,m} + S_{y,i,m}Ks_{y,i,m})l_{y,i,m}\right)\right) \tag{30}$$

where N_P is the total number of plants in the canopy, and $n_{y,l}$ and $n_{y,c}$ are the number of layers and subcylinders in plant individual y, respectively. The flux of sky-diffuse PFD is calculated for both sky bands and azimuth directions to account for differential placement of neighboring plants. Sky-diffuse flux (analogous to Eq. 25) is calculated as:

$$D_{x,p,w,a} = D_{sky}A_{w,a}$$
$$\times \exp\left(\sum_{y=1}^{N_P}\left(\sum_{i=1}^{n_l}\sum_{m=1}^{n_c}(L_{y,i,m}Kl_{y,i,m} + S_{y,i,m}Ks_{y,i,m})l_{y,i,m}\right)\right) \tag{31}$$

where $A_{w,a}$ is the view factor including azimuth orientation (a) of skyband subsections.

As with the uniform canopy model, reflected and transmitted diffuse within diverse canopies is difficult to model. One approach is to calculate average reflected and transmitted diffuse for horizontal layers of points within the canopy using formulations similar to Eqs. 26 and 27 (Norman and Welles 1983, Ryel et al. 1993). Cescatti (1997) used a similar approach, but used a weighted average for the flux at each point within a layer.

E. Parameterization

A lengthy treatise of data collection methods is too voluminous for this chapter, but an overview is provided. The reader is referred to the cited literature for further details. Estimating model parameters is often challenging and time consuming because of the difficulty involved in measuring flux rates and structural parameters. Since parameter estimates can affect model outputs and study conclusions, parameter estimation must be performed carefully. Assessing model sensitivity to parameter estimates (Forrester 1961, Steinhorst et al. 1978) should be conducted to evaluate the robustness of model outputs.

1. Model Parameters for Single-Leaf Photosynthesis

Parameters for single-leaf photosynthesis and stomatal models are often derived from gas exchange measurements (Harley et al. 1992, Falge et al. 1996). For the model of Farquhar et al. (1980), the initial slope of the A versus c_i relationship characterizes the activity status of Rubisco, and A at saturating light and CO_2

relates to the maximum RuBP regeneration rate of the Calvin cycle (von Caemmerer and Farquhar 1981). The initial slope of the relationship A versus incident PFD at saturating CO_2 is an estimate of the maximum quantum-use efficiency of the light reaction of photosynthesis, and can also be measured with chlorophyll fluorescence (Schreiber et al. 1994). Biochemical parameters are often obtained from the literature, but some may vary considerably within or among species (Evans and Seemann 1984, Keys 1986).

2. Canopy Structural Parameters

Both destructive and nondestructive methods are used to estimate density and distribution of foliage. Destructive methods often involve harvesting foliage from portions of the plant canopy (Caldwell et al. 1986, Beyschlag et al. 1994), with foliage area estimated with analytical devices such as the leaf-area meter. Within

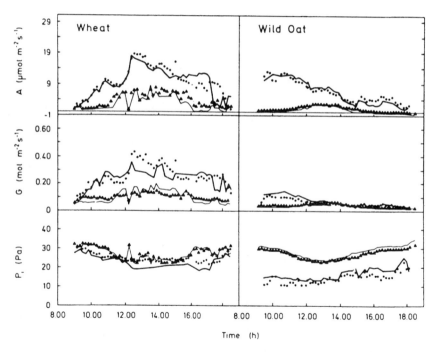

Figure 8 Model validation conducted for single-leaf photosynthesis model of Tenhunen et al. (1987) for wheat (*Triticum aestirum*) and wild oat (*Avena fatua*). Model output for the course of a single day was compared with data measured by gas exchange for net photosynthesis (A), stomatal conductance (G), and intercellular CO_2 (P$_i$). (*Source*: Beyschlag et al. 1990.)

canopy light measurements are the most commonly used nondestructive methods for estimating canopy structure (Norman and Campbell 1989), and are often combined with inverted light extinction equations to estimate foliage area in many types of canopies (Lang et al. 1985, Perry et al. 1988, Walker et al. 1988). This estimation procedure has been automated with the LAI-2000 plant canopy analyzer (LI-COR, Lincoln, NE), which efficiently calculates foliage area and average leaf inclination despite some limitations (Gower and Norman 1991, Deblonde et al. 1994, Stenberg et al. 1994). High-contrast fisheye photography has also been used to estimate foliage area (Anderson 1971, Bonhomme and Chartier 1972). An additional nondestructive method, the linear probe or inclined-point-quadrat method, uses the frequency of contacts of foliage obtained from the repeated movement of a rod oriented at a fixed angle through the canopy (Warren Wilson 1960, Caldwell et al. 1983). Foliage orientation can be measured in situ

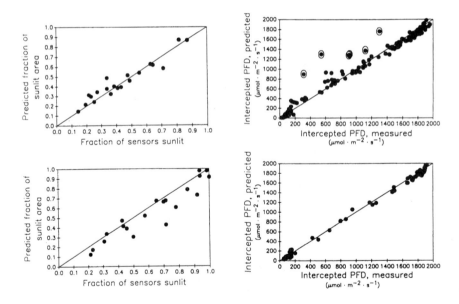

Figure 9   Model validation for light flux in uniform canopy as attenuated through mixtures of wheat and wild oat (upper left) and through monocultures of wheat or wild oat (lower left). Measured data were collected with photodiodes mounted on a long stick inserted into the canopy at defined canopy depths. Fluxes less than 400 μmol m² s⁻¹ are for shaded leaves and sensors. The fraction of foliage sunlit was also compared between model and measured data for mixtures (upper right) and monocultures (lower right). Measured flux data was bimodal in nature and allowed for estimating the portion of fully sunlit sensors. Data points from measurements made deep in the canopy where individual photodiodes may not have been fully sunlit are circled. (*Source*: Ryel et al. 1990.)

(Caldwell et al. 1986) with the LAI-2000 analyzer or by linear probe (Warren Wilson 1960, 1963). Leaf reflectance and transmittance of PFD, total shortwave and longwave radiation are often measured with a spectroradiometer (Brunner and Eller 1977) or obtained from suitable literature values (Sinclair and Thomas 1970, Ross 1975).

F. Model Validation

Model validation is necessary to assess whether model simulations mimic the dynamics of the modeled system. Model validation is conducted by comparing model output with independently measured variables (Forrester 1961, Innis 1974) to evaluate model performance. Comparisons are often made graphically, but statistical evaluation methods have been proposed (Kitanidis and Bras 1980).

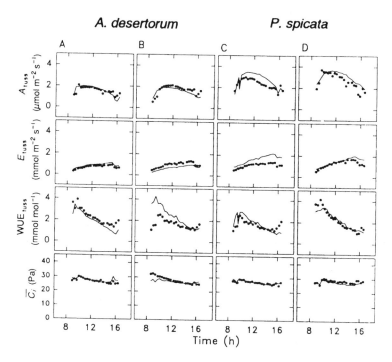

Figure 10 Validation for cylinder model of Ryel et al. (1994) conducted for whole-plant net photosynthesis (A), transpiration (E), water-use efficiency (WUE), and average intercellular CO_2 concentration (C_i) for tussocks of crested wheat grass (*Agropyron desertorum*) and blue-bunch wheat grass (*Pseudoroegneria spicata*). Measured data were from whole-plant gas-exchange cuvettes. (*Source*: Ryel et al. 1994.)

Simulations extrapolating beyond the range of validation run the risk of producing faulty results, but are often performed to investigate system dynamics in response to hypothetical conditions. However, validations are conducted under a range of conditions to increase the confidence of extrapolations using the model.

Diurnal course measurements of single-leaf net photosynthesis, transpiration, intercellular CO_2, and stomatal conductance for H_2O are usually compared with simulated results for photosynthesis and conductance models (Figure 8; Tenhunen et al. 1987, Ryel et al. 1993). Validation of light interception models are conducted by comparing model outputs to light sensor measurements taken within the canopy (Figure 9; Norman and Welles 1983, Beyschlag et al. 1994). Model predictions of whole-plant or canopy gas fluxes can be compared with several types of field measurements for model validation, including whole-plant gas exchange measured within large gas-exchange cuvettes (Figure 10), sap-flow measurements (Cermak et al. 1973, Ishida et al. 1991) of a suitable sample of plants within a canopy (Dye and Olbrich 1993), and eddy-correlation measurements (Baldocchi et al. 1986, Dugas et al. 1991, Verma 1990) of gases within and above the canopy (Baldocchi and Harley 1995, Aber et al. 1996).

ACKNOWLEDGMENTS

Part of this work was funded by the Deutsche Forschungsgemeinschaft, Bonn (SFB 251, University of Würzburg). We would also like to thank Mr. K. Weigel, Bielefeld, for the artwork in Figures 2–4 and 7.

REFERENCES

Aber, D.A., P.B. Reich, and M.L. Goulden. 1996. Extrapolating leaf CO_2 exchange to the canopy: a generalized model of forest photosynthesis compared with measurements by eddy correlation. Oecologia 106: 257–267.

Amthor, J.S. 1994. Scaling CO_2-photosynthesis relationships from the leaf to the canopy. Photosynthesis Research 39: 321–350.

Anderson, M.C. 1971. Radiation and crop structure. In: Sestak, Z., J. Catsky, and P.G. Jarvis, eds. Plant Photosynthetic Production. Manual of Methods. The Hague, the Netherlands: W. Junk, Publishers, pp 412–466.

Baldocchi, D.D., and P.C. Harley. 1995. Scaling carbon dioxide and water vapour exchange form leaf to canopy in a deciduous forest. II. Model testing and application. Plant, Cell and Environment 18: 1157–1173.

Baldocchi, D.D., S.B. Verma, D.R. Matt, and D.E. Anderson. 1986. Eddy-correlation measurements of carbon dioxide efflux from the floor of a deciduous forest. Journal of Applied Ecology 23: 967–976.

Baldocchi, D.D., and S. Collineau. 1994. The physical nature of solar radiation in heterogeneous canopies: spatial and temporal attributes. In: Caldwell, M.M., and R.W. Pearcy, eds. Exploitation of Environmental Heterogeneity by Plants. San Diego, CA: Academic Press, pp 21–71.

Ball, J.T., I.E. Woodrow, and J.A. Berry. 1987. A model predicting stomatal conductance and its contribution to the control of photosynthesis under different environmental conditions. In: Biggens, J., ed. Progress in Photosynthesis Research, vol. IV. Dordrecht: Martinus Nijhoff, pp 221–224.

Barnes, P.B., W. Beyschlag, R.J. Ryel, S.D. Flint, and M.M. Caldwell. 1990. Plant competition for light analyzed with a multispecies canopy model. III. Influence of canopy structure in mixtures and monocultures of wheat and wild oat. Oecologia. 82: 560–566.

Beyschlag, W., and J. Eckstein. 1998. Stomatal patchiness. In: Behnke, H.D., U. Lüttge, K. Esser, J.W. Kadereits, and M. Runge, eds. Progress in Botany, vol. 59. Berlin, Heidelberg, New York: Springer-Verlag, pp 283–298.

Beyschlag, W., P.W. Barnes, R.J. Ryel, M.M. Caldwell, and S.D. Flint. 1990. Plant competition for light analyzed with a multispecies canopy model. II. Influence of photosynthetic characteristics on mixtures of wheat and wild oat. Oecologia 82: 374–380.

Beyschlag, W., R.J. Ryel, and M.M. Caldwell. 1994. Photosynthesis of vascular plants: assessing canopy photosynthesis by means of simulation models. In: Schulze, E.-D., and M.M. Caldwell, eds. Ecophysiology of Photosynthesis. Ecological Studies Vol. 100. Berlin, Heidelberg, New York: Springer-Verlag, pp 409–430.

Beyschlag, W., R.J. Ryel, and C. Dietsch. 1994. Shedding of older needle age classes does not necessarily reduce photosynthetic primary production of Norway spruce. Analysis with a three-dimensional canopy photosynthesis model. Trees Structure and Function 9: 51–59.

Beyschlag, W., R.J. Ryel, and I. Ullmann. 1992. Experimental and modeling studies of competition for light in roadside grasses. Botanica Acta 105: 285–291.

Beyschlag, W., R.J. Ryel, I. Ullmann, and J. Eckstein. 1996. Experimental studies on the competitive balance between two central European roadside grasses with different growth forms. 2. Controlled experiments on the influence of soil depth, salinity and allelopathy. Botanica Acta 109: 449–455.

Bonhomme, R., and P. Chartier. 1972. The interpretation and automatic measurement of hemispherical photographs to obtain sunlit foliage area and gap frequency. Israel Journal of Agricultural Research 22: 53–61.

Brown, P.S., and J.P. Pandolfo. 1969. An equivalent obstacle model for the computation of radiative flux in obstructed layers. Agricultural Meteorology 6: 407–421.

Brunner, U., and B.M. Eller. 1977. Spectral properties of juvenile and adult leaves of *Piper betle* and their ecological significance. Physiologia Plantarum 41: 22–24.

Burt, J.E., and F.M. Luther. 1979. Effect of receiver orientation on erythema dose. Photochemistry and Photobiology 29: 85–91.

Buwalda, J.G. 1991. A mathematical model of carbon acquisition and utilization by kiwifruit vines. Ecological Modelling 57: 43–64.

Caldwell, M.M., G.W. Harris, and R.S. Dzurec. 1983. A fiber optic point quadrat system for improved accuracy in vegetation sampling. Oecologia 59: 417–418.

Caldwell, M.M., H.P. Meister, J.D. Tenhunen, and O.L. Lange. 1986. Canopy structure, light microclimate and leaf gas exchange of *Quercus coccifera* L. in a Portuguese macchia: measurements in different canopy layers and simulations with a canopy model. Trees Structure and Function 1: 25–41.

Campbell, G. 1986. Extinction coefficients for radiation in plant canopies calculated using an ellipsoidal inclination angle distribution. Agricultural and Forest Meteorology 36: 317–321.

Cermak, J., M. Deml, and M. Penka. 1973. A new method of sap flow rate determination in trees. Biologia Plantarum 15: 171–178.

Cescatti, A. 1997. Modelling the radiative transfer in discontinuous canopies of asymmetric crowns. I. Model structure and algorithms. Ecological Modelling 101: 263–274.

Charles-Edwards, D.A., D. Doley, and G.M. Rimmington. 1986. Modelling Plant Growth and Development. Sydney: Academic Press.

Charles-Edwards, D.A., and J.H.M. Thornley. 1973. Light interception by an isolated plant. A simple model. Annals of Botany 37: 919–928.

Chen, D.-X., M.B. Coughenour, A.K. Knapp, and C.E. Owensby. 1994. Mathematical simulation of C_4 grass photosynthesis in ambient and elevated CO_2. Ecological Modelling 73: 63–80.

Collatz, G.J., M. Ribas-Carbo, and J.A. Berry. 1992. Coupled photosynthesis-stomatal conductance model for leaves of C_4 plants. Australian Journal of Plant Physiology 19: 519–538.

Czaran, T., and S. Bartha 1989. The effect of spatial pattern of community dynamics: a comparison of simulated and field data. Vegetatio 83: 229–239.

Deblonde, G., M. Penner, and A. Royer. 1994. Measuring leaf area index with the LI-COR LAI-2000 in pine stands. Ecology 75: 1507–1511.

De Pury, D.G.G., and G.D. Farquhar. 1997. Simple scaling of photosynthesis from leaves to canopies without the errors of big-leaf models. Plant, Cell and Environment 20: 537–557.

Dougherty, R.L., J.A. Bradford, P.I. Coyne, and P.L. Sims. 1994. Applying an empirical model of stomatal conductance to three C-4 grasses. Agricultural and Forest Meteorology 67: 269–290.

Dugas, W.A., L.J. Fritschen, L.W. Gay, A.A. Held, A.D. Matthias, D.C. Reicosky, P. Steduto, and J.L. Steiner. 1991. Bowen ratio, eddy correlation, and portable chamber measurements of sensible and latent heat flux over irrigated spring wheat. Agricultural and Forest Meteorology 56: 1–20.

Duncan, W.C., R.S. Loomis, W.A. Williams, and R. Hanau. 1967. A model for simulating photosynthesis in plant communities. Hilgardia 38: 181–205.

Dwyer, M.J., E.G. Patton, and R.H. Shaw. 1997. Turbulent kinetic energy budgets from a large-eddy simulation of airflow above and within a forest canopy. Boundary Layer Meteorology 84: 23–43.

Dye, P.J., and B.W. Olbrich. 1993. Estimating transpiration from 6-year-old *Eucalyptus grandis* trees: development of a canopy conductance model and comparison with independent sap flux measurements. Plant, Cell and Environment 16: 45–53.

Eckstein, J., W. Beyschlag, K.A. Mott, and R.J. Ryel. 1996. Changes in photon flux can induce stomatal patchiness. Plant, Cell and Environment 19: 1066–1074.

Evans, J.R., and J.R. Seemann. 1984. Differences between wheat genotypes in specific activity of ribulose-1,5-bisphosphate carboxylase and the relationship to photosynthesis. Plant Physiology 74: 759–765.

Falge, E., R.J. Ryel, M. Alsheimer, and J.D. Tenhunen. 1997. Effects of stand structure and physiology on forest gas exchange: a simulation study for Norway spruce. Trees Structure and Function 11: 436–448.

Falge, E., W. Graber, R. Siegwolf, J.D. Tenhunen. 1995. A model of the gas exchange response of *Picea abies* to habitat conditions. Trees Structure and Function 10: 277–287.

Fan, S.-M., M.L. Goulden, J.W. Munger, B.C. Daube, P.S. Bakwin, S.C. Wofsy, J.S. Amthor, D.R. Fitzjarrald, K.E. Moore, and T.R. Moore. 1995. Environmental controls on the photosynthesis and respiration of a boreal lichen woodland: a growing season of whole-ecosystem exchange measurements by eddy correlation. Oecologia 102: 443–452.

Farquhar, G.D., S. von Caemmerer, and J.A. Berry. 1980. A biochemical model of photosynthetic CO_2 assimilation in leaves of C_3 species. Planta 149: 78–90.

Farquhar, G.D., and S. von Caemmerer. 1982. Modelling of photosynthetic response to environmental conditions. In: Lange, O.L., P.S. Nobel, C.B. Osmond, and H. Ziegler, eds. Encyclopedia of Plant Physiology. New Series, vol. 12B. Berlin, Heidelberg, New York: Springer-Verlag, pp 549–587.

Field, C.B., J.A. Gamon, and J. Peñelas. 1994. Remote sensing of terrestrial photosynthesis. In: Schulze, E.-D., and M.M: Caldwell, eds. Ecophysiology of Photosynthesis. Ecological Studies, vol. 100. Berlin, Heidelberg, New York: Springer-Verlag, pp 511–527.

Forrester, J.W. 1961. Industrial Dynamics. Cambridge, MA: MIT Press.

Fukai, S., and R.S. Loomis. 1976. Leaf display and light environments in row-planted cotton communities. Agricultural Meteorology 17: 353–379.

Gates, D.M. 1980. Biophysical Ecology. Berlin, Heidelberg, New York: Springer-Verlag.

Gower, S.T., and J.M. Norman. 1991. Rapid estimation of leaf area index in conifer and broad-leaf plantations. Ecology 72: 1896–1900.

Grace, J.C., P.G. Jarvis, and J.M. Norman. 1987a. Modelling the interception of solar radiant energy in intensively managed stands. New Zealand Journal of Forestry Science 17: 193–209.

Grace, J.C., D.A. Rook, and P.M. Lane. 1987b. Modelling canopy photosynthesis in *Pinus radiata* stands. New Zealand Journal of Forestry Science 17: 204–228.

Harley, P.C., J.D. Tenhunen, and O.L. Lange. 1986. Use of an analytical model to study limitations on net photosynthesis in *Arbutus unedo* under field conditions. Oecologia 70: 393–401.

Harley, P.C., R.B. Thomas, J.F. Reynolds, and B.R. Strain. 1992. Modelling photosynthesis of cotton grown in elevated CO_2. Plant, Cell and Environment 15: 271–282.

Hoffmann, F. 1995. FAGUS, a model for growth and development of beech. Ecological Modelling 83: 327–348.

Hogeweg, P., B. Hesper, C.P. van Schail, and W.G. Beeftink 1985. Patterns in vegetation succession, an ecomorphological study. In: White, J., ed. The Population Structure of Vegetation. Dordrecht: Dr. W. Junk Publishers, pp 637–666.

Innis, G.S. 1974. A spiral approach to ecosystem simulation, I. In: Innis, G.S., and R.V.

O'Neill, eds. Systems Analysis of Ecosystems. Fairland, MD: International Cooperative Publishing House, pp 211–386.

Ishida, T., G.S. Campbell, and C. Calissendorff. 1991. Improved heat balance method for determining sap flow rate. Agricultural and Forest Meteorology 56: 35–48.

Jarvis, P.G. 1976. The interpretation of the variations in leaf water potential and stomatal conductance found in canopies in the field. Philosophical Transactions of the Royal Society of London B 273: 593–610.

Jarvis, P.G. 1995. Scaling processes and problems. Plant, Cell and Environment 18: 1079–1089.

Johnson, F., H. Eyring, and R. Williams. 1942. The nature of enzyme inhibitions in bacterial luminescence; sulfanilamide, urethane, temperature, and pressure. Journal of Cell Comparative Physiology 20: 247–268.

Johnson, I.R., and J.H.M. Thornley. 1985. Dynamic model of the response of a vegetative grass crop to light temperature and nitrogen. Plant, Cell and Environment 8: 485–499.

Johnson, I.R., A.J. Parsons, and M.M. Ludlow. 1989. Modelling photosynthesis in monocultures and mixtures. Australian Journal of Plant Physiology 16: 501–516.

Jones, H.G., and K.H. Higgs. 1989. Empirical models of the conductance of leaves in apple orchards. Plant, Cell and Environment 12: 301–308.

Jordan, D.B., and W.L. Ogren. 1984. The CO_2/O_2 specificity of ribulose 1,5-bisphosphate carboxylase/oxygenase: dependence on ribulose-bisphosphate concentration, pH and temperature. Planta 161: 308–313.

Keys, A.J. 1986. Rubisco: its role in photorespiration. Proceedings of the Royal Society of London B 313: 325–336.

Kitanidis, P.K., and R.L. Bras. 1980. Real-time forecasting with a conceptual hydrologic model. 2. Applications and results. Water Resource Research 16: 1034–1044.

Kull, O., and P.G. Jarvis. 1995. The role of nitrogen in a simple scheme to scale up photosynthesis from leaf to canopy. Plant, Cell and Environment 18: 1174–1182.

Küppers, M., C. Giersch, H. Schneider, and M.U.F. Kirschbaum. 1997. Leaf gas exchange in light and sunflecks: response patterns and simulations. In: Rennenberg, H., W. Eschrich, and H. Ziegler, eds. Trees—Contributions to Modern Tree Physiology. Leiden, The Netherlands: Backhuys Publishers, pp 77–96.

Kuuluvainen, T., and T. Pukkala. 1987. Effect of crown shape and tree distribution on the spatial distribution of shade. Agricultural and Forest Meteorology 40: 215–231.

Lang, A.R., G.Y. Xiang, and J.M. Norman 1985. Crop structure and the penetration of direct sunlight. Agricultural and Forest Meteorology 35: 83–101.

Leegood, R.C., M.D. Adcock, and H.D. Doncaster. 1989. Analysis of the control of photosynthesis in C_4 plants by changes in light and carbon dioxide. Philosophical Transactions of the Royal Society of London B 323: 339–355.

Lemeur, R. 1973. A method for simulating the direct solar radiation regime of sunflower, Jerusalem artichoke, corn and soybeans using actual stand structural data. Agricultural Meteorology 12: 229–247.

Leuning, R. 1983. Transport of gases into leaves. Plant, Cell and Environment 6: 181–194.

Leuning, R. 1995. A critical appraisal of a combined stomatal-photosynthesis model for C₃ plants. Plant, Cell and Environment 18: 339–355.

Leuning, R., F.M. Kelliher, D.G.G. dePury, and E.-D. Schulze. 1995. Leaf nitrogen, photosynthesis, conductance and transpiration: scaling from leaves to canopies. Plant, Cell and Environment 18: 1183–1200.

Lloyd, J. 1991. Modelling stomatal responses to environment in *Macadamia integrifolia*. Australian Journal of Plant Physiology 18: 649–660.

Mann, J.E., G.L. Curry, and P.J.H. Sharp. 1979. Light penetration by isolated plants. Agricultural Meteorology 20: 205–214.

McNaughton, K.G. 1994. Effective stomatal and boundary-layer resistances of heterogeneous surfaces. Plant, Cell and Environment 17: 1061–1068.

Mott, K.A., and D.F. Parkhurst. 1991. Stomatal responses to humidity in air and helox. Plant, Cell and Environment 14: 509–515.

Niinemets, Ü. 1997. Distribution patterns of foliar carbon and nitrogen as affected by tree dimensions and relative light conditions in the canopy of *Picea abies*. Trees Structure and Function 11: 144–154.

Nobel, P. 1983. Biophysical Plant Physiology and Ecology. New York: W.H. Freeman and Company.

Norman, J.M. 1979. Modeling the complete crop canopy. In: Barfield, B.J., and J.F. Gerber, eds. Modification of the Aerial Environment of Crops. Joseph, MI: American Society of Agricultural St. Engineers, pp 249–277.

Norman, J.M., and G.S. Campbell. 1989. Canopy structure. In: Pearcy, R.W., J.R. Ehleringer, H.A. Mooney, and P.W. Rundel, eds. Plant Physiological Ecology. Field Methods and Instrumentation. London, New York: Chapman and Hall, pp 301–325.

Norman, J.M., and J.M. Welles. 1983. Radiative transfer in an array of canopies. Agronomy Journal 75: 481–488.

Oker-Blom, P. 1985. The influence of penumbra on the distribution of direct solar radiation in a canopy of scots pine. Photosynthetica 19: 312–317.

Oker-Blom, P., and S. Kellomäki 1982a. Effect of angular distribution of foliage on light absorption and photosynthesis in the plant canopy: theoretical computations. Agricultural Meteorology 26: 105–116.

Oker-Blom, P., and S. Kellomäki. 1982b. Theoretical computations on the role of crown shape in the absorption of light by forest trees. Mathematical Biosciences 59: 291–311.

Oker-Blom, P., T. Pukkala, and T. Kuuluvainen. 1989. Relationship between radiation interception and photosynthesis in forest canopies: effect of stand structure and latitude. Ecological Modelling 49: 73–87.

Pearcy, R.W., L.J. Gross, and D. He. 1997. An improved dynamic model of photosynthesis for estimation of carbon gain in sunfleck light regimes. Plant, Cell and Environment 20: 411–424.

Pearcy, R.W., and W.A. Pfitsch. 1994. The consequences of sunflecks for photosynthesis and growth of forest understory plants. In: Schulze, E.-D., and M.M. Caldwell, eds. Ecophysiology of Photosynthesis. Ecological Studies, vol. 100. Berlin, Heidelberg, New York: Springer-Verlag, pp 343–359.

Pearcy, R.W., and D.A. Sims. 1994. Photosynthetic acclimation to changing light environments: scaling from the leaf to the whole plant. In: Caldwell, M.M., and R.W. Pearcy, eds. Exploitation of Environmental Heterogeneity by Plants. San Diego, CA: Academic Press, pp 145–174.

Peisker, M., and S.A. Henderson. 1992. Carbon: terrestrial C_4 plants. Plant, Cell and Environment 15: 987–1004.

Perry, S.G., A.B. Fraser, D.W. Thompson, and J.M. Norman. 1988. Indirect sensing of plant canopy structure with simple radiation measurements. Agricultural and Forest Meteorology 42: 255–278.

Pospíšilová, J., and J. Šantruč-ek. 1994. Stomatal patchiness. Biologia Plantarum. 36: 421–453.

Reynolds, J.F., B. Acock, R.L. Dougherty, and J.D. Tenhunen. 1987. A modular structure for plant growth simulation models. In: Pereira, J.S., and J.J. Landsberg, eds. Biomass Production by Fast Growing Trees. Dordrecht: Kluver Academic Publishers, pp 123–134.

Reynolds, J.F., J.L. Chen, P.C. Harley, D.W. Hilbert, R.L. Dougherty, and J.D. Tenhunen. 1992. Modeling the effects of elevated CO_2 on plants: extrapolating leaf response to a canopy. Agricultural and Forest Meteorology 61: 69–94.

Roden, J.S., and R.W. Pearcy. 1993. The effect of flutter on the temperature of poplar leaves and its implications for carbon gain. Plant, Cell and Environment 16: 571–577.

Ross, J. 1975. Radiative transfer in plant communities. In: Monteith, J.W., ed. Vegetation and the Atmosphere. London, New York, San Francisco: Academic Press, pp 13–55.

Ruimy, A., P.G. Jarvis, D.D. Baldocchi, and B. Saugier. 1995. CO_2 fluxes over plant canopies and solar radiation: a review. Advances in Ecological Research 26: 1–68.

Ryel, R.J., P.W. Barnes, W. Beyschlag, M.M. Caldwell, and S.D. Flint. 1990. Plant competition for light analyzed with a multispecies canopy model. I. Model development and influence of enhanced UV-B conditions on photosynthesis in mixed wheat and wild oat canopies. Oecologia 82: 304–310.

Ryel, R.J., and W. Beyschlag. 1995. Benefits associated with steep foliage orientation in two tussock grasses of the American intermountain west. A look at water-use-efficiency and photoinhibition. Flora 190: 251–260.

Ryel, R.J., W. Beyschlag, and M.M. Caldwell. 1993. Foliage orientation and carbon gain in two tussock grasses as assessed with a new whole-plant gas-exchange model. Functional Ecology 7: 115–124.

Ryel, R.J., W. Beyschlag, and M.M. Caldwell. 1994. Light field heterogeneity among tussock grasses: theoretical considerations of light harvesting and seedling establishment in tussocks and uniform tiller distributions. Oecologia 98: 241–246.

Ryel, R.J., W. Beyschlag, B. Heindl, and I. Ullmann. 1996. Experimental studies of the competitive balance between two central European roadside grasses with different growth forms. 1. Field experiments on the effects of mowing and maximum leaf temperatures on competitive ability. Botanica Acta 109: 441–448.

Sage, R.F. 1990. A model describing the regulation of ribulose-1,5-bisphosphate carboxylase, electron transport, and triose-phosphate use in response to light intensity and CO_2 in C_3 plants. Plant Physiology 94: 1728–1734.

Schreiber, U., W. Bilger, and C. Neubauer. 1994. Chlorophyll fluorescence as a nonintrusive indicator for rapid assessment of in vivo photosynthesis. In: Schulze, E.-D., and M.M. Caldwell, eds. Ecophysiology of Photosynthesis. Ecological Studies, vol. 100. Berlin, Heidelberg, New York: Springer-Verlag, pp 49–70.

Schuepp, P.H. 1993. Leaf boundary layers. Tansley Review No. 59. New Phytologist 125: 477–507.

Sellers, P.J., J.A. Berry, G.J. Collatz, C.B. Field, and F.G. Hall. 1992. Canopy reflectance, photosynthesis and transpiration. III. A reanalysis using improved leaf models and a new canopy integration scheme. Remote Sensing Environment 42: 187–216.

Sharkey, T.D. 1985. Photosynthesis in intact leaves of C_3 plants: physics, physiology and rate limitations. The Botanical Review 51: 53–105.

Silvertown, J., S. Holtier, J. Johnson, and P. Dale. 1992. Cellular automaton models of interspecific competition for space—the effect of pattern on process. Journal of Ecology 80: 527–534.

Sinclair, R., and B.A. Thomas. 1970. Optical properties of leaves of some species in arid south Australia. Australian Journal of Botany 18: 261–273.

Smith, E. 1937. The influence of light and carbon dioxide on photosynthesis. General Physiology 20: 807–830.

Spitters, C.J.T., H.A.J.M. Toussaint, and J. Goudriaan. 1986. Separating the diffuse and direct component of global radiation and its implications for modeling canopy photosynthesis. Part I. Components of incoming radiation. Agricultural and Forest Meteorology 38: 217–229.

Steinhorst, R.K., H.W. Hunt, G.S. Innis, and K.P. Haydock. 1978. Sensitivity analyses of the ELM model. In: Innis, G.S., ed. Grassland Simulation Model. Ecological Studies, vol. 26. Heidelberg, New York: Springer-Verlag, Berlin, pp 231–255.

Stenberg, P., S. Linder, H. Smolander, and J. Flowerellis. 1994. Performance of the LAI-2000 plant canopy analyzer in estimating leaf area index of some Scots pine stands. Tree Physiology 14: 981–995.

Tenhunen, J.D., P.C. Harley, W. Beyschlag, and O.L. Lange. 1987. A model of net photosynthesis for leaves of the sclerophyll *Quercus coccifera*. In: Tenhunen, J.D., F. Catarino, O.L. Lange, and W.C. Oechel, eds. Plant Response to Stress—Functional Analysis in Mediterranean Ecosystems. Berlin, Heidelberg, New York: Springer-Verlag, pp 339–354.

Tenhunen, J.D., R.A. Siegwolf, and S.F. Oberbauer, 1994. Effects of phenology, physiology and gradient in community composition, structure and microclimate on tundra ecosystem CO_2 exchange. In: Schulze, E.-D., and M.M. Caldwell, eds. Ecophysiology of Photosynthesis. Ecological Studies, vol. 100. Heidelberg, New York: Springer-Verlag, Berlin, pp 431–460.

Thornley, J.H.M., and I.R. Johnson. 1990. Plant and Crop Modelling. A Mathematical Approach to Plant and Crop Physiology. Oxford: Clarendon Press.

Ustin, S.L., M.O. Smith, and J.B. Adams. 1993. Remote sensing of ecological processes: a strategy for developing and testing ecological models using spectral misture analysis. In: Ehleringer, J.R., and C.B. Field, eds. Scaling Physiological Processes: Leaf to Globe. San Diego, CA: Academic Press, pp 339–357.

Van Tongeren, O., and I.C. Prentice. 1986. A spatial simulation model for vegetation dynamics. Vegetatio 65: 163–173.

Verma, S.B. 1990. Micrometeorological methods for measuring surface fluxes of mass and energy. In: Goel, N.S., and J.M. Norman, eds. Instrumentation for Studying Vegetation Canopies for Remote Sensing in Optical and Thermal Infrared Regions. Remote Sensing Reviews, vol. 5. Chur, Switzerland: Harwood Academic Publishers, pp 99–115.

Von Caemmerer, S., and G.D. Farquhar. 1981. Some relationships between the biochemistry of photosynthesis and the gas exchange of leaves. Planta 153: 376–387.

Walker, G.K., R.E. Blackshaw, and J. Dekker. 1988. Leaf area and competition for light between plant species using direct sunlight transmission. Weed Technology 2: 159–165.

Wang, Y.P., and P.G. Jarvis. 1990. Description and validation of an array model—MAESTRO. Agricultural and Forest Meteorology 51: 257–280.

Warren Wilson, J. 1960. Inclined point quadrats. New Phytologist 58: 1–8.

Warren Wilson, J. 1963. Estimation of foliage denseness and foliage angle by inclined point quadrats. Australian Journal of Botany 11: 95–105.

Webb, W.L. 1991. Atmospheric CO_2, climate change, and tree growth: a process model I. Model structure. Ecological Modelling 56: 81–107.

Whitehead, D., D.V.W. Okali, and F.E. Fasehun. 1981. Stomatal response to environmental variables in two tropical rainforest species during the dry season in Nigeria. Journal of Applied Ecology 18: 581–587.

Wolfram, S. 1983. Statistical mechanics of cellular automata. Reviews of Modern Physics 55: 601–643.

23

Ecological Applications of Remote Sensing at Multiple Scales

John A. Gamon and Hong-lie Qiu
California State University–Los Angeles, Los Angeles, California

I. INTRODUCTION

We are restrained by our senses, which often limit our view to a particular favorite or accessible geographic location. Consequently, much ecological research has traditionally focused on specific organisms, populations, and communities defined by geographic region, and ecology is largely a collection of case histories in search of unifying principles. However, the growing human pressures on the planet's resources are altering ecosystem function at regional to global scales and placing a new urgency on studies that integrate information across spatial and temporal scales. Ecologists are now faced with the challenge of measuring and understanding ecological processes at these multiple scales. Answering this challenge will necessarily involve the use of remote sensing, which has the unique ability to provide an objective, synoptic view of the earth and its atmosphere.

In this chapter, remote sensing will primarily be defined as the measurement of electromagnetic radiation (reflectance, fluorescence, or longwave emission) with noninvasive sampling (Figure 1). *Remote* implies measurements from a distance (e.g., from an airborne or satellite platform), and many of the most spectacular examples have been from these great distances. However, the fundamental tools and principles of radiation measurement can be applied at almost any scale. Indeed, one of the strengths of optical sampling is that, unlike many other measurement methods, it is eminently "scaleable"–it can be applied at many spatial scales and used to analyze how information content changes with scale (Ustin et al. 1993, Foody and Curran 1994, Quattrochi and Goodchild 1997, Wessman 1992). This ability to bridge scales allows us to extend our otherwise limited perception to new domains and to reach a new understanding of complex ecological phenomena. Remote sensing has additional virtues, including the ability to sample nondestructively and without direct contact, thus avoiding the problem common to many measurement techniques of disturbing or destroying the object of measurement. Because digital remote sensing provides a consistent data format, it provides a degree of objectivity that is lacking in many other methods of data collection.

Remote sensing per se is not new. Properly positioned, the eye is a powerful (if subjective) remote sensing device. Aerial photography has been used for many decades for mapping and reconnaissance. Continued advances in digital and optical technology are providing ever more powerful tools for collecting and analyzing remotely sensed data, and a vast array of sensors are now available for use by the ecological community. These technological advances are redefining the questions that can be addressed by ecologists and are contributing to paradigm shifts in both the concepts and methodology of ecology. Much of what is currently "new" about remote sensing is the way in which investigators are finding innovative ways of using these tools, often in combination with other methods, to address ecological questions at multiple scales.

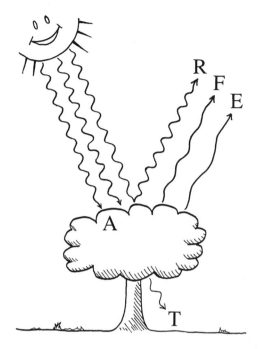

Figure 1 Schematic illustrating passive remote sensing, which is dependent on solar radiation. Radiation can either be absorbed (A), transmitted (T), or reflected (R). Absorbed radiation can be released rapidly as fluorescence (F) or more slowly as longwave, thermal emission (E). Remotely positioned sensors can infer information about surface features by detecting patterns of reflected, transmitted, fluoresced, or emitted radiation. In a real landscape, multiple scattering and signals from multiple scene components (not illustrated) significantly influence the pattern of radiation detected.

The rapid pace of advance in this field, combined with the uncertain future of many current and planned sensors, makes a comprehensive review a difficult, if not impossible, task. Remote sensing is now being used to infer a dizzying array of earth surface and atmospheric properties and processes, including surface temperature, moisture, topography, albedo, mineral composition, vegetation cover and type, vegetation dynamics and land use change, atmospheric composition, irradiance, and surface–atmosphere fluxes. Some of these (e.g., surface–atmosphere fluxes) require the incorporation of remotely sensed measurements with models, whereas others (e.g., surface temperature and vegetation cover) can be determined more directly. Clearly, a full survey of these applications is well beyond the scope of this review. For a more thorough coverage of remote sensing technology and applications, the interested reader might pursue current information on the World Wide Web or refer to any of a number of recent references and reviews on the topic (Lillesand and Kiefer 1987, Hobbs and Mooney 1990, Richards 1993, Ustin et al. 1993, Foody and Curran 1994, Danson and Plummer 1995, Gholz et al. 1997, Kasischke et al. 1997, Quattrochi and Goodchild 1997, Sabins 1997). Instead of a comprehensive survey, we have chosen to present a few specific applications that illustrate the range, power, and potential of optical (visible and infrared) remote sensing for addressing ecological questions at several levels of inquiry. We selected a particular focus on carbon stocks and fluxes associated with terrestrial vegetation. This topic is of critical concern today because human activities are perturbing vegetation cover, composition, and the carbon cycle in significant and measurable ways (Vitousek 1994, Amthor 1995, Houghton 1995). A central goal of global ecology is to clearly define vegetation–atmosphere fluxes in the context of these perturbations.

The chapter begins with a short presentation of a few remote sensing fundamentals. It then illustrates recent examples of remote sensing as a mapping tool, perhaps the most obvious and traditional of remote sensing applications. Newer extensions of this mapping capability are now explicitly considering spectral and temporal aspects of landscapes, and linking this information to process models. We also examine some examples of the application of remote sensing to models of terrestrial carbon flux and consider how new developments offer to improve our understanding of carbon stocks and fluxes. An implicit message is that the relevant technology and appropriate data are increasingly becoming available, but are only slowly being adopted by the ecological community at large. The chapter ends by considering the removal of barriers to the use of remote sensing. The goal is to encourage ecologists to continue exploring innovative ways of applying this powerful and exciting technology.

II. FUNDAMENTALS

A. Sensors

Remote sensing typically involves noncontact measurement of electromagnetic radiation and the inference of patterns or processes from this radiation. Sensors can be classified as imaging and nonimaging instruments (Table 1). Imaging sensors view a given ground area (''scene'') with a characteristic array of detectors or pixels, each covering a specific ground area (''instantaneous field of view,'' or IFOV). Sensors can be further divided into passive or active sensors, both of which have advantages and disadvantages. Passive sensors rely on external (typically solar) energy sources, and generally sample during daylight (Figure 1). Passive sensors can, in principle, sample any wavelength of radiation, provided the energy level is high enough for an adequate signal-to-noise ratio. However, in practice, a given detector is usually restricted to a specific wavelength

Table 1 Selected Sensors Useful for Ecological Applications[a]

Platform	Sensor	Status	No. of bands	Approximate IFOV (m)	Wavelength range	Source
Satellite	AVHRR	In orbit	5	1100	0.58–12.5 μm	NOAA (USA)
	Landsat Multispectral Scanner (MSS)	In orbit	4 vis-NIR	79	0.5–1.1 μm	EOSAT, Lanham, MD
	Landsat Thematic Mapper (TM)	In orbit	7 vis-mid-IR 1 thermal	30 120	0.45–2.35 μm & 10.5–12.5 μm	EOSAT, Lanham, MD
	SPOT, panchromatic	In orbit	1	10	0.51–0.73 μm	SPOT Corporation Reston, VA
	SPOT, multispectral		3	20	0.50–0.89 μm	
	ERS-1 and ERS-2 (radar)	In orbit	1	218	5.7 cm	European Space Agency
	Radarsat (radar)	In orbit	1	Variable	5.6 (C-Band)	Radarsat International Richmond, BC (Canada)
	Moderate Resolution Imaging Spectrometer (MODIS)	Planned (1999)	36	250–1000	0.4–14.4	NASA (USA)
Space Shuttle	Shuttle Imaging Radar (SIR)	Flew 1994	Variable	Variable	3.0 cm 6.0 cm 24.0 cm	NASA (USA)
Airborne	AVIRIS	Operational	224	20	400–2500 nm	NASA (USA)
	CASI (spectral mode)	Commercially available	288	Variable	430–870 nm	ITRES, Calgary, Alberta (Canada)
	Scanning Lidar Imager of Canopies by Echo Recovery (SLICER, laser altimeter)	Operational	1	Variable	1.06 μm	NASA (USA)
Personal spectrometers (diode array, nonimaging)	FieldSpec FR	Commercially available	1512–1582	N/A	350–2500 nm	Analytical Spectral Devices, Inc., Boulder, CO
	GER 1500	Commercially available	512	N/A	300–1050 nm	Geophysical and Environmental Research Corp, Milbrook, NY
	GER 2600/3700	Commercially available	640/704	N/A	300–2500 nm	Spectron Engineering, Denver, CO
	SE-590 (with CE390WB-R)	Commercially available	256	N/A	400–1100 nm	
	UniSpec (VIS/NIR model)	Commercially available	256	N/A	300–1100 nm	PP Systems, Haverhill MA

[a] Except for SLICER and the personal spectrometers, all sensors are imaging instruments.
Sources: Lillesand and Kiefer (1987), Richards (1993), Vane et al. (1993), Jensen (1996), Sabins (1997), World Wide Web, and vendor literature.

range. For example, silicon photodiodes are generally sensitive to the near-UV to near-IR range (approximately 350–1100 nm; Pearcy 1989), which corresponds well to the region of strongest solar irradiance (Nobel 1991). Atmospheric absorption, primarily by water vapor and other atmospheric gases in specific infrared wavelengths, and atmospheric scattering, particularly by aerosols in the blue and near-UV wavelengths, can further reduce the usable spectral region when sampling from aircraft or satellite. Because passive detectors are directly dependent on the solar source, they work poorly in some circumstances (e.g., low light or cloudy conditions). By contrast, active sensors can operate independently of light conditions because they sample the return of a signal originating from the instrument. Laser altimeters (light detecting and ranging, or LIDAR) and active microwave ("radar") sensors provide good examples of this approach (Table 1). Radar sensors offer the additional advantage of being able to penetrate clouds, and thus are largely independent of light or weather conditions, and are now emerging as useful sensors for a variety of ecological, hydrological, and topographical studies (Hess and Melack 1994, Kasischke 1997, Smith 1997).

B. The Concept of Scale

Scale in remote sensing has several dimensions, including the spatial, spectral, and temporal dimensions, each of which provides a rich source of ecologically relevant information (Figure 2). Sensor look angle and radiation polarization provide additional sampling dimensions that may be relevant in specific cases (Barnsley 1994). For example, certain structural features of vegetation are sensitive to polarization of optical radiation (Vanderbilt et al. 1990) or microwave radiation (Hess and Melack 1994). Issues of scale must be carefully considered when applying remote sensing to biological or ecological processes. Because ecological processes occur over definable time frames and geographic regions, often with characteristic spectral "signatures," detectibility of a given process will vary with the temporal, spatial, and spectral scale of measurement. Although the principles of detection operate similarly at all scales, the actual information content changes with scale, and consequently the methods or tools of interpretation often have to change. For example, changes in certain key biochemical processes involving photosynthetic regulation at the leaf scale can be detected over seconds to minutes only with narrow spectral bands, whereas changes in green vegetation cover at regional scales can be detected over seasons to years with broad spectral bands. Fortunately, the technology of remote sensing is sufficiently mature that a wide range of sensors now exist for ready usage at a wide range of scales (Table 1).

C. Vegetation Indices

In terrestrial systems, vegetated and nonvegetated terrestrial areas can be readily distinguished based on their contrasting reflectance patterns in the red and near-

Spatial Scale   Temporal Scale

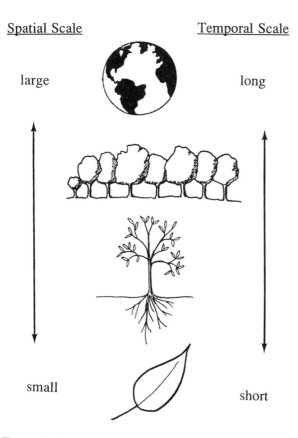

large   long

small   short

Figure 2 Remote sensing involves multiple dimensions, including spatial, temporal, and spectral dimensions, each of which must be considered when designing a sampling protocol or interpreting remotely sensed data. For a variety of physical and biological reasons, the scales of these three dimensions tend to be interdependent. At coarse geographic scales (large grain size or IFOV), processes tend to be detectable over larger time frames, whereas many processes at a fine spatial scale (small grain size or IFOV) can often be detected over short time periods. Not shown in this figure is the scale of the spectral dimension, which tends to match those of the spatial and temporal dimensions. Many geographically large-scale features and processes (e.g., changing vegetation cover) can be detected adequately with broad spectral bands, whereas detection of small features (e.g., biochemical properties or processes) often requires narrow spectral bands. Ground-based sampling is generally restricted to limited areas, but is ideally suited for fine spectral (narrow-band) or fine temporal (rapid and continuous) sampling. Global coverage with satellite sensors tends to involve large grain sizes (IFOV) using broad spectral bands (although new sensor designs are improving the spectral and spatial resolution of satellite sensors). Consequently, a range of sensors and platforms (see Table 1) are needed to link information across scales.

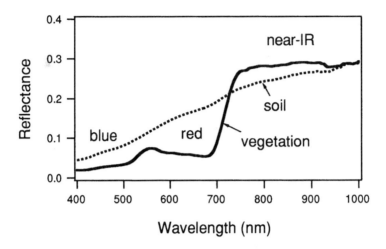

Wavelength (nm)

Figure 3 Typical spectral reflectance patterns for green vegetation and bare soil in the visible (400–700 nm) to near-infrared (> 700 nm) region. Chlorophyll absorbs well in the visible (particularly the blue and red), but not in the infrared, resulting in the characteristic vegetation ''red edge'' at 700 nm. This red edge feature is generally lacking in nonphotosynthetic surfaces. The contrasting red and near-infrared reflectance is the basis for many vegetation indices depicting relative amounts of green vegetation cover. Note also the slight water absorption feature (dip) between 900 and 1000 nm, which can serve as an indicator of water content.

infrared (NIR) spectral regions (Figure 3). This contrast results from differential chlorophyll absorption in red and NIR wavebands, and reveals the radiation requirements for photosynthesis. Photosynthetic pigments absorb most effectively in the visible region (400–700 nm), where energy is most abundant and strong enough to drive electron transport, yet weak enough to avoid excessive damage to biological molecules. By contrast, there is insufficient energy in the NIR to drive photosynthesis; consequently, photosynthetic pigments cannot use or absorb these wavelengths, and vegetation canopies effectively scatter (reflect and transmit) most NIR radiation.

Numerous vegetation indices have been developed that characterize this contrasting reflectance on either side of this ''red edge'' boundary, including the Normalized Difference Vegetation Index (NDVI), the simple ratio (SR), the soil-adjusted vegetation index and its derivatives (SAVI, TSAVI, SARVI), the greenness vegetation index (GVI), and the perpendicular vegetation index (PVI) (Perry and Lautenschlager 1984, Baret and Guyot 1991, Wiegand et al. 1991, Huete et al. 1997). Other indices are based on the slope or inflection point of the ''red edge'' near 700 nm (Curran et al. 1990, Gitelson et al. 1996, Johnson et al. 1994).

Essentially, all of these indices collapse a full spectrum into a single, readily usable value that scales with green canopy cover or related measures of vegetation abundance. Of all these indices, the most commonly used index is undoubtedly the normalized difference vegetation index:

$$NDVI = (R_{NIR} - R_{RED})/(R_{NIR} + R_{RED}) \tag{1}$$

where R_{NIR} is the reflectance (or radiance) in the NIR region, and R_{RED} is the reflectance (or radiance) in the red region (Figure 3).

Each of these indices suffers from certain well-discussed limitations (Holben 1986, Myneni and Asrar 1994, Myneni and Williams 1994, Sellers et al. 1996b). Many of these limitations arise because reflectance of any spectral region is necessarily influenced by multiple factors that can confound a simple interpretation of a given reflectance signature. Complex landscapes contain many components that can have multiple, overlapping effects on any single index, and the concept of a pure index becomes elusive, particularly at large spatial scales where many components contribute to the measured signal. Additional errors arise from atmospheric effects (absorption or scattering), variation in solar angle, and sensor calibration drifts. Consequently, it is often necessary to correct indices for known confounding factors, and this is now a common practice with global NDVI datasets (Goward et al. 1994, James and Kalluri 1994, Los et al. 1994, Sellers et al. 1994, Townshend 1994, Sellers et al. 1996b). Despite these flaws, all of these indices provide some measure of light absorption by green vegetation, and thus can be used to distinguish relative levels of energy capture for photosynthetic processes.

Some of the most potent and spectacular applications of remote sensing in ecology derive from the ability to detect spectral features associated with specific, biologically important compounds (Table 2). Some of these compounds, notably photosynthetic pigments (chlorophylls and carotenoids) and photoprotective pigments (carotenoids and anthocyanins), are positioned for capturing solar energy; consequently, they are well suited for detection from above. Because photosynthetic pigments control light absorption for photosynthetic carbon uptake, their detection can be linked to photosynthetic production models. NDVI provides a good example of an operational index that is influenced by both vegetation structure and chlorophyll content (Figure 3). Global NDVI dynamics are significantly correlated with seasonally and latitudinally changing atmospheric CO_2 content (Tucker et al. 1986, Fung et al. 1987), demonstrating a strong influence of terrestrial photosynthesis on atmospheric CO_2 levels. NDVI is also strongly linked to primary production at continental scales (Goward et al. 1985). This realization of strong links between NDVI, CO_2 fluxes, and primary production, along with the increasing availability of standardized global NDVI datasets (Goward et al. 1994, James and Kalluri 1994, Los et al. 1994, Sellers et al. 1994, Townshend

Table 2 Examples of Biologically Important Compounds Detectable with Spectral Reflectance

Compounds	Function	References
Chlorophylls	Photosynthesis (PAR absorption)	Gitelson and Merzlyak (1994, 1996), Gitelson et al. (1996), Carter (1994), Johnson et al. (1994), Peñuelas et al. (1994, 1995a), Gamon et al. 1995, Yoder and Pettigrew-Crosby (1995)
Carotenoids Carotenes Xanthophylls	Accessory photosynthetic pigments Photosynthesis (PAR absorption) Photoprotection	Peñuelas et al. (1994, 1995a, 1997a), Gamon et al. (1990, 1992, 1997), Filella et al. (1996)
Flavonoids Anthocyanins	Protection (pathogen deterrent and sunscreen)	Gould et al. (1995), Coley and Barone (1996)
Water	Essential for structural support and most metabolic processes	Bull (1991), Carter (1991), Peñuelas et al. (1993, 1994, 1997b), Gao and Goetz (1995), Roberts et al. (1997) Zhang et al. (1997), Gamon et al. (1998), Ustin et al. (1998)
Lignin	Plant cell wall structure, resists decomposition	Peterson et al. (1988), Wessman et al. (1988), Johnson et al. (1994), Gastellu-Etchegorry et al. (1995), Martin and Aber (1997)
Nitrogen-containing compounds (e.g., proteins and pigments)	Needed for many metabolic processes	Peterson et al. (1988) Peñuelas et al. (1994) Filella et al. (1995), Gamon et al. (1995), Gastellu-Etchegorry et al. (1995), Johnson et al. (1994), Yoder and Pettigrew-Crosby (1995), Martin and Aber (1997)
Cellulose	Plant cell wall structure	Gastellu-Etchegorry et al. (1995), Martin and Aber (1997)

1994, Wharton and Myers 1997), has spurred tremendous progress in global-scale studies of carbon flux.

D. New Opportunities with Hyperspectral Sensors

The advent of hyperspectral sensors—those with many, narrow, adjacent bands—is now allowing exploration of many additional biologically active compounds with spectral reflectance. Operational methods for reliably quantifying many of these compounds are currently under development (Table 2), and some have been discussed in recent reviews (Curran 1989, Wessman 1990, Peñuelas and Filella 1998). A number of pigments, notably carotenoids (accessory photosynthetic pigments; Young and Britton 1993) and anthocyanins (phenolic pigments; Strack 1997), may serve as indicators of stress or leaf senescence in many plant species. Additionally, distinct absorption features for water vapor and liquid water are now remotely detectable (Gao and Goetz 1990, Green et al. 1991 and 1993, Roberts et al. 1997, Ustin et al. 1998, Gamon et al. 1998). Water vapor features are now being used to remove confounding atmospheric effects in the derivation of apparent surface reflectance from uncorrected radiance signals (Green et al. 1991, 1993, Roberts et al. 1997). Liquid water absorption features may eventually provide useful indices of leaf area index and vegetation cover (Roberts et al. 1997, Gamon et al. 1998, Ustin et al. 1998) or canopy moisture status (Peñuelas et al. 1993, 1997b, Zhang et al. 1997, Gamon et al. 1998, Ustin et al. 1998). Particularly in combination with data from alternate sensors (e.g., radar and LIDAR), passive remote detection of water status may provide useful insight into physiological and hydrological processes. It also may be possible to detect or infer canopy nitrogen and lignin content remotely, which could provide useful inputs for models of ecosystem production and N cycling (Wessman et al. 1988, Aber et al. 1990, Matson et al. 1994, Martin and Aber 1997, Peterson et al. 1988; however, see cautions by Curran 1989, Curran and Kupiec 1995, Grossman et al. 1996).

The emergence of hyperspectral sensors is also spurring the application of new analytical procedures. In contrast to simple indices, which are typically limited to information from one or two broad spectral bands, these methods often take advantage of the additional information present in the shape of narrow-band spectra, and are often able to resolve subtle features not detectable with broad-band indices. These alternative approaches include derivative spectra (Demetriades-Shah 1990), continuum removal (Clark and Roush 1984), hierarchical foreground/background analysis (Pinzon et al. submitted), and a variety of other feature fitting methods (e.g., Gao and Goetz 1994, 1995). Another promising example is provided by spectral mixture analysis, which models a spectrum as a mix of component "endmember" spectra (Ustin et al. 1993, Roberts et al. 1993, 1997). These endmembers can be spectra of pure components (e.g., species, canopy components, minerals, or soil types; Roberts et al. 1993, 1998), or can be spectra derived from recognizable landscape components present in the image itself (Tompkins et al. 1997). Because spectral mixture analysis uses the additional leverage of multiple spectral bands (instead of the few bands used in most spectral indices), it can be less sensitive to confounding factors, assuming all of the appropriate spectral endmembers are properly included in the model. Because it "unmixes" a spectrum to fundamental components (e.g., areas of bare ground and vegetation), it is particularly well suited to situations where the components are smaller than the IFOV. On the other hand, spectral mixture analysis is computationally intensive, and the resulting product can be highly dependent on the particular endmembers selected (Roberts et al. 1998), making quantitative interpretations difficult. While many challenges to interpetation and validation of hyperspectral data remain, the continued emergence of hyperspectral sensors combined with improved computational capabilities will undoubtedly lead to increased refinement and application of new analytical methods applicable to ecological studies.

III. REMOTE SENSING AS A MULTIDIMENSIONAL MAPPING TOOL

The traditional strength of remote sensing lies in its ability to make objective maps of regions much larger than can be sampled from the ground. When image features are linked to understood objects or events, these maps allow us to greatly extend our view of ecologically significant patterns and processes. Historically, aerial photography has provided a useful tool for local- to regional-scale mapping, and this application has been well described elsewhere (Knipling 1969, Yost and Wenderoth 1969, Salerno 1976, Lillesand and Kiefer 1987). In contrast to the high spatial resolution (fine detail) of aerial photography, many current imaging sensors operate at coarser spatial resolution (typically, an IFOV of several meters to several thousand meters), and this resolution typically degrades with sampling distance (Figure 2). However, current imaging spectrometers offer added benefits of a digital data format and supplementary spectral information that can be thought of as multiple data layers for a given image. Multitemporal coverage, now provided by many satellite and some aircraft sensors (Table 1), adds an additional dimension that is particularly useful for examining time-dependent processes, including phenological, successional, and land-use change (Justice et al. 1985, Malingreau et al. 1989, Hobbs et al. 1990, Hall et al. 1991, DeFries and Townshend 1994a, 1994b). Exploration of spectral and temporal information in a spatial context offers powerful new ways of distinguishing features and functional states. This combination of spatial, temporal, and spectral dimensions in a structured, multidimensional data volume can readily reveal ecologically significant patterns and processes. The uniform data format provides potent inputs for mapping, statistical, and digital modeling tools (e.g., Geographic Information Systems; Jensen 1996) that are now revolutionizing the field of ecology. Landscape ecology (Turner and Gardner 1991) and global ecology (Gates 1993, Solomon and Shugart 1993, Walker and Steffen 1996) provide examples of new fields emerging from this synthesis of pattern and process.

A simple illustration of the power of combined spectral, temporal, and spatial information can be illustrated by imagery from the Santa Monica mountains region of southern California. Ongoing research in this area is seeking improved ways to map vegetation patterns and landscape processes, in part to improve models of fire behavior and assist resource managers in this increasingly urbanized and disturbed region. Remote sensing is providing maps of real-time vegetation cover and land usage not attainable from ground surveys due to the mountainous terrain, complex land ownership, and varying successional states and community composition caused by frequent fire and other disturbances (Figure 4). Normalized difference vegetation index (NDVI) images, derived from reflectance in two bands, the red and NIR (Figure 3), readily depict relative levels of green vegetation cover across this landscape. However, a single overpass during spring, the peak biomass period for this vegetation, fails to clearly distinguish several native vegetation types (Figure 5A), which for this landscape include chaparral, coastal sage scrub, grassland, southern oak woodland, and riparian woodland (Raven et al. 1986). In the spring, these vegetation types have similar reflectance spectra (Figure 6A) and are not readily distinguished by NDVI. However, these vegetation types have contrasting seasonal patterns of green leaf display that can be readily captured by multitemporal NDVI sampling (Gamon et al. 1995). This contrasting phenology is readily revealed in the second image at the end of the summer drought (fall 1994), illustrating a general decline in NDVI (darkening across the image) relative to the spring image, with distinct patches emerging that reveal different vegetation types (Figure 5B). In this season, distinct spectra emerge for these different vegetation types due to the seasonal contrasts in canopy greenness (Figure 6B), allowing NDVI to more fully separate these vegetation types (compare Figure 5A and 5B). In this way, the addition of multitemporal sampling can reveal contrasting seasonal trajectories of canopy greenness that are not evident in a single overpass.

In this landscape, key vegetation types can also be readily separated by more fully using the rich spectral dimension present in hyperspectral imagery. With spectral mixture analysis, which provides additional leverage to the problem of separating spectrally similar landscape components, an individual vegetation type (e.g., coastal sage scrub) can be readily depicted as an endmember fraction map, where varying intensity represents varying quantities of that vegetation type

Pacific Ocean

(Figure 5C). Recently developed elaborations of spectral mixture analysis that employ many spectral endmembers suggest that further separation of vegetation types into dominant species may be possible (Roberts et al. 1998). Spectral mixture analysis is also able to depict relative levels of green or dead biomass associated with vegetation composition, varying seasonal or successional states, or contrasting disturbance regimes (Gamon et al 1993, Wessman et al. 1997, Roberts et al. 1998). These improvements in the ability to distinguish vegetation types and functional states are possible with a new generation of narrow-band imaging spectrometers now available as aircraft sensors or under development for new satellite platforms (Table 1). The ability to accurately depict dominant species or functional vegetation types will undoubtedly improve ecosystem models that require spatially explicit vegetation inputs.

At the global scale, narrow-band imagery is not yet commonly available, but multitemporal, broad-band satellite imagery capturing seasonal variation in NDVI is now being used to develop improved global vegetation classifications (DeFries and Townshend 1994a, 1994b, DeFries et al. 1995, Sellers et al. 1996b, Nemani and Running 1997). NDVI dynamics detected from satellite are providing valuable insights into changing vegetation activity at global and decadal scales. For example, multiyear observations of AVHRR-derived NDVI for the continent of Africa have documented long-term vegetation dynamics in the Sahel and have demonstrated a surprising resiliency to Sahel vegetation (Tucker et al. 1991). This study is particularly notable because it counters the often-stated (but poorly documented) view that desertification in this region is an irreversible and inevitable process. Recent satellite NDVI evidence suggests that early "fingerprints" of global warming are now detectable as increased vegetation activity in northern latitudes (Myneni et al. 1997). Although the exact interpretation of this and similar observations are subject to question due to the difficulties of direct validation at the global scale, it is clear that remote sensing will play an increasingly critical role in elucidating global-scale vegetational change.

Figure 4 Western end (Ventura County) of the Santa Monica Mountains, California. This scene depicts the complex justaposition of urban, agricultural, and natural habitats that makes extensive ground-based sampling difficult. Frequent disturbances present additional challenges to mapping the vegetation of this region. The vegetation of the mountainous area at the eastern end of the image is in an early successional state due to a wildfire in the previous year. This image depicts 608 nm radiance collected with the AVIRIS sensor (Table 1) in fall 1994.

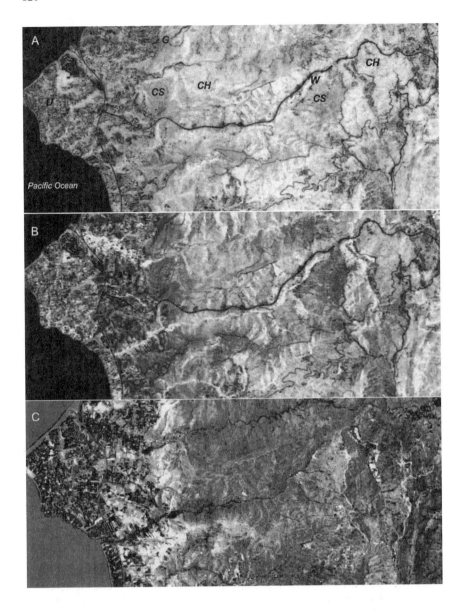

IV. LINKING REMOTE SENSING TO MODELS OF PHOTOSYNTHETIC PRODUCTION

A. APAR and Radiation-Use Efficiency

Remote sensing has emerged as a critical tool for spatially explicit models of terrestrial photosynthesis and net primary production (NPP), and a wide variety of model formulations exist (Heimann and Keeling 1989, Aber and Federer 1992, Potter et al. 1993, Running and Hunt 1993, Prince and Goward 1995, Ruimy et al. 1996, Sellers et al. 1996a). In several of these photosynthesis models, carbon gain by vegetation can be represented by the following relationship:

$$\text{photosynthetic rate} = \text{efficiency} \times \text{APAR} \tag{2}$$

where the photosynthetic rate is the instantaneous photosynthetic rate, APAR indicates the quantity of absorbed photosynthetically active radiation, and efficiency represents photosynthetic radiation-use efficiency, the efficiency of converting absorbed radiation to fixed (organic) carbon. Although not explicitly discussed here, this equation is most appropriately applied to gross photosynthetic rate, ignoring respiratory losses. To derive net photosynthesis, the effects of respiratory processes (photorespiration and mitochondrial respiration) could be incorporated into the "efficiency" coefficient. Alternatively, a separate respiratory term or coefficient could be added to Eq. 2.

If integrated over time (typically a growing season) and space (typically, canopies, stands, or regions), equation 2 is often expressed as:

$$\text{primary productivity} = \varepsilon \times \Sigma \text{ APAR} \tag{3}$$

where primary productivity is usually expressed as NPP, often estimated by aboveground biomass accumulated in a growing season, Σ APAR is the annual integral of radiation absorbed by vegetation, and ε represents the efficiency with which absorbed radiation is converted to biomass (Monteith 1977). Again, the

---

Figure 5 Images of the Pt. Dume region of the Santa Monica Mountains of southern California. (A) The NDVI in spring 1995; (B) the NDVI in fall 1994 (brighter areas indicate higher NDVI values, signifying more green vegetation). Different native vegetation types, which include chaparral, coastal sage scrub, annual grassland, and riparian areas are difficult to separate in the spring (A), but are more readily distinguished in fall (B) due to the contrasting seasonal patterns of these vegetation types. Vegetation types can also be readily separated with spectral mixture analysis, which models a landscape in terms of "endmember" fractions, where each endmember represents a spectral type (vegetation type in this case). (C) An example of an endmember fraction image for coastal sage scrub (brighter areas indicate areas with higher coastal sage scrub content). Images derived from NASA's AVIRIS sensor (see Table 1). In these images, north is to the right.

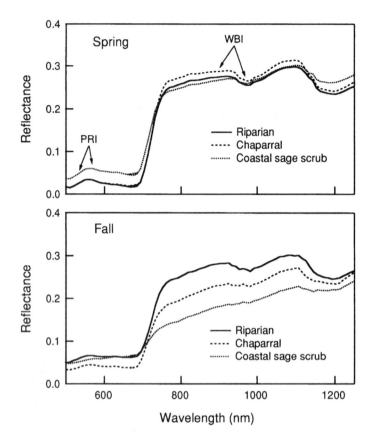

Figure 6 Spectra for representative vegetation types depicted in the AVIRIS image shown in Figure 5. Because vegetation types are spectrally similar in spring, they are hard to separate from a single overpass during that season. However, the contrasting spectral patterns emerging by fall allow ready separation of distinct vegetation types. Note how the water absorption band (near 970 nm), which is clearly visible in the spring spectra, becomes less apparent in the fall spectra, indicating a progressive drying of the landscape (see also Figure 11). Arrows indicate wavebands used for the photochemical reflectance index (PRI) and the water band index (WBI), discussed in Section IV.B. (*Source*: adapted from Gamon et al. 1998.)

respiratory contribution of nonphotosynthetic organs (e.g., stems, branches, and roots) can be incorporated into ε, or more explicitly included as a separate respiratory term or coefficient (Prince and Goward 1995, Ruimy et al. 1996, Landsberg et al. 1997).

Equations 2 and 3 can be viewed as a simple conceptual model, or can be further elaborated as a more explicit, mechanistic model. The APAR and efficiency (ε) terms determine the maximum attainable photosynthetic rate, often called the potential photosynthetic rate. Environmental or physiological factors that reduce photosynthetic rate through a variety of mechanisms impact either APAR or light-use efficiency, or both. For example, water stress, temperature extremes, or nutrient limitations can all reduce effective leaf-area index (LAI), and thus absorbed radiation (APAR). These same conditions can reduce light-use efficiency through down-regulation of photosynthetic processes by stomatal closure, enzyme inactivation, and altered light energy distribution involving photoprotective mechanisms (Björkman and Demmig-Adams 1994, Gamon et al. 1997; Figure 7).

This simple model can be applied at several temporal and spatial scales and can be readily linked to remote sensing, which is also scaleable, allowing application and validation at multiple spatial and temporal levels. The remote sensing link is usually provided by the APAR term, which can be further defined as the product of irradiance of photosynthetically active radiation (PAR) and the fraction of that irradiance that is absorbed by photosynthetic canopy elements (F_{APAR}).

$$APAR = F_{APAR} \times PAR \qquad (4)$$

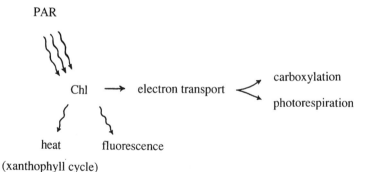

Figure 7 Schematic depicting possible fates of photosynthetically active radiation (PAR) absorbed by photosynthetic pigments (Chl). This absorbed energy can be used to drive photosynthetic electron transport and carbon uptake via carboxylation. Under conditions of reduced photosynthetic light-use efficiency, the photosynthetic system down-regulates, and an increased proportion of absorbed energy is dissipated by fluorescence or heat production. The operation of the xanthophyll cycle is linked to this heat dissipation (Pfündel and Bilger 1994, Demmig-Adams and Adams 1996, Demmig-Adams et al. 1996). Both xanthophyll pigment conversion and chlorophyll fluorescence provide useful means for optically detecting reductions in photosynthetic efficiency.

Both PAR irradiance (Frouin and Pinker 1995) and F_{APAR} can be determined remotely. F_{APAR} can be derived from NDVI or other vegetation indices, and this relationship has been well tested both with theoretical studies (Kumar and Monteith 1981, Asrar et al. 1984, Sellers 1987, Prince 1991) and empirical measurements at many spatial and temporal scales (Daughtry et al. 1983, Goward et al. 1985, Bartlett et al. 1990, Steinmetz et al. 1990, Gamon et al. 1995, Joel et al. 1997). Because of the strong links between vegetation indices and radiation absorbed by photosynthetic canopy elements, a number of models explicitly or implicitly use some form of Eq. 2 and 3 to derive photosynthetic fluxes or primary production. These models vary widely in complexity and detail, but can be roughly divided as follows: (1) models that assume a uniform efficiency for all vegetation types (Heimann and Keeling 1989, Myneni et al. 1995); and (2) models that allow efficiency to vary, either according to biome type (Ruimy et al 1994) or according to dynamic environmental conditions, notably temperature and water availability (Potter et al. 1993, Field et al. 1995, Prince and Goward 1995).

1. Models Assuming Constant Efficiency

Models in the first category assume that efficiency (ε) is relatively constant across time and space, that variation in photosynthetic flux (or NPP) can be primarily described by light absorption, and that environmental stresses act through reducing light absorption rather than light-use efficiency. This view appears to be supported from a number of studies, largely on unstressed crops (Monteith 1977, Kumar and Monteith 1981, Russell et al. 1989). This assumption of constant ε greatly simplifies the derivation of photosynthetic fluxes and primary production from satellite-derived NDVI and allows the derivation of potential photosynthetic flux or production without explicit consideration of photosynthetic down-regulation (Myneni et al. 1995). Global patterns of NDVI significantly correlate with spatial and temporal patterns of atmospheric CO_2 measured independently (Tucker et al. 1986, Fung et al. 1987), indicating that NDVI patterns capture much of the spatial and temporal variation in CO_2 flux and production at the global scale. However, it is important to remember that potential fluxes or production derived in this way do not necessarily agree with actual production or fluxes under conditions of photosynthetic down-regulation, and that direct validation is not readily available at this large scale.

This approach to deriving potential photosynthesis from NDVI by assuming a constant efficiency (ε) has now been carefully examined at the stand and landscape levels, using a combination of approaches that include remote sensing, modeling, biomass harvesting, APAR sampling, and flux measurements (Bartlett et al. 1990, Gamon et al. 1993, 1995, Running and Hunt 1993, Runyon et al. 1994, Valentini et al. 1995, Joel et al. 1997, Landsberg et al. 1997). The general conclusion that can be drawn from these studies is that the ability to predict photosynthetic fluxes from vegetation indices (and APAR) varies widely with vegetation type, condition, and temporal scale. For example, in certain annual grasslands, well-managed crops, and possibly deciduous forests, where physiological activity closely tracks canopy greenness (and light absorption), it is often possible to model photosynthesis and NPP accurately from APAR alone by assuming a constant efficiency (Monteith 1977, Russell et al. 1989, Gamon et al. 1993). By contrast, in water- or nutrient-stressed canopies (Joel et al. 1997, Landsberg 1997) or in evergreens exposed to periodically unfavorable environmental conditions (Running and Nemani 1988, Hunt and Running 1992, Running and Hunt 1993, Runyon et al. 1994, Gamon et al. 1995), it is inappropriate to assume a constant efficiency, particularly over short time periods (hours to months). In these cases, photosynthetic down-regulation can be significant, making it difficult to accurately predict fluxes from APAR alone. Because photosynthetic down-regulation can exert a significant feedback effect on atmospheric processes (Sellers et al. 1996a), varying efficiency should be considered in land–surface parameterizations of global circulation models. Clearly, efficiency must be closely considered if accurate flux predictions are to be made from remote sensing at a global scale.

2. Models with Varying Efficiency

In response to the realization of varying efficiency (ε), a number of photosynthesis or production models now include explicit parameters that consider the impacts of environmental stressors and attendant photosynthetic down-regulation on modeled fluxes. Some of these models now attempt to incorporate varying values for ε, either by biome type (Ruimy et al. 1994) or by linking ε to varying physiological status or environmental conditions (Potter et al. 1993, Prince and Goward 1995). Some of these models are considerably more complex than Eq. 2 and 3, but all share the common theme of defining a potential flux linked to absorbed radiation, scaled down by a term (or terms) that incorporate limiting environmental or physiological factors. In their current formulation, most of these models suffer from the fact that some of the information needed to run or validate them is not readily available at the same scale as remotely sensed inputs; consequently, some critical model parameters must be assumed or derived indirectly from measurements often made at inappropriate scales (Hall et al. 1995, Sellers et al. 1995). The challenge is to develop sensors and algorithms that provide all the necessary model inputs at the appropriate scale, without reference to external assumptions or excessive model tuning. In an attempt to resolve this problem, Prince and Goward (1995) developed a global scale model (GLO-PEM) with many variables derived entirely from 8 $\times$ 8-km resolution satellite data. This model yields reasonable estimates of gross and net primary production that are within the range of other global models, which demonstrates that deriving global fluxes entirely from remote sensing remains a reasonable goal. Advances in satellite sensors and improvements in data availability are likely to make such internally consistent global models more attainable. However, direct validation of global scale flux models remains an elusive goal, largely due to the shortage of large-scale field measurements (Hall et. al. 1995, Prince and Goward 1995, Sellers et al. 1995).

Several contrasting modeling approaches were recently compared by Field et al. (1995), who reported that several models converged on surprisingly similar results, despite their varying formulations and assumed or calculated ε values. In part, this convergence results from the retrospective nature of remote sensing, which largely captures what has already happened (Demetriades-Shah 1992, Field et al. 1995). To the extent that current or future conditions match past conditions, remote sensing driven models should do a good job of describing large-scale flux and production patterns, even if they do not fully capture the mechanisms that lead to those patterns. Model convergence also results because efficiency and APAR are clearly correlated; reductions in efficiency eventually lead to reductions in APAR, largely through altered canopy structure or reduced canopy development. Consequently, there are multiple ways to partition environmental stresses using the two aforementioned terms, and the literature abounds with apparently workable variations of this model (Potter et al. 1993, Prince and Goward 1995, Ruimy et al. 1996, Landsberg et al. 1997).

Clearly, the exact significance of down-regulation and varying efficiency remains a fundamental question, particularly at the global scale, where we lack direct mechanisms for validating efficiency, productivity, and fluxes (Hall et al. 1995, Prince and Goward 1995, Sellers et al. 1996). Consequently, it remains fundamentally unclear whether current models can adequately predict changing conditions due to successional state, climate change, or other disturbance. Several possible approaches for providing the missing information through remote sensing are addressed below.

B. New Approaches To Resolving the Question of Efficiency

The wide variety of current and emerging sensors (Table 1) provides a range of potential tools for evaluating radiation-use efficiency. New indices using narrow reflectance and fluorescence bands can provide direct assessment of physiological processes linked to photosynthetic light-use efficiency. For example, both chlorophyll fluorescence (at approximately 685 and 735 nm) and reflectance at 531 nm provide indicators of fundamental photoregulatory processes linked to carboxylation (Figures 7 and 8). The fluorescence index $\Delta F/Fm'$ (Genty et al. 1989) is a widely used measure of photosystem II radiation-use efficiency. However, as

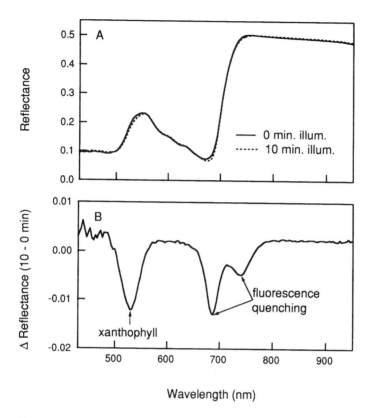

Figure 8 (A) Reflectance of a single Douglas fir (*Pseudotsuga menziesii*) needle sampled immediately upon illumination (0 min. illum.) and 10 minutes after illumination (10 min. illum.) with irradiance equivalent to full sun. (B) When plotted as a difference spectrum (10 min. minus 0 min.), subtle changes in apparent reflectance appear that can be attributed to xanthophyll pigment conversion (feature near 531 nm) and chlorophyll fluorescence quenching (double feature near 685 and 735 nm). Optical indices of xanthophyll pigment activity and chlorophyll fluorescence can be used to monitor changing photosynthetic light-use efficiency (Gamon et al. 1997). This Douglas fir needle spectrum was sampled from the Wind River Canopy Crane with a prototype "reflectometer" similar in design to the UniSpec (see Table 1).

with most commercial instruments, this fluorescence index is best applied at very close range (mm to cm), in part due to a requirement for saturating light pulses (Bolhar-Nordenkampf et al. 1989). New algorithms and advances in fluorescence technology, including laser-induced fluorescence, are beginning to relax this limitation (Günther et al. 1994). However, at this time, quantitative applications of chlorophyll fluorescence to photosynthesis are most easily applied at the leaf scale.

In contrast to fluorescence, reflectance is now being applied at several spatial scales to monitor the activity of xanthophyll cycle pigments. One common expression of this signature is the photochemical reflectance index (PRI):

$$PRI = (R_{531} - R_{REF})/(R_{531} + R_{REF}) \tag{5}$$

where R_{531} represents reflectance at 531 nm (the xanthophyll cycle band) and R_{REF} indicates reflectance at a reference wavelength (typically 570 nm; Peñuelas et al. 1995b, Gamon et al. 1997). This index requires a narrow-band detector, with bandwidths of approximately 10 nm full-width half maximum (FWHM) or finer, which includes the CASI and AVIRIS sensors, and any of the personal spectrometers listed in Table 1.

Because the xanthophyll cycle pigments function in photosynthetic light regulation (Pfündel and Bilger 1994, Demmig-Adams and Adams 1996, Demmig-Adams et al. 1996), and because interconversion of these pigments is detectable with spectral reflectance (Gamon et al. 1990), PRI provides a measure of radiation-use efficiency at the level of the fundamental photosynthetic light reactions. Because these reactions are closely linked to photosynthetic carbon uptake, they can also provide a near-direct index of photosynthetic light-use efficiency (Gamon et al. 1992 and 1997, Peñuelas et al. 1995b, 1997a, Filella et al. 1996).

Recently, relationships between PRI, light-use efficiency, and actual photosynthetic rates were examined for multiple species and functional types, and were found to be remarkably robust, provided sampling was restricted to leaves from upper-canopy regions under high light (Figure 9). Because this is the canopy portion and light exposure typically sampled with remote sensing, it is possible to detect this signal remotely at several spatial scales, at least for uniform, closed stands (Figure 10). Furthermore, this index can be applied to the efficiency term of Eq. 2 to derive real-time photosynthetic rates (Gamon et al. in preparation). However, more study is needed to test this index for landscapes and larger geographic regions, where the confounding influences of vegetation and landscape structure, atmospheric effects, and potential calibration errors can potentially impact this relatively small physiological signal. In particular, accurate spectral and radiometric calibration, as well as proper correction for atmospheric scattering in the derivation of surface reflectance, appear to be critical challenges in deriving a valid PRI from high-altitude platforms (Gamon et al. 1998). While a number of aircraft sensors exist that can potentially detect this narrow-band signal (Table 1), there clearly remain a number of technical challenges to sampling and validating this subtle signal, particularly from great distances. Currently, we lack satellite sensors with the appropriate bands for this index, and there are no direct ways for validating optical estimates of gross photosynthesis at scales larger than

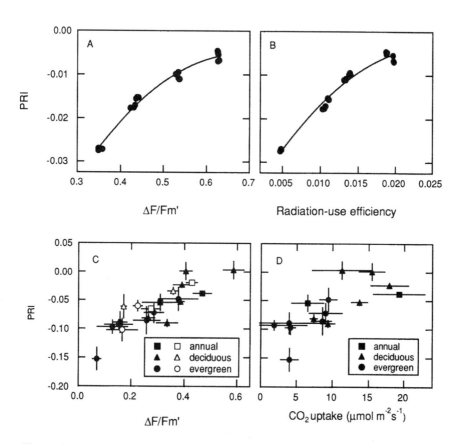

Figure 9 The photochemical reflectance index (PRI), an index of xanthophyll cycle activity, can be used to detect changing light-use efficiency. (A) PRI vs. $\Delta F/Fm'$ (a fluorescence-based index of light-use efficiency) for a single cotton (*Gossypium barbadense*) leaf exposed to varying light levels and CO_2 concentrations. (B) PRI vs. radiation-use efficiency (calculated as gross CO_2 assimilation rate divided by incident photon flux density) for the same cotton leaf shown in (A). (C) PRI vs. $\Delta F/Fm'$ for top-canopy leaves from 20 species all sampled at midday in full sun in Los Angeles, CA. Species include different functional types (annual, deciduous, or evergreen), and were either well fertilized (solid symbols) or given varying fertilizer treatments (open symbols). (D) Midday PRI vs. midday net CO_2 uptake rates for 15 of the 20 species shown in (C). CO_2 uptake was measured up to several days later under similar midday light conditions as PRI. Squared correlation coefficients (r^2 values) were (A) 0.93, (B) 0.94, (C) 0.75, and (D) 0.54. In all cases correlations were highly significant ($P < .01$). (*Source*: data from Gamon et al. 1997.)

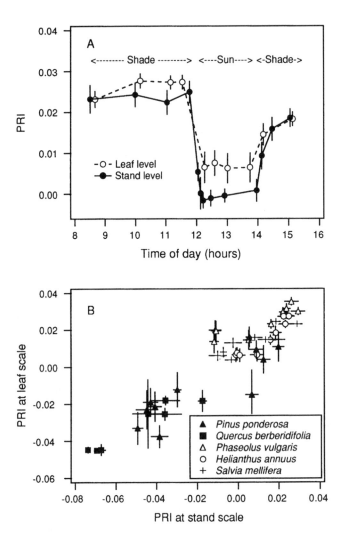

Figure 10 (A) The photochemical reflectance index (PRI) sampled at both the leaf and stand scales for sunflower (*Helianthus annuus*) exposed to alternating shade (PPFD approximately 60 μmol photons $m^{-2}s^{-1}$) and sun (PPFD approximately 1600 μmol photons $m^{-2}s^{-1}$). Similar kinetics of PRI, indicating xanthophyll cycle pigment conversion linked to changing photosynthetic light-use efficiency, are detectable at both the leaf and stand scales. (B) Results of five scaling experiments combining measurements for 5 species of varying photosynthetic rate, phenology, and structure (low PRI values indicate plants with low midday photosynthetic rates; Gamon et al. 1997). Drought-tolerant evergreens with low photosynthetic rates and PRI values are indicated with solid symbols, and annual crop plants with high photosynthetic rates and PRI values are indicated with open symbols. Pluses (+) represent *Salvia mellifera*, a drought-deciduous species native to California's coastal sage scrub communities. Data were collected with a prototype ''reflectometer'' similar in design to the UniSpec (see Table 1).

the single stand or canopy. Consequently, in the immediate future, this optical index of efficiency will continue to be most readily applicable at the leaf to stand scales.

Another approach to determining radiation-use efficiency takes advantage of the multidimensional nature of remotely sensed imagery to define a set of ecophysiological syndromes or functional vegetation types by combining multiple remotely sensed parameters sampled simultanously. Partly because of the confounding link between efficiency and light absorption previously discussed, strong correlations often exist between multiple reflectance indices. Contrasting patterns of ecophysiological function associated with contrasting functional types can be revealed as subtle perturbations in the relationship between structural and physiological indices. This approach is represented in Figure 11, which depicts seasonal changes in the relationship between the NDVI and liquid water content for the southern California landscape depicted in Figure 5. In this Mediterranean-climate ecosystem, water is a limiting resource during much of the year, and

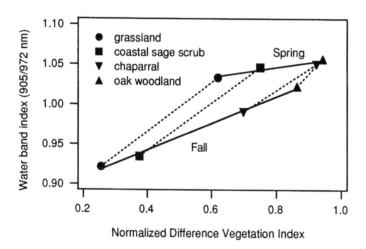

Figure 11 Values for the water band index (WBI) vs. the normalized difference vegetation index (NDVI) for different vegetation types, derived from AVIRIS images of Point Dume (Santa Monica Mountains, CA). The WBI was expressed as the ratio of reflectance at 905 and 972 nm. Each point represents the mean of 17 to 22 AVIRIS pixels (standard error bars are too small to see). Solid lines represent regression lines for samples selected from a single date in spring 1995 (top regression) or fall 1994 (bottom regression). Dashed lines represent hypothetical seasonal trajectories for each vegetation type in NDVI-WBI space. The progressive drying of the vegetation can also be visualized as the flattening of the 970-nm liquid water absorption feature (see Figure 6). (*Source*: data derived from the AVIRIS images depicted in Figure 5; adapted from Gamon et al. 1998.)

reductions in water availability are associated with reduced photosynthetic efficiency (Mooney et al. 1975, Tenhunen et al. 1985, Gamon et al. 1995). In this case, an index of liquid water absorption (the water band index, or WBI) and the NDVI are both derived from the same sensor (NASA's AVIRIS sensor), ensuring a common sampling scale. In both spring and fall, the WBI is strongly correlated with the NDVI, suggesting that the WBI may function as an alternate index of green leaf area. Recently, other investigators have reported similar correlations between measures of liquid water absorption and canopy greenness (Roberts et al. 1997, Ustin et al. 1998). However, the seasonal change in the NDVI–water relationship indicates a combination of changing canopy structure and loss of leaf water content, mirroring functional changes linked to seasonally varying photosynthetic efficiency in this drought-prone landscape (Gamon et al. 1998).

The tendency for functionally distinct vegetation types to occupy distinct trajectories in NDVI-WBI space suggests that functional classifications of vegetated landscapes can be derived entirely from imaging spectrometry using this approach. A similar example at the continental scale is provided by Nemani and Running (1997), who compared the NDVI to surface temperature, both derived from the AVHRR sensor (Table 1). In that study, different trajectories on the plot of the NDVI versus surface temperature indicate different vegetation types, presumably having different functional behavior. The elegance of these approaches is that they are entirely derived from multiple bands on a single sensor, thus ensuring that the inputs are at identical spatial scales. However, a remaining challenge lies in obtaining quantitative expressions of efficiency from these seasonally changing trajectories. Furthermore, existing efforts to derive efficiency entirely from multiple remotely sensed signals remain difficult to validate, particularly at large scales, due to the shortage of appropriate field measurements (Prince and Goward 1995). Although more work is clearly needed to fully understand the functional significance of the seasonally changing stuctural and physiological patterns discussed here, it is likely that further investigation of multiple dimensions in remotely sensed data will lead to an improved understanding of photosynthetic light-use efficiency at progressively larger spatial and temporal scales. The synergistic combination of multiple simultaneous bands and sensors is a principle goal of NASA's planned Earth Observing Stations (EOS) program (Ustin et al. 1991, Wickland 1991, Running et al. 1994). The emergence of new narrow-band sensors (Table 1) will provide new opportunities to explore ecosystem function using multiple indices.

V. REMOVING THE BARRIERS

Although remote sensing has considerable power and promise, there remain substantial barriers to the use of remote sensing for ecological studies. These limitations are both technical and cultural. On the technical side, many processes of interest to ecologists are currently inaccessible or poorly described by remote sensing. Remote sensing is particularly useful for depicting the production and energy input end of ecosystem fluxes, but it does not readily depict many important processes associated with allocation, respiration, or belowground processes. Similarly, most remote sensing approaches cannot readily describe species composition, distribution, and other demographic variables at the scales traditionally studied by population ecologists. Because remote sensing alone is insufficient to depict many processes critical to ecology, it is likely that examination of these critical processes will continue to require multiple approaches.

Because not all variables can be directly derived from remote sensing, we must often evaluate whether it is possible to do without information that was previously considered critical. Furthermore, because remote sensing measures differently from most field sampling methods, it forces us to rethink how we structure our concepts and models. One example is the concept of leaf area index, which is generally thought to be an important parameter for ecosystem and global photosynthesis models (Running and Hunt 1993, Sellers et al. 1996b). This view derives, in part, from the fact that leaf-level gas exchange is typically expressed on a leaf area basis (Field et al. 1989). Furthermore, leaf area is readily measurable at the canopy and stand scale, and ecologists, foresters, and agronomists now have a wide array of tools and procedures for determining leaf area (Norman and Campbell 1989, Daughtry 1990, Welles 1990, Chen and Cihlar 1995). Many remote sensing studies attempt to translate vegetation indices into leaf area index, as if they were necessarily closely related, and this often requires varying calibrations and assumptions for different vegetation types (Sellers et al. 1996b). However, as previously described, remote vegetation indices sample light absorption, or "effective" leaf area index, more directly, which incorporates clumping at the leaf and branch scales and can differ substantially from leaf area index determined by traditional methods (Chen 1996). Consequently, when using remote sensing, we must reconsider the functional significance and relevance of leaf area index as it has traditionally been applied. Perhaps it is time to more clearly define a new parameter, effective leaf area index, that is more closely linked to remote vegetation indices, light absorption, and photosynthetic production at the ecosystem level.

Similarly, ecologists have long considered the relative importance of species versus functional types in ecosystem function, and this topic has received considerable attention recently as new experimental evidence continues to emerge linking functional categories to ecosystem performance (Grime 1997, Hooper and Vitousek 1997, Tilman et al. 1997, Wardle et al. 1997). Although species will continue to be a fundamental unit for many branches of ecology, recent publications suggest that the concept of functional types will gradually emerge as a more accessible unit of ecosystem function (Chapin 1993, Schulze and Mooney 1994, Chapin et al. 1996, 1997, Lavorel et al. 1997, Smith et al. 1997). In many cases, remote sensing may be more able to depict functional types than species, and it is likely that the use of remote sensing will further spur the development of the concept and application of functional types.

Clearly, remote sensing presents a number of challenges to traditional views and practices, and this leads to the issue of culture and how we perform science. The scientific method, with experimental treatments, controls, and independent replicates, is not easily followed with the tools of remote sensing, particularly at the larger scales. One solution to this dilemma lies in the use of ''natural experiments,'' where temporal or spatial dimensions can be substituted for independent treatments. Another approach lies in the linking of remotely sensed parameters to process models that can be run as a series of ''virtual'' experiments, which is a common procedure in global modeling (Sellers et al. 1996a). In either case, remote sensing frequently requires that we redefine what is considered an acceptable level of evidence, and will force us to continually reevaluate the process of science itself, leading to new paradigms.

Reluctance to adopt remote sensing is often due to the ''fuzzy'' nature of data derived from great distances; we often are more comfortable with the exactness of readily or directly measurable quantities that match the scales best perceived by our senses. However, fine-scale observations may not be directly relevant to critical, large-scale processes. Moreover, there is great benefit in examining patterns and processes at multiple scales. This concern with how information transcends (or fails to transcend) scales can yield valuable insight into complex ecological processes, and remote sensing will undoubtedly continue to provide a principle tool for exploring the influence of scale on these processes (Wessman 1992, Ehleringer and Field 1993, Quattroch and Goodchild 1997). It is likely that ecologists will gain greater comfort with remote sensing as they gain familiarity with issues of scaling. The redefinition of traditional ecological questions to more closely match scales appropriate for remote sensing will undoubtedly lead to further progress in ecology.

Historically, remote sensing has been an engineering-driven field disproportionately influenced by military, intelligence, and commercial (e.g., mining) industries; consequently, many of the instruments of remote sensing have been designed in ignorance of the needs of ecology. As a community, we have inherited the tools designed for other purposes and have been struggling to adapt these tools to new uses. There is a great need for ecologists to influence the design and use of remote sensing tools, so that the instrumentation more closely reflects the evolving needs of the ecological community. This can most effectively be done if the larger scientific community works across disciplinary boundaries and if ecologists gain familiarity with the ''view from above.'' This need to combine varied disciplines and develop new tools to address key ecological issues has been a primary focus of a number of recent initiatives sponsored by NASA's

Mission to the Planet Earth (recently renamed the "Earth Science Enterprise") (Ustin et al. 1991, Wickland 1991, Asrar and Dozier 1994, Running et al. 1994). These initiatives have included large-scale field studies (Sellers et al. 1990, 1992, Peterson and Waring 1994, Prince et al. 1995, Margolis and Ryan 1997), new multisensor platforms designed with input from ecologists (Running et al. 1994), and increasing attempts to create standardized data products accessible to the ecological community (Goward et al. 1994, James and Kalluri 1994, Los et al. 1994, Sellers et al. 1994, Townshend 1994, Wharton and Myers 1997). These interdisciplinary efforts will undoubtedly lead to many new advances in this rapidly evolving field.

ACKNOWLEDGMENTS

Thanks to D. Roberts and M. Gardner for provision of atmospheric correction for AVIRIS imagery (Figure 5), and to J. Surfus, D. Garcia, L. Lee, E. Yi, and D. Ponciano for assistance with the experimental results depicted in Figure 10. The spectra in Figure 8 were collected with assistance from personnel at the Wind River Canopy Crane Research Facility located within the T. T. Munger Research Natural Area in Washington state. This facility is a cooperative scientific venture among the University of Washington, the USFS Pacific Northwest Research Station, and the USFS Gifford Pinchot National Forest. Thanks also to C. J. Fotheringham, S. D. Prince, J. Randerson, D. Roberts, and A. Ruimy, who reviewed the manuscript and provided many valuable suggestions for the final draft. Many of the research findings presented here were supported by grants from NASA, NSF, and the U.S. EPA. This is publication number 5 of the Center for Spatial Analysis and Remote Sensing (CSARS) at California State University, Los Angeles.

REFERENCES

Aber, J.D., C.A. Wessman, D.L. Peterson, J.M. Melillo, and J.H. Fownes. 1990. Remote sensing of litter and soil organic matter decomposition in forest ecosystems. In: Hobb, R.J., and H.A. Mooney, eds. Remote Sensing of Biosphere Functioning. New York: Springer-Verlag, pp 87–103.

Aber, J.D., and C.A. Federer. 1992. A generalized, lumped-parameter model of photosynthesis, evaporation, and net primary production in temperate and boreal forest ecosystems. Oecologia 92: 463–474.

Amthor, J.S. 1995. Terrestrial higher-plant response to increasing atmospheric [CO_2] in relation to the global carbon cycle. Global Change Biology 1: 243–274.

Asrar, G., M. Fuchs, E.T. Kanemasu, and J.L. Hatfield. 1984. Estimating absorbed photosynthetic radiation and leaf area index from spectral reflectance in wheat. Agronomy Journal 76: 300–306.

Asrar, G., and J. Dozier. 1994. EOS Science Strategy for the Earth Observing System. Woodbury, NY: American Institute of Physics Press.

Baret, F., and G. Guyot. 1991. Potentials and limits of vegetation indices for LAI and APAR assessment. Remote Sensing of Environment 35: 161–173.

Barnsley, M.J. 1994. Environmental monitoring using multiple-view-angle (MVA) remotely-sensed data. In: Foody, G., and P. Curran, eds. Environmental Remote Sensing from Regional to Global Scales. Chichester: Wiley, pp 181–201.

Bartlett, D.S., G.H. Whiting, and J.M. Hartman. 1990. Use of vegetation indices to estimate intercepted solar radiation and net carbon dioxide exchange of a grass canopy. Remote Sensing of Environment 30: 115–128.

Björkman, O., and B. Demmig-Adams. 1994. Regulation of photosynthetic light energy capture, conversion, and dissipation in leaves of higher plants. In: Schulze, E.-D., and M.M. Caldwell, eds. Ecophysiology of Photosynthesis. Berlin: Springer-Verlag, pp 17–47.

Bohlar-Nordenkampf, H.R., S.P. Long, N.R. Baker, G. Oquist, U. Schreiber, and E.G. Lechner. 1989. Chlorophyll fluorescence as a probe of the photosynthetic competence of leaves in the field: a review of current instrumentation. Functional Ecology 3: 497–514.

Bull, C.R. 1991. Wavelength selection for near-infrared reflectance moisture meters. Journal of Agricultural Engineering Research 49: 113–125.

Carter, G.A. 1991. Primary and secondary effects of water concentration on the spectral reflectance of leaves. American Journal of Botany 78: 916–924.

Carter, G.A. 1994. Ratios of leaf reflectances in narrow wavebands as indicators of plant stress. International Journal of Remote Sensing 15: 697–703.

Chapin, F.S. III 1993. Functional role of growth forms in ecosystem and global processes. In: Ehleringer, J.R., and C.B. Field, eds. Scaling Physiological Processes: Leaf to Globe. San Diego, CA: Academic Press, pp 287–312.

Chapin, F.S. III, B.H. Walker, R.J. Hobbs, D.U. Hooper, J.H. Lawton, O.E. Sala, and D. Tilman. 1997. Biotic control over the functioning of ecosystems. Science 277: 500–504.

Chapin, F.S. III, H.L. Reynolds, C.M. D'Antonio, and V.M. Eckhart. 1996. The functional role of species in terrestrial ecosystems. In: Walker, B., and W. Steffen, eds. Global Change and Terrestrial Ecosystems. Cambridge: Cambridge University Press, pp 403–428.

Chen, J.M. 1996. Optically-based methods for measuring seasonal variation of leaf area index in boreal conifer stands. Agricultural and Forest Meteorology. 80: 135–163.

Chen, J.M., and J. Cihlar. 1995. Plant canopy gap-size analysis theory for improving optical measurements of leaf area index. Applied Optics 34: 6211–6222.

Clark, R.N., and T.L. Roush. 1984. Reflectance spectroscopy: quantitative analysis techniques for remote sensing applications. Journal of Geophysical Research 89: 6329–6340.

Coley, P.D. and J.A. Barone. 1996. Herbivory and plant defenses in tropical forests. Annual Review of Ecology and Systematics 27: 305–335.

Curran, P.J. 1989. Remote sensing of foliar chemistry. Remote Sensing of Environment 30: 271–278.

Curran, P.J., and J.A. Kupiec. 1995. Imaging spectrometry: a new tool for ecology. In: Danson, F.M., and S.E. Plummer, eds. Advances in Environmental Remote Sensing. Chichester: Wiley, pp 71–88.

Curran, P.J., J.L. Dungan, and H.L. Gholz. 1990. Exploring the relationship between reflectance red edge and chlorophyll content in slash pine. Tree Physiology 7: 33–48.

Danson, F.M., and S.E. Plummer, eds. 1995. Advances in Environmental Remote Sensing. Chichester: Wiley.

Daughtry, C.S.T. 1990. Direct measurements of canopy structure. Remote Sensing Reviews. 5: 45–60.

Daughtry, C.S.T., K.P. Gallo, and M.E. Bauer. 1983. Spectral estimates of solar radiation intercepted by corn canopies. Agronomy Journal 75: 527–531.

DeFries, R.S., and J.R.G. Townshend. 1994a. NDVI-derived land cover classification at a global scale. International Journal of Remote Sensing 15: 3567–3586.

DeFries, R.S., and J.R.G. Townshend. 1994b. Global land cover: comparison of ground-based data sets to classifications with AVHRR data. In: Foody, G., and P. Curran, eds. Environmental Remote Sensing from Regional to Global Scales. Chichester: Wiley, pp 84–110.

DeFries, R.S., C.B. Field, I. Fung, C.O. Justice, S. Los, P.A. Matson, E. Matthews, H.A. Mooney, C.S. Potter, K. Prentice, P.J. Sellers, J.R.G. Townshend, C.J. Tucker, S.L. Ustin, and P.M. Vitousek. 1995. Mapping the land surface for global atmosphere-biosphere models: toward continuous distributions of vegetation's functional properties. Journal of Geophysical Research—Atmospheres 100: 867–882.

Demetriades-Shah T.H., M. Fuchs, E.T. Kanemasu, and I. Flitcroft. 1992. A note of caution concerning the relationship between cumulative intercepted solar radiation and crop growth. Agricultural Forest Meteorology 58: 193–207.

Demetriades-Shah, T.H., M.D. Steven, and J.A. Clark. 1990. High resolution derivative spectra in remote sensing. Remote Sensing of Environment 33: 55–64.

Demmig-Adams, B., A.M. Gilmore, and W.W. Adams III. 1996. In vivo functions of carotenoids in higher plants. FASEB Journal 10: 403–412.

Demmig-Adams, B., and W.W. Adams III. 1996. The role of xanthophyll cycle carotenoids in the protection of photosynthesis. Trends in Plant Science 1: 21–26.

Ehleringer, J.R., and C.B. Field, eds. 1993. Scaling Physiological Processes: Leaf to Globe. San Diego, CA: Academic Press.

Field, C.B., J.T. Ball, and J.A. Berry. 1989. Photosynthesis: principles and field techniques. In: Pearcy, R.W., J. Ehleringer, H.A. Mooney, and P.W. Rundel, eds. Plant Physiological Ecology: Field Methods and Instrumentation. London: Chapman and Hall, pp 209–253.

Field, C.B., J.T. Randerson, and C.M. Malmstrom. 1995. Global net primary production: combining ecology and remote sensing. Remote Sensing of Environment 51: 74–88.

Filella, I., L. Serrano, J. Serra, and J. Peñuelas. 1995. Evaluating wheat nitrogen status with canopy reflectance indices and discriminant analysis. Crop Science 35: 1400–1405.

Filella, I., T. Amaro, J.L. Araus, and J. Peñuelas. 1996. Relationship between photosynthetic radiation-use efficiency of barley canopies and the photochemical reflectance index (PRI). Physiologia Plantarum 96: 211–216.

Foody, G., and P. Curran, eds. 1994. Environmental Remote Sensing from Regional to Global Scales. Chichester: Wiley.

Frouin, R., and R.T. Pinker. 1995. Estimating photosynthetically active radiation (PAR) at the earth's surface from satellite observations. Remote Sensing of Environment 51: 98–107.

Fung, I.Y., C.J. Tucker, and K.C. Prentice. 1987. Application of advanced very high resolution radiometer vegetation index to study atmosphere-biosphere exchange of CO_2. Journal of Geophysical Research 92: 2999–3015.

Gamon, J.A., C.B. Field, D.A. Roberts, S.L. Ustin, and R. Valentini. 1993. Functional patterns in an annual grassland during an AVIRIS overflight. Remote Sensing of Environment 44: 239–253.

Gamon, J.A., C.B. Field, M.L. Goulden, K.L. Griffin, A.E. Hartley, G. Joel, J. Peñuelas, and R. Valentini. 1995. Relationships between NDVI, canopy structure, and photosynthesis in three Californian vegetation types. Ecological Applications 5: 28–41.

Gamon, J.A., C.B. Field, W. Bilger, O. Björkman, A.L. Fredeen, and J. Peñuelas. 1990. Remote sensing of the xanthophyll cycle and chlorophyll fluorescence in sunflower leaves and canopies. Oecologia 85: 1–7.

Gamon, J.A., J. Peñuelas, and C.B. Field. 1992. A narrow-waveband spectral index that tracks diurnal changes in photosynthetic efficiency. Remote Sensing of Environment 41: 35–44.

Gamon, J.A., L.-F. Lee, H.-L. Qiu, S. Davis, D.A. Roberts, and S.L. Ustin. 1998. A multiscale sampling strategy for detecting physiologically significant signals in AVIRIS imagery. In: Summaries of the Seventh Annual JPL Earth Science Workshop, January 12–16, 1988, Pasadena, CA [website] (http://makula.jpl.nasa.gov/docs/workshops/98_docs/toc.htm).

Gamon, J.A., L. Serrano, and J.S. Surfus. 1997. The photochemical reflectance index: an optical indicator of photosynthetic radiation use efficiency across species, functional types, and nutrient levels. Oecologia. 112: 492–501.

Gao, B.-C., and A.F.H. Goetz. 1990. Column atmospheric water vapor and vegetation liquid water retrievals from airborne imaging spectrometer data. Journal of Geophysical Research 95: 3549–3564.

Gao, B.-C., and A.F.H. Goetz. 1994. Extraction of dry leaf spectral features from reflectance spectra of green vegetation. Remote Sensing of Environment 47: 369–374.

Gao, B.-C., and A.F.H. Goetz. 1995. Retrieval of equivalent water thickness and information related to biochemical components of vegetation canopies from AVIRIS data. Remote Sensing of Environment 52: 155–162.

Gastellu-Etchegorry, J.P., F. Zagolski, E. Mougin, G. Marty, and G. Giordano. 1995. An assessment of canopy chemistry with AVIRIS—a case study in the Landes Forest, south-west France. International Journal of Remote Sensing 16: 487–501.

Gates, D.M. 1993. Climate Change and its Biological Consequences. Sunderland: Sinauer Associates.

Genty, B., J.-M. Briantais, and N.R. Baker. 1989. The relationship between the quantum yield of photosynthetic electron transport and quenching of chlorophyll fluorescence. Biochimica et Biophysica Acta 990: 87–92.

Gholz, H.L., K. Nakane, and H. Shimoda, eds. 1997. The Use of Remote Sensing in the Modeling of Forest Productivity. Dordrecht: Kluwer Academic Publishers.

Gitelson, A., and M.N. Merzlyak. 1994. Spectral reflectance changes associated with autumn senescence of *Aesculus hippocastanum* L. and *Acer platanoides* L. leaves: spectral features and relation to chlorophyll estimation. Journal of Plant Physiology 143: 286–292.

Gitelson, A., and M.N. Merzlyak. 1996. Signature analysis of leaf reflectance spectra: algorithm development for remote sensing of chlorophyll. Journal of Plant Physiology 148: 494–500.

Gitelson, A., M.N. Merzlyak, and H.K. Lichtenthaler. 1996. Detection of red edge position and chlorophyll content by reflectance measurements near 700 nm. Journal of Plant Physiology 148: 501–508.

Gould, K.S., D.N. Kuhn, D.W. Lee, and S.F. Oberbauer. 1995. Why leaves are sometimes red. Nature 378: 241–242.

Goward, S.N., C.J. Tucker, and D.G. Dye. 1985. North American vegetation patterns observed with the NOAA-7 advanced very high resolution radiometer. Vegetatio 64: 3–14.

Goward, S.N., S. Turner, D.G. Dye, and S. Liang. 1994. The University of Maryland improved Global Vegetation Index product. International Journal of Remote Sensing 15: 3365–3395.

Green, R.O., J.E. Conel, and D.A. Roberts. 1993. Estimation of aerosol optical depth, pressure elevation, water vapor and calculation of apparent surface reflectance from radiance measured by the Airborne Visible-Infrared Imaging Spectrometer (AVIRIS) using MODTRAN2. SPIE Proceedings 1937: 2–12.

Green, R.O., J.E. Conel, J.S. Margolis, C.J. Brugge, and G.L. Hoover. 1991. An inversion algorithm for retrieval of atmospheric and liquid water absorption from AVIRIS radiance with compensation for atmospheric scattering. Proceedings of the Third AVIRIS Workshop, May 20–21, 1991, JPL 91-28: 51–61.

Grime, J.P. 1997. Biodiversity and ecosystem function: the debate deepens. Science 277: 1260–1261.

Grossman, Y.L., S.L. Ustin, S. Jacquemoud, and E.W. Sanderson. 1996. Critique of stepwise multiple linear regression for the extraction of leaf biochemistry information from leaf reflectance data. Remote Sensing of Environment 56: 182–193.

Günther, K.P., H.-G. Dahn, and W. Lüdeker. 1994. Remote sensing vegetation status by laser-induced fluorescence. Remote Sensing of Environment 47: 10–17.

Hall, F.G., D.B. Botkin, D.E. Strebel, K.D. Woods, and S.J. Goetz. 1991. Large-scale patterns of forest succession as determined by remote sensing. Ecology 72: 628–640.

Hall, F.G., J.R. Townshend, and E.T. Engman. 1995. Status of remote sensing algorithms for estimation of land surface state parameters. Remote Sensing of Environment 51: 138–156.

Heimann, M., and C.D. Keeling. 1989. A three-dimensional model of atmospheric CO_2 transport based on observed winds: 2. Model description and simulated tracer experiments. In: Peterson, D.H., ed. Aspects of Climate Variability in the Pacific and the Western Americas. AGU Monograph, 55. Washington, D.C.: American Geophysical Union, pp 237–275.

Hess, L.L., and J.M. Melack. 1994. Mapping wetland hydrology and vegetation with synthetic aperture radar. International Journal of Ecology and Environmental Sciences 20: 197–205.

Hobbs, R.J. 1990. Remote sensing of spatial and temporal dynamics of vegetation. In: Hobbs, R.J., and H.A. Mooney, eds. Remote Sensing of Biosphere Functioning. New York: Springer-Verlag. pp 203–219.

Hobbs, R.J., and H.A. Mooney, eds. 1990. Remote Sensing of Biosphere Functioning. New York: Springer-Verlag.

Holben, B.N. 1986. Characteristics of maximum value composite images from temporal AVHRR data. International Journal of Remote Sensing 7: 1417–1434.

Hooper, D.U., and P.M. Vitousek. 1997. The effects of plant composition and diversity on ecosystem processes. Science 277: 1302–1305.

Houghton, R.A. 1995. Land-use change and the carbon cycle. Global Change Biology 1: 275–287.

Huete, A.R., H.Q. Liu, K. Batchily, and W. van Leeuwen. 1997. A comparison of vegetation indices over a global set of TM Images for EOS-MODIS. Remote Sensing of Environment 59: 440–451.

Hunt, E.R., and S.W. Running. 1992. Simulated dry matter yield for aspen and spruce stands in the North American boreal forest. Canadian Journal of Remote Sensing 18: 126–133.

James, M.E., and S.N.V. Kalluri. 1994. The Pathfinder AVHRR land data set: an improved course resolution data set for terrestrial monitoring. International Journal of Remote Sensing 15: 3347–3363.

Jensen, J.R. 1996. Introductory Digital Image Processing: A Remote Sensing Perspective, 2nd Ed. Upper Saddle River, NJ: Prentice Hall.

Joel, G., J.A. Gamon, and C.B. Field. 1997. Production efficiency in sunflower: the role of water and nitrogen stress. Remote Sensing of Environment 62: 176–188.

Johnson, L.F., C.A. Hlavka, and D.L. Peterson. 1994. Multivariate analyis of AVIRIS data for canopy biochemical estimation along the Oregon Transect. Remote Sensing of Environment 47: 216–230.

Justice, C.O., J.R.G. Townshend, B.N. Holben, and C.J. Tucker. 1985. Analysis of the phenology of global vegetation using meteorological satellite data. International Journal of Remote Sensing 6: 1271–1318.

Kasischke, E.S., J.M. Melack, M.C. Dobson. 1997. The use of imaging radars for ecological applications—a review. Remote Sensing of Environment 59: 141–156.

Knipling, E.B. 1969. Leaf reflectance and image formation on color infrared film. In: Johnson, P.L., ed. Remote Sensing in Ecology. Athens, GA: University of Georgia Press, pp 17–29.

Kumar, M., and J.L. Monteith. 1981. Remote sensing of crop growth. In: Smith, H., ed. Plants and the Daylight Spectrum. London: Academic Press, pp 133–144.

Landsberg, J.J., S.D. Prince, P.G. Jarvis, R.E. McMurtrie, R. Luxmoore, and B.E. Medlyn. 1997. Energy conversion and use in forests: an analysis of forest production in terms of radiation utilisation efficiency (ε). In: Gholz, H.L., K. Nakane, and H.

Shimoda, eds. The Use of Remote Sensing in the Modeling of Forest Productivity. Dordrecht: Kluwer Academic Publishers, pp 273–298.

Lavorel, S., S. McIntyre, J. Landsberg, and T.D.A. Forbes. 1997. Plant functional classifications: from general groups to specific groups based on response to disturbance. Trends in Ecology and Evolution 12: 474–478.

Lillesand, T.M., and R.W. Kiefer. 1987. Remote Sensing and Image Interpretation, 2nd Ed. New York: Wiley.

Los, S.O., C.O. Justice, and C.J. Tucker. 1994. A global 1° by 1° NDVI data set for climate studies derived from the GIMMS continental NDVI data. International Journal of Remote Sensing 15: 3493–3518.

Malingreau, J.P., C.J. Tucker, and N. Laporte. 1989. A VHRR for monitoring global tropical deforestation. International Journal of Remote Sensing 10: 855–867.

Margolis, H.A., and M.G. Ryan. 1997. A physiological basis for biosphere-atmosphere interactions in the boreal forest: an overview. Tree Physiology 17: 491–499.

Martin, M.E., and J.D. Aber. 1997. High spectral resolution remote sensing of forest canopy lignin, nitrogen, and ecosystem processes. Ecological Applications 72: 431–443.

Matson, P.A., L. Johnson, C. Billow, J. Miller, and R. Pu. 1994. Seasonal patterns and remote spectral estimation of canopy chemistry across the Oregon Transect. Ecological Applications 4: 280–298.

Monteith, J.L. 1977. Climate and the efficiency of crop production in Britain. Philosophical Transactions of the Royal Society of London 281: 277–294.

Mooney, H.A., A.T. Harrison, and P.A. Morrow. 1975. Environmental limitations of photosynthesis on a California evergreen shrub. Oecologia (Berlin) 19: 293–301.

Myneni, R.B., and D.L. Williams. 1994. On the relationship between FAPAR and NDVI. Remote Sensing of Environment 49: 200–211.

Myneni, R.B., and G. Asrar. 1994. Atmospheric effects and spectral vegetation indices. Remote Sensing of Environment 47: 390–402.

Myneni, R.B., C.D. Keeling, C.J. Tucker, G. Asrar, and R.R. Nemani. 1997. Increased plant growth in the northern high latitudes from 1981–1991. Nature 386: 698–702.

Myneni, R.B., S.O. Los, and G. Asrar. 1995. Potential gross primary productivity of terrestrial vegetation from 1982–1990. Geophysical Research Letters 22: 2617–2620.

Nemani, R., and S. Running. 1997. Land cover characterization using multitemporal red, near-IR, and thermal-IR data from NOAA/AVHRR. Ecological Applications 7: 79–90.

Nobel, P.S. 1991. Physicochemical and Environmental Plant Physiology. San Diego, CA: Academic Press.

Norman, J.M, and G.S. Campbell. 1989. Canopy structure. In: Pearcy, R.W., J. Ehleringer, H.A. Mooney, and P.W. Rundel, eds. Plant Physiological Ecology: Field Methods and Instrumentation. London: Chapman and Hall, pp 301–325.

Pearcy, R.W. 1989. Radiation and light measurements. In: Pearcy, R.W., J. Ehleringer, H.A. Mooney, and P.W. Rundel, eds. Plant Physiological Ecology: Field Methods and Instrumentation. London: Chapman and Hall, pp 97–116.

Peñuelas, J., J.A. Gamon, A.L. Fredeen, J. Merino, and C.B. Field. 1994. Reflectance indices associated with physiological changes in nitrogen- and water-limited sunflower leaves. Remote Sensing of Environment 48: 135–146.

Peñuelas, J., F. Baret, and I. Filella. 1995a. Semi-empirical indices to assess carotenoids/ chlorophyll a ratio from leaf spectral reflectance. Photosynthetica 31: 221–230.

Peñuelas, J., I. Filella, and J.A. Gamon. 1995b. Assessment of photosynthetic radiation-use efficiency with spectral reflectance. New Phytologist 131: 291–296.

Peñuelas, J., I. Filella. 1998. Visible and near-infrared reflectance techniques for diagnosing plant physiological status. Trends in Plant Science 3: 151–156.

Peñuelas, J., I. Filella, C. Biel, L. Serrano, and R. Save. 1993. The reflectance at the 950-970 nm region as an indicator of plant water status. International Journal of Remote Sensing 14: 1887–1905.

Peñuelas, J., J. Llusia, J. Piñol, and I. Filella. 1997a. Photochemical reflectance index and leaf photosynthetic radiation-use efficiency assessement in Mediterranean trees. International Journal of Remote Sensing 18: 2863–2868.

Peñuelas, J., J. Piñol, R. Ogaya, and I. Filella. 1997b. Estimation of plant water concentration by the reflectance Water Index WI (R900/R970). International Journal of Remote Sensing 18: 2869–2875.

Perry, C.R., and L.F. Lautenschlager. 1984. Functional equivalence of spectral vegetation indices. Remote Sensing of Environment 14: 169–182.

Peterson, D.L., and R.H. Waring. 1994. Overview of the Oregon Transect Ecosystem Research project. Ecological Applications 4: 211–225.

Peterson, D.L., J.D. Aber, P.A. Matson, D.H. Card, N. Swanberg, C. Wessman, and M. Spanner. 1988. Remote sensing of forest canopy and leaf biochemical content. Remote Sensing of Environment 24: 85–108.

Pfündel, E., and W. Bilger. 1994. Regulation and possible function of the violaxanthin cycle. Photosynthesis Research 42: 89–109.

Pinzon, J.E., S.L. Ustin, C.M. Castenada, and M.O. Smith. 1998. Investigation of leaf biochemistry by hierarchical foreground/background analysis. IEEE Trans. Geosci. and Remote Sensing 36: 1–15.

Potter, C.S., J.T. Randerson, C.B. Field, P.A. Matson, P.M. Vitousek, H.A. Mooney, and S.A. Klooster. 1993. Terrestrial ecosystem production: a process model based on global satellite and surface data. Global Biogeochemical Cycles 7: 811–841.

Prince, S.D, and S.N. Goward. 1995. Global primary production: a remote sensing approach. Journal of Biogeography 22: 815–835.

Prince, S.D. 1991. A model of regional primary production for use with coarse-resolution satellite data. International Journal of Remote Sensing 12: 1313–1330.

Prince, S.D., Kerr Y.H., J.-P. Goutorbe, T. Lebel, A. Tinga, P. Bessemoulin, J. Brouwer, A.J. Dolman, E.T. Engman, J.H.C. Gash, M. Hoepffner, P. Kabat, B. Monteny, F. Said, P. Sellers, and J. Wallace. 1995. Geographic, biological, and remote sensing aspects of the Hydrologic Amospheric Pilot Experiment in the Sahel (HAPEX-Sahel). Remote Sensing of Environment 51: 215–234.

Quattrochi, D.A., and M.A. Goochild, eds. 1997. Scale in Remote Sensing and GIS. Boca Raton, FL: CRC Press.

Raven, P.H., H.J. Thompson, and B.A. Prigge. 1986. Flora of the Santa Monica Mountains, California, 2nd Ed. Southern California Botanists, Special Publication No. 2. Los Angeles: University of California.

Richards, J.A. 1993. Remote Sensing Digital Image Analysis. Berlin: Springer-Verlag.

Roberts, D.A., J.B. Adams, and M.O. Smith. 1993. Discriminating green vegetation non-photosynthetic vegetation and soils in AVIRIS data. Remote Sensing of Environment 44: 255–270.

Roberts, D.A., M. Gardner, R. Church, S. Ustin, G. Scheer, and R.O. Green. 1998. Mapping chaparral in the Santa Monica Mountains using multiple endmember spectral mixture models. Remote Sensing of Environment 65: 267–279.

Roberts, D.A., R.O. Green, and J.B. Adams. 1997. Temporal and spatial patterns in vegetation and atmospheric properties from AVIRIS. Remote Sensing of Environment 62: 223–240.

Ruimy, A., G. Dedieu, and B. Saugier. 1996. TURC: a diagostic model of continental gross primary productivity and net primary productivity. Global Biogeochemical Cycles 10: 269–285.

Ruimy, A., B. Saugier, and G. Dedieu. 1994. Methodology for the estimation of terrestrial primary production from remotely sensed data. Journal of Geophysical Research 99: 5263–5283.

Running, S.W., and E.R. Hunt, Jr. 1993. Generalization of a forest ecosystem process model for other biomes, BIOME-BGC, and an application for global-scale models. In: Ehleringer, J.R., and C.B. Field, eds. Scaling Physiological Processes: Leaf to Globe. San Diego, CA: Academic Press, pp 141–158.

Running, S.W., and R.R. Nemani. 1988. Relating seasonal patterns of the AVHRR vegetation index to simulated photosynthesis and transpiration of forests in different climates. Remote Sensing of Environment 24: 347–367.

Running, S.W., C.O. Justice, V. Salomonson, D. Hall, J. Barker, Y.J. Kaufmann, A.H. Strahler, A.R. Huete, J.-P. Muller, V. Vanderbilt, Z.M. Wan, P. Teillet, and D. Carneggie. 1994. Terrestrial remote sensing science and algorithms planned for EOS/MODIS. International Journal of Remote Sensing 15: 3587–3620.

Runyon, J., R.H. Waring, S.N. Goward, and J.M. Welles. 1994. Environmental limits on net primary production and light-use efficiency across the oregon transect. Ecological Applications 4: 226–237.

Russell, G., P.G. Jarvis, and J.L. Monteith. 1989. Absorption of radiation by canopies and stand growth. In: Russell, G., B. Marshall, and P.G. Jarvis, eds. Plant Canopies: Their Growth, Form and Function. Cambridge: Cambridge University Press, pp 21–39.

Sabins, F.F. 1997. Remote Sensing: Principles and Interpretation, 3rd ed. New York: W.H. Freeman and Company.

Salerno, A.E. 1976. Aerospace photography. In: Schanda, E., ed. Remote Sensing for Environmental Sciences. Berlin: Springer-Verlag, pp 11–83.

Schulze, E.-D., and H.A. Mooney, eds. 1994. Biodiversity and Ecosystem Function. Berlin: Springer-Verlag.

Sellers, P.J. 1987. Canopy reflectance, photosynthesis, and transpiration. II. The role of biophysics in the linearity of their interdependence. Remote Sensing of Environment 21: 143–183.

Sellers, P.J., B.W. Meeson, F.G. Hall, G. Asrar, R.E. Murphy, R.A. Schiffer, F.P. Bretherton, R.E. Dickinson, R.G. Ellingson, C.B. Field, K.F. Huemmrich, C.O. Justice, J.M. Melack, N.T. Roulet, D.S. Schimel, and P.D. Try. 1995. Remote sensing of the land surface for studies of global change: models–algorithms–experiments. Remote Sensing of Environment 51: 3–26.

Sellers, P.J., C.J. Tucker, G.J. Collatz, S.O. Los, C.O. Justice, D.A. Dazlich, and D.A. Randall. 1994. A global 1° by 1° NDVI data set for climate studies. Part 2: the generation of global fields of terrestrial biophysical parameters from the NDVI. International Journal of Remote Sensing 15: 3519–3545.

Sellers, P.J., F.G. Hall, D.E. Strebel, G. Asrar, and R.E. Murphy. 1990. Satellite remote sensing and field experiments. In: Hobbs, R.J., and H.A. Mooney, eds. Remote Sensing of Biosphere Functioning. New York: Springer-Verlag, pp 169–201.

Sellers, P.J., F.G. Hall, G. Asrar, D.E. Strebel, and R.E. Murphy. 1992. An Overview of the First ISLSCP Field Experiment. Journal of Geophysical Research 97: 18345–18372.

Sellers, P.J., L. Bounoua, G.J. Collatz, D.A. Randall, D.A. Dazlich, S.O. Los, J.A. Berry, I. Fung, C.J. Tucker, C.B. Field, and T.G. Jensen. 1996a. Comparison of radiative and physiological effects of doubled atmospheric CO_2 on climate. Science 271: 1402–1406.

Sellers, P.J., S.O. Los, C.J. Tucker, C.O. Justice, D.A. Dazlich, G.J. Collatz, and D.A. Randall. 1996b. A revised land surface parameterization (SiB2) for atmospheric GCMs. Part II: the generation of global fields of terrestrial biophysical parameters from satellite data. Journal of Climate 9: 706–737.

Smith, T.M., H.H. Shugart, and F.I. Woodward, eds. 1997. Plant Functional Types: Their Relevance to Ecosystem Properties and Global Change. Cambridge: Cambridge University Press.

Smith, L.C. 1997. Satellite remote sensing of river inundation area, stage, and discharge: a review. Hydrological Processes 11: 1427–1440.

Solomon, A.M., and H.H. Shugart, eds. 1993. Vegetation Dynamics and Global Change. New York: Chapman and Hall.

Steinmetz, S., M. Guerif, R. Delecolle, and F. Baret. 1990. Spectral estimates of the absorbed photosynthetically active radiation and light-use-efficiency of a winter wheat crop subjected to nitrogen and water deficiencies. International Journal of Remote Sensing 11: 1797–1808.

Strack, D. 1997. Phenolic metabolism. In: Dey, P.M., and J.B. Harborne, eds. Plant Biochemistry. San Diego, CA: Academic Press, pp 387–416.

Tenhunen, J.D, F.M. Catarino, O.L. Lange, and W.C. Oechel, eds. 1985. Plant Responses to Stress: Functional Analysis in Mediterranean Ecosystems. Berlin: Springer-Verlag.

Tilman, D., J. Knops, D. Wedin, P. Reich, M. Ritchie, and E. Siemann. 1997. The influence of functional diversity and composition on ecosystem processes. Science 277: 1300–1302.

Tompkins, S., J.F. Mustard, C.M. Pieters, and D.W. Forsyth. 1997. Optimization of endmembers for spectral mixture analysis. Remote Sensing of Environment 59: 472–489.

Townshend, J.R.G. 1994. Global data sets for land applications from the Advanced Very High Resolution Radiometer: an introduction. International Journal of Remote Sensing 15: 3319–3332.

Tucker, C.J., H.E. Dregne, and W.W. Newcomb. 1991. Expansion and contraction of the Sahara Desert from 1980 to 1990. Science 253: 299–301.

Tucker, C.J., I.Y. Fung, C.D. Keeling, and R.H. Gammon. 1986. Relationship between atmospheric CO_2 variations and a satellite-derived vegetation index. Nature 319: 195–199.

Turner, M.G., and R.H. Gardner, eds. 1991. Quantitative Methods in Landscape Ecology. New York: Springer-Verlag.

Ustin, S.L., C.A. Wessman, B. Curtis, E. Kasischke, J.B. Way, and V. Vanderbilt. 1991. Opportunities for using the EOS imaging spectrometers and synthetic aperture radar in ecological models. Ecology 72: 1934–1945.

Ustin, S.L., D.A. Roberts, J.E. Pinzon, S. Jacquemoud, G. Scheer, C.M. Castenada, and A. Palacios. 1998. Estimating canopy water content of chaparral shrubs using optical methods. Remote Sensing of Environment 65: 280–291.

Ustin, S.L., M.O. Smith, and J.B. Adams. 1993. Remote sensing of ecological processes: a strategy for developing and testing ecological models using spectral mixture analysis. In: Ehleringer, J.R., and C.B. Field, eds. Scaling of Physiological Processes: Leaf to Globe. San Diego, CA: Academic Press, pp 339–357.

Valentini, R., J.A. Gamon, and C.B. Field. 1995. Ecosystem gas exchange in a California grassland: seasonal patterns and implications for scaling. Ecology 76: 1940–1952.

Vanderbilt, V.C., L. Grant, S.L. Ustin. 1990. Polarization of light by vegetation. In: Ross, J., and R.B. Myneni, eds. Photon-Vegetation Interactions: Applications in Optical Remote Sensing and Plant Ecology. Berlin: Springer-Verlag, pp 194–228.

Vane, G., R.O. Green, T.G. Chrien, H.T. Enmark, E.G. Hansen, and W.M. Porter. 1993. The airborne visible/infrared imaging spectrometer (AVIRIS). Remote Sensing of Environment 44: 127–143.

Vitousek, P.M. 1994. Beyond global warming: ecology and global change. Ecology 75: 1861–1876.

Walker, B., and W. Steffen, eds. 1996. Global Change and Terrestrial Ecosystems. Cambridge: Cambridge University Press.

Wardle, D.A., O. Zackrisson, G. Hörnberg, and C. Gallet. 1997. The influence of island areas on ecosystem properties. Science 277: 1296–1299.

Welles, J.M. 1990. Some indirect methods of estimating canopy structure. Remote Sensing Reviews 5: 31–43.

Wessman, C.A. 1990. Evaluation of canopy biochemistry. In: Hobbs, R.J., and H.A. Mooney, eds. Remote Sensing of Biosphere Functioning. New York: Springer-Verlag, pp 135–156.

Wessman, C.A. 1992. Spatial scales and global change: bridging the gap from plots to GCM grid cells. Annual Review of Ecology and Systematics 23: 175–200.

Wessman, C.A., C.A. Bateson, and T.L. Benning. 1997. Detecting fire and grazing patterns in tallgrass prairie using spectral mixture analysis. Ecological Applications 7: 493–511.

Wessman, C.A., J. Aber, D. Peterson, and J. Melillo. 1988. Remote sensing of canopy chemistry and nitrogen cycling in temperate forest ecosystems. Nature 335: 154–156.

Wharton; S.W., and M.F. Myers. 1997. MTPE EOS Data Products Handbook. Greenbelt, MD: NASA/Goddard Space Flight Center.

Wickland, D.E. 1991. Mission to planet earth: the ecological perspective. Ecology 72: 1923–1933.

Wiegand C.L., A.J. Richardson, D.E. Escobar, and A.H. Gerbermann. 1991. Vegetation indices in crop assessments. Remote Sensing of Environment 35: 105–119.

Yoder, B.J., and R.E. Pettigrew-Crosby. 1995. Predicting nitrogen and chlorophyll content and concentration from reflectance spectra (400–2500 nm) at leaf and canopy scales. Remote Sensing of Environment 53: 199–211.

Yost, E., and S. Wenderoth. 1969. Ecological applications of multispectral color aerial photography. In: Johnson, P.L., ed: Remote Sensing in Ecology. Athens, GA: University of Georgia Press, pp 46–62.

Young A., and G. Britton, eds. 1993. Carotenoids in Photosynthesis. London: Chapman and Hall.

Zhang, M., S.L. Ustin, E. Rejmankova, and E.W. Sanderson. 1997. Remote sensing of salt marshes: potential for monitoring. Ecological Applications 7: 1039–1053.

24

Generalization in Functional Plant Ecology: The Species Sampling Problem, Plant Ecology Strategy Schemes, and Phylogeny

Mark Westoby
Macquarie University, New South Wales, Australia

I. INTRODUCTION

This chapter discusses how we generalize across species, and how we choose species for study. If a process is universal, it should not matter in which species we study it. An expectation that the processes under study should be universal is implicit in much physiological or developmental genetic work. Correspondingly, a single species is studied, chosen for experimental convenience, perhaps tobacco or *Arabidopsis*. In functional ecology, on the other hand, most generalizations will be conditional and comparative. Functional ecology aims to work out how things operate differently in pioneer versus shade-tolerant species, on low-nutrient versus high-nutrient soils, among monocots versus among dicots, and in rare compared with widespread species.

Functional ecology makes progress through the struggle to generalize, to understand what is similar and what is different about species and situations. Progress is manifested at two levels in publication. In primary journal publications that report fresh data, comparisons may be between as few as two species up to tens of species, depending on which traits were studied and how difficult and time consuming it was to characterize them. But later, in literature reviews or meta-analyses, numbers of such primary studies are gathered together and generalizations are sought. It is at the stage of compiling literature reviews that knowledge can be said to become consolidated and reliable. At this stage, evidence may be available about tens to hundreds of species (e.g., edited review books about ecophysiology such as Lange et al. 1984, Lambers et al. 1989, Roy and Garnier. 1994, Schulze and Caldwell 1994, Mulkey et al. 1996). Similarly, many hundreds of field experiments have now been conducted on the major interactions between species such as competition, herbivory, and predation (Connell 1983, Crawley 1983, Schoener 1983, Sih et al. 1985, Price et al. 1986, Hairston

1989, Goldberg and Barton 1992, Gurevitch et al. 1992, Wilson and Agnew 1992, Goldberg 1996). Generalizing across the many primary studies has now become an outstanding problem for ecology.

This chapter argues in the following sequence. First, the reader is briefly reminded of the rules for inferring generalizations from samples. These rules are very familiar to ecologists in contexts such as vegetation sampling, but curiously, they have been almost completely ignored in the context of selecting species for study. Some problems in applying the generalization rules to species selection are acknowledged, and the distinction between primary data reports and subsequent reviews and meta-analyses becomes important here. Then, to arrive at a conditional generalization, species have to be categorized in some way. The three main types of categorization are habitat, ecological strategy (traits of the species itself), and phylogeny. For primary data collection, anyone could use their own basis for categorizing species, but during subsequent meta-analyses, only widely adopted categorization schemes can come into play. Hence, achieving consensus on at least some categorization schemes should be an important part of the research agenda of comparative functional ecology. A new ecological strategy scheme is proposed for this purpose. Finally, phylogeny is a potential categorization scheme. Species with recent common ancestors are more likely to have similar traits. A brief outline is provided of the recent debate about whether phylogeny should be regarded as an alternative explanation to present-day functionality. More constructively, phylogeny provides a potential tool for choosing study species to arrive more efficiently at generalizations, and to place better-defined boundaries around the generalizations.

II. INFERENCE RULES FOR GENERALIZING FROM A SAMPLE

The rules for inferring a generalization from a sample have been well appreciated by ecologists for many years, for example, in the context of describing vegetation by means of quadrat sampling. The rules are:

1. Quadrats are placed at random, i.e., each possible location within some category should have an equal chance of being sampled.
2. Accordingly, the scope and boundaries of the category itself need to be defined. Is it a particular patch of forest that is being described, or is it all patches of that forest type within a continent? In addition to conceptual decisions about the scope of the category, there usually need to be some exclusion rules—perhaps sites are not sampled on cliffs because of the practical inconvenience, or in the middle of forest tracks because these are not thought interesting.

3. There needs to be some replication of the samples to give a sense of the range of variation. Alternative indices of the range of variation are not discussed here; biometry texts provide recipes for calculating these and discuss their merits and assumptions. The point is that all recipes require replication, because the logic by which we generalize about a category requires some indication of the range of variation within the category.

These rules come into play during the design of a sampling procedure. Generalization rules forge the link between the sampling procedure and the scope of the generalizations that can subsequently be inferred. The design of the study is shaped by contemplating what conclusions one might be seeking to draw. The decision of what category to sample, together with the exclusion rules, draws boundaries on the conclusions that can be drawn. The legitimacy of the conclusions depends on the equal-chance-of-being-sampled principle, and the replication allows the strength of the conclusions to be assessed.

III. APPLICATION OF THE INFERENCE RULES TO GENERALIZATION ACROSS SPECIES: AN ISSUE THAT HAS BEEN UNDERESTIMATED

In contexts such as vegetation sampling and manipulative field experiments, the inference rules have been thoroughly assimilated into the practices of ecological researchers. This contrasts sharply with the problem of choosing a set of species to study. Even today, choosing species seems rarely to be perceived as a sampling problem, where the rules for generalization need to be applied. Some researchers may legitimately argue that they have chosen particular species because they are only interested in those species, and they have no interest in generalizing. Alternatively, they may believe that the mechanism they are studying is universal; therefore, it does not matter in which species they study it. But most studies of functional ecology or ecophysiology of plants seek to interpret their results by references to categories of plants—shade-tolerant versus pioneer, low versus high altitude, annual versus perennial, inhabitants of low-nutrient versus more fertile soils, rare and endangered versus widespread, and constitutively slow-growing versus capable of rapid growth. In other words, the species are to be interpreted as samples from some category. Nevertheless, these studies rarely address explicitly the representativeness and replication criteria. If the inference rules were being followed, the methodology of these reports would first define categories of species (e.g., listing all the low-nutrient species in New South Wales) and then indicate exclusion rules that had to be applied (e.g., only species available from seed merchants could be considered, and among those, some had to be eliminated because they could not be persuaded to germinate in reasonable numbers within a few days of each other). Then, within those boundaries, they would confirm that the species studied were chosen at random. But in reality, most reports about comparative ecophysiology hardly comment on the choice of species, and it is common for two categories to be compared by means of two species, with no replication. Imagine if a manuscript were submitted in which two vegetation types were compared by means of one quadrat in each, placed at the location the investigator thought most representative or convenient; it would get short shrift from referees and editors. But many reports that have been successfully published and are respectfully cited in comparative ecophysiology do exactly this with regard to choice of species.

How have these differences in research culture, in expectations about sampling, come about? Probably one factor has been the physiologist's expectation that truly interesting processes occur reliably in whatever species is studied. The other main factor must be the sheer difficulty and laboriousness of some types of measurement. Remember, also, that for an investigator seeking to dissect mechanisms, there are always strong incentives to measure more processes and at more frequent intervals within one species, rather than to include further species. Consider investigations of complete carbon budgets, for example, including dissecting root respiration into growth and maintenance components. Within the framework of a Ph.D., it is unreasonable to expect more than a couple of species to be studied. Then, because there are too few species for replication within categories, it must seem unimportant to apply the equal-chance rule within each category. There are also strong reasons to use species for which seeds are available and are known to germinate readily, which can lead to a few species being used repeatedly by successive investigators. In summary, the tradition of using few species, with those used not chosen according to explicit rules of sampling, is entirely understandable within ecophysiology and other areas such as genetics and demography of rare species.

Nevertheless, in a subsequent literature review or meta-analysis, the following issue must be faced: is species coverage in the primary research reports representative, and of what categories? As more and more primary research reports accumulate, this issue is becoming more pressing. One way of defining the problem in species selection is to say that species should be selected in such a way as to make better generalizations possible to the literature review 5–10 years later.

How might this come about? At one extreme, a grand overall species-selection design could be decided by a well-meaning dictator or committee. Each laboratory worldwide would then be instructed on which species to select. I feel confident this is not going to happen (and, indeed, should not be allowed to, because the slackening of competitive energy would surely outweigh the benefits of coordination). At the other extreme, and far more likely, is that primary research publications will move toward discussing their species selection explicitly in relation to the inference rules. Explicit discussion could be very valuable. Even when there can be little replication, and when many exclusion rules have been required so that there is only a small choice of species, compilers of literature reviews will nevertheless be better placed if the primary studies have spelled out their species-selection criteria explicitly. In addition to explicit discussion, investigators should be encouraged to make available as much background information as possible about their species: information that is not referred to in the report for the purpose of proving a conclusion, but that might subsequently be useful to reviewers or meta-analysts seeking to investigate different questions. The limited page allotment in journals should no longer discourage contributors and editors from reporting background information about traits and habitats of the study species, since such material can now be placed at websites.

IV. CRITERIA ON WHICH TO COMPARE SPECIES

If species sampling is one side of a coin, the other side is categories into which species are grouped for comparison. Criteria on which species might be compared fall under three main headings: (1) habitat; (2) attributes of the species themselves—ecological strategies; and (3) phylogeny (or in practice, usually taxonomy).

The distinction between primary research reports and reviews or meta-analyses is important here again. Individual research groups can and do categorize species according to whatever criteria they think are meaningful for the question in hand. But at the stage when knowledge is consolidated across many primary reports, the reviewer can use only categories that were adopted in common by all of the primary reports, or alternatively, those that are simple enough for species to be attributed to them by the reviewer. As a consequence, most reviews group species into rather simple categories: herbaceous versus woody, deciduous versus evergreen, temperate versus tropical, and so forth. For categories to be used at review that captured a subtler degree of difference between species, consensus on the categorization scheme would need to be achieved in advance. Developing wider consensus on more expressive categorization schemes is thus a major priority for improved generalization in functional ecology.

A. Habitat

Habitat descriptors will be discussed very briefly. Rainfall and temperature are not too much of a problem to attribute to plant species. Maps or climate-interpolating software can produce estimates of yearly means or various seasonal patterns or extremes for a given location where a species is known to occur.

Shading under canopies often used to be described in qualitative categories, but measurements on an absolute scale of photosynthetically active radiation (PAR) have become more typical over the last couple of decades and have been important in clarifying seedling performance in the shade.

Soil nutrients remain an unresolved problem. Many reports have compared species from infertile versus fertile soils within particular landscapes. But synthesis across these reports is very difficult because they do not share common measurements that describe soil fertility of the sites where the species is successful. There is some reason to believe that relatively fertile soils in Australia might fall into an infertile category in northwest Europe, for instance, and this complicates any attempt to relate Australian to European studies. There are certainly many complexities in measuring soil fertility, but it would still be a step forward if even one or two lowest-common-denominator measures could be agreed upon.

B. Traits of the Species Themselves: Ecological Strategy Schemes

The literature on plant ecological strategy schemes can be summarized into three main strands of thinking. One strand categorizes species by reference to distribution (realized niche) on one or more gradients (e.g., Dyksterhuis 1949 for grazing, Noble and Slatyer 1980 for time after disturbance, Ellenberg 1988 for soil and other habitat features). A second strand categorizes species according to physiognomy (e.g., Raunkiaer 1934, Dansereau 1951, Mueller-Dombois and Ellenberg 1974, Box 1981, Sarmiento and Monasterio 1983, Barkman 1988, Orshan 1989, Prentice et al. 1992), and has been active especially within plant geography. In a third strand, axes or categories are named according to concepts (as distinct from naming them according to traits or realized niche). Examples include the r-K spectrum (Cody 1966, MacArthur and Wilson 1967) and several schemes that have developed this spectrum into a three-cornered arrangement (Greenslade 1972, 1983, Grime 1974, Whittaker 1975, Southwood 1977). The three-cornered schemes add a category of opportunities where the physical environment permits only slow acquisition of resources. This situation is called "stress" in Grime's CSR triangle (1974, 1979, Grime et al. 1988), the best developed three-cornered scheme for plants. The CSR triangle has two dimensions: the C-S axis, reflecting adaptation to opportunities for rapid growth versus continuing enforcement of slow growth (competitors to stress tolerators), and the R axis, reflecting adaptation to disturbance (ruderals).

Among conceptual strategy schemes, the CSR triangle is the most widely cited in textbooks (Begon et al. 1996, Cockburn 1991, Colinvaux 1993, Crawley 1996, Ingrouille 1992), reflecting wide acceptance that exploiting opportunities for fast versus slow growth, and coping with disturbance, are two of the most important forces shaping the ecologies of plants within landscapes. Yet reports in functional ecology do not routinely report CSR axis scores for the species studied, because there is no explicit quantitative protocol for scoring a species from anywhere worldwide (see qualitative and partly subjective keys in Grime 1984, Grime et al. 1988). In other words, the CSR scheme is widely cited for conceptual discussion, but not widely adopted for practical comparisons. The only general-purpose scheme that has been widely adopted is Raunkiaer's life-form categorization (1907, English translation 1934), based on the location of the buds where regrowth arises after the unfavorable season of the year. Raunkiaer life-forms are very easily attributed for most species, which is why the scheme is widely adopted, but the scheme conveys only a modest amount of information about differences between species.

In summary, existing schemes that can easily be applied worldwide capture very few of the differences between species, especially with regard to how they exploit different opportunities within multispecies vegetation and between different sites in a landscape. On the other hand, schemes that seek to be more expressive about these differences between species are not designed so that species anywhere can readily be categorized. Consequently, schemes such as the CSR triangle have not been able to be used to group species worldwide during literature reviews or meta-analyses. In this context, I have recently (Westoby 1998) proposed a new leaf-height-seed (LHS) scheme (Appendix 1) designed to express at least some of the differences between species addressed by the CSR triangle and related schemes, while using axes readily measured on the plant itself, and therefore offering the potential for worldwide comparison.

As outlined in Appendix 1, the LHS scheme captures a substantial part of the same spectra of strategy variation as the CSR scheme, while resolving some difficulties with it. SLA variation (the L dimension) is crucial to the CS axis (Grime et al. 1988, 1997), i.e., to leaf longevity, mean residence time of nutrients, soil nutrient adaptation, and potential relative growth rate (RGR). Canopy height at maturity (the H dimension) is arguably the most central single trait that needs to be adjusted to the duration of the growth opportunity between disturbances (R axis); it is also treated by Grime et al. (1988) as a significant predictor of C versus S strategy. The LHS scheme does not prejudge what parts of the LHS volume will be occupied, compared with the CSR triangle, which decides a priori that the high-S-high-R quadrant is not a viable strategy. By separating seed mass (S dimension) as a distinct axis, it expresses something about dispersal to new growth opportunities, independently of what is expressed by canopy height about the duration of the growth opportunity between disturbances. Seed mass also expresses some significant differences about seedling establishment between species. Most important, because the LHS axes chosen require modest effort, experimentalists may be willing to report them for their species with a view to subsequent meta-analysis by others, even though they have no immediate use for the data themselves.

C. Phylogeny

Regrettably, phylogenetic relatedness is often interpreted as an alternative reason (sometimes called "phylogenetic constraint") why species should be similar (Hodgson and Mackey 1986, Kelly and Purvis 1993, Silvertown and Dodd 1996, Harvey 1996). Common ancestry or phylogeny is seen as a source of confounding or error that requires controlling for, in competition with explanations that invoke natural selection or functionality continuing into the present day.

This competing-explanation approach is incorrect with regard to explaining present-day ecological function. Phylogenetic niche conservatism is commonplace; hence, species can often have similar trait combinations both because they are phylogenetically related and because they are subject to similar continuing forces of natural selection. The issues of interpretation have been debated elsewhere, for example, in a forum in the *Journal of Ecology* (Westoby et al. 1995a, 1995b, 1995c, 1996, 1997, Harvey et al. 1995a, 1995b, Ackerly and Donoghue 1995, Rees 1995, Price 1997) and are summarized in Appendices 2–4.

Precisely because functionally important traits are sometimes phylogenetically conservative, phylogeny can and should be seen not as a source of confounding, as a technical difficulty to be overcome, but more positively as a basis on which to select species for study. Through better species selection, we might arrive at generalizations more efficiently and with better-defined boundaries on the generalization. People concerned with methods of phylogenetic analysis have mostly been using datasets already in existence, and have not as yet paid much attention to species-selection designs. But for experimentalists who collect new data, because substantial effort is involved for each species, it is worth thinking carefully about how that effort should be allocated.

Species-selection design, like any other aspect of design, depends on the question under study. It is important to be careful about the exact formulation of questions invoking phylogeny. A range of different question formulations and corresponding species-selection designs are discussed by Westoby et al. (1997). Here a more general overview is provided.

A traditional idea is that one should compare species within a genus rather than more distantly related species, because other unmeasured attributes are less likely to vary in such a comparison. This idea is actually not a very good compromise. On the one hand, it is not a safe means of controlling for the influence of third variables. Other unmeasured variables are capable of varying within genera as well as between genera. On the other hand, there is no way to tell how far a generalization from a within-genus study extends to other lineages. The within-genus study sacrifices all power to assess generality across lineages, without decisively controlling for third variables.

To assess the consistency of a pattern across lineages, typical designs are based on phylogenetically independent contrasts (PICs). A phylogenetic contrast, or radiation, is a branch-point in the phylogeny with the set of branches descending from it (Felsenstein 1985, Grafen 1989, Harvey and Pagel 1991). In the most simple case, it is a pair of species descended from a common ancestor. (Using pairs maximizes the number of PICs in the design relative to the number of species required.) *Independence* refers here to the set of contrasts within a particular study being independent of each other, representing separate divergences or radiations in the phylogenetic tree. Each PIC provides one replicate for testing whether a divergence in attribute X has consistently been associated with a divergence in Y across separate evolutionary divergences.

Suppose the phylogenetic tree being adopted is simply the existing taxonomy, and the aim is to select 20 PICs from a pool of candidate species. A simple rule for obtaining further PICs is to include new genera rather than more than two species within a genus, new families rather than more than two genera within a family, new orders rather than more than two families within an order, and so forth. The effect of this rule is to spread sampling across the phylogenetic tree, so that any PIC-based design will have some degree of breadth of coverage of different lineages. Nevertheless, unless there are a very large number of PICs, some lineages may go unrepresented, and nothing in the simple rule described allocates equal representation to different major branches. To achieve these design aims, one might spread PICs through the phylogenetic tree more systematically, for example, by treating major branches of the angiosperm tree, such as rosids, asterids, and palaeoherbs, as blocks. To my knowledge, there is no study to date that has implemented such a design.

PIC-based designs are good for assessing consistency of a relationship across many lineages, but they have countervailing disadvantages (several complications of using PICs are discussed in more detail in Westoby et al. 1997). Probably the most important complication is that they will usually not satisfy the "equal chance of being selected" rule in the species selected from a particular habitat or strategy. Suppose, for example, we wish to contrast species from infertile soils with species from more fertile soils (Cunningham et al. in press, Wright and Westoby in press). For each species chosen from an infertile soil, a related species will be sought on fertile soil to form a phylogenetically independent contrast. This means that from the list of all species on fertile soil, species are more likely to be chosen if they belong to genera or families that are also present on infertile soil. The species chosen according to a PIC-based design will not give a fair representation of the overall shift in the frequency distribution of, e.g., species leaf sizes between habitats. Specifically, they will tend to underestimate the contribution from families and orders that are present in one habitat, but not the other.

One can select PICs branching across higher as well as lower taxonomic levels; therefore, in principle it might be possible to select species in such a way that they both constitute a set of PICs between two habitats and also are proportionately representative of the phylogenetic species mixture occurring within each habitat. However, such a design has never been attempted to my knowledge. This is on the premise that a PIC between orders within a superorder is constructed by selecting one species at random from each order. In some designs, such a PIC can also be constructed by estimating each order's trait value from species within that order that have also been used to build PICs between families, genera, or species. But this is only possible when the species have been randomly sampled from the phylogeny within each order, not when the PICs have deliberately been contrasted for, e.g., leaf size or soil habitat (Westoby et al. 1997). There is no doubt that species selection on the basis of phylogeny is an area in which many further developments and improvements can be expected over the next few years.

V. CONCLUSION

The current situation in functional plant ecology is that large numbers of detailed field experiments and ecophysiological studies on one or a few species have accumulated, more than have been satisfactorily digested, interpreted, and generalized. Emerging wide-area applied problems, notably global change of climate and land use, are creating urgent demand for plant functional type classifications that might permit worldwide generalizations (Steffen et al. 1992, Körner 1993, Woodward and Cramer 1996, Smith et al. 1997). The gradual accumulation of comparative information in electronic databases is reaching critical mass, allowing patterns to have their generality quantified much more widely and quickly than a decade ago. Together, these trends mean that generalization across species and the associated topic of ecological strategy schemes will become keys to research progress in functional plant ecology over the next 10–20 years.

In this context, the selection of species for study is an issue deserving closer attention than it has received up to the present. The maxim is to be explicit. This means explicitly describing the boundaries on categories of species that are to be compared in any given study. Ideally, one would then select replicate species at random within those categories. This is a counsel of perfection that will be difficult to meet in practice, but again, investigators should be encouraged to think about and list explicitly whatever exclusion rules they have found necessary to use, which prevented them from choosing at random from the whole list of species within a particular category. The work of subsequent literature review and generalization must surely become more rigorous and powerful once reviewers have available to them a clearer knowledge of what sorts of species have been studied and what sorts have been avoided.

APPENDIX 1: A PROPOSED LHS (LEAF-HEIGHT-SEED) PLANT ECOLOGY STRATEGY SCHEME

The LHS scheme (Westoby 1998) would consist of three axes: (1) typical specific leaf area (SLA) of mature leaves, developed in full light, or the fullest light the species naturally grows in; (2) typical height of the canopy of the species at maturity; and (3) typical seed mass.

The strategy of a species would be characterized in the scheme by a position in a three-dimensional volume. Each dimension is known to vary widely between species at any given level of the other two; thus, the volume occupied by present-day species extends considerably in all three dimensions. Each of these traits is correlated with a number of others, but they have not been chosen only as conveniently measured indicators. Rather, it is believed that they themselves are fundamental trade-offs controlling plant strategies. They are fundamental because it is ineluctable that a species cannot both deploy a large light-capturing area per gram and also build strongly reinforced leaves that may have long lives; cannot support leaves high above the ground without incurring the expense of a tall stem; and cannot produce large, heavily provisioned seeds without producing fewer of them per gram of reproductive effort.

As would be expected for traits of such ecological importance, plants have some capacity to shift trait values in response to the circumstances in which they find themselves. In other words, none of SLA, height at maturity, or seed mass are absolute constants within species. Nevertheless, variation between species is much greater than within species, and many previous investigators have seen no insuperable difficulty in recording characteristic species values for comparative purposes (e.g., for height at maturity; Hubbell and Foster 1986, Grime et al. 1988, Keddy 1989, Bugmann 1996, Chapin et al. 1996). All three axes would be log-scaled, reflecting the fact that the difference between 30 and 31 m (to take canopy height at maturity as an example) is not nearly so important as the difference between 30 cm and 130 cm.

A. Specific Leaf Area

Specific leaf area is the light-catching area deployed per unit of previously photosynthesized dry mass allocated to the purpose. SLA is like an expected rate of return on investment. High SLA permits (given favorable growth conditions) a shorter payback time on a gram of dry matter invested in a leaf (Poorter 1994). At first glance, it might appear that a low rate of return on investment would not be evolutionarily competitive, but low SLA species achieve greater leaf life span (Reich et al. 1992, 1997), through extra structural strength and sometimes through allocation to tannins, phenols, or other defensive compounds. Therefore, light capture per gram invested can be at least as great in a low-SLA species when considered through the whole life of the investment. Reich et al. (1997) have shown across six biomes that SLA is closely correlated with mass-based net photosynthetic capacity and mass-based leaf N, as well as leaf life span. Higher leaf water content and reduced lamina depth can contribute to higher SLA (Witkowski and Lamont 1991, Garnier and Laurent 1994, Cunningham et al. in press). Grime et al. (1997) found SLA to be among the major contributors to the "primary axis of specialization" they identified by ordination of 67 traits among 43 species, corresponding to the C-S axis of the CSR scheme.

Potential relative growth rate (RGR), measured on exponentially growing seedlings given plentiful water and nutrients, has been seen as an indicator of responsiveness to favorable conditions (Grime and Hunt 1975, Leps et al. 1982, Loehle 1988, Poorter 1989, Reich et al. 1992, Chapin et al. 1993, Aerts and van der Peijl 1993, van der Werf et al. 1993, Turner 1994). Because potential RGR is made up of net assimilation rate × leaf fraction × SLA, variation in SLA necessarily influences potential RGR. Indeed, in most comparative studies, SLA has been the largest of the three sources of variation in potential RGR (Poorter 1989, Poorter and Remkes 1990, Poorter and Lambers 1991, Lambers and Poorter 1992, Reich et al. 1992, Garnier and Freijsen 1994, Saverimuttu and Westoby 1996, Cornelissen et al. 1996, Grime et al. 1997, Hunt and Cornelissen 1997, Poorter and van der Werf 1998). High SLA species can have strategies associated with rapid production of new leaf during early life. Faster turnover of plant parts also permits a more flexible response to the spatial patchiness of light and soil resources (Grime 1994b). On the other hand, species with low SLA and long-lived leaves can eventually accumulate a much greater mass of leaf and capture a great deal of light in that way; and the long mean residence time of nutrients made possible by leaf longevity permits a progressively larger share of nitrogen pools to be sequestered (Aerts and van der Pijl 1993).

B. Canopy Height at Maturity

Height obviously conditions how plants make a living, in different ways depending on vegetation dynamics. In some vegetation types, a characteristic vertical profile of leaf area and light attenuation persists over time despite turnover of individual plants. Species with canopies at different depths in this profile are operating at different light incomes, heat loads, wind speeds, humidities, and with different capital costs for supporting leaves and lifting water to the leaves. In other vegetation types, disturbances, or the death of large individual trees, destroy canopy cover, and daylight becomes available near the ground. The successional process that ensues can be understood as a race upward for the light. Because light descends from above, the leading species at a given time have a considerable advantage. In this race, unlike a standard athletic contest, there is not a single winner determined after a fixed distance. Rather, any species that is among the leaders at some stage during the race is a winner, in that being among the leaders for a reasonable period permits a sufficient carbon profit to be accumulated for the species to ensure it runs again in subsequent races. The entry in subsequent races may occur via vegetative regeneration, via a stored seed bank, or via dispersal to other locations, but the prerequisite for any of these is sufficient carbon accumulation at some stage during vegetative growth. Races are restarted when a new disturbance destroys the accumulated stem height. The duration of an individual race can be measured in years, or ideally, in units of biomass accumulation, calibrating intervals between disturbances to the productivity of a site. But within a race series having some typical race duration, one finds successful growth strategies that have been designed by natural selection to be among the leaders early in a race, and other successful strategies that join the leaders at various later stages. Species that achieve most of their lifetime photosynthesis with leaves deployed at 10–50 cm have different stem tissue properties from those designed for 1–5 m, and those, in turn, are different from species that achieve 30–40 m. The canopy height that species have been designed to achieve by natural selection is the simplest measure of this spectrum of strategies.

C. Seed Mass

Seed mass variation expresses a species' chance of successfully dispersing a seed into an establishment opportunity from a given area of ground already occupied by a species. Seed mass is also a useful indicator of a cotyledon-stage seedling's ability to survive various hazards.

Species having smaller seed mass can produce more seeds from within a given reproductive effort, and seed mass is therefore the best easy predictor of seed output per square meter of canopy cover. It might be thought that distance of dispersal would be the major influence on a species' chance of dispersing a seed to a forest gap or another establishment opportunity. However, dispersal distances have not proved tidily related to dispersal morphology, to seed mass, or to any other plant attribute (Hughes et al. 1994). Among unassisted species, larger seeds do not travel as far from a given height of release; on the other hand, larger seeds tend to have wings, arils, etc., or to be released from a greater height. Similarly, among wind-assisted species, larger seeds tend to have larger wings or longer pappuses. Because reduced dispersal associated with larger seed mass tends to be counteracted by extra investment in dispersal-assisting structures, or sometimes by being released from a taller plant, the net effect is that dispersal distance is not tidily related to any of these attributes. At present, seed mass (as a surrogate for seed output per ground area occupied) is the best predictor of the chance that an occupied site will disperse a propagule to an establishment opportunity.

Species having larger seed mass have been shown experimentally to survive better under a variety of different seedling hazards (Westoby et al. 1996), including drought, removal of cotyledons, and dense shade below the compensation point. The tendency to survive longer applies only during the cotyledon phase, while seed reserves are being deployed into the fabric of the seedling (Saverimuttu and Westoby 1996). Capacity to continue growth into later seedling life under a low light level is determined more by canopy architecture and leaf properties (Kitajima 1994). It seems likely that tolerance of seedling hazards is endowed not by seed mass as such, but by a tendency for larger seeds to retain more metabolic reserves uncommitted to the fabric of the seedling over a longer period, and therefore available to support respiration while in carbon deficit (Westoby et al. 1996).

D. The LHS Scheme in Relation to Grime's CSR Triangle

Where each axis of the CSR scheme implies a complex of plant traits (Grime et al. 1997), the LHS scheme has axes defined by single quantitative traits. The benefit of the LHS scheme's simple protocol for positioning a species outweighs any loss of information in the LHS axes compared with the CSR axes, for the purpose of facilitating worldwide comparisons of species.

The CSR scheme has been made triangular rather than rectangular because the most stressful and most frequently disturbed corner is said not to be occupied (Grime et al. 1988), or because ineluctable trade-offs are said to prevent a species from being highly adapted to more than one of the three "primary strategies": C, S, or R (Grime 1994a). The idea that a whole quadrant is missing due to the combination of high stress and high disturbance has been criticized (Grubb 1985), and experiments with crossed gradients of fertility and disturbance (Campbell and Grime 1992, Burke and Grime 1996) have not produced wholly unoccupied space at the low-fertility high-disturbance corner. The LHS scheme avoids prejudging the question of whether any particular corner of the LHS volume is not viable.

Another difficulty in the CSR scheme is the ruderality axis. Adaptation to disturbance might, in principle, include adaptations for surviving individual disturbances, together with adaptations for completing life history within a short interval between disturbances and with adaptations for dispersing through space or time to freshly disturbed locations. Grubb (1985) criticized the CSR scheme for not distinguishing continuing from episodic disturbance. According to Grime

(Grime et al. 1988, Grime and Hillier 1992), the scheme is for adults, not juveniles: a given adult strategy can occur in combination with several different juvenile strategies, which has the effect of separating dispersal and seed bank strategies from the main CSR categorization of a species. The LHS scheme
disentangles these disparate elements to some extent. The canopy height at maturity axis reflects adaptation to the interval between disturbances (calibrated in
units of height growth rather than time). The seed mass axis (more exactly its
inverse, seed number per mass allocated to seed production) reflects the potential for dispersal to freshly disturbed locations. Adaptations for continuing the
lineage through particular types of disturbance (e.g., lignotubers for resprouting
after fire, soil seed banks with a light requirement for germination after soil turnover, basal tillering in graminoids for grazing tolerance) have deliberately been
left out of the LHS scheme because they do not lend themselves to any simple
generalization.

APPENDIX 2: FREQUENTLY ASKED QUESTIONS (FAQs)
ABOUT PHYLOGENY AND FUNCTIONAL ECOLOGY

FAQ1: *When related species tend to be similar (for example, seed mass more
similar within than between genera), should this be attributed to "phylogenetic
constraint"?*
Answer: No. The term "constraint" clearly implies that the trait has been under
directional selection toward different values, but nevertheless has failed to respond to selection. (Remaining unchanged due to absence of directional selection
or due to continuing convergent selection cannot usefully be called constraint,
nor inertia, an alternative sometimes seen.) There are two reasons why similarity
of species with a common ancestor should not be regarded as positive evidence
for constraint or failure to respond to directional selection.

First, given that differences between species, genera, or families are under
consideration, the hypothesized constraint needs to have applied over millions,
perhaps tens of millions, of years. Thus, features of genetic architecture that might
be measured in a present-day population and might restrict response to selection
over tens of generations, such as low heritability or genetic correlations between
traits, could only account for constraint in this context if the low heritability or
the genetic correlations survived a million years of mutation and genetic re-
arrangement under directional selection. This would be sufficiently surprising
that it certainly should not be accepted as a null hypothesis, especially not for
quantitative traits such as seed mass. Rather, it is a decidedly strong biological
hypothesis: a definite mechanism for the constraint should be proposed, and
means sought to test it.

Second, there are alternative well-established mechanisms through which species could tend to maintain similar traits over time after diverging from a common ancestor. Therefore, correlation of a trait with phylogeny cannot be regarded as evidence for constraint rather than for continuing functionality. Phylogenetic niche conservatism is a process whereby, because ancestors have a particular constellation of traits, their descendants tend to be most successful using similar ecological opportunities; therefore, natural selection tends to maintain the same traits among most, if not all, descendant lineages. Niche conservatism is at least as likely a cause of similarity among related species as constraint—more likely, for quantitative traits—and explicitly invokes ecological functionality continuing into the present day.

Sometimes the term "phylogenetic effect" is used to refer to the tendency for phylogenetically related species to have similar traits. This term is defensible provided it means "effect" only in a purely statistical sense, a label for variation correlated with phylogeny. However, the temptation seems strong to see a phylogenetic effect as an alternative causal interpretation to ecological functionality, and this is wrong. The term "phylogenetic effect" were better eschewed (Westoby et al. 1995c). If constraint is being inferred, a specific mechanism should be proposed. If not, one might refer to phylogenetic conservatism, identifying the pattern in the outcome without hinting at any particular mechanism.

FAQ2: Is it true that phylogenetically related species are not independent as evidence for present-day ecological function?
Answer: Yes, in part, but mainly no. Actually, formulating the question around the term "independence" is not helpful (see Appendix 3). The grain of truth in this idea is that a correlation across present-day species between traits X and Y might be caused through a cross-correlation with Z rather than reflecting a direct functional relationship between X and Y. Related species are more likely to have similar values for Z. But the argument usually connected to the claim about nonindependence is that radiations, separate divergences on the phylogenetic tree, are independent events, and that therefore a test for correlated divergence deconfounds the X-Y correlation "to a large extent" (Harvey et al. 1995a) from third-variable influences (see Appendix 4 for further discussion of the sense in which correlated-change analysis deconfounds or partials out third variables). This implication that analyzing divergences rather than present-day species is an improved, "phylogenetically corrected" method for assessing ecological function is unsafe for two reasons.

First, the problem of cross-correlation with third variables is not confined to related species. Hence, analyzing for correlations in divergence rather than correlations across present-day species does not overcome the well-known problems of inferring causation from correlation.

Second, just because an X-Y correlation is cross-correlated with Z, this does not necessarily mean Z is the true cause. It remains just as likely that the true mechanism runs from X to Y, and Z is a secondary correlate, so far as anyone can tell from the correlation pattern alone. In such cross-correlated situations, it is not conducive to sensible interpretation to deconfound X-Y from any influence of Z without at the same time looking at the raw X-Y correlation. This is all the more so when the third, fourth, etc., variables from which X-Y is being deconfounded are not explicitly identified, but rather are an aggregate of all variables that have been conservative down the phylogenetic tree.

In summary, evolution by natural selection has given rise to cross-correlated patterns of traits among present-day species. Selection for ecological functionality has inherently been confounded with phylogeny during the history of evolution, and statistical "corrections" are not capable of converting that inherently confounded history into the ideal experiment in which phylogeny is orthogonally crossed with present-day function. In this situation, the credibility of a hypothesis connecting traits to ecological functions cannot be judged according to the pattern of correlation and cross-correlation alone, but must rest also on whether the physiological or morphological mechanism is convincing and the outcomes are well tested in field experiments. In short, everyone should be aware that correlation cannot prove causation; it is also important to remember that disappearance of correlation after correction or partialling does not disprove causation.

The qualification "as evidence for present-day ecological function" in FAQ2 is important. If the issues under study were to do with the historical process of evolutionary divergence, then naturally data about present-day species should be transformed by "hanging on the phylogenetic tree" (Grafen 1989) to give rise to inferred data about the radiations.

FAQ3: Is it obligatory to "correct" for "effects" of phylogeny?
Answer: No (when concerned with present-day function). Although advocates of phylogenetic correction or correlated-divergence analysis (Kelly and Purvis 1993, Rees 1993, Harvey 1996, Silvertown and Dodd 1996) have taken the view that cross-species correlation analysis has been superseded, correlated-divergence analysis cannot be considered obligatory because: (1) tests for correlated evolutionary divergence (phylogenetic correction procedures) do not reliably control for all potentially confounding third variables (see FAQ2); and (2) phylogenetic correction does remove from consideration correlations that have been phylogenetically conservative, many of which may also reflect present-day function (phylogenetic niche conservatism; see also FAQ2). A trait can be perfectly functional, but have arisen in only one or a few separate radiations, so that a correlated-divergence analysis would never show statistical significance. Conversely, a trait can be repeatedly correlated with an ecological outcome across many radiations or phylogenetically independent contrasts, but nevertheless not be the true cause.

APPENDIX 3: MEANINGS OF INDEPENDENCE AND ADAPTATION

The debate over phylogenetic correction has (like most debates) mixed issues of how we obtain reliable knowledge together with issues of semantics. The following key words require comment.

A. Independence

Arguments for phylogenetic correction typically begin from the formulation by Felsenstein (1985), that species do not represent independent data points on the grounds that related species will have similarities by reason of common ancestry. To claim that species lack independence purely because they have similarities cannot be justified. If correlation with another trait were sufficient to vitiate independence, then two species that both occurred in Europe could not be considered independent, nor could two species that both had alternate leaves. Carried to its logical conclusion, evidence could never be found for anything, because some correlate could always be found that would be regarded as vitiating the independence.

In general, independence is not an absolute property, but makes sense only in the context of a particular model of causation. The issue is whether two species represent separate items of evidence for that causation process. The model connected with the Felsenstein formulation of nonindependence focuses on the process of change in a trait. The present-day trait value is viewed as caused by the past process of change, rather than the process of change being caused by an attraction toward the present-day trait value, an attraction arising from ecological functionality. The claim that species are not independent items of evidence, but rather, the change along each phylogenetic branch is an independent item, makes sense only in the context of this particular model of the generating process. Price (1997) gave an example of a model in which the evolutionary process positions species in trait space according to the present-day ecological context, and shows formally that a better test of that process is obtained by considering each species as an independent case rather than by considering each radiation as an independent case.

B. Adaptation

A sector of the scientific community wishes to reserve ''adaptation'' to refer only to the natural selection under which a trait first emerged, excluding natural selection that may be maintaining it at present (Gould and Vrba 1982, Harvey and Pagel 1991). Although others continue with a broader useage of adaptation that can refer also to ecological functionality in the present day (see Williams

1992, Reeve and Sherman 1993 for balanced discussion), advocates of phylogenetic correction have chosen to insist that tests for adaptation must exclude trait
maintenance (Harvey et al. 1995a). In practical effect, this definition of adaptation
insists that questions about the emergence of traits are legitimate, whereas questions about trait maintenance are not.

Under these circumstances, the word "adaptation" is best avoided for the
present. Throughout this chapter, traits have been referred to as functional, or
having ecological significance, to avoid being sidetracked by this issue of the
definition of adaptation.

APPENDIX 4: RELATIONSHIP BETWEEN CORRELATED
DIVERGENCE ANALYSIS (PHYLOGENETIC CORRECTION)
AND PARTIALLING OUT THE CROSS-CORRELATION WITH
A THIRD VARIABLE

According to people who believe correlated-divergence analysis should be obligatory (Harvey 1996), one of its major benefits is in deconfounding an X-Y correlation, partialling out potential influences of whatever third traits Z1, Z2, etc., may
be phylogenetically conservative. Westoby et al. (1995a) described phylogenetic
correction as extracting variation in this sense and discarding it from consideration as potentially related to ecological function. In response, Harvey et al.
(1995a) asserted that correlated-change procedures should not be regarded as
extracting any component from the cross-species dataset. What, then, are the
similarities and differences between correlated-divergence analyses and partial
correlation analysis?

Correlated-divergence analysis transforms a species × traits data table by
"hanging it on the tree" (Grafen 1989), producing a new dataset where each
row is a radiation or node in the phylogenetic tree, and each column is a measure
of divergence in a trait at the radiation in question. In the simplest case, the
measure of divergence would simply be the difference in the trait between the
two species descended from a branch point. (There are various complications
where three or more branches descend from a node, or where branch lengths are
not assumed equal, but the essential logic is the same for these more complicated
cases.) The question is whether divergence in Y is correlated with divergence in
X, and it is tested by fitting a regression through the origin to the data points
derived one from each radiation.

Thus, correlated-divergence analysis has similarities to a paired design,
such as if one set out to study 1000 biology students and paired each one with
a humanities student, matched for age, gender, and university. Then, if one
wished to analyze for a relationship between biology versus humanities and at-
tending live drama, the number of plays attended during the preceding year for

each biology student would be subtracted from the number attended for the corresponding humanities student, and one would test whether the difference in plays attended was significantly different from zero. In correlated-divergence analysis, species are similarly matched into pairs, using the criterion of common ancestry, which has the effect of pairing them according to any number of phylogenetically conservative traits. Depending on the species-selection design, pairs may be deliberately contrasted on some attribute, e.g., soil habitat, or may simply be random species descended from a branch point in the phylogenetic tree. In any event, the point of subtracting trait values between pair members is to remove from consideration trait variation associated with matters for which pairs have been matched, such as age, gender, and university. Although this has advantages for some questions, analyzing only the differences also has distinct disadvantages. Suppose there was some tendency for humanities students to attend more live drama, but the tendency of females rather than males to attend live drama was much stronger—this second fact, putting the first in perspective, would be rendered invisible by the pairing and subtraction process. Furthermore, if one then drew the conclusion that the average humanities student attended more plays than the average biology student, this might be wrong if a greater proportion of biology students were female.

REFERENCES

Ackerly, D.D., and M.J. Donoghue. 1995. Phylogeny and ecology reconsidered. Journal of Ecology 83: 730–732.

Aerts, R., and M.J. van der Peijl. 1993. A simple model to explain the dominance of low-productive perennials in nutrient-poor habitats. Oikos 66: 144–147.

Barkman, J.J. 1988. New systems of plant growth forms and phenological plant types. In: Werger, M.J.A., P.J.M. van der Aart, H.J. During, and J.T.A. Verhoeven, eds. Plant Form and Vegetation Structure. Adaptation, Plasticity and Relation to Herbivory. The Hague: SPB Academic Publishing, pp 9–44.

Begon, M., J.L. Harper, and C.R. Townsend. 1996. Ecology, 3rd ed. Oxford: Blackwell Scientific.

Box, E.O. 1981. Macroclimate and Plant Forms. The Hague: Dr W. Junk.

Bugmann, H. 1996. Functional types of trees in temperate and boreal forests: classification and testing. Journal of Vegetation Science 7: 359–370.

Burke, M.J.W., and J.P. Grime. 1996. An experimental study of plant community invasibility. Ecology 77: 776–790.

Campbell, B.D., and J.P. Grime. 1992. An experimental test of plant strategy theory. Ecology 73: 15–29.

Chapin, F.S. III, K. Autumn, and F. Pugnaire. Evolution of suites of traits in relation to environmental stress. American Naturalist 139: 1293–1304.

Chapin, F.S. III, M. Bret-Harte, M. Syndonia, S.E. Hobbie, and H. Zhong. 1996. Plant functional types as predictors of transient responses of arctic vegetation to global change. Journal of Vegetation Science 7: 347–358.

Cockburn, A. 1991. An Introduction to Evolutionary Ecology. Oxford: Blackwell Scientific.

Cody, M.L. 1966. A general theory of clutch size. Evolution 20: 174–184.

Colinvaux, P. 1993. Ecology 2. New York: Wiley.

Connell, J.H. 1983. On the prevalence and relative importance of interspecific competition: evidence from field experiments. American Naturalist 122: 661–696.

Cornelissen, J.H.C., P. Castro Diez, and R. Hunt. 1996. Seedling growth, allocation and leaf attributes in a wide range of woody plant species and types. Journal of Ecology 84: 755–765.

Crawley, M.J. 1983. Herbivory. Oxford: Blackwell Scientific.

Crawley, M.J., ed. 1996. Plant Ecology, 2nd ed. Oxford: Blackwell Scientific.

Cunningham, S.A., B. Summerhayes, and M. Westoby. 1999. Evolutionary divergences in leaf structure and chemistry, comparing gradients of rainfall and soil nutrients. Ecology, in press.

Dansereau, P. 1951. Description and recording of vegetation upon a structural basis. Ecology 32: 172–229.

Dyksterhuis, E.J. 1949. Condition and management of rangeland based on quantitative ecology. Journal of Range Management 2: 104–115.

Ellenberg, H. 1988. Vegetation of Central Europe, 4th ed. Berlin: Springer-Verlag.

Felsenstein, J. 1985. Phylogenies and the comparative method. American Naturalist 125: 1–15.

Garnier, E., and G. Laurent. 1994. Leaf anatomy, specific mass and water content in congeneric annual and perennial grasses. New Phytologist 128: 725–736.

Garnier, E., and A.H.J. Freijsen. 1994. On ecological inference from laboratory experiments conducted under optimum conditions. In: Roy, J., and E. Garnier, eds. A Whole-Plant Perspective on Carbon-Nitrogen Interactions. The Hague: SPB Publishing, pp 267–292.

Goldberg, D.E. 1996. Competitive ability: definitions, contingency and correlated traits. Philosophical Transactions of the Royal Society of London B 351: 1377–1385.

Goldberg, D.E., and A.M. Barton. 1992. Patterns and consequences of interspecific competition in natural communities: a review of field experiments with plants. American Naturalist 139: 771–801.

Gould, S.J., and E. Vrba. 1982. Exaptation—a missing term in the science of form. Paleobiology 8: 4–15.

Grafen, A. 1989. The phylogenetic regression. Philosophical Transactions of the Royal Society of London, Series B 205: 581–598.

Greenslade, P.J.M. 1972. Evolution in the Staphylinid genus *Priochirus* (Coleoptera). Evolution 26: 203–220.

Greenslade, P.J.M. 1983. Adversity selection and the habitat templet. American Naturalist 122: 352–365.

Grime, J.P. 1974. Vegetation classification by reference to strategies. Nature 250: 26–31.

Grime, J.P. 1977. Evidence for the existence of three primary strategies in plants and its relevance to ecological and evolutionary theory. American Naturalist 111: 1169–1194.

Grime, J.P. 1979. Plant Strategies and Vegetation Processes. Chichester: Wiley.

Grime, J.P. 1984. The ecology of species, families and communities of the contemporary British flora. New Phytologist 98: 15–33.

Grime, J.P. 1994a. Defining the scope and testing the validity of CSR theory: a response to Midgley, Laurie and Le Maitre. Bulletin of the South African Institute of Ecologists 13: 4–7.

Grime, J.P. 1994b. The role of plasticity in exploiting environmental heterogeneity. In: M.M. Caldwell, and R. Pearcy, eds. Exploitation of Environmental Heterogeneity in Plants. San Diego, CA: Academic Press, pp 1–18.

Grime, J.P., and S.H. Hillier. 1992. The contribution of seedling regeneration to the structure and dynamics of plant communities and larger units of landscape. In: Fenner, M., ed. Seeds: The Ecology of Regeneration in Plant Communities. Wallingford, UK: CAB International, pp 349–364.

Grime, J.P., and R. Hunt. 1975. Relative growth rate: its range and adaptive significance in a local flora. Journal of Ecology 63: 393–342.

Grime, J.P., J.G. Hodgson, and R. Hunt. 1988. Comparative Plant Ecology. London: Unwin-Hyman.

Grime, J.P., et al. 1997. Integrated screening validates a primary axis of specialization in plants. Oikos 79: 259–281.

Grubb, P.J. 1985. Plant populations and vegetation in relation to habitat, disturbance and competition: problems of generalization. In: White, J., ed. The Population Structure of Vegetation. Dordrecht: Junk, pp 595–621.

Gurevitch, J., L.L. Morrow, A. Wallace, and J.S. Walsh. 1992. A meta-analysis of field experiments on competition. American Naturalist 140: 539–572.

Hairston, N.G. Sr. 1989. Ecological Experiments: Purpose, Design and Execution Cambridge: Cambridge University Press.

Harvey, P.H. 1996. Phylogenies for ecologists. Journal of Animal Ecology 65: 255–263.

Harvey, P.H., and M.D. Pagel. 1991. The Comparative Method in Evolutionary Biology. Oxford: Oxford University Press.

Harvey, P.H., A.F. Read, and S. Nee. 1995a. Why ecologists need to be phytogenetically challenged. Journal of Ecology, 83: 535–536.

Harvey, P.H., A.F. Read, and S. Nee. 1995b. Further remarks on the role of phylogeny in comparative ecology. Journal of Ecology 83: 735–736.

Hubbell, S.P., and R.B. Foster. 1986. Commonness and rarity in a tropical forest: implications for tropical tree diversity. In: Soule, M.E., ed. Conservation Biology. Sunderland, MA: Sinauer, pp 205–231.

Hughes, L., M. Dunlop, K. French, M. Leishman, B. Rice, L. Rodgerson, and M. Westoby. 1994. Predicting dispersal spectra: a minimal set of hypotheses based on plant attributes. Journal of Ecology 82: 933–950.

Hunt, R., and J.H.C. Cornelissen. 1997. Physiology, allocation and growth rate: a reexamination of the Tilman model. American Naturalist 150: 122–130.

Ingrouille, M. 1992. Diversity and Evolution of Land Plants. London: Chapman and Hall.

Keddy, P.A. 1989. Competition. London: Chapman and Hall.

Kelly, C.K., and A. Purvis. 1993. Seed size and establishment conditions in tropical trees: on the use of taxonomic relatedness in determining ecological patterns. Oecologia 94: 356–360.

Kitajima, K. 1994. Relative importance of photosynthetic traits and allocation pattern as correlates of seedling shade tolerance of 13 tropical trees. Oecologia 98: 419–428.

Körner, Ch. 1993. Scaling from species to vegetation: the usefulness of functional groups. In: Schulze, E.-D., and H.A. Mooney, eds. Biodiversity and Ecosystem Function. Berlin: Springer-Verlag, pp 117–140.

Lambers, H., and H. Poorter. 1992. Inherent variation in growth rate between higher plants: a search for physiological causes and ecological consequences. Advances in Ecological Research 23: 188–261.

Lambers, H., H. Konings, M.L. Cambridge, and T.L. Pons, eds. 1989. Causes and Consequences of Variation in Growth Rate and Productivity of Higher Plants. The Hague: SPB Academic Publishing.

Lange, O.L., P.S. Nobel, C.B. Osmond, and H. Ziegler, eds. 1984. Encyclopaedia of Plant Physiology, New Series, vol 12B. Physiological Plant Ecology II. Water Relations and Carbon Assimilation. New York: Springer-Verlag.

Leps, J., J. Osborna-Kosinova, and K. Rejmanek. 1982. Community stability, complexity and species life-history strategies. Vegetatio 50: 53–63.

Loehle, C. 1988. Tree life history strategies: the role of defenses. Canadian Journal of Forest Research 18: 209–222.

MacArthur, R.H., and E.O. Wilson. 1967. The Theory of Island Biogeography. Princeton, NJ: Princeton University Press.

Mueller-Dombois, D., and H. Ellenberg. 1974. Aims and Methods of Vegetation Ecology. New York: Wiley.

Mulkey, S.S., R.L. Chazdon, and A.P. Smith, eds. 1996. Tropical Forest Plant Ecophysiology. London: Chapman and Hall.

Noble, I.R., and R.O. Slatyer. 1980. The use of vital attributes to predict successional changes in plant communities subject to recurrent disturbances. Vegetatio 43: 5–21.

Orshan, G. 1989. Plant Pheno-Morphological Studies in Mediterranean Ecosystems. Geobotany Volume 12. The Hague: Junk.

Poorter, H. 1989. Interspecific variation in relative growth rate: on ecological causes and physiological consequences. In: Lambers, H., et al., eds. Causes and Consequences of Variation in Growth Rate and Productivity in Higher Plants. The Hague: SPB Academic Publishing, pp 45–68.

Poorter, H. 1994. Construction costs and payback time of biomass: a whole plant perspective. In: Roy, J., and E. Garnier, eds. A Whole-Plant Perspective on Carbon-Nitrogen Interactions. The Hague: SPB Publishing, pp 111–127.

Poorter, H., and H. Lambers. 1991. Is interspecific variation in relative growth rate positively correlated with biomass allocation to the leaves? American Naturalist 138: 1264–1268.

Poorter, H., and C. Remkes. 1990. Leaf area ratio and net assimilation rate of 24 wild species differing in relative growth rates. Oecologia 83: 553–559.

Poorter, H., and A. van der Werf. 1998. Is inherent variation in RGR determined by LAR at low irradiance and by NAR at high irradiance? A review of herbaceous species. In: Lambers, H., H. Poorter, and M. van Vuuren, eds. Inherent Variation in Plant Growth. Physiological Mechanisms and Ecological Consequences. Leiden: Backhuys, pp 309–336.

Prentice, I.C., W. Cramer, S.P. Harrison, R. Leemans, R.A. Monserud, and A.M. Solomon. 1992. A global biome model based on plant physiology and dominance, soil properties and climate. Journal of Biogeography 19: 117–134.

Price, T. 1997. Correlated evolution and independent contrasts. Philosophical Transactions of the Royal Society of London, Series B 352: 519–529.

Price, P.W., M. Westoby, B. Rice, P.R. Atsatt, R.S. Fritz, J.N. Thompson, and K. Mobley. 1986. Parasite mediation in ecological interactions. Annual Reviews of Ecology and Systematics 17: 487–505.

Raunkiaer, C. 1934. The Life Forms of Plants and Statistical Plant Geography. Oxford: Clarendon Press.

Rees, M. 1993. Trade-offs among dispersal strategies in British plants. Nature 366: 150–152.

Rees, M. 1995. EC-PC comparative analyses? Journal of Ecology 83: 891–892.

Reeve, H.K., and P.W. Sherman. 1993. Adaptation and the goals of evolutionary research. Quarterly Review of Biology 68: 1–32.

Reich, P.B., M.B. Walters, and D.S. Ellsworth. 1992. Leaf life-span in relation to leaf, plant and stand characteristics among diverse ecosystems. Ecological Monographs 62: 365–392.

Reich, P.B., M.B. Walters, and D.S. Ellsworth. 1997. From tropics to tundra: global convergence in plant functioning. Proceedings of the National Academy of Science 94: 13730–13734.

Roy, J., and E. Garnier, eds. 1994. A Whole-Plant Perspective on Carbon-Nitrogen Interactions. The Hague-SPB Publishing.

Sarmiento, G., and M. Monasterio. 1983. Life forms and phenology. In: Bourliere, F., ed. Ecosystems of the World 13: Tropical Savannas. Amsterdam: Elsevier, pp 79–108.

Saverimuttu, T., and M. Westoby. 1996. Seedling longevity under deep shade in relation to seed size. Journal of Ecology 84: 681–689.

Schoener, T.W. 1983. Field experiments on interspecific competition. American Naturalist 122: 240–285.

Schulze, E.-D., and M.M. Caldwell. eds. 1994. Ecophysiology of Photosynthesis. Berlin: Springer-Verlag.

Sih, A., P. Crowley, M. McPeek, J. Petranka, and K. Strohmeier. 1985. Predation, competition and prey communities. Annual Review of Ecology and Systematics 16: 269–311.

Silvertown, J., and M. Dodd. 1996. Comparing plants and connecting traits. Philosophical Transactions of the Royal Society of London, Series B 351: 1233–1239.

Smith, T.M., H.H. Shugart, and F.I. Woodward, eds. 1997. Plant Functional Types: Their Relevance To Ecosystem Properties and Global Change. Cambridge: Cambridge University Press.

Southwood, T.R.E. 1977. Habitat, the templet for ecological strategies. Journal of Animal Ecology 46: 337–365.

Steffen, W.L., B.H. Walker, J.S. Ingram, and G.W. Koch, eds. 1992. Global Change and Terrestrial Ecosystems: The Operational Plan. Stockholm: International Geosphere-Biosphere Program, IGBP Report No. 21.

Turner, I.M. 1994. A quantitative analysis of leaf form in woody plants from the world's major broadleaved forest types. Journal of Biogeography 21: 413–419.

van der Werf, A., M. van Nuenen, A.J. Visser, and H. Lambers. 1993. Contribution of physiological and morphological plant traits to a species' competitive ability at high and low nitrogen supply. Oecologia 94: 434–440.

Westoby, M. 1998. A leaf-height-seed (LHS) plant ecology strategy scheme. Plant and Soil 199: 213–227.

Westoby, M., M.R. Leishman, and J.M. Lord. 1995a. On misinterpreting the "phylogenetic correction." Journal of Ecology 83: 531–534.

Westoby, M., M.R. Leishman, and J.M. Lord. 1995b. Further remarks on phylogenetic correction. Journal of Ecology 83: 727–730.

Westoby, M., M.R. Leishman, and J.M. Lord. 1995c. Issues of interpretation after relating comparative datasets to phylogeny. Journal of Ecology, 83: 892–893.

Westoby, M., M.R. Leishman, and J.M. Lord. 1996. Comparative ecology of seed size and seed dispersal. Philosophical Transactions of the Royal Society of London, Series B 351: 1309–1318.

Westoby, M., S.A. Cunningham, C.R. Fonseca, J.McC. Overton, and I.J. Wright. 1998. Phylogeny and variation in light capture area deployed per unit investment in leaves: designs for selecting study species with a view to generalizing. In: Lambers, H., H. Poorter, and M. van Vuuren, eds. Inherent Variation in Plant Growth. Physiological Mechanisms and Ecological Consequences. Leiden: Backhuys, pp 539–566.

Whittaker, R.H. 1975. Communities and Ecosystems. New York: Macmillan.

Williams, G.C. 1992. Natural Selection: Domains, Levels, and Challenges. Oxford: Oxford University Press.

Wilson, J.B., and A.D.Q. Agnew. 1992. Positive-feedback switches in plant communities. Advances in Ecological Research 23: 263–336.

Witkowski, E.T.F., and B.B. Lamont. 1991. Leaf specific mass confounds leaf density and thickness. Oecologia 88: 486–493.

Woodward, F.I., and W. Cramer. 1996. Plant functional types and climatic changes: introduction. Journal of Vegetation Science 7: 306–308.

Wright, I.J., and M. Westoby. 1999. Differences in seedling growth behavior among species: trait correlation, and trait shifts along nutrient compared to rainfall gradients. Journal of Ecology, in press.

Index